Recent Titles in This Serie

147 Neil Robertson and Paul Seymour, Edit
146 Martin C. Tangora, Editor, Algebraic topology, 1993
145 Jeffrey Adams, Rebecca Herb, Stephen Kudla, Jian-Shu Li, Ron Lipsman, Jonathan Rosenberg, Editors, Representation theory of groups and algebras, 1993
144 Bor-Luh Lin and William B. Johnson, Editors, Banach spaces, 1993
143 Marvin Knopp and Mark Sheingorn, Editors, A tribute to Emil Grosswald: Number theory and related analysis, 1993
142 Chung-Chun Yang and Sheng Gong, Editors, Several complex variables in China, 1993
141 A. Y. Cheer and C. P. van Dam, Editors, Fluid dynamics in biology, 1993
140 Eric L. Grinberg, Editor, Geometric analysis, 1992
139 Vinay Deodhar, Editor, Kazhdan-Lusztig theory and related topics, 1992
138 Donald St. P. Richards, Editor, Hypergeometric functions on domains of positivity, Jack polynomials, and applications, 1992
137 Alexander Nagel and Edgar Lee Stout, Editors, The Madison symposium on complex analysis, 1992
136 Ron Donagi, Editor, Curves, Jacobians, and Abelian varieties, 1992
135 Peter Walters, Editor, Symbolic dynamics and its applications, 1992
134 Murray Gerstenhaber and Jim Stasheff, Editors, Deformation theory and quantum groups with applications to mathematical physics, 1992
133 Alan Adolphson, Steven Sperber, and Marvin Tretkoff, Editors, p-Adic methods in number theory and algebraic geometry, 1992
132 Mark Gotay, Jerrold Marsden, and Vincent Moncrief, Editors, Mathematical aspects of classical field theory, 1992
131 L. A. Bokut', Yu. L. Ershov, and A. I. Kostrikin, Editors, Proceedings of the International Conference on Algebra Dedicated to the Memory of A. I. Mal'cev, Parts 1, 2, and 3, 1992
130 L. Fuchs, K. R. Goodearl, J. T. Stafford, and C. Vinsonhaler, Editors, Abelian groups and noncommutative rings, 1992
129 John R. Graef and Jack K. Hale, Editors, Oscillation and dynamics in delay equations, 1992
128 Ridgley Lange and Shengwang Wang, New approaches in spectral decomposition, 1992
127 Vladimir Oliker and Andrejs Treibergs, Editors, Geometry and nonlinear partial differential equations, 1992
126 R. Keith Dennis, Claudio Pedrini, and Michael R. Stein, Editors, Algebraic K-theory, commutative algebra, and algebraic geometry, 1992
125 F. Thomas Bruss, Thomas S. Ferguson, and Stephen M. Samuels, Editors, Strategies for sequential search and selection in real time, 1992
124 Darrell Haile and James Osterburg, Editors, Azumaya algebras, actions, and modules, 1992
123 Steven L. Kleiman and Anders Thorup, Editors, Enumerative algebraic geometry, 1991
122 D. H. Sattinger, C. A. Tracy, and S. Venakides, Editors, Inverse scattering and applications, 1991
121 Alex J. Feingold, Igor B. Frenkel, and John F. X. Ries, Spinor construction of vertex operator algebras, triality, and $E_8^{(1)}$, 1991
120 Robert S. Doran, Editor, Selfadjoint and nonselfadjoint operator algebras and operator theory, 1991
119 Robert A. Melter, Azriel Rosenfeld, and Prabir Bhattacharya, Editors, Vision geometry, 1991
118 Yan Shi-Jian, Wang Jiagang, and Yang Chung-chun, Editors, Probability theory and its applications in China, 1991

(Continued in the back of this publication)

CONTEMPORARY MATHEMATICS

147

Graph Structure Theory

Proceedings of the AMS-IMS-SIAM Joint Summer
Research Conference on Graph Minors
held June 22 to July 5, 1991, with support
from the National Science Foundation
and the Office of Naval Research

Neil Robertson
Paul Seymour
Editors

American Mathematical Society
Providence, Rhode Island

EDITORIAL BOARD

Craig Huneke, managing editor
Clark Robinson   J. T. Stafford
Linda Preiss Rothschild   Peter M. Winkler

The AMS-IMS-SIAM Joint Summer Research Conference in the Mathematical Sciences on Graph Minors was held at the University of Washington, Seattle, from June 22 to July 5, 1991, with support from the National Science Foundation, Grant DMS-8918200, and the Office of Naval Research, Grant N00014-91-J-1496.

This work relates to Department of Navy Grant N00014-91-J-1496, issued by the Office of Naval Research. The United States Government has a royalty-free license throughout the world in all copyrightable material contained herein.

1991 *Mathematics Subject Classification.* Primary 05C75, 05C40, 05C10; Secondary 05C15, 05C38, 05B35, 05C85, 68R10, 06A06, 57M15, 57M25.

Library of Congress Cataloging-in-Publication Data

AMS-IMS-SIAM Joint Summer Research Conference on Graph Minors (1991: University of Washington)
  Graph structure theory: proceedings of an AMS-IMS-SIAM Joint Summer Research Conference on Graph Minors, held June 22 to July 5, 1991, at the University of Washington, Seattle, with support from the National Science Foundation and the Office of Naval Research/Neil Robertson, Paul Seymour, editors.
  p. cm.—(Contemporary mathematics; v. 147)
  ISBN 0-8218-5160-8
  1. Graph theory—Congresses. I. Robertson, Neil, 1938– . II. Seymour, Paul D. III. National Science Foundation (U.S.) IV. United States. Office of Naval Research. V. Title. VI. Series: Contemporary mathematics (American Mathematical Society); v. 147.
QA166.J65 1991   93-18553
511$'$.5—dc20   CIP

Copying and reprinting. Individual readers of this publication, and nonprofit libraries acting for them, are permitted to make fair use of the material, such as to copy an article for use in teaching or research. Permission is granted to quote brief passages from this publication in reviews, provided the customary acknowledgment of the source is given.

Republication, systematic copying, or multiple reproduction of any material in this publication (including abstracts) is permitted only under license from the American Mathematical Society. Requests for such permission should be addressed to the Manager of Editorial Services, American Mathematical Society, P.O. Box 6248, Providence, Rhode Island 02940-6248.

The appearance of the code on the first page of an article in this book indicates the copyright owner's consent for copying beyond that permitted by Sections 107 or 108 of the U.S. Copyright Law, provided that the fee of $1.00 plus $.25 per page for each copy be paid directly to the Copyright Clearance Center, Inc., 27 Congress Street, Salem, Massachusetts 01970. This consent does not extend to other kinds of copying, such as copying for general distribution, for advertising or promotional purposes, for creating new collective works, or for resale.

Copyright ©1993 by the American Mathematical Society. All rights reserved.
The American Mathematical Society retains all rights except those granted
to the United States Government.
Printed in the United States of America.
The paper used in this book is acid-free and falls within the guidelines
established to ensure permanence and durability. ∞
This volume was printed directly from copy
typeset by the authors using $\mathcal{A}_{\mathcal{M}}\mathcal{S}$-TEX,
the American Mathematical Society's TEX macro system.

10 9 8 7 6 5 4 3 2 1   98 97 96 95 94 93

Contents

Preface	ix
Alphabetical list of authors	xi
Polynomials W. T. TUTTE	1
Tutte invariants for 2-polymatroids JAMES OXLEY AND GEOFF WHITTLE	9
Extremal matroid theory JOSEPH P. S. KUNG	21
Subexponentially computable truncations of Jones-type polynomials TERESA M. PRZYTYCKA AND JÓZEF H. PRZYTYCKI	63
Knots and braids: Some algorithmic questions D. J. A. WELSH	109
A survey of linkless embeddings NEIL ROBERTSON, P. D. SEYMOUR, AND ROBIN THOMAS	125
On a new graph invariant and a criterion for planarity YVES COLIN DE VERDIÈRE	137
Four problems on plane graphs raised by Branko Grünbaum OLEG BORODIN	149
Counterexamples to a conjecture of Las Vergnas and Meyniel BRUCE REED	157
An extremal function for the achromatic number BÉLA BOLLOBÁS, BRUCE REED, AND ANDREW THOMASON	161
The asymptotic structure of H-free graphs H. J. PRÖMEL AND A. STEGER	167
Induced minors and related problems MICHAEL FELLOWS, JAN KRATOCHVÍL, MATTHIAS MIDDENDORF, AND FRANK PFEIFFER	179

Induced circuits in graphs on surfaces
ALEXANDER SCHRIJVER   183

Tree-representation of directed circuits
ANDRÁS FRANK AND TIBOR JORDÁN   195

Intercyclic digraphs
WILLIAM MCCUAIG   203

Eulerian trails through a set of terminals in specific, unique and all orders
JØRGEN BANG-JENSEN AND SVATOPLUK POLJAK   247

2-reducible cycles containing two specified edges in $(2k+1)$-edge-connected graphs
HARUKO OKAMURA   259

Edge-disjoint cycles in n-edge-connected graphs
ANDREAS HUCK   279

Finding disjoint trees in planar graphs in linear time
B. A. REED, N. ROBERTSON, A. SCHRIJVER, AND P. D. SEYMOUR   295

Surface triangulations without short noncontractible cycles
TERESA M. PRZYTYCKA AND JÓZEF H. PRZYTYCKI   303

Representativity and flexibility on the projective plane
R. P. VITRAY   341

On non-null separating circuits in embedded graphs
XIAOYA ZHA AND YUE ZHAO   349

Projective-planar graphs with even duals II
SEIYA NEGAMI   363

2-factors, connectivity, and graph minors
NATHANIEL DEAN AND KATSUHIRO OTA   381

A conjecture in topological graph theory
JOHN PHILIP HUNEKE   387

On the closed 2-cell embedding conjecture
XIAOYA ZHA   391

Cycle cover theorems and their applications
CUN-QUAN ZHANG   405

Cones, lattices and Hilbert bases of circuits and perfect matchings
LUIS A. GODDYN   419

Regular maps from voltage assignments
PAVOL GVOZDJAK AND JOZEF ŠIRÁŇ   441

The infinite grid covers the infinite half-grid
BOGDAN OPOROWSKI   455

Dominating functions and topological graph minors
REINHARD DIESTEL   461

Notes on rays and automorphisms of locally finite graphs
H. A. JUNG   477

Quasi-ordinals and proof theory
L. GORDEEV   485

Minor classes :Extended abstract
DIRK VERTIGAN   495

Well-quasi-ordering finite posets
JENS GUSTEDT   511

The immersion relation on webs
GUOLI DING   517

Structural descriptions of lower ideals of trees
NEIL ROBERTSON, P. D. SEYMOUR, AND ROBIN THOMAS   525

Finite automata, bounded treewidth, and well-quasiordering
KARL ABRAHAMSON AND MICHAEL FELLOWS   539

Graph grammars, monadic second-order logic and the theory of graph minors
BRUNO COURCELLE   565

Graph reductions, and techniques for finding minimal forbidden minors
ANDRZEJ PROSKUROWSKI   591

An upper bound on the size of an obstruction
JENS LAGERGREN   601

An obstruction-based approach to layout optimization
MICHAEL A. LANGSTON   623

Decomposing 3-connected graphs
COLLETTE R. COULLARD AND DONALD K. WAGNER   631

Graph planarity and related topics
A. K. KELMANS   635

Excluding a graph with one crossing
NEIL ROBERTSON AND PAUL SEYMOUR   669

Open problems
NATHANIEL DEAN   677

Preface

This volume contains the proceedings of a conference on Graph Minors, held at the University of Washington, Seattle, between June 22 and July 5, 1991. The topics of the talks included algorithms on tree-structured graphs, well-quasi-ordering, logic, infinite graphs, disjoint path problems, surface embeddings, knot theory, graph polynomials, matroid theory, and combinatorial optimization. We have tried to organize the volume so that related papers are near one another.

This volume also contains an English translation of a paper by Yves Colin de Verdière, which was first published in a refereed journal in French. It seems worthwhile to make this paper more easily accessible, and we are grateful to the American Mathematical Society and Academic Press for permission to include the translation. All the other papers in this volume have been refereed.

This conference was one in the series of AMS-IMS-SIAM Joint Summer Research Conferences. It was funded by the Office of Naval Research and by the National Science Foundation; we would like to express our thanks to the ONR and NSF for their generous support. Thanks are also due to the other members of the organizing committee: Harvey Friedman and Bruce Reed; Carole Kohanski, who handled the detailed organization of the workshop; and Donna Harmon, Christine Thivierge, and Alison Buckser at the American Mathematical Society, who shepherded us successfully through the publication process.

Neil Robertson
Paul Seymour

ALPHABETICAL LIST OF AUTHORS

Karl Abrahamson and **Mike Fellows**
Finite automata, bounded treewidth and well-quasiordering

Jørgen Bang-Jensen and **Svatopluk Poljak**
Eulerian trails through a set of terminals in specific, unique and all orders

Béla Bollobás, Bruce Reed and **Andrew Thomason**
An extremal function for the achromatic number

Oleg Borodin
Four problems on plane graphs raised by Branko Grünbaum

Yves Colin de Verdière
On a new graph invariant and a criterion for planarity

Collette Coullard and **Donald Wagner**
Decomposing 3-connected graphs

Bruno Courcelle
Graph grammars, monadic second-order logic and the theory of graph minors

Nathaniel Dean
Open problems

Nathaniel Dean and **Katsuhiro Ota**
2-factors, connectivity and graph minors

Reinhard Diestel
Dominating functions and topological graph minors

Guoli Ding
The immersion relation on webs

Michael Fellows, Jan Kratochvil, Matthias Middendorf and Frank Pfeiffer
Induced minors and related problems

András Frank and Tibor Jordán
Tree-representation of directed cycles

Luis Goddyn
Cones, lattices and Hilbert bases of circuits and perfect matchings

L. Gordeev
Quasi-ordinals and proof theory

Jens Gustedt
Well quasi ordering finite posets

Pavol Gvozdjak and Jozef Sirán
Regular maps from voltage assignments

Andreas Huck
Edge-disjoint cycles in n-edge-connected graphs

Philip Huneke
A conjecture in topological graph theory

H. A. Jung
Notes on rays and automorphisms of locally finite graphs

A. K. Kelmans
Graph planarity and related topics

Joseph Kung
Extremal matroid theory

Jens Lagergren
An upper bound on the size of an obstruction

Mike Langston
An obstruction-based approach to layout optimization

William McCuaig
Intercyclic digraphs

Seiya Negami
Projective-planar graphs with even duals II

Haruko Okamura
2-reducible cycles containing two specified edges in
$(2k+1)$-edge-connected graphs

Bogdan Oporowski
The infinite grid covers the infinite half-grid

James Oxley and **Geoff Whittle**
Tutte invariants for 2-polymatroids

H. J. Prömel and **A. Steger**
The asymptotic structure of H-free graphs

Andrzej Proskurowski
Graph reductions, and techniques for finding minimal
forbidden minors

Teresa Przytycka and **Józef Przytycki**
Subexponentially computable truncations of Jones-type
polynomials

Teresa Przytycka and **Józef Przytycki**
Surface triangulations without short noncontractible cycles

Bruce Reed
Counterexamples to a conjecture of Las Vergnas and Meyniel

B.A. Reed, N. Robertson, A. Schrijver and **P. D. Seymour**
Finding disjoint trees in planar graphs in linear time

Neil Robertson and **Paul Seymour**
Excluding a graph with one crossing

Neil Robertson, P. D. Seymour and **Robin Thomas**
A survey of linkless embeddings

Neil Robertson, P. D. Seymour and **Robin Thomas**
Structural descriptions of lower ideals of trees

Alexander Schrijver
Induced circuits in graphs on surfaces

W. T. Tutte
Polynomials

Dirk Vertigan
Minor classes

R. P. Vitray
Representativity and flexibility in the projective plane

D. J. A. Welsh
Knots and braids: some algorithmic questions

Xiaoya Zha
On the closed 2-cell embedding conjecture

Xiaoya Zha and **Yue Zhao**
On non-null separating circuits in embedded graphs

Cun-Quan Zhang
Cycle cover theorems and their applications

POLYNOMIALS

W. T. TUTTE

1. Kirchhoff's Equations.

The graphs of this paper are of the general kind. That is they can have loops and multiple edges.

It was in a study of electrical networks that I first encountered graph-polynomials. There was a variable called the "conductance" associated with each edge of a graph G. It was important to consider the sum $C(G)$ of the conductance-products over all the spanning trees of G. This sum could be expressed as a determinant. There were other determinants derived from conductance-products over spanning double trees. Some standard problems about electrical flows in G could be solved in terms of these determinants, the determinants of "Kirchhoff's Equations" [1]. Note that the loops of G are irrelevant in the computation of $C(G)$, and that $C(G)$ is zero for a disconnected graph.

At the time I was interested only in the special case in which each conductance had the value 1, and perhaps I did not appreciate the polynomial C(G) in its full beauty. For that special case, in which $C(G)$ is the number of spanning trees of G, I was very interested in the recursion formula

$$(1) \qquad C(G) = C(G_A) + C(G^A).$$

Here A is any link of G, that is any edge of G having two distinct ends. G_A is the graph obtained from G be deleting that edge, and G^A is the graph derived from G by contracting A, with its two ends, into a single vertex. I looked for other functions of graphs that satisfied this recursion.

2. Chromatic Polynomials.

I found a recursion similar to (1), attributed to R. M. Foster, in a paper of Hassler Whitney [3]. For any positive integer l the number of vertex-colourings of G in λ colours was denoted by $P(G, \lambda)$. It was understood that in a "vertex-colouring" each edge must have two ends of different colours. The function

---

1991 *Mathematics Subject Classification.* Primary 05C15

This paper is in final form and no version of it will be submitted for publication elsewhere.

$P(G,\lambda)$ was found to take the form of a polynomial in λ. Once this was recognized it became possible to define the value of the polynomial at any real or complex value of λ. $P(G,\lambda)$ was called the "chromatic polynomial" of G. This term is now often abbreviated to "chromial". Note that $P(G,\lambda)$ is identically zero whenever G has a loop.

Foster's recursion formula was

$$(2) \qquad P(G,l) = P(G_A,l) - P(G^A,l).$$

The analogy with (1) is not perfect, there being a difference instead of a sum on the right. However this discrepancy can be regarded as trivial; it can be adjusted by multiplying the function $P(G,\lambda)$ by a power of -1, the index of -1 being the number of vertices of G.

Another difference between $C(G)$ and $P(G,\lambda)$ is that $C(G)$ is zero for a disconnected G whereas $P(G,\lambda)$ is "multiplicative", that is the chromial of G is the product of the chromials of the components of G. As a rule I prefer graph-polynomials to be multiplicative. Sometimes it is possible to work with connected graphs only, and then it does not matter whether the polynomial is multiplicative or not.

3. The Dichromate.

It will be convenient to have names for three very simple graphs. First we name the "vertex-graph" which has a single vertex and no edges. Then the "loop-graph", consisting of a single vertex and a single incident loop. Last the "link-graph", defined by a single link and its two ends. We shall not use the null graph.

It will also be convenient to have the following notation. We write $\alpha_0(G)$ for the number of vertices of G, $\alpha_1(G)$ for the number of edges and $p_0(G)$ for the number of components. We write also

$$(3) \qquad p_1(G) = \alpha_1(G) - \alpha_0(G) + p_0(G).$$

We define a graph-polynomial $\chi(G;x,y)$ in two variables x and y. To begin with we assume that there is such a polynomial satisfying the following rules.
  (i) If G is the vertex-graph, then $\chi(G;x,y) = 1$.
  (ii) $\chi(G;x,y)$ is multiplicative.
  (iii) If G has j isthmuses, k loops and no other edge, then

$$(4) \qquad \chi(G;x,y) = x^j y^k.$$

  (iv) If G has an edge A that is neither a loop nor an isthmus then the recursion (1) applies. That is

$$(5) \qquad \chi(G;x,y) = \chi(G_A;x,y) + \chi(G^A;x,y).$$

We call $\chi(G;x,y)$, the polynomial assumed to be defined by these rules, the "dichromate" of G. Clearly if such a polynomial exists it is uniquely determined.

We leave the justification of the assumption to the next Section. Meanwhile we take note of some of its consequences.

3.1. *The adjunction of an isolated vertex to G makes no difference to the dichromate, by Rules (i) and (ii).*

3.2. *The dichromate of a link-graph is x, and that of a loop-graph is y, by Rule (iii).*

3.3. *If G is connected then $\chi(G; 1, 1)$ is the number of spanning trees of G.*

To prove this we define $m(G)$ as the number of edges of G that are neither loops nor isthmuses. If $m(G) = 0$ then G has just one spanning tree, obtained from G by deleting any loops. But $\chi(G; 1, 1) = 1$ by Rule (iii). So the theorem is true in this case.

We complete the proof by induction on $m(G)$, using the two recursions (1) and (5).

3.4. *The dichromate of G is related to the chromial by the equation*

$$(-1)^{\alpha_0(G)}(-\lambda)^{p_0(G)} P(G, \lambda) = \chi(G; 1 - \lambda, 0). \tag{6}$$

This can be proved by showing that the expression on the left satisfies the four rules for the appropriate values of x and y.

If G is a connected planar graph we denote its dual by G^*. It is well-known that in the change from G to G^* any isthmus is transformed into a loop and any loop into an isthmus. Moreover the dual of G_A is G^{*A} and the dual of G^A is G^*_A. Using these observations we can deduce from the rules that

$$\chi(G^*; x, y) = \chi(G; y, x). \tag{7}$$

From (6) and (7) we can relate the dichromate of G to the chromatic polynomial of G^*.

Consider the special case in which G consists of two vertices joined by two links. Then G_A is a link-graph and G^A a loop-graph. By Rule (iv)

$$\chi(G; x, y) = x + y. \tag{8}$$

We can regard this result as a simple example of the rule that the dichromate of a self-dual graph must be symmetrical in the two variables. This rule is a consequence of (7).

Now let G be a 3-circuit. If A is one of its edges we find that each of the two edges of G_A is an isthmus, and that G^A is the graph of Equation (8). So, by Rule (iv),

$$\chi(G; x, y) = x^2 + x + y. \tag{9}$$

It is not difficult to continue in this way until we get to the complete graph K_4 of 4 vertices. We then find that

$$\chi(K_4; x, y) = 2x + 3x^2 + x^3 + 2y + 4xy + 3y^2 + y^3. \tag{10}$$

Again we have a self-dual planar graph and a symmetrical dichromate.

4. An Existence Proof for the Dichromate.

In this Section we define a graph-polynomial $R(G;x,y)$ in two variables by an explicit formula. We then show that it satisfies our four rules for the dichromate. We then know that there is indeed a polynomial $\chi(G;x,y)$ satisfying the four rules. The rules themselves guarantee the uniqueness of this polynomial, by an inductive argument over the number $m(G)$ of edges of G that are neither loops nor isthmuses.

Let S be any subset of the edge-set $E(G)$ of G. Let us write $G:S$ for the spanning subgraph of G corresponding to S, that is the subgraph whose vertices are those of G and whose edges are the members of S.

We define $R(G;x,y)$ by the following equation

(11) $$R(G;x,y) = \sum_S (x-1)^{p_0(G:S)-p_0(G)}(y-1)^{p_1(G:S)}.$$

It is easy to verify Rule (i) for $R(G;x,y)$. For the vertex-graph S can only be the null set. Then both indices on the right of (11) are zero and we must put $R=1$. For the link-graph or loop-graph with edge A the set S must be $\{A\}$ or the null set. So for the link-graph the first index is 1 and the second is zero for the non-null S and both indices are zero when S is null, whence $R=x$. For the loop-graph the first index is zero and the second 1 for the non-null S, whence $R=y$.

To prove Rule (ii) for R suppose G to be the union of two disjoint subgraphs H and K. Then the possible choices for S are the unions, each of an arbitrary subset S_H of $E(H)$ with an arbitrary subset S_K of $E(K)$. For each such choice we can verify that the first index on the right of (11) is

$$p_0(G:S_H) + p_0(G:S_K) - p_0(H) - p_0(K)$$

and that the second is

$$p_1(G:S_H) + p_1(G:S_K).$$

We deduce that

(12) $$R(G;x,y) = R(H;x,y).R(K;x,y).$$

Repeated application of this result shows that $R(G;x,y)$ is the product of the R-polynomials of the components of G. Thus R is multiplicative in the sense of Rule (ii).

The same argument, leading to Equation (12), can be applied when G is the union of two subgraphs H and K having just one vertex v in common. Repeated application of (12), in this interpretation, then shows that $R(G;x,y)$ is the product of the R-polynomials of the blocks of G, that is the maximal non-separable subgraphs. Rule (iii) follows from this result since each loop and each isthmus of G defines a one-edged block. We state the property of this paragraph by saying that the polynomial R is "block-multiplicative".

To prove Rule (iv) for R we split the sum on the right of (11) into two sums, the first over subsets S containing A and the second over subsets S not containing A. Since A is neither a loop nor an isthmus of G the first sum is readily identified as $R(G^A; x, y)$ and the second as $R(G_A; x, y)$.

We have now verified the existence and uniqueness of $\chi(G; x, y)$. We have also found two new theorems for it, which we state below as 4.1 and 4.2. The second is a strengthening of Rule (iv).

4.1. *$\chi(G; x, y)$ is given explicitly by the sum on the right of (11).*
4.2. *$\chi(G; x, y)$ is block-multiplicative.*

5. The Dichromate as a Sum Over Spanning Trees.

The sum of the numerical coefficients in $\chi(G; x, y)$ is the number of spanning trees of G, by 3.3. This suggested that each spanning tree should be associated with a product of the form $x^a y^b$, and that $\chi(G; x, y)$ should be the sum of this product over all the spanning trees.

The 3-circuit seemed to be a counter-example to this conjecture. It has three spanning trees and they are all equivalent under the symmetry of the circuit. So does not the conjecture imply that the dichromate of the 3-circuit is of the form $3x^a y^b$, contrary to (9)?

This argument from symmetry was circumvented by enumerating the edges of G as a sequence

$$U = (e_1, e_2, e_3, \ldots, e_m),$$

so destroying the symmetry, and then defining a and b in terms of U, as well as of the particular spanning tree under consideration. Unexpectedly but conveniently the final sum over all the spanning trees was found to be independent of the enumeration U. Consider any spanning tree T of a connected graph G. Let A be an edge of T. Its deletion from T splits the tree into two subtrees T_1 and T_2. We write $C(T, A)$ for the set of all edges of G, other than A, which have one end in T_1 and the other in T_2. Clearly no member of $C(T, A)$ belongs to T. We say that the members of $C(T, A)$ are the edges of G "covered by A", with respect to T.

Now let B be an edge of G not in T. Its ends are joined by an arc $L(B)$ in T. If B is a loop then $L(B)$ reduces to a vertex-graph. The edges of $L(B)$ constitute a set $D(T, B)$. We say they are the edges of G "covered by B", with respect to T.

Let A and B be distinct edges of G. It is clear that if A covers B with respect to T then B covers A with respect to T.

An edge e_j of G is said to be "dominant" with respect to T if it covers no edge with a suffix exceeding j. We define $a(T)$ as the number of dominant edges in T, and $b(T)$ as the number of dominant edges not in T, for the fixed enumeration U. In [2] $a(T)$ is called the "internal activity" and $b(T)$ the "external activity" of T.

We now define a graph-polynomial $F(G; x, y)$. If G is connected we define it as a sum over the spanning trees T of G, thus:

$$(13) \qquad F(G; x, y) = \sum_T x^{a(T)} y^{b(T)}.$$

If G is not connected we define $F(G; x, y)$ as the product of the F-polynomials of the components of G, taking the enumeration induced by U in each component.

5.1. *The polynomials $F(G; x, y)$ and $\chi(G; x, y)$ are identical for each enumeration U of $E(G)$.*

Since both $\chi(C; x, y)$ and $F(G; x, y)$ are multiplicative it is only necessary to prove this theorem for the connected case. We proceed by induction over $\alpha_1(G)$. We note first that if G is edgeless then it can only be a vertex-graph. There is just one spanning tree T, and for it the indices $a(T)$ and $b(T)$ are both zero. We therefore write $F(G; x, y) = 1$. So the theorem is true in this case, by Rule (i).

Assume as an inductive hypothesis that the theorem holds whenever $\alpha_1(G)$ is less than some positive integer q, and consider the case $\alpha_1(G) = q$. Then let A be the edge e_1 of U.

Suppose first that A is a loop of G. Then it belongs to no spanning tree and is dominant for every spanning tree. Its deletion converts G into G_A, and the spanning trees of G become the spanning trees of G_A. Moreover the deletion does not affect the dominance or otherwise of any other edge with respect to any spanning tree. (For G_A we use the enumeration U_1 derived from U by deleting e_1 and then reducing each remaining suffix by 1). Hence

$$\begin{aligned} F(G; x, y) &= y F(G_A; x, y) \\ &= y \chi(G_A, x, y), \quad \text{by the induction hypothesis,} \\ &= \chi(G; x, y), \quad \text{by Rule (i) and 4.2.} \end{aligned}$$

Suppose next that A is an isthmus of G. Then it belongs to every spanning tree and is dominant for every spanning tree. Its contraction converts G into G^A and the spanning trees of G become the spanning trees of G^A. Moreover the deletion does not affect the dominance or otherwise of any other edge with respect to any spanning tree. (We use U_1 for G^A). Hence

$$\begin{aligned} F(G; x, y) &= x F(G^A; x, y) \\ &= x \chi(G^A; x, y), \quad \text{by the induction hypothesis,} \\ &= \chi(G; x, y), \quad \text{by Rule (i) and 4.2.} \end{aligned}$$

In the remaining case A is neither a loop nor an isthmus of G. Then A is not dominant for any spanning tree. We split the sum on the right of (12) into two sums, the first over the spanning trees containing A and the second over the spanning trees not containing A. Let us contract A to change G into G^A, for which we use the enumeration U_1. Then the spanning trees of the first sum are changed into the spanning trees of G^A. With respect to any one of them the dominance or otherwise of any edge other than A is unchanged. So the first sum is $F(G^A; x, y)$. We treat the second sum by deleting A to change G into G_A, and then using U_1. The spanning trees of the second sum are those of G_A, and with respect to any one of them the dominance or otherwise of any edge other than A is unchanged. Accordingly the second sum is $F(G_A; x, y)$. Hence

$$\begin{aligned} F(G; x, y) &= F(G_A; x, y) + F(G^A; x, y) \\ &= c(G_A; x, y) + c(G^A; x, y), \quad \text{by the induction hypothesis,} \\ &= c(G; x, y), \quad \text{by Rule (iv).} \end{aligned}$$

This completes the proof of the theorem.

6. A Conjecture.

Using 5.1 we can rewrite (13), for a connected G, as

$$\chi(G; x, y) = \Sigma \chi(i, j) x^i y^j. \tag{14}$$

Here $\chi(i, j)$ is the number of spanning trees T of G for which exactly i edges of T and exactly j edges not in T are dominant in the chosen U. The sum is over all values of i from 0 to $\alpha_0(G) - 1$ and over all values of j from 0 to $\alpha_1(G) - \alpha_0(G) + 1 = p_1(G)$.

The coefficients are conveniently regarded as the entries in a matrix C of $\alpha_0(G) - 1$ rows and $p_1(G)$ columns, the entry in the ith row and jth column being $\chi(i, j)$.

Until recently it could be said that in all known cases the matrix C shows a simple regularity. In each row and each column: no entry lies between two larger ones. Accordingly it was conjectured that the regularity holds for every G. However news was received at the Seattle meeting that a counterexample had been discovered by Werner Schwärtzler of Bonn University. I have learned that it is to be published in the Journal of Combinatorial Theory.

7. Some Associated Polynomials.

Let G be a connected graph with n vertices. For each spanning tree T of G let $q(G, T; i, j)$ be the number of enumerations U of $E(G)$ for which $a(T) = i$ and $b(T) = j$. Then from the results of Section 6 we have

$$n! \chi(i, j) = \sum_T q(G, T; i, j). \tag{15}$$

For each T we can define the two-variable polynomial

$$J(G, T; x, y) = \sum q(G, T; i, j) x^i y^j, \tag{16}$$

the sum being over all relevant values of i and j. Then by (15) we have

$$n! \chi(G; x, y) = \sum_T J(G, T; x, y). \tag{17}$$

We now define a polynomial $Q(H;x)$ for an arbitrary graph H. In it the coefficient of x^k is the number of enumerations of the vertex-set $V(H)$ of H in which exactly k vertices are dominant. A "dominant" vertex is one that is joined to no vertex coming later in the enumeration.

In a generalization $Q(H, R, S; x, y)$ of $Q(H, x)$ we split $V(H)$ into two complementary subsets R and S. We then define the coefficient of the product $x^i y^j$ to be the number of enumerations of $V(G)$ with exactly i dominant vertices in R and exactly j in S.

Going back to G and T we can define a graph H and a partition (R, S) of $V(H)$ as follows. The vertices of H are the edges of G. Two vertices of H are joined by an edge if and only if each covers the other with respect to T in G. The set S is made up of the edges of T, and R of the edges of G not in T. Evidently

$$(18) \qquad J(G, T; x, y) = Q(H, S, T; x, y).$$

In this special case H is bipartite.

It is natural to ask if $Q(H, x)$ and $Q(H, S, T; x, y)$ each have a property analogous to that once conjectured for $\chi(G; x, y)$, as explained in Section 6. But these polynomials have been calculated for too few graphs to justify a formal conjecture.

In Reconstruction Theory a "vertex-deleted" subgraph of H is a subgraph obtained by deleting one vertex and its incident edges. Thus there are as many vertex-deleted subgraphs of H as H has vertices. It is easy to show that for a connected graph H having at least two vertices the polynomial $Q(H, x)$ is the sum of the polynomials $Q(K, x)$ over all the vertex-deleted subgraphs of H. Thus the polynomial is trivially reconstructible.

References.

[1] R. L. Brooks, C. A. B. Smith, A. H. Stone and W. T. Tutte, The dissection of rectangles into squares, Duke Math. J., 7 (1940), 312-340.

[2] W. T. Tutte, A contribution to the theory of chromatic polynomials, Can. J. Math., 6 (1954), 80-91.

[3] Hassler Whitney, The coloring of graphs, Ann. Math., 33 (1932), 688-718.

DEPARTMENT OF COMBINATORICS AND OPTIMIZATION, UNIVERSITY OF WATERLOO, WATERLOO, ONTARIO N2L 3G1, CANADA

Tutte Invariants for 2-Polymatroids

JAMES OXLEY and GEOFF WHITTLE

Introduction

This paper describes a theory of Tutte invariants for 2-polymatroids that parallels the corresponding theory for matroids. The paper is a slightly informal exposition of the main results of [13] and contains no proofs. In particular, it shows that 2-polymatroid invariants obeying deletion-contraction recursions arise in the enumeration of many combinatorial structures including matchings and perfect matchings in graphs, colourings in hypergraphs, and common bases in pairs of matroids. The main result is that, just as for matroids, there is a two-variable polynomial that is essentially the universal Tutte invariant for 2-polymatroids.

Section 1 of the paper presents some basic facts about polymatroids. Section 2 summarizes the theory of Tutte-Grothendieck invariants for matroids which we are seeking to generalize, and Section 3 describes this generalization. The graph and matroid terminology used throughout will follow Bondy and Murty [1] and Oxley [11], respectively.

1. Polymatroids

We begin with an example. Let M be the rank-3 matroid that is represented geometrically in Figure 1. Now pick some set of flats of M, say the lines that are labelled 1,2,3, and 4 and the points labelled 5,6,7,8, and 9. Let $E = \{1, 2, \ldots, 9\}$ and, for each subset X of E, let $f(X)$ be the rank in M of the union of the flats that are labelled by members of X. So, for example,

---

1991 Mathematics Subject Classification. Primary 05B35.

The authors' research was partially supported by grants from the Louisiana Education Quality Support Fund through the Board of Regents and from the Commonwealth of Australia through the Australian Research Council.

This paper is a summary of the main results of [13], which has been submitted for publication elsewhere.

$$f(\{1\}) = f(\{2\}) = f(\{3\}) = f(\{4\}) = 2,$$

$$f(\{5\}) = f(\{6\}) = f(\{7\}) = f(\{8\}) = f(\{9\}) = 1,$$

$$f(\{1,5\}) = f(\{1,5,6\}) = 2, f(\{1,2\}) = 3, \text{ and so on.}$$

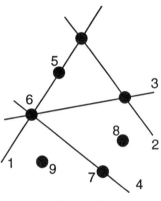

Figure 1

Then the pair (E, f) is an example of a polymatroid.

Next consider slightly modifying this example by allowing the set E to be a multiset of flats of M. This amounts geometrically to adding flats in parallel to existing flats as shown, for instance, in Figure 2. In that case, $E = \{1, 1', 2, 3, 3', 4, 5, 5', 5'', 6, 7, 7', 8, 9\}$ and, for example, $f(\{1, 1'\}) = 2$,

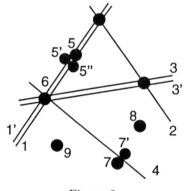

Figure 2

$f(\{1, 5, 5'\}) = 2$, and so on. Again, (E, f) is an example of a polymatroid.

Formally, a *polymatroid* (E, f) consists of a finite set E and a function $f : 2^E \to \mathbb{Z}$ such that

(i) $f(\emptyset) = 0$;

(ii) if $X \subseteq Y$, then $f(X) \leq f(Y)$; and

(iii) if $X, Y, \subseteq E$, then $f(X) + f(Y) \geq f(X \cup Y) + f(X \cap Y)$.

But we find it easier to think of polymatroids geometrically as in the above examples. Indeed, every polymatroid arises as a multiset of flats of some matroid in the manner described there [7, 8, 10].

This paper will focus on 2-polymatroids, where, for a positive integer k, a *k-polymatroid* is a polymatroid (E, f) such that $f(\{e\}) \leq k$ for all e in E. Thus a 1-polymatroid is just the rank function of a matroid, and both of the examples looked at earlier are 2-polymatroids. Geometrically, every 2-polymatroid can be viewed as consisting of a multiset of lines, points, and loops of some matroid.

Two well-known classes of 2-polymatroids will receive the most attention here. The members of the first class arise from graphs in the following way. Let G be a graph having edge set E and, for all $X \subseteq E$, let $f_G(X)$ be $|V(X)|$, the number of vertices of G that are incident with some edge in X. It is not difficult to check that (E, f_G) is indeed a 2-polymatroid. Comparing this 2-polymatroid with the more familiar cycle matroid $M(G)$ of G, we note that the rank of X in $M(G)$ is $|V(X)| - k(X)$ where $k(X)$ is the number of components of the induced graph $G[X]$. Moreover, unlike $M(G)$, the 2-polymatroid (E, f_G) uniquely determines G up to the possible presence of isolated vertices.

Our second fundamental class of examples of 2-polymatroids arises from matroids. Let M_1 and M_2 be matroids on a common ground set E and, for all $X \subseteq E$, let $f(X) = r_1(X) + r_2(X)$ where r_i is the rank function of M_i. Since each of (E, r_1) and (E, r_2) is a 1-polymatroid, it is easy to show that $(E, r_1 + r_2)$ is a 2-polymatroid.

2. Tutte-Grothendieck invariants for matroids

Much of the motivation for our results on 2-polymatroid isomorphism invariants derives from the well-established theory of Tutte-Grothendieck invariants for matroids. This theory, which grew out of work of Tutte [14] for graphs, is reviewed in detail in [3]. We now briefly summarize some of the relevant aspects of the theory.

Let $\mathfrak{M}$ be a class of matroids that is closed under isomorphism and the taking of minors. A function ψ on $\mathfrak{M}$ taking values in a field K is an *isomorphism invariant* if $\psi(M) = \psi(N)$ whenever $M \cong N$.

Several numbers that one can associate with a matroid M such as its number of bases, its number of independent sets, and its number of spanning sets obey the following two basic recursions:

(2.1) $\psi(M) = \psi(M \backslash e)\psi(M|\{e\})$ *if e is a separator (a loop or coloop) of M; and*

(2.2) *for some fixed non-zero members σ and τ of K,*

$$\psi(M) = \sigma\psi(M\backslash e) + \tau\psi(M/e)$$

if e is not a separator of M.

An isomorphism invariant on $\mathfrak{M}$ that obeys (2.1) and (2.2) is called a *generalized T-G invariant* on $\mathfrak{M}$. There are many well-known important examples of such invariants; for instance, the chromatic polynomial is a generalized T-G invariant on the class of graphic matroids as is the flow polynomial. One particularly attractive feature of these invariants is that they are all evaluations of a certain universal invariant. To state this result formally, we shall need another definition. For a matroid M having ground set E and rank function r, the *(matroid) rank generating function* is given by

(2.3) $s(M; u, v) = \sum_{X \subseteq E} u^{r(E)-r(X)} v^{|X|-r(X)}$,

or, equivalently,

(2.4) $s(M; u, v) = \sum_{X \subseteq E} u^{r(E)-r(X)} v^{r^*(E)-r^*(E-X)}$.

It is not difficult to show that this function is a generalized T-G invariant with $\sigma = \tau = 1$. Moreover, for the two single-element matroids, $U_{1,1}$ and $U_{0,1}$, which consist of a single coloop and a single loop, respectively,

$$s(U_{1,1}; u, v) = u + 1 \text{ and } s(U_{0,1}; u, v) = v + 1.$$

These matroids are distinguished here because they are the only irreducible matroids with respect to the operations in (2.1) and (2.2).

Extending a result of Brylawski [2], Oxley and Welsh [12] proved that every generalized T-G invariant is easily expressible in terms of the rank-generating function:

(2.5) THEOREM. *Let $\mathfrak{M}$ be a class of matroids that is closed under isomorphism and the taking of minors. If ψ is a generalized T-G invariant from $\mathfrak{M}$ into a field K such that $\psi(U_{1,1}) = x$ and $\psi(U_{0,1}) = y$, then, for all M in $\mathfrak{M}$,*

$$\psi(M) = \sigma^{|E(M)|-r(M)} \tau^{r(M)} s(M; \frac{x}{\tau} - 1, \frac{y}{\sigma} - 1).$$

This result is more usually stated in terms of the Tutte polynomial $t(M; x, y)$, where

$$t(M; x, y) = s(M; x - 1, y - 1).$$

However, the above form of the result extends more naturally to 2-polymatroids. Some well-known basic evaluations of the rank generating function are as follows:

(2.6) $s(M; 1, 0)$ is the number of independent sets of M;

(2.7) $s(M;0,0)$ is the number of bases of M; and

(2.8) $s(M;0,1)$ is the number of spanning sets of M.

3. Isomorphism invariants for 2-polymatroids

Our approach to developing a theory of isomorphism invariants for 2-polymatroids will be to try to mimic the corresponding theory for matroids. But there are several potential problems that one needs to solve.

Firstly, what does it mean for an element e to be a separator in a 2-polymatroid (E,f)? Here we follow Cunningham [4] and define e to be a *separator* if $f(e)+f(E-e)=f(E)$. It should be noted that whereas the separators in a matroid are of just two types, loops and points, those in a 2-polymatroid are of three types: loops, points, and lines.

Next we need to define deletion and contraction in a 2-polymatroid (E,f). Deletion is straightforward; we define it in terms of restriction of f: if $A \subseteq E$ and $X \subseteq E - A$, then

$$(f \backslash A)(X) = f(X).$$

For contraction, we again look to matroids. If r_M is the rank function of a matroid M on E and $A \subseteq E$, then the rank function of M/A is defined by

$$r_{M/A}(X) = r_M(X \cup A) - r_M(A)$$

for all $X \subseteq E - A$. This suggests defining contraction in a 2-polymatroid (E,f) analogously, that is,

$$(f/A)(X) = f(X \cup A) - f(A)$$

for all $X \subseteq E - A$ [6]. It is not difficult to show that $(E-A, f/A)$ is indeed a 2-polymatroid. Moreover, this definition of contraction is consistent with the matroid definition in another sense. If the polymatroid f is represented as a multiset E of flats of a matroid M and $A \subseteq E$, then f/A has a natural representation as a multiset of flats of $M/(\cup_{a \in A} a)$.

We now have analogues for 2-polymatroids of two of the three fundamental matroid operations of deletion, contraction, and the taking of duals. The basic link between these operations in the matroid case is

(3.1) $$M^* \backslash e = (M/e)^*.$$

Whittle [15] proposed that duality for 2-polymatroids should be an idempotent operation for which the analogue of (3.1) always holds. Moreover, he showed that if this occurs, that is, if, for all 2-polymatroids (E,f) and all e in E, $(f^*)^* = f$ and $f^* \backslash e = (f/e)^*$, then the dual (E, f^*) of (E, f) must be defined by

(3.2) $$f^*(X) = 2|X| + f(E - X) - f(E)$$

for all $X \subseteq E$. The last equation should be compared with the formula for the usual dual of a matroid rank function r, which is given by

$$r^*(X) = |X| + r(E - X) - r(E).$$

Next we consider the elements in a 2-polymatroid that are not separators. Such elements obey one of the following three conditions, where the dual f^* of f is its 2-polymatroid dual:

(i) $f(E - e) = f(E)$ and $f^*(E - e) = f^*(E) - 1$;

(ii) $f(E - e) = f(E) - 1$ and $f^*(E - e) = f^*(E)$; and

(iii) $f(E - e) = f(E)$ and $f^*(E - e) = f^*(E)$.

Note that if f is a matroid rank function and f^* denotes the rank function of the dual matroid, then those elements obeying (i) and (ii) above are precisely the loops and coloops, respectively, of the matroid. Conditions (i), (ii), and (iii) are equivalent to (i)', (ii)', and (iii)', respectively, where (i)' – (iii)' are as follows:

(i)' $f(E - e) = f(E)$ and $f(e) = 1$;

(ii)' $f(E - e) = f(E) - 1$ and $f(e) = 2$; and

(iii)' $f(E - e) = f(E)$ and $f(e) = 2$.

In view of the existence of these three different types of nonseparator elements in 2-polymatroids, the definition of a generalized Tutte invariant for 2-polymatroids, which we shall give next, will allow three distinct variants on the fundamental deletion-contraction recursion.

Let $\mathfrak{N}$ be a class of 2-polymatroids that is closed under isomorphism, deletion, and contraction. Assume that $\mathfrak{N}$ contains $U_{0,1}, U_{1,1}$, and $U_{2,1}$, the single-element 2-polymatroids of ranks zero, one, and two which correspond to a loop, a point, and a line. An isomorphism invariant ψ on $\mathfrak{N}$ is a *generalized Tutte invariant* for $\mathfrak{N}$ if, whenever f is a member of $\mathfrak{N}$ having ground set E and $e \in E$, $\psi(f) \in \mathbb{C}[x, y, z, a, b, c, d, m, n]$ where

(3.3) $$\psi(U_{2,1}) = x, \quad \psi(U_{0,1}) = y, \quad \text{and} \quad \psi(U_{1,1}) = z;$$

(3.4) $\psi(f) = \psi(f \backslash (E - e))\psi(f \backslash e)$ if e is a separator of f;

and

(3.5)
$$\psi(f) = \begin{cases} a\psi(f \backslash e) + b\psi(f/e) & \text{if } f(E - e) = f(E) \text{ and } f(e) = 1; \\ c\psi(f \backslash e) + d\psi(f/e) & \text{if } f(E - e) = f(E) - 1 \text{ and } f(e) = 2; \\ m\psi(f \backslash e) + n\psi(f/e) & \text{if } f(E - e) = f(E) \text{ and } f(e) = 2. \end{cases}$$

An important example of such an invariant is the 2-*polymatroid rank generating function* which is defined as follows:

(3.6) $$S(f;u,v) = \sum_{X \subseteq E} u^{f(E)-f(X)} v^{2|X|-f(X)}$$

or, equivalently,

(3.7) $$S(f;u,v) = \sum_{X \subseteq E} u^{f(E)-f(X)} v^{f^*(E)-f^*(E-X)}.$$

Indeed, it is not difficult to check that, $S(f;u,v)$ is a generalized Tutte invariant on the class of all 2-polymatroids having

$x = 1 + u^2$, $y = 1 + v^2$, $z = u + v$, $m = n = 1$, $a = d = 1$, $b = v$, and $c = u$.

The reader should note the similarity between (2.4) and (3.7), the difference being that, in the first, the duality is matroid duality while, in the second, it is 2-polymatroid duality. One striking difference occurs here between $S(f;u,v)$ and an arbitrary generalized T-G invariant for matroids. For the latter, the four parameters, x, y, σ, and τ are independent. But, for the former the nine parameters, x, y, z, m, n, a, b, c, and d, are far from being independent. A natural question here is whether some such dependence is forced for all generalized Tutte invariants for 2-polymatroids. Our main result will answer this. Before presenting it, we look at certain interesting evaluations of $S(f;u,v)$ for the two special classes of 2-polymatroids distinguished earlier. We begin with the analogues of (2.6)–(2.8).

Recall that, for a graph G, $f_G(X) = |V(X)|$ for all $X \subseteq E(G)$. It is not difficult to see that

(3.8) $S(f_G; 1, 0)$ is the number of matchings of G.

Moreover, if G has no isolated vertices, then

(3.9) $S(f_G; 0, 0)$ is the number of perfect matchings of G; and

(3.10) $S(f_G; 0, 1)$ is the number of subsets of $E(G)$ that cover $V(G)$.

Now suppose that r_1 and r_2 are the rank functions of matroids M_1 and M_2 on E. Then

(3.11) $S(r_1 + r_2; 1, 0)$ is the number of common independent sets of M_1 and M_2;

(3.12) $S(r_1 + r_2; 0, 0)$ is the number of common bases of M_1 and M_2; and

(3.13) $S(r_1 + r_2; 0, 1)$ is the number of common spanning sets of M_1 and M_2.

Generalizing the above, we note that, for an arbitrary 2-polymatroid (E, f),

(3.14) $S(f; 1, 0)$ is the number of matchings of (E, f), and

(3.15) $S(f; 0, 1)$ is the number of spanning sets of (E, f),

where a *matching* is a set X such that $f(X) = 2|X|$, while a *spanning set* is a set Y for which $f(Y) = f(E)$.

The 2-polymatroid rank generating function is closely related to the matroid rank generating function. Indeed, if $s(f; u, v)$ is defined for a 2-polymatroid f by simply replacing r by f in (2.3), then

(3.16) $S(f; u, v) = v^{f(E)} s(f; uv^{-1}, v^2)$ provided $v \neq 0$.

The last observation may suggest that $S(f; u, v)$ contains little more information than $s(f; u, v)$. In practice, however, several of the more interesting evaluations of $S(f; u, v)$ arise when $v = 0$. For instance, if G is a graph, then

(3.17) $u^{f_G(E)/2} S\left(f_G; u^{-1/2}, 0\right)$ is the matching generating polynomial $\sum_{k \geq 0} m_k u^k$ of G where m_k is the number of k-edge matchings of G.

If the graph G has n isolated vertices and $i = \sqrt{-1}$, then

(3.18) $u^n \, i^{f_G(E)} S\left(f_G; -iu, 0\right)$ is the matching defect polynomial of G (Lovász and Plummer [9]).

Among the interesting properties of $S(f; u, v)$ that are easily proved are the following:

(3.19) $S(f^*; u, v) = S(f; v, u)$;

(3.20) $S(f; 1, 1) = 2^{|E|}$;

(3.21) $S(f; -u, -v) = (-1)^{f(E)} S(f; u, v)$; and

(3.22) $S(f; \frac{1}{u}, u) = (1 + u^2)^{|E|} u^{-f(E)}$ provided $u \neq 0$.

The matching generating and matching defect polynomials are just two examples of several single-variable polynomials that arise as special cases of $S(f; u, v)$. For example, if G is a graph without isolated vertices and $\omega(G)$ is a random subgraph of G obtained by independently deleting each edge of G with probability $1 - p$, then

(3.23) $(1-p)^{|E|-f_G(E)/2} p^{f_G(E)/2} S\left(f_G; 0, p^{1/2}(1-p)^{-1/2}\right)$ is the probability that $\omega(G)$ has no isolated vertices.

Another evaluation of $S(f; u, v)$ is the stability polynomial $A(G; p)$ of a graph G, a polynomial that has been studied by a number of authors (see, for example, Farr [5]). For G having no isolated vertices, $A(G; p)$ is defined as follows. Suppose that the vertices of G are chosen independently, each with probability p. Then $A(G; p)$ is the probability that the chosen set of vertices is stable. Farr showed that $A(G; p) = \sum_{X \subseteq E} (-1)^{|X|} p^{f_G(X)}$. Hence

(3.24) $A(G;p) = (ip)^{f_G(E)} S(f_G; -ip^{-1}, i)$.

Finally we note that $S(f; u, v)$ has several important applications to hypergraphs. For instance,

(3.25) $i^{f(E)} S(f; -i\lambda, i) = \sum_{X \subseteq E} (-1)^{|X|} \lambda^{f(E)-f(X)} = P(f; \lambda)$ where $P(f; \lambda)$
is the characteristic polynomial of f (Helgason [7]), which enumerates colourings of a hypergraph in the same way that the chromatic polynomial enumerates colourings in a graph.

Evidently 2-polymatroid rank generating functions arise in a wide variety of contexts. The next theorem, the main result of the paper, indicates why these functions are so pervasive by noting that $S(f; u, v)$ is essentially the universal Tutte invariant for 2-polymatroids.

(3.26) THEOREM. *Let ψ be a generalized Tutte invariant on the class of all 2-polymatroids and suppose that at most two of x, y, z, a, b, c, d, m, and n are identically zero. Then one of the following occurs:*
  (i) $a = m; d = n; mx = mn + c^2; ny = mn + b^2; z = b + c; m \neq 0; n \neq 0;$ *and for all 2-polymatroids f,* $\psi(f) = m^{|E|-f(E)/2} S\left(f; \frac{c}{(mn)^{1/2}}, \frac{b}{(mn)^{1/2}}\right);$
  (ii) $z^2 = xy = ax + bz = cz + dy = mx + ny;$ $yz = az + by;$ $xz = cx + dz;$ *and, for all 2-polymatroids f, $\psi(f) = Q(f)$ where*

$$Q(f) = \begin{cases} y^{|E|-f(E)} z^{f(E)} & \text{if } f(E) \leq |E|; \\ x^{f(E)-|E|} z^{f^*(E)} & \text{if } f(E) \geq |E|. \end{cases}$$

Of the two functions arising here, $S(f; u, v)$ is, by now, quite familiar. The other function Q is basically trivial. The only information it conveys about (E, f) is the cardinality of E and the value of $f(E)$. Thus, in the 2-polymatroid case, just as in the matroid case, there is essentially a unique universal Tutte invariant.

The proof of the theorem involves looking at a number of small 2-polymatroids. For each of these, one evaluates ψ in two different ways. For instance, if (E, f) is represented geometrically by a single point on a line, then, on deleting and contracting the line, we get

$$\psi(\phi) = c\psi(\bullet) + d\psi(U_{0,1})$$
$$= cz + dy.$$

On the other hand, deleting and contracting the point gives

$$\psi(\phi) = a\psi(/) + b\psi(\bullet)$$
$$= ax + bz.$$

Thus, for ψ to be well-defined,

$$cz + dy = ax + bz.$$

By looking at several other examples, one obtains a number of other relations between the nine variables involved. A detailed case analysis of these relations leads eventually to the result. In fact, one can drop the restriction that at most two of $x, y, z, a, b, c, d, m,$ and n are zero for, in so doing, one merely admits six more trivial invariants each of which is a monomial conveying very limited information.

References

1. J. A. Bondy and U. S. R. Murty, "Graph Theory with Applications", Macmillan, London; Elsevier, New York, 1976.

2. T. H. Brylawski, A decomposition for combinatorial geometries, *Trans. Amer. Math. Soc.* **171** (1972), 235-282.

3. T. H. Brylawski and J. G. Oxley, The Tutte polynomial and its applications, in "Matroid Applications" (N. White, Ed.), pp. 123-225, Cambridge University Press, Cambridge, 1991.

4. W. H. Cunningham, Decomposition of submodular functions, *Combinatorica* **3** (1983), 53-68.

5. G. E. Farr, A correlation inequality involving stable sets and chromatic polynomials, submitted.

6. S. Fujishige, Lexicographically optimal base of a polymatroid with respect to a weight vector, *Math. Oper. Res.* **5** (1980), 186-196.

7. T. Helgason, Aspects of the theory of hypermatroids, in "Hypergraph Seminar" (C. Berge and D. K. Ray-Chaudhuri, Eds.), pp. 191-214, Lecture Notes in Math. 411, Springer-Verlag, Berlin, 1974.

8. L. Lovász, Flats in matroids and geometric graphs, in "Combinatorial Surveys", (P. Cameron, Ed.), pp. 45-86, Academic Press, London, 1977.

9. L. Lovász and M. D. Plummer, "Matching Theory", Ann. Discrete Math. 29, North-Holland, Amsterdam, 1986.

10. C. J. H. McDiarmid, Rado's Theorem for polymatroids, *Math. Proc. Camb. Phil. Soc.* **78** (1975), 263-281.

11. J. G. Oxley, "Matroid Theory", Oxford University Press, Oxford, to appear.

12. J. G. Oxley and D. J. A. Welsh, The Tutte polynomial and percolation, in "Graph Theory and Related Topics" (J. A. Bondy and U. S. R. Murty, Eds.), pp. 329-339, Academic Press, New York, 1979.

13. J. G. Oxley and G. P. Whittle, A characterization of Tutte invariants of 2-polymatroids, submitted.

14. W. T. Tutte, A ring in graph theory, *Proc. Camb. Phil. Soc.* **43** (1947), 26-40.

15. G. P. Whittle, Duality in polymatroids and set functions, submitted.

MATHEMATICS DEPARTMENT, LOUISIANA STATE UNIVERSITY, BATON ROUGE, LA 70803, USA
E-mail address: oxley @ marais.math.lsu.edu

MATHEMATICS DEPARTMENT, UNIVERSITY OF TASMANIA, HOBART, TASMANIA, 7001, AUSTRALIA E-mail address: whittle @ hilbert.maths.utas.edu.au

Extremal matroid theory

JOSEPH P. S. KUNG

1. INTRODUCTION

As matroid theory is the common generalization of graph theory and projective geometry, one can, in an expansionist mood, classify every extremal problem in either area under extremal matroid theory. But, more plausibly, the reverse is the case: extremal matroid theory is motivated by and derives many – but not all – of its problems and methods from graphical and geometric extremal problems. In this survey, we shall begin by discussing several classical extremal theorems and problems connected with matroids. Next, in §3, we shall present results about excluding submatroids. In §§4, 5, and 6, results about excluding minors will be discussed. Finally, in §7, we shall discuss the matroid version of the direction problem in real and complex space. It goes without saying that any survey reflects the philosophy and research work of its author. This survey concentrates on what I personally find interesting.

2. EXTREMAL PROBLEMS

2.1. Extremal graph theory

We begin with two well-known theorems. The first is a corollary of Euler's formula for polyhedra and is perhaps the oldest result in extremal combinatorics. Let $v(\Gamma)$ be the number of vertices and $e(\Gamma)$ the number of edges in a graph Γ.

(2.1) THEOREM. *Let Γ be a simple planar graph with at least 3 vertices. Then*

$$e(\Gamma) \leq 3v(\Gamma) - 6.$$

The graphs attaining this bound are the planar triangulations.

1991 *Mathematics Subject Classification.* Primary 05B35; Secondary 05B20, 05C35, 05D99, 06C10, 51M04.

The author was supported by the National Security Agency under Grant MDA904-91-0030. The following information is included at the request of *Mathematical Reviews*: This is a survey paper with several new results. These include bounds for binary matroids with no k-wheel-minors (§5.4) and a proof of the polynomial-exponential gap for size functions of minor-closed classes (§6.1). This paper is in its final form and will not be published elsewhere.

(2.2) DIRAC'S THEOREM [**25**]. *Let Γ be a simple graph with at least 2 vertices not containing the complete graph K_4 as a minor (or subcontraction). Then*

$$e(\Gamma) \leq 2v(\Gamma) - 3.$$

The bounds on $e(\Gamma)$ are both *linear* in $v(\Gamma)$. This is not an accident and is due to the fact that the two classes of graphs under consideration are closed under minors.

(2.3) THEOREM. *Let m be an integer greater than 2. There exists a constant c_m such that for any simple graph Γ with no K_m-minor,*

$$e(\Gamma) < c_m v(\Gamma).$$

This theorem was first proved by Mader in [**60**] with $c_m = 2^{m-3}$. Mader's proof is one of the sources of the method of cones (see §6) used in extremal matroid theory. His idea is to do induction on m using the neighborhood graph of a vertex. The case $m = 3$ is easy. Now let $m \geq 4$. Let u be a vertex of a minor-minimal graph Γ satisfying $e(\Gamma) \geq 2^{m-3}v(\Gamma)$. By minimality, there are more than 2^{m-3} edges incident on u and every such edge is on at least 2^{m-3} graphical triangles. Hence, the graph Γ_u induced by the vertices adjacent but not equal to u has minimum degree at least 2^{m-3} and hence, satisfies $e(\Gamma_u) \geq 2^{m-4}v(\Gamma_u)$. By induction, Γ_u contains a K_{m-1}-minor Δ. The minor induced by u and Δ is a K_m-minor in Γ.

The bound given in Mader's proof is not the best possible. Using random graphs, Thomason [**78**] showed that up to $o(1)$,

$$0.265 m \sqrt{\log_2 m} \leq c_m \leq 2.68 m \sqrt{\log_2 m}.$$

(2.3) implies that if one excludes any graph as a minor, then there is a drastic drop in the maximum number of edges, from $\binom{v(\Gamma)}{2}$, quadratic in $v(\Gamma)$, to a number bounded by a linear function in $v(\Gamma)$.

When subgraphs are excluded, the situation is qualitatively different: the maximum number of edges is quadratic in $v(\Gamma)$, with some degenerate exceptions. This is illustrated by one of the best-known results in extremal graph theory, Turán's theorem [**82**].

(2.4) TURÁN'S THEOREM. *Let Γ be a simple graph on v vertices not containing K_m as a subgraph. Then Γ contains at most $t_{m-1}(v)$ edges, where $t_{m-1}(v)$ is the number of edges in the complete $(m-1)$-partite graph $T_{m-1}(v)$ on v vertices with $\lfloor (v+k-1)/(m-1) \rfloor$ vertices in the kth class. In addition, $T_{m-1}(v)$ is the unique graph on v vertices with no K_m-subgraph having $t_{m-1}(v)$ edges.*

Surveys of extremal graph theory can be found in [**13,14,15**].

2.2. Matrices

A matrix M with integer entries is said to be *totally unimodular* if all its subdeterminants equal -1, 0, or 1. Totally unimodular matrices arose in algebraic topology and linear programming. See [**9**] and [**73**, p. 299].

(2.5) HELLER'S THEOREM [**36**]. *Let M be an $n \times c$ totally unimodular matrix with no repeated columns. Then*

$$c \leq 2\binom{n}{2} + 2n + 1.$$

A totally unimodular matrix M with n rows has $2\binom{n}{2} + 2n + 1$ distinct columns if and only if M is the matrix made up from the zero column, the $n \times n$ identity matrix and its negative, and the $n \times 2\binom{n}{2}$ vertex-edge incidence matrix of the complete directed graph on n vertices.

Besides putting restrictions on subdeterminants, one can also exclude certain submatrices. The first theorem of this type is due to Sauer [**71**] and Shelah [**76**]. They proved the case $q = 2$ of the following theorem due to Anstee and Murty [**8**]. Let $S_m(q)$ be the $m \times q^m$ matrix consisting of all possible columns on m rows with entries from the set $\{0, 1, 2, \cdots, q-1\}$.

(2.6) THEOREM. *Let M be a $n \times c$ matrix with entries from $\{0, 1, 2, \cdots, q-1\}$ and no repeated columns. Suppose that no submatrix of M is a row or column permutation of $S_m(q)$. Then*

$$c \leq \sum_{i=n-m+1}^{n} \binom{n}{i} (q-1)^i.$$

The bound in (2.6) is sharp but a characterization of the matrices with the maximum number of columns is not known. Because every matrix with m rows, no repeated columns, and entries in $\{0, 1, \cdots, q-1\}$ can be obtained from a submatrix of $S_m(q)$ by row or column permutations, (2.6) yields a bound (not necessarily sharp, of course) for matrices not containing any matrix in a given set of matrices as a submatrix. For further work on excluding submatrices, see the papers [**2,3,4,5,7,8**].

When $q = 2$ and $m = 3$, the bounds in (2.5) and (2.6) are identical. This is not accidental because one can prove Heller's theorem from (2.6) [**12**].

2.3. Directions

In his 1970 note [**72**], Scott proposed the problem of finding the minimum number of directions or slopes determined by a finite set or *configuration* of points in the real Euclidean plane $\mathbb{R}^2$. This problem was solved by Ungar [**87**]. (See also [**21,22,44**].)

(2.7) UNGAR'S THEOREM. *Let S be a configuration of n points in $\mathbb{R}^2$. Then S determines at least n directions if n is even and $n - 1$ directions if n is odd.*

The *direction-critical* configurations for which equality occurs have not been determined. There are four known infinite families for odd n, one of which consists of sets of vertices of a regular $(n-1)$-gon together with its center. See [**42,43,44**].

2.4. Matroids and their minors

We shall assume familiarity with the basic ideas of matroid theory [24,51,70, 91,95,96,98]. Because our interest is in counting the number of points (or rank-1 flats), we shall use the word "matroid" to mean "simple matroid" or "geometry". The *size* $|G|$ of a matroid G is the number of points in G. We shall use G to denote both the matroid and its set of points.

The *contraction* G/U of a matroid G by a set U of points is defined to be the (simple) matroid induced on the flats covering the closure $\bar{U}$ of U by the upper interval $[\bar{U}, \hat{1}]$ in the lattice $L(G)$ of flats of G. In particular, if a is a point, G/a is a matroid on the set of lines (or rank-2 flats) containing a with a lattice of flats isomorphic to the $[a, \hat{1}]$. Thus, G/U is the simplification of the contraction as it is usually defined and $G/U = G/\bar{U}$. If Γ is a simple graph and a an edge in the graph, then the contraction, as defined here, is the graphic matroid $M(\Gamma/a)$, where Γ/a is the graph obtained from Γ by contracting a and removing loops and all but one edge from each set of multiple edges.

When a matroid G is represented by the column vectors of a matrix M_G over a field, contraction by a point a has the following explicit interpretation. Take the column c_a representing a. By row operations on M_G, reduce c_a so that all but one of its entries, say, the entry at row u, are zero. Now delete row u and column c_a. The resulting matrix represents a matroid whose simplification is isomorphic to G/a.[1] A representation for G/a can be obtained by removing all but one column from each set of projectively equivalent columns.

A *minor* of a matroid G is a matroid which can be obtained from G by a sequence of deletions or contractions of points.

We shall use the standard designations for frequently occurring matroids. In particular, $U_{r,s}$ is the uniform matroid with s points and rank r, $U_{2,s}$ is the s-point line, $PG(n,q)$ is the rank-$(n+1)$ projective geometry of dimension n over the finite field $GF(q)$, $AG(n,q)$ is the rank-$(n+1)$ affine geometry of dimension n over $GF(q)$ [obtained by deleting a hyperplane from $PG(n,q)$], F_7 is the Fano plane, F_7^- is the non-Fano plane, and R_{10} is the regular rank-5 matroid on 10 points that is neither graphic nor cographic [10,75]. In addition, we shall denote the (orthogonal) dual of a matroid G by $G^\perp$. The *nullity* of G is the rank of its dual $G^\perp$. The *corank* of a flat U in G is the difference $\text{rank}(G) - \text{rank}(U)$.

2.5. Extremal matroid theory

Given the examples in §§2.1 and 2.2, it is reasonable to say that extremal matroid theory is concerned with the following question:

Let $\mathcal{C}$ be a class of matroids satisfying given properties. Determine the size function

$$h(\mathcal{C};n) = \max\{|G| : G \in \mathcal{C} \text{ and } \text{rank}(G) = n\}$$

and characterize the matroids of maximum size.

---
[1] If a row is simply deleted, then one obtains a "quotient" of G.

Of course, we may have to settle for upper or lower bounds for $h(\mathcal{C}; n)$. The rank is the correct parameter because for a connected graph Γ on v vertices, the rank of the cycle matroid $M(\Gamma)$ equals $v - 1$. In addition, the rank of a matrix is at most the number of rows. Because most matrices of maximum size contain a full identity submatrix, the rank is a good substitute for the number of rows.

Historically, the first theorem in extremal matroid theory is Heller's theorem published in 1957. At about the same time, Tutte [84] showed that a matroid is regular (that is, representable over every field) if and only if it can be represented over the integers by a totally unimodular matrix. Hence, Heller's theorem says:[2]

$$h(\mathcal{R}; n) = \binom{n+1}{2},$$

where $\mathcal{R}$ is the class of regular matroids. Heller's work was not followed up until the 1970's, when Murty [61] and Baclawski and White [9] extended it (see (6.4)).

In 1980, I was led, while considering characteristic sets [38], to ask the question: What is the maximum size of a matroid representable over $GF(p)$ and $GF(q)$, where p and q are different primes? Heller's theorem indicated that it should be quadratic, and this in fact is the case (see §6.4). Somewhat earlier, Seymour's decomposition theorem [75] showed another direction in which Heller's theorem can be extended. Much of this survey will be concerned with work done since then.

Besides the class of regular matroids, other important classes of matroids are:

$\mathcal{L}(q)$, the class of $GF(q)$-representable matroids,

$\mathcal{G}$, the class of graphic matroids, and

$\mathcal{G}^\perp$, the class of cographic matroids.

Their size functions are given by the following formulas:

$$h(\mathcal{L}(q); n) = \frac{q^n - 1}{q - 1};$$

$$h(\mathcal{G}; n) = \binom{n+1}{2};$$

$$h(\mathcal{G}^\perp; n) = 3n - 3.$$

The first two formulas are elementary. The formula for cographic matroids can be proved[3] by observing that if G is a connected cographic matroid, then there exists a 3-edge-connected graph Γ (possibly having multiple edges) such that G is the dual of the cycle matroid $M(\Gamma)$ of Γ. Because Γ is connected, $M(\Gamma)$ has rank $v - 1$ and nullity $e - v + 1$, where $v = v(\Gamma)$ and $e = e(\Gamma)$. Moreover, every vertex of Γ has degree at least 3 and hence $3v \leq 2e$. We conclude that

$$|G| \leq e \leq e + [2e - 3v] = 3[e - v + 1] - 3 = 3\text{rank}(G) - 3.$$

---

[2]Note that the zero column is never used when representing a simple matroid. Moreover, a column and its negative represent the same projective point. Hence, the bound in Heller's theorem yields the bound of $\binom{n}{2} + n = \binom{n+1}{2}$ for regular matroids.

[3]This proof is implicit in Jaeger [39]; this explicit version of the proof was communicated to me by Lindström (see [49]). Related papers by Jaeger are [40,41].

2.6. The critical problem

The critical problem for matroids was posed by Crapo and Rota [**24**, Chap. 15] as a geometric analogue of coloring and flow problems for graphs. Let G be a set of points in projective space $\mathrm{PG}(m,q)$. The *critical exponent* $c(G;q)$ is the minimum corank of a subspace U in $\mathrm{PG}(m,q)$ such that $U \cap G = \emptyset$. The critical exponent of G can be calculated from the *characteristic polynomial*

$$\chi_G(\lambda) = \sum_{X: X \in L(G)} \mu(\hat{0}, X) \lambda^{\mathrm{rank}(G) - \mathrm{rank}(X)}.$$

Here, μ is the Möbius function of the lattice $L(G)$ of flats of G.

(2.8) THEOREM (CRAPO AND ROTA). *Let G be a set of points in $\mathrm{PG}(m,q)$. Then $c(G;q)$ is the minimum positive integer c such that $\chi_G(q^c) \neq 0$.*

An important consequence of this theorem is that $c(G;q)$ depends only on q and $L(G)$. It does not depend on the representation $G \hookrightarrow \mathrm{PG}(m,q)$; in particular, it is independent of the rank $m+1$ of the ambient projective space. Hence, the critical exponent $c(G;q)$ over $\mathrm{GF}(q)$ can be defined for any matroid G in $\mathcal{L}(q)$ by defining it to be the critical exponent of any one of its $\mathrm{GF}(q)$-representations.

From the definition, it is immediate that if H is a submatroid of G, where G is a (simple) $\mathcal{L}(q)$-matroid, then $c(H;q) \leq c(G;q)$. The critical exponent behaves badly under contraction and duality. For example, the affine geometry $\mathrm{AG}(n,q)$ has critical exponent 1; however, all of its one-point contractions are isomorphic to $\mathrm{PG}(n-1,q)$ and have critical exponent n. In the same vein, $\mathrm{PG}(n,q)$ has critical exponent $n+1$ but its dual $\mathrm{PG}(n,q)^{\perp}$ is affine and has critical exponent 1.[4]

The critical exponent of a set G of points in $\mathrm{PG}(m,q)$ can also be defined to be the minimum number c of hyperplanes $H_1, H_2, \cdots, H_c$ in $\mathrm{PG}(m,q)$ such that for every point a in G, there exists a hyperplane H_i such that $a \notin H_i$. Thus, $c(G;q)$ equals the minimum number c such that G can be partitioned into c affine subsets of points.

A graphic matroid $M(\Gamma)$ is affine as a binary matroid if and only if Γ is bipartite (see [**24**,**91**]). Hence, for a graphic matroid $M(\Gamma)$, the critical exponent over $\mathrm{GF}(2)$ is the minimum number c such that the edge set of Γ can be partitioned into c subsets, $\Delta_1, \Delta_2, \cdots, \Delta_c$, each of which induces a bipartite subgraph. Let (A_i, B_i) be a bipartition of the graph induced by the edge set Δ_i. If v is a vertex of Γ, let $b_i(v) = 0$ if $v \in A_i$ and 1 if $v \in B_i$. Then the assignment $v \mapsto (b_1(v), b_2(v), \cdots, b_c(v))$ is a proper coloring of Γ with colors from $\{0,1\}^c$. We deduce that the chromatic number of a graph Γ is at most $2^{c(M(\Gamma);2)}$. From this, we see that for graphic matroids, the critical exponent over $\mathrm{GF}(2)$ approximates the chromatic number up to the nearest power of 2.

---

[4] To see this, use the fact [**98**] that if G is represented by the column vectors of a $n \times s$ matrix M over $\mathrm{GF}(q)$, then $G^{\perp}$ is represented by the column vectors of any matrix N whose row vectors span the orthogonal complement of the subspace spanned by the row vectors of M in $[\mathrm{GF}(q)]^s$.

2.7. Gain-graphic matroids

Gain-graphic matroids are important examples which come up naturally when studying minor-closed classes of matroids.[5] In this section, we briefly describe these matroids.[6]

The rank-n *Dowling matroid* $Q_n(\mathrm{GF}(q)^\times)$, or briefly, $Q_n(q)$, over the multiplicative group $\mathrm{GF}(q)^\times$ of the finite field $\mathrm{GF}(q)$ is the $\mathcal{L}(q)$-matroid of size $(q-1)\binom{n}{2}+n$ consisting of the following points in projective space $\mathrm{PG}(n-1,q)$:

$$e_1, e_2, \cdots, e_n, \text{ and } e_i + \alpha e_j, \text{ for all pairs } i \neq j \text{ and all } \alpha \in \mathrm{GF}(q)^\times.$$

Here e_i is the *standard basis vector* with the i-coordinate 1 and the other coordinates 0. The points in $Q_n(2)$ are the column vectors in the vertex-edge incidence matrix of the complete graph K_{n+1} over $\mathrm{GF}(2)$ with the last row deleted. Similarly, the points in $Q_n(3)$ can be thought of as column vectors in the incidence matrix of a complete signed graph on n vertices. Thus, $Q_n(q)$ is a "q-analogue" of the cycle matroid $M(K_{n+1})$. It can be represented graphically by the complete graph K_n with "half-edges" at every vertex and $q-1$ multiple edges labelled by $\mathrm{GF}(q)^\times$ between every pair of vertices.

It turns out that because the points in $Q_n(q)$ are linear combinations of one or two basis vectors, the dependencies can be specified graphically without using the additive structure of $\mathrm{GF}(q)$ or the commutativity of multiplication. Hence, one can define the rank-n Dowling matroid $Q_n(A)$ over any group A for all n. Moreover, when $n=3$, the dependencies can be specified without using associativity: hence, $Q_3(A)$ can be defined over any quasi-group or Latin square A. A *gain-graphic matroid over the group A* is a submatroid of $Q_n(A)$ for some n. In particular, the gain-graphic matroids over the group of order 1 are the graphic matroids and those over the (cyclic) group of order 2 are the signed-graphic matroids [**100**]. We shall denote by $\mathcal{Z}(A)$ the class of gain-graphic matroids over the group A. By definition, the rank-n Dowling matroid $Q_n(A)$ is the rank-n matroid of maximum size in $\mathcal{Z}(A)$; hence,

$$h(\mathcal{Z}(A); n) = |A|\binom{n}{2} + n.$$

For our purposes, Dowling matroids have two important properties [**27**]. Firstly, they are *upper homogeneous*: if a is a point in $Q_n(A)$, then $Q_n(A)/a \cong Q_{n-1}(A)$. It follows from this that contractions of gain-graphic matroids over a group A are gain-graphic over A. Secondly, they have the following representability property.

---

[5] Historically, gain graphs arose from two different sources. In response to a problem of Rota to find a "q-analogue" of the partition lattice, Dowling [**27**,**28**] defined lattices of group-labelled partitions which turned out to be the lattices of flats of complete gain-graphic matroids. (See also [**26**].) Zaslavsky [**99**] discovered a connection between signed graphs and arrangements of hyperplanes defined by root systems. These two lines of thought were brought together by Zaslavsky, who defined and studied gain graphs and their generalization, biased graphs, in [**100**,**101**,**102**,**103**] and other papers. Gain graphs were called "voltage graphs" from the time of their discovery until the middle 1980's, when their name was changed to avoid confusion with systems subjected to Kirchhoff's voltage laws.

[6] A brief but complete description, based on Zaslavsky's work, is in [**46**, p. 492].

(2.9) THEOREM (DOWLING). *When $n \geq 3$, the Dowling matroid $Q_n(A)$ over the group A is representable over a skew field $\mathbb{K}$ if and only if A is a subgroup of the multiplicative group of $\mathbb{K}$.*

A *co-gain-graphic* matroid over A is the dual of a matroid in $\mathcal{Z}(A)$. Let $\mathcal{Z}(A)^\perp$ be the class of co-gain-graphic matroid over A. Using the method in §2.5 for computing the size function of the class of cographic matroids, one can show that
$$h(\mathcal{Z}(A)^\perp; n) = 3n - 3.$$
It is noteworthy that the size function of $\mathcal{Z}(A)^\perp$ does not depend on A.

3. EXCLUDING SUBMATROIDS

In this section, we consider results about excluding submatroids. A class $\mathcal{C}$ is said to be *submatroid-closed* if every matroid that is isomorphic to a submatroid of a member of $\mathcal{C}$ is also in $\mathcal{C}$. If $\{M_\alpha\}$ is a collection of matroids, we denote by $\mathcal{EX}_{sub}(M_\alpha)$ the submatroid-closed class of matroids not containing any of the matroids M_α as submatroids. If $\mathcal{C}$ is submatroid-closed, then $\mathcal{C} = \mathcal{EX}_{sub}(M_\alpha)$, where $\{M_\alpha\}$ is the collection of matroids not in $\mathcal{C}$.

3.1. Excluding non-affine matroids

In this section, we consider the question of bounding the number of points in matroids in $\mathcal{L}(q)$ without a given matroid as a submatroid. There are surprisingly tight bounds on the size function of many submatroid-closed classes.

(3.1) LEMMA. *Let $\{M_\alpha\}$ be a collection of matroids in $\mathcal{L}(q)$ and let c be the minimum $\min\{c(M_\alpha; q)\}$ of their critical exponents. Then,*
$$h(\mathcal{EX}_{sub}(M_\alpha) \cap \mathcal{L}(q); n) \geq q^{n-1} + q^{n-2} + \cdots + q^{n-c+1} = \frac{q^n - q^{n-c+1}}{q-1}.$$

PROOF. Let $E_{c-1}(n,q)$ be the rank-n matroid $PG(n-1,q)\backslash H$ obtained by removing a subspace H of codimension $c-1$ from the rank-n projective geometry $PG(n-1,q)$. By construction, $E_{c-1}(n,q)$ has critical exponent $c-1$. Because the critical exponent of a submatroid of $E_{c-1}(n,q)$ is at most $c-1$, $E_{c-1}(n,q)$ does not contain any of the matroids M_α as submatroids. The number of points in $E_{c-1}(n,q)$ equals
$$\frac{q^n-1}{q-1} - \frac{q^{n-c+1}-1}{q-1} = \frac{q^n - q^{n-c+1}}{q-1} = \sum_{l=n-c+1}^{n-1} q^l.$$
This yields the lower bound for the size function. □

Because $PG(m,q)$ has critical exponent $m+1$, (3.1) yields a lower bound of $(q^n - q^{n-c+1})/(q-1)$ for the size function of the class of $\mathcal{L}(q)$-matroids with no $PG(m,q)$-submatroid. This bound is in fact sharp. This is a consequence of the following theorem due to Bose and Burton [16].[7]

---
[7] Bose and Burton used this theorem to show that certain linear codes are unique.

(3.2) THEOREM (BURTON AND BOSE). *Let G be a matroid of rank n in $\mathcal{EX}_{sub}(\mathrm{PG}(m,q)) \cap \mathcal{L}(q)$. Then*

$$|G| \leq \frac{q^n - q^{n-m}}{q-1} = q^{n-1} + q^{n-2} + \cdots + q^{n-m+1}$$

points. $E_m(n,q)$ is the unique rank-n matroid containing $(q^n - q^{n-m})/(q-1)$ points and no $\mathrm{PG}(m,q)$-submatroid.

SKETCH OF PROOF. The main step in the proof is to show that for a set F of points in $\mathrm{PG}(r-1,q)$,

(a) $|F| < |\mathrm{PG}(k-1,q)| = (q^k - 1)/(q-1)$ implies that $c(F;q) \leq k - 1$, and
(b) $|F| = (q^k - 1)/(q-1)$ and $\mathrm{rank}(F) > k$ imply that $c(F;q) \leq k - 1$.

This is done by considering the average number of points in F on a subspace of a given rank.[8] □

The full projective geometries $\mathrm{PG}(m,q)$ are geometric analogues of the complete graphs K_{m+1}. Hence, (3.2) is an analogue in $\mathcal{L}(q)$ of Turán's theorem and the matroids $E_m(n,q)$ are analogues of the m-partite graphs $T_m(n+1)$. We remark that the *density* $|E_m(n;q)|/|\mathrm{PG}(n-1,q)|$ of $E_m(n,q)$ in $\mathrm{PG}(n-1,q)$ has the following approximation:

$$|E_m(n;q)|/|\mathrm{PG}(n-1,q)| = 1 - \frac{q^{n-m} - 1}{q^n - 1} \approx 1 - q^{-m}.$$

This suggests that there should be an extension of (3.2) along the lines of the Erdős-Stone theorem [**31**] in extremal graph theory. (3.1) and (3.2) yield the following corollary.

(3.3) COROLLARY. *Let $\{M_\alpha\}$ be a collection of matroids in $\mathcal{L}(q)$, $c = \min\{c(M_\alpha;q)\}$, and $r = \min\{\mathrm{rank}(M_\alpha)\}$. Then,*

$$\frac{q^n - q^{n-c+1}}{q-1} \leq h(\mathcal{EX}_{sub}(M_\alpha) \cap \mathcal{L}(q); n) \leq \frac{q^n - q^{n-r+1}}{q-1}.$$

In particular, if $m > q$, the class $\mathcal{EX}_{sub}(M(K_m)) \cap \mathcal{L}(q)$ satisfies the bounds given in (3.3) with $c = \lceil \log_q m \rceil$ and $r = m - 1$.

The lower bounds in (3.1) and (3.3) are non-trivial (and exponential) only if $c > 1$, that is, only if all the matroids M_α are non-affine. Not much is known about excluding affine matroids. Note that for any positive integer k, there exists a submatroid-closed class with size function a polynomial in n of degree k. For example, the class of binary matroids which can be represented over $\mathrm{GF}(2)$ with vectors having at most k non-zero coordinates has size function $\binom{n}{k} + \binom{n}{k-1} + \cdots + \binom{n}{0}$. There might be a difference between excluding a finite and an infinite set of affine submatroids.

---

[8]The bound in (3.2) is very similar to the bound in (2.4). (2.4) is about excluding submatrices, that is, submatroids of quotients obtained from a specific representation of a matroid. If the matrix contains a full identity submatrix, then one is excluding a minor from a specific representation of a matroid. Thus, (2.4) yields bounds for matroids not containing minors obtainable using a specific recipe.

3.2. Excluding graphs from regular matroids

One can exclude submatroids from subclasses of $\mathcal{L}(q)$. For example, excluding submatroids from $\mathcal{Z}(A)$ would yield an "extremal gain-graph theory".[9] Another natural subclass is the class $\mathcal{R}$ of regular matroids. Lee [58] obtained the first extremal regular-matroid theorem by extending Turán's theorem for $m = 3$. Roberta Tugger [81] then observed that every excluded-subgraph theorem with a bound quadratic in the number of vertices extends readily to an extremal regular-matroid theorem. To show how this is done, we shall prove the following extension of all the cases of Turán's theorem.

(3.4) TUGGER'S THEOREM. (a) *For $m \geq 4$,*
$$h(\mathcal{EX}_{sub}(M(K_m)) \cap \mathcal{R}; n) = t_{m-1}(n+1).$$

(b)
$$h(\mathcal{EX}_{sub}(M(K_3)) \cap \mathcal{R}; n) = \begin{cases} t_2(n+1) & \text{if } n \neq 5 \\ 10 & \text{if } n = 5. \end{cases}$$

SKETCH OF PROOF. By Seymour's decomposition theorem [75], a regular matroid can be obtained by taking 1-, 2-, or 3-sums of graphic or cographic matroids and copies of the matroid R_{10}. Now observe that

(a) Any cographic matroid of rank n has at most $3n - 3$ points,
(b) R_{10} has rank 5 and 10 points, and
(c) The size function is quadratic: more precisely,
$$t_{r-1}(n+1) \approx \frac{(r-2)}{2(r-1)}(n+1)^2.$$

Hence, it is respected as an upper bound when 1-, 2-, and 3-sums are taken. □

With minor adjustments, (3.4) also holds when $\mathcal{R}$ is replaced by the classes $\mathcal{EX}(F_7) \cap \mathcal{L}(2)$ or $\mathcal{EX}(F_7^\perp) \cap \mathcal{L}(2)$.

4. MINOR-CLOSED CLASSES AND THE GROWTH RATE CONJECTURE

At present, most of the results in extremal matroid theory are about excluding minors.[10] In §§4, 5, and 6, we shall describe all the known results in a unified way.

4.1. Minor-closed classes

A *minor-closed* class $\mathcal{C}$ of matroids is a class satisfying the conditions:

(MC1) If G is a matroid in $\mathcal{C}$ and H is isomorphic to a minor of G, then H is also in $\mathcal{C}$;
(MC2) $\mathcal{C}$ contains a rank-n matroid for every non-negative integer n.

---

[9]This theory is perhaps too similar to extremal graph theory to be worth extensive study.
[10]One can also exclude series minors. This is analogous to excluding topological subgraphs or subdivisions in extremal graph theory. Two papers about series minors are [10,11].

The second condition (MC2) is there simply to avoid unnecessary pedantry when stating theorems; it is equivalent to $\mathcal{C}$ containing all the free matroids $U_{n,n}$ of rank n on n points. If $\{M_\alpha\}$ is a collection of matroids, the class $\mathcal{EX}(M_\alpha)$ of matroids not having any of the matroids M_α as minors satisfies (MC1). Conversely, any class $\mathcal{C}$ satisfying (MC1) has this form: take $\{M_\alpha\}$ to be the collection of all the matroids not in $\mathcal{C}$. The simplest minors to exclude are lines. We denote by $\mathcal{U}(q)$ the class $\mathcal{EX}(U_{2,q+2})$ of matroids with no $(q+2)$-point-line-minor.

Unions and intersections of arbitrary families of minor-closed classes are also minor-closed classes. If $\mathcal{B}$ is a class of matroids, there is a unique smallest minor-closed class containing it. This class is the *completion* $\overline{\mathcal{B}}$ of $\mathcal{B}$ defined by:

$$\overline{\mathcal{B}} = \bigcap \{\mathcal{C} : \mathcal{B} \subseteq \mathcal{C}, \mathcal{C} \text{ minor-closed}\}.$$

The size function of $\overline{\mathcal{B}}$ may be larger than the size function of $\mathcal{B}$. However, if the contraction of every matroid in $\mathcal{B}$ is a submatroid of a matroid in $\mathcal{B}$, then $\overline{\mathcal{B}}$ is the class of all submatroids of matroids in $\mathcal{B}$; in this case, the size functions of $\mathcal{B}$ and $\overline{\mathcal{B}}$ are equal.

Now suppose that $\mathcal{C}$ and $\mathcal{D}$ are two minor-closed classes with the same size function $h(n)$. Then the union $\mathcal{C} \cup \mathcal{D}$ also has size function $h(n)$. Hence, if $h(n)$ is the size function of a minor-closed class, there exists a unique *maximum* minor-closed class $\mathcal{M}$ such that $h(\mathcal{M};n) = h(n)$; this class $\mathcal{M}$ is obtained by taking the union of all the minor-closed classes with size function $h(n)$. If $\{M_\alpha\}$ is an antichain under the minor order, the class $\mathcal{EX}(M_\alpha)$ is a maximum class if and only if for all α, $|M_\alpha| > h(\mathcal{EX}(M_\alpha); \text{rank}(M_\alpha))$ and for all proper minors N of M_α, $|N| \leq h(\mathcal{EX}(M_\alpha); \text{rank}(N))$. We conjecture that *a maximum class in* $\mathcal{U}(q)$ *can be characterized by a finite set of forbidden minors*.

Two functions h and k defined on the non-negative integers are *asymptotically equal* if there exists an integer N such that for all $n \geq N$, $h(n) = k(n)$. When this is the case, we write $h(n) \sim k(n)$.

When studying minor-closed classes, it is often technically more convenient to work with the first difference of the size function. The *growth rate* $g(\mathcal{C};n)$ of a class $\mathcal{C}$ of matroids is the difference

$$h(\mathcal{C};n) - h(\mathcal{C};n-1).$$

The *maximum growth rate* $g(\mathcal{C})$ is defined to be $\max\{g(\mathcal{C};n) : 1 \leq n < \infty\}$ if this maximum exists; it is said to be *infinite* otherwise.

A line is said to be *long* if it contains at least three points. Long lines in binary matroids are 3-point lines. Long lines are important because they determine the number $|G| - |G/a|$ of points *destroyed* when a point a is contracted in a matroid G. More precisely, because points in G/a are lines in G containing a,

$$|G| - |G/a| = 1 + \sum_{\ell : a \in \ell} (|\ell| - 2),$$

where the sum is over all the long lines incident on a. The *cone* C_a *at the point* a is the union of the point sets of all the long lines incident on a.

(4.1) LEMMA. *Let a be a point in a matroid G.*
(a) *The number of points in the cone C_a equals*

$$1 + \sum_{\ell: a \in \ell} (|\ell| - 1) = |G| - |G/a| + \#\text{long lines on } a.$$

(b) *Suppose G is in $\mathcal{U}(q)$ and $|G| - |G/a| \geq m + 1$. Then a is incident on at least $\lceil \frac{m}{q-1} \rceil$ long lines and*

$$|C_a| \geq m + \lceil \frac{m}{q-1} \rceil + 1.$$

PROOF. To prove (b), observe that $|\ell| - 2 \leq q - 1$ for every line in a $\mathcal{U}(q)$-matroid. □

Let $\mathcal{C}$ be a class of matroids. A matroid E in $\mathcal{C}$ is said to be *extremal* if any simple proper extension of E of the same rank is not in $\mathcal{C}$. Matroids in $\mathcal{C}$ having maximum size are extremal but the converse is not necessarily the case.

(4.2) LEMMA. *Let G be a matroid in a minor-closed class $\mathcal{C}$. Suppose X is a subset of points in G such that the restriction $G|X$ is an extremal matroid in $\mathcal{C}$. Then X is a modular flat of G.*

PROOF. See [**49**, (3.1)]. □

An easy and frequently used case of (4.2) is: a $(q+1)$-point line is modular in a $\mathcal{U}(q)$-matroid.

4.2. Excluding lines

In this section, we shall consider the size functions of the minor-closed classes $\mathcal{U}(q)$ and $\mathcal{L}(q)$. Since every matroid containing a circuit contains a $U_{2,3}$-minor, $\mathcal{U}(1)$ is the class $\{U_{n,n} : 0 \leq n < \infty\}$ of free matroids. When q is a prime power, $U_{2,q+2}$ is not representable over $\mathrm{GF}(q)$; hence, $\mathcal{L}(q) \subseteq \mathcal{U}(q)$. The next theorem[11] clarifies somewhat the relationship between the size functions of $\mathcal{U}(q)$ and $\mathcal{L}(q)$.

(4.3) THEOREM. *Suppose q is a positive integer greater than 1. Then,*

$$h(\mathcal{U}(q); n) \leq \frac{q^n - 1}{q - 1},$$

with equality for $n \geq 4$ if and only if q is a prime power. If q is a prime power and $n \geq 4$, a rank-n matroid in $\mathcal{U}(q)$ contains the maximum number of points if and only if it is a projective geometry $\mathrm{PG}(n, q)$. A rank-3 matroid in $\mathcal{U}(q)$ contains the maximum number of points if and only if it is a projective plane of order q.

PROOF. We begin with a technical lemma.

---
[11]Weaker versions of this theorem can be found in [**44**,**47**].

(4.4) LEMMA. *Let G be a rank-n matroid in $\mathcal{U}(q)$ and let X be a copoint in G. Then $|G\backslash X| \leq q^{n-1}$.*

PROOF. We proceed by induction on n, the rank of G. The lemma is evident when n equals 1 or 2. Now, let U be a coline contained in X. The upper interval $[U, \hat{1}]$ is a line in $\mathcal{U}(q)$ and contains at most $q+1$ copoints. By induction, if Y is such a copoint, $|Y\backslash U| \leq q^{n-2}$. Because $G\backslash X$ is the union of $Y\backslash U$, where Y ranges over all the copoints in $[U, \hat{1}]$ not equal to X,

$$|G\backslash X| \leq q(q^{n-2}) = q^{n-1}. \quad \square$$

From the lemma and induction, we conclude that if G is in $\mathcal{U}(q)$, X is a copoint in G, and U is a coline contained in X, then

$$|G| \leq |X| + tq^{n-2} \leq 1 + q + q^2 + \cdots + q^{n-2} + q^{n-1},$$

where $t+1$ is the number of copoints in the upper interval $[U, \hat{1}]$. If equality is attained, then

(a) $|X| = 1 + q + q^2 + \cdots + q^{n-2}$, that is, X has maximum size in $\mathcal{U}(q)$, and
(b) $[U, \hat{1}]$ contains $q+1$ copoints.

In particular, $G|X$ is extremal; hence, by (4.2), X is a modular copoint. We conclude that if $|G| = 1 + q + q^2 + \cdots + q^{n-1}$, then every copoint in G is modular and hence G is a modular matroid. Moreover, every upper interval of rank 2, and hence every line (by modularity), has $q+1$ points. Therefore, G is a projective geometry of order q. When rank$(G) \geq 4$, this forces q to be a prime power [**1,74**]. In this case, G is isomorphic to $\mathrm{PG}(n-1, q)$. When rank$(G) = 3$, G is a projective plane of order q. Such projective planes *may* exist[12] when q is not a prime power. $\square$

(4.5) COROLLARY. *Let $\mathcal{C}$ be a minor-closed class. The following are equivalent:*

(a) *The size function $h(\mathcal{C}; n)$ is defined and finite for all n;*
(b) *$h(\mathcal{C}; 2)$ is defined and finite;*
(c) *$\mathcal{C} \subseteq \mathcal{U}(q)$, where $q + 1 = h(\mathcal{C}; 2)$.*

It seems likely that when $q \geq 2$,

$$h(\mathcal{U}(q); n) \sim \frac{q_*^n - 1}{q_* - 1},$$

where q_* is the largest prime power less than or equal to q. Some caution is necessary here since $\mathcal{U}(q)$ is known not to be well-quasi-ordered when $q \geq 3$. An interesting problem is to *determine all the extremal matroids in $\mathcal{U}(3)$*. The known extremal matroids are: the projective geometries $\mathrm{PG}(n, 2)$ and $\mathrm{PG}(n, 3)$, $U_{3,5}$, certain non-representable self-dual matroids obtained by "relaxing" circuit-hyperplanes, and their parallel connections (at points).[13] Are these all the extremal matroids?

---

[12]None is actually known to exist at present.
[13]See [**70**, Chap. 14] for more details.

4.3. Varieties

A *variety* of matroids is a class closed under minors *and direct sums* in which there is exactly one extremal matroid for each rank. The minor-closed classes $\mathcal{L}(q)$ and $\mathcal{Z}(A)$ are varieties. In fact, they motivated the definition! Because of this, they are called the *classical* varieties. There are three other kinds of varieties:

(a) $\mathcal{U}(1)$, the variety of free matroids,
(b) $\mathcal{M}(s)$, the variety of *matchstick matroids*, consisting of submatroids of direct sums of $(s+1)$-point lines, and
(c) $\mathcal{O}(s)$, the variety of *origami matroids*, consisting of submatroids of the matroids O_n defined as follows: Take an infinite sequence of $(s+1)$-point lines $\ell_2, \ell_3, \cdots$. Choose two distinct points x_i and y_i on the line ℓ_i. Define O_n inductively by setting $O_2 = \ell_2$ and defining O_n to be the parallel connection of O_{n-1} and ℓ_n in which the points y_{n-1} and x_n are identified.

These three kinds of varieties are said to be *degenerate*. Rather surprisingly, the classical and degenerate varieties are all the varieties of finite matroids [46].[14]

(4.6) CLASSIFICATION OF VARIETIES. *Let $\mathcal{V}$ be a variety of finite matroids. Then $\mathcal{V}$ equals $\mathcal{U}(1)$, $\mathcal{M}(s)$ for some positive integer s, $\mathcal{O}(s)$ for some positive integer s, $\mathcal{Z}(A)$ for some finite group A, or, $\mathcal{L}(q)$ for some prime power q.*

Related results can be found in [35,49].

4.4. The growth rate conjecture

There are many natural questions one can ask about minor-closed classes. For example, one can ask whether the classical varieties are well-quasi-ordered under minors. Another question is whether these varieties can be characterized by a finite set of forbidden minors.[15] A less well-known problem is to prove the growth rate conjecture [52]. This conjecture was inspired by the classification of varieties. Roughly speaking, this conjecture says that there are exactly three possible behavior for the size function of a minor-closed class contained in $\mathcal{U}(q)$: exponential, quadratic, and linear. We first state the technical version of this conjecture for binary matroids.

(4.7) CONJECTURE.
(a) *Let $h(n)$ be the size function of the class $\mathcal{EX}(\mathrm{PG}(m,2)) \cap \mathcal{L}(2)$ of binary matroids with no $\mathrm{PG}(m,2)$-minor, where m is a fixed integer greater than 1.*

---

[14]The condition that $\mathcal{V}$ be closed under direct sums is essential. For example, since a minor of a wheel (see §5.1) is a subgraph of a smaller wheel, the completion of the class of all cycle matroids of wheels is a graphic minor-closed class having exactly one extremal matroid for each rank. Other examples can be obtained by taking completions of suitable classes of uniform matroids.

[15]It is easy to see that the degenerate varieties are well-quasi-ordered. The degenerate varieties, as well as $\mathcal{L}(2)$, $\mathcal{L}(3)$, and $\mathcal{G}$, can also be characterized by finite sets of forbidden minors. The problem of finding the forbidden minors for $\mathcal{Z}(\{+,-\})$, the class of signed-graphic matroids, deserves more attention.

Then $h(n)$ is quadratic, that is,

$$\binom{n+1}{2} \leq h(n) \leq u(n),$$

where $u(n)$ is a quadratic polynomial in n whose coefficients depend only on m.

(b) Let $h(n)$ be the size function of the class $\mathcal{EX}(M(K_m)) \cap \mathcal{L}(2)$ of binary matroids with no $M(K_m)$-minor, where m is a fixed integer greater than 2. Then $h(n)$ is linear, that is,
$$h(n) \leq cn - d,$$
where c and d are constants depending only on m.

(4.7) implies the following striking conjecture.

(4.8) THE GROWTH RATE CONJECTURE FOR BINARY MATROIDS. *Let $h(n)$ be the size function of a minor-closed class $\mathcal{C}$ of binary matroids. Then*

(a) $h(n) = 2^n - 1$ and $\mathcal{C}$ equals $\mathcal{L}(2)$,
(b) $h(n)$ is quadratic and $\mathcal{C}$ contains $\mathcal{G}$, or
(c) $h(n)$ is linear.

To see that (4.7) implies (4.8), suppose that $\mathcal{C} \neq \mathcal{L}(2)$. Then, for some m, $\mathrm{PG}(m, 2)$ is not in $\mathcal{C}$ and $\mathcal{C} \subseteq \mathcal{EX}(\mathrm{PG}(m, 2)) \cap \mathcal{L}(2)$. Hence, $h(n)$ is bounded above by some quadratic polynomial in n. If $\mathcal{C}$ contains the class $\mathcal{G}$ of graphic matroids, then $h(n) \geq \binom{n+1}{2}$. If not, then, for some m, $M(K_m)$ is not in $\mathcal{C}$, $\mathcal{C} \subseteq \mathcal{EX}(M(K_m)) \cap \mathcal{L}(2)$, and $h(n)$ is linear.

For $\mathcal{U}(q)$-matroids, the technical version of the growth rate conjecture is more complicated because of the existence of proper subfields and subgroups in finite fields and their multiplicative groups. In addition, the exact size function of $\mathcal{U}(q)$ is not known when q is not a prime power. In the remainder of this section, I shall describe what seems to be true in *my* current state of ignorance.

(4.9) CONJECTURE. *Let $\{M_\alpha\}$ be a collection of matroids in $\mathcal{U}(q)$ and let $h(n)$ be the size function of the class $\mathcal{EX}(M_\alpha) \cap \mathcal{U}(q)$.*

(a) *Suppose that for some prime power r, $r \leq q$, $\{M_\alpha\} \cap \mathcal{L}(r) = \emptyset$. Let $\hat{r}$ be the maximum such prime power. Then,*

$$h(n) \sim \frac{\hat{r}^n - 1}{\hat{r} - 1}.$$

(b) *Suppose that for every prime power r, $r \leq q$, there exists a matroid M_α which is in $\mathcal{L}(r)$. Then,*
$$h(n) \leq u(n),$$
where $u(n)$ is a quadratic polynomial in n whose coefficients depend only on q and $\{M_\alpha\}$.

(c) *Suppose that for every group A having order at most $q - 1$, there exists a matroid M_α which is in $\mathcal{Z}(A)$. Then,*
$$h(n) \leq cn - d,$$

where c and d are constants depending on only q and $\{M_\alpha\}$.

As in the binary case, (4.9) implies that there is a discrete hierarchy amongst size functions of minor-closed classes in $\mathcal{U}(q)$. This is best shown using an example. We emphasize that the last four size-function estimates are conjectural.

$$h(\mathcal{U}(3); n) = \tfrac{1}{2}(3^n - 1)$$
$$\downarrow$$
$$h(\mathcal{EX}(U_{3,5}) \cap \mathcal{U}(3); n) = \tfrac{1}{2}(3^n - 1)$$
$$\downarrow$$
$$h(\mathcal{EX}(U_{3,5}, \mathrm{PG}(4,3)) \cap \mathcal{U}(3); n) \sim 2^n - 1$$
$$\downarrow$$
$$2\binom{n}{2} + n \leq h(\mathcal{EX}(U_{3,5}, \mathrm{PG}(4,3), \mathrm{PG}(3,2)) \cap \mathcal{U}(3); n) \leq u_1(n)$$
$$\downarrow$$
$$\binom{n+1}{2} \leq h(\mathcal{EX}(U_{3,5}, \mathrm{PG}(4,3), \mathrm{PG}(3,2), Q_3(3)) \cap \mathcal{U}(3); n) \leq u_2(n)$$
$$\downarrow$$
$$h(\mathcal{EX}(U_{3,5}, \mathrm{PG}(4,3), \mathrm{PG}(3,2), Q_3(3), M(K_5)) \cap \mathcal{U}(3); n) \leq cn - d.$$

Here, $u_1(n)$ and $u_2(n)$ are quadratic polynomials in n, and, c and d are constants. This hierarchical behavior can be summarized in the following way: *The size function of a minor-closed class in $\mathcal{U}(q)$ can only be exponential, quadratic, or linear.* This conjecture, which we call the *growth rate conjecture for $\mathcal{U}(q)$*, is probably quite difficult; indeed, it is not even particularly plausible. In §6.1, we shall make this conjecture more credible by showing that for a minor-closed class in $\mathcal{U}(q)$, the size function is either exponential or polynomial. We can also restrict the growth rate conjecture to $\mathcal{L}(q)$; the restricted conjecture may well be much easier to settle.

We end by remarking that Mader's proof [60] of (2.3) extends easily to gain-graphs. Therefore, if A is a finite group, the size function of a minor-closed class in $\mathcal{Z}(A)$ is either linear or quadratic.

4.5. Framed gain-graphic matroids

It is tempting to conjecture that if $\mathcal{C} \subseteq \mathcal{U}(q)$ has quadratic size function, then $h(\mathcal{C}; n) \sim h(\mathcal{Z}(A); n)$ for some group A, $|A| \leq q - 1$. But this is false.

We first describe a useful construction. Let G be a rank-n $\mathcal{L}(q)$-matroid represented as a set of points in rank-$(n+1)$ projective space $\mathrm{PG}(n, q)$. Let ω be a point not in the hyperplane spanned by the points representing G. The *framed matroid $F_q(G)$* is the rank-$(n+1)$ matroid represented by the following set of points:

$$\omega + \alpha a, \text{ for all } a \in G \text{ and all } \alpha \in \mathrm{GF}(q).$$

The point ω is called the *apex*. Four easy facts about framing are:

(a) If H is a hyperplane in $\mathrm{PG}(n, q)$ not containing ω, then $H \cap F_q(G)$ represents a matroid isomorphic to G;

(b) If b is a point in $F_q(G)$ not equal to ω, then $F_q(G)/b \cong F_q(G/b)$, where the right-hand side is the matroid obtained as follows: Intersect $F_q(G)$ with a hyperplane H containing b but not ω to obtain a matroid $\hat{G}$ isomorphic to G in H. Contract $\mathrm{PG}(n,q)$ by b and frame the contraction $\hat{G}/b$ using the apex $\omega \vee b$ in the rank-n projective geometry $\mathrm{PG}(n,q)/b$.

(c) $F_q(G)/\omega \cong G$;

(d) $|F_q(G)| = q|G| + 1$.

Given a minor-closed class $\mathcal{C}$ in $\mathcal{L}(q)$, we define $\mathcal{F}_q(\mathcal{C})$ to be the class of all submatroids of matroids of the form $F_q(G)$, where G is a matroid in $\mathcal{C}$. By (b),(c), and (d), this class is minor-closed and

$$h(\mathcal{F}_q(\mathcal{C}); n+1) = qh(\mathcal{C}; n) + 1.$$

Now let A be a subgroup of the multiplicative group $\mathrm{GF}(q)^\times$. Let $F_{q,n+1}(A)$ be the rank-$(n+1)$ framed Dowling matroid $F_q(Q_n(A))$. Framed Dowling matroids are almost "upper homogeneous."

(4.10) LEMMA. *Let a be a point not equal to the apex ω in the framed Dowling matroid $F_{q,n+1}(A)$. The contraction $F_{q,n+1}(A)/a$ is isomorphic to $F_{q,n}(A)$. The contraction $F_{q,n+1}(A)/\omega$ by the apex is isomorphic to $Q_n(A)$.*

A framed (A,q)-gain-graphic matroid is a matroid in the class $\mathcal{F}_q(\mathcal{Z}(A))$. Note that

$$\mathcal{Z}(A) \subset \mathcal{F}_q(\mathcal{Z}(A)) \subset \mathcal{EX}(\mathrm{PG}(2,q)) \cap \mathcal{L}(q)$$

and for $n \geq 1$,

$$h(\mathcal{F}_q(\mathcal{Z}(A)); n) = q\left[|A|\binom{n-1}{2} + n - 1\right] + 1.$$

Therefore, $\mathcal{F}_q(\mathcal{Z}(A))$ has quadratic size function strictly greater than $h(\mathcal{Z}(A); n)$ for all $n \geq 3$. Starting with $\mathcal{Z}(A)$, the framing construction can be iterated k times to obtain a minor-closed class in $\mathcal{L}(q)$ with size function equal to

$$q^k\left[|A|\binom{n-k}{2} + n - k\right] + q^{k-1} + q^{k-2} + \cdots + q + 1$$

when $n \geq k$.

4.6. Critical exponents

Growth rates and critical exponents are closely related by the following result.[16]

---

[16]The basic idea is in Jaeger's proof of the 8-flow theorem [**39**]. It is stated in terms of growth rates in [**50**]. The result is stated here in a version due to Oxley [**65**,**67**].

(4.11) THEOREM. *Let $\mathcal{C}$ be a submatroid-closed class in $\mathcal{L}(q)$. Suppose that there exists a constant c such that for every matroid G in $\mathcal{C}$, $|G| \leq \operatorname{crank}(G)$. Then, for every matroid G in $\mathcal{C}$, the critical exponent $c(G; q)$ over $\operatorname{GF}(q)$ is at most c.*

PROOF. Because $|H| \leq \operatorname{crank}(H)$ for all submatroids H in G, G can be partitioned into c independent sets by the matroid partition theorem [30]. Since independent sets are affine, G has critical exponent at most c. □

From the proof, it is evident that the constant c is very rarely a sharp bound because independent sets are the smallest possible affine submatroids. For example, (1.1) and (4.11) imply that the critical exponent over $\operatorname{GF}(2)$ of the class of planar-graphic matroids is at most 3. However, by the 4-color theorem, it is in fact 2. On the other hand, the constant can be sharp. For example, the class of cographic matroids has size function $3n - 3$ and critical exponent 3 [39]. There should be a more refined version of this theorem.

The *critical exponent* $c(\mathcal{C}; q)$ of a class $\mathcal{C}$ of matroids in $\mathcal{L}(q)$ is defined to be the maximum $\max\{c(G; q) : G \in \mathcal{C}\}$ if it exists and is said to be *infinite* otherwise. (4.11) implies that the critical exponent of a submatroid-closed class $\mathcal{C}$ is bounded above by its maximum growth rate.

(4.12) CONJECTURE. *Let $\mathcal{C}$ be a minor-closed class in $\mathcal{L}(q)$. Then its critical exponent $c(\mathcal{C}; q)$ is finite if and only if its maximum growth rate is finite.*

This conjecture is implied by technical versions of the growth rate conjecture. It is inspired partly by Wagner's theorem [88]: There exists a constant $\phi(m)$ such that, if the chromatic number of a graph Γ is at least $\phi(m)$, then Γ contains a K_m-minor.[17] A stronger conjecture is the following.

(4.13) CONJECTURE. *There exists a constant $\psi(m)$ depending only on m such that for all prime powers q,*

$$c(\mathcal{EX}(M(K_m)) \cap \mathcal{L}(q); q) \leq \psi(m).$$

Brylawski [20] has conjectured that the best constant is 2 when $m = 4$. Whittle [97] has proved that $c(\mathcal{TR} \cap \mathcal{L}(q); q) = 2$ for every q, where $\mathcal{TR}$ is the class of transversal matroids.

The *chromatic number* $\pi(\mathcal{C})$ of a class $\mathcal{C}$ of matroids is defined in [62,90,92,93] to be the minimum positive integer k such that the value $\chi_G(i)$ of the characteristic polynomial evaluated at i is non-zero for every matroid G in $\mathcal{C}$ and every integer $i \geq k$, if such an integer exists; it is defined to be *infinite* otherwise. Because there is no known analogue of (4.11) for chromatic numbers, bounds for them are much harder to obtain. All the results so far rely on decomposition theory. Two examples from [90] are:

$$\pi(\mathcal{EX}(F_7, M(K_{3,3})^\perp, M(K_{3,3})) \cap \mathcal{L}(2)) = 5;$$
$$\pi(\mathcal{EX}(F_7^\perp, M(K_{3,3})^\perp, M(K_5)) \cap \mathcal{L}(2)) = 5.$$

---

[17]Note that (2.3) and (4.11) imply Wagner's theorem. However, Wagner's proof yields the better constant.

Two attractive conjectures (also in [90]) are:

$$\pi(\mathcal{EX}(F_7, M(K_5)) \cap \mathcal{L}(2)) = 5;$$
$$\pi(\mathcal{EX}(F_7, M(K_{3,3})) \cap \mathcal{L}(2)) = 5.$$

Both conjectures are equivalent to Tutte's 5-flow conjecture.

5. EXCLUDING GRAPHIC MINORS

Most of the published theorems about excluding graphic minors are about excluding small wheels. In this section, we shall describe these and other related results. We shall conclude with a hitherto unpublished theorem about excluding k-wheels from binary matroids. We begin by recalling some elementary facts about wheels and whirls.

5.1. Wheels and whirls

For $k \geq 2$, the k-wheel is the graph obtained from a k-cycle by adding a new vertex h called the *hub* and all possible edges between h and the vertices in the cycle. The edges in the cycle are called *rim* edges and the new edges are called *spokes*. The k-whirl is the (non-binary) matroid obtained by "relaxing" the circuit formed by the rim edges: that is, the k-whirl is the matroid on the same set with the same independent set as the k-wheel, with the exception that the circuit formed by the rim edges is declared independent. In [86], Tutte proved that k-wheels and k-whirls are the minimal 3-connected rank-k matroids.

The cycle matroid of $M(W_k)$ is represented in rank-k projective space over any field by the following $2k$ points:

$$e_1, e_2, \cdots, e_k, e_1 - e_2, e_2 - e_3, \cdots, e_{k-1} - e_k, e_k - e_1,$$

where e_i is the standard basis vector. The vectors e_i represent the spokes and the vectors $e_i - e_{i+1}$ the rim. When $k \geq 3$, the k-whirl can be represented over any field except GF(2) by changing, say, the vector $e_k - e_1$ to $e_k + \alpha e_1$, where the non-zero scalar α is chosen so that the rim is independent.

This representation of the k-wheel motivates the following definition. A *ring R of k long lines* is a matroid with a distinguished subset of points $x_1, x_2, \cdots, x_k$ such that (a) the lines $x_1 \vee x_2, x_2 \vee x_3, \cdots, x_{k-1} \vee x_k$, and $x_k \vee x_1$ are long, and (b) the point set of R is the union of these long lines. An easy but useful fact is: *A ring R of k long lines contains a k-wheel or a k-whirl if and only if R has rank k.*

One way to attempt to produce wheels or whirls is to first produce an eel. A matroid E is said to be an *eel of length l* if there exists a *distinguished basis* $\{x_1, x_2, ..., x_l\}$ satisfying: (a) the lines $x_1 \vee x_2, x_2 \vee x_3, \cdots, x_{l-2} \vee x_{l-1}$, and $x_{l-1} \vee x_l$ (but not necessarily $x_l \vee x_1$) are long, and (b) if a point y in E is not on one of the long lines $x_i \vee x_{i+1}$, $1 \leq i \leq l-1$, then the line $y \vee x_l$ is a long line. The point x_1 is called the *head* and the point x_l is called the *tail* of E. The *width* of the eel E is the number $|C_{x_l}|$ of points in the cone at the tail x_l.

(5.1) LEMMA.
(a) *Let E be an eel in $\mathcal{U}(q)$ of length l and width w. Then*

$$l \geq \lceil \log_q[(q-1)w + 1] \rceil.$$

(b) *Suppose that G is a matroid in which the cone C_a at every point a contains at least w points. Then G contains an eel-submatroid having width at least w. In particular, if $G \in \mathcal{U}(q)$ and $|G| - |G/a| \geq m + 1$ for every point a in G, then G contains a eel-submatroid E having width at least*

$$m + \lceil \frac{m}{q-1} \rceil + 1.$$

PROOF. (a) The eel E has rank l and contains at least w points. Hence, by (4.3), $(q^l - 1)/(q - 1) \geq w$.

(b) We construct an eel E in G as follows: Start by choosing a point x_1, the head of E. Choose a long line ℓ_1 on x_1 and a point x_2 on ℓ_1 distinct from x_1. Supposing that $\ell_1, \ell_2, \cdots, \ell_{i-1}$ and $x_1, x_2, \cdots, x_i$ have already be chosen, choose ℓ_i to be a long line on x_i such that $\text{rank}(\ell_1 \vee \ell_2 \vee \cdots \vee \ell_{i-1} \vee \ell_i) = i + 1$ if such a line exists. If such a line does not exist, then all the long lines on x_i are in the closure of $\{x_1, x_2, \cdots, x_i\}$. Let E be the union of the lines ℓ_i and all the long lines incident on x_i. Because $\{x_1, x_2, \cdots, x_i\}$ is a basis for E and $|C_{x_i}| \geq w$, E is an eel with tail x_i having width at least w. The second assertion now follows from (4.1b). □

5.2. Excluding 3-wheels and 3-whirls

The 3-wheel W_3 is isomorphic to the complete graph K_4. It is a natural candidate for an excluded minor. Results about excluding 3-wheels and 3-whirls from matroids give a good overview of the two known general methods for studying size functions of minor-closed classes: using decomposition theory and using long lines.

We begin with an extension of Dirac's theorem to binary matroids.

(5.2) THEOREM.

$$h(\mathcal{EX}(M(K_4)) \cap \mathcal{L}(2); n) = 2n - 1.$$

There are three very different ways to prove this theorem. The first is to observe that a binary matroid with no $M(K_4)$-minor is graphic. This follows from the fact that $M(K_4)$ is a minor of all four binary forbidden minors, $F_7, F_7^\perp, M(K_5)^\perp,$ and $M(K_{3,3})^\perp$, in Tutte's characterization of graphic matroids [85]. Hence, (5.2) is basically Dirac's theorem.

The *second* is to use Duffin's theorem [29] that graphs without $M(K_4)$-minors are series-parallel networks. Using Tutte's characterization of graphic matroids as in the first proof, we can restate Duffin's theorem as a decomposition theorem.

(5.3) THEOREM. *A matroid is 3-connected, binary, and has no $M(K_4)$-minor if and only if it is isomorphic to the 3-point line $U_{2,3}$ or the single point $U_{1,1}$.*

Using (5.3) and induction, it is easy to show that a rank-n binary matroid with no $M(K_4)$-minor has maximum size if and only if it is a parallel connection of $n-1$ copies of 3-point lines. Hence, such matroids have $2(n-1)+1$ points.

The first two proofs rely strongly on properties of $M(K_4)$ and $\mathcal{L}(2)$. The third uses long lines and is more "robust". We shall prove the following theorem due to Hipp [37] using eels.

(5.4) THEOREM.

$$h(\mathcal{EX}(M(W_3), \mathcal{W}^3) \cap \mathcal{U}(q); n) = q(n-1) + 1.$$

A matroid in $\mathcal{EX}(M(W_3), \mathcal{W}^3) \cap \mathcal{U}(q)$ has the maximum number of points if and only if it is a parallel connection of $(q+1)$-point lines.

Both parts of this theorem can be handled with one technical lemma.

(5.5) LEMMA. *Let G be a $\mathcal{U}(q)$-matroid. Suppose that every point a in G satisfies the conditions: (a) $|G| - |G/a| \geq q$, and, (b) a is on at least two long lines. Then G contains a 3-wheel or 3-whirl as a minor.*

PROOF. By (4.1a), $|C_a| \geq q+2$ for every point a in G. Hence, by (5.1b), G contains an eel having width at least $q+2$. Let E be a minimal eel-minor in G having width at least $q+2$; let l be its length, x_1 its head, and x_l its tail. Since E/x_1 is an eel-minor of G, it has strictly smaller width. Hence, there are two points y and z on long lines through the tail x_l that are identified when x_1 is contracted: that is, x_1, y, and z are collinear in E. If y and z are on the same line ℓ,[18] then x_1, y, z, and x_l are collinear and E contains an l-wheel or l-whirl, where $l \geq 3$. Contracting if necessary, we conclude that $\mathcal{C}$ contains a 3-wheel or 3-whirl. If y and z are on distinct long lines ℓ_1 and ℓ_2, then $x_1 \vee \ell_1$ is a rank-3 flat containing the long lines ℓ_1 and ℓ_2 through x_l as well as the long line $x_1 \vee y \vee z$. These lines form a 3-wheel or 3-whirl and we are done. □

To prove the first part of (5.4), suppose $\mathcal{C}$ is a minor-closed class in $\mathcal{U}(q)$ and there exists an integer N such that $g(\mathcal{C}; N) \geq q+1$. Let G be a rank-N matroid of maximum size in $\mathcal{C}$. Then, for every point a in G, $|G| - |G/a| \geq q+1$. Moreover, by (4.1b), a is on at least two long lines. Hence, by (5.5), G and hence $\mathcal{C}$ contains a 3-wheel or a 3-whirl. Therefore, $g(\mathcal{EX}(M(W_3), \mathcal{W}^3) \cap \mathcal{U}(q)) \leq q$. From this and the fact that parallel connections of $(q+1)$-point lines are in this class, we conclude that the size function is $q(n-1) + 1$.

Now suppose that H is a rank-n matroid in $\mathcal{EX}(M(W_3), \mathcal{W}^3) \cap \mathcal{U}(q)$ with $q(n-1)+1$ points. Then there exists a point b in H on exactly one long line ℓ. [Otherwise, H satisfies both conditions in (5.5); hence, H contains $M(W_3)$ or $\mathcal{W}^3$ as a minor.] Since $|H| - |H/b| \geq q$, ℓ is a $(q+1)$-point line and is modular. Next, we show that $H \backslash \ell$ does not span. If not, $H \backslash \ell$ contains a basis B. If $|B| = 3$, then $B \cup \ell$ contains a 3-wheel because ℓ is modular. Otherwise, contract points of B not on long lines in the submatroid $B \cup \ell$ to reduce rank($B \cup \ell$) to 3.

---
[18]In this case, x_l may equal y or z.

Because $\text{rank}(H\backslash\ell) \leq n-1$, the partition $(H\backslash\ell, \ell)$ is a 2-separation of H and H is a parallel connection one of whose parts is the $(q+1)$-point line ℓ. By induction, H is a parallel connection of $(q+1)$-point lines. This completes the proof of the second part of (5.4). Since the 3-whirl is not binary, the case $q = 2$ of (5.4) is (5.2).

It is now natural to ask what the size function is when $M(K_4)$ is excluded from $\mathcal{U}(q)$ or $\mathcal{L}(q)$. A somewhat crude bound is given in the next theorem [53].

(5.6) THEOREM.
(a) *Let s be a positive integer. Then*

$$h(\mathcal{EX}(M(K_4)) \cap \mathcal{U}(q); n) \leq [6q^{q-1} + 8q - 1]n.$$

(b) *Let q be a prime power. Then*

$$h(\mathcal{EX}(M(K_4)) \cap \mathcal{L}(q); n) \leq [6q^3 - 1]n.$$

Similar bounds [53] hold for classes of matroids not containing $M(K_3*[r])$, where $K_3*[r]$ is the graph obtained by taking a graphical triangle K_3 on the vertices c_1, c_2, c_3 and r other vertices $d_1, d_2, \cdots, d_r$ and adding all possible edges of the form $\{c_i, d_j\}$. These bounds yield linear size functions for the classes $\mathcal{EX}(M(K_{3,3})) \cap \mathcal{U}(q)$ and $\mathcal{EX}(M(K_{3,3})) \cap \mathcal{L}(q)$. In addition, the minor-closed class $\mathcal{BO}$ of base-orderable matroids and the minor-closed class $\mathcal{GA}$ of gammoids are contained in $\mathcal{EX}(M(K_4))$; hence, the submatroid-closed class of transversal matroids $\mathcal{TR}$ is also contained in $\mathcal{EX}(M(K_4))$.[19] Thus, (5.6) implies that intersections of these classes with $\mathcal{U}(q)$ or $\mathcal{L}(q)$ have linear size functions. The exact size functions of $\mathcal{GA} \cap \mathcal{U}(q)$, $\mathcal{GA} \cap \mathcal{L}(q)$, $\mathcal{TR} \cap \mathcal{U}(q)$, and $\mathcal{TR} \cap \mathcal{L}(q)$ are known. They all equal $q(n-1)+1$; moreover, for all four classes, the rank-n matroids of maximum size are parallel connections of $n-1$ $(q+1)$-point lines.

(5.7) THEOREM.

$$h(\mathcal{GA} \cap \mathcal{U}(q); n) = (q-1)n + 1.$$

The rank-n matroids in $\mathcal{GA} \cap \mathcal{U}(q)$ having maximum size are parallel connections of $n-1$ $(q+1)$-point lines.

PROOF. We first show that $h(\mathcal{GA} \cap \mathcal{U}(q); n) \leq (q-1)n + 1$. We claim that in any gammoid G, there exists a point a on at most one long line. To prove this, let Γ be a graph and B and G subsets of the vertex set V of Γ such that the independent sets of G are precisely those subsets in G linked to B. Three points a, b, and c form a circuit if and only if there exists a cut-set $\{e, f\}$ with two vertices disconnecting $\{a, b, c\}$ from B. In the strict gammoid on V, all five points a, b, c, e, f are on the same line. From this, we see that if a is a point having maximum graphical distance from B, then a is on at most one long line. Now suppose that the growth rate $g(\mathcal{GA} \cap \mathcal{U}(q); n) > q$ for some n. Then by (4.1b), there exists a rank-n gammoid in which every point is on two long lines, a contradiction. We can now complete the proof as in (5.4). □

---
[19]See [17] or [91, Chap. 14] for details.

The bounds given in (5.6) are probably far from exact. In fact, it is not unreasonable to conjecture that

$$h(\mathcal{EX}(M(K_4)) \cap \mathcal{U}(q); n) \leq \lambda(q)n,$$

where $\lambda(q)$ is a linear function of q. Oxley has obtained a decomposition theorem for $\mathcal{EX}(M(K_4)) \cap \mathcal{L}(3)$. From this, he obtained the exact size function for this class [65].

(5.8) THEOREM (OXLEY).

$$h(\mathcal{EX}(M(K_4)) \cap \mathcal{L}(3); n) = \begin{cases} 4n - 3 & \text{if } n \text{ is odd} \\ 4n - 4 & \text{if } n \text{ is even.} \end{cases}$$

The matroids in $\mathcal{EX}(M(K_4)) \cap \mathcal{L}(3)$ attaining these bounds are parallel connections of copies of the affine ternary plane $AG(2,3)$ if n is even, and parallel connections of copies of $AG(2,3)$ and a 4-point line if n is odd.

In the same paper [65], Oxley also obtained a decomposition theorem for the class $\mathcal{EX}(M(K_4), \mathcal{W}^3) \cap \mathcal{L}(4)$.

Another way to look at matroids with no $M(K_4)$-minors is to use the Crapo beta invariant [23]. The beta invariant is defined on all matroids – simple or non-simple – by the formula:

$$\beta(M) = \sum_{X : X \subseteq E(M)} (-1)^{\operatorname{rank}(M)+|X|} \operatorname{rank}(X),$$

the sum being over all subsets X of the set $E(M)$ of elements of the matroid M. Crapo proved that if M is a series-parallel extension of N, then $\beta(M) = \beta(N)$. In particular, if M does not have loops, M and its simplification have the same beta invariant. The beta invariant also satisfies the Tutte recursion:[20] if a is neither a loop nor an isthmus,

$$\beta(G) = \beta(G \backslash a) + \beta(G/a).$$

In addition, the beta invariant equals 0 for a rank-0 matroid and 1 for a loopless rank-1 matroid. From this, we see that $\beta(G) \geq 0$ and $\beta(G) \geq \beta(H)$ if H is a minor of G. Hence, the class $\mathcal{C}_{\beta \leq k}$ of matroids having beta invariant at most k is minor-closed. Because $\beta(U_{2,k+2}) = k$, $\mathcal{C}_{\beta \leq k} \subseteq \mathcal{U}(k+1)$. Brylawski [18] showed that $\mathcal{C}_{\beta \leq 1}$ equals $\mathcal{EX}(M(K_4)) \cap \mathcal{L}(2)$. Oxley [63] has studied the structure of matroids in $\mathcal{C}_{\beta \leq k}$. Roughly speaking, the matroids in $\mathcal{C}_{\beta \leq k}$ are series-parallel extensions of a small set of matroids in $\mathcal{C}_{\beta \leq k}$. From this, one can show that the size function $h(\mathcal{C}_{\beta \leq k}; n)$ is linear in n.

---

[20]See [19,83] for the general theory of Tutte (-Grothendieck) invariants.

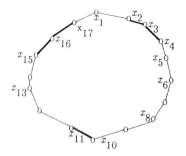 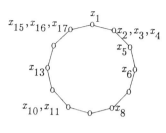

FIGURE 1. Truing a bent wheel. The even circuit is $\{x_2, x_4, x_{10}, x_{11}, x_{15}, x_{17}\}$. The edges represent 3-point lines. The interior points in the lines represented by thicker edges are contracted.

5.3. Excluding the complete graph on 5 vertices

Except for the following theorem [**52**] (see also [**50**]), not much is known when one excludes the complete graphic matroids $M(K_m)$ for $m > 4$.

(5.9) THEOREM. *The maximum growth rate of $\mathcal{EX}(M(K_5)) \cap \mathcal{L}(2)$ is at most 8.*

(4.11) and (5.9) imply that the critical exponent of $\mathcal{EX}(M(K_5)) \cap \mathcal{L}(2)$ over $GF(2)$ is at most 8. Walton and Welsh [**90**] (see also [**94**]) have observed that there are no counterexamples to the conjecture that it is 3.[21]

5.4. Excluding wheels from binary matroids

We conclude this section by presenting a new result about excluding wheels from binary matroids.

(5.10) THEOREM. *The maximum growth rate of the class $\mathcal{EX}(M(W_k)) \cap \mathcal{L}(2)$ is at most 2^{2k-3}.*

To prove (5.10), we need to consider certain binary matroids which are "almost" wheels. A binary matroid U is said to be a *bent wheel of rank l* if there exist a basis $\{x_1, x_2, \cdots, x_l\}$ and an *extra* point x_{l+1} distinct from the points in the basis satisfying: (a) the fundamental circuit of x_{l+1} relative to $\{x_1, x_2, \cdots, x_l\}$ contains an even number of points, and (b) the remaining points in U are

$$x_1 + x_2, x_2 + x_3, \cdots, x_{l-1} + x_l, x_l + x_{l+1}, x_{l+1} + x_1.$$

In particular, the lines $x_1 \vee x_2, x_2 \vee x_3, \cdots, x_{l+1} \vee x_1$ are 3-point lines and the bent wheel U is a ring of $l + 1$ 3-point lines of rank l. Bent wheels can be made into true wheels by suitable contractions.

---

[21]It is known [**50,90**] that $c(\mathcal{EX}(F_7, M(K_5)) \cap \mathcal{L}(2); 2) = c(\mathcal{EX}(F_7^\perp, M(K_5)) \cap \mathcal{L}(2); 2) = 3$.

(5.11) LEMMA. *A bent wheel of rank l contains a $M(W_k)$-minor, where*

$$k = \lceil \frac{l+1}{2} \rceil.$$

PROOF. The idea of the proof is illustrated in Figure 1. Let U be the bent wheel of length l described earlier. Arrange the points $x_1, x_2, \cdots, x_{l+1}$ clockwise on a circle. If $j \equiv i + k \pmod{l+1}$, where $1 \leq k \leq l+1$, then the arc $[i, j]$ is defined to be the set $\{x_i, x_{i+1}, \cdots, x_{j-1}, x_j\}$. The set I of *interior* points is the set $\{x_1 + x_2, x_2 + x_3, \cdots, x_{l-1} + x_l, x_l + x_{l+1}, x_{l+1} + x_1\}$. If $[i,j]$ is an arc, let $\text{Int}[i,j]$ be the set $\{x_i + x_{i+1}, x_{i+1} + x_{i+2}, \cdots, x_{j-1} + x_j\}$. Using the fact that in a binary matroid, symmetric differences of circuits are disjoint unions of circuits, it is easy to check that the set

$$\{x_i\} \cup \text{Int}[i,j] \cup \{x_j\} = \{x_i, x_i + x_{i+1}, x_{i+1}\} \triangle \{x_{i+1}, x_{i+1} + x_{i+2}, x_{i+2}\}$$
$$\triangle \cdots \triangle \{x_{j-1}, x_{j-1} + x_j, x_j\}$$

is a circuit.

Let $\{x_{i_1}, x_{i_2}, \cdots, x_{i_m}\}$, where $i_1 < i_2 < \cdots < i_m = l+1$ and m is even, be the fundamental circuit of x_{l+1} relative to the basis $\{x_1, x_2, \cdots, x_l\}$. Let

$$S = \text{Int}[i_1, i_2] \cup \text{Int}[i_3, i_4] \cup \cdots \cup \text{Int}[i_{m-1}, i_m].$$

Consider the set S and its complement $I \backslash S$. One of these sets contains at most $\lfloor \frac{l+1}{2} \rfloor$ points.

Relabelling if necessary,[22] we can assume that $|S| \leq \lfloor \frac{l+1}{2} \rfloor$. The set S is the mod-2 sum of the circuits $\{x_{i_1}, x_{i_2}, \cdots, x_{i_m}\}$, $\{x_{i_1}\} \cup \text{Int}[i_1, i_2] \cup \{x_{i_2}\}$, $\{x_{i_3}\} \cup \text{Int}[i_3, i_4] \cup \{x_{i_4}\}$, $\cdots$, and $\{x_{i_{m-1}}\} \cup \text{Int}[i_{m-1}, i_m] \cup \{x_{i_m}\}$; it is in fact a circuit. Hence, $\text{rank}(S) = |S| - 1$ and the contraction U/S has rank t, where

$$t = l - |S| + 1 \geq \lceil \frac{l+1}{2} \rceil.$$

Because U/S is a binary ring of t 3-point lines having rank t, U/S is a t-wheel. Because $t \geq k$, U/S, and hence, U contains a k-wheel minor. □

We shall now prove (5.10) by showing that a minor-closed class $\mathcal{C}$ of binary matroids having maximum growth rate at least $2^{2k-3} + 1$ contains a k-wheel. Let $g(\mathcal{C}; N) \geq 2^{2k-3} + 1$ and let G be a matroid of maximum size in $\mathcal{C}$. Then for every point a in G, $|G| - |G/a| \geq 2^{2k-3} + 1$. By (5.1b), G, and hence $\mathcal{C}$, contains an eel having width at least $2^{2k-2} + 1$. Among all the eels having width at least $2^{2k-2} + 1$ in $\mathcal{C}$, let E be one having minimum length l. By (5.1a),

$$l \geq \lceil \log_2[(2^{2k-2} + 1) + 1]] = 2k - 1.$$

---

[22]Change the basis to $\{x_{i_m+1}, \cdots, x_l, x_{l+1}, x_1, x_2, \cdots, x_{i_m-1}\}$ and let x_{i_m} be the extra point.

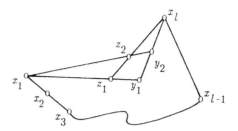

FIGURE 2. A bent wheel in an eel.

Let $\{x_1, x_2, \cdots, x_l\}$ be the distinguished basis, x_1 the head and x_l the tail of E. Since the contraction E/x_1 by the head is an eel of smaller length, E/x_1 has width strictly less than $2^{2k-3} + 2$. Hence, two points y_1 and z_1 on long lines through the tail x_l are identified when x_1 is contracted: that is, the points x_1, y_1, and z_1 are collinear in E. There are two possible cases.

Case 1. The tail x_l is on the line $x_1 \vee y_1 \vee z_1$. Then, $x_1 \vee x_l$ is a 3-point line and the points $x_1, x_2, \cdots, x_l$ form the spokes of an l-wheel. Since $l \geq k$, E contains a k-wheel minor.

Case 2. The tail x_l is *not* on the line $x_1 \vee y_1 \vee z_1$. Let ℓ_1 and ℓ_2 be the 3-point lines $x_l \vee y_1$ and $x_l \vee z_1$ respectively. Let Π be the plane spanned by ℓ_1 and ℓ_2. Note that $x_1 \in \Pi$. Since E is binary, the 3-point lines ℓ_1 and ℓ_2 are modular; hence, they intersect the copoint X of E spanned by $x_1, x_2, \cdots, x_{l-1}$ at two distinct points. These two points are collinear with x_1 because they are in the rank-2 intersection $\Pi \cap X$. Now relabel the points in Π as shown in Figure 2 so that $\Pi \cap X = \{x_1, y_1, z_1\}$. Note that x_1, y_2 and z_2 are also collinear.

Suppose that y_1 is a point on one of the lines $x_i \vee x_{i+1}$, where $1 \leq i \leq l$. If $i \geq k$, then the lines $x_1 \vee x_2, x_2 \vee x_3, \cdots, x_i \vee x_{i+1}$, and $\{x_1, y_1, z_1\}$ form a i-wheel, where $i \geq k$; if $i < k$, then the lines $x_l \vee x_{l+1}, x_{l-1} \vee x_l, \cdots, x_{i+1} \vee x_{i+2}$, $x_i \vee x_{i+1}$, and $\{x_1, y_1, z_1\}$ form a i-wheel, where $i \geq k$; in both cases, E contains a k-wheel minor. The same argument applies to z_1. We can now suppose that neither y_1 nor z_1 is on one of the lines $x_i \vee x_{i+1}$.

Consider the fundamental circuit C of y_1 relative to the basis $\{x_1, x_2, \cdots, x_l\}$. Since y_1 is in the copoint X, x_l is not in C. If C is even, then E contains a bent wheel of rank l. (See Figure 2.) Otherwise, C is odd. Taking the symmetric difference of C with the 3-element circuit $\{y_1, y_2, x_l\}$, we obtain $(C \backslash \{y_1\}) \cup \{y_2, x_l\}$. This set is in fact the fundamental circuit of y_2 relative to the basis $\{x_1, x_2, \cdots, x_l\}$. This circuit is even and hence E contains a bent wheel of rank l. In both cases, E contains a k-wheel minor by (5.11). This completes the proof of (5.10).

Since the rank of $\mathrm{PG}(k-2, 2)$ is $k-1$, it does not contain a k-wheel minor. Hence, the maximum growth rate of $\mathcal{EX}(M(W_k)) \cap \mathcal{L}(2)$ is at least 2^{k-2} and it is probable that it is exactly that.[23] With more attention to the small rank cases, it is possible to improve the bound in (5.10), but not enough to attain sharpness.

From his decomposition theorem for the class $\mathcal{EX}(M(W_4)) \cap \mathcal{L}(2)$, Oxley [67] proved that

$$h(\mathcal{EX}(M(W_4)) \cap \mathcal{L}(2); n) = \begin{cases} 3n - 2 & \text{if } n \text{ is odd} \\ 3n - 3 & \text{if } n \text{ is even.} \end{cases}$$

He also characterized the matroids having maximum size. Another result [68] of Oxley along similar lines is:

$$h(\mathcal{EX}(M(W_5)) \cap \mathcal{R}; n) = \begin{cases} 3n - 2 & \text{if } n \equiv 1 \pmod{3} \\ 3n - 3 & \text{otherwise.} \end{cases}$$

The matroids of maximum size are characterized. Other than this, no other exact result about excluding higher wheels is known. Indeed, the exact size functions of the *graphic* classes $\mathcal{EX}(M(W_k)) \cap \mathcal{G}$ are not known for $k \geq 6$.[24] Using long lines and carefully analysing the rank-4, 5 and 6 cases, Hipp [37] obtained the following related result: For $n \geq 3$,

$$h(\mathcal{EX}(M(W_4), \mathcal{W}^4) \cap \mathcal{L}(3); n) = 6n - 5.$$

The matroids of maximum size are not known in this case. Matroids attaining this bound are parallel connections of copies of $PG(2,3)$ and the matroid V_{19}, the rank-4 ternary matroid obtained by taking the points in two $PG(2,3)$'s in $PG(3,3)$ and removing three points from their line of intersection.

6. QUADRATIC GROWTH

6.1. Cones

A matroid C is said to be a *cone* if there exists a point ω in C such that every other point in C is on a long line incident on ω, or, equivalently, C equals the cone C_ω at ω. The vertex ω is called an *apex* of C. Examples of cones are the Fano plane F_7 and, more generally, any projective geometry. Gain-graphic matroids, however, are never cones unless they are parallel connection of long lines at a common basepoint. This is one reason why cones are useful for studying the growth rate of classes excluding non-gain-graphic matroids. The following easy lemma is useful when doing induction with cones.

(6.1) LEMMA. *Let C be a cone with apex ω and let U be a flat of C not containing ω. Then the contraction C/U is a cone with apex $U \vee \omega$.*

Cones were *implicitly* used by Heller [36] and Mader [60]. We shall illustrate how they are used by proving an extension of Heller's theorem.

---

[23] Note that the maximum growth rate does not give sufficient information for calculating the size function exactly.

[24] See (6.16) for a partial result. Gubser [34] has recently obtained a sharp bound of $e(\Gamma) \leq \lceil (14v(\Gamma) - 27)/5 \rceil$ for a simple planar graph Γ with no 6-wheel-minor.

(6.2) THEOREM. *Let $\mathcal{C}$ be a minor-closed class in $\mathcal{U}(q)$. Suppose that for some integer n,*

$$g(\mathcal{C}; n) > (q-1)^k \binom{n}{k}.$$

Then $\mathcal{C}$ contains a matroid of rank $k+2$ containing at least $2^{k+2} - 1$ points.

PROOF. Let $g(\mathcal{C}; N) > (q-1)^k \binom{N}{k}$, G a rank-N matroid in $\mathcal{C}$ of maximum size, and ω a point in G. Then $|G| - |G/\omega| \geq (q-1)^k \binom{N}{k} + 1$. Let C be the cone at the point ω. By (4.1b), the point ω is on at least $(q-1)^{k-1}\binom{N}{k}$ long lines and the number of points in C is at least

$$(q-1)^k \binom{N}{k} + (q-1)^{k-1}\binom{N}{k} + 1 = q(q-1)^{k-1}\binom{N}{k} + 1$$
$$\geq q(q-1)^{k-1}\binom{\text{rank}(C)}{k} + 1.$$

Thus, to prove (6.2), it suffices to show: *if C is a cone satisfying*

(∗) $$|C| > q(q-1)^{k-1}\binom{\text{rank}(C)}{k},$$

then C contains a rank-$(k+2)$ minor with at least $2^{k+2} - 1$ points.

We shall first tackle the induction step. Suppose $k \geq 2$. Let C be a minor-minimal cone satisfying (∗), ω its apex, r_C its rank, and a a non-apex point on C. Because the contraction C/a is a cone (by (6.1)) and C is minor-minimal,

$$|C/a| \leq q(q-1)^{k-1}\binom{r_C - 1}{k}$$

and

$$|C| - |C/a| > q(q-1)^{k-1}\left[\binom{r_C}{k} - \binom{r_C - 1}{k}\right] = q(q-1)^{k-1}\binom{r_C - 1}{k-1}.$$

As in an earlier step, the point a is the apex of a cone D with at least

$$q(q-1)^{k-1}\binom{r_D - 1}{k-1} + q(q-1)^{k-2}\binom{r_D - 1}{k-1} + 1 = q^2(q-1)^{k-2}\binom{r_D - 1}{k-1} + 1$$

points, where $r_D = \text{rank}(D)$. Hence, the image of D/ω in the contraction C/ω is a cone having rank $r_{D/\omega} = r_D - 1$ and containing at least

$$q(q-1)^{k-2}\binom{r_{D/\omega}}{k-1} + 1$$

points. By induction, D/ω contains a minor E of rank $k+1$ and size at least $2^{k+1} - 1$. Since a minor of D/ω is also a minor of D, $E = F/X$, where F and X are sets of points of D, $X \subseteq F$, and $\omega \in X$. Let $\bar{F}$ and $\bar{X}$ be the closure of F and

$P_7 \qquad G_7 \qquad T_7$

FIGURE 3. Fans.

X in C and consider the interval $[\bar{X}, \bar{F}]$ in the lattice $L(C)$ of flats of C. Let U be a complement of the point ω in $[\hat{0}, \bar{X}]$: that is, U is a flat contained in $\bar{X}$, $\omega \notin U$, and $U \vee \omega = \bar{X}$. The minor $\bar{F}/U$ of C is a cone with apex $U \vee \omega$. Moreover, $\bar{F}/X$ is a submatroid of $\bar{F}/\bar{X}$ and $\bar{F}/\bar{X} = \bar{F}/(U \vee \omega)$; hence, $|\bar{F}/(U \vee \omega)| \geq 2^{k+1} - 1$. Because every point in $\bar{F}/(U \vee \omega)$ corresponds to a long line incident on $U \vee \omega$ in $\bar{F}/U$, $\bar{F}/U$ is a minor of C containing at least $2[2^{k+1} - 1] + 1 = 2^{k+2} - 1$ points.

The proof for the case $k = 1$ is similar, except that we can take advantage of the ranks being small to obtain better inequalities. Let C be a minor-minimal cone satisfying $|C| > q\text{rank}(C)$ and let ω be its apex. As in the general case, there exists a non-apex point a in C such that $|C| - |C/a| > q$. By (4.1a), a is on at least two long lines. Let ℓ be a long line containing a but not ω. Then the plane $\omega \vee \ell$ has rank 3 and contains at least 7 points. This allows us to start the induction.

This completes the proof of (6.2). Note that this proof only used the fact that $|G| - |G/\omega| > (q-1)^k \binom{N}{k}$ for *one* point ω; the full power of $g(\mathcal{C}; N) > (q-1)^k \binom{N}{k}$ is not used. $\square$

A matroid F is said to be a *fan* if (a) $\text{rank}(F) = 3$, (b) F is the union of at least three long lines with a common point ω, and (c) there exists a long line in F not incident on ω. The proof of the case $k = 1$ actually yields: *if C is a cone satisfying $|C| > q\text{rank}(C)$, then C contains a fan.* There are five minimal fans: the Fano plane F_7, the non-Fano plane F_7^-, and the three matroids G_7, P_7, and T_7 shown in Figure 3. Thus, we have proved the following theorem:

(6.3) THEOREM.

$$h(\mathcal{EX}(F_7, F_7^-, G_7, P_7, T_7) \cap \mathcal{U}(q); n) \leq (q-1)\binom{n+1}{2}.$$

Because only F_7 is binary, (6.3) for the case $q = 2$ and the fact that $|M(K_{n+1})| = \binom{n+1}{2}$ imply the following extension of Heller's theorem [**9,61,75**].[25]

(6.4) COROLLARY.

$$h(\mathcal{EX}(F_7) \cap \mathcal{U}(2); n) = \binom{n+1}{2}.$$

---

[25] Murty's proof in [**61**] is essentially Heller's. Baclawski and White [**9**] started by proving a version of (6.16) and then used multilinear algebra to finish their proof. The theorem also follows from Seymour's decomposition theorem [**75**] for $\mathcal{EX}(F_7) \cap \mathcal{L}(2)$.

Because $|U_{2,4}| = \binom{3}{2} + 1$, $|F_7| = \binom{4}{2} + 1$, and all the rank-2 minors of F_7 are submatroids of $U_{2,3}$, (6.4) implies that $\mathcal{EX}(F_7) \cap \mathcal{U}(2)$ is the maximum class with size function $\binom{n+1}{2}$.

Observing that F_7 is representable only over fields of characteristic two [98] and that the four minimal fans not equal to F_7 all contain the 3-whirl, we obtain the following corollary.

(6.5) COROLLARY. *If q is an odd prime power,*

$$h(\mathcal{EX}(\mathcal{W}^3) \cap \mathcal{L}(q); n) \leq (q-1)\binom{n+1}{2}.$$

Once again, the bound in (6.5) is not exact. For $q = 3$, the exact size function can be deduced from work of Oxley [66] (see also [89]):[26] $h(\mathcal{EX}(\mathcal{W}^3) \cap \mathcal{L}(3); n) = \binom{n+1}{2}$ when $n \geq 4$, and $3(n-1)+1$ when $n \leq 3$. This result makes the following conjecture plausible: when q is an odd prime power, $h(\mathcal{EX}(\mathcal{W}^3) \cap \mathcal{L}(q); n) \sim \binom{n+1}{2}$. Using another result of Oxley [69] and (6.4), we can deduce the following exact results:[27] (a) $h(\mathcal{EX}(\mathcal{W}^3) \cap \mathcal{L}(4); n) = 2^n - 1$ when $n \geq 4$, and $4(n-1)+1$ when $n \leq 3$, and, (b) $h(\mathcal{EX}(F_7, \mathcal{W}^3) \cap \mathcal{L}(4); n) = \binom{n+1}{2}$ when $n \geq 6$, and $4(n-1)+1$ when $n \leq 6$. We shall let the reader conjecture the obvious extensions to $\mathcal{L}(2^r)$.

We conclude this section with a result promised in §4.4. This result offers significant supporting evidence for the growth rate conjecture.[28]

(6.6) POLYNOMIAL OR EXPONENTIAL GROWTH. *Let $\mathcal{C}$ be a minor-closed class in $\mathcal{U}(q)$. Either there exists a positive integer k such that, for all n,*

$$h(\mathcal{C}; n) \leq (q-1)^k \binom{n+1}{k}$$

and $h(\mathcal{C}; n)$ is bounded above by a polynomial in n, or, for all n,

$$h(\mathcal{C}; n) \geq 2^n - 1$$

and $h(\mathcal{C}; n)$ is bounded below by a function exponential in n.

The lower bound of $2^n - 1$ seems crude, but it is sharp since $\mathcal{C}$ can be the class $\mathcal{L}(2)$.[29] For a class $\mathcal{C}$ in $\mathcal{L}(q)$, where q is a power of the prime p, the sharp

---

[26]We use (3.1) in [66]: A ternary but non-binary 3-connected matroid having rank and nullity at least 3 has a $\mathcal{W}^3$-minor. Hence, if G is a matroid in $\mathcal{EX}(\mathcal{W}^3) \cap \mathcal{L}(3)$, one of the following holds:
  (a) $G \in \mathcal{L}(2) \cap \mathcal{L}(3)$, and hence G is regular,
  (b) G has nullity at most 2,
  (c) G has rank 2 and is a line with at most four points, or
  (d) G is not 3-connected and hence is a 2-sum or is disconnected.

The result now follows from induction and Heller's theorem.

[27]Use (1.5) in [69].

[28]When $q = 2$, (6.6) can also be deduced from (2.6).

[29]If one wants a minor-closed class in $\mathcal{U}(q)$ but not in $\mathcal{U}(q-1)$, then $\mathcal{L}(2) \cup \{U_{2,q+1}\}$ is an example. In this case, the bound will only be sharp for $n \geq 3$.

lower bound of $(p^n - 1)/(p - 1)$ can be obtained using methods in [54] and the following technical version of (6.2) obtained by examining the proof: *Let $\mathcal{C}$ be a minor-closed class in $\mathcal{U}(q)$ satisfying the hypothesis in (6.2). Then $\mathcal{C}$ contains a matroid G with rank $k + 2$ satisfying the property: there exist points ω_{k+2}, $\omega_{k+1}, \cdots, \omega_2$ in G such that G is a cone with apex ω_{k+2}, G/ω_{k+2} is a cone with apex $\omega_{k+2} \vee \omega_{k+1}$, $G/\omega_{k+2} \vee \omega_{k+1}$ is a cone with apex $\omega_{k+2} \vee \omega_{k+1} \vee \omega_k$, $\cdots$, and $G/\omega_{k+2} \vee \omega_{k+1} \vee \cdots \vee \omega_3$ is a cone with apex $\omega_{k+2} \vee \omega_{k+1} \vee \cdots \vee \omega_3 \vee \omega_2$.*

6.2. Excluding the dual Fano configuration

The results in this section are motivated by two exact results obtained by decomposition theory. Let $s(n)$ be the function defined as follows:

$$s(n) = \begin{cases} \binom{n+1}{2} & \text{if } n \neq 3 \\ 7 & \text{if } n = 3. \end{cases}$$

The first result can be read off Seymour's decomposition theorem [75] for $\mathcal{EX}(F_7^\perp) \cap \mathcal{L}(2)$. It contains (6.4) as an easy corollary.

(6.7) THEOREM.
$$h(\mathcal{EX}(F_7^\perp) \cap \mathcal{L}(2); n) = s(n).$$

Since $F_7^\perp$ is a one-point deletion of the rank-4 affine binary geometry AG(3, 2), it is natural to ask for a decomposition theorem for $\mathcal{EX}(\text{AG}(3, 2)) \cap \mathcal{L}(2)$. This problem is open. The following bounds for the size function are not hard to obtain, but the exact result is elusive. Let V_{11} be the matroid obtained by taking the union of two Fano planes in PG(3, 2), or, equivalently, removing a 4-point circuit from PG(3, 2).

(6.8) THEOREM.

$$\binom{n+1}{2} + 1 \leq h(\mathcal{EX}(\text{AG}(3, 2)) \cap \mathcal{L}(2); n)$$
$$\leq h(\mathcal{EX}(V_{11}) \cap \mathcal{L}(2); n) \leq \binom{n+1}{2} + n.$$

The lower bound follows from the observation that the binary rank-n matroid represented by the $\binom{n+1}{2} + 1$ vectors $e_1 + e_2 + e_3$, e_i, $1 \leq i \leq n$, and $e_i + e_j$, $1 \leq i, j \leq n$, does not have a AG(3, 2)-minor. The upper bound can be obtained using cones and a connectivity argument.

The second exact result is due to Oxley [67] and follows from his decomposition theorem for binary matroids excluding the matroid P_9 and its dual; here, P_9 is the binary matroid obtained by adding the point $e_1 + e_2 + e_3$ to the GF(2)-representation of the 4-wheel given in §5.1.

(6.9) THEOREM (OXLEY).

$$h(\mathcal{EX}(P_9, P_9^\perp) \cap \mathcal{L}(2); n) = s(n).$$

Both P_9 and $P_9^\perp$ contain $F_7^\perp$; hence, (6.9) extends (6.7). The next result is the "best" possible such extension. To state it, we first list all three non-isomorphic binary rank-4 matroids on 11 points. They are

(a) V_{11},
(b) $M(K_5)^+$, obtained by adding one point in $PG(3,2)$ to $M(K_5)$, or, equivalently, removing a basis from $PG(3,2)$, and
(c) C_{11}, obtained by removing a 3-point line and a point outside it from $PG(3,2)$.

(6.10) THEOREM.

$$h(\mathcal{EX}(V_{11}, M(K_5)^+, C_{11}) \cap \mathcal{L}(2); n) = s(n).$$

The proof is similar to that of (6.8). Both proofs are somewhat lengthy and are omitted. It is easy to check that $\mathcal{EX}(V_{11}, M(K_5)^+, C_{11}) \cap \mathcal{L}(2)$ is a maximum class. Hence, (6.10) cannot be extended further.

6.3. Geometric algebra

A *Reid matroid*[30] R is a rank-3 matroid consisting of three long lines ℓ_1, ℓ_2, and ℓ_3, such that the intersection $\ell_1 \cap \ell_2 \cap \ell_3$ is a point ω and the line ℓ_3 is a 3-point line. Let ω, α, and β be the three points on ℓ_3. The *incidence graph* $I(R)$ is the bipartite graph defined as follows. The vertex set is the union $(\ell_1 \backslash \{\omega\}) \cup (\ell_2 \backslash \{\omega\})$. A vertex u in $\ell_1 \backslash \{\omega\}$ is connected to a vertex v in $\ell_2 \backslash \{\omega\}$ by an edge if u, v, and α, or u, v, and β are collinear. The vertices in $I(R)$ have degree at most 2 and $I(R)$ is a disjoint union of paths and cycles of even length.

Let $R_{cycle}[k]$ be the Reid matroid whose incidence graph is a cycle of length $2k$ and let $R_{path}[k]$ be the Reid matroid whose incidence graph is a path of length $2k - 1$. For example, $R_{cycle}[2]$ is the Fano plane F_7 and $R_{path}[2]$ is the non-Fano plane F_7^-. The Reid matroids code equations of the form $k = 0$ or $k \neq 0$. More precisely, $R_{cycle}[k]$ is representable over a field $\mathbb{K}$ if and only if k is prime and $\mathbb{K}$ has characteristic k; $R_{path}[k]$ is representable over a field $\mathbb{K}$ if and only if $2 \neq 0$, $3 \neq 0, \cdots, k-1 \neq 0$, and $k \neq 0$ in $\mathbb{K}$. In particular, $R_{cycle}[k]$ is not representable over the rationals $\mathbb{Q}$.[31] These facts are not hard to show using determinants [33,54,98].[32]

6.4. Matroids representable over two different characteristics

Because $R_{cycle}[2] \cong F_7$, the following theorem [54] is yet another extension of Heller's theorem.

---

[30]Reid's work, done around 1971, was unpublished. An account can be found in [33].

[31]Except when $k = 2$, it has not been proved that the matroid $R_{cycle}[k]$ are non-orientable. But this is almost certainly true.

[32]The idea of doing algebra with geometrical configurations goes back to the Greeks (see Euclid's *Elements*). The idea used here first appeared in von Staudt [77]. We remark that the equation $\zeta^k = 1$ can be coded by a submatroid of a rank-3 Dowling matroid, *cf.* (2.9). Gordon [32] has shown that $R_{cycle}[k]$ is algebraic over a field if and only if that field has characteristic k.

(6.11) THEOREM. *Let q be a positive integer greater than 2. Then*

$$h(\mathcal{EX}(R_{cycle}[s] : 2 \leq s \leq q) \cap \mathcal{U}(q); n) \leq (q^\nu - q^{\nu-1})\binom{n+1}{2} - n,$$

where $\nu = 2^{q-1} - 1$.

Even though the constant is extremely crude, the degree of n is exact. When $q = 3$, an exact result is known [55,56].

(6.12) THEOREM.

$$h(\mathcal{EX}(F_7, R_{cycle}[3]) \cap \mathcal{U}(3); n) = 2\binom{n}{2} + n = n^2.$$

When $n \geq 4$, a rank-n matroid G in $\mathcal{EX}(F_7, R_{cycle}[3]) \cap \mathcal{U}(3)$ has n^2 points if and only if G is the Dowling matroid $Q_n(3)$. A rank-3 matroid G in this class has 9 points if and only if $G \cong Q_3(3)$ or $G \cong \mathrm{AG}(2,3)$.

Because of the representability properties of Reid matroids, (6.11) yields a quadratic bound on the size function of the intersection of two incomparable varieties of representable matroids.[33]

(6.13) COROLLARY. *Let q and r be coprime prime powers. Then*

$$h(\mathcal{L}(q) \cap \mathcal{L}(r); n) \leq (q^\nu - q^{\nu-1})\binom{n+1}{2} - n,$$

where $\nu = 2^{q-1} - 1$.

(2.9), (6.12), and (6.13) inspired the following rash conjecture [54].

(6.14) CONJECTURE. *Let p and q be distinct primes. Then*

$$h(\mathcal{L}(p) \cap \mathcal{L}(q); n) \sim \gcd(p-1, q-1)\binom{n}{2} + n.$$

For sufficiently large n, there is a unique rank-n matroid of maximum size, the Dowling matroid $Q_n(A)$, where A is the cyclic group of order $\gcd(p-1, q-1)$.

Because it is an asymptotic result, (6.14) is probably quite difficult at present. Part of the difficulty lies at the rank-3 level and suggests the following problem of interest in its own right: For a given prime p, find the minimum number $\gamma(p)$ with the property: If G is a rank-3 $\mathcal{L}(p)$-matroid and $|G| \geq \gamma(p)$, then G can only be represented over a field of characteristic p. The two known cases are: $\gamma(2) = 7$ [98] and $\gamma(3) = 10$ [55]. It is possible, however, to prove the following related result.[34]

---

[33]Other consequences of (6.11) can be found in [54]. These include quadratic bounds on the maximum number of projective inequivalent columns in totally T-adic matrices. Totally T-adic matrices are integer matrices all of whose subdeterminants are in a given subset T of integers. Non-quadratic bounds for T-adic matrices can be found in [6,57,59].

[34]The proof will appear elsewhere if enough people are interested in it.

(6.15) THEOREM. *Let q be an integer and let M_1, M_2, $\cdots$ be all the rank-3 matroids in $\mathcal{U}(q)$ not contained in any rank-3 Dowling matroid $Q_3(A)$, where A is a quasi-group of order $q-1$. Then*

$$h(\mathcal{EX}(M_1, M_2, \cdots) \cap \mathcal{U}(q); n) = (q-1)\binom{n}{2} + n,$$

The rank-n matroids of maximum size are precisely the Dowling matroids $Q_n(A)$, where $|A| = q - 1$.

6.5. Binary classes not containing Fano planes

Reasonably sharp bounds on size functions of minor-closed classes of binary matroids not containing a Fano plane can be obtained using the method in §3.2. This method uses Seymour's decomposition theorem for $\mathcal{EX}(F_7) \cap \mathcal{L}(2)$. A more elementary approach is to use the following technical version of (6.4) due essentially to Heller [36].

(6.16) LEMMA (HELLER). *Let C be a binary cone with apex ω such that the contraction C/ω contains a circuit. Then C contains a F_7-minor. In particular, a binary rank-n cone with at least n 3-point lines incident on its apex contains a F_7-minor.*

To show how this lemma is used, we shall prove the following variation on (5.10).[35]

(6.17) THEOREM. *The maximum growth rate of $\mathcal{EX}(F_7, M(W_k)) \cap \mathcal{L}(2)$ is exactly k.*

PROOF. Let $\mathcal{C}$ be a minor-closed class in $\mathcal{EX}(F_7) \cap \mathcal{L}(2)$ having maximum growth rate at least $k+1$. We shall show that $\mathcal{C}$ contains F_7 or $M(W_k)$.

By (5.1b), $\mathcal{C}$ contains an eel having width at least $2k+1$. Among all the eels having width at least $2k+1$ in $\mathcal{C}$, let E be one having minimum length l. Let $x_1, x_2, \cdots, x_l$ be the distinguished basis, x_1 the head, x_l the tail of E. In addition, let x'_1 be the third point on the line spanned by x_1 and x_2. Because the cone C_{x_l} at the tail contains at least k 3-point lines, either E has a F_7-minor, or, $\operatorname{rank}(C_{x_l}) \geq k+1$. Hence, we can suppose that $l \geq k + 1$.

By minimality, the contractions E/x_1 and E/x'_1 have widths strictly less than $k+1$. Hence, two points y_1 and z_1 in the cone C_{x_l} at the tail are identified when x_1 is contracted: that is, the points $x_1, y_1,$ and z_1 are collinear in E. Similarly, x'_1 is collinear with two points y'_1 and z'_1 in C_{x_l}.

If the tail x_l is on one of the lines $x_1 \vee y_1$ or $x'_1 \vee y'_1$, then E contains a $M(W_k)$-minor. Now suppose that x_l is on neither $x_1 \vee y_1$ nor $x'_1 \vee y'_1$. Let y_2 and z_2 be the third point on the lines $x_l \vee y_1$ and $x_l \vee z_1$. Then the six points x_1, $y_1, y_2, z_1, z_2,$ and x_l form a submatroid isomorphic to $M(K_4)$. Similarly, $x'_1, y'_1,$ and x_l span a plane containing an $M(K_4)$. There are now two cases to consider, depending on whether the two $M(K_4)$-submatroids intersect. These two cases

---

[35] Other results of the same type can be found in [50].

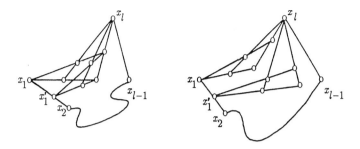

FIGURE 4. Two configurations containing F_7-minors.

are shown in Figure 4. It is easy to check that, in both cases, E contains an F_7-minor. □

Since graphic matroids have no F_7-minor, we obtain the following result in extremal graph theory.[36]

(6.18) COROLLARY. *Let $e(\Gamma^v)$ be the maximum number of edges in a simple graph Γ^v on v vertices with no k-wheel minor. Then*

$$\binom{k}{2}\lfloor\frac{v-1}{k-1}\rfloor \leq e(\Gamma^v) \leq kv - \binom{k+1}{2}.$$

The lower bound is derived from the fact that parallel connections of K_k contain no W_k-minors.

6.6. Beyond cones

As observed at the end of the proof of (6.2), the method of cones does not make full use of the growth rate being large. One way to go further is look at "cones" with two or more layers. With this idea, I have proved that the classes $\mathcal{EX}(\mathrm{PG}(3,2))\cap\mathcal{L}(2)$ and $\mathcal{EX}(\mathrm{PG}(2,p))\cap\mathcal{L}(q)$, where q is a p-power, have quadratic size function. I hope that this will lead to a proof of the quadratic-exponential gap for $\mathcal{L}(q)$.

7. DIRECTIONS AND MODULAR COPOINTS

Not all of extremal matroid theory is concerned with size functions. In fact, the chapter entitled "Extremal Problems" in Welsh's book [91] has empty intersection with this survey and is devoted to "internal" extremal problems. We shall discuss one such problem which originated in the direction problem (§2.3) in classical combinatorial geometry.

---

[36] This result is not hard to prove graph-theoretically. It is stated using the notation in [13].

A *direction* in affine real space $\mathbb{R}^n$ may be defined as an equivalence class of parallel lines, or, equivalently, as a point in the hyperplane at infinity in projective space $\mathrm{PG}(n, \mathbb{R})$. Scott's direction problem (§2.3) can be reformulated over any field $\mathbb{K}$ as follows [21,22,44]. Let S be a set of s points in affine n-space $\mathrm{AG}(n, \mathbb{K})$. A point a in the hyperplane H at infinity is said to be *determined by* G if a is the intersection of H with a line ℓ spanned by two or more points in G. Given n, s and $\mathbb{K}$, an extension of Scott's problem is to find the minimum number $\delta_n(s)$ of points on the hyperplane H at infinity determined by a set G of s points in $\mathrm{AG}(n, \mathbb{K})$. Note that if all the points H_G determined by G on H are added to G, then we obtain a matroid $G \cup H_G$ in $\mathrm{PG}(n, \mathbb{K})$ containing a modular copoint H_G.

The only known result about this problem is Ungar's theorem (2.7). There does not seem to be a reasonable conjecture for $\delta_n(s)$ over $\mathbb{R}$ when $n \geq 3$. However, for complex n-space $\mathbb{C}^n$, there is a natural conjecture because there are candidates for the direction-critical matroids.[37]

(7.1) CONJECTURE. *Let G be a spanning set of points in $\mathbb{C}^n$ containing a modular flat H, s be the number of points outside H, and δ the number of points in H. Then*
$$\delta \geq (q-1)\binom{n}{2} + n,$$
where
$$q = \lceil \frac{s-1}{n} \rceil.$$

The critical configurations are isomorphic to the rank-$(n+1)$ Dowling matroid $Q_{n+1}(C_q)$, where C_q is the cyclic subgroup of $\mathbb{C}^\times$ consisting of all the qth roots of unity.

Because $Q_{n+1}(C_q)$ is $\mathbb{K}$-representable whenever $\mathbb{K}$ contains all the qth roots of unity, (7.1) should hold over any field containing *all* the roots of unity. Note that this conjecture has not been verified for the complex plane.

8. CONCLUDING REMARKS

It might be appropriate to end with some informal observations. Results about size functions proved using long lines are almost never exact; on the other hand, the proofs are robust: they generally work for all q for both $\mathcal{U}(q)$ and $\mathcal{L}(q)$. Thus, one trades exactness for generality. Two developments which may lead to sharper results are possible. The first is in decomposition theory. Here, bounds on size functions can be useful in indicating the form of the decomposition theorem. For example, with hindsight, Heller's theorem that $h(\mathcal{R}; n) = \binom{n+1}{2}$ suggests that graphic matroids "predominate" in the class $\mathcal{R}$ of regular matroids and points the way to Seymour's decomposition theorem. Analogously, but alas, much more speculatively, (6.12) indicates that a similar decomposition theory, with signed graphs instead of graphs, should exist for the class of ternary matroids with no $R_{cycle}[3]$-minor. Similar speculative ideas are suggested by the results in §6.2.

---

[37] Our conjecture was partly inspired by Jamison's work in [45]

Because the size function of $\mathcal{EX}(V_{11}, M(K_5)^+, C_{11}) \cap \mathcal{L}(2)$ equals $\binom{n+1}{2}$ when $n \geq 4$, it is probable that for such integers n, the rank-n matroids of maximum size are graphic. A decomposition theorem for this class should follow the same pattern as the theorem for $\mathcal{EX}(F_7^{-1}) \cap \mathcal{L}(2)$. However, the matroids of maximum size for $\mathcal{EX}(AG(3,2)) \cap \mathcal{L}(2)$ are not graphic. One would expect, therefore, that a decomposition theorem for this class would have some new features. Although there are deep results[38] and a lot of current activity in decomposition theory, *basic* questions, such as what is the right notion of k-sum and k-connectivity in this context, remain to be answered. It is also evident that because many small examples need to be examined, some computer assistance will be needed. This important area has not been adequately surveyed in this paper, partly because it is still in flux and mostly because I do not know it well enough. We refer the reader to two forthcoming books [70,80].

The second development is in random matroid theory.[39] Random graphs are useful for finding lower bounds in extremal graph theory [15] and random matroids should perform the same rôle. However, there is a major hurdle for anyone doing random matroid theory – or *any* matroid theory: the absence of vertices. It is easy, for example, to write down the probability that a random graph does not have a K_m-minor (see, for example, [78]) because vertices exist; to write down the probability that a random subset of points in $PG(n,2)$ does not contain a F_7-minor is hard (at least for me). Perhaps a theory of random rank functions may work better.

There are obviously more *fundamental* questions in extremal matroid theory than there are known methods. This is an exciting state for a subject to be in.

Acknowledgement. I would like to thank the referee for carefully reading the manuscript and suggesting many improvements. I would also like to thank Jeff Kahn, James Oxley, Paul Seymour, and Tom Zaslavsky for many discussions about extremal matroid theory over the past twelve years.

References

1. A. A. Albert and R. Sandler, *An introduction to finite projective planes*, Holt, Rinehart, and Winston, New York, 1968.
2. N. Alon, *On the density of sets of vectors*, Discrete Math. **46** (1983), 199–202.
3. R. P. Anstee, *Properties of (0,1)-matrices with no triangles*, J. Combin. Theory Ser. A **29** (1980), 186–198.
4. R. P. Anstee, *General forbidden configuration theorems*, J. Combin. Theory Ser. A **40** (1985), 108–124.
5. R. P. Anstee, *A forbidden configuration theorem of Alon*, J. Combin. Theory Ser. A **47** (1988), 16–27.
6. R. P. Anstee, *Forbidden configurations, discrepancy and determinants*, Europ. J. Combin. **11** (1990), 15–19.
7. R. P. Anstee and Z. Füredi, *Forbidden submatrices*, Discrete Math. **62** (1986), 225–243.

---

[38]To the papers already quoted, we should add the series [79] of papers by Truemper.

[39]Some work has been done on random subsets of points in projective space [47,48,64] from the matroidal point of view.

8. R. P. Anstee and U. S. R. Murty, *Matrices with forbidden subconfigurations*, Discrete Math. **54** (1985), 113–116.
9. K. Baclawski and N. L. White, *Higher order independence in matroids*, J. London Math. Soc. (2) **19** (1979), 193–202.
10. R. E. Bixby, *A strengthened form of Tutte's characterization of regular matroids*, J. Combin. Theory Ser. B **20** (1976), 216–221.
11. R. E. Bixby, *Kuratowski's and Wagner's theorems for matroids*, J. Combin. Theory Ser. B **22** (1977), 31–53.
12. R. E. Bixby and W. H. Cunningham, *Short cocircuits in binary matroids*, Europ. J. Combin. **8** (1987), 213–225.
13. B. Bollobás, *Extremal graph theory*, Academic Press, London and New York, 1978.
14. B. Bollobás, *Graph theory*, Springer-Verlag, Berlin and New York, 1979.
15. B. Bollobás, *Extremal graph theory with emphasis on probabilistic methods*, Amer. Math. Soc., Providence, Rhode Island, 1986.
16. R. C. Bose and R. C. Burton, *A characterization of flat spaces in finite geometry and the uniqueness of the Hamming and the MacDonald codes*, J. Combin. Theory **1** (1966), 244–257.
17. R. A. Brualdi and E. B. Scrimger, *Exchange systems, matchings, and transversals*, J. Combin. Theory **5** (1968), 244–257.
18. T. H. Brylawski, *A combinatorial model for series-parallel networks*, Trans. Amer. Math. Soc. **154** (1971), 1–22.
19. T. H. Brylawski, *A decomposition for combinatorial geometries*, Trans. Amer. Math. Soc. **171** (1972), 235–282.
20. T. H. Brylawski, *An affine representation for transversal geometries*, Stud. Appl. Math. **54** (1975), 143–160.
21. G. R. Burton and G. B. Purdy, *The directions determined by n points in the plane*, J. London Math. Soc. (2) **20** (1979), 109–114.
22. R. Cordovil, *The directions determined by n points in the plane, a matroidal generalization*, Discrete Math. **43** (1983), 131–137.
23. H. H. Crapo, *A higher invariant for matroids*, J. Combin. Theory **2** (1967), 406–417.
24. H. H. Crapo and G.-C. Rota, *On the foundations of combinatorial theory: Combinatorial geometries*, M. I. T. Press, Cambridge, Mass., 1970.
25. G. A. Dirac, *In abstrakten Graphen vorhandene vollständige 4-Graphen und ihre Unterteilungen*, Math. Nachr. **22** (1960), 61–85.
26. P. Doubilet, G.-C. Rota, and R. P. Stanley, *On the foundations of combinatorial theory (VI): The idea of generating function*, in *Proceedings of the Sixth Berkeley Symposium on Mathematical Statistics and Probability*, Vol. 2, Probability theory, University of California Press, Berkeley, California, 1972, pp. 267–318.
27. T. A. Dowling, *A q-analog of the partition lattice*, in *A survey of combinatorial theory*, J. N. Srivastava, ed., North-Holland, Amsterdam, 1973, pp. 101–115.
28. T. A. Dowling, *A class of geometric lattices based on finite groups*, J. Combin. Theory Ser. B **14** (1973), 61–86; erratum, ibid. **15** (1973), 211.
29. R. J. Duffin, *Topology of series-parallel networks*, J. Math. Anal. Appl. **10** (1965), 303–318.
30. J. Edmonds, *Minimal partition of a matroid into independent sets*, J. Res. Nat. Bur. Standards Sect. B **69** (1965), 67–77.
31. P. Erdős and A. H. Stone, *On the structure of linear graphs*, Bull. Amer. Math. Soc. **52** (1946), 1087–1091.
32. G. Gordon, *Algebraic characteristic sets of matroids*, J. Combin. Theory Ser. B **44** (1988), 64–74.
33. C. Greene, *Lectures in combinatorial geometries*, Notes taken by D. Kennedy from the National Science Foundation Seminar in Combinatorial Theory, Bowdoin College, Maine, unpublished, 1971.
34. B. S. Gubser, *Planar graphs having no minor isomorphic to the 6-wheel*, preprint.
35. H. Groh, *Varieties of topological geometries*, Trans. Amer. Math. Soc., to appear.

36. I. Heller, *On linear systems with integral valued solutions*, Pacific J. Math. **7** (1957), 1351–1364.
37. J. W. Hipp, *The maximum size of combinatorial geometries excluding wheels and whirls as minors*, Doctoral dissertation, University of North Texas, Denton, Texas, August, 1989.
38. A. W. Ingleton, *Representations of matroids*, in *Combinatorial mathematics and its applications*, D. J. A. Welsh, ed., Academic Press, London and New York, 1971, pp. 149–169.
39. F. Jaeger, *Flows and generalized coloring theorems in graphs*, J. Combin. Theory Ser. B **26** (1979), 205–216.
40. F. Jaeger, *Interval matroids and graphs*, Discrete Math. **27** (1979), 331–336.
41. F. Jaeger, *Two characterizations of cographic matroids*, Unpublished research report, Institut I. M. A. G., Grenoble, France, 1979.
42. R. E. Jamison, *Planar configurations which determine few slopes*, Geom. Dedicata **16** (1984), 17–34.
43. R. E. Jamison, *Structure of slope-critical configurations*, Geom. Dedicata **16** (1984), 249–277.
44. R. E. Jamison, *A survey of the slope problem*, in *Discrete geometry and convexity (New York, 1982)*, Ann. New York Acad. Sci. 440, New York Acad. Sci., New York, 1985, pp. 34–51.
45. R. E. Jamison, *Modular hyperplanes in complex configurations*, preprint.
46. J. Kahn and J. P. S. Kung, *Varieties of combinatorial geometries*, Trans. Amer. Math. Soc. **271** (1982), 485–499.
47. D. G. Kelly and J. G. Oxley, *Asymptotic properties of random subsets of projective spaces*, Math. Proc. Cambridge Philos. Soc. **91** (1982), 119–130.
48. D. G. Kelly and J. G. Oxley, *Threshold functions for some properties of random subsets of projective spaces*, Quart. J. Math. Oxford (2) **33** (1982), 463–469.
49. J. P. S. Kung, *Numerically regular hereditary classes of combinatorial geometries*, Geom. Dedicata **21** (1986), 85–105.
50. J. P. S. Kung, *Growth rates and critical exponents of minor-closed classes of binary geometries*, Trans. Amer. Math. Soc. **293** (1986), 837–857.
51. J. P. S. Kung (ed.), *A sourcebook in matroid theory*, Birkhäuser, Boston and Basel, 1986.
52. J. P. S. Kung, *Excluding the cycle geometries of the Kuratowski graphs from binary geometries*, Proc. London Math. Soc. **55** (1987), 209–242.
53. J. P. S. Kung, *The long-line graph of a combinatorial geometry. I. Excluding $M(K_4)$ and the $(q+2)$-point line as minors*, Quart. J. Math. Oxford (2) **39** (1988), 223–234.
54. J. P. S. Kung, *The long-line graph of a combinatorial geometry. II. Geometries representable over two fields of different characteristics*, J. Combin. Theory Ser. B **50** (1990), 41–53.
55. J. P. S. Kung, *Combinatorial geometries representable over $GF(3)$ and $GF(q)$. I. The number of points*, Discrete Comput. Geom. **5** (1990), 84–95.
56. J. P. S. Kung and J. G. Oxley, *Combinatorial geometries representable over $GF(3)$ and $GF(q)$. II. Dowling geometries*, Graphs and Combin. **4** (1988), 323–332.
57. J. Lee, *Subspaces with well-scaled frames*, Ph. D. thesis, Cornell University, Ithaca, New York, 1986.
58. J. Lee, *Turán's triangle theorem and binary matroids*, Europ. J. Combin. **10** (1989), 85–90.
59. J. Lee, *Subspaces with well-scaled frames*, Linear Algebra Appl. **114/115** (1989), 21–56.
60. W. Mader, *Homomorphieeigenschaften und mittlere Kantendichte von Graphen*, Math. Ann. **174** (1967), 265–268.
61. U. S. R. Murty, *Extremal matroids with forbidden restrictions and minors (synopsis)*, Proceedings of the Seventh Southeastern Conference on Combinatorics, Graph Theory, and Computing, Louisiana State University, Baton Rouge, LA, 1976, *Congressus Numerantium*, No. 17, Utilitas Math., Winnipeg, Manitoba, 1976, pp. 463–468.
62. J. G. Oxley, *Colouring, packing and the critical problem*, Quart. J. Math. Oxford (2) **29** (1978), 11–22.

63. J. G. Oxley, *On Crapo's beta invariant for matroids*, Stud. Appl. Math. **66** (1982), 267–277.
64. J. G. Oxley, *Threshold distribution functions for some random representable matroids*, Math. Proc. Cambridge Philos. Soc. **95** (1984), 335–347.
65. J. G. Oxley, *A characterization of the ternary matroids with no $M(K_4)$-minor*, J. Combin. Theory Ser. B **42** (1987), 212–249.
66. J. G. Oxley, *On nonbinary 3-connected matroids*, Trans. Amer. Math. Soc. **300** (1987), 663–679.
67. J. G. Oxley, *The binary matroids with no 4-wheel minor*, Trans. Amer. Math. Soc. **301** (1987), 63–75.
68. J. G. Oxley, *The regular matroids with no 5-wheel minor*, J. Combin. Theory Ser. B **46** (1989), 292–305.
69. J. G. Oxley, *On an excluded-minor class of matroids*, Discrete Math. **82** (1990), 35–52.
70. J. G. Oxley, *Matroid Theory*, Oxford Univ. Press, Oxford, to appear.
71. N. Sauer, *On the density of families of sets*, J. Combin. Theory Ser. A **13** (1972), 145–147.
72. P. R. Scott, *On the sets of directions determined by n points*, Amer. Math. Monthly **77** (1970), 502–505.
73. A. Schrijver, *Theory of linear and integer programming*, Wiley, Chichester, 1986.
74. A. Seidenberg, *Lectures on projective geometry*, van Nostrand, Princeton, 1962.
75. P. D. Seymour, *Decomposition of regular matroids*, J. Combin. Theory Ser. B **28** (1980), 305–359.
76. S. Shelah, *A combinatorial problem: stability and order for models and theories in infinitary languages*, Pacific J. Math. **41** (1972), 247–261.
77. G. K. C. von Staudt, *Beiträge zur Geometrie der Lage*, Nuremberg, 1856.
78. A. G. Thomason, *An extremal function for contractions of graphs*, Math. Proc. Cambridge Philos. Soc. **95** (1984), 261–265.
79. K. Truemper, *A decomposition theory for matroids. I. General results*, J. Combin. Theory Ser. B **39** (1985), 43–76; *II. Minimal violation matroids*, ibid. **39** (1985), 282–297; *III. Decomposition conditions*, ibid. **41** (1986), 275–305; *IV. Decompositions of graphs*, ibid. **39** (1988), 259–292; *V. Testing of matrix total unimodularity*, ibid. **49** (1990), 241–281; *VI. Almost regular matroids*, ibid., to appear; *VII. Analysis of minimal violation matrices*, ibid., to appear.
80. K. Truemper, *Matroid decomposition*, Academic Press, New York, to appear.
81. R. T. Tugger, *Private communication*, June, 1990.
82. P. Turán, *On an extremal problem in graph theory*, in Hungarian, Mat. Fiz. Lapok **48** (1941), 436–452.
83. W. T. Tutte, *A ring in graph theory*, Proc. Cambridge Philos. Soc. **43** (1947), 26–40.
84. W. T. Tutte, *A homotopy theorem for matroids, I.*, Trans. Amer. Math. Soc. **88** (1958), 144–160; *II.*, ibid. **88** (1958), 161–174.
85. W. T. Tutte, *Matroids and graphs*, Trans. Amer. Math. Soc. **90** (1959), 527–552.
86. W. T. Tutte, *Connectivity in matroids*, Canad. J. Math. **18** (1966), 1301–1324.
87. P. Ungar, *$2N$ noncollinear points determine at least $2N$ directions*, J. Combin. Theory Ser. A **33** (1982), 343–347.
88. K. Wagner, *Beweis einer Abschwächung der Hadwiger-Vermutung*, Math. Ann. **153** (1964), 139–141.
89. P. N. Walton, *Some topics in combinatorial theory*, D. Phil. thesis, Oxford University, Oxford, 1981.
90. P. N. Walton and D. J. A. Welsh, *On the chromatic number of binary matroids*, Mathematika **27** (1980), 1–9.
91. D. J. A. Welsh, *Matroid theory*, Academic Press, London and New York, 1976.
92. D. J. A. Welsh, *Colouring problems and matroids*, in *Surveys in combinatorics (Proceedings of the Seventh British Combinatorial Conference, Cambridge 1979)*, Cambridge Univ. Press, Cambridge, 1979, pp. 229–257.
93. D. J. A. Welsh, *Colourings, flows, and projective geometry*, Nieuw Arch. Wisk. (3) **28** (1980), 159–176.

94. D. J. A. Welsh, *Matroids and combinatorial optimisation*, in *Matroid theory and its applications*, A. Barlotti, ed., Liguori, Naples, 1982, pp. 323–416.
95. N. L. White (ed.), *Theory of matroids*, Cambridge Univ. Press, Cambridge, 1986.
96. N. L. White (ed.), *Combinatorial geometries*, Cambridge Univ. Press, Cambridge, 1987.
97. G. P. Whittle, *On the critical exponent of transversal matroids*, J. Combin. Theory Ser. B **37** (1984), 94–95.
98. H. Whitney, *On the abstract properties of linear dependence*, Amer. J. Math. **57** (1935), 509–533.
99. T. Zaslavsky, *The geometry of root systems and signed graphs*, Amer. Math. Monthly **88** (1981), 88–105.
100. T. Zaslavsky, *Signed graphs*, Discrete Appl. Math. **4** (1982), 47–74; *erratum, ibid.* **5** (1983), 248.
101. T. Zaslavsky, *The biased graphs whose matroids are binary*, J. Combin. Theory Ser. B **42** (1987), 337–347.
102. T. Zaslavsky, *Biased graphs. I. Bias, balance, and gains*, J. Combin. Theory Ser. B **47** (1989), 32–52; *II. The three matroids, ibid.* **51** (1991), 46–72; *III. Chromatic and dichromatic invariants*, to appear; *IV. Geometric realizations*, to appear.
103. T. Zaslavsky, *Biased graphs whose matroids are special binary matroids*, Graphs Combin. **6** (1990), 77–93.

DEPARTMENT OF MATHEMATICS, UNIVERSITY OF NORTH TEXAS, DENTON, TEXAS 76203, U. S. A.

Subexponentially Computable Truncations of Jones-type Polynomials

TERESA M.PRZYTYCKA and JÓZEF H. PRZYTYCKI

ABSTRACT. We show that an essential part of the new (Jones-type) polynomial link invariants can be computed in subexponential time. This is in a sharp contrast to the result of Jaeger, Vertigan and Welsh that computing the whole polynomial and most of its evaluations is $\#P$-hard.

1 Introduction

The discovery by V. Jones, in 1984, of a new powerful knot invariant led to a rapid growth of research in knot theory and elevated the theory of knots and links from its relative isolation. In particular, it has been noted that the "objects" similar to the Jones polynomial were studied in graph theory (the dichromatic polynomial) and statistical mechanics (e.g. the partition function for the Pott's model of anti-ferromagnetism). Jones type invariants of knots are widely used: from solving old problems in topology to applications in physics, chemistry, and biology (compare [11,37,40]).

The roots of this paper lie in a practical need for computing polynomial invariants for knots and links that have a large number of crossings. All currently known algorithms for the new knot polynomials have exponential time complexity with respect to the number of crossings (we consider a function $f(n)$ to have exponential growth if $f(n) = 2^{n^{\Theta(1)}}$). In practice, computing such a polynomial for a knot with more than 35 crossings is unrealistic.

1991 Mathematics Subject Classification. 57M25, 05C15; secondary 68R10.

This paper is in final form and no version of it will be submitted for publication elsewhere.

One can even argue that the main difference between old and new invariants of links is their computational complexity: practically all old link invariants (e.g. the Alexander polynomial, and the classical signature) can be computed in polynomial time while the Jones type invariants (i.e. the Jones, skein and Kauffman polynomials) are $\mathcal{NP}$-hard and therefore are believed not to have polynomial time algorithms. In fact it is believed that problems that are $\mathcal{NP}$-hard have at least exponential complexity (compare [18]).

It has been shown by Jaeger [25], in 1987, that the problem of computing the skein (Homfly) polynomial is $\mathcal{NP}$-hard. This result has been strengthened by Vertigan [77,78] in 1990 by proving that computing most of the evaluations of the Jones polynomial is #P-hard. Thus computing Jones type polynomials might still be intractable even if $\mathcal{P} = \mathcal{NP}$. In fact there are only a few evaluations of the Jones polynomial that can be computed in polynomial time. Interestingly all these evaluations have been well understood before.

In this paper we describe other restrictions of the Jones type polynomials which can be computed in subexponential time but still leave new and useful link invariants.

The paper is organized as follows:

In the second section, we describe two methods of translating graphs to knots and vice versa. We introduce also the important notion of a *matched diagram*.

In the third section, we describe polynomials of graphs (including chromatic or weighted graphs) emphasizing the dichromatic polynomial.

In the fourth section, we sketch the constructions of Jones type polynomials and describe the evaluations of these polynomials which can be computed in polynomial time.

In the fifth section, we analyze relations between link and graph polynomials.

In the sixth section, we show how to compute useful "quotients" of polynomial invariants in polynomial time.

In the seventh section, we show how to compute a part of the dichromatic polynomial in subexponential time. Then we describe the analogous result for the skein polynomial of links.

In the eighth section, we illustrate how our results are used for practical problems in knot theory (e.g. for computing braid index, and checking amphicheirality and periodicity).

In the last section, we discuss possible generalizations.

2 Dichromatic graphs and link diagrams

We consider two methods of assigning a link diagram to a plane graph. The first method is a classical one and was discovered by Tait more than a hundred years ago (see [5,9,33]). The second method is based on the idea of Jaeger [25].

First, we establish some notation and terminology. We should stress that the definition of a dichromatic graph which we give below has been chosen to reflect the dependency between knots and graphs.

By a graph $G = (V(G), E(G))$ we understand a finite graph (we allow multiedges and loops). By $|V(G)|$ (resp., $|E(G)|$) we denote the number of nodes (resp., edges) in G. The number of connected components is denoted by p_0 and the cyclomatic number by p_1 (i.e. $p_1 = |E(G)| - |V(G)| + p_0$).

An edge e is called an *isthmus* of a graph G if its removal disconnects a component of G. A *loop* is an edge (v, w) such that $v = w$.

We use $G - e$ to denote the graph obtained from graph G by removing edge e, and G/e to denote the graph obtained from $G - e$ by identifying the endvertices of edge e (contracting e).

The notion of a *chromatic* (or weighted) graph has been considered already by Kirchhoff [34]. Our definition is motivated by a connection between graphs and link diagrams, which will be explained later.

A *chromatic graph* is a graph with a function c on the edges, where $c : E(G) \mapsto Z \times \{d, l\}$. The first element of the pair $c(e)$ is called the *color* and the second the *attribute* (d - for dark, l - for light) of the edge e. Note that chromatic graphs are extension of signed graphs where the attribute of an edge corresponds to its sign (plus or minus).

A *plane graph* is a planar graph with its embedding on the plane.

The *dual* to a connected chromatic plane graph G is the graph $G^* = (V(G^*), E(G^*))$ where $V(G^*)$ and $E(G^*)$ are defined as for non-chromatic graphs and the edge e^* dual to e has assigned the same color as e and the opposite attribute. It is convenient to consider a plane graph G to lie on the sphere S^2, then the dual G^* is uniquely embedded in S^2. The dual of a non-connected graph is the disjoint sum of duals of its connected components.

By a *link* (resp., *oriented link*) we understand several circles (resp., oriented circles) embedded in S^3. Two links L_1 and L_2 are (ambient) *isotopic* if and only if there exists an isotopy $F : S^3 \times I \mapsto S^3$ such that $F_0 = Id$ and $F_1(L_1) = L_2$. If the links are oriented the isotopy must preserve the orientation. Informally two links are isotopic if and only if one can be continuously transformed to the other.

By a *link invariant* we mean a link isotopy class invariant.

A *diagram* D of a link L is a regular projection of L into the plane (*i.e.* a 4-valent graph) together with an overcrossing-undercrossing structure denoted as in Figure 2.1.

Figure 2.1

We can analyze properties of links up to isotopy by considering their diagrams exclusively. This is the case because of the following classical theorem of Reidemeister [60,2] (see also [9]; Proposition 1.14).

Theorem 2.1 (Reidemeister) *Two link diagrams (resp., oriented link diagrams) represent isotopic links if and only if they can be connected by a finite sequence of Reidemeister moves $\Omega_i^{\pm 1}$ ($i = 1, 2, 3$) shown in Figure 2.2 (in an oriented case we allow any consistent orientation of diagrams).*

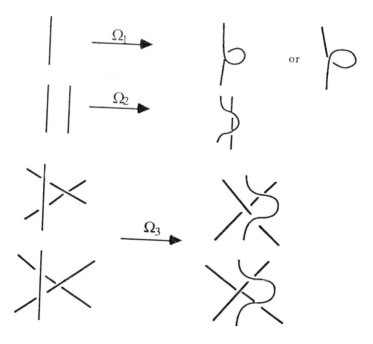

Figure 2.2

Now, we describe the classical correspondence between graphs and links.

Consider chromatic plane graphs with one color and attributes denoted by A and B. We call such graphs A, B-*graphs*. We can associate with an A, B-graph an unoriented link diagram, $D_N(G)$, together with a chessboard-like coloring of the plane, according to the rules presented in Figure 2.3.

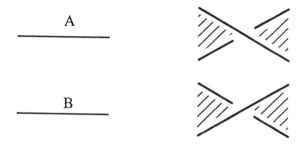

Figure 2.3

Observe that $D_N(G) = D_N(G^*)$ (as diagrams on S^2) with the role of white and black colors switched in the chessboard-like coloring.

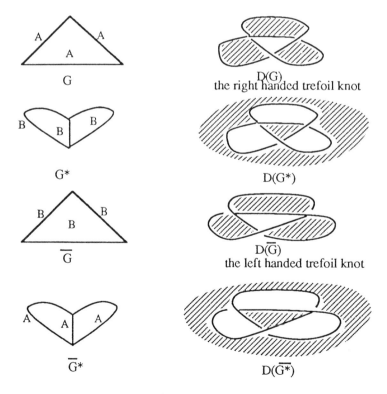

Figure 2.4

Let $\overline{G}$ be the graph obtained from an A, B-graph G by switching attributes of edges. Then $D_N(\overline{G}) = \overline{D_N(G)}$, where $\overline{D}$ denotes the mirror image of D (i.e overcrossings are changed to undercrossings and vice versa); compare Figure 2.4.

We use D_N, D_0, D_∞ to denote three link diagrams that are identical, except near one crossing point where they look like in Figure 2.5.

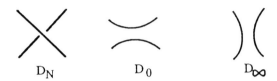

Figure 2.5

Let G be a plane A, B-graph and $e \in E(G)$ be an edge that is not a loop. If the crossing of D_N (see Figure 2.5) corresponds to the edge e then

(i) if the attribute of e is A then $D_N(G-e) = D_\infty(G)$ and $D_N(G/e) = D_0(G)$;

(ii) if the attribute of e is B then $D_N(G-e) = D_0(G)$ and $D_N(G/e) = D_\infty(G)$.

Finally, note that if we ignore the crossing structure of $D(G)$ we get the medial graph of G.

Now, we describe a correspondence between graphs and oriented links that is based on an idea of Jaeger [25].

Consider a chromatic graph with two colors. We call such a graph a *dichromatic graph*. The two colors of a dichromatic graph are denoted by $+$ and $-$. We use d^+ (resp., d^-) to denote an arbitrary dark edge of color $+$ (resp.$-$). By l^+ (resp., l^-) we an arbitrary light edge of color $+$ (resp. $-$). Let G be a dichromatic plane graph. We can associate with G an oriented link diagram, $D(G)$, and a chessboard-like coloring of the plane, according to the rules shown in Figure 2.6. (*i.e.* an edge (v, w) is replaced by one of the diagrams shown in Figure 2.6 in such a way that v and w remain in the black regions).

2.2

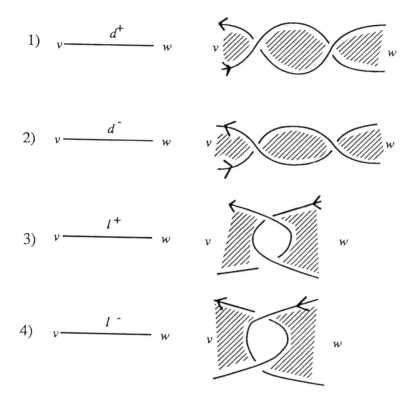

Figure 2.6

In particular we have:

Figure 2.7

We use the convention $D(\cdot) = o$ (*i.e.* the trivial knot diagram, o, corresponds to the one vertex graph). Diagrams of links obtained in the above manner are called *matched diagrams* [4] [1]. Matched diagrams have the following duality property:

Property 2.3 *If G^* is the dual to the dichromatic plane graph G then $D(G) = -D(G^*)$ with black and white regions exchanged, where $-D$ is obtained form D by reversing the orientation of each component of D.*

Let D_+, D_-, D_0 denote diagrams of links that are identical, except near one crossing point where they look like in Figure 2.8. We associate a sign (+ or -) to each crossing according to the convention presented in Figure 2.8.

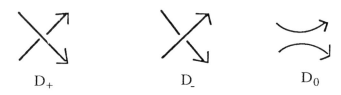

Figure 2.8

If one considers link diagrams up the second Reidemeister move then we have the following useful correspondence (edges d^+, l^+, d^-, l^- correspond to crossings of D_+, D_- in Figure 2.8; we choose (either) one of the (two) crossings in $D_\pm(G)$):

Property 2.4 *For any edge that is not a loop we have*

(i) $\quad D(G - d^+) = D_0(G) \qquad\qquad D(G/d^+) = D_-(G)$

(ii) $\quad D(G - d^-) = D_0(G) \qquad\qquad D(G/d^-) = D_+(G)$

(iii) $\quad D(G - l^+) = D_-(G) \qquad\qquad D(G/l^+) = D_0(G)$

(iv) $\quad D(G - l^-) = D_+(G) \qquad\qquad D(G/l^-) = D_0(G).$

It is an open question whether every link has a matched diagram. It is very unlikely that it is the case. Nevertheless any 2-bridge link [9] possesses a matched diagram.

Consider the following move on an oriented link diagram (called in [54] a

---

[1] A concept similar to matched diagrams was also considered by Conway in late 1960's (personal communication, Seattle, July 1991).

t_3-move).

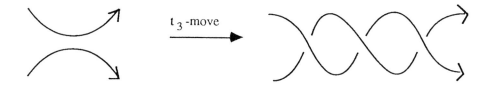

Figure 2.9

The following conjecture can be stated:

Conjecture 2.5 ([58]) *Any oriented link has a diagram t_3-equivalent to a matched diagram (i.e. any oriented link has a diagram that can be transformed to a matched diagram by t_3 moves, their inverses, and Reidemeister moves).*

For us, the importance of t_3 moves is implied by the following observation. If G' is obtained from G by changing an edge d^- into l^+ (or any of the following: d^+ into l^-, l^+ into d^-, l^- into d^+) then $D(G)$ and $D(G')$ are t_3-equivalent as illustrated in Figure 2.10.

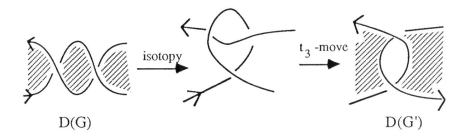

Figure 2.10

This idea was used in [58] to prove partially the conjecture 14C in [42].

We use t_3-equivalence of link diagrams to obtain the following commutative diagram between graphs and link diagrams. We will refer to Figure 2.11 when comparing various invariants of graphs and links.

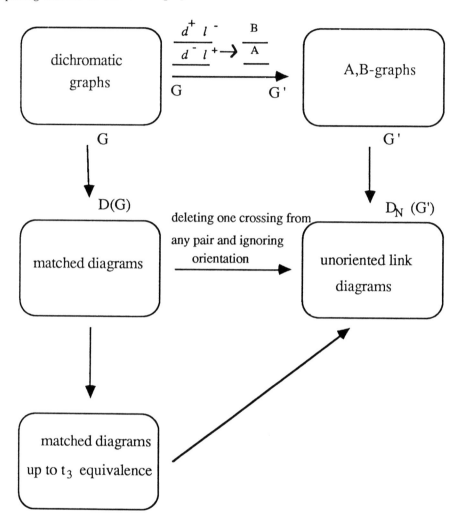

Figure 2.11

3 Polynomial invariants of graphs

Let $\tau(G)$ denote the *complexity* of graph G, that is, the number of its spanning trees. Invariant τ was introduced and studied by Kirchhoff [34]. It has been noted in [7] that if e is an edge of G that is not a loop then

$$\tau(G) = \tau(G - e) + \tau(G/e) \tag{1}$$

As noted by Tutte ([76]; p51), this equality had been long familiar to the authors of [7]. It was the equality (1) that inspired Tutte to investigate graphs invariants, $W(G)$, which satisfy the following identity

$$W(G) = W(G - e) + W(G/e) \tag{2}$$

This led to the discovery of the dichromatic polynomial and its variant, the Tutte polynomial [73,75]. [2]

Fortuin and Kastelyn generalized the dichromatic polynomial to chromatic (weighted) graphs [16] (compare also Heilmann [19]). The research of [16,19] was motivated by "statistical mechanics" considerations. One should stress here that only slightly earlier Temperley discovered that the partition function for the Potts model is equivalent to the dichromatic polynomial of the underlying graph [13,65].

The dichromatic polynomial for chromatic graphs gained new importance after the Jones discovery of new polynomial invariants of links and the observation of Thistlethwaite that the Jones polynomial of links is closely related to the Tutte polynomial of graphs. Several researches rediscovered the dichromatic polynomial and analyzed its properties [32,46,72,51,85,88].

The following version of the dichromatic polynomial is motivated by connections between graphs and links.

Theorem 3.1 *There exists an invariant of chromatic graphs $R(G) = R(G; \mu, r_1, r_2, A_i, B_i)$ which is uniquely defined by the following properties:*
1. $R(T_n) = \mu^{n-1}$; where T_n is the n-vertex graph with no edges,
2.
$$R(G) = (\frac{r_1}{\mu})^{\epsilon(d_i)} B_i R(G - d_i) + r_2^{\delta(d_i)} A_i R(G/d_i)$$

$$R(G) = (\frac{r_1}{\mu})^{\epsilon(l_i)} A_i R(G - l_i) + r_2^{\delta(l_i)} B_i R(G/l_i)$$

where

$$\epsilon(e) = \begin{cases} 0 & \text{if } e \text{ is not an isthmus} \\ 1 & \text{if } e \text{ is an isthmus} \end{cases}$$

$$\delta(e) = \begin{cases} 0 & \text{if } e \text{ is not a loop} \\ 1 & \text{if } e \text{ is a loop} \end{cases}$$

---

[2] H.Whitney [80,81] was considering graph invariants $m_{i,j}$ which are essentially the coefficients of the dichromatic polynomial. He also analyzed closer the topological graph invariants m_i which corresponds to the coefficient of the flow polynomial [83]. R.M.Foster noticed [81] that $m_{i,j}$ invariants satisfy $m_{i,j}(G) = m_{i,j}(G - e) + m_{i-1,j}(G/e)$.

Our variables have been chosen in such a way that the invariants for a plane graph G and its dual G^* are symmetric in the following sense:

Lemma 3.2 *If G is a plane graph then*

$$R(G) = R(G; \mu, r_1, r_2, A_i, B_i) = R(G^*; \mu, r_2, r_1, A_i, B_i)$$

Note, that $R(G)$ is a 2-isomorphism invariant of connected chromatic graphs. Generally, when G is not necessarily connected and $\mu \neq 1$ then the polynomial measures also the number of connected components of the graph. If we put $\mu = 1$ then the dichromatic polynomial, R, and its property described in Lemma 3.2 can be extended to matroids (see [88] for a full analysis of the Tutte polynomial of colored matroids) or more generally to colored Tutte set systems (see [53]).

Let S denote a subset of edges of a graph G. By $(G : S)$ we denote the subgraph of G which includes all the vertices of G but only edges in S. The polynomial $R(G)$ has the following "state model" expansion:

Lemma 3.3

$$R(G; \mu, r_1, r_2, A_i, B_i) =$$
$$\mu^{p_0(G)-1} \sum_{S \in 2^{E(G)}} r_1^{p_0(G:S)-p_0(G)} r_2^{p_1(G:S)} \left(\prod_{i=1}^{n} A_i^{\alpha_i + \alpha'_i} \cdot B_i^{\beta_i + \beta'_i} \right)$$

where the sum is taken over all subsets of $E(G)$, and α_i is the number of dark edges in S of the ith color, α'_i is the number of light edges in $E(G) - S$ of the ith color, β_i is the number of dark edges in $E(G) - S$ of the ith color, and β'_i is the number of light edges in S of the ith color.

In the above lemma we consider a subset S of the set of edges to be the state of G in the sense that edges in S are marked to be contracted and the edges in $E(G) - S$ are marked to be deleted.

Below we list a few easy but useful properties of $R(G)$.

Lemma 3.4
(i) $R(\overline{G}; \mu, r_1, r_2, A_i, B_i) = R(G; \mu, r_1, r_2, B_i, A_i)$
(ii) *For any i, the number of ith colored edges of G is equal to $\alpha_i + \alpha'_i + \beta_i + \beta'_i$ which is equal to the highest power of A_i in $R(G)$*
(iii) *If $G_1 * G_2$ is a one vertex product of G_1 and G_2 and $G_1 \sqcup G_2$ is a disjoint sum of G_1 and G_2 then*

$$R(G_1 \sqcup G_2) = \mu R(G_1 * G_2) = \mu R(G_1) R(G_2)$$

(iv) *If G is a loop or isthmus then we have*

$$R(\mathcal{O}_{d_i}) = B_i + r_2 A_i$$

$$R(\mathcal{O}_{l_i}) = A_i + r_2 B_i$$

$$R(\multimap_{d_i}) = A_i + r_1 B_i$$

$$R(\multimap_{l_i}) = B_i + r_1 A_i$$

where d_i (resp.,l_i) denotes a dark (resp., a light) edge of the ith color.

(v) *If $Q(G;t,z)$ is the Traldi's version of the dichromatic polynomial [72] then*

$$Q(G;t,z) = \frac{tR(G;\mu,r_1,r_2,A_i,B_i)}{\prod_i B_i^{E_i(G)}}$$

where $r_1 = \mu = t, r_2 = z$, $E_i(G)$ denotes the number of ith colored edges in G, and the weight, $w(e)$, of an edge e of G is defined by:

$$w(e) = \begin{cases} \frac{A_i}{B_i} & if\ e\ is\ a\ d_i\ edge \\ \frac{B_i}{A_i} & if\ e\ is\ an\ l_i\ edge \end{cases}$$

Note that both versions of the dichromatic polynomial are equivalent because, by Lemma 3.3 (ii), $E_i(G)$ is determined by $R(G)$. Furthermore $Q(G;t,z)$ determines $E_i(G)$ and $p_0(G)$.

4 Polynomial invariants of links

4.1

In 1928 J.W.Alexander found the Laurent polynomial invariant of oriented links and proved that polynomials of L_+, L_-, L_0 (compare Figure 2.8) are linearly related [1]. In early 1960's, J. Conway rediscovered Alexander's formula and normalized the Alexander polynomial, $\Delta_L(t) \in Z[t^{\pm 1/2}]$, defining it recursively as follows:

(i) $\Delta_o(t) = 1$, where o denotes a knot isotopic to a simple circle

(ii)
$$\Delta_{L_+} - \Delta_{L_-} = (\sqrt{t} - \frac{1}{\sqrt{t}})\Delta_{L_0}$$

It was already known to Alexander and Conway that computing $\Delta_L(t)$ has polynomial time complexity. This follows from the fact that the computation of this polynomial can be reduced to the computation of the determinant of a certain matrix.

4.2

In the spring of 1984, V.Jones discovered his invariant of links, $V_L(t) \in Z[t^{\pm 1/2}]$, and in July 1984 Jones [27], and Lickorish and Millett proved that the Jones polynomial is defined recursively as follows:

(i) $V_o = 1$,

(ii) $\frac{1}{t}V_{L_+}(t) - tV_{L_-}(t) = (\sqrt{t} - \frac{1}{\sqrt{t}})V_{L_o}(t)$.

Jaeger, Vertigan and Welsh [26] proved that computing $V_L(t)$ is #P-hard. This result has been strengthened by Vertigan [77,78], who showed that computing $V_L(c)$ is #P-hard for any non-zero complex number c except for $c = \pm 1, \pm i, \pm e^{2\pi i/3}, \pm e^{4\pi i/3}$ where the time complexity is polynomial.

4.3

In the summer and the fall of 1984, the Alexander and the Jones polynomials were generalized to the skein (named also Flypmoth, Homfly, generalized Jones, 2-variable Jones, Jones-Conway, Thomflyp, twisted Alexander) polynomial, $P_L \in Z[a^{\pm 1}, z^{\pm 1}]$, of oriented links. This polynomial is defined recursively as follows [15,59]:

(i) $P_o = 1$;

(ii) $aP_{L_+} + a^{-1}P_{L_-} = zP_{L_o}$.

In particular $\Delta_L(t) = P_L(i, i(\sqrt{t} - \frac{1}{\sqrt{t}}))$, $V_L(t) = P_L(it^{-1}, i(\sqrt{t} - \frac{1}{\sqrt{t}}))$. Jaeger [25] proved that computing $P_L(a, z)$ is $\mathcal{NP}$-hard. This result was generalized by Vertigan [77,78] to show that computing $P_L(a_0, z_0)$ is #P-hard for any non-zero complex numbers a_0, z_0 except

1. $a_0 = \pm i$, (then P_L reduces to the Alexander polynomial),
2. $z_0 = \pm(a_0 + a_0^{-1})$
3. $(a_0, z_0) = (\pm 1, \pm\sqrt{2})$
4. $(a_0, z_0) = (\pm 1, \pm 1)$
5. $(a_0, z_0) = (\pm e^{\pm \pi i/6}, \pm 1)$

where $\pm$'s are independent.

4.4

In 1985 L.Kauffman found another approach to the Jones polynomial. It starts from an invariant, $<D> \in Z[\mu, A, B]$, of an unoriented link diagram D called the Kauffman bracket [31]. The Kauffman bracket is defined recursively by:

(i)
$$<\underbrace{o\ldots o}_{i}> = \mu^{i-1}$$

(ii)
$$<\!\asymp\!> = B<\!\smile\!> + A<\!)(\!>$$

(iii)
$$<\!\times\!> = A<\!\smile\!> + B<\!)(\!>$$

where $\asymp$, $\times$, $\smile$ and $)($ denote four diagrams that are identical except near one crossing as shown on the diagrams, and $<\underbrace{o\ldots o}_{i}>$ denotes a diagram of i trivial components (i simple circles).

The Kauffman bracket is a source of our version of the dichromatic polynomial of chromatic graphs.

If we assign $B = A^{-1}$ and $\mu = -(A^2 + A^{-2})$ then the Kauffman bracket gives a variant of the Jones polynomial for oriented links. Namely, for $A = t^{-\frac{1}{4}}$ and D being an oriented diagram of L we have

$$V_L(t) = (-A^3)^{-w(D)} <D> \qquad (3)$$

where $w(D)$ is the *planar writhe* (*twist* or *Tait number*) of D equal to the algebraic sum of signs of crossings.

It should be noted, as observed first by Kauffman, that bracket $<\ >_{\mu,A,B}$ is an isotopy invariant of alternating links (and their connected sums) under the assumption that the third Tait conjecture (recently proven by Menasco and Thistlethwaite [40]) holds.

4.5

In the summer of 1985, L.Kauffman discovered another invariant of links, $F_L(a,z) \in Z[a^{\pm 1}, z^{\pm 1}]$, generalizing the polynomial discovered at the beginning of 1985 by Brandt, Lickorish, Millett and Ho [6,20]. To define the Kauffman polynomial we first introduce the polynomial invariant of link diagrams $\Lambda_D(a,z)$. It is defined recursively by:

(i) $\Lambda_o(a,z) = 1$,

(ii) $\Lambda_{\curvearrowright}(a,z) = a\Lambda_{\mid}(a,z)$; $\Lambda_{\curvearrowleft}(a,z) = a^{-1}\Lambda_{\mid}(a,z)$,

(iii) $\Lambda_{\times}(a,z) + \Lambda_{\times}(a,z) = z(\Lambda_{\smile}(a,z) + \Lambda_{)(}(a,z))$.

The Kauffman polynomial of oriented links is defined by

$$F_L(a,z) = a^{-w(D)} \Lambda_D(a,z)$$

where D is any diagram of an oriented link L. It was observed in [66] that computing $F_L(a,z)$ is NP-hard. This result has been strengthened by Vertigan [77,78,79] to show that computing $F_L(a_0, z_0)$ is $\#P$-hard with some exceptions. Most of these exceptions were already studied and understood before (see [37], and [29,64] for the additional case $(a_0, z_0) = (1, 2\cos\frac{2\pi}{5})$). The exception $(a_0, z_0) = (-q^{\pm 1}, q + q^{-1})$ where $q^8 = -1$ seems to be noted in [77,78,79] for the first time.

5 Relations between link and graph polynomials

The two correspondences between graphs and link diagrams described in Section 2 led to relations between polynomial invariants of links and graphs (described in sections 3 and 4). Such a relation was observed for the first time by Thistlethwaite [66].

We take Figure 2.11 as the base for our comparison.

I. Starting from the skein polynomial of links (in particular of matched diagrams) $P_D(a, z)$, we can define the dichromatic plane graphs invariant $W(G)$ by putting

$$W(G) = P_{D(G)}(a, z).$$

Then the following holds (see Section 3 for notation):

5.1

$$W(G) = a^{-1}zW(G - d^+) - a^{-2}W(G/d^+)(\frac{a+a^{-1}}{z})^{\delta(d^+)}$$

$$W(G) = -a^{-2}W(G - l^+) + a^{-1}zW(G/l^+)(\frac{a+a^{-1}}{z})^{\delta(l^+)}$$

$$W(G) = azW(G - d^-) - a^2W(G/d^-)(\frac{a+a^{-1}}{z})^{\delta(d^-)}$$

$$W(G) = -a^2W(G - l^-) + azW(G/l^-)(\frac{a+a^{-1}}{z})^{\delta(l^-)}$$

$$W(T_i) = (\frac{a+a^{-1}}{z})^{i-1}.$$

One can easily recognize that $W(G)$ is a special case of the dichromatic polynomial $R(G, \mu, r_1, r_2, A_\pm, B_\pm)$, for $r_1 = r_2 = \mu = \frac{a+a^{-1}}{z}$, $A_+ = -a^{-2}, B_+ =$

$a^{-1}z, A_- = -a^2, B_- = az$. The equality $r_1 = r_2$ reflects the fact that $D(G) = -D(G^*)$ (see Property 2.3)

II. A relation of the skein polynomial of $D(G)$ to the Kauffman bracket is presented in the following lemma:

Lemma 5.2 *Let G be a dichromatic plane graph, let G' be the corresponding A, B-graph (see Figure 2.11), let $D_N(G')$ be the unoriented link diagram obtained from G' by rules from Figure 2.3 and let $D(G)$ be the oriented link diagram obtained from G by rules described in Figure 2.6. Let in_+ (resp., in_-) be the number of positive (resp., negative) edges of G. Then for $B^{-1} = A = a^{1/2}, \mu = -A^2 - A^{-2}$ the following holds:*

$$<D_N(G')> = (-a^{\frac{3}{2}})^{in_+ - in_-} P_{D(G)}(a, -1) = (-a^{\frac{3}{2}})^{\frac{1}{2}w(D(G))} P_{D(G)}(a, -1).$$

Proof: Note that for $z = -1$ we can rewrite 5.1 as follows (see formulas (i) - (iv) of Section 2)

$$-a^{\frac{3}{2}} P_{D(G)}(a, -1) = a^{\frac{1}{2}} P_{D(G-d^+)}(a, -1) + a^{-\frac{1}{2}} P_{D(G/d^+)}(a, -1)(-(a+a^{-1}))^{\delta(d^+)}$$

$$-a^{\frac{3}{2}} P_{D(G)}(a, -1) = a^{-\frac{1}{2}} P_{D(G-l^+)}(a, -1) + a^{\frac{1}{2}} P_{D(G/l^+)}(a, -1)(-(a+a^{-1}))^{\delta(l^+)}$$

$$-a^{-\frac{3}{2}} P_{D(G)}(a, -1) = a^{-\frac{1}{2}} P_{D(G-d^-)}(a, -1) + a^{\frac{1}{2}} P_{D(G/d^-)}(a, -1)(-(a+a^{-1}))^{\delta(d^-)}$$

$$-a^{-\frac{3}{2}} P_{D(G)}(a, -1) = a^{\frac{1}{2}} P_{D(G-l^-)}(a, -1) + a^{-\frac{1}{2}} P_{D(G/l^-)}(a, -1)(-(a+a^{-1}))^{\delta(l^-)}$$

$$P_{D(T_i)}(a, -1) = [-(a+a^{-1})]^{i-1}.$$

The lemma follows by comparing the above formulas with corresponding formulas for the Kauffman bracket $<D_N(G')>$. □

If we orient $D_N(G')$ to get $D_N^{or}(G')$ in such a way that the orientation agrees with that of $D(G)$ (i.e. positive crossings of $D_N^{or}(G')$ correspond to positive crossings of $D(G)$) then the following holds:

Corollary 5.3 *For $a = t^{-\frac{1}{2}}$,*

$$V_{D_N^{or}(G')}(t) = P_{D(G)}(a, -1).$$

In [36] there are given necessary and sufficient conditions for existence of $D_N^{or}(G')$.

We have mentioned before that it is unlikely that every link has a matched diagram (*i.e.* has a diagram of a form $D(G)$). One can hope to use Lemma 5.2 and some properties of skein polynomials and the Kauffman bracket to find a link without a matched diagram.

One observation should be made. If $\tilde{G}$ is obtained by changing an edge l^- to d^+ in a planar graph, G, then the associated nonoriented diagrams are equal (*i.e.* $D_N(G') = D_N(\tilde{G}')$) and therefore $< D_N(G') > = < D_N(\tilde{G}') >$. Now, by Lemma 5.2, $P_{D(\tilde{G})}(a,-1) = a^{-3} P_{D(G)}(a,-1)$. The last equality can be put in a more general context. Namely, $D(G)$ and $D(\tilde{G})$ are t_3-equivalent (see Figure 2.9) and it can be easily checked that for any t_3 move on a diagram D, $P_{t_3(D)}(a,-1) = a^{-3} P_D(a,-1)$ (compare [54]).

6 Polynomial algorithms

We express the time complexity of our algorithms as a function of the number of crossings, m, and we assume that the number of link components, $\text{com}(L)$, of a link L is less than or equal to the number of crossings [3].

The main goal of the paper is to show that a substantial part of the skein (Homfly) polynomial $P_L(a,z)$ can be computed in subexponential time (more precisely in $O(m^{c \ln m})$). We will discuss the corresponding algorithm in the next section. In this section we show that "essential parts" of the new polynomial link invariants can still be computed in polynomial time (so faster than in $O(m^{c \ln m})$ time) where "essential" can be interpreted as useful for nontrivial applications, as discussed in Section 8.

Because of the recursive definition of the new link invariants, the computation of such an invariant can be visualized with the help of a *resolving tree*. In a resolving tree of a link polynomial (resp., of a graph polynomial) each internal node corresponds to one application of the recursive formula and each leaf is labeled with a polynomial of a trivial link (resp., of a graph whose polynomial can be immediately computed).

Theorem 6.1 *Let $s \geq 2$ be a natural number and let $((t-1)^s)$ be the ideal in $Z[t^{\pm \frac{1}{2}}]$ generated by $(t-1)^s$. If $V_L(t)$ is the Jones polynomial of a link diagram L then*

$$V_L(t) \bmod ((t-1)^s)$$

can be computed in $O(m^s)$ time, where m denotes the number of crossings of L.

---

[3] This inequality is implied by an assumption that no component of a diagram is a simple circle, so there is no real loss of generality.

Proof: The theorem can be derived from the analogous theorem for the skein polynomial (Theorem 6.2). However, as a warm-up, we give a separate proof, using the Kauffman bracket $<L> \in Z[A^{\pm 1}]$ and unoriented link diagrams.

From the identities

$$<\!\!\times\!\!> = A <\!\!\asymp\!\!> + A^{-1} <\!\!)(\!\!>$$

and

$$<\!\!\times\!\!> = A^{-1} <\!\!\asymp\!\!> + A <\!\!)(\!\!>$$

we get

$$A <\!\!\times\!\!> - A^{-1} <\!\!\times\!\!> = (A^2 - A^{-2}) <\!\!\asymp\!\!> \qquad (4)$$

This formula can be used to build a binary *resolving tree* for L, leaves of which can be used to compute $<L>$. An internal node of this tree looks as in Figure 6.1 and leaves represent *descending diagrams* (see [22,39,59] for the definition of a descending diagram; the important fact is that descending diagrams represent trivial links) for which

$$<D> = (-A^3)^{w(D)}(-A^2 - A^{-2})^{com(D)-1}$$

where $w(D)$ is the writhe (or Tait) number of D and $com(D)$ is the number of (link) components of D. One can easily show (compare [22,39,59]) that the depth of the resolving tree is less than m.

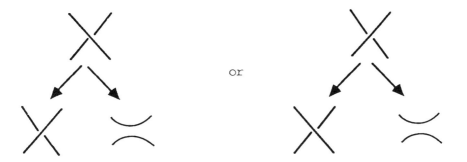

Figure 6.1

The crucial observation is that if D^l is a leaf of the resolving tree such that the path joining the root with this leaf goes at least s times to the right then the polynomial introduced to $<D>$ by D^l is divisible by $(A^2 - A^{-2})^s$. Therefore if we are interested in $<D> \bmod ((A^2 - A^{-2})^s)$ we can ignore such a leaf. Thus to find $<D> \bmod ((A^2 - A^{-2})^s)$ it suffices to consider at most $\sum_{i=0}^{s-1} \binom{m}{i} =$

$O(m^{s-1})$ leaves. Therefore the number of nodes of the resolving tree required by the computation of $<D> \mod((A^2 - A^{-2})^s)$ is $O(m^s)$.

Furthermore all possible polynomials for the leaves can be tabulated in $O(m^2)$ time. For each leaf we compute, based on the path from that leaf to the root, the polynomial by which the given leaf is multiplied in the process of computing the whole polynomial. To keep the algorithm within $O(m^s)$ time complexity we tabulate values of polynomials $(A^2 - A^{-2})^i, (-A^2 - A^{-2})^j, i = 1\ldots m, j = 1, \ldots, s$. Then, for every leaf, we can compute the corresponding polynomial in $O(sm)$ time and order the sum of the leaves in $O(m^s)$ time using the classic linear time integer sorting algorithm.

Now, the theorem follows from the fact that

$$V_D(t) = (-A^3)^{-w(D)} <D> \text{ for } A = t^{-\frac{1}{4}}.$$

□

Our result should be compared with that of Jaeger, Vertigan, and Welsh [26,77,78] (see Section 4). Namely, Theorem 3 of [26] (see 4.2) can be interpreted as an answer to the question: For which polynomial $w \in Z[t]$, without multiple roots, can the Jones polynomial modulo the ideal generated by w ($V_L(t) \mod(w)$) be computed in polynomial time. We allow w to have multiple root. Then Theorem 6.1 gives an example of w with multiple root ($t = 1$) for which a polynomial time algorithm exists. The simplest open question is whether $V_D(t) \mod((t+1)^s)$ can be computed in polynomial time for $s \geq 2$.

Theorem 6.2 *Let $\mathcal{R}$ be the subring of the ring $Z[a^{\pm 1}, z^{\pm 1}]$ generated by $a^{\pm 1}$, $\frac{a+a^{-1}}{z}$ and z, then for any oriented link L:*
(a) *the skein polynomial $P_L(a, z) \in \mathcal{R}$, and*
(b) $P_L(a, z) \mod (z^s) \in \mathcal{R}/(z^s)$
can be computed in polynomial time (more precisely in $O(m^s)$ time) where (z^s) is the ideal of $\mathcal{R}$ generated by z^s and m denotes the number of crossings of L.

Proof:
(a) The proof of this part follows by an easy induction on the depth of the binary resolving tree for L (see [57]: Lemma 1.1).
(b) We proceed as in the proof of Theorem 6.1. The relation $aP_{L_+}(a, z) + a^{-1}P_{L_-}(a, z) = zP_{L_0}(a, z)$ allows us to build, for any link diagram L, a binary resolving tree with leaves representing descending diagrams. As in Theorem 6.1, the depth of the tree is less than m and the number of leaves

which contribute to $P_L(a,z)\mod(z^s)$ is no more than $\sum_{i=0}^{s-1} \binom{m}{i}$.
Therefore the number of internal nodes of the resolving tree introduced in the computation of $P_L(a,z)\mod(z^s)$ is $O(m^s)$.

□

The following useful criterion (which is a slight modification of Lemma 1.5 of [57]) can be used as a tool to see how much of the skein polynomial survives if we consider it modulo (z^n) in $\mathcal{R}$.

Lemma 6.3 *Let $w(a,z) \in \mathcal{R}$ and $w(a,z) = \sum_i v_i(a) z^i$ where $v_i(a) \in Z[a^{\pm 1}]$. Then $w(a,z)$ is in the ideal (z^s) if and only if for all $i \leq s$, $v_i(a)$ is in the ideal $((a+a^{-1})^{s-i})$ of $Z[a^{\pm 1}]$.*

Proof: Let $\mathcal{R}'$ denote the ring $\mathcal{R}$ treated as a $Z[a^{\pm 1}]$ module. Let I_s be the ideal in $\mathcal{R}'$ generated by the elements $z^i(a+a^{-1})^{s-i}, (i \leq s)$ and $z^i, (i > s)$. Because I_s was chosen so that $w(a,z) \in I_s$ if and only if $v_i(a) \in ((a+a^{-1})^{s-i})$, where $i \leq s$, it therefore suffices to prove that $I_s = (z^s)$.

First note that $I_s \subset (z^s)$ because $z^i((a+a^{-1})^{s-i}) = z^s(\frac{a+a^{-1}}{z})^{s-i} \in (z^s)$. On the other hand I_s, which is an ideal in $\mathcal{R}'$, is also an ideal in $\mathcal{R}$ and contains z^s. This is the case because $z(z^i(a+a^{-1})^{s-i}) = z^{i+1}((a+a^{-1})^{s-i-1})(a+a^{-1}) \in I_s (i < s), z(z^i) = z^{i+1} \in I_s$ $(i \geq s)$ and $\frac{a+a^{-1}}{z}(z^i(a+a^{-1})^{s-i}) = z^{i-1}(a+a^{-1})^{s-i+1} \in I_s (i \leq s)$. Therefore $I_s = (z^s)$. □

As noted in 4.3 (point 1), $P_L(a,z)\mod(a^2+1)$ can be computed in polynomial time. This follows from the fact that this formula is equivalent to the Alexander polynomial. It is however an open question of great interest whether $P_L(a,z)\mod((a^2+1)^s)$ can be computed in polynomial time. Similarly, from the condition 4.3 4. we deduce that $P_L(a,z)\mod(a^2-1, z^2-1)$ can be computed in polynomial time. Now considering the map from oriented matched diagrams to unoriented diagrams, $D(G) \mapsto D_N(G')$ (see Figure 2.11), and using the relation of Lemma 5.2, we can show that for matched diagrams $P_L(a,z)\mod((a^2-1)^s, z^2-1)$ can be computed in polynomial time. Theorem 6.2 and Lemma 6.3 are related to a recent work of P. Traczyk on the Poincaré Conjecture [71].

Namely, Traczyk essentially shows that the third skein module [4] of a homotopy 3-sphere, $S_3(\Sigma^3)$, is generated by the unknot if considered modulo (z^s). Traczyk is testing various available homotopy spheres to check whether the un-

---

[4] A skein module is a generalization of the skein polynomial to any oriented 3-manifold. For an overview of skein modules see [56,23].

knot is a free generator of $S_3(\Sigma^3)$mod (z^s) (if it is not, a counterexample to the Poincaré Conjecture would be found).

More generally, if a 3-manifold, M, is simply connected (not necessarily compact) then $S_3(M)\mod(z^s)$ is generated by the unknot. This can be used to prove that for uncountably many Whitehead type manifolds the third skein module $S_3()$ is not free (compare [24]).

The main idea behind theorem 6.2 was to see how much of the skein polynomial of the link is preserved if we ignore nodes which can be reached from the root by going s or more times to the right. In terms of link diagrams this is equivalent to saying that we perform at most $s-1$ smoothings on a link diagram. We can weaken this condition in two ways: (i) by allowing at most $s-1$ smoothings of selfcrossings of a link diagram (this is the idea behind Section 7), or (ii) by allowing at most $s-1$ smoothings between different components of a link. The later idea awaits exploration; however, for a fixed number of link components we get a polynomial time algorithm.

Theorem 6.4 *The Kauffman polynomial $F_L(a,z)$ is an element of the ring $\mathcal{R}$ and it can be computed $mod(z^s)$ in a polynomial time (more precisely in $O(m^s)$ time).*

Proof: It is convenient to work with $\Lambda_L(a,z)$ which is an invariant of unoriented link diagrams (up to Reidemeister moves II and III). The fact that $\Lambda_L(a,z) \in \mathcal{R}$ follows by an easy inductive argument: this is obviously true when L is a trivial link diagram and from the fact that this is true for three diagrams ╳,≍,)(or ╳,≍,)(of Figure 6.2 it follows that it is true for the fourth diagram of Figure 6.2.

Figure 6.2

To compute $\Lambda_L(a,z)\mod(z^s)$, we build a trinary resolving tree for L, leaves of which represent descending diagrams (so their polynomial can be immediately found). The internal nodes of the tree look as in the Figure 6.3.

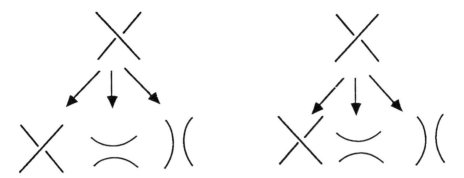

Figure 6.3

As in Theorem 6.1 and 6.2, the depth of the tree is less than m, where m is the number of crossings. Furthermore the leaves of the tree that can be reached from the root by going at least s times right (*i.e* following the right or the center branch of a tree presented in Figure 6.3), not necessarily in consecutive steps, introduce to $\Lambda_L(a,z)$ a polynomial divisible in $\mathcal{R}$ by z^s. Thus these leaves can be ignored in computing $\Lambda_L(a,z) \bmod(z^s)$. Therefore to find $\Lambda_L(a,z) \bmod (z^s)$ we need to consider at most $\sum_{i_1+i_2<s} \begin{pmatrix} m \\ i_1, i_2 \end{pmatrix}$ leaves, where $\begin{pmatrix} m \\ i_1, i_2 \end{pmatrix} = \frac{m!}{i_1! i_2! (m-i_1-i_2)!}$. Therefore the number of internal nodes of the resolving tree involved in the computation is $O(m^s)$. □

7 The $O(m^{c \ln m})$ algorithm

As we mentioned in the introduction, computing Jones type invariants is #P-hard. However two non-isotopic knots (resp., non-isomorphic graphs) can be often distinguished by computing only a part of an invariant. One way of doing this is to compute some evaluations of a polynomial of a graph or knot. However by the result of Vertigan [77,78], this approach does not lead to efficient algorithms (except for few specific evaluations). In the previous section we presented polynomial time algorithms to compute various link invariants modulo some ideals. In this section we take an alternative approach. Namely we discuss complexity of computing the s first coefficients of the dichromatic polynomial for chromatic planar graphs.

We assume that the number of connected components in the graph is $O(m)$. The algorithm which we present has an intriguing complexity of $O(m^{c \ln m})$

where m is the number of edges. Thus it is subexponential but not polynomial. The proof, we present, translates directly to matched diagrams of links.

A number of useful properties of $R(G)$ were given in Lemmas 3.3 and 3.4

We will change the variables in the dichromatic polynomial of chromatic graphs so that the variable μ corresponds to the variable z^{-1} of the skein polynomial.

Let $B'_i = B_i\mu$, $A'_i = A_i$, and $r'_1 = r_1\mu^{-1}$, $r'_2 = r_2\mu^{-1}$. Then by rewriting the recursive definition of $R(G)$ (see 3.1) we obtain the invariant $\hat{R} \in Z[\mu^{\pm 1}, r'_1, r'_2, A'_i, B'_i]$ with the following recursive properties:

1. $\hat{R}(T_n) = \mu^{n-1}$; where T_n is the n-vertex graph with no edges,

2.
$$\hat{R}(G) = (r'_1)^{\epsilon(d_i)}\mu^{-1}B'_i\hat{R}(G - d_i) + (r'_2\mu)^{\delta(d_i)}A'_i\hat{R}(G/d_i)$$
$$\hat{R}(G) = (r'_1)^{\epsilon(l_i)}A'_i\hat{R}(G - l_i) + (r'_2\mu)^{\delta(l_i)}\mu^{-1}B'_i\hat{R}(G/l_i)$$

Denote by G_d (resp.G_l) the graph obtained from G by removing all light (resp. dark) edges (but retaining all vertices) and by $\deg_\mu P$ the maximal degree of μ in the polynomial P. Write

$$\hat{R}(G) = \sum_{j \leq \deg_\mu \hat{R}(G)} q_j(r'_1, r'_2, A'_i, B'_i)\mu^j$$

We abbreviate $q_j(r'_1, r'_2, A'_i, B'_i)$ to q_j.

The following two lemmas will be proven by a joint inductive argument.

Lemma 7.1 $\deg_\mu \hat{R}(G) = p_0(G_d) + p_1(G_d) - 1$.

Lemma 7.2 *Let l (resp., d) be a light (resp., dark) edge, then*

$$deg_\mu \hat{R}(G - l) = deg_\mu \hat{R}(G)$$

$$deg_\mu \hat{R}(G/d) + \delta(d) = deg_\mu \hat{R}(G)$$

$$deg_\mu \hat{R}(G/l) + \delta(l) = \begin{cases} deg_\mu \hat{R}(G) + 1 & \text{if both endverticces of } l \text{ are in the} \\ & \text{same connected component of } G_d \\ deg_\mu \hat{R}(G) - 1 & \text{otherwise} \end{cases}$$

$$deg_\mu \hat{R}(G - d) = \begin{cases} deg_\mu \hat{R}(G) + 1 & \text{if } d \text{ is an isthmus of } G_d \\ deg_\mu \hat{R}(G) - 1 & \text{otherwise} \end{cases}$$

Proof: (of Lemma 7.1 and Lemma 7.2 - sketch) We prove both lemmas by induction on the number of edges in G. For a graph without edges Lemma 7.1 is obvious and Lemma 7.2 is an empty statement. The inductive step for Lemma 7.2 follows from the inductive hypothesis for Lemma 7.1. Then the inductive step for Lemma 7.1 is an immediate consequence of the inductive step for Lemma 7.2 and the defining recurrence for $\hat{R}$ (note that all components of the sums are positive, and thus we do not have cancellations). □

The above lemma and the recursive definition of $\hat{R}$ give the following:

Corollary 7.3 *All powers of μ with non-zero coefficients have the same parity.*

Let $b(G) = p_0(G_d) + p_1(G_d) - 1$. Note that in Lemmas 7.1 and 7.2 only the attribute of an edge was important, not its color. In particular if $G = G_d$ (or $G = G_l$) (i.e all edges have the same attribute) then, using the argument from the previous section, one shows that the cost of computing $q_{b(G)-2s}$ is polynomial of order $O(m^{s+1})$ (provided s is a constant). Namely, we need to consider only these paths of the resolving tree of the full polynomial that go at most s times right.

Our algorithm to compute the sth highest coefficient, $q_{b(G)-2s}$, is also based on the observation that the resolving tree for computing such a coefficient can be made considerably smaller than the resolving tree for the full polynomial. In fact we will show that this tree belongs to a certain family of trees and we show an upper bound on the number of leaves for a tree in this family.

We consider a family of trees in which each non-leaf node has one distinguished child called the *left child*. A child which is not a left child is called a *right child*. An edge from a node to its left child is called a *left edge* and an edge from a node to any of its right children is called a *right* edge. If u and v are two nodes of a tree, T, then the *right distance* between u and v is defined to be the number of right edges on the unique path in T between u and v. Each internal node has associated with it a *type* that is an element from a finite set. The *type of an edge* is equal to the type of the corresponding parent node.

Before we describe the family of trees which contains resolving trees constructed by our algorithm, we define a simpler tree for which the number of leaves can be easily approximated.

Definition 7.4 *Let s_1, s_2, m, k be non-negative integers such that $k \geq 1$. A rooted tree T is called (s_1, s_2, m, k)-unbalanced if and only if it satisfies the following conditions:*

1. *If $m = 0$ then T is the one vertex tree.*
2. *If $m \geq 1$ then the root of T has as the left child $(s_1, s_2, m-1, k)$-unbalanced*

tree, and the set of right children depends on the type of the root in the following way:

(i) $s_1 \geq 1$ and there is one right child which is a $(s_1 - 1, s_2, m - 1, k)$-unbalanced tree, or

(ii) $s_2 \geq 1$ and there are k right children which are $(s_1, s_2 - 1, m - 1, k)$-unbalanced trees, or

(iii) $s_1 = s_2 = 0$ and there is no right child.

An example of an $(1, 1, 3, 2)$-unbalanced tree is given in Figure 7.1

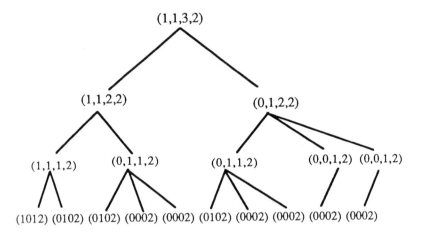

Figure 7.1 An example of an $(1, 1, 3, 2)$-unbalanced tree

Informally, an (s_1, s_2, m, k)-unbalanced tree is a tree of height m such that the right distance from the root to any leaf is bounded by $s_1 + s_2$. Furthermore, a path from the root to a leaf contains at most s_2 right edges of type (ii).

We should stress that in our main theorem (Theorem 7.10) we use unbalanced trees with $k = 1$, that is, binary trees. This simplifies some of the considerations. In particular, Lemma 7.5 is simpler for $k = 1$. We keep the option $k > 1$ for possible improvements of the main theorem (one such improvement is presented in Theorem 7.11).

Lemma 7.5 *Let $F(s_1, s_2, m, k)$ denote the maximal number of leaves in an (s_1, s_2, m, k)-unbalanced tree. Then*

$$F(s_1, s_2, k, m) = \sum_{i=0}^{s_2} k^i \binom{m}{i} + k^{s_2} \sum_{i=1}^{s_1} \binom{m}{s_2 + i}$$

In particular:

(a) If $s_1 \leq s'_1$, $s_2 \leq s'_2$, $m \leq m'$, $k \leq k'$ then $F(s_1, s_2, k, m) \leq F(s'_1, s'_2, k', m')$

(b) $F(s_1, s_2, k, m) \leq k^{s_2} m^{s_1+s_2} + 1$ and equality holds if and only if either $s_1 + s_2 = 1$ or $m = k = 1, s_1 + s_2 \geq 1$, or $m = 1, s_2 = 1, s_1 \geq 1$.

Proof: Given any m, we build an (s_1, s_2, m, k)-unbalanced tree with the maximal number of leaves in a top down fashion, following the recursive definition, creating type (ii) nodes whenever it is possible. For a tree obtained in this way we can compute precisely the number of leaves for which the right subpaths from the root to a given leaf has length j (compare Section 6). Namely we get: if $j \leq s_2$ then the number of such leaves is $k^j \binom{m}{j}$, and if $s_2 \leq j \leq s_2 + s_1$ then it is $k^{s_2} \binom{m}{j}$. The point (b) of the second part of the lemma follows from the observation that for $m \geq j > 1$, $\binom{m}{j} < m^{j-1}(m-1)$. □

We will describe now a family of rooted trees, called α-trees, which strictly generalize the computational tree used in our algorithm. We will approximate the number of leaves of an α-tree by the number of leaves in an appropriate unbalanced tree.

Definition 7.6 *Let s, m, m_d, h, k be non-negative integers such that $k \geq 1$ and α be a real number, where $0 < \alpha < 1$. A rooted tree T is called an $(s, m, m_d, k, h, \alpha)$-tree (or shortly an α-tree) if and only if it satisfies the following conditions:*

1. *If $m = 0$ then T is the one vertex tree.*

2. *If $m \geq 1$ then the root of T is a node of type (i), (ii), (iii) or (iv) described below. It has as the left child an $(s, m-1, m''_d, k, h, \alpha)$-tree where $m''_d \leq m_d$, or an empty tree; and either*

    (i) *there is one right child which is an $(s-1, m-1, m'_d, k, h, \alpha)$-tree and $m'_d \leq m_d$, $h' \leq h$ or*

    (ii) *$m_d \geq 1$ and there are at most k right children each of them being an $(s, m', m'_d, k, h, \alpha)$-tree, where $m'_d \leq \alpha m_d$, $m' < m$ or*

    (iii) *$m_d \geq 1$ and there is one right child which is an $(s, m-1, m'_d, k, h, \alpha)$-tree where $m'_d \leq m_d$, or*

    (iv) *$s = 0$ and there is no right child.*

3. *Let u be any node of T, whose right distance from the root is greater than h and let u be the root of an $(s', m', m'_d, k, h, \alpha)$-tree. Then either*

(a) $s' < s$, or

(b) $m'_d \leq \alpha m_d$.

Lemma 7.7 *The number of leaves of an $(s, m, m_d, k, h, \alpha)$-tree is less than or equal to $F(s_1, s_2, m, k)$ where $s_1 = (h+1)s + h\lfloor \log_{\frac{1}{\alpha}} m_d \rfloor + h$, $s_2 = \lfloor \log_{\frac{1}{\alpha}} m_d \rfloor + 1$. In particular an $(s, m, m_d, k, h, \alpha)$-tree has at most*

$$k^{\lfloor \log_{\frac{1}{\alpha}} m_d \rfloor + 1} m^{(h+1)(s + \lfloor \log_{\frac{1}{\alpha}} m_d \rfloor + 1)} + 1$$

leaves.

Proof: It suffices to show that the number of leaves of our tree is at most the maximum number of leaves of an (s_1, s_2, m, k)-unbalanced tree with s_1, s_2 defined as in the lemma. To show this we consider a path for the root to a leaf and will approximate the maximal number of right edges in such a path. The number of right edges of type (ii) is at most $s_2 = \lfloor \log_{\frac{1}{\alpha}} m_d \rfloor + 1$. By Condition 3, the total number of right edges is bounded by $s_1 + s_2 = (h+1)(s + \lfloor \log_{\frac{1}{\alpha}} m_d \rfloor + 1)$. Thus our tree has no more leaves than an (s_1, s_2, m, k)-unbalanced tree with s_1, s_2 defined as in the lemma. □

Now we state and prove our key lemma.

Lemma 7.8 *Let G be a planar chromatic graph with m edges and m_d dark edges. Then one can construct a resolving tree, T, to compute $q_{b(G)-2s}$, such that T is an $(s, m, m_d, 1, 1, \frac{4}{5})$-tree where each node of T has associated with it a planar chromatic graph and the following conditions are satisfied:*

- *the root of T has associated with it the graph G,*
- *each leaf of T has associated with it a set of isolated vertices,* [5]
- *a coefficient of the graph associated with an internal node can be computed from the corresponding coefficients of the graphs associated with its children using a constant number of algebraic operations on polynomials,*
- *the graph associated with a non-root node can be computed from the graph associated with the parent node in $O(m)$ time.*

Proof: We proceed by induction on the number of edges, m, of the graph G.

If $m = 0$ then the lemma follows immediately from the definition of an α-tree. Thus assume that the lemma holds for any m', $0 \leq m' < m$.

We consider the following cases:

---

[5] Note that we can reduce slightly the size of a computational resolving tree if we allow, in the leaves of the resolving tree, forests with selfloops and compute polynomials for them using a direct formula.

Case 1: G has a dark edge, d, not being an isthmus of G_d. By Lemma 7.2 and recursive formulas for $\hat{R}$ it follows that the degree of $\hat{R}(G-d)$ is equal to the degree of $\hat{R}(G)$ minus two. More generally, in this case $q_{b(G)-2s} = B_i' q_{b(G-d)-2s+2} + (r_2')^{\delta(d)} A_i' q_{(G/d)-2s}$ where d_i is a dark edge of ith color.

Thus to compute $q_{b(G)-2s}$ we need to compute $q_{b(G/d)-2s}$ and, if $s \geq 1$, $q_{b(G-d)-2s+2}$. By the inductive hypothesis one can construct a resolving tree T_1 to compute $q_{b(G/d)-2s}$ such that T_1 is an $(s, m-1, m_d-1, 1, 1, \frac{4}{5})$-tree. Furthermore, if $s \geq 1$ then one can construct a resolving tree T_2 for $G-b$ such that T_2 is an $(s-1, m-1, m_d-1, 1, 1, \frac{4}{5})$-tree. Thus if $s \geq 1$ we can let the root of the resolving tree for G be a node of type (i) and let the roots of T_1 and T_2 be left and right children respectively. If $s=0$ then we let the root of the resolving tree be of type (iv) with T_1 as the left child. By the definition of an α-tree, the resulting resolving tree for G is an $(s, m, m_d, 1, 1, \frac{4}{5})$-tree.

Case 2: There exist a light edge, l, joining different components of G_d. Then the degree of $\hat{R}(G/l)$ is equal to the degree of $\hat{R}(G)$ minus two or more generally $q_{(b(G)-2s} = (r_1')^{\epsilon(l)} A_i' q_{b(G-l)-2s} + B_i' q_{b(G/l)-2s+2}$ where l is a light edge of the ith color. Thus we let T_1 be the resolving tree for $G-l$, which by the inductive hypothesis is an $(s, m-1, m_d, 1, 1, \frac{4}{5})$-tree, and (if $s \geq 1$) we let T_2 be the resolving tree for G/l, which by the inductive hypothesis is an $(s-1, m-1, m_d, 1, 1, \frac{4}{5})$-tree. Again, by definition, the resulting tree is an $(s, m, m_d, 1, 1, \frac{4}{5})$-tree with the root of type (i) or (iv).

Case 3: G_d is a forest, no light edge joins different components of G_d, and there exist a dark edge, d, such that $G_d - d$ can be partitioned into two groups of connected components each having at most $\frac{4}{5} m_d$ edges.

We let the root be of type (iii) and we associate with its left child the graph G/d and with the right child the graph $G-d$. The resolving tree for G/d is constructed inductively to be an $(s, m-1, m_d-1, 1, 1, \alpha)$ tree and the resolving tree for $G-d$ is constructed as follows:

Let $l_1, \ldots l_k$, where $k \geq 0$, be the set of light edges connecting two trees from different groups of the partition. We iteratively delete/contract edges l_i associating the graph obtained by removing l_i with the left subtrees and the graph obtained by contracting l_i with the root of the right subtree (see Figure 7.2). The resolving tree for the right subtree is then found inductively (we lower the value of the parameter s). The resolving tree for the left subtree is a subject of the next iteration. If all edges l_i are removed then, by the definition of d, the connected components of of the resulting graph can be partitioned into two groups each of which has at most $\frac{4}{5}m_d$ dark edges. We associate the subgraph with larger number of dark edges, say G_a, with the left subtree and the subgraph with smaller number of dark edges, say G_b, with the right subtree. The resolving trees for these two graphs are known by the inductive assumption.

To see that Condition 3 of the definition of an α-tree is satisfied note that at any step of the above construction, the graph associated with a right subtree is obtained from the graph associated with the parent node either by contracting a light edge connecting two different connected component of the dark forest, or (in the last iteration) by dividing the graph into two subgraphs each of which has at most $\frac{4}{5}m_d$ dark edges.

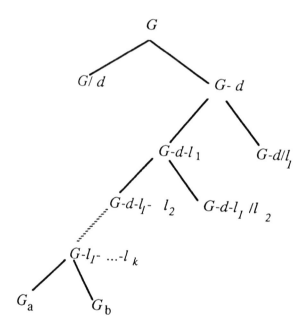

Figure 7.2. Construction of the resolving tree in Case 3

Case 4: Cases 1-3 do not hold.

For simplicity of the presentation we assume that G_d (and therefore G) is connected. The generalization to non-connected G is straightforward.

We will use the following well known fact:

Lemma 7.9 *If G_d is a tree then either*

(i) *there is an edge d in G_d such that no component of $G_d - d$ has more than $(1 - \xi)m_d$ edges, where m_d is the number of edges of G_d, and $0 < \xi < \frac{1}{2}$ or*

(ii) *for any plane embedding of G_d there is a vertex v of G_d and three pairwise disjoint (except v) curves γ_1, γ_2 and γ_3 from v to infinity ($\gamma_i \cap G_d = v$) which split G_d into three connected subgraphs, G_1, G_2 and G_3, each of which has no less than $\beta = (\frac{1}{3} - \frac{2}{3}\xi)m_d$ edges.*

Let G be a chromatic plane graph. Assume that the case (ii) of Lemma 7.9 holds for G_d. Let w be a point in the unique region of $R^2 - G$ which touches $G_i - v$ for any i. We can assume that γ_i goes from v to w, cuts light edges perpendicularly and any light edge is cut at most once. It is useful to think of $\gamma_1 \cup \gamma_2 \cup \gamma_3$ as a subgraph H of the graph dual to G, composed of edges which are dual to light edges joining different components G_i (and disjoint from v). Let w be the vertex of H corresponding to the face containing point w. If w has valency less than 2 then G can be split along v into two subgraphs no one of which has more than $(1 - \beta)m_d$ dark edges. Otherwise w has valency 2 or 3. Consider the case of valency 3 (the case of valency 2 being analogous). Let e_1, e_2 and e_3 be light edges of G dual to edges adjacent to w in H. Let c_i be the unique cycle of G_d plus e_i, and D_i the region bounded by c_i (including the boundary) with w outside of it (see Figure 7.3). Let $G'_i = G_d \cap D_i$ and $G''_i = G_i - \bigcup_{i=1,2,3} G'_i$. Then G'_i has no more than $(1 - \beta)m_d$ edges and there is some i, say $i = 1$, such that $G'_1 \cup \bigcup_{i=1,2,3} G''_i$ has at least βm_d edges.

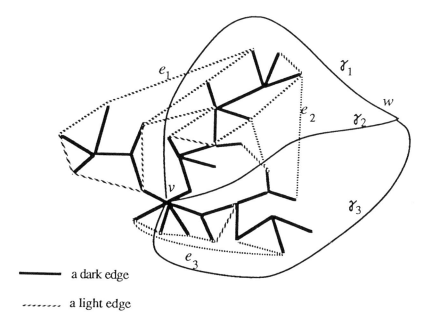

—— a dark edge

-------- a light edge

Figure 7.3 Splitting a graph in a vertex

From this there follows immediately the following fact of great importance for our algorithm. Let G/c_1 be a plane graph obtained from G by contracting all edges of the cycle c_1. Then the 2-connected components of G/c_1 can be grouped into five subgraphs of G/c_1 none of which has more than $(1-\beta)m_d$ dark edges; namely, $(G \cap D_1)/c_1$, G_1'', G_2'', G_3'' and G''' containing all other edges of the graph G/c_1. It follows from this that the 2-connected components of G/c_1 can be grouped into two subgraphs G_a and G_b none of which has more than $(1-\beta)m_d$ dark edges. Notice that $\hat{R}(G/c_1) = \hat{R}(G_a)\hat{R}(G_b)$.

We use the above facts to build a resolving tree for G satisfying the conditions of Case 4.

We let $\xi = \frac{1}{5}$ which gives $\beta = \frac{1}{5}$ (this leads to $\alpha = \frac{4}{5} = 1 - \xi = 1 - \beta$). We consider the following cases depending on the valency of w:

Case a: w has valency at most 1. This means that at most one γ_i cuts light edges of G. Thus the remaining two curves partition the biconnected components of G into two groups each having at most $\frac{4}{5}m_d$ edges. We associate one group of these components with the left child of G and the other with the right child of G.

Case b: w has valency 3 (the case when w has valency 2 is analogous and is omitted). Let, as before, e_1 be the light edge of G and c_1 the cycle of $G_d \cup e_1$ such that the 2-connected components of G/c_1 can be grouped into two subgraphs of G/c_1 each of which has at most $\frac{4}{5}m_d$ dark edges. We associate the graph $G - e_1$ with the root of the left subtree and the graph G/e_1 with the root of the right subtree. As in the Case 3 we construct a resolving tree of G/e_1 that guarantees Condition 3 of the definition of an α-tree iteratively.

Let $d_1, \ldots d_k$, where $k \geq 0$ be the dark edges of the dark cycle, c_1/e_1, introduced by contraction of e_1. We iteratively delete/contract edges d_i associating the graph obtained by removing d_i with the root of the right subtree and the graph obtained by contracting d_i with the root of the left subtree. (See Figure 7.4) If all edges d_i are contracted then by the definition of c_1, the biconnected components of the resulting graph can be grouped into two subgraphs neither of which has more than $\frac{4}{5}m_d$ dark edges. We let the subgraph with the larger number of dark edges be in the left subtree and the subgraph with the smaller number of dark edges be in the right subtree. Using an argument similar to the one used in Case 3 we can argue that Condition 3 of the definition of an α-tree is satisfied.

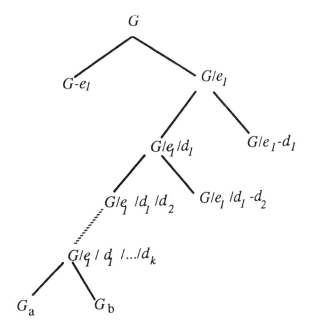

Figure 7.4 Construction of the resolving tree, Case 4.

Finally, using standard graph algorithms for planar graphs (see, for example [12,21]) we conclude than each step of the construction can be carried in $O(m)$ time. In particular a plane embedding of a planar graph can be found in $O(m)$ time. We represent an embedding by cyclic ordering of edges adjacent to any vertex of the graph. □

Now we are ready to prove our main theorem.

Theorem 7.10 *Let G be a planar chromatic graph with m edges, $m_d > 0$ dark edges, and $O(m)$ connected components. Then the coefficient $q_{b(G)-2s}$ can be computed in $O((s+1)^2 m^4 m^{2(s+\lfloor \log_{\frac{5}{4}} m_d \rfloor)})$ time.*

Proof: By Lemma 7.8, one can construct a resolving tree, T, to compute $q_{b(G)-2s}$ which is an $(s, m, m_d, 1, 1, \frac{4}{5})$-tree. By Lemma 7.7 this tree has at most most $t = m^{2(s+\lfloor \log_{\frac{5}{4}} m_d \rfloor +1)} + 1 = m^2 m^{2(s+\lfloor \log_{\frac{5}{4}} m_d \rfloor)} + 1$ leaves.

Each path from the root to a leaf has length at most m. Thus T has at most $mt + 1$ nodes. Computing a coefficient in a leaf takes $O(1)$ time. Since we are interested in s^{th} highest coefficient we need to compute at most s highest coefficients in every node. Thus computing coefficients in one node takes $O((s+1)^2)$ algebraic operations on polynomials. Since for any internal node of the resolving tree the graphs associated with the left and right subtree of the resolving tree can be found in $O(m)$ time, the result follows.

We need to note that the resulting polynomial is not presented in ordered form. □

One can slightly improve the above theorem. We sketch this improvement for $s = 0$.

Theorem 7.11 (a) *There is a resolving computational tree for $q_{b(G)}$ of depth no more than m and number of leaves of order $O(2^{\log_{\frac{5}{4}} m_d} m^{\lfloor \log_{\frac{5}{4}} m_d \rfloor + 1})$*
(b) *$q_{b(G)}$ can be computed in $O(m^3 2^{\log_{\frac{5}{4}} m_d} m^{\lfloor \log_{\frac{5}{4}} m_d \rfloor)}$ time*

Proof:(sketch) A careful analysis of our algorithm allows us, for $s = 0$, to build an $(0, m, m_d, 2, 0, \frac{4}{5})$ resolving tree for G. Therefore by Lemma 7.7, the number of leaves of this tree is bounded by $2^{\lfloor \log_{\frac{5}{4}} m_d \rfloor + 1} m^{(\lfloor \log_{\frac{5}{4}} m_d \rfloor + 1)} + 1$. As in the proof of Theorem 7.10 we get that the time complexity for $q_{b(G)}$ is $O(m^3 2^{\log_{\frac{5}{4}} m_d} m^{\lfloor \log_{\frac{5}{4}} m_d \rfloor})$. □

Remark. Using a more involved computation we can improve further Theorem 7.11 by approximating the number of leaves in a $(0, m, m_d, k, 0, \alpha)$-tree slightly tighter. Namely we get that, for sufficiently large m_d, the number of leaves of the tree is bounded from above by the function $\phi_{k,\alpha}(m, m_d)$ where

$$\phi_{k,\alpha}(m, m_d) = k^{\log_{\frac{1}{\alpha}} m_d} m^{\log_{\frac{1}{\alpha}} m_d} m_d^{-\log_{\frac{1}{\alpha}} \ln m_d + a} e^{\ln^t m_d}$$

where $a = \frac{1}{\ln \frac{1}{\alpha}}(\ln \ln \frac{1}{\alpha} + 1)$ and t is any real number greater than 0.

To accomplish this we perform computation in three steps:

STEP 1. We note that the number of leaves in a $(0, m, m_d, k, 0, \alpha)$-tree is bounded from above by the function $f(m, m_d)$ defined recursively as follows: $f(m, 0) = f(m, m) = 1$ and $f(m, m_d) = f(m - 1, m_d) + kf(m - 1, \lfloor \alpha m_d \rfloor)$.

STEP 2. Let $g_{\alpha,k}(x, y) = g(x, y)$ be a continuous (smooth with respect to y) function defined for $x \geq y \geq 1$ by the initial condition

$$g(x, 1) = 1 + k$$

the partial differential equation

$$\frac{\partial g(x, y)}{\partial x} = kg(x, \alpha y)$$

and the condition that $g(x, y)$ is non-decreasing with respect to y. [6] Then for

---

[6] If we consider the region $x \geq 1, y \geq 1$ and $g(1, y) = 1+k$, $g(x, \frac{1}{\alpha}) = k(1+k)x - (k-1)(k+1)$ and assume that g is linear with respect to y, for $1 \leq y \leq \frac{1}{\alpha}$ then g, satisfying our assumptions, is uniquely determined.

any natural m_d,
$$g(m, m_d) \geq f(m, m_d).$$

STEP 3. Let $\phi(x,y) = k^{\log_{\frac{1}{\alpha}} y} x^{\log_{\frac{1}{\alpha}} y} y^{-\log_{\frac{1}{\alpha}} \ln y + a} e^{\ln^t y}$ where a, t are as before. Then we show that for y large enough, $\frac{\partial \phi(x,y)}{\partial x} > k\phi(x, \alpha y)$ and finally

$$\lim_{y \mapsto \infty} \frac{\phi(x,y)}{g(x,y)} = \infty.$$

Now, combining the results of steps 1-3 we obtain the desired bound for the number of leaves of a $(0, m, m_d, k, 0, \alpha)$-tree.

Theorems 7.10 and 7.11 have an immediate application to knot theory. We use the fact that every dichromatic planar graph defines a matched diagram of an oriented link. So we can translate Theorem 7.10 to polynomials of links as follows:

Corollary 7.12 *Consider a matched diagram D of an oriented link L and its skein polynomial $P_L(a,z) = \sum_{i=m}^{M} P_i(a) z^i$ where $P_m(a), P_M(a) \neq 0$. Then*

(i) *[39] $m = 1 - com(L)$ where $com(L)$ is the number of components of the link L,*

(ii) *$P_{m+2i}(a)$ can be computed in time $O(n(D)^{c \ln n(D) + 2i})$ where c is some constant and $n(D) > 0$ denotes the number of crossings of the matched diagram D.*

Proof: Point (i) follows from Lemma 7.1. Namely, if G is a planar dichromatic graph and $D(G)$ is its matched diagram then $p_0(G_d) + p_1(G_d)$ is equal to the number of components, $com(D(G))$, of the link with matched diagram $D(G)$. The variable z in $P_{D(G)}(a, z)$ is proportional to the variable μ^{-1} in $\hat{R}(G)$, more exactly: $\hat{R}(G) = P_{D(G)}(a, z)$ for $\mu = (a + a^{-1})z^{-1}$, $r'_1 = r'_2 = 1$, $A'_+ = -a^{-2}$, $A'_- = -a^2$, $B'_+ = 1 + a^{-2}$, $B'_- = 1 + a^2$. Therefore $m \geq 1 - com(D(G))$. Now if we consider $P_L(a, z)$ in the ring $\mathcal{R}$ (as in Theorem 6.2), and then take $P_L(a, z)$ mod (z) we get $P_L(a, z) \equiv ((a + a^{-1})z^{-1})^{com(L)-1}$ mod (z). Therefore $m \leq 1 - com(D(G))$.

Point (ii) follows from Theorem 7.10 and the relation between the bracket $\hat{R}(G)$ of a planar dichromatic graph G and the skein polynomial $P_{D(G)}(G)(a, z)$ of the corresponding matched diagram. □

It remains an open question whether Corollary 7.12 (ii) holds for any oriented link diagram. Computations performed so far support the positive answer.

8 Applications

As we mentioned at the beginning, this research is motivated by a practical need to compute the new polynomial invariants of links (for links with many crossings).

We refer the reader to [9,33] or [63] for the definitions of some of the terms used in this section.

Our applications can be divided into three groups:

A. The recognition problem (*i.e.* the problem whether two link diagrams represent the same link). In particular we are concerned with the following questions:

1. Is a given link amphicheiral, (*i.e.* isotopic to its mirror image)?
2. Let K and K' be a pair of mutant knots. Do the (k,m)-cables about these knots have the same polynomial invariants?

B. Applications of the new link polynomial to classical properties of links. In particular:

1. Periodicity of links (*i.e.* for a link L and a number n, is there a Z_n action on S^3 with a circle as a fixed point set which maps L onto itself and such that L is disjoint from the fixed point set?)
2. Computing the braid index of a link L (*i.e.* the minimal number n such that L can be realized as a closed n braid).

C. Recognition problem for 3-manifolds using Witten-Reshetikchin-Turaev [84, 62] invariants.

Before we analyze the above problems with more detail we should explain why the links we are interested in may have so many crossings (> 35) while the original questions can be asked for relatively small links (e.g links that are already tabulated). The potential of the method we use was first recognized by H.Morton and H.Short [43,44]. Assume, for example, that we try to distinguish two links L_1 and L_2 that have the same polynomial invariants. If the links are in fact isotopic then the same is true for the cables about them. Therefore if some polynomial invariant distinguishes cables about L_1 and L_2 then L_1 and L_2 are not isotopic. However if $n(L)$ is the number of crossings of a link L then the natural diagram of the k-cable about L has at least $k^2 n(L)$ crossings and thus may exceed the number 35 below which the computation of the full polynomial is feasible.

Now we explain problems A(1)(2), B(1)(2), and C.

A(1) If $\overline{K}$ is the mirror image of a knot K then

$$P_K(a,z) = P_{\overline{K}}(a^{-1},z).$$

Therefore if K is an amphicheiral knot then

$$P_K(a,z) = P_K(a^{-1},z).$$

This gives a criterion for amhicheirality. However a knot may not be amphicheiral even when the above equality holds. The following, stronger criterion may also be used. If K is amphicheiral, so is the $(k,0)$ cable about K. This method was used for the first time by Morton and Short [43] for the knot 9_{42} (in the notation of Rolfsen [63]). Namely $P_{9_{42}}(a,z) = P_{9_{42}}(a^{-1},z)$ but the polynomial of the $(2,0)$ cable about 9_{42} is not symmetric with respect to a. Morton and Short had to face the problem that the natural diagram of the $(2,0)$ cable about 9_{42} has 37 crossings. In their solution, they used the fact that this knot has a presentation as an 8-braid and they found a quadratic time algorithm to compute $P_L(a,z)$ under the assumption that the number of braid strings, s, is fixed. However, due to a preprocessing step that consists of building of a data base of polynomials of a certain family of links, the storage requirement of their algorithm grows exponentially with s. Thus the practical limit of this algorithm is when s is equal to 9. In contrast, if we use an algorithm suggested by Theorem 7.11 for the first two terms of the skein polynomial of the $(2,0)$ cable about 9_{42} then we can exclude amphicheirality of this knot in less than one second.

A(2) It has been proven in [38] and [55] that if two knots K and K' are mutants of each other then their 2-cables have the same skein polynomials. Morton and Traczyk [40] proved that any cable about K and K' have the same Jones polynomial. It was conjectured that this is not true for the skein polynomial. The simplest pair of mutants which proves this conjecture are the Conway and Kinoshita-Terasaki knots. They have 11 the crossings each and braid index 4. Thus a 3-cable about each of them has at least $9 \cdot 11$ crossings and the braid index is expected to be 12. Morton and Traczyk used a special trick to distinguish these knots. Our algorithm allows to distinguish them by a relatively short computation. The same observation is true for the Kauffman polynomial.

B(1) There are various criteria for periodicity of links for which our method can be used In particular it can be used for two criteria introduced by Murasugi and Traczyk [45,68] that use the Jones polynomial. We focus our attention on the criterion described in [57] that uses the skein polynomial $P_L(a, z) \in \mathcal{R}$, as this criterion is the motivation for our Theorem 6.2. Combining this theorem and Theorem 1.2 of [57] we obtain:

Theorem 8.1 *Let L be an r-periodic link, where r is a prime number. Then the skein polynomial $P_L(a, z)$ satisfies*

$$P_L(a, z) \equiv P_L(a^{-1}, z) \bmod (r, z^r)$$

where (r, z^r) is the ideal in $\mathcal{R}$ generated by r and z^r. Furthermore $P_L(a, z) \bmod (r, z^r)$ can be computed in polynomial time.

The power of Theorem 8.1 lies in the fact that if K is an r-periodic link then, for properly chosen m, the (k, m) cable about K is also r-periodic.

The criterion of Theorem 8.1 was generalized by Traczyk [70] and Yokota [87] as follows:

Theorem 8.2 *Let $P_K(a, z) = \sum_{i=0}^{M} P_{2i}(a) z^{2i}$ and K be an r-periodic knot ($r \neq 2$) such that the linking number of K with the axis of rotation is equal to k. Then*

(a) *[70] The polynomial $P_0(a) = \sum_{i=0}^{M} d_{2i} a^{2i}$ satisfies $d_{2i} \equiv d_{2i+2} \bmod r$ except possibly when $2i + 1 \equiv \pm k \bmod r$.*

(b) *[87] $P_{2i}(a) \equiv c_{2i} P_0(a) \bmod r$ for some constant c_{2i} and $2i \leq r - 3$.*

Traczyk wrote a computer program that, using ideas of our $O(m^{c \log m})$ algorithm, computes the first few terms of P_{2i}. He ran this program on 165 ten-crossing prime knots listed in [63]. Two of them are known to have period 5. Traczyk found [70] that for the remaining 163 knots, the criterion of Theorem 8.2 (a) immediately excludes 149 knots; using 2-cables excludes 9 knots of the remaining 14 and using 3-cables excludes the remaining 5. Note that 3-cables about a 10-crossing knot are expected to have at least 90 crossings.

B(2) The theorem of Morton [41] and Franks-Williams [17] says that the braid index, $b(L)$ of a link L satisfies

$$b(L) \geq \frac{1}{2} \operatorname{span}_a P_L(a, z) + 1 \tag{5}$$

For alternating links tabulated in knot tables this inequality becomes equality. Thus it was conjectured that if L is an alternating link then (5) becomes an equality. The conjecture was disproved in [48]. To check a possible knot counterexample the authors of [48] had to use equality (5) for a certain 72-crossing knot. Using ideas described in Section 7, J.Hoste wrote a program (in Spring 1989) to compute the first terms of $P_L(a,z)$. The smallest link counterexample found in [48] has 15 crossings and the smallest knot counterexample has 18 crossings (see Figure 8.1, where the graph G is drawn together with the diagram $D = D_N^0(G)$ which describes our oriented knot K_D).

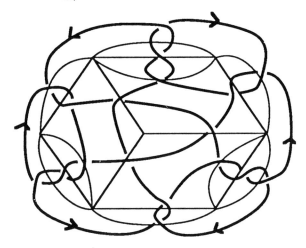

Figure 8.1

One can easily find that $b(K_D) \leq 6$ [48]. On the other hand, $\frac{1}{2}\mathrm{span}_a P_{K_D}(a,z) + 1 = 5$. To prove that $b(K_D) > 5$ the authors of [48] use the idea of Morton and Short of considering a 2-cable K_D^2 about K_D (if $b(K_D) = 5$ then $b(K_D^2) \leq 10$). J.Hoste computed the first 2 terms of $P(a,z)$ of the simplest 2-cable (with $4 \cdot 18 = 72$ crossings) and found that $\mathrm{span}_a P_{K_D^2}(a,z) \geq 20$, and so $b(K_d^2) \geq \frac{1}{2}20 + 1 = 11$. Therefore $b(K_d) > 5$. The computation took 37 minutes on a Mac plus personal computer.

C To compute the Witten–Reshetikchin–Turaev invariant of a 3-manifold given by a link surgery description, one has to find the Jones polynomial of the link and its parallel copies at p^{th} root of unity. So we are essentially interested in $V(t) \bmod (t^p - 1)$. By Theorem 6.1 we can compute $V(t)$ mod $(t-1)^p$ in polynomial time. For prime p, $t^p - 1 \equiv (t-1)^p \bmod p$. Therefore we should be able to compute the Witten–Reshetikchin–Turaev invariant modulo p (i.e. an element of $Z[t^{\pm 1}]/(p, t^p - 1)$) in polynomial time. In fact, there is a good chance that an invariant analogous to the Witten-Reshetikchin-Turaev invariant of a 3-manifold can be defined with values in the ring $Z[t^{\pm 1}]/(t - 1)^p$. These ideas are currently being developed in [30].

9 Generalizations of the polynomial R(G)

We propose in [53] two ways of generalizing the dichromatic polynomial. The first of them is to define polynomials on supergraphs (called also set systems or shortly setoids). The notion of a supergraph is more general than that of a matroid. However, because of the generality of this notion, the deletion-contraction rule cannot be always used as a computation method alternative to the state model formula.

The second generalization is based on the deletion-contraction rule but instead of linear formulas as in Theorem 3.1 we use operations in an abstract algebra.

With the above generalizations essential parts of corresponding invariants can be still computed in subexponential time.

Acknowledgments. We thank the referee and the editors for helpful comments and suggestions on the earlier version of the paper.

Added in proof. D. Vertigan has announced a polynomial time algorithm to compute $P_{m+2i}(a)$ for any link diagram (letter of March 26, 1992).

References

[1] J.W.Alexander. Topological invariants of knots and links. *Trans. Amer. Math. Soc.* 30 (1928) 275-306.

[2] J.W.Alexander, G.B. Briggs. On types of knotted curves. *Ann. of Math* 28 (1926/27), 562-586.

[3] R.P. Anstee. Lectures on graph invariants, UBC, Vancouver, 1986.

[4] R.P.Anstee, J.H.Przytycki, D.Rolfsen. Knot polynomials and generalized mutation. *Topology and Applications* 32 (1989) 237-249.

[5] C.Bankwitz. Über die Torsinszahlen der alternierenden Knoten. *Math. Ann.* 103 (1930) 145-161.

[6] R.D.Brandt, W.B.R.Lickorish, K.C.Millett. A polynomial invariant for unoriented knots and links. *Invent. Math.* 84 (1986), 563-573.

[7] R.I.Brooks, C.A.B.Smith, A.H.Stone, W.T.Tutte. The Dissection of rectangles into squares. *Duke Math. Jour.* (1940) 312 -340.

[8] T.Brylawski, J.Oxley. The Tutte polynomial and its application; draft of a chapter for the third volume of the matroid theory series edited by N.White, 1988.

[9] G.Burde, H.Zieschang. *Knots*, De Gruyter (1985).

[10] J.H.Conway. An enumeration of knots and links. *Computational problems in abstract algebra* (ed. J.Leech), Pergamon Press (1969) 329 - 358.

[11] H. de la Harpe. Introduction to knot and link polynomials. *Fractals, quasicrystals, chaos, knots and algebraic quantum mechanics* (Maratena, 1987), NATO Adv.Sci.Ser.C:Math.Phys.Sci.235,Kluwer Acad.Publ.,Dordrecht (1988), 233-263.

[12] S.Even. *Graph Algortithms* Computer Science Press, Inc. (1979)

[13] J.W.Essam. Graph theory and statistical physics. *Discrite Math.* 1 (1971/72) 83-112.

[14] E.F.Farell. In introduction to matching polynomials. *Journal of Combinatorial Theory*, B27 (1979) 75-86.

[15] P.Freyd, D.Yetter, J.Hoste, W.B.R.Lickorish, K.Millett, A.Ocneanu. A new polynomial invariant of knots and links. *Bull. Amer. Math. Soc.*, 12 (1985) 239-249.

[16] C.M Fortuin, P.W.Kasteleyn. On the random-cluster model I. Introduction and relation to other models. *Physica* 57 (1972), 536 - 564.

[17] J.Franks, R.Williams. Braids and the Jones polynomial. *Trans. Amer. Math. Soc* 303 (1987) 97-108.

[18] M.R.Gary, D.S.Johnson. *Computers and Intractability: a Guide to Theory of NP Completeness.* Ed. V.Klee, New York (1979).

[19] O.J.Heilmann. Location of the zeros of the grand partition function of certain classes of lattice gases. *Studies in Applied Math.* 50 (1971) 385-390.

[20] C.F.Ho. A new polynomial for knots and links; preliminary report. *Abstracts AMS* 6(4) (1985), p 300.

[21] J.Hopcroft, R.E.Tarjan. Efficient planarity tesitng. *J. Assoc. Comput. Math*, 21 (1974) 549-568.

[22] J.Hoste. A polynomial invariant of knots and links. *Pacific J. Math.* 124 (1986) 295-320.

[23] J.Hoste, J.H.Przytycki. A survey of skein modules of 3-manifolds. To appear in Proc. of "Knots" conference, Japan 1990.

[24] J.Hoste, J.H.Przytycki. The $(2,\infty)$-skein module of Whitehead manifolds. Preprint 1991.

[25] F.Jaeger. On Tutte polynomials and link polynomials. *Proc. Amer, Math. Soc.*, 103 (2) (1988) 647-654.

[26] F.Jaeger, D.L.Vertigan, D.J.A.Welsh. On the Computational Complexity of the Jones and Tutte Polynomials. *Math. Proc. Camb. Phil. Soc.* 108, (1990), 35-53.

[27] V.Jones. A polynomial invariant for knots via Von Neuman algebras. *Bull. Amer. Math. Soc* 12 (1985) 103-111.

[28] V.Jones. Hecke algebra representations of braid groups and link polynomials, *Ann. of Math.* (1987) 335-388.

[29] V.Jones. On a certain value of the Kauffman polynomial. *Comm. Math. Phys.* to appear.

[30] J. Kania-Bartoszyńska, J.H.Przytycki. Work in progress.

[31] L.Kauffman. State models and the Jones polynomial. *Topology 26* (1987) 395-407.

[32] L.Kauffman. A Tutte polynomial for signed graphs. *Discrete Appl. Math* 25 (1989) 105-127.

[33] L.Kauffman. *On Knots*. Princeton University Press, 1987.

[34] G.R.Kirchhoff. Über die Auflosung der Gleichungen, auf welche man bei der Untersuchung der linearen Vertheilung galvanischer strome geführt wird. *Annalen d. Physik und Chemie* 72 (1847) 497-508.

[35] B.Korte, L.Lovász. Greedoids, a structural framework for the Greedy algorithm. in *Progress in combinatorial Optimization*, Proc. of the Silver Jubilee Conference on Combinatorics, Waterloo, June 1982, (W.R. Pullyblank, Ed.), 221-243, Academic Press, London/New York, 1984.

[36] K.Kobayashi. Coded graph of oriented links and Homfly polynomials. *Topology and Computer Science*, Kinokuniyu Company Ltd., (1987) 277-294.

[37] W.B.Lickorish. Polynomials for links.*Bull. London Math. Soc.* 20 (1998) 558-588.

[38] W.B.Lickorish, A.S.Lipson. Polynomials of 2-cable-like links. *Proc. Amer. Math. Soc* 100 (1987) 355-361.

[39] W.B.R.Lickorish, K.Millett. A polynomial invariant of oriented links. *Topology* 26 (1) (1987) 107-141.

[40] W.M. Menasco, M.B.Thistlethwaite. The Tait Flyping Conjecture. *Bull. Amer. Math. Soc.* 25 (2) (1991) 403-412.

[41] H.R.Morton. Seifert circles and knot polynomials. *Math. Proc. Cambridge Phil. Soc.* 99 (1986) 107-109.

[42] H.R.Morton. Problems. In *Braids*, ed. J.S.Birman and A.Libgober, Contemporary Math. Vol. 78 (1988) 557-574.

[43] H.R.Morton, H.B.Short. The 2-variable polynomial of cable knots. *Math. Proc. Cambridge Phil. Soc* 101 (1987) 267-278.

[44] H.R.Morton, H.B.Short. Calculating the 2-variable polynomial for knots presented as closed braids. *J.Algorithms* 11(1) (1990) 117-131.

[45] K.Murasugi. Jones polynomial of periodic links. *Pacific J. Math.* 131 (2) (1988) 319 - 329.

[46] K.Murasugi. On invariants of graphs with application to knot theory. *Trans. Amer. Math. Soc.*, 314 (1989) 1-49.

[47] K.Murasugi. On the braid index of alternating links. *Trans. Amer. Math. Soc.* 326, (1991) 237-260.

[48] K.Murasugi, J.H.Przytycki. An index of a graph with applications to knot theory. Submitted to *Trans. Amer. Math. Soc.*.

[49] S.Negami, Polynomial invariants of graphs, *Trans. Amer. Math.*, Soc. Vol 299,2 (1987) 601-622.

[50] J.G.Oxley, D.J.A. Welsh. The Tutte polynomial and percolation. *Graph Theory and Related Topics*, 329-339 (1979) Academic Press Inc.

[51] T.M.Przytycka, J.H.Przytycki. Signed dichromatic graphs of oriented link diagrams and matched diagrams. Manuscript, University of British Columbia, (July 1987).

[52] T.M.Przytycka, J.H.Przytycki. Invariants of chromatic graphs. Dept. of Computer Science, The University of British Columbia, Technical Report 88-22.

[53] T.M.Przytycka, J.H.Przytycki. Algebraic structures underlying the Tutte polynomial of graphs. Manuscript 1988.

[54] J.H.Przytycki. t_k-moves on links. In *Braids*, ed. J.S.Birman and A.Libgober, Contemporary Math. Vol. 78 (1988) 615-656.

[55] J.H.Przytycki. Equivalence of cables of mutants of knots. *Canad. J. Math.* 26 (2) 1989, 250-478.

[56] J.H.Przytycki. Skein modules of 3-manifolds. *Bull. Polon. Acad. Sci.: Math* 39 (1991).

[57] J.H.Przytycki. On Murasugi's and Traczyk's criteria for periodic links. *Math Ann.*, 283 (1989) 465-478.

[58] J.H.Przytycki. t_3 and $\bar{t}_4$ moves conjecture for oriented links with matched diagrams. *Math. Proc. Camb. Phil. Soc.* 108 (1990) 55-61.

[59] J.H.Przytycki, P.Traczyk. Invariants of links of Conway type. *Kobe J. Math.* 4 (1987) 115-139.

[60] K.Reidemeister. Elementare Begründung der Knotentheorie. *Abh. Math. Sem. Univ. Hamburg* 5 (1927), 24-32.

[61] K.Reidemeister. *Knotentheorie*. Ergebn. Math. Grenzgeb., Bd.1; Berlin: Springer-Verlag (1932) (English translation: Knot theory, BSC Associates, Moskow, Idaho, USA, 1983).

[62] N.Y.Reshetikchin, V.Turaev. Invariants of three manifolds via link polynomials and quantum groups. *Invent. Math.* 103 (1991) 547-597.

[63] D.Rolfsen. *Knots and links*. Publish or Perish, 1976.

[64] Y.Rong. The Kauffman Polynomial and the two Fold Cover of a Link. Preprint 1989.

[65] H.N.V.Temperley. E.H. Lieb. Relations between the "percolation" and "coloring" problem and other graph-theoretical problems associated with regular planar lattices: some exact results for the "percolation" problem. *Proc. Roy. Soc.* Lond. A 322 (1971) 251-280.

[66] M.B.Thistlethwaite. A spanning tree expansion for the Jones polynomial, *Topology* 26 (1987) 297-309.

[67] A.Thompson. A polynomial invariant of graphs in 3-manifolds. Preprint 1989.

[68] P.Traczyk. 10_{101} has no period 7: a criterion for periodic links. *Proc. Amer. Math. Soc.* 180 (1990) 845-846.

[69] P.Traczyk. A criterion for knots of period 3. *Topology and its applications* 36(3) (1990) 275-281.

[70] P.Traczyk. Periodic knots and the skein polynomial. *Invent. Math.* 106(1) (1991), 73-84.

[71] P.Traczyk, Preprint 1990.

[72] L.Traldi. A dichromatic polynomial for weighted graphs and link polynomials. *Proc, Amer. Math. Soc.*, 106 (1) (1989) 279-286.

[73] W.T.Tutte. A ring in graph theory. *Proc. Cambridge Phil. Soc.* 43 (1947) 26-40.

[74] W.T.Tutte. Rotors in graph theory. *Annals of Discrete Math.* 6 (1980) 343-347.

[75] W.T.Tutte. *Graph theory*. Encyclopedia of Mathematics and its Applications 21 (Cambridge University Press, 1984).

[76] W.T.Tutte, *Selected papers of W.T.Tutte*. D.McCarthy and R.G.Stanton, eds. Charles Babbage Research Center, St.Pierre, Man., Canada (1979).

[77] D.L.Vertigan. On the computational complexity of Tutte, Jones, Homfly and Kauffman invariants. PhD. thesis.

[78] D.L.Vertigan. The computational complexity of Jones, Homfly and Kauffman invariants for links. Preprint.

[79] D.L.Vertigan. Private communications Seattle, July 1991.

[80] H.Whitney. A logical expansion in mathematics. *Bull. Amer. Math. Soc.* 38 (1932) 572-579.

[81] H.Whitney. The coloring of graphs. *Ann. of Math.* 33(1932) 688-718.

[82] H.Whitney. 2-isomorphic graphs. *Amer. J.Math.* 55 (1933) 245- 254.

[83] H.Whitney. A set of topological invariants of graphs. *Amer. J. Math.* 55(2) (1932) 231-235.

[84] E.Witten. Quantum field theory and the Jones polynomial. *Comm. Math. Phis.* 121 (1989) 351-399.

[85] D.Yetter. On graph invariants given by linear recurrence relation. *J.Com. Th.* Ser. B. 48(1) (1990) 6-18.

[86] Y.Yokota. The Jones polynomial of periodic knots. *Proc. Amer. Math. Soc.* to appear.

[87] Y.Yokota. The skein polynomial of periodic knots. *Math. Ann.* 291 (1991), 281-291.

[88] T. Zaslavsky. Strong Tutte Functions of Matroids and Graphs. Preprint 1990.

Teresa M.Przytycka
Department of Computer Science, University of California, Riverside, CA 92521
On leave from
Instytut Informatyki, Uniwersytet Warszawski, Warsaw, Poland.

Józef H.Przytycki
Department of Mathematics, University of California, Riverside, CA 92521
On leave from
Department of Mathematics, Uniwersytet Warszawski, Warsaw, Poland.

Knots and Braids: Some Algorithmic Questions

D.J.A. Welsh
University of Bonn and University of Oxford

1 Introduction

In this lecture I shall discuss some problems with an algorithmic flavour which arise in combinatorial knot theory.

A *link* with $c(L)$ components in the three-sphere S^3 is a smooth submanifold that consists of $c(L)$ disjoint simple closed curves. A *knot* is a link with one component. Two links K, L are *ambient isotopic* if there exists a homotopy $h_t : S^3 \to S^3$ ($0 \le t \le 1$), such that each h_t is a homeomorphism, $h_0 = 1$ and $h_1(K) = L$. We restrict attention to tame links and thus we may assume that for each link L considered, the projection $\pi[L]$ of L to $\mathbf{R}^2$ is a finite 4-regular plane graph. The *link diagram* $D(L)$ of L arising from $\pi[L]$ is obtained by indicating at each crossing which one of the two curve segments goes over the other.

Let D be any link diagram. The underlying 4-regular plane graph G is Eulerian and the dual plane graph, whose vertices are the faces of G, is bipartite. Thus the faces of G can be 2-coloured. We colour the boundary faces black, and if two black faces share a crossing we join them by a signed edge according to the convention shown in Figure 1. In this way, given any link diagram D we get a plane signed graph $G(D)$ in which each edge corresponds to a crossing in D.

Conversely, given any plane signed graph G we can associate with it, in a canonical way, a link diagram $D(G)$ such that $G(D(G)) = G$. The construction is easy; draw the medial graph $m(G)$ of G (see for example [9]) and this will be the link diagram where the over/under nature of the crossings is determined by the sign of the appropriate edge in G.

---

[0]1991 Mathematics Subject Classification. Primary 05C15, 57M25.
This paper is in final form and no version of it will be submitted for publication elsewhere.

© 1993 American Mathematical Society
0271-4132/93 $1.00 + $.25 per page

+ ve   Figure 1:   − ve

We use $L(D)$ to denote the link having diagram D, and if G is any signed plane graph, then $L(G)$ denotes the link $L(G(D))$.

The terms *unknot* and *unlink* have their obvious meaning. A link diagram D is *alternating*, if the crossings are alternately over/under/over.... This corresponds exactly to the associated graph $G(D)$ having edges of only one sign.

The fundamental theorem of Reidemeister [15] states:

(1.1) THEOREM. *Two links K and L are ambient isotopic if and only if a link diagram of K can be transformed into a link diagram of L by a finite sequence of the moves (R1), (R2), (R3) and their inverses.*

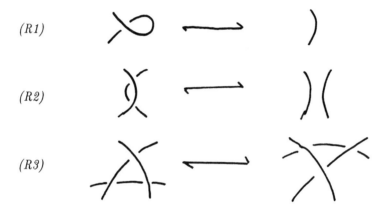

Figure 2:

These moves, known as the *Reidemeister moves*, are applied locally. In each case, away from the crossings to which the move is being applied, the diagram remains unchanged.

It is an easy exercise to check that the Reidemeister moves (R1)—(R3) on a link diagram D correspond to the following moves and their inverses on the signed graph $G(D)$.

(I) Delete a loop which bounds a face and contract an edge which has an endpoint of degree 1.

(II) Delete any pair of oppositely signed parallel edges which bound a face.

(II') If u is a vertex of degree 2 and xu and uy are opposite signs, contract xu and uy.

(III) For any triangle which bounds a face and is signed as shown perform the signed star-triangle interchange shown in Figure 3.

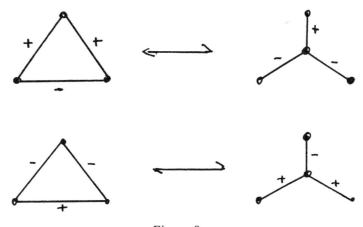

Figure 3:

We will use $\mathcal{SG}$ to denote the class of signed graphs, and $\mathcal{PG}$ to denote the set of planar graphs.

A fundamental algorithmic question already raised in [21] is:

(1.2) PROBLEM. *Find a function f such that any diagram D on n crossings which is isotopic to the unknot can be shown to be so by a sequence of not more than $f(n)$ Reidemeister moves.*

This question is readily transformed into a question about $\mathcal{SG}$ and the moves (I—III). We shall return to this problem in §5. First we consider the problem from an alternative perspective.

2 Braids and the braid group

A *braid* on m strings is constructed as follows.

Take m distinct points $P_1, \ldots, P_m$ in a horizontal line and link them to n distinct points $Q_1, \ldots, Q_m$ lying in a parallel line by m disjoint simple arcs (strings) f_i in $\mathbf{R}^3$, with f_i starting at P_i and ending at $Q_{\pi(i)}$ and where π is a permutation of $(1, 2, \ldots, m)$. The f_i are required to "run downwards" as illustrated in the example shown in Figure 4a.

The collection of strings constitutes an *m-braid*. The map $i \mapsto \pi(i)$ is the *permutation* of the braid. The braid will be *closed* by joining the points $P_i Q_i$ as illustrated in Figure 4b.

(2.1) *Each closed braid defines a link of μ components, where μ is the number of cycles in the permutation π.*

The oriented link formed by closing the braid α will be denoted by $\hat{\alpha}$. The trivial m-braid is a configuration in which no crossing of strings occurs. For example Figure 4b shows a braid on 3 strings representing a link of two components and having crossing number ≤ 5. Hence trivially every closed braid is a link. The converse also holds.

(2.2) THEOREM (Alexander). *Every link can be represented as a closed braid.*

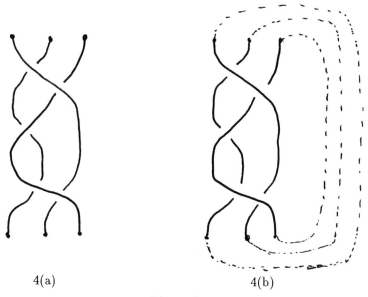

4(a)   4(b)

Figure 4:

There is an obvious way in which braids on the same number of strings can be composed. Namely, if z is a braid having end points $Q_1, \ldots, Q_m$ and z' is a braid having initial points $P'_1, \ldots, P'_m$, their composition zz' is obtained by identifying Q_i with P'_i for $1 \leq i \leq m$; the resulting braid has initial points $P_1, \ldots, P_m$ and endpoints $Q'_1, \ldots, Q'_m$. It is straightforward to check:

(2.3) *Under the above composition the isotopy classes of m-braids form a group, called the braid group B_m.*

It is also clear that the braid group B_m is generated by the *elementary braids* σ_i, σ_i^{-1} $(1 \leq i \leq m-1)$ representing simple interchanges.

Figure 5:

Defining relations for B_m were proved by Artin to be

$$\sigma_i \sigma_j = \sigma_j \sigma_i \quad \text{if} \quad |i - j| \geq 2$$

and

$$\sigma_i \sigma_{i+1} \sigma_i = \sigma_{i+1} \sigma_i \sigma_{i+1}.$$

Also, there exists a "Reidemeister type" theorem for braids due to A.A. Markov. This is in terms of moves of the following kind:

Markov Moves

TYPE I: Replace braid $\alpha \in B_m$ by a conjugate $\gamma \alpha \gamma^{-1} \in B_m$ with $\gamma \in B_m$.

TYPE II: Replace $\alpha \in B_m$ by $\alpha \sigma_m \in B_{m+1}$ or $\alpha \sigma_m^{-1} \in B_{m+1}$.

TYPE II^{-1}: Replace a braid of the form $\alpha \sigma_m \in B_{m+1}$, respectively $\alpha \sigma_m^{-1} \in B_{m+1}$, by $\alpha \in B_m$, provided α is a word in the generators $\sigma_1, \ldots, \sigma_{m-1}$ only.

(2.4) THEOREM. *Two braids have closures which are equivalent as links if and only if they are connected by a finite sequence of elementary moves of type I, II and II^{-1}.*

A given link can be represented in infinitely many different ways over many different braid groups and if $\alpha \in B_m$ and $\beta \in B_n$ have isotopic closures then the sequence of Markov moves transforming α to β may be long and go through several different braid groups.

A proof of Markov's theorem is given in the book of Birman [2].

Pictorially, Markov's moves are easy to understand. The conjugacy relation represented by a type I move is nothing more than the observation that the closure of the braid $\gamma \alpha \gamma^{-1}$ is isotopic to $\hat{\alpha}$ since closing the braid allows γ^{-1} to cancel out the effect of γ.

The type II moves are just the moves representing the introduction of a new string (or its inverse). However it is these moves which cause difficulties, because they change the number of strings and thus stop the problem being a "simple conjugacy problem"! For although deciding conjugacy in a given braid group is difficult, Makanin [10] and Garside [6] independently give algorithms which decide whether two given braids are conjugate or not. However these algorithms are complicated and Paterson and Razborov [12] have recently proved the following interesting result which may have some implications for the knot equivalence question. Consider the following computational problem:

NON-MINIMAL BRAIDS

Instance: A braid group B and a word w, in the standard generators of B.

Question: Is there a shorter word w' which is equivalent to w in B?

(2.5) THEOREM: NON-MINIMAL BRAIDS is NP-complete.

Note: Unlike the majority of NP-completeness results, perhaps the more surprising aspect of this result is that NON-MINIMAL BRAIDS belongs to NP. Artin's original algorithm for the word problem in the braid group involves generating a canonical form which is exponential in the length of the original word. However, due to recent (as yet unpublished) work of Thurston [17] there is a polynomial time algorithm for the word problem in B. Hence this can be used to show that NON-MINIMAL BRAIDS belongs to NP.

To show it is NP-complete, Paterson and Razborov use a reduction from

NON-MINIMAL FEEDBACK ARC SET

Instance: Digraph G and subset A of edges such that each circuit in G contains some edge of A (i.e. A is a *feedback arc set*).

Question: Is there a feedback arc set A' with $|A'| < |A|$?

The analogue of (1.2) can equally well be posed for braids. Specifically we can ask:

(2.6) PROBLEM: *Find a function g such that given a word ω, of length n in the standard generators of the braid group B_m, if its closure is isotopic to the unlink then it can be demonstrated in at most $g(n)$ Markov moves.*

A more precise form of (2.6) would be the following

(2.7) PROBLEM: *Does there exist a polynomial g satisfying the conditions of (2.6)?*

It would be reasonable to expect some relationship between the number of Markov moves and the number of Reidemeister moves needed to show equivalence to the unknot.

For example the following question does not seem quite as difficult as some we have been considering.

(2.8) PROBLEM: *Does there exist a polynomial time algorithm which will transform a link diagram to an isotopic closed braid with the same number of crossings?*

The constructions obtained by adapting the proofs of Alexander [1] and Yamada [23] can give closed braids with substantially more crossings than the original link diagram. However in a recent paper Vogel [20] gives a polynomial time construction which only adds 2 crossings to the diagram.

3 The braid index and the Seifert graph of a link

As we have seen, each link in 3-space has many different representations as a closed braid. The minimum number of strings in any braid representation of L is known as the *braid index* of L and is denoted by $\beta(L)$. In other words $\beta(L)$ is the smallest m for which there exists $\alpha \in B_m$ with $\hat{\alpha}$ isotopic to L.

The braid index characterises the unknot in the following sense:

(3.1) K is the unknot iff $\beta(K) = 1$.

Thus any polynomial time algorithm which determines the braid index would be of great interest.

A classical theorem about links is the following.

(3.2) Any oriented link L is the boundary of a compact connected orientable surface.

A canonical way of constructing such a surface, was given by H. Seifert in (1934) and the resulting surface is known as a *Seifert surface*. The key step in the construction of a Seifert surface from an oriented link diagram D is to "split" each crossing of D in the obvious way shown in Figure 6 and then to glue the resulting set of disjoint discs together using twisted bands to preserve orientability. For more details see [8 Chapter V].

Figure 6:

The *Seifert graph* $\Gamma(D)$ of an oriented link diagram D is a signed graph whose vertices are the Seifert circles (or discs) constructed in the above splitting process and with signed edges joining two circles whenever they share a crossing. The sign of the crossing is determined by the following convention.

Figure 7:

An easy property of $\Gamma(D)$ is that:

(3.3) Any Seifert graph is planar and bipartite.

PROOF (sketch) Planarity is obvious; to show that it is bipartite assume there is an odd circuit. This forces a contradiction on the clockwise/anticlockwise orientations of the Seifert circles. □

Other properties of the Seifert graph pointed out in [11] are the following:

(3.4) If the Seifert graph is nonseparable it uniquely determines the underlying link.

(3.5) The Seifert graph of a closed m-braid is the block sum of $m-1$ graphs, each consisting of parallel edges. Thus the "natural diagram" of a closed m-braid has exactly m Seifert circles.

An immediate consequence of this is:

(3.6) Every link L has at least one diagram D_0 for which the number of Seifert circles $s(D_0)$ equals the braid index $\beta(L)$.

In 1987, S. Yamada proved the following striking result:

(3.7) THEOREM. *For any diagram D of L, $s(D) \geq \beta(L)$.*

Combining this with (3.6) gives

(3.8) $$\beta(L) = \min_D s(D)$$

where the minimum is taken over all diagrams D representing L.

An extension of Yamada's theorem has recently been obtained by Murasugi and Przytycki [11] using the following combinatorial concept. Define the *cycle index* or *index*, $\text{ind}(G)$, of a graph G by:

(3.9) $$\text{ind}(G) = \max_{X \subseteq E(G)} |X| \; : \; |X \cap C| < |C|/2, \text{ for all circuits } C \text{ of } G.$$

The *index* of a link diagram D is the index of the unsigned version of $\Gamma(D)$ and is denoted by $\text{ind}(D)$.

This is not the original definition of Murasugi and Przytycki but for bipartite graphs it is equivalent to it by a theorem of Traczyk [18].

In [11] it is shown that

(3.10) THEOREM. *For any link diagram D of a link L,*

$$\beta(L) \leq s(D) - \text{ind}(D).$$

Since $\text{ind}(D)$ is nonnegative, this extends Yamada's theorem, and moreover it is conjectured in [11] that

(3.11) CONJECTURE. *For an alternating link diagram D of an (alternating) link L,*

$$\beta(L) = s(D) - \text{ind}(D).$$

From the complexity point of view this is interesting, for if we suppose that the conjecture is true, then determining the braid index of an alternating link L *given* an alternating diagram representing L, reduces to the problem of finding the index of a planar bipartite graph. Having tried unsuccessfully to develop such an algorithm it did not seem unreasonable to make

(3.12) CONJECTURE: *Finding the index of a bipartite planar graph is NP-hard.*

Since I first raised this question in [22] Fraenkel [3] has shown that finding the index is NP-hard for bipartite (not necessarily planar) graphs, and he and Loebl [4] now have a proof of the full conjecture.

If both conjectures (3.11) and (3.12) are true it would mean that:

(3.13) Determining the braid index of an alternating link L, even when presented with an alternating diagram, is NP-hard.

However, there is a curiosity here, in that Frank [5] had previously found a polynomial time algorithm which will solve the following problem

(3.14) $\max|X| : |X \cap C| \leq |C|/2$ for all circuits C of G.

Thus this is an interesting example of a situation where a very small change in the constraints has a drastic effect on the difficulty of an optimisation problem.

4 Classes of linklessly embeddable graphs

An *embedding* of a graph G in $\mathbf{R}^3$ is a mapping ϕ of the vertex set and edge set such that
(a) each vertex is mapped to a distinct point in $\mathbf{R}^3$;

(b) the edges are mapped to piecewise linear non-self-intersecting curves such that edges meet only at the images of vertices and such that if (u,v) is an edge, its image is a piecewise-linear-curve joining $\phi(u)$ to $\phi(v)$.

A pair of vertex disjoint cycles of G is said to be *linked* in the embedding if there do not exist disjoint topological balls in $\mathbf{R}^3$ containing them.

A graph G is *linklessly embeddable* if it has an embedding in $\mathbf{R}^3$ in which no pair of cycles is linked. We denote the class of such graphs by $\mathcal{LE}$.

In this section we shall not be concerned with $\mathcal{LE}$ (for a recent survey see [15]) but with two distinct subsets of it which are of interest in their own right.

We call G an *apex graph* if by the deletion of at most one vertex it becomes planar. If $\mathcal{A}$ denotes this class then it is easy to see:

(4.1) $\mathcal{A}$ is a proper subset of $\mathcal{LE}$.

Call G a *delta-wye* graph if G can be reduced to a graph with no edges by a finite sequence of the following operations:

I Delete loops and contract isthmuses.

II Replace a pair of parallel edges by a single edge

II' If u is a vertex of degree 2 and xu, uy are its incident edges, delete them and insert a single edge xy

III Perform any star-triangle interchange as illustrated in Figure 8.

Figure 8:

It is known that:

(4.2) If $\mathcal{PG}$ denotes the class of planar graphs and $\mathcal{DY}$ the class of delta-wye graphs then $\mathcal{PG} \subseteq \mathcal{DY} \subseteq \mathcal{LE}$ and in each case the inclusion is strict.

It is also well known that:

(4.3) Both $\mathcal{A}$ and $\mathcal{DY}$ are closed under the taking of minors.

Thus by Robertson-Seymour theory there is a polynomial time, $O(n^3)$, algorithm for deciding membership. However, as far as I know, the (finite) list of forbidden minors for membership of either class is not known.

The relationship between $\mathcal{A}$ and $\mathcal{DY}$ is not clear.

If $K_{5,5}\backslash M$ denotes the graph obtained from $K_{5,5}$ by deleting the edges of a perfect matching then in [14] and [19] it was observed that:

(4.4) $K_{5,5}\backslash M$ is linklessly embeddable but it is not even wye-delta reducible to an apex graph..

It is easier to find examples of delta-wye graphs which are not apex graphs.

Thus we have two reasonably attractive, minor closed, classes of graphs lying between the two sets $\mathcal{PG}$ and $\mathcal{LE}$. Moreover we know that neither class can contain as a minor K_6 or the Petersen graph P_{10}.

It is tempting therefore to make

(4.5) CONJECTURE: *If $G \in \mathcal{DY}$ and has no loops then G is 5-vertex colourable.*

(4.6) CONJECTURE: *If $G \in \mathcal{DY}$ and has no isthmuses then G has a 4-flow.*

Note: any counterexample to these conjectures would provide a counterexample to Hadwiger's conjecture and Tutte's 4-flow conjecture respectively.

Assuming the 4-colour theorem, it is clear that all apex graphs are 5-colourable. However the analogue of (4.6) for $\mathcal{A}$ is very appealing.

(4.7) CONJECTURE: *Every apex graph having no isthmuses has a 4-flow.*

However Paul Seymour has pointed out to me that Conjecture (4.7) turns out to be equivalent to the following long standing conjecture of Grötzsch.

(4.8) CONJECTURE: *A planar graph with all vertex degrees at most 3 and with no subgraph in which one vertex has degree 2 and all others have degree 3 is 3 edge-colourable.*

This suggests that proving (4.7) may be quite difficult. Accordingly, the following weakening of (4.7) (which is even more certainly true), and which was suggested by a question of U.S.R. Murty at this meeting, is the following:

(4.9) PROBLEM *Prove that every apex graph with no isthmus has a 5-flow.*

5 Reidemeister graphs

We close with a brief description of some recent joint work with W. Schwärzler [16].

For $G, H \in \mathcal{SG}$ we say that G is *R-equivalent* to H, written $G \sim H$, if G can be transformed to H by some finite sequence of the moves (I), (II), (II*), (III) defined on page 3 together with their inverses, where (II*) is the following transformation.

(II*) Contract any pair of oppositely signed edges which form an edge-cutset.

Note that (II*) is the exact matroid dual of (II) and is a slightly more powerful version of (II′) which is the move corresponding to a Reidemeister move for link equivalence. Accordingly it follows that

(5.1) If D and D' are different link diagrams representing the same link then $G(D) \sim G(D')$.

Call $G \in \mathcal{SG}$ a *Reidemeister graph (R-graph)* if $G \sim K_1$. Thus R-graphs are the graphic equivalents of the unknot. They include many non-planar signed graphs; for example 260 different signed versions of K_5 are R-graphs.

It is easy to see that $\sim$ is an equivalence relation on $\mathcal{SG}$; formally it bears some resemblance to ΔY equivalence discussed earlier. However, algorithmically it is much harder to handle. For example we know of no algorithm which will decide whether, for $G, H \in \mathcal{SG}$, $G \sim H$.

Accordingly we have to rely on partial invariants which at least are able to detect non-equivalence. The most useful of these turns out to be the extension of Kauffman's bracket polynomial which we show in [23] can be defined for any signed matroid (E, M) by

$$\langle M; A \rangle = A^{|E^-|-|E^+|-2r(M)} \sum_{X \subseteq E} A^{4(r(X)-|X^-|)}(-A^4-1)^{r(M)+|X|-2r(X)}$$

where r is the usual rank function.

(M is a *signed matroid* if its groundset E is partitioned into *positive* and *negative* elements, and $X^+(X^-)$ denotes the subset of positive (negative) elements of X.)

Then if G is a graph with $k(G)$ components,

$$\langle G; A \rangle = (-A^{-2} - A^2)^{k(G)-1} \langle M(G); A \rangle.$$

When D is a link diagram representing the unlink with μ components, we know that

(5.2) $\qquad \langle D \rangle = \langle G(D) \rangle = A^\alpha(-A^2 - A^{-2})^{\mu-1}$

where α is an integer.

A major open question about the bracket (or equivalently the Jones) polynomial of a link is the following:

(5.3) PROBLEM. *Does there exist a link L which is not the unlink but which has bracket polynomial satisfying (5.2)?*

The bracket polynomial itself is not invariant under $\sim$. However we show in [16] that if

$$\text{span} \langle M \rangle = \text{max degree} \langle M \rangle - \text{min degree} \langle M \rangle$$

then:

(5.4) THEOREM. *If G and H are R-equivalent then $\text{span}\langle G \rangle = \text{span}\langle H \rangle$.*

Another key property of $\langle G \rangle$ is that:

(5.5) When G is planar and μ denotes the number of components of the link $L(G)$, then for all signings

$$|\langle G : 1 \rangle| = 2^{\mu-1}.$$

Now when G is planar it always has a signing under which $L(G)$ is the unlink with μ components, where μ is determined by

$$2^{\mu-1} = |T(G; -1, -1)|,$$

and where T is its Tutte polynomial, (see for example [7]). It follows that for planar graphs

(5.6) $\qquad \text{min span}\langle G \rangle \leq 4(\mu(L(G)) - 1)$

where the minimum is taken over all possible signings. However many other non-planar graphs also satisfy (5.6).

A curious feature of our investigation into the bracket polynomial is that so far, computer search for minimal graphs not satisfying (5.6) has yielded just the 7 graphs which are the known minor minimal graphs not linklessly embeddable in $\mathbf{R}^3$, namely K_6, the Petersen graph P_{10}, and the 5 other graphs obtainable from these by triangle to star and star to triangle unsigned transformations.

As yet we have no satisfactory explanation of this.

ACKNOWLEDGEMENT

I am very grateful to the referee, A. Chalcraft, A.S. Fraenkel, F. Jaeger, M. Loebl, and particularly to Paul Seymour, for their comments on a first draft of this paper.

REFERENCES

[1] Alexander, J.W., A lemma on systems of knotted curves, Proc. Nat. Acad. Sci. USA **9** (1923), 93-95.

[2] Birman, J.S., *Braids links and mapping class groups*, Ann. Math. Studies 82, Princeton N.J., Princeton Univ. Press (1974).

[3] Fraenkel, A.S., The complexity of circuit intersection in graphs (to appear).

[4] Fraenkel, A.S. and Loebl, M. Complexity of circuit intersections in graphs (to appear).

[5] Frank, A., Conservative weightings and ear-decompositions of graphs (to appear).

[6] Garside, F.A., The braid group and other groups, Quart. J. Math. Oxford (2) **20** (1969), 235-254.

[7] Jaeger, F., Vertigan, D.L. and Welsh, D.J.A., On the computational complexity of the Jones and Tutte polynomials, Math. Proc. Camb. Phil. Soc. **108** (1990), 35-53.

[8] Kauffman, L.H., *On Knots*, Princeton Univ. Press, Ann. Math. Stud. (1987).

[9] Kauffman, L.H., A Tutte polynomial for signed graphs, Discrete Applied Math. **25** (1989), 105-127.

[10] Makanin, G.S., The conjugacy problem in the braid groups, Soviet Math. Doklady **9** (1968), 1156-1157.

[11] Murasugi, K. and Przytycki, J.H., The index of a graph with applications to knot theory (preprint).

[12] Paterson, M.S. and Razborov, A.A., The set of minimal braids is CO-NP complete, J. Algorithms **12** (1991), 393-408.

[13] Reidemeister, K., Homotopieringe und Linsenräume, Abh. Math. Sem. Hamburg II (1935), 102-109.

[14] Robertson, N., (private communication).
[15] Robertson, N., Seymour, P.D. and Thomas R., A survey of linkless embeddings. This proceedings.
[16] Schwärzler, W. and Welsh, D.J.A., Knots matroids and the Ising model Math. Proc. Camb. Phil. Soc. (to appear)
[17] Thurston, W.P., Finite state algorithms for the braid groups, preliminary draft (1988).
[18] Traczyk, P., On the index of graphs: Index versus cycle index (preprint).
[19] Vertigan, D.L. and Welsh, D.J.A., Delta-wye graphs (unpublished manuscript).
[20] Vogel, P., Representation of links by braids: A new algorithm, Comment. Math. Helvetici **65** (1990) 104-113.
[21] Welsh, D.J.A., The complexity of knots, Ann. Disc. Math, (to appear).
[22] Welsh, D.J.A., Knots, colourings and the complexity of counting, ARIDAM Lectures (1991).
[23] Yamada, S., The minimal number of Seifert circles equals the braid index of a link, Inv. Math. **89** (1987), 347-356.

Current address   Merton College, University of Oxford, Oxford, England.
E-mail address   dwelsh @ uk.ac.ox.vax

A SURVEY OF LINKLESS EMBEDDINGS

NEIL ROBERTSON, P. D. SEYMOUR and ROBIN THOMAS

ABSTRACT. We announce results about flat (linkless) embeddings of graphs in 3–space. A piecewise-linear embedding of a graph in 3–space is called *flat* if every circuit of the graph bounds a disk disjoint from the rest of the graph. We have shown that:

(i) An embedding is flat if and only if the fundamental group of the complement in 3–space of the embedding of every subgraph is free.

(ii) If two flat embeddings of the same graph are not ambient isotopic, then they differ on a subdivision of K_5 or $K_{3,3}$.

(iii) Any flat embedding of a graph can be transformed to any other flat embedding of the same graph by "3–switches," an analog of 2–switches from the theory of planar embeddings. In particular, any two flat embeddings of a 4–connected graph are either ambient isotopic, or one is ambient isotopic to a mirror image of the other.

(iv) A graph has a flat embedding if and only if it has no minor isomorphic to one of seven specified graphs. These are the graphs that can be obtained from K_6 by means of $Y\Delta$- and ΔY-exchanges.

1991 Mathematics Subject Classification. Primary 05C10, 05C75, 57M05, 57M15, 57M25

Research of the first author was performed under a consulting agreement with Bellcore and was supported by NSF under Grant No. DMS-8903132 and by ONR under Grant No. N00014-91-J-1905.

Research of the third author was was supported by NSF under Grant No. DMS-8903132 and by DIMACS Center, Rutgers University, New Brunswick, New Jersey 08903, USA.

This paper is a preliminary version, and the detailed version will be published elsewhere.

1. INTRODUCTION

All spatial embeddings are assumed to be piecewise linear. If C, C' are disjoint simple closed curves in S^3, then their *linking number*, $\text{lk}(C, C')$, is the number of times (mod 2) that C crosses over C' in a regular projection of $C \cup C'$. In this paper graphs are finite, undirected and may have loops and multiple edges. Every graph is regarded as a topological space in the obvious way. We say that an embedding of a graph G in S^3 is *linkless* if every two disjoint circuits of G have zero linking number. The following is a result of Sachs [16] and Conway and Gordon [4].

(1.1) *The graph K_6 has no linkless embedding.*

Proof. Let ϕ be an embedding of K_6 into S^3. By studying the effect of a crossing change in a regular projection, it is easy to see that the mod 2 sum $\sum \text{lk}(\phi(C_1), \phi(C_2))$, where the sum is taken over all unordered pairs of disjoint circuits C_1, C_2 of K_6, is an invariant independent of the embedding. By checking an arbitrary embedding we can establish that this invariant equals 1. □

A graph is a *minor* of another if the first can be obtained from a subgraph of the second by contracting edges. Our main result is a theorem that a graph is linklessly embeddable if and only if it has no minor isomorphic to K_6 or six other closely related graphs. However, we find it much easier to work with the following stronger concept, suggested by Böhme [1] and Saran [18]. We say that an embedding ϕ of a graph G in S^3 is *flat* if for every circuit C of G there exists an open disk in S^3 disjoint from $\phi(G)$ whose boundary is $\phi(C)$. Clearly every flat embedding is linkless, but the converse is false. However, we shall see later that a graph admits a linkless embedding if and only if it admits a flat embedding, and so the classes of embeddable graphs are the same. The reason why we prefer flat embeddings is that they work better. For instance, there is a uniqueness theory parallel to the theory of planar embeddings, and a theorem which characterizes flat embeddings in terms of the fundamental group of the complement.

If G is a graph and X is a vertex or a set of vertices, we denote by $G \backslash X$ the graph obtained from G by deleting X. A graph G is *nearly-planar* if there exists a vertex v of G such that $G \backslash v$ is planar. It may be helpful to notice the following fact.

(1.2) *Every nearly-planar graph admits a flat embedding.*

Proof. Let G be nearly-planar, and let v be such that $G \backslash v$ is planar. We may assume that G is simple, because it is easy to construct a flat embedding of a graph given a flat embedding of its underlying simple graph. We embed $G \backslash v$ in the xy-plane in $R^3 \subseteq S^3$, embed v anywhere not in this plane, and embed all edges from v to the planar graph as straight line segments. It is easy to check that this defines a flat embedding. □

The following lemma was proved by Böhme [1] (see also [18]).

(1.3) Let ϕ be a flat embedding of a graph G into S^3, and let $C_1, C_2, \ldots, C_n$ be a family of circuits of G such that for every $i \neq j$, the intersection of C_i and C_j is either connected or null. Then there exist pairwise disjoint open disks $D_1, D_2, \ldots, D_n$, disjoint from $\phi(G)$ and such that $\phi(C_i)$ is the boundary of D_i for $i = 1, 2, \ldots, n$.

An embedding ϕ of a graph G in S^3 is *spherical* if there exists a surface $\Sigma \subseteq S^3$ homeomorphic to S^2 such that $\phi(G) \subseteq \Sigma$. Clearly if ϕ is spherical then G is planar. We illustrate the use of (1.3) with the following, which is a special case of a theorem of Wu [22].

(1.4) Let ϕ be an embedding of a planar graph G in S^3. Then ϕ is flat if and only if it is spherical.

Proof. Clearly if ϕ is spherical then it is flat. We prove the converse only for the case when G is 3–connected. Let $C_1, C_2, \ldots, C_n$ be the collection of face–boundaries in some planar embedding of G. These circuits satisfy the hypothesis of (1.3). Let $D_1, D_2, \ldots, D_n$ be the disks as in (1.3); then $\phi(G) \cup D_1 \cup D_2 \cup \cdots \cup D_n$ is the desired sphere. □

The paper is organized as follows. In Section 2 we present a characterization of flat embeddings in terms of the fundamental group of the complement, in Section 3 we discuss a uniqueness theory of flat embeddings, in Section 4 we state our main result, an excluded minor characterization of linklessly embeddable graphs, and finally in Section 5 we discuss three conjectures and some algorithmic aspects of flat embeddings.

2. THE FUNDAMENTAL GROUP

The following is a result of Scharlemann and Thompson [19].

(2.1) Let ϕ be an embedding of a graph G in S^3. Then ϕ is spherical if and only if
(i) G is planar, and
(ii) for every subgraph G' of G, the fundamental group of $S^3 - \phi(G')$ is free.

The "only if" implication is easy to see. The point of the theorem is the converse. It is easy to see that (ii) cannot be replaced by the weaker condition that the fundamental group of $S^3 - \phi(G)$ is free. We use (2.1) to prove the following generalization.

(2.2) Let ϕ be an embedding of a graph G in S^3. Then ϕ is flat if and only if for every subgraph G' of G, the fundamental group of $S^3 - \phi(G')$ is free.

Proof. Here we only prove "only if." Let G' be a subgraph of G such that $\pi_1(S^3 - \phi(G'))$ is not free. Choose a maximal forest F of G' and let G'' be obtained from G' by contracting all edges of F, and let ϕ'' be the induced embedding of G''. Then $\pi_1(S^3 - \phi''(G'')) = \pi_1(S^3 - \phi(G'))$ is not free, but

G'' is planar, and so ϕ'' is not flat by (2.1) and (1.4). Hence ϕ is not flat, as desired. □

Let G be a graph, and let e be an edge of G. We denote by $G\backslash e$ (G/e) the graph obtained from G by deleting (contracting) e. If ϕ is an embedding of G in S^3, then it induces embeddings of $G\backslash e$ and (up to ambient isotopy) of G/e in the obvious way. We denote these embeddings by $\phi\backslash e$ and ϕ/e, respectively.

(2.3) *Let ϕ be an embedding of a graph G in S^3, and let e be a nonloop edge of G. If both $\phi\backslash e$ and ϕ/e are flat, then ϕ is flat.*

Proof. Suppose that ϕ is not flat. By (2.2) there exists a subgraph G' of G such that $\pi_1(S^3 - \phi(G'))$ is not free. If $e \notin E(G')$, then $\phi\backslash e$ is not flat by (2.2). If $e \in E(G')$ then ϕ/e is not flat by (2.2), because $\pi_1(S^3 - (\phi/e)(G'/e)) = \pi_1(S^3 - \phi(G'))$ is not free. □

We say that a graph G is a *coforest* if every edge of G is a loop. The following follows immediately from (2.3).

(2.4) *Let ϕ be an embedding of a graph G in S^3. Then ϕ is flat if and only if the induced embedding of every coforest minor of G is flat.*

3. UNIQUENESS

We begin this section by recalling the following two classical results. The first is Kuratowski's theorem [8]. (A graph H is a *subdivision* of a graph G if H can be obtained from G by replacing edges by internally–disjoint paths.)

(3.1) *A graph is planar if and only if it has no subgraph isomorphic to a subdivision of K_5 or $K_{3,3}$.*

Let ϕ be an embedding of a graph G in S^2. Let P be a simple closed curve in S^2 meeting $\phi(G)$ in a set A containing at most two points. Let D be a chord of P (that is, a simple curve with only its distinct endpoints in common with P) and assume that every member of A is on D. Let B be the open disk of $S^2 - P$ containing the interior of D. Let ϕ' be an embedding obtained from ϕ by taking a reflection through D in B, and by leaving ϕ unchanged in $S^2 - B$. We say that ϕ' was obtained from ϕ by a *2–switch*. The second classical result is a theorem of Whitney [21], perhaps stated in a slightly unusual way.

(3.2) *Let ϕ_1, ϕ_2 be two embeddings of a graph in S^2. Then ϕ_1 can be obtained from ϕ_2 by a series of 2–switches.*

We shall see in (3.10) that a similar theorem holds for flat embeddings.

Let ϕ_1, ϕ_2 be two embeddings of a graph G in S^3. We say that ϕ_1, ϕ_2 are *ambient isotopic* if there exists an orientation preserving homeomorphism h of S^3 onto S^3 such that $\phi_1 = h\phi_2$. (We remark that by a result of Fisher [5] h can be realized by an ambient isotopy.) The following follows from (1.4) and (3.2).

(3.3) *Any two flat embeddings of a planar graph are ambient isotopic.*

(3.4) *The graphs K_5 and $K_{3,3}$ have exactly two non-ambient isotopic flat embeddings.*

Sketch of proof. Let G be $K_{3,3}$ or K_5, let e be an edge of G, and let H be $G\backslash e$. Notice that H is planar. From (1.3) it follows that if ϕ is a flat embedding of G, then there is an embedded 2-sphere $\Sigma \subseteq S^3$ with $\phi(G) \cap \Sigma = \phi(H)$. If ϕ_1 and ϕ_2 are flat embeddings of G, we may assume (by replacing ϕ_2 by an ambient isotopic embedding) that this 2-sphere Σ is the same for both ϕ_1 and ϕ_2. Now ϕ_1 is ambient isotopic to ϕ_2 if and only if $\phi_1(e)$ and $\phi_2(e)$ belong to the same component of $S^3 - \Sigma$. □

As a curiosity we deduce from (3.1), (3.3) and (3.4) that a graph has a unique flat embedding if and only if it is planar.

Our next objective is to determine the relation between different flat embeddings of a given graph. We denote by $f|X$ the restriction of a mapping f to a set X.

(3.5) *Let ϕ_1, ϕ_2 be two flat embeddings of a graph G that are not ambient isotopic. Then there exists a subgraph H of G isomorphic to a subdivision of K_5 or $K_{3,3}$ for which $\phi_1|H$ and $\phi_2|H$ are not ambient isotopic.*

A question arises if there is any analogue of (3.5) when the embeddings are not necessarily flat. The following follows immediately from (2.4).

(3.6) *Let ϕ_1, ϕ_2 be two embeddings of a graph G such that they are not ambient isotopic and exactly one of them is flat. Then G has a coforest minor H such that the embeddings of H induced by ϕ_1 and ϕ_2 are not ambient isotopic.*

We do not know if (3.6) remains true when none of ϕ_1, ϕ_2 is flat.

We denote the vertex-set and edge-set of a graph G by $V(G)$ and $E(G)$ respectively. Let G be a graph and let H_1, H_2 be subgraphs of G isomorphic to subdivisions of K_5 or $K_{3,3}$. We say that H_1 and H_2 are 1-*adjacent* if there exist $i \in \{1, 2\}$ and a path P in G such that P has only its endpoints in common with H_i and such that H_{3-i} is a subgraph of the graph obtained from H_i by adding P. We say that H_1 and H_2 are 2-*adjacent* if there are seven vertices $u_1, u_2, \ldots, u_7$ of G, and thirteen paths L_{ij} of G ($1 \leq i \leq 4$ and $5 \leq j \leq 7$, or $i = 3$ and $j = 4$), such that

(i) each path L_{ij} has ends u_i, u_j,
(ii) the paths L_{ij} are mutually vertex-disjoint except for their ends,
(iii) H_1 is the union of L_{ij} for $i = 2, 3, 4$ and $j = 5, 6, 7$, and
(iv) H_2 is the union of L_{ij} for $i = 1, 3, 4$ and $j = 5, 6, 7$.

(Notice that if H_1 and H_2 are 2-adjacent, then they are both isomorphic to subdivisions of $K_{3,3}$, and that L_{34} is used in neither H_1 nor H_2.) We denote by $\mathcal{K}(G)$ the simple graph with vertex-set all subgraphs of G isomorphic to subdivisions of K_5 or $K_{3,3}$ in which two distinct vertices are adjacent if they are either 1-adjacent or 2-adjacent. The following is easy to see, using (3.4).

(3.7) *Let ϕ_1, ϕ_2 be two flat embeddings of a graph G, and let H, H' be two adjacent vertices of $\mathcal{K}(G)$. If $\phi_1|H$ is ambient isotopic to $\phi_2|H$, then $\phi_1|H'$ is ambient isotopic to $\phi_2|H'$.*

We need the following purely graph-theoretic lemma.

(3.8) *If G is a 4-connected graph, then $\mathcal{K}(G)$ is connected.*

We prove (3.8) in [12] by proving a stronger result, a necessary and sufficient condition for $H, H' \in V(\mathcal{K}(G))$ to belong to the same component of $\mathcal{K}(G)$ in an arbitrary graph G. The advantage of this approach is that it permits an inductive proof using the techniques of deleting and contracting edges.

If ϕ is an embedding of a graph G in S^3 we denote by $-\phi$ the embedding of G obtained by composing ϕ with the antipodal map. The following is our uniqueness theorem.

(3.9) *Let G be a 4-connected graph and let ϕ_1, ϕ_2 be two flat embeddings of G. Then ϕ_1 is ambient isotopic to either ϕ_2 or $-\phi_2$.*

Proof. If G is planar then ϕ_1 is ambient isotopic to ϕ_2 by (3.3). Otherwise there exists, by (3.1), a subgraph H of G isomorphic to a subdivision of K_5 or $K_{3,3}$. By replacing ϕ_2 by $-\phi_2$ we may assume by (3.4) that $\phi_1|H$ is ambient isotopic to $\phi_2|H$. From (3.7) and (3.8) we deduce that $\phi_1|H'$ is ambient isotopic to $\phi_2|H'$ for every $H' \in V(\mathcal{K}(G))$. By (3.5) ϕ_1 and ϕ_2 are ambient isotopic, as desired. □

Actually, the 4-connectedness is not necessary for (3.9). It turns out that what is necessary and sufficient for the conclusion of (3.9) is, roughly, that no two subgraphs isomorphic to subdivisions of K_5 or $K_{3,3}$ are "separated" by a separation of order at most 3. Let us call such graphs *Kuratowski 4-connected*.

We now state a generalization of (3.9). Let ϕ be a flat embedding of a graph G, and let $\Sigma \subseteq S^3$ be a surface homeomorphic to S^2 meeting $\phi(G)$ in a set A containing at most three points. In one of the open balls into which Σ divides S^3, say B, choose an open disk D with boundary a simple closed curve ∂D such that $A \subseteq \partial D \subseteq \Sigma$. Let ϕ' be an embedding obtained from ϕ by taking a reflection of ϕ through D in B, and leaving ϕ unchanged in $\Sigma - B$. We say that ϕ' is obtained from ϕ by a *3-switch*. The following analog of (3.2) generalizes (3.9).

(3.10) *Let ϕ_1, ϕ_2 be two flat embeddings of a graph G in S^3. Then ϕ_2 can be obtained from ϕ_1 by a series of 3-switches.*

4. THE PETERSEN FAMILY

Let G be a graph and let v be a vertex of G of valency 3 with distinct neighbors. Let H be obtained from G by deleting v and adding an edge between

every pair of neighbors of v. We say that H was obtained from G by a $Y\Delta$-exchange and that G was obtained from H by a ΔY-exchange. We say that two graphs are $Y\Delta$-equivalent if one can be obtained from a graph isomorphic to the other by a sequence of the following operations and their inverses:
(i) Deleting a vertex of valency ≤ 1,
(ii) suppressing a vertex of valency 2 (that is, contracting an edge incident to it),
(iii) deleting a parallel edge or a loop,
(iv) $Y\Delta$-exchange.

(4.1) *If G, H are $Y\Delta$-equivalent, then G has a flat embedding if and only if H does.*

It follows from (4.1) and (1.2) that if a graph is $Y\Delta$-equivalent to a nearly-planar graph, then it admits a flat embedding. The converse is false, because $K_{5,5}$ minus a perfect matching is a counterexample.

The *Petersen family* is the set of all graphs that can be obtained from K_6 by doing $Y\Delta$- and ΔY-exchanges. There are (up to isomorphism) exactly seven such graphs, one of which is the Petersen graph. The Petersen family is depicted in Figure 1. The following is our main theorem.

(4.2) *For a graph G, the following conditions are equivalent.*
(i) G has a flat embedding,
(ii) G has a linkless embedding,
(iii) G has no minor isomorphic to a member of the Petersen family.

Here (i) $\Rightarrow$ (ii) is trivial. Sachs [16] has in fact shown that no member of the Petersen family has a linkless embedding, from which (ii) $\Rightarrow$ (iii) follows because the property of having a linkless embedding is closed under taking minors. (Sachs stated his result in a weaker form, but the proof is adequate.) The hard part is that (iii) $\Rightarrow$ (i), which we now briefly sketch.

Sketch of the proof that in (4.2), (iii) $\Rightarrow$ (i). Suppose that G is a minor-minimal graph with no flat embedding. It can be shown that G is "basically 5-connected", which is a certain weaker form of 5-connectivity (see the next section for a precise definition). From (4.1) we may assume that G has no triangles. Suppose that there are edges e, f of G and an end v of e not adjacent to either end of f such that $G\backslash v, G\backslash e/f, G/e/f$ are all Kuratowski 4-connected. Since G is minor-minimal with no flat embedding, there are flat embeddings ϕ_1, ϕ_2, ϕ_3 of $G\backslash e, G/e, G/f$, respectively. By (3.9), since $\phi_3\backslash e$ and ϕ_1/f are both flat embeddings of the Kuratowski 4-connected graph $G\backslash e/f$, we may assume that $\phi_3\backslash e \simeq \phi_1/f$, and similarly that $\phi_3/e \simeq \phi_2/f$. (Here and later $\simeq$ means "ambient isotopic to.") From the first equation there is a 1-edge uncontraction of $\phi_3\backslash e$ which yields an embedding ambient isotopic to ϕ_1, and similarly there is a 1-edge uncontraction of ϕ_3/e yielding ϕ_2. These two uncontractions can be viewed as "local" operations at a vertex common to $\phi_3\backslash e$ and ϕ_3/e, and it

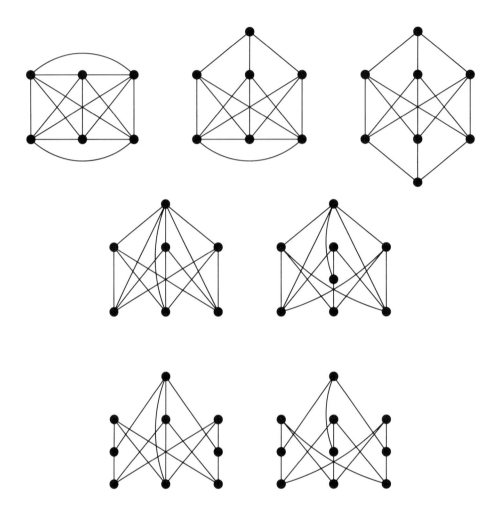

Figure 1: The Petersen Family

can be argued (the details are quite complicated, see [14]) that they are the "same" uncontraction operation. Let ϕ be obtained from ϕ_3 by performing this uncontraction; then $\phi \backslash e \simeq \phi_1$ and $\phi/e \simeq \phi_2$. Since ϕ_1 and ϕ_2 are flat, so is ϕ by (2.3), a contradiction since G has no flat embedding. Thus no two such edges e, f exist. But now a purely graph-theoretic argument [13] (using the non-existence of such edges e, f, the high connectivity of G and that G is not nearly-planar) implies that G has a minor in the Petersen family. □

There have been a number of other attempts [10, 18, 2] at proving (iii) ⇒ (i) and (iii) ⇒ (ii). However, none of them is correct. The question whether (iii) ⇒ (i) was first raised by Sachs [16], and that (i) and (ii) are equivalent was conjectured by Böhme [1].

We mention the following corollary, which is vaguely related to the so-called "strong embedding conjecture." Let ψ be an embedding of a graph G in a surface (=compact 2–manifold without boundary) Σ. We say that ψ is k-representative if every non-null-homotopic closed curve in Σ meets $\psi(G)$ at least k times. The strong embedding conjecture states that every 2–connected graph has a 2-representative embedding in some surface. It is also possible that every 3–connected graph has such an embedding in a nonorientable surface. From (4.2) we deduce the following.

(4.3) *If a graph G admits a 3–representative embedding into some nonorientable surface, then G has a minor isomorphic to a member of the Petersen family other than $K_{4,4}^-$ ($K_{4,4}$ minus an edge).*

Proof. Let ψ be a 3-representative embedding of G in a nonorientable surface Σ. By [15, Proposition 7.3] we may assume (by taking a minor of G) that G is 3–connected. We first show that G has a minor isomorphic to a member of the Petersen family. By (4.2) it suffices to show that G has no flat embedding. Suppose for a contradiction that G has a flat embedding ϕ into S^3. Let $C_1, C_2, \ldots, C_n$ be the collection of face–boundaries in the embedding ψ; since G is 3–connected and ψ is 3-representative, $C_1, C_2, \ldots, C_n$ are circuits and satisfy the hypothesis of (1.3). Let $D_1, D_2, \ldots, D_n$ be the disks as in (1.3). Then $\psi(G) \cup D_1 \cup D_2 \cup \ldots \cup D_n$ is homeomorphic to Σ, a contradiction because Σ has no embedding in S^3.

Thus G has a minor isomorphic to a member of the Petersen family, and so we may assume that it has a minor isomorphic to $K_{4,4}^-$ and to no other member of the Petersen family. Now it is easy to show, using the splitter theorem [20] of the second author, that G is isomorphic to $K_{4,4}^-$. But $K_{4,14}^-$ has no 3-representative embedding in any nonorientable surface, and the theorem follows. □

Conversely, every member of the Petersen family except $K_{4,4}^-$ admits a 3-representative embedding in the projective plane.

5. REMARKS

It would be nice to have a structural description of all linklessly embeddable graphs. Let us say that a graph G has a *hamburger structure* if either $|V(G)| \leq 4$ or there are vertices $v_1, v_2, \ldots, v_5$ of G and three subgraphs G_1, G_2, G_3 of G such that

(i) $G_1 \cup G_2 \cup G_3 = G$,
(ii) $V(G_i) \cap V(G_j) = \{v_1, v_2, \ldots, v_5\}$ for $i \neq j$, and
(iii) each G_i can be embedded in a closed disk with vertices $v_1, v_2, \ldots, v_5$ (in this order) on the boundary.

It is not difficult to see that if G has a hamburger structure, then it has a flat embedding. We say that a graph G is *basically 5–connected* if G is simple, 3–connected, and cannot be expressed as a union of two subgraphs G_1 and G_2, where $E(G_1) \cap E(G_2) = \emptyset$, and either
(i) $|V(G_1) \cap V(G_2)| = 3$ and $|E(G_1)|, |E(G_2)| \geq 4$, or
(ii) $|V(G_1) \cap V(G_2)| = 4$ and $|E(G_1)|, |E(G_2)| \geq 7$.

(5.1) Conjecture. *Let G be a basically 5–connected, triangle–free linklessly embeddable graph. Then either there are two vertices u, v of G such that $G \backslash \{u, v\}$ is planar, or else G has a hamburger structure.*

From (4.1) we see that the requirement that G be triangle–free is not restrictive. One can also modify the definition of "hamburger structure" so that (5.1) could be true for all basically 5–connected linklessly embeddable graphs. The point of (5.1) is that if $G \backslash \{u, v\}$ is planar then there is a simple polynomial–time algorithm to test if G has a flat embedding. The algorithm is based on a study of homotopy of paths joining the neighbors of u and v in $G \backslash \{u, v\}$.

A second relevant conjecture is the following, due to Jorgensen [7].

(5.2) Conjecture. *Let G be a 6–connected graph with no minor isomorphic to K_6. Then G is nearly–planar.*

This was motivated by Hadwiger's conjecture [6]. One case of the latter states that every loopless graph with no minor isomorphic to K_6 is 5–colorable. Mader [9] showed that every minor-minimal counterexample G is 6–connected, in which case (5.2) and the Four Color Theorem would imply that G is 5–colorable, a contradiction. However, we believe that we have now obtained a proof of this case of Hadwiger's conjecture, without proving (5.2). We do not even know if (5.2) holds for linklessly embeddable graphs.

Our third conjecture relates linklessly embeddable graphs and a graph parameter $\mu(G)$ introduced by Colin de Verdière in [3]. We refer the reader to that paper for a definition of $\mu(G)$ (an English translation appears in this volume), which is in terms of the multiplicities of the second largest eigenvalues of certain matrices associated with G.

(5.3) Conjecture. *A graph G has a flat embedding if and only if $\mu(G) \leq 4$.*

The "if" part of (5.3) follows from our main result, and so the problem is about the converse.

Finally, let us mention two algorithmic aspects of flat embeddings. In [19] Scharlemann and Thompson describe an algorithm to test if a given embedding is spherical. Using their algorithm, (1.4) and (2.4), we can test if a given embedding is flat, by testing the flatness of all coforest minors. At the moment there is no known *polynomial-time* algorithm to test if an embedding of a given coforest is flat, because it includes testing if a given knot is trivial. On the other hand, we can test if a given graph G has a flat embedding in time $O(|V(G)|^3)$. This is done by testing the absence of minors isomorphic to members of the Petersen family, using (4.2) and the algorithm [11] of the first two authors.

REFERENCES

1. T. Böhme, On spatial representations of graphs, *Contemporary Methods in Graph Theory*, R. Bodendieck ed., Mannheim, Wien, Zürich (1990), 151–167.
2. T. Böhme, Lecture at the AMS Summer Research Conference on Graph Minors, Seattle, WA, June 1991.
3. Y. Colin de Verdière, Sur un nouvel invariant des graphes et un critère de planarité, *J. Combin. Theory Ser. B* 50 (1990), 11–21.
4. J. H. Conway and C. McA. Gordon, Knots and links in spatial graphs, *J. Graph Theory*, 7 (1983), 445–453.
5. G. M. Fisher, On the group of all homeomorphisms of a manifold, *Trans. Amer. Math. Soc.* 97 (1960), 193–212.
6. H. Hadwiger, Über eine Klassifikation der Streckenkomplexe, *Vierteljschr. Naturforsch. Gessellsch. Zürich*, 88 (1943), 133–142.
7. L. Jorgensen, Contraction to K_8, to appear.
8. C. Kuratowski, Sur le problème des courbes gauches en topologie, *Fund. Math.* 15 (1930), 271–283.
9. W. Mader, Über trennende Eckenmengen in homomorphiekritische Graphen, *Math. Ann.* 175 (1968), 245–252.
10. R. Motwani, A. Raghunathan and H. Saran, Constructive results from graph minors: Linkless embeddings, *Proc. 29th Symposium on the Foundations of Computer Science*, Yorktown Heights, 1988.
11. N. Robertson and P. D. Seymour, Graph minors. XIII. The disjoint paths problem, submitted.
12. N. Robertson, P. D. Seymour and R. Thomas, Kuratowski chains, submitted.
13. N. Robertson, P. D. Seymour and R. Thomas, Petersen family minors, submitted.
14. N. Robertson, P. D. Seymour and R. Thomas, Sachs' linkless embedding conjecture, manuscript.
15. N. Robertson and R. Vitray, Representativity of surface embeddings, In: *Paths, flows, and VLSI layout*, B. Korte, L. Lovász, H. J. Prömel, and A. Schrijver, eds., Springer-Verlag, Berlin, Heidelberg, 1990.
16. H. Sachs, On spatial representation of finite graphs, Proceedings of a conference held in Łagów, Poland, February 10–13, 1981, *Lecture Notes in*

Mathematics, Vol. 1018, Springer-Verlag, Berlin, Heidelberg, New York, Tokyo, 1983.

17. H. Sachs, On spatial representation of finite graphs, Colloquia Mathematica Societatis János Bolyai, *37. Finite and infinite sets*, A. Hajnal, L. Lovász and V. T. Sós, eds., North-Holland, Budapest 1984, 649–662.
18. H. Saran, *Constructive Results in Graph Minors: Linkless Embeddings*, PhD thesis, University of California at Berkeley, 1989.
19. M. Scharlemann and A. Thompson, Detecting unknotted graphs in 3-space, *J. Diff. Geometry*, 34 (1991), 539–560.
20. P. D. Seymour, Decomposition of regular matroids, *J. Combin. Theory Ser. B* 28 (1980), 305–359
21. H. Whitney, 2-isomorphic graphs, *Amer. J. Math.* 55 (1933), 245–254.
22. Y.-Q. Wu, On planarity of graphs in 3-manifolds, to appear in *Comment. Math. Helv.*

DEPARTMENT OF MATHEMATICS, OHIO STATE UNIVERSITY, COLUMBUS, OH 43210, USA
E-mail: robertso@function.mps.ohio-state.edu

BELLCORE, 445 SOUTH STREET, MORRISTOWN, NJ 07962, USA
E-mail: pds@bellcore.com

SCHOOL OF MATHEMATICS, GEORGIA INSTITUTE OF TECHNOLOGY, ATLANTA, GA 30332, USA
E-mail: thomas@math.gatech.edu

On a New Graph Invariant and a Criterion for Planarity

YVES COLIN de VERDIÈRE

A finite graph is said to be planar if it may be embedded in the plane without its edges crossing. A natural problem, solved by Kuratowski [KI], is to find a characterization of planar graphs. For other references on this subject consult [BE, WE, TE]. In our previous articles [C–C,[CV$_i$]$_{1 \le i \le 4}$] we developed methods which enable us to give a global invariant associated with a finite graph which is apparently new. This integer invariant $\mu(\Gamma)$ satisfies:

Theorem. Γ *is planar if and only if* $\mu(\Gamma) \le 3$.

The aims of this article are the definition of $\mu(\Gamma)$, the study of the property of monotonicity of μ with respect to the operations of deletion and contraction (in the sense of [H–T]) of a graph; and the proof of the above theorem and others concerning the embedding of graphs in surfaces. The underlying question which seems to be of greatest importance is: is $\mu(\Gamma)$ related to the chromatic number $C(\Gamma)$? Known examples and the preceding theorem lead us to propose

Conjecture. *For each* Γ, $\mu(\Gamma) \ge C(\Gamma) - 1$.

The study of this conjecture, which implies the 4 colour theorem, could lead us to a new proof of that theorem!!

1991 Mathematics Subject Classification. Primary 05C10, 05C75.
Translated from J. Combin. Theory Ser. B 50 (1990), 11–21 by Neil Calkin.

It should also be noted that the conjecture is weaker than Hadwiger's conjecture [TE p. 52; OE p. 146].

1. Construction of $\mu(\Gamma)$

This construction is based on a property of *transversality* introduced by Arnold [AD] and called in [CV3] the Strong Arnold Hypothesis (SAH).

We start by defining some terms:

Γ is a finite, connected, undirected graph without loops;

$V(\Gamma)$ or V is the set of vertices, of size v_Γ or v;

$E(\Gamma)$ or E is the set of edges, of size e_Γ or e;

$\mathcal{S}_v$ is the set of symmetric real $v \times v$ matrices.

We denote by $\mathcal{O}_\Gamma$ the set of matrices in $\mathcal{S}_v$ such that if $A = (a_{ij}) \in \mathcal{O}_\Gamma$, we have

(i) $a_{ij} < 0$ if $\{i,j\} \in E$

(ii) $a_{ij} = 0$ if $\{i,j\} \notin E$ and $i \neq j$.

To every measure $\nu = \sum_{i \in V} V_i \delta(i)$ $(V_i > 0)$ on V we associate a bijection $A \mapsto q_A$ from $\mathcal{O}_\Gamma$ to the set Q_Γ of quadratic forms on $\mathbb{R}^V = L^2(V,\nu)$ of the form

$$q((x_i)) = \sum_{i \in V} c_i x_i^2 + \sum_{\{i,j\} \in E} c_{\{i,j\}} (x_i - x_j)^2,$$

where the $c_{\{i,j\}}$ are > 0; this bijection is defined by

$$\langle Ax \mid y \rangle_{L^2(\nu)} = q_A(x,y).$$

As Γ is connected, it is well known and easy to check that the spectrum of $A \in \mathcal{O}_\Gamma$ is of the form $\lambda_1 < \lambda_2 \leq \cdots \leq \lambda_v$ where the values are repeated according to their multiplicities (usual convention).

The Strong Arnold Hypothesis. Let $\lambda_0 \in \mathbb{R}$, $n_0 \geq 0$ an integer, and consider the submanifold $W_{\lambda_0, n_0} \subset \mathcal{S}_v$ of symmetric matrices having λ_0 as an eigenvalue of multiplicity n_0; we say that the eigenvalue λ_0 of multiplicity n_0 of $A_0 \in \mathcal{O}_\Gamma$ satisfies the SAH if $\mathcal{O}_\Gamma$ and W_{λ_0, n_0} intersect *transversally* in A_0. Since the codimension of W_{λ_0, n_0} is $n_0(n_0+1)/2$, this is only possible if $v+e \geq \frac{1}{2} n_0(n_0+1)$.

Let $L : T_{A_0} \mathcal{O}_\Gamma \to Q(E_0)$ (with $E_0 = \mathrm{Ker}(A_0 - \lambda_0 I)$ and $Q(E_0)$ the space of quadratic forms on E_0) be defined by

$$L(dA) = \langle dA \cdot \mid \cdot \rangle_{|E_0}$$

where the scalar product is that of $L^2(V,\nu)$. Then we have

Criterion (∗). *SAH is equivalent to the surjectivity of L.*

Now let us define:

Definition. $\mu(\Gamma)$ is the greatest integer n_0 for which there exists $A_0 \in \mathcal{O}_\Gamma$ for which the second eigenvalue λ_2 is of multiplicity n_0 and satisfies SAH. Such an A_0 is said to be optimal for Γ.

Some examples:

1. If K_N is the complete graph on N vertices, $\mu(K_N) = N - 1$. Indeed $T_A \mathcal{O}_{K_N} = \mathcal{S}_N$ and then L is surjective for every A_0. It suffices to take for A_0 the matrix each of whose entries is equal to -1 and having spectrum $-N < 0 = 0 = 0 = \cdots = 0$. Conversely, if λ_2 is of multiplicity $v_\Gamma - 1$, Γ is the complete graph K_v: indeed, the eigenspace E_{λ_2} is the space orthogonal to the eigenfunction $\phi_0 \in E_{\lambda_1}$. For $f \in E_{\lambda_2}$ we have

$$\sum_{i=2}^{v} a_{1i} f(i) = \mu_0 f(1);$$

then if there exists an i such that $a_{1i} = 0$, there is another relation (in addition to the orthogonality to ϕ_0) between the values of $f(i)$: thus for all i, $a_{1i} \neq 0$, so we see that Γ is complete.

2. If $K_{3,3}$ is the complete bipartite graph on 6 vertices, we have $\mu(K_{3,3}) = 4$ (Figure 1).

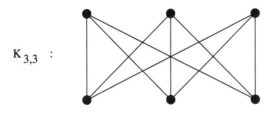

$K_{3,3}$:

Figure 1

Indeed, $\mu(K_{3,3}) \leq 4$ by the preceding remark. Let A_0 be such that $a_{ii} = 0$ and $a_{ij} = -1$ if and only if $\{i, j\} \in E$. The spectrum of A_0 is $-3 < 0 = 0 = 0 = 0 < 3$. The eigenspace $E_0 = \text{Ker } K_0$ is defined by

$$E_0 = \{(x_i) \mid x_1 + x_2 + x_3 = 0, x_4 + x_5 + x_6 = 0\}.$$

Let $L_i(x) = x_i$; then L_1, L_2, L_4, L_5 form a basis for E_0^*. It is clear that $L_1^2, L_2^2, L_4^2, L_5^2, L_1 L_4, L_1 L_5, L_2 L_4$ and $L_2 L_5$ are in the image of L; furthermore, $L_1 L_2$ and $L_4 L_5$ are obtained by the restriction of $x_3^2 = (x_1 + x_2)^2$ and $x_6^2 = (x_4 + x_5)^2$ to E_0.

3. If I_N is a *path* on N vertices ($N \geq 2$) then $\mu(I_N) = 1$.
4. If C_N is a *cycle* on N vertices ($N \geq 3$) then $\mu(C_N) = 2$.
5. If Γ is a *star* with 3 branches, $\mu(\Gamma) = 2$.

2. Properties of $\mu(\Gamma)$ Relative to Reductions and Contractions

2a. Reduction (Deletion). A reduction Γ_1 of Γ is a connected graph defined in the following manner: $E(\Gamma_1) \subset E(\Gamma)$ and $V(\Gamma_1)$ are the vertices in $V(\Gamma)$ which are the endpoints of at least one edge of $E(\Gamma_1)$. We delete the edges of $\Gamma \backslash \Gamma_1$ and then we delete those vertices of Γ which are isolated by this operation (Figure 2).

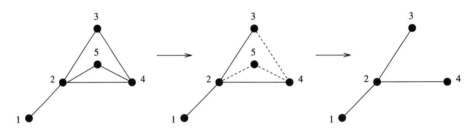

Figure 2

Theorem 2.1. *If Γ_1 is obtained from Γ by reduction (connected) then we have $\mu(\Gamma_1) \leq \mu(\Gamma)$.*

Proof. Let $n = \mu(\Gamma_1)$ and A_0 be optimal for Γ_1. We define, for each $A \in \mathcal{O}_{\Gamma_1}$ and for each $\epsilon > 0$, a quadratic form of Q_Γ by

$$q_{\epsilon,A}((x_i)) = C\Sigma' x_i^2 + \epsilon\Sigma''(x_i - x_j)^2 + q_A((x_i))$$

where Σ' runs over $V(\Gamma)\backslash V(\Gamma_1)$, Σ'' over $E(\Gamma)\backslash E(\Gamma_1)$ and q_A is the quadratic form associated with a measure μ_0 on $V(\Gamma_1)$. We choose on $V(\Gamma)$ the measure $\nu_0 = \mu_0 + \Sigma'\delta(i)$ and C larger than all the eigenvalues of q_A for A close to A_0.

For $\epsilon = 0$, the spectrum of q_0, A_0 relative to $L^2(V(\Gamma), \nu_0)$ has $\lambda_2 = \lambda_2(A_0)$ as second eigenvalue with multiplicity n. We also have the SAH for this eigenvalue relative to deformations of $\mathcal{O}_\Gamma$. This property is clearly true for q_{ϵ,A_ϵ} ($\epsilon > 0$) and A_ϵ chosen sufficiently close to A_0. But for $\epsilon > 0$, $q_{\epsilon,A_\epsilon} \in Q_\Gamma$.

2b. Contractions. Let Γ be a connected graph. We shall say that Γ_0 is a *contraction* of Γ if Γ_0 may be defined in the following fashion: Let $V(\Gamma) = \bigcup_{i=1}^N A_i$ be a partition of $V(\Gamma)$ into connected non-empty pieces (Figure 3); then

(i) $V(\Gamma_0) = \{1, 2, \ldots, N\}$,
(ii) $\{i, j\} \in E(\Gamma_0)$ if and only if there exists an edge $\{\alpha, \beta\} \in E(\Gamma)$ such that $\alpha \in A_i$ and $\beta \in A_j$.

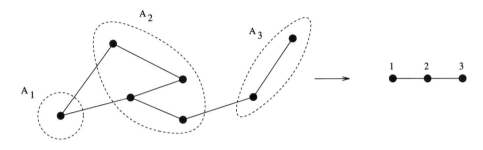

Figure 3

We denote by:

$E_{i,j} \subset E(\Gamma)$ the edges which join vertices of A_i to vertices of A_j;
$n_{i,j} = \# E_{i,j}$;
Γ_i the graph such that $V(\Gamma_i) = A_i$ and $E(\Gamma_i) = E_{i,i}$ ($i \geq 1$);
p the natural projection of $V(\Gamma)$ onto $V(\Gamma_0)$;
$\nu = \sum_{\alpha \in V(\Gamma)} \delta(\alpha)$ and ν_0 its image under p: $\nu_0(\{i\}) = n_i$.

Then we have

Theorem 2.2. *If Γ_0 is a contraction of Γ, then $\mu(\Gamma) \geq \mu(\Gamma_0)$.*

Proof. Let $A_0 \in \mathcal{O}_{\Gamma_0}$ be optimal, $q_{A_0} \in \mathcal{Q}_{\Gamma_0}$ the associated quadratic form relative to ν_0, $\lambda_2(A_0)$, $F_0 = \text{Ker}(A_0 - \lambda_2 I)$, and $m_0 = \mu(\Gamma_0) = \dim(F_0)$. The space $L^2(V(\Gamma_0), \nu_0)$ is naturally identified with, and isometric by $f \mapsto f \circ p$ to the subspace E_0 of functions $L^2(V(\Gamma), \nu)$ constant on each A_i.

To each quadratic form $q \in \mathcal{Q}_{\Gamma_0}$ it is then natural to associate a lifting, $p^*(q)$, a quadratic form on $L^2(V(\Gamma), \nu)$ satisfying $p^*(q)(f \circ p) = q(f)$. For example, we may define p^* by extending linearly the formulae

$$p^*(x_i^2) = \frac{1}{n_i} \sum_{\alpha \in A_i} x_\alpha^2$$

$$p^*((x_i - x_j)^2) = \frac{1}{n_{i,j}} \sum_{\alpha, \beta \in E_{i,j}} (y_\alpha - y_\beta)^2.$$

Now let $q_i \in \mathcal{Q}_{\Gamma_i}$ ($i \geq 1$) be defined by

$$q_i(y) = \sum_{(\alpha, \beta) \in E(\Gamma_i)} (y_\alpha - y_\beta)^2.$$

To each $A \in \mathcal{O}_{\Gamma_0}$ with quadratic form $q_A \in Q_{\Gamma_0}$, we associate, for each $\epsilon > 0$, the quadratic form $q_{\epsilon,A} \in Q_\Gamma$ defined by

$$q_{\epsilon,A} = \sum_{i=1}^{N} q_i + \epsilon p^*(q_A).$$

When $\epsilon = 0$, the spectrum of $q_{0,A}$ consists of the eigenvalue 0 of multiplicity N and eigenvalues > 0 (those of q_i on Γ_i). Since $q_{\epsilon,A}$ is an analytic function of ϵ and of A, we may apply the theory of analytic perturbations of Kato [KO]; if we denote by $E_{\epsilon,A}$ the sum of the eigenspaces of $q_{\epsilon,A}$ corresponding to the N smallest eigenvalues of $q_{\epsilon,A}$ where ϵ is small, $E_{\epsilon,A}$ is close to E_0 and we may denote by $U_{\epsilon,A}$ the "small" canonical isometry of E_0 onto $E_{\epsilon,A}$ and $\widetilde{q}_{\epsilon,A} = U^*_{\epsilon,A}(q_{\epsilon,A|B_{\epsilon,A}})$. The family $\widetilde{q}_{\epsilon,A}$ of quadratic forms on E_0 is analytic in ϵ and A and has as eigenvalues N least eigenvalues of $q_{\epsilon,A}$. Further, $\widetilde{q}_{0,A} = 0$ and so $r_{\epsilon,A} = \left(\frac{1}{\epsilon}\right)\widetilde{q}_{\epsilon,A}$ is also analytic in ϵ and A.

We have $r_{0,A} = q_A$. Indeed $r_{0,A}$ is the derivative of $\widetilde{q}_{\epsilon,A}$ with respect to ϵ at $\epsilon = 0$ which is thus equal (see [CV2] for a similar calculation) to the derivative of $q_{\epsilon,A|B_0}$; that is to say $p^*(q_A)|_{E_0} = q_A$ identifying E_0 and $L_2(V(\Gamma_0); V_0)$.

Thus by SAH for A_0, for sufficiently small $\epsilon > 0$ there exists an operator $A_\epsilon \in \mathcal{O}_{\Gamma_0}$ close to A_0 such that r_{ϵ,A_ϵ} has λ_2 as its second eigenvalue with multiplicity m_0, and then $q_{\epsilon,A}$ has $\epsilon\lambda_2$ as its second eigenvalue with the same multiplicity. The SAH for this operator is shown using the uniqueness of deformations of $\mathcal{O}_{\Gamma_0}$ at the origin; indeed the linear map L_ϵ of criterion (*) of §1, $L_\epsilon : T_{A_\epsilon}\mathcal{O}_{\Gamma_i} \to Q_{F_\epsilon}$ depends continuously on ϵ and is surjective for $\epsilon = 0$ since A_0 satisfies SAH.

2c. Topological invariance. Looking at the preceding results, it is natural to ask whether $\mu(\Gamma)$ is a topological invariant of Γ. Recall that two graphs are said to be homeomorphic if we may subdivide their edges to obtain isomorphic graphs. That $\mu(\Gamma)$ is not a topological invariant may be seen from the following: $\mu(\Gamma_1) = 3$ and $\mu(\Gamma_2) = 2$ (Figure 4).

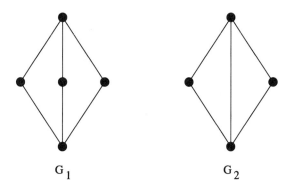

Figure 4

3. Relations Between $\mu(\Gamma)$ and the Embeddings of Γ in a Surface

Recall [CV1, CV4] that we may associate with every compact manifold X an integer invariant $m(X)$ defined in the following fashion:

$m(X)$ is the maximal multiplicity of the second eigenvalue of a positive elliptic second order symmetric differential operator having real coefficients acting on $C^\infty(X; \mathbb{R})$. Known results on $m(X)$ are the following: $m(S^2) = 3$ [CG], $m(p^2(\mathbb{R})) = 5$ and $m(\mathbb{R}^2/\mathbb{Z}^2) = 6$ [CG, BN], if B is the Klein bottle, $m(B) = 5$ [CV4], and if X is the orientable surface of genus g, $m(X) \leq 4g + 3$ [CG, BN]. On the other hand, if $\dim(X) \geq 3$, $m(X) = +\infty$ [CV2]. We have

Theorem 3.1. *If Γ is an (injective) embedding into X, $\mu(\Gamma) \leq m(X)$.*

Theorem 3.2. *Γ is planar if and only if $\mu(\Gamma) \leq 3$.*

Theorem 3.1 is proved in [CV4, Theorem 7.1 and Corollary 7.3] by constructing a Schrödinger operator given that the multiplicity of the second eigenvalue is $\mu(\Gamma)$, to give an embedding of Γ on X. Since $\mu(S^2) = 3$, we see in particular that if Γ is planar, $\mu(\Gamma) \leq 3$; in particular neither K_5 nor $K_{3,3}$ are planar, following the calculations in §1.

It remains to prove that if Γ is not planar then $\mu(\Gamma) \geq 4$. This is a consequence of the results of §2 and the version of Harary and Tutte of Kuratowski's theorem.

Theorem 3.3. *([H–T]) If Γ is non–planar, there exists a graph Γ_1, isomorphic to $K_{3,3}$ or K_5 which is a reduction of a contraction of Γ.*

The proof (that planar $\Leftrightarrow \mu(\Gamma) \leq 3$) is complete with the remark of §1:

$$\mu(K_{3,3}) = \mu(K_5) = 4.$$

It would be of interest to have generalisations of Theorem 3.2 to other surfaces than the sphere.

4. Variants of $\mu(\Gamma)$

Let Γ be such that $\mu(\Gamma) = m$ and let A_0 be optimal for Γ. Let $X \subset V(\Gamma)$; we shall say that X is *generic* if the linear forms, $L_\alpha : E_0 \to \mathbb{R}$ defined by $L_\alpha(\phi) = \phi(\alpha)$ for $\alpha \in X$ generate the dual E_0^* of E_0. We shall say that X is *positive generic* if X is generic and there exists a linear relation $\sum_{\alpha \in X} a_\alpha L_\alpha = 0$ in E_0^* with $a_\alpha > 0$. Necessarily $\# X \geq m + 1$.

Now let $\Gamma_0 = S_X(\Gamma)$ be defined by adding a vertex 0 to Γ, together with the edges $\{0, \alpha\}$ for $\alpha \in X$; then we have

Theorem 4.1. *If X is positive generic then*

$$\mu(S_X(\Gamma)) = \mu(\Gamma) + 1.$$

Proof. 1st *Step.* Let A_0 be optimal for Γ; construct an operator $B_0 \in \mathcal{O}_{\Gamma_0}$ having $\lambda_2 = \lambda_2(A_0)$ as its second eigenvalue of multiplicity $m + 1$ with SAH. Let $q_{\epsilon,A}$ be a quadratic form on $L^2(V(\Gamma_0), \nu_0)$ defined for $\epsilon = (\epsilon_0, (\epsilon_\alpha)_{\alpha \in X})$ and $A \in \mathcal{O}_\Gamma$ by

$$q_{\epsilon,A}(x_0, (x_i)) = (\lambda_2 + \epsilon_0)x_0^2 - \sum_{\alpha \in X} \epsilon_\alpha x_0 x_\alpha + q_A(x_i).$$

Then let $\epsilon_\alpha > 0$, $q_{\epsilon,A} \in Q_{\Gamma_0}$. For $\epsilon = 0$, q_{0,A_0} has λ_2 as its second eigenvalue, with multiplicity $m + 1$. Further, this eigenvalue satisfies SAH relative to deformations $A \in \mathcal{O}_\Gamma$, $\epsilon \in \mathbb{R}^{1 + \# X}$; indeed the linear map L used in criterion $(*)$ is $L(dA, d\epsilon) = (d\epsilon \cdot x_0^2 - \sum_{\alpha \in X} d\epsilon_\alpha \cdot x_0 \cdot x_\alpha + dq_A)_{|F_0}$, where $F_0 = \mathbb{R}v_0 \oplus E_0$ with $v_0(i) = \delta_{0i}$, which is surjective on $Q(F_0)$ since X is generic.

Since there is a linear relation between the $(L_\alpha)_{\alpha \in X}$ we have in fact the existence of a germ of the submanifold of $W_{\lambda_2, m+1} \cap S_{\nu+1}$ in the neighborhood of q_{0,A_0}: the tangent space to this manifold contains the vector given by

$$d\epsilon_\alpha = a_\alpha,$$
$$d\epsilon_0 = 0,$$
$$dA = 0,$$

which is in $\mathrm{Ker}(L)$. As the $a_\alpha > 0$, this manifold meets $\mathcal{O}_{\Gamma_0}$.

2nd *Step.* The second step depends upon:

Theorem 4.2. *If Γ is obtained from Γ_0 by the deletion of the edges adjacent to the vertex 0, then $\mu(\Gamma_0) = \mu(\Gamma)$ or $\mu(\Gamma_0) = \mu(\Gamma) + 1$.*

Proof. We know already from §1 that $\mu(\Gamma) \leq \mu(\Gamma_0)$. Let A_0 be optimal with A_0 and let E_0 be the corresponding eigenspace. Let $F_0 \subset E_0$ be the set of $\phi \in E_0$ such that $\phi(0) = 0$. F_0 is clearly the second eigenspace of an operator B_0 of $\mathcal{O}_\Gamma$ (in the matrix A_0 suppress all the entries a_{i0} and a_{0i}) and thus satisfies the SAH, as we see using the criterion $(*)$ since $L(L_0 L_i)|_{F_0} = 0$ (i a vertex adjacent to 0).

We may also obtain an interesting corollary: denote by $cr(\Gamma)$ the minimum number of crossings in a planar embedding of Γ; then

Corollary 4.3. $\mu(\Gamma) \leq 3 + cr(\Gamma)$.

In particular, if $cr(\Gamma) = 1$, we see that $\mu(\Gamma) = 4$. This is the case for $K_{3,3}$ and K_5, as the following embeddings show (Figure 5).

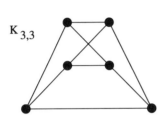
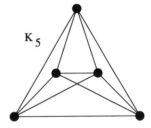

Figure 5

5. n–Critical Graphs

Recall [TE, p. 32 and following] that there is a natural order relation on (the isomorphism classes of) graphs given in the following way:

Definition 5.1. Γ_1 is a minor of Γ if Γ_1 is a reduction of a contraction of Γ.

It is natural then to make

Definition 5.2. Γ is an n–critical graph if $\mu(\Gamma) = n$ and for every minor Γ_1 of Γ, not isomorphic to Γ, $\mu(\Gamma_1) < \mu(\Gamma)$.

Then we have:

Theorem 5.3. For a graph Γ, $\mu(\Gamma) \geq n$ if and only if there is a minor Γ_1 of Γ which is n–critical.

General (and difficult) results in graph theory [R-S] imply that, for each n, there are only finitely many n–critical graphs. These are known for $n \leq 4$.

Theorem 5.4. The n–critical graphs for $n \geq 4$ are the following

for $n = 0$, K_1;

for $n = 1$, K_2;

for $n = 2$, K_3 and $K_{3,1}$;
for $n = 3$, K_4 and $K_{3,2}$;
for $n = 4$, K_5 and $K_{3,3}$.
For every n, K_{n+1} is n-critical.

Proof. The cases $n = 0, 1$ are trivial. The case $n = 4$ has already been dealt with (Theorem 3.2). It remains to deal with $n = 2$ and $n = 3$: for $n = 3$ we use the notion of *outer-planar* graphs [C-H].

Definition 5.5. Γ is outer–planar if Γ is planar and there exists an embedding j of Γ in $\mathbb{R}^2$ such that the vertices of Γ are all in the closure of the unbounded connected component of $\mathbb{R}^2 \setminus j(\Gamma)$. Then

Theorem 5.6. *Γ is not outer–planar if and only if Γ has a minor isomorphic to K_4 or $K_{3,2}$.*

Since $\mu(K_4) = \mu(K_{3,3}) = 3$ we have the more precise result.

Theorem 5.7. *Γ is outer–planar if and only if $\mu(\Gamma) \leq 2$.*

Proof. Indeed, if Γ is not outer–planar, $\mu(\Gamma) \geq 3$ from 5.6 and the calculations of μ for the graphs K_4 and $K_{3,2}$. Conversely, if Γ is outer–planar, let $\Gamma_1 = S_X(\Gamma)$ where $X = V(\Gamma)$; then Γ_1 is planar (Figure 6) and $\mu(\Gamma_1) = \mu(\Gamma) + 1$ from 4.1, since $V(\Gamma)$ is positive generic. Thus $\mu(\Gamma) + 1 \leq 3$.

From the above we see that the 3–critical Γ are K_4 and $K_{3,2}$. The 2–critical case is clear: $\mu(\Gamma) \geq 2$ if and only if Γ is not a path; that is, if it contains K_3 or $K_{3,1}$ as a minor (Figure 6).

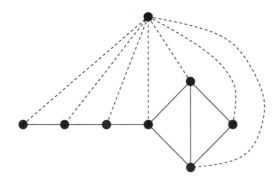

Figure 6

Acknowledgements

I am indebted to my colleagues from Grenoble (C. Benzaken, F. Jaeger, C. Payan, and N. H. Xuong) for discussions during the preparation of this work, especially about minors and k–critical graphs. In particular, the result that graphs with $\mu \leq 2$ are outer-planar graphs uses an idea of F. Jaeger.

References

[AD] V. Arnold, Modes and quasi–modes, *J. Funct. Anal.* **6** (1972), 94–101.

[BE] D. Barnette, Map coloring, polyhedra, and the four–color problem, *Dolciani Math. Exp.* **8** (1983).

[BN] G. Besson, Sur la multiplicité de la première valeur propre des surfaces riemanniennes, *Ann. Inst. Fourier* **30** (1980), 109–128.

[C–C] B. Colbois and Y. Colin de Verdière, Multiplicité de la première valeur propre du laplacien des surfaces à courbure constante, *Comment. Math. Helv.* **63** (1988), 194–208.

[CG] S. Cheng, Eigenfunctions and nodal sets, *Comment. Math. Helv.* **51** (1979), 43–55.

[C–H] C. Chartrand and Harary, Planar permutation graphs, *Ann. Inst. H. Poincaré B* **3** (1967), 433–438.

[CV1] Y. Colin de Verdière, Spectres de variétés riemanniennes et spectres de graphes, *Proc. Intern. Congress of Math.*, Berkeley, 1986, 522-530.

[CV2] Y. Colin de Verdière, Sur la multiplicité de la première valeur propre non nulle du laplacien, *Comment. Math. Helv.* **61** (1986), 254–270.

[CV3] Y. Colin de Verdière, Sur une hypothèse de transversalité d'Arnold, *Comment. Math. Helv.* **63** (1988), 184–193.

[CV4] Y. Colin de Verdière, Constructions de laplaciens dont une partie finie du spectre est donné, *Ann. Sci. École Norm. Sup.* **20** (1987), 599–615.

[H–T] F. Harary and W. Tutte, A dual form of Kuratowski's theorem, *Canad. Math. Bull.* **8** (1965), 17–20.

[KI] K. Kuratowski, Sur le problème des courbes gauches en topologie, *Fund. Math.* **15** (1930), 271–283.

[KO] T. Kato, "Perturbation Theory for Linear Operators," Springer–Verlag, Berlin/New York, 1976.

[OE] O. Ore, "The Four–color Problem," Academic Press, New York, 1967.

[R–S] Robertson and Seymour, Graphs minors I., *J. Combin. Theory Ser. B* **35** (1983), 39–61.

[TE] W. Tutte, Graph theory, *Encyclopedia Math.* **21** (1984).

[WE] A. White, "Graphs, Groups and Surfaces," North–Holland, Amsterdam, 1984.

Institut Fourier, B.P. 74, 38402 St. Martin d'Heres, Cedex, France

Four Problems on Plane Graphs Raised by Branko Grünbaum

OLEG BORODIN

ABSTRACT. The following four problems on coloring and structure of plane graphs raised by Branko Grünbaum and contributed to or solved by the author are discussed: (a) acyclic coloring, (b) the three color problem, (c) cyclic connectivity, and (d) the number of light edges.

Gruñbaum's problems on colorings and structural properties of plane graphs have attracted me from my very first steps as a mathematician. Therefore, being here in the University of Seattle where Branko Grünbaum has been working more than a quarter of a century, I would like to discuss some of these problems. Graphs considered in the paper are assumed to have no loops or multiple edges.

1. Acyclic coloring

In 1973, Grünbaum introduced [13] a new class of vertex colorings in which various restrictions were placed on the structure of all 1-, 2-, and more chromatic induced subgraphs. Acyclic coloring is that one among them in which every 1-chromatic subgraph is an independent set of vertices, and every 2-chromatic subgraph is an acyclic graph (i.e., a forest). The main justification for introducing this concept was (see [13,p.390]) the following.

CONJECTURE 1. *Every planar graph is acyclically 5-colorable.*

It should be observed that a slightly stronger kind of coloring in which there should be no 1-chromatic K_2, no 2-chromatic C_{2t}, and no 3-chromatic outerplanar subgraph, may require arbitrarily many colors for some planar graphs [20].

Conjecture 1 led to a number of intermediate results. First, Grünbaum proved [13] that 9 colors suffice, then Mitchem [24], Albertson and Berman [4], and Kostochka [19] step by step reduced this upper bound to 6. At last, in [6,7] I succeeded in proving Conjecture 1: in [6] there is only a sketch of the proof

---

1991 Mathematics Subject Classification. Primary 05C10, 05C15.
This paper is in final form and no version of it will be submitted for publication elsewhere.

with the main ideas and precise formulations of all lemmas, but [7] contains the whole proof. The proof consists of constructing an unavoidable set of 450 reducible configurations. It should be noticed that at that time I was unaware of the Four Color Theorem by Appel and Haken [5], which appeared about the same time.

Now I would like to describe one feature of the proof in [6,7] which may turn out to be of use elsewhere. Some of these 450 configurations are related to each other, in the sense that they all may be generated from the simplest configuration "a 4-vertex (i.e., a vertex of degree 4) adjacent to three 6-vertices" by replacing two of these 6-vertices by some "weak" vertices. The latter are defined to be major vertices adjacent to sufficiently many minor vertices (i.e. those of degree 4 or 5). The point is that the reducibility proof was done for the whole family at once, and it was 6 or 7 pages long. Who knows, perhaps some such trick could be used to simplify the proof of the Four Color Theorem [5] too?

Next I wish to formulate some unsolved problems concerning acyclic coloring. In 1976, Albertson and Berman [3], and independently myself, conjectured that for each closed surface S^N except for the plane P, its acyclic chromatic number is the same as its usual chromatic number:

$$\text{if } S^N \neq P \text{ then } x_a(S^N) = x(S^N) = \lfloor (7 + \sqrt{49 - 24N})/2 \rfloor.$$

As far as I know, this conjecture still remains open.

Recall that a graph is called $k - degenerate$ if it can be destroyed by consecutively removing vertices with current degree less then k. For example, the 1-degenerate graphs are exactly those without edges, the 2-degenerated ones are the acyclic graphs, while a proper subset of the 3-degenerate graphs is constituted by the outerplanar graphs. I conjectured [6,7] that for each planar graph there exists a 5-coloring in which every k-chromatic subgraph is k-degenerate for all $1 \leq k \leq 4$. (If $1 \leq k \leq 2$, this coincides with a result of [6,7].) To the best of my knowledge, the following two much easier problems are also open, namely those concerning the existence of a similar 10^{10}-coloring for $k = 10^{10} - 1$ and even for $k = 3$. Note that, if in the last problem one demands that all 3-chromatic subgraphs be outerplanar instead of being 3-degenerate, then the answer is negative (see [20]).

2. The three color problem

The natural question of which plane graphs are 3-colorable cannot have a simple answer (assuming P$\neq$NP): as proved by Holyer in 1981, the problem is NP-hard [17].

However, as early as 1959, Grötzsch established [11] that if a planar graph does not contain 3-cycles, then it is 3-colorable. In 1963, Grünbaum generalized [12] this theorem to all planar graphs with at most three 3-cycles. (The example of K_4 shows that the result is the best possible with respect to the number of

3-cycles.) Unfortunately, the proof in [12], based on the idea of the extendability of an arbitrary 3-coloring of any short cycle to a 3-coloring of the whole graph, had some gaps, which were more recently filled in by Aksenov [1]. In fact, [1] contains the only correct proof of Grünbaum's theorem; it further develops the coloring extension idea due to Grötzsch [11].

Here I would like to announce a proof of Grünbaum's three color theorem which is free of extension arguments and is therefore simpler than that in [1]. The proof makes use of the idea tof searching for an unavoidable set of reducible configurations inside an appropriate short cycle if any, instead of in the whole graph.

Another direction in the three color problem is known as Grünbaum's and Havel's problem (see [26]) and consists of determining whether there exists a finite d_0 such that if the minimal distance between triangles in a plane graph is not less than d_0, then the graph is 3-colorable. In 1963-78, Grünbaum, Havel, Aksenov, Melnikov and Steinberg raised the lower bound for d_0 up to 4 (for details see [2]). In other words, there were constructed 4-chromatic planar graphs in which the minimal distance between triangles was 3.

Now I would like to preliminarily announce the following related result. There exist many planar graphs with large distances between triangles (10 or so), which do not contain those 3-color reductions due to Grötzsch and Grünbaum and in which clenching each 4-face draws some triangles together. I strongly believe that the same construction may be extended to arbitrarily large d_0: a skeleton and certain blocks to be inserted into its holes are already prepared for arbitrary d_0, and it just remains to solve some boundary problems, which at least I can solve for some specific d_0. (The work was laid aside because of more urgent occupations.)

A practical consequence of these constructions is that at least, one acquires a conviction that the positive solution of Grünbaum-Havel's problem "hardly" exists. Moreover, if among these irreducible graphs there were found 4-chromatic ones for arbitrary d_0, then the problem of Grünbaum and Havel would be solved in the negative.

3. Cycle connectivity

We recall that the cycle connectivity (c.c.) of a graph is defined to be the least number of edges whose removal results in at least two connected components each of which contains a cycle. As it is well known, there exist graphs with connectivity three and arbitrarily large c.c. For planar 4-connected graphs, the c.c. is also unbounded as shown by the double n-pyramid. Note that each planar graph is obviously at most 5-connected. In 1975, Grünbaum raised

CONJECTURE 2.[14]. *The cyclic connectivity of each 5-connected planar graph is at most 11.*

A similar intermediate result with 13 instead of 11 was proved a few years earlier by Plummer [25]. The bound 11 is attained, for example, by inserting a new vertex into each face of a dodecahedron and then joining it with all the boundary vertices (each face of the resulting triangulation is clearly incident with a 5-vertex and two 6-vertices).

Let δ denote the minimum degree of the graph under consideration. Recently I proved [8] a stengthening Conjecture 2 under weaker assumptions, as follows:

THEOREM 3. *In each plane graph with $\delta = 5$ there is a triangle such that the degree sum of its vertices is at most 17.*

The concept of the weight of a triangle or of an edge defined as the sum of the degrees of its vertices was introduced by Kotzig [21]. In 1963, Kotzig proved that in each plane triangulation with $\delta = 5$, the minimum weight of a face was at most 18, and conjectured that in fact it was at most 17 (see [22]). A weaker upper bound, 19, under the same assumption was obtained by Lebesgue [23] as early as 1940. Although Grünbaum's [14] and Kotzig's [22] conjectures are formulated in somewhat different terms, Theorem 3 clearly settles them both.

Let f_{ijk} be the number of triangles incident with an i-vertex, a j-vertex, and a k-vertex. Then Theorem 3 may be stated as: if $\delta \geq 5$, then

$$f_{555} + f_{556} + f_{557} + f_{566} > 0.$$

More recently, I have found constructions which show that no term of this inequality may be removed without upsetting it. In particular, I have found a plane graph having

$$\delta = 5, \quad \text{and} \quad f_{555} = f_{556} = f_{557} = 0, \ f_{566} = 36.$$

This graph confirms, on the other hand, that the fourth coefficient in the following theorem is the best possible:

THEOREM 4([9]). *For each plane graph with $\delta = 5$, we have*

$$18f_{555} + 9f_{556} + 5f_{557} + 4f_{566} \geq 144.$$

All coefficients except possibly the third are the best possible. Theorem 4 may be regarded as a non-superfluons version of the following inequality of Lebesgue [23] for plane triangulations with $\delta = 5$:

$$f_{555} + \frac{2}{3}f_{556} + \frac{3}{7}f_{557} + \frac{1}{4}f_{558} + \frac{1}{9}f_{559} + \frac{1}{3}f_{566} + \frac{2}{21}f_{567} \geq 20.$$

PROBLEM 5. *What is the optimal value of the third coefficient in Theorem 4?*

4. The number of light edges.

At the end of the preceding Section, we have seen that the number of triangles whose weight does not exceed the upper bound (for this reason named *light* triangles) is relatively large. The same happens for edges as well. Let e_{ij} be the number of edges joining a vertex of degree i with one of degree j. In 1955, Kotzig proved

THEOREM 6([21]). *In each 3-polytope, there exists an edge of weight at most 13.*

This bound is the best possible, as shown by inserting a new vertex into each face of a plane graph with $\delta = 5$ and joining it with the boundary vertices of the corresponding face. Kotzig's theorem may be also stated as

$$\sum_{i+j \leq 13} e_{ij} > 0.$$

In 1973, Grünbaum [15] conjectured, for specific constants a_{ij} and M, that each plane triangulation satisfies

$$\sum_{i+j \leq 13} a_{ij} e_{ij} \geq M.$$

Jucovič obtained [18] a similar relation with coefficients different from those in [15]. The complete solution to this problem is given by my recent result:

THEOREM 7([10]). *Each plane triangulation satisfies*

$$20e_{33} + 25e_{34} + 16e_{35} + 10e_{36} + 6\tfrac{2}{3}e_{37} + 5e_{38} + 2\tfrac{1}{2}e_{39} + 2e_{3,10}$$
$$+ 16\tfrac{2}{3}e_{44} + 11e_{45} + 5e_{46} + 1\tfrac{2}{3}e_{47}$$
$$+ 5\tfrac{1}{3}e_{55} + 2e_{56} \geq 120.$$

Each coefficient is the best possible. For arbitrary planar graphs with $\delta = 3$, it was only known that

$$\sum_{i+j \leq 13} e_{ij} \geq 3$$

(Jucovič [18]), and the question was raised by Grünbaum [15] and Jucovič [18] of finding the best possible relation of the type

$$\sum_{i+j \leq 13} a_{ij} e_{ij} \geq M.$$

A complete answer to this problem now is: to obtain the best possible relation for planar graphs with $\delta = 3$, just replace $a_{33} = 20$ by $a_{33} = 40$ in Theorem 7 (see[10]).

In [18], Jucovič also asked about similar inequalities for planar graphs (1) with $\delta = 4$, and $\delta = 3$ but without 4-vertices. It is remarkable that complete answers to these two questions by Jucovič also follow easily from the proof of Theorem 7 (with $a_{33} = 40$): to obtain the best possible relation in each of the two cases, one should merely erase the terms containing subscripts 3 or 4, respectively.

A similar qestion about the case $\delta = 5$ has a much longer history and has not been completely settled yet. In 1904, Wernicke [27] proved that $e_{55} + e_{56} > 0$. In 1973, Grünbaum [13] proved that $e_{55} = 0$ implies $e_{56} > 60$ and conjectured that $2e_{55} + e_{56} \geq 60$. However, a counterexample to the last conjecture was found by Fisk (see [16, p.133]) which has $e_{55} = 28$, $e_{56} = 0$. Grünbaum and Shephard proved [16] a weaker relation $4e_{55} + e_{56} \geq 60$. Theorem 7 already implies a better bound $\frac{8}{3} e_{55} + e_{56} \geq 60$, but again it is not the best possible, as follows from my recent result:

THEOREM 8([9]). *Each planar graph with $\delta = 5$ satisfies*

$$\frac{18}{7} e_{55} + e_{56} \geq 60.$$

The second coefficient is the best possible. As for the first coefficient in Theorem 8, it is only known to be between $\frac{15}{7}$ and $\frac{18}{7}$.

PROBLEM 9. *What is the best value for the first coefficient in Theorem 8?*

PROBLEM 10. *Describe the class of exceptions for the conjecture*

$$2e_{55} + e_{56} \geq 60.$$

Finally, we remark that there exist [9] rather weak sufficient conditions for $2e_{55} + e_{56} \geq 60$: for example, $e_{57} = 0$, or merely that no 7-vertex is adjacent to seven or six 5-vertices.

References

1. V.A.Aksenov, *On extension of 3-colorings of plane graphs*, Diskret. Analis 26 (1974), 3–19 (in Russian).
2. V.A.Aksenov and L.S.Melnikov, *Essay on the theme: the three-color problem*, in: Combinatorics, Coll. Math. Soc. János Bolyai 18, edited by A.Hajnal and V.T.Sós (North-Holland, 1976), pp.23–24.
3. M.O.Albertson and D.M.Berman, *The acyclic chromatic number*, in: Proceedings of the 7th S-E Conference on Combinatorics, Graph Theory and Computing, Baton Rouge 1976, Congr. Num. 17 (1976), 51–60.
4. M.O.Albertson and D.M.Berman, *Every planar graph has an acyclic 7-coloring*, Israel J.Math. 28 (1977), 169–174.
5. K.Appel and W.Haken, *The existence of unavoidable sets of geographically good configurations*, Illinois J.Math. 20 (1976), 218–297.
6. O. V. Borodin, *A proof of B.Grünbaum's conjecture on the acyclic 5-colourability of planar graphs*, Dokl. Akad. Nauk SSSR 231 (1976), 18–20 (in Russian).
7. O. V. Borodin, *On acyclic colorings of planar graphs*, Discrete Math. 25 (1979), 211–236.
8. O. V. Borodin, *Solutions of Kotzig's and Grünbaum's problems on the separability of a cycle in planar graphs*, Matem. zametki 46, No.5 (1989), 9–12 (in Russian).
9. O. V. Borodin, *Structural properties of planar maps with the minimal degree 5*, Math. Nachrichten (submitted).
10. O. V. Borodin, *A structural property of planar graphs and the simultaneous colouring of their edges and faces*, Math. Slovaca 40,No. 2 (1990), 113–116.
11. H.Grötzsch, *Ein Dreifarbensatz für dreikreisfrei Netze auf der Kugel*, Wiss. Z. Martin-Luther-Univ. Halle-Wittenberg, Math.-Nat. Reihe 8 (1959),109–120.
12. B.Grünbaum, *Grötzsch's theorem on 3-coloring*, Michigan Math. J. 10 (1963), 303–310.
13. B.Grünbaum, *Acyclic colorings of planar graphs*, Israel J. Math. 14 (1973), 390–408.
14. B.Grünbaum, *Polytopal graphs*, Math. Assos. of America Studies in Mathematics 12 (1975), 201–224.
15. B.Grünbaum, *New views on some old questions of combinatorial geometry*, in: Proc. Int. Colloq. Rome, 1973, Accad. nac. dei lincei Rome (1976), 451–468.
16. B.Grünbaum and G.C.Shephard, *Analogues for tilings of Kotzig's theorem on minimal weight of edges*, Ann. Discrete Math. 12 (1981), 129–140.
17. I.J.Holyer, *The NP-completeness of edge coloring*, SIAM J. Comput. 10 (1981), 718–720.
18. E.Jucovič, *Strengthening of a theorem about 3-polytopes*, Geom. Dedic. 13 (1974), 20–34.

19. A.V.Kostochka, *Acyclic 6-coloring of planar graphs*, Diskret. Analiz 28 (1976), 40–56 (in Russian).
20. A.V.Kostochka and L.S.Melnikov, *Note to the paper of Grünbaum on acyclic colorings*, Discrete Math. 14 (1976), 403–406.
21. A.Kotzig, *Contribution to the theory of eulerian polyhedra*, Mat. čas. 5 (1955), 101–103 (in Russian).
22. A.Kotzig, *From the theory of Euler's polyhedrons*, Mat. čas. 13 (1963), 20–34 (in Russian).
23. Lebesgue, *Quelques conséquences simple de la formule d'Euler*, J. de Math. 9 (1940), 27–43.
24. J.Mitchem, *Every planar graph has an acyclic 6-coloring*, Duke Math. J. 41 (1974), 177–181.
25. M.D.Plummer, *On the cyclic connectivity of planar graphs*, Graph Theory and Applications, Berlin (1972), 235–242.
26. B.Toft, *Graph colouring problems*, I, prepr. Odense University No. 2 (1987).
27. P.Wernicke, *Über den Kartographischen Vierfarbensatz*, Math. Ann. 58 (1904), 413–426.

INSTITUTE OF MATHEMATICS, NOVOSIBIRSK, 630090, RUSSIA

Counterexamples to a Conjecture of Las Vergnas and Meyniel

Bruce Reed

ABSTRACT: In 1981, Las Vergnas and Meyniel conjectured that if G contains k disjoint sets of vertices $S_1, ..., S_k$ such that $S_i \cup S_j$ is connected for all $i \neq j$ then G has a K_k minor. Jorgensen proved this conjecture for k at most 8 and Duchet independently proved it for k at most 9 (unpublished). In this paper a probabilistic argument is presented which shows that the conjecture is false for all sufficiently large k.

We assume that the reader is familiar with standard graph theory terminology and the fundamentals of probabilistic graph theory. For any undefined terms see Bollobas[1]. For our purposes, a *clique minor* of order k in a graph G consists of a family of k vertex disjoint connected subgraphs of G called *nodes*, such that for any two nodes M and N, there is an edge of G with one endpoint in M and the other in N. By a *clique pseudo-minor* of order k, we mean a family of k disjoint subsets of vertices of G called *pseudo-nodes* such that for any two pseudo-nodes M and N the graph induced by $M \cup N$ is connected (we note that any clique minor corresponds to a clique pseudo-minor).

In 1943, Hadwiger [3] conjectured that, for all k, any loopless graph which does not have a vertex colouring with k colours contains a clique minor of order $k+1$. This is a difficult conjecture as a proof for the case $k = 4$ would imply the four colour theorem.

In 1981, Las Vergnas and Meyniel [4] made a number of conjectures related to Hadwiger's conjecture; in particular they conjectured that any graph which contains a clique pseudo-minor of order k contains a clique minor of order k.

Subsequently, this was proved for $k \leq 8$ by Jorgensen [5]; Duchet [2] independently obtained a proof (unpublished) for $k \leq 9$. The purpose of this note is to prove that the conjecture is false for all sufficiently large k. We do this by showing that a random choice from an appropriately defined probability space is almost surely a counterexample.

Theorem: *For k sufficently large, there is a graph G with a clique pseudo-minor of order k and no clique minor of order greater than $\frac{2k}{3} + 2000009 log(k)$.*

1991 Mathematics Subject Classification. Primary 05C80, 05C78
Supported by a grant from The Alexander Von Humboldt Stiftung
This paper is in final form and no version of it will be submitted elsewhere

Proof: Fix an integer k. By a *candidate* G we mean a graph on $2k$ vertices consisting of k pairs $P_1, ..., P_k$ of non-adjacent vertices of G and, for each $1 \leq i < j \leq k$, exactly three of the possible four edges between P_i and P_j.

By a *pair* of G, we mean one of $P_1, ..., P_k$. We generate a random candidate G on vertices $v_1, ..., v_{2k}$ by letting the pairs be $\{v_1, v_2\}, ..., \{v_{2k-1}, v_{2k}\}$ and for any two pairs choosing randomly three of the four edges between the pairs where each of the four possibilities is equally likely and the choices are made independently.

Lemma 1: *Almost surely a random candidate contains no clique with more than $8\log(k)$ vertices.*

Proof. Let $m = \lceil 8\log(k) \rceil$. Consider the probability that an arbitrary set $W = \{v_1, ..., v_m\}$ of m vertices of a random candidate G forms a clique in G. If any two vertices of W form a pair in G then W obviously does not induce a clique. Otherwise, for every two vertices of W, the probability that there is an edge between the two vertices is $\frac{3}{4}$ and these probabilities are independent. Thus, the probability that W forms a clique is $\frac{3}{4}^{\binom{m}{2}}$. Since there are $\binom{2k}{m}$ sets of m vertices in G, it follows that the probability that some such set induces a clique is at most $\binom{2k}{m}(\frac{3}{4})^{\binom{m}{2}}$ which goes to 0 as k goes to infinity, as required. □

Lemma 2: *Almost surely a random candidate contains no clique minor with more than $2000001\log(k)$ nodes such that each node of the minor consists of an edge.*

Proof: Let $m = \lceil 2000001\log(k) \rceil$. Let $N_1 = (a_1, b_1), ..., N_m = (a_m, b_m)$ contain $2m$ distinct vertices of G and let us examine the probability that there is a clique minor of order m in G with nodes $N_1, ..., N_m$. Note that no N_i consists of two vertices from the same pair, as each N_i forms an edge. Thus, we can find a subset $\{N'_1, ..., N'_{\lceil \frac{m}{3} \rceil}\}$ of $\{N_1, ..., N_m\}$ such that there is at most one vertex of each pair of G in $\bigcup_{i=1}^{\lceil \frac{m}{3} \rceil} N'_i$. Then, for any $1 \leq i < j \leq \lceil \frac{m}{3} \rceil$, the probability that there is an edge between N'_i and N'_j is $(1 - (\frac{1}{4})^4) = \frac{255}{256}$ and these event probabilities are all independent. Thus the probability that the desired minor exists is $(\frac{255}{256})^{\binom{\lceil \frac{m}{3} \rceil}{2}} < \frac{255}{256}^{\frac{m^2}{19}}$.

Now, there are less than $\frac{2k^{2m}}{2^m m!}$ choices for $\{N_1, ..., N_m\}$ so the probability that some such choice yields the desired clique minor is at most $\frac{2k^{2m}}{2^m m!} \frac{255}{256}^{\frac{m^2}{19}}$ which goes to zero as k goes to infinity as required. □

Now, by Lemmas 1 and 2, almost surely all but $2000009\log(k)$ nodes of any clique minor in a random candidate G contain at least three vertices of G. Thus, for almost every candidate G with k pairs, the largest clique minor contains at most $\frac{2k}{3} + 2000009\log(k)$ nodes. However, the pairs of G are the nodes of a clique pseudo-minor of order k. This completes the proof of our theorem. □

We have shown that there are graphs with a pseudo-minor of order k and no clique minor of order greater than $\frac{2k}{3} + O(\log(k))$. On the other hand, it is easy to see that any pseudo-minor of order k yields a clique minor of order $\lfloor \frac{k}{2} \rfloor$ (simply partition the pseudo-nodes of the pseudo-minor into disjoint pairs). It would be of interest to know the largest constant c such that every graph containing a pseudo-minor of order n contains a clique minor of order $\lfloor cn \rfloor$. It would also be of interest to know the smallest k for which the conjecture of Las Vergnas and Meyniel fails.

References

1. Bollobas B., *Random Graphs*, Academic Press, London, 1985.

2. Duchet P., Private Communication.

3. Hadwiger H., Uber Eine Klassifikation der Streckenkomplexe, *Viertelsjahr. Naturforsch. Ges. Zurich* 88(1943), 133-142.

4. Las Vergnas M. and Meyniel H., Kempe Classes and the Hadwiger Conjecture, *Journal of Combinatorial Theory, Series B* 31(1981), 95-104.

5. Jorgensen L. K., Contractions to Complete Graphs, *Annals of Discrete Mathematics* 41(1989), 307-310.

Forschungsinstitut fur Diskrete Mathematik, Universitat Bonn, Bonn 1, Germany.
email: breed@watserv1.uwaterloo.ca

An extremal function for the achromatic number

BÉLA BOLLOBÁS, BRUCE REED AND ANDREW THOMASON

November 12, 1991

ABSTRACT. The achromatic number $\psi(G)$ of a graph G is defined to be the maximum number of parts into which the vertex set may be partitioned so that between any two parts there is at least one edge. We show $\max\{e(G) \; ; \; |G| = n \text{ and } \psi(G) \leq k\} = (k-1)n - \binom{k}{2}$ if n is large, and investigate the extremal function for n not large.

1. Introduction

Given a graph G of order n, (all graphs considered in this paper are of course *simple*, that is, have no loops or multiple edges), we say two disjoint subsets A, B of the vertex set $V(G)$ are *joined* if there is an edge ab with $a \in A$ and $b \in B$. A k-partition $V_1 \cup V_2 \cup \ldots V_k$ of $V(G)$ is *complete* if each pair V_i, V_j is joined, $1 \leq i,j \leq k$, otherwise it is *incomplete*. The maximum value of k for which G has a complete k-partition is called here the *achromatic number* of G, denoted $\psi(G)$. Our purpose in this note is to investigate the extremal function $\max\{e(G) \; ; \; |G| = n \text{ and } \psi(G) \leq k\}$.

The parameter $\psi(G)$ has been called the *pseudo-achromatic* number by some authors, who reserve the term achromatic number for the case in which each part of the partition is required to be an independent set; we shall denote the maximum value of k for which G has a complete k-partition into independent sets by $\psi_i(G)$. Note that $\psi_i(G)$ is at least $\chi(G)$, the chromatic number of G. The corresponding extremal function $\max\{e(G) \; ; \; |G| = n \text{ and } \psi_i(G) \leq k\}$ is easily shown to equal $t_k(n)$, the size of the complete k-partite Turán graph $T_k(n)$. Indeed $\psi_i(T_k(n)) \leq k$, and if $|G| = n$ and $e(G) > t_k(n)$ then (by Turán's theorem) $K_{k+1} \subset G$ so $\chi(G) \geq k+1$.

1991 *Mathematics Subject Classification*. Primary 05C35, 05C15.
The second author was supported by a grant from the Von Humboldt Stiftung.
This paper is in final form and no version of it will be submitted for publication elsewhere.

Another commonly imposed restriction on the complete partition is that each part span a connected subgraph. The maximum value of k for which G has a complete k partition into connected parts is the *Hadwiger* number of G, namely the largest k for which K_k is a minor of G. The maximum size of a graph with no K_{k+1} minor was determined by Kostochka [5] and by Thomason [6]; if we define
$$c(k) = \inf\{\ c\ ;\ e(G) \geq c|G|\ \text{implies}\ G\ \text{has a}\ K_{k+1}\ \text{minor}\ \}$$
then $0.265k\sqrt{\log_2 k} \leq c(k) \leq 2.68k\sqrt{\log_2 k}$ for large k, as shown in [6].

We shall impose no restriction on the subgraphs spanned by the parts of our partitions. For extremely small graphs, the behaviour of the extremal function is much like that described in the previous paragraph. However for large graphs the situation is very different.

2. Extremal graphs

Bollobás, Catlin and Erdős [2] investigated $\psi(G)$ for $G \in \mathcal{G}(n,p)$, the probability space of random labelled graphs of order n where edges are chosen independently with probability p. They showed

$$\psi(G) \leq n\Big/\sqrt{\log_b n - d\log_b \log_e n} = n\sqrt{\log b}\Big/\sqrt{\log n}\,(1 + o(1))$$

almost surely, where $d > 2$ is fixed and $b = 1/(1-p)$. The proof holds for any value of p; hence by choosing $p = p(n)$ so that $\psi(G) \leq k$ almost surely, we obtain a lower bound on the function $\max\{e(G)\ ;\ |G| = n\ \text{and}\ \psi(G) \leq k\}$ for large k. For a given constant $C > 0$ there is a constant $c > 0$, such that if $n < Ck\sqrt{\log k}$ then $p(n) > c$, so the lower bound obtained is necessarily within a constant factor of the actual maximum value. On the other hand, for larger values of n, $p(n)$ is small, $\log b \approx p$, $\log n \approx \log k$ and G has size approximately $k^2 \log k/2$.

The size of random graphs with $\psi(G) = k$ therefore does not (significantly) increase with $|G|$, so random graphs provide useful examples only when their order is not large compared with k. The graph $G = K_{k-1} + \overline{K}_{n-k+1}$ (notation is that of [1]) satisfies $\psi(G) = k$, and its size, $(k-1)n - \binom{k}{2}$, increases with n. For $n \gg k \log k$ its size is greater than the random examples.

It seems likely that the graph $K_{k-1} + \overline{K}_{n-k+1}$ and random graphs between them describe the extremal function well. We prove in (3.2) that $K_{k-1} + \overline{K}_{n-k+1}$ is the unique extremal graph for large n, and in (4.2) that even for $n \gg k \log^2 k$ its size is within a constant factor of being best possible. Unfortunately we are unable to show that the size of random graphs comes within a constant factor of best possible for extremely small n ($n < k \log k$). The methods of [6], which at first sight ought to provide information in this range, fail to work. (It would be needed to show that a graph of order $n < k \log k$ with size $p^2 n = k^2 \log k$ has $\psi \geq k + 1 \approx n\sqrt{p/\log n}$; the methods of [6] yield only $\psi \geq pn\sqrt{p/\log n}$.)

3. The extremal function for large graphs

It is not difficult to verify that the graph $K_{k-1} + \overline{K}_{n-k+1}$ is extremal for $n \geq k^3$, and by means of the following lemma we can do even better.

(3.1). *Let $X_1 \cup X_2 \cup \ldots \cup X_k$ be a k-partition of a subset $X \subset V(G)$. If $\Gamma_{G-X}(X_i) \geq 2(k+1)\log(k+1)$ for $1 \leq i \leq k$, then $\psi(G) \geq k+1$.*

PROOF. From a random $(k+1)$-partition $Y_0 \cup Y_1 \cup \ldots Y_k$ of $V(G) - X$ we may form the $(k+1)$-partition $Y_0 \cup \bigcup_{i=1}^{k}(X_i \cup Y_i)$ of $V(G)$. It is enough to show that the probability of this partition being incomplete is less than one.

For a fixed i and j, the probability that there are no X_i–Y_j edges is at most $(k/(k+1))^{2(k+1)\log(k+1)} < (k+1)^{-2}$. Hence the expected number of pairs of parts which are not joined is at most $\frac{1}{2}k(k+1)(k+1)^{-2} < 1$. □

(3.2). *Let $\psi(G) \leq k$ and $|G| \geq 2(k+1)^2[1+\log k]$. If $e(G) \geq (k-1)|G| - \binom{k}{2}$ then $G = K_{k-1} + \overline{K}_{n-k+1}$.*

PROOF. We proceed by induction on k, the case $k = 1$ being trivial. Let G be a graph satisfying the conditions of the theorem. If for some vertex $v \in G$ it happens that $\psi(G-v) \leq k-1$, then $e(G-v) \geq e(G) - (n-1) \geq (k-2)(n-1) - \binom{k-1}{2}$ and by the induction hypothesis $G - v = K_{k-2} + \overline{K}_{n-k+1}$. Therefore G must itself be $K_{k-1} + \overline{K}_{n-k+1}$ (or else $e(G)$ would be smaller than stated), and so the theorem follows in this case.

Consequently we may assume that, for each $v \in G$, $\psi(G-v) \geq k$, and it will be enough to show $|G| < 2(k+1)^2[1+\log k]$. For a given v let H be a minimal subgraph of $G - v$ with $\psi(H) \geq k$. Since only one edge need be selected from each joined pair of parts in a complete k-partition of $V(G) - v$ to form H, we have $|H| \leq k(k-1)$ and each part of the partition of $V(H)$ has order at most $k-1$. If now $d(v) \geq (k-1)^2 + 1$ we may add, to any part of H not containing a neighbour of v, some vertex of $\Gamma(v) \setminus V(H)$, and so obtain a complete $(k+1)$-partition of $\{v\} \cup \Gamma(v) \cup V(H)$ with $\{v\}$ as one of the parts. This contradicts $\psi(G) \leq k$, so we have $d(v) \leq (k-1)^2$ for any $v \in G$.

Therefore $\Delta(G) \leq (k-1)^2$. Let ℓ be the maximum size of a set of independent edges in G. Then $\ell < \binom{k+1}{2}$ else there would be a complete $(k+1)$-partition of the endvertices of these edges. Since $e(G) \leq 2\ell\Delta(G)$ we have at once that $|G| \leq k^3$. Moreover by (3.1) there cannot be k vertices of degree at least $2(k+1)\log(k+1)$, so

$$\begin{aligned}e(G) &\leq (k-1)\Delta(G) + (2\ell-k+1)2(k+1)\log(k+1) \\ &\leq (k-1)^3 + 2(k^2-1)(k+1)\log(k+1),\end{aligned}$$

which together with the lower bound on $e(G)$ stated in the conditions of the theorem implies $|G| < 2(k+1)^2[1+\log k]$, as required. □

4. The extremal function for small graphs

The most interesting part of this investigation is the attempt to reduce the bound $2(k+1)^2[1+\log k]$ stated in (3.2) to $Ck\log k$, for some positive constant C. Our best attempt is described in (4.2) below, where we consider graphs of order as low as $Ck\log^2 k$ at the cost of increasing the size to $3k|G|$. The proof relies on (4.1), which is of independent interest.

(4.1). *Let G be an n by m bipartite graph, with bipartition (U,V), with each vertex in U having degree at least r. Then V has a partition $V = \bigcup_{i=1}^r V_i$ such that each V_i is joined to at least $n/5$ vertices of U.*

PROOF. We in fact prove the stronger result in which the lower bound $n/5$ is replaced by cn, where $c = (1-e^{-1})/3$. We proceed by induction on r, the case $r=1$ being trivial. First delete edges so each vertex in U has degree exactly r. If now any vertex $v \in V$ has degree greater than cn, take $V_r = \{v\}$ and apply the induction hypothesis with $r-1$ to the graph $G-v$. Hence we assume no vertex in V has degree greater than cn.

Consider random partitions of V into parts $V_1, \ldots, V_r$. The probability that a given vertex $u \in U$ is not in $\Gamma(V_i)$ is $(1-1/r)^r < e^{-1}$. Hence the expected value of $|\Gamma(V_i)|$ is $(1-e^{-1})n$, and the expected value of $\sum_{i=1}^r |\Gamma(V_i)|$ is $r(1-e^{-1})n$. Thus we may take a partition $V = \bigcup_{i=1}^r V_i$ such that $\sum_{i=1}^r |\Gamma(V_i)| \geq r(1-e^{-1})n$. Let $A = \{V_i;\ |\Gamma(V_i)| < cn\}$ and $B = \{V_i;\ |\Gamma(V_i)| \geq (1-e^{-1})n\}$. Then

$$0 \leq \sum_{i=1}^r \left[|\Gamma(V_i)| - (1-e^{-1})n\right] \leq |B|e^{-1}n - 2|A|cn$$

so $|B| \geq |A|$. Hence it is enough to show we can split each $V_i \in B$ into two parts V_i' and V_i'' so that both $|\Gamma(V_i')| \geq cn$ and $|\Gamma(V_i'')| \geq cn$. But this is clearly possible, since no vertex in V has degree more than cn and so

$$\min\{|\Gamma(V_i')|, |\Gamma(V_i'')|\} \geq (|\Gamma(V_i)| - cn)/2 \geq cn,$$

thus completing the proof. □

The constant 5 in the statement of (4.1) is clearly not the smallest possible value for which the theorem would be true, which is probably 2. An improvement in the constant in (4.1) would improve somewhat the constant in the next theorem.

(4.2). *If G is a graph of order at least $200(k+1)\log^2(k+1)$ and size at least $3k|G|$ then $\psi(G) \geq k+1$.*

PROOF. Let G be a graph with $|G| \geq 200(k+1)\log^2(k+1)$ and $e(G) \geq 3k|G|$. Repeatedly delete vertices of degree less than $2k$ until there remains a graph H with $\delta(H) \geq 2k$. At most $(2k-1)|G|$ edges will be removed by this process, so at least $(k+1)|G|$ are left. It follows that $|H| \geq \sqrt{2(k+1)|G|} \geq 20(k+1)\log(k+1)$.

A well-known argument (see for example Erdős [3]) implies that H contains a spanning bipartite subgraph B with $\delta(B) \geq k$. Let the vertex classes of B be X and Y. We may assume $|Y| \geq |B|/2 \geq 10(k+1)\log(k+1)$. By (4.1) there is a partition $X_1 \cup X_2 \cup \ldots \cup X_k$ of X such that $\Gamma_Y(X_i) \geq |Y|/5 \geq 2(k+1)\log(k+1)$. It now follows from (3.1) that $\psi(G) \geq k+1$. $\square$

5. Remarks

The bound $C_1 k \log^2 k$ in (4.2) could be brought down to the best possible $C_2 k \log k$ if more were known about our extremal function for very small graphs. In particular if it were known that graphs of order at most $C_3 k \log k$ and size $k^2 \log k$ had $\psi \geq k+1$ (for appropriate constants C_*), this would be achieved by considering graphs G of size $2k|G| + k^2 \log k$, and selecting a minimal subgraph H of size at least $2k|H| + k^2 \log k$. If $|H|$ were less than $C_3 k \log k$ we would be done, otherwise $\delta(H) \geq 2k$ and we would proceed as in the proof of (4.2).

Karabeg and Karabeg [4] state that $\max\{e(G) \; ; \; \psi(G) \leq k\} = (k-1)|G| - \binom{k}{2}$ for all $|G|$ if $k \leq 4$. They are also interested in extremely small values of $|G|$, and state $\max\{e(G) \; ; \; \psi(G) \leq k\} = \binom{|G|}{2} - 2(|G| - k + 2)$ if $|G| \leq 4k/3$.

We are informed by a referee that a form of (3.2) has been obtained independently by Z. Füredi.

References

1. B. Bollobás, *Extremal Graph Theory*, Academic Press (1978).
2. B. Bollobás, P. Catlin and P. Erdős, Hadwiger's conjecture is true for almost every graph, *Europ. J. Combin. Theory*, **1** (1980) 195–199.
3. P. Erdős, On some extremal problems in graph theory, *Israel J. Math.* **3** (1965) 113–116.
4. A. Karabeg and D. Karabeg, Graph Compaction, *Graph Theory Notes of New York* **XXI** (J.W. Kennedy and L.V. Quintas, eds.), New York Academy of Sciences (1991) 44–51.
5. A.V. Kostochka, A lower bound for the Hadwiger number of graphs by their average degree, *Combinatorica* **4** (1984), 307–316.
6. A. Thomason, An extremal function for contractions of graphs, *Math. Proc. Cambridge Phil. Soc.* **95** (1984), 261–265.

DEPARTMENT OF PURE MATHEMATICS AND MATHEMATICAL STATISTICS, 16 MILL LANE, CAMBRIDGE CB2 1SB, ENGLAND
  E-mail address: B.Bollobas@pmms.cam.ac.uk

FORSCHUNGSINSTITUT FÜR DISKRETE MATHEMATIK, UNIVERSITÄT BONN, NASSESTRASSE 2, BONN, GERMANY
  E-mail address: OR788@dbnuor1.bitnet

DEPARTMENT OF PURE MATHEMATICS AND MATHEMATICAL STATISTICS, 16 MILL LANE, CAMBRIDGE CB2 1SB, ENGLAND
  E-mail address: A.G.Thomason@pmms.cam.ac.uk

The Asymptotic Structure of H-free Graphs

H.J. PRÖMEL AND A. STEGER

November 8, 1991

ABSTRACT. In this paper we focus on two questions.
   (1) How many edges can an H-free graph have?
   (2) How does a "typical" H-free graph look like?
For excluded *weak* subgraphs both problems have been well-studied for a long time. We survey some of the important results. The main stress of the paper however lies on explaining the recent results on excluding *induced* subgraphs which complement the results in the weak case in an unexpected way.

We also indicate how the methods envolved in proving these asymptotic results can be used in other branches of graph theory. In particular we comment on applications to graph coloring and to perfect graphs.

1. Introduction

A graph $G = (V, E)$ is called triangle-free, if it does not contain a triangle, i.e. a complete graph on three vertices, as a subgraph. The question of how many edges a triangle-free graph on n vertices may have, was already answered by Mantel in 1907.

THEOREM 1.1 (MANTEL [18]). *Every graph on n vertices with more than $t_2(n) = \lfloor \frac{n}{2} \rfloor \cdot \lceil \frac{n}{2} \rceil$ edges contains a triangle and the complete bipartite graph with parts as equal as possible is the only triangle-free graph on n vertices with $t_2(n)$ many edges.*

This result of Mantel shows that the extremal graph, and therefore also all of its subgraphs, are bipartite. However, as observed by Tutte in 1948, publishing under a pseudonym, there are also triangle-free graphs of arbitrary high chromatic number.

1991 *Mathematics Subject Classification.* Primary 05C80, 05C35; Secondary 68R10, 05C30.
The final version of this paper will be submitted for publication elsewhere.

THEOREM 1.2 (DESCARTES [3]). *For every $r \in \mathbb{N}$ there exists a triangle-free graph $G = (V, E)$ such that $\chi(G) > r$.*

While this shows that there are triangle-free graphs of arbitrary high chromatic number, a natural question is to ask: What chromatic number do we expect, if we pick a triangle-free graph "at random"? That is, what is the chromatic number of a *typical* triangle-free graph? As it turned out the answer is quite simple, namely 2. This was proved in 1976 by Erdős, Kleitman and Rothschild. More precisely, they showed:

THEOREM 1.3 (ERDŐS, KLEITMAN, ROTHSCHILD [9]). *Let $\mathcal{F}orb_n(K_3)$ denote the class of all triangle-free graphs on n vertices and let $\mathcal{C}ol_n(2)$ denote the class of all bipartite graphs on n vertices. Then*

$$\lim_{n \to \infty} \frac{|\mathcal{F}orb_n(K_3)|}{|\mathcal{C}ol_n(2)|} = 1.$$

Obviously, the type of questions we just considered is not limited to triangle-free graphs. Whenever H is some fixed graph we may ask: How many edges can an H-free graph have and what is the chromatic number of a typical H-free graph, or more generally, how does a typical H-free graph look like?

In fact, if H is not a clique we have to distinguish two cases, namely excluding H as a weak or as an induced subgraph. It turns out that the answers for the weak case are a lot easier to obtain and the fundamental results here are known for a long time. They essentially state that the number of edges in an extremal graph as well as the asymptotic structure of H-free graphs is basically determined by the chromatic number $\chi(H)$.

In the induced case, however, answers to the corresponding problems were found only during the last two years (cf. [21], [22], [23]). It was shown that in the induced case, like in the weak case, the asymptotic properties of H-free graphs are mainly determined by a single parameter, say $\tau(H)$, which is a common generalization of the chromatic number and the clique covering number. (To prevent any misunderstandings, let us add at this point, that asking for the maximum number of edges a graph without induced H-subgraph can have, is obviously not very sensible. In fact one has to modify the definition of an *extremal graph* in a natural way. For a precise definition the reader is refered to section three.)

The aim of this paper is to survey these results. In particular we try to point out the interaction between extremal graph theory and asymptotic properties on the one hand and the similarities between the weak case and the induced case on the other hand. We also indicate how the methods envolved in proving these asymptotic results can be used in other branches of graph theory. In particular we comment on applications to graph coloring and to perfect graphs.

In this paper graphs are always simple, i.e. loops and multiple edges are not allowed. If H is a fixed graph we denote by $\mathcal{F}orb_n(H)$ the set of all graphs on

n vertices which do not contain H as a weak subgraph and by $\mathcal{F}orb_n^*(H)$ the set of all graphs on n vertices which do not contain H as an induced subgraph. We often write $G = (V_n, E)$ to indicate that the graph G has n vertices. All logarithms in this paper are to base two.

2. Excluding weak subgraphs

Let $ex_n(H)$ denote the maximum number of edges which a graph not containing H as a (weak) subgraph may have. The problem of determining $ex_n(H)$ and that of analysing the structure of H-free graphs with $ex_n(H)$ many edges is a well studied problem in extremal graph theory (cf. Bollobás [2], Simonovits [28]).

The asymptotic structure of H-free graphs was investigated only much later. By today, however, a fairly complete picture is known. In this section we survey some of the results. We start with the case that H is a clique.

2.1. Cliques. A classical result of Turán (1941) states that the (unique) extremal K_{l+1}-free graph is the complete l-partite graph where all parts are chosen as equal as possible. For $l = 2$ this implies Mantel's result.

THEOREM 2.1 (TURÁN [29]). *Denote by $t_n(l)$ the number of edges of a complete l-partite graph $T_n(l)$ on n vertices with parts as equal as possible. Then $ex_n(K_{l+1}) = t_n(l)$ and $T_n(l)$ is the unique graph on n vertices with $t_n(l)$ many edges that does not contain a clique of size $l + 1$.*

While Tutte's result (1.2) states that there exist K_{l+1}-free graphs (i.e., in fact triangle-free graphs) with arbitrary high chromatic number, Kolaitis, Prömel and Rothschild [16] extended the result (1.3) of Erdős, Kleitman and Rothschild to show that almost all K_{l+1}-free graphs have chromatic number l.

THEOREM 2.2 (KOLAITIS, PRÖMEL, ROTHSCHILD [16]). *Let $\mathcal{F}orb_n(K_{l+1})$ denote the class of all K_{l+1}-free graphs and let $\mathcal{C}ol_n(l)$ denote the class of all l-colorable graphs on n vertices. Then*

$$\lim_{n \to \infty} \frac{|\mathcal{F}orb_n(K_{l+1})|}{|\mathcal{C}ol_n(l)|} = 1.$$

In particular, (2.2) allows to determine the number of K_{l+1}-free graphs via counting the number of (labelled) l-colorable graphs. This gives (cf. [26]) that

$$|\mathcal{F}orb_n(K_{l+1})| = \Theta(2^{\frac{l-1}{2l}n^2 + n \log l - \frac{l-1}{2} \log n}).$$

The general idea behind the proof of Theorem (2.2) is to partition the set $\mathcal{F}orb_n(K_{l+1})$ suitably into a finite family of sets, so that one can show inductively that all except one of them are asymptotically negligible. Then structural properties of these negligible sets are used to prove that all graphs in the set remaining have the desired property, i.e., in case of Theorem (2.2) they are all l-colorable. A similar method was first used by D. Kleitman and B. Rothschild

[14] to show properties of almost all partially ordered sets. Therefore this method is sometimes called the "Kleitman-Rothschild method".

2.2. Color-critical graphs. In the last section we have seen that the class $\mathcal{F}orb_n(K_{l+1})$ of K_{l+1}-free graphs has the property that on the one hand the extremal graph is the complete l-partite graph $T_n(l)$ and on the other hand almost all graphs in this class have chromatic number l. So one might ask conversely, for what graphs H is it true that the class $\mathcal{F}orb_n(H)$ has these properties. In this section we give a complete characterization of those graphs.

An edge e of a graph H is called *color-critical*, if the deletion of e from H reduces the chromatic number, i.e. $\chi(H \setminus e) < \chi(H)$. Observe that in particular every edge of a clique is color-critical.

THEOREM 2.3 (SIMONOVITS [27]). *Let H be a graph with chromatic number $l+1$. Then for all sufficiently large n one has that $ex_n(H) = t_n(l)$ if and only if H contains a color-critical edge.*

The condition that H contains a color-critical edge is easily seen to be necessary. Otherwise add an arbitrary edge to the Turán graph $T_n(l)$. The new graph has now more edges and all its $(l+1)$-chromatic subgraphs contain a color-critical edge. The sufficiency was originally proven in a more general context. A direct proof by progressive induction can be found in [28].

Concerning the asymptotic structure Erdős [6] conjectured that the result (2.2) generalizes to the same class of graphs for which Simonovits' theorem (2.3) is true. This turned out to be the case.

THEOREM 2.4 (PRÖMEL, STEGER [20]). *Let H be a graph with chromatic number $l + 1$. Then*

$$\lim_{n\to\infty} \frac{|\mathcal{F}orb_n(H)|}{|\mathcal{C}ol_n(l)|} = 1$$

if and only if H contains a color-critical edge.

Again it is easy to see that the condition that H contains a color-critical edge is necessary. Compare $|\mathcal{C}ol_n(l)|$ with the number of graphs on n vertices with chromatic number at most $l + 1$ and parts $V_0, \ldots, V_l$ such that $|V_0| = 1$ and $|\Gamma(V_0) \cap V_1| = 1$. As $|\mathcal{C}ol_n(l)|$ is asymptotically well-understood, it is easy to see that for large n the second number dominates the first.

The opposite direction was first shown in Prömel [19] for $l = 2$. Both this and the final proof are based on the Kleitman-Rothschild method.

2.3. The general case. The graphs we considered so far, viz. cliques and graphs containing a color-critical edge, are up to a few special cases the only examples of graphs for which the extremal graph and the asymptotic structure are precisely known. However due to the work of Erdős, Stone and Simonovits and of Erdős, Frankl and Rödl it is known that the behaviour of these properties is closely connected to the chromatic number of the excluded graph.

THEOREM 2.5 (ERDŐS, SIMONOVITS [10], ERDŐS, STONE [11]). *For all graphs H we have that*

$$ex_n(H) = (1 - \frac{1}{\chi(H)-1})\frac{n^2}{2} + o(n^2).$$

In fact one knows considerably more about the structure of extremal graphs than just an asymptotic for the number of their edges. Erdős and Simonovits showed that every extremal graph can be obtained from a Turán graph by changing at most $o(n^2)$ many edges (the Asymptotic Structure Theorem) and furthermore that if an H-free graph has sufficiently many edges than it can already be obtained from a Turán graph by changing $o(n^2)$ many edges (the First Stability Theorem). For precise statements of theses results and many more we refer the reader to Bollobás [2] and Simonovits [28].

Observe that (2.5) determines the leading term in $ex_n(H)$, if the chromatic number of H is at least three. If one excludes bipartite graphs H (known as the degenerate case in Extremal Graph Theory), very little is known. In the case of even cycles for example good answers are only known for C_4, C_6 and C_{10}.

A question of Erdős and Simonovits, worth a considerable amount of money, states

PROBLEM 2.1 (ERDŐS, SIMONOVITS, CF. [5]). *Let H be a bipartite graph. Is it true that there exists $0 \leq \alpha = \alpha(H) < 1$ such that*

$$\lim_{n \to \infty} \frac{ex_n(H)}{n^{1+\alpha}} = c_H \qquad 0 < c_H < \infty.$$

Structural results complementing the Erdős, Stone, Simonovits results in extremal graph theory were obtained by Erdős, Frankl and Rödl. They incorporate the characterization (2.2) in order to show that H-free graphs are "close" to being $(\chi(H) - 1)$-colorable.

THEOREM 2.6 (ERDŐS, FRANKL, RÖDL [8]). *If $\chi(H) = l+1$, then the graphs in $\mathcal{F}orb_n(H)$ are "almost" l-colorable, i.e., every graph can be made $K_{\chi(H)}$-free by removing at most $o(n^2)$ edges. In particular,*

$$|\mathcal{F}orb_n(H)| = 2^{(1-\frac{1}{\chi(H)-1})\frac{n^2}{2}+o(n^2)}$$

and therefore, for $\chi(H) \geq 3$

$$|\mathcal{F}orb_n(H)| = 2^{ex_n(H)\cdot(1+o(1))}.$$

Again this result degenerates to $|\mathcal{F}orb_n(H)| = 2^{o(n^2)}$ whenever H is bipartite. Even for the simplest case, a cycle of length four, the best known upper bound is $2^{cn^{3/2}}$ with $c \approx 1.08$ due to Kleitman and Winston [15]. This contrasts the lower bound given by $ex_n(C_4) \sim \frac{1}{2}n^{3/2}$. Nevertheless it is believed that also for bipartite graphs (different from trees) the subgraphs of the extremal graph already represent "most" H-free graphs. More precisely, Erdős conjectured:

CONJECTURE 2.1 (ERDŐS [**7**]). *If H is bipartite and not a tree then*

$$|\mathcal{F}orb_n(H)| = 2^{ex_n(H)\cdot(1+o(1))}.$$

2.4. Application: fast coloring algorithms. It is well-known that the determination of the chromatic number of a graph is NP-hard. In fact, even deciding for any fixed $l \geq 3$, whether a given graph is l-colorable or not, is NP-complete. It is therefore quite unlikely that there exist algorithms for this problem whose worst-case running times are polynomially bounded. The importance of the chromatic number within Graph Theory as well as in applications (e.g. in scheduling and assignment problems) led several researchers to look for algorithms that are, while exponential in the worst case, at least fast "on average".

Wilf [**31**] was the first to investigate the average time behaviour of coloring algorithms. He showed that for every natural number l there exists an algorithm which decides whether $\chi(G) \leq l$ for a given graph G and if yes, colors it with $\chi(G)$ many colors – and whose expected running time is constant(!), assuming equal distribution on the class of all graphs. The reason for this at first sight surprising result is that a randomly chosen graph contains with very high probability a clique of size $l+1$ and is therefore in particular not l-colorable. Even more can be shown: every such randomly chosen graph does not only contain a K_{l+1}-clique, but it contains such a clique with very high probability already on an initial piece of constant size. In these cases the algorithms therefore simply reports correctly "the graph is not l-colorable". Otherwise it colors the graph optimally with an exponential branch and bound algorithm, reports whether $\chi(G) \leq l$, and outputs a coloring with $\chi(G)$ many colors. As the probability that the graph does not contain a K_{l+1} is sufficiently small, the exponential running time in this case does not destroy the overall constant expected running time.

If one excludes however the K_{l+1} as a pathological reason for a graph of not being l-colorable, i.e. if one only considers graphs in $\mathcal{F}orb_n(K_{l+1})$, then the difficulty of the problem increases considerably. Going one step further and considering only l-colorable graphs Turner [**30**] developed an algorithm which colors almost every l-colorable graph with l colors and whose running time is polynomially bounded. This result was strengthened by Dyer and Frieze [**4**]. Their algorithm colors every l-colorable graph with l colors and has linear expected running time assuming equal distribution on the class of all l-colorable graphs.

Based on a further improvement of the characterization (2.2) the remaining gap between Wilf's algorithm and the algorithm of Dyer and Frieze was recently closed in Prömel, Steger [**25**]. There an algorithm is described which colors every graph in $\mathcal{F}orb_n(K_{l+1})$ with $\chi(G)$ many colors (regardless how large $\chi(G)$ might be) and whose expected running time is still linear. Roughly speaking the improvement of the characterization allows to play off "structure" against "convergence". That is, if one requires for example l-colorability not for the whole graph, but for the whole graph except some "bad" vertices, then it can be shown that the fraction of K_{l+1}-free graphs not having this property decreases proportionally to the number of "bad" vertices. More precisely, the following result was shown.

THEOREM 2.7 (PRÖMEL, STEGER [25]). *Let $f(n)$ be an integral function such that $1 \leq f(n) \leq n/\log^3 n$. Then there exists a set $\mathcal{X}_n(f) \subseteq \mathcal{F}orb_n(K_{l+1})$ such that*
$$Prob(G \in \mathcal{X}_n(f)) = 2^{-\Omega(n \cdot f(n))}$$
and such that for every $G = (V_n, E)$ in $\mathcal{F}orb_n(K_{l+1}) \setminus \mathcal{X}_n(f)$ there exists a set $A \subseteq V_n$ of "bad" vertices of size $|A| \leq f(n) - 1$ in such a way that the graph $G[V_n \setminus A]$ induced by $V_n \setminus A$ is l-colorable and can be l-colored in linear time.

While for $f(n) \equiv 1$ this result just implies the characterization (2.2) of Kolaitis, Prömel and Rothschild, the general case allows to construct the coloring algorithm described above.

3. Excluding induced graphs

If one excludes induced subgraphs instead of weak subgraphs, the picture changes completely. While classes of graphs defined by excluding a whole sequence of induced subgraphs, as for example the class of triangulated graphs or the class of Berge graphs, played an important rôle in graph theory, the question of what happens if one excludes a single graph remained unanswered until recently.

In this section we describe some analogues to the classical theorem (2.5) of Erdős, Stone and Simonovits and to the asymptotic result (2.6) of Erdős, Frankl and Rödl, which were discovered not long ago. In order to motivate the necessary notions and definitions we start with an example.

3.1. A special case: quadrilaterals. Perhaps the simplest graph for which the answers to the weak and induced problems do not coincide is the quadrilateral, a cycle of length four. While in the weak case, determining the asymptotic structure or just the number $|\mathcal{F}orb_n(C_4)|$ of C_4-free graphs is one of the challenging open problems (cf. section 2.3), surprisingly the related questions for graphs without *induced* quadrilaterals could be solved.

Let $Split(n)$ denote the class of all split graphs on n vertices, i.e., the class of graphs whose vertex set can be partitioned into a clique and a stable set. While it is easy to see that no split graph contains an induced C_4, it was shown in Prömel, Steger [21] that asymptotically also the contrary is true.

THEOREM 3.1 (PRÖMEL, STEGER [21]). *Almost all graphs without induced C_4 are split graphs, i.e.*
$$\lim_{n \to \infty} \frac{|\mathcal{F}orb_n^*(C_4)|}{|Split(n)|} = 1.$$

This generalizes a result of Bender, Richmond and Wormald [1] who showed that a similar result holds, if one replaces the class $\mathcal{F}orb_n^*(C_4)$ by the class of all triangulated graphs, that is, if one excludes all induced cycles of length at least four, instead of just the induced cycles of length exactly four.

Using an asymptotic formula for the number of bipartite graphs, (3.1) implies that there exist constants c_0 and c_1 such that

$$|\mathcal{F}orb_n^*(C_4)| = (c_r + o(1)) \cdot 2^{\frac{n^2}{4} + n - \frac{1}{2}\log n},$$

where $r = n \bmod 2$.

3.2. Extremal graphs. In order to succeed in finding counterparts to the classical theorem (2.5) of Erdős, Stone and Simonovits and to the asymptotic result (2.6) of Erdős, Frankl and Rödl one first had to develop an appropriate notion of an "extremal graph" for the class $\mathcal{F}orb_n^*(H)$ of graphs not containing the graph H as an induced subgraph. Obviously, asking just for a graph with maximum number of edges does not lead to a meaningful concept: in this case the complete graph would be the (unique) extremal graph for all graphs H different from a clique. That is, one had to generalize some other notion inherent to extremal graphs.

Recall that in case of excluding weak subgraphs every subgraph of an H-free graph is also H-free, i.e., the property of being H-free is *hereditary*. As a consequence extremal graphs in the class $\mathcal{F}orb_n(H)$ are exactly those graphs which give rise to a maximum number of H-free graphs by taking subgraphs. In fact (2.6) states, that even the number of graphs in $\mathcal{F}orb_n(H)$ is essentially given by the number of subgraphs of the extremal graph, i.e., that

$$\log|\mathcal{F}orb_n(H)| = (1 + o(1)) \cdot ex_n(H).$$

In the previous section we have seen that a similar result also holds in the case of excluded quadrilaterals. Namely, denote by $K^*(\lfloor\frac{n}{2}\rfloor, \lceil\frac{n}{2}\rceil)$ the complete bipartite graph $K(\lfloor\frac{n}{2}\rfloor, \lceil\frac{n}{2}\rceil)$ on n vertices in which the first class is replaced by a clique. Observe, that $K^*(\lfloor\frac{n}{2}\rfloor, \lceil\frac{n}{2}\rceil)$ as well as all subgraphs arising from $K^*(\lfloor\frac{n}{2}\rfloor, \lceil\frac{n}{2}\rceil)$ by deleting edges from the original $K(\lfloor\frac{n}{2}\rfloor, \lceil\frac{n}{2}\rceil)$ do not contain an induced C_4. On the other hand the logarithm of the number of these subgraphs is $\lfloor\frac{n}{2}\rfloor\lceil\frac{n}{2}\rceil$ which indeed corresponds to the order of $\log|\mathcal{F}orb_n^*(C_4)|$. This motivates the following definition (cf. Prömel, Steger [22]):

DEFINITION 3.1. *Let $ex_n^*(H)$ denote the maximum number of edges a graph $G = (V_n, E)$ may have such that there exists a graph $G_0 = (V_n, E_0)$ with $E \cap E_0 = \emptyset$ such that $(V_n, E_0 \cup X)$ does not contain an induced H-subgraph for all $X \subseteq E$. We call such a graph G an extremal graph for the class $\mathcal{F}orb_n^*(H)$ and G_0 a supplemental graph with respect to the extremal graph G.*

With this definition at hand it is not difficult to see that in our example of a quadrilateral one indeed has $ex_n^*(C_4) = \lfloor \frac{n}{2} \rfloor \cdot \lceil \frac{n}{2} \rceil$ and that for $n \geq 6$ the $K(\lfloor \frac{n}{2} \rfloor, \lceil \frac{n}{2} \rceil)$ is the unique extremal graph.

3.3. A new parameter. In the study of classes of graphs defined by excluding weak subgraphs the chromatic number played a central rôle for the extremal graph as well as for asymptotic properties of the class. As we will see results of a similar flavour are true in the induced case as well — if one replaces the chromatic number by an appropriately defined parameter. As a motivation for the definition of this new parameter $\tau(H)$ consider the following definitions of the chromatic number $\chi(H)$ and of the clique covering number $\sigma(H)$:

$\chi(H)$ is the largest integer k such that no $(k-1)$-partite graph contains H as an induced subgraph,

$\sigma(H)$ is the largest integer k such that no $(k-1)$-partite graph, in which *each* class is replaced by a clique, contains H as an induced subgraph.

As a common generalization define:

$\tau(H)$ is the largest integer k for which there exists an integer $\tilde{k}$ between 0 and $k-1$ such that no $(k-1)$-partite graph, in which $\tilde{k}$ classes are replaced by a clique, contains H as an induced subgraph.

One easily observes that $\max\{\chi(H), \sigma(H)\} \leq \tau(H) < \chi(H) + \sigma(H)$, which in particular implies that $\tau(H)$ is well-defined.

The counterpart of (2.5) for induced subgraphs now reads as follows.

THEOREM 3.2 (PRÖMEL, STEGER [22]). *For all graphs H we have that*

$$ex_n^*(H) = (1 - \frac{1}{\tau(H) - 1})\frac{n^2}{2} + o(n^2).$$

Theorem (3.2) is easily proved by combining the proof of (2.5) with Ramsey's theorem.

It can moreover be shown that also the structural theorems of Extremal Graph Theory (like the Asymptotic Structure Theorem or the First Stability Theorem) remain true for induced subgraphs after straightforward modifications.

3.4. A general asymptotic. The problem of determining the structure of a typical graph in $\mathcal{F}orb_n^*(H)$, or at least estimating the number of graphs in this class, is considerably more difficult. However, after generalizing Szemerédi's powerful uniformity lemma to hypergraphs an analogue to Theorem (2.6) was proven.

THEOREM 3.3 (PRÖMEL, STEGER [23]). *If $\tau(H) = l + 1$, then the graphs in $\mathcal{F}orb_n^*(H)$ are "almost" l-partite split-graphs, i.e., there exists a partition $V_n = A_1 \cup \ldots \cup A_l$ of the vertex set into l classes such that after changing at most $o(n^2)$ edges all the classes A_i form either cliques or stable sets. In particular,*

$$|\mathcal{F}orb_n^*(H)| = 2^{(1 - \frac{1}{\tau(H)-1})\frac{n^2}{2} + o(n^2)}$$

and therefore, for $\tau(H) \geq 3$

$$|\mathcal{F}orb_n^*(H)| = 2^{ex_n^*(H)\cdot(1+o(1))}.$$

As in (2.6) this result degenerates to $|\mathcal{F}orb_n^*(H)| = 2^{o(n^2)}$, whenever $\tau(H) = 2$, i.e. H is "bipartite". However, in contrary to the weak case, there are only five such "bipartite" graphs H. For all of them the class $\mathcal{F}orb_n^*(H)$ is well understood.

3.5. The strong perfect graph conjecture. An important motivation after studying $\mathcal{F}orb_n^*(C_4)$ to investigate asymptotic properties of $\mathcal{F}orb_n^*(H)$ in general, and of $\mathcal{F}orb_n^*(C_5)$ in particular, was one of the most prominent problems in graph theory.

A graph G is called *perfect*, if G and each of its induced subgraphs have the property that the chromatic number χ equals the size of a maximum clique ω. Since their introduction by C. Berge in 1960 various striking results on perfect graphs have been proven. Many "natural" classes of graphs were identified as perfect graphs (e.g. triangulated graphs, interval graphs, unimodular graphs, and many more) and several interesting structural properties were discovered (as for example that with G also the complement of G is perfect, Lovász [**17**]). Perfect graphs are also of interest from an algorithmic point of view. While determining the chromatic number of a graph is in general a difficult problem, more precisely, an NP-hard problem, Grötschel, Lovász and Schrijver [**12**] showed that for perfect graphs the chromatic number as well as several other optimization problems are solvable in polynomial time.

Despite this progress, a possible characterization of perfect graphs in terms of forbidden subgraphs, also posed by Berge in 1960, remained open and is one of the outstanding open problems in graph theory.

THE STRONG PERFECT GRAPH CONJECTURE. *A graph $G = (V,E)$ is perfect if and only if neither G nor its complement $\overline{G}$ contains an odd cycle of length at least five as an induced subgraph.*

Let $Perf(n)$ denote the set of all perfect graphs on n vertices and let $Berge(n)$ denote the set of all graphs on n vertices which contain neither an odd cycle nor the complement of such a cycle as an induced subgraph (nowadays these graphs are also known as *Berge* graphs). Combining a suitable generalization of the Kleitman-Rothschild method on the one hand with Szemerédi's uniformity lemma on the other hand we were able to show that almost all graphs in $\mathcal{F}orb_n^*(C_5)$ are so-called generalized split graphs and therefore by a result of Hayward [**13**] are in particular perfect. As trivially $Berge(n) \subseteq \mathcal{F}orb_n^*(C_5)$, this shows that the strong perfect graph conjecture is at least almost always true.

THEOREM 3.4 (PRÖMEL, STEGER [**24**]). *Almost all Berge graphs are perfect, i.e.*

$$\lim_{n\to\infty} \frac{|Perf(n)|}{|Berge(n)|} = 1.$$

REFERENCES

1. E.A. Bender, L.B. Richmond, and N.C. Wormald, *Almost all chordal graphs split*, J. Austral. Math. Soc. (Series A) **38** (1985), 214–221.
2. B. Bollobás, *Extremal Graph Theory*, Academic Press, New York, London, 1978.
3. B. Descartes, *A three color problem*, Eureka, 1948.
4. M.E. Dyer and A.M. Frieze, *The solution of some random NP-hard problems in polynomial expected time*, J. Algorithms **10** (1989), 451–489.
5. P. Erdős, *On the combinatorial problems which I would most like to see solved*, Combinatorica **1** (1981), 25–42.
6. P. Erdős, Private Communication, Oberwolfach, June 1988.
7. P. Erdős, *Some of my old and new combinatorial problems*, Paths, Flows, and VLSI-Layout (B. Korte, L. Lovász, H.J. Prömel and A. Schrijver, eds.), Algorithms and Combinatorics 9, Springer-Verlag Berlin Heidelberg, 1990, pp. 35-45.
8. P. Erdős, P. Frankl, and V. Rödl, *The asymptotic number of graphs not containing a fixed subgraph and a problem for hypergraphs having no exponent*, Graphs Combin. **2** (1986), 113–121.
9. P. Erdős, D.J. Kleitman, and B.L. Rothschild, *Asymptotic enumeration of K_n-free graphs*, International Colloquium on Combinatorial Theory, Atti dei Convegni Lincei **17**, Vol. 2, Rome, 1976, pp. 19-27.
10. P. Erdős and M. Simonovits, *A limit theorem in graph theory*, Studia Sci. Math. Hung. **1** (1966), 51–57.
11. P. Erdős and A.H. Stone, *On the structure of linear graphs*, Bull. Amer. Math. Soc. **52** (1946), 1089–1091.
12. M. Grötschel, L. Lovász, and A. Schrijver, *The ellipsoid method and its consequences in combinatorial optimization*, Combinatorica **1** (1981), 169–197.
13. R. Hayward, *Weakly triangulated graphs*, J. Combin. Theory, Series B **39** (1985), 200–209.
14. D.J. Kleitman and B.L. Rothschild, *Asymptotic enumeration of partial orders on a finite set*, Trans. Amer. Math. Soc **205** (1975), 205–220.
15. D.J. Kleitman and K.J. Winston, *On the number of graphs without 4-cycles*, Discrete Math. **41** (1982), 167–172.
16. Ph.G. Kolaitis, H.J. Prömel, and B.L. Rothschild, *K_{l+1}-free graphs: asymptotic structure and a 0 − 1 law*, Trans. Amer. Math. Soc. **303** (1987), 637–671.
17. L. Lovász, *Normal hypergraphs and the perfect graph conjecture*, Discrete Math. **2** (1972), 253–267.
18. W. Mantel, *Problem 28, soln. by H. Gouwentak, W. Mantel, J. Teixeira de Mattes, F. Schuh and W.A. Wythoff*, Wiskundige Opgaven **10** (1907), 60–61.
19. H.J. Prömel, *Almost bipartite-making graphs*, Random graphs '87 (M. Karoński, J. Jaworski and A. Ruciński, eds.), 1991, pp. 275-282.
20. H.J. Prömel and A. Steger, *The asymptotic number of graphs not containing a fixed color-critical subgraph*, Combinatorica (to appear).
21. H.J. Prömel and A. Steger, *Excluding induced subgraphs: quadrilaterals*, Random Struct. Alg. **2** (1991), 55–71.
22. H.J. Prömel and A. Steger, *Excluding induced subgraphs II: extremal graphs*, Discrete Appl. Math. (to appear).
23. H.J. Prömel and A. Steger, *Excluding induced subgraphs III: a general asymptotic*, Random Struct. Alg. **3** (1992), 19–31.
24. H.J. Prömel and A. Steger, *Almost all Berge graphs are perfect*, Combinatorics, Probability & Computing, 1992, in print.
25. H.J. Prömel and A. Steger, *Coloring clique-free graphs in linear expected time*, Random Struct. Alg., 1992, in print.
26. H.J. Prömel and A. Steger, *Random l-colorable graphs*, Forschungsinstitut für Diskrete Mathematik, Universität Bonn, 1992.
27. M. Simonovits, *Extremal graph problems with symmetrical extremal graphs. Additional chromatic conditions*, Discrete Math. **7** (1974), 349–376.
28. M. Simonovits, *Extremal graph theory*, Selected Topics in Graph Theory 2 (L.W. Beineke,

R.J. Wilson, eds.), Academic Press, London, 1983, pp. 161-200.
29. P. Turán, *Egy gráfelméleti szélsőértékfeladatról*, Mat. Fiz. Lapok **48** (1941), 436–452.
30. J.S. Turner, *Almost all k-colorable graphs are easy to color*, J. Algorithms **9** (1988), 63–82.
31. H.S. Wilf, *Backtrack: and O(1) expected time algorithm for the graph coloring problem*, Information Processing Letters **18** (1984), 119–121.

RESEARCH INSTITUTE OF DISCRETE MATHEMATICS, UNIVERSITY OF BONN, NASSESTR. 2, D-5300 BONN 1, GERMANY
E-mail address: PROEMEL@dbnuor1.bitnet

RESEARCH INSTITUTE OF DISCRETE MATHEMATICS, UNIVERSITY OF BONN, NASSESTR. 2, D-5300 BONN 1, GERMANY
E-mail address: OR210@dbnuor1.bitnet

Induced Minors and Related Problems

MICHAEL FELLOWS, JAN KRATOCHVÍL,
MATTHIAS MIDDENDORF AND FRANK PFEIFFER

ABSTRACT. Many well-studied and natural families of graphs are lower ideals in the induced minor order. That is, these families of graphs are closed under the operations (1) vertex deletion and (2) edge contraction. Intersection graphs of topological structures provide many examples, including interval graphs, string graphs, permutation graphs and co-comparability graphs. One of our principal motivations has been to develop the basic theory of induced minors, and to explore the similarities and differences of minors and induced minors.

We also study the computational complexity of determining the presence of various induced structures in a graph. Our results show that the decision problems concerning induced structures in a graph are usually more difficult than the corresponding problems for non-induced structures. The effect on these problems of restricting the input to planar graphs is also considered.

1. Notation

In this paper, we consider finite graphs without loops or multiple edges. The graphs are undirected, if not explicitely stated otherwise. The notation used is standard, we just stress the notions of *induced subgraph* and *induced minor*.

A subgraph $H \subset G$ is called an *induced subgraph* if in addition to $E(H) \subset E(G)$, $E(H) = \{xy | x, y \in V(H)\ \&\ xy \in E(G)\}$.

A graph H is called an *induced minor* of G (denoted by $H \prec_i G$) if a graph isomorphic to H can be obtained by edge contractions from an induced subgraph of G. In other words, H is an induced minor of G if there are disjoint subsets

---

1991 *Mathematics Subject Classification.* Primary 05C55, 68Q25; Secondary 05C15, 68R10;.

The first author is supported in part by the U.S. Office of Naval Research under contract N00014-88-K-0456, by the U.S. National Science Foundation under grant MIP-8919312, and by the National Science and Engineering Research Council of Canada.

The second author acknowledges the support of the U.S. Office of Naval Research when visiting the University of Idaho in the spring of 1990, and the hospitality of the University of Cologne in the fall of 1990.

The detailed version of this paper will appear elsewhere

V_u of $V(G)$ indexed by vertices $u \in V(H)$ such that for every $u, v \in V(H)$, there are $x \in V_u$ and $y \in V_v$ such that $xy \in E(G)$ if and only if $uv \in E(H)$.

2. Induced minors

Robertson and Seymour have established the following fundamental results concerning the minor order [**RS**].

THEOREM A. *Every lower ideal in the minor order has a finite obstruction set.*

THEOREM B. *For every fixed graph H, the problem H-MINOR TESTING which takes as input a graph G of order n and determines if $G \geq H$ in the minor order is solvable in time $O(n^3)$.*

These two theorems have the consequence that every lower ideal in the minor order is recognizable in time $O(n^3)$.

The situation for the induced minor order is quite different. There are lower ideals in the induced minor order having infinite obstruction sets, even when attention is restricted to planar graphs [**Th**]. In [**MNT**] it is shown that there are induced minor lower ideals that are undecidable. On the other hand, an analog of Theorem A does hold for series-parallel graphs [**Th**].

We show that though an analogue of Theorem B does not hold for induced minors in general, it does hold for planar inputs. We prove the following results:

THEOREM 2.1. *There is a graph graph H, such that the problem H-INDUCED MINOR TESTING which takes as input a graph G of order n and determines if $H \prec_i G$ is NP-complete.*

PROOF. Rather technical and will appear elsewhere.

THEOREM 2.2. *For every planar graph H there is a constant $k = k(H)$ such that every planar graph G of tree width $w(G) \geq k$ contains an induced minor isomorphic to H.*

SKETCH OF THE PROOF. We use the following result of [**RS**]: For every h and p there is a k such that every graph G of tree width $w(G) \geq k$ contains a minor isomorphic to K_p or a subgraph isomorphic to a wall of size h. If $p \geq 5$, a planar graph cannot contain K_p as a minor. Thus if a planar graph G has sufficiently large tree width, it contains a wall of large height. Such a wall is contained in G as a subgraph, not necessarily induced, and may have undesirable edges. However, since G is assumed to be planar, additional edges may lie only inside the cells of the wall. It follows that the wall, and hence G, contains all 'small' planar graphs as induced minors. □

THEOREM 2.3. *For every fixed planar graph H, the problem H-INDUCED MINOR TESTING which takes as input a planar graph G of order n and determines if $H \prec_i G$ is solvable in time $O(n \log^2 n)$.*

PROOF. From Theorem 2.2, it is straightforward to apply the techniques of second-order monadic graph properties for bounded tree width [**Co**]. Let k be such as in Theorem 2.2 for the graph H. In time $O(n \log^2 n)$ using the algorithm of Lagergren [**La**] we can either discover that the tree width of G is greater than k, so that H is necessarily an induced minor of G, or obtain a tree decomposition of G of width at most $3k$. Since the property of having an induced minor isomorphic to H is easily expressible in second order monadic logic, we can determine in linear time from this tree decomposition whether H is an induced minor of G. □

This has the consequence that every induced minor order lower ideal having a finite obstruction set is recognizable in polynomial time for planar graphs.

3. NP-hard induced substructure problems

In this section we present a number of combinatorial reductions which collectively demonstrate that induced substructure problems tend to be more difficult than their non-induced counterparts. We have the following results:

THEOREM 3.1. *The problem INDUCED MATCHING is NP-complete even when restricted to planar graphs.*

PROOF. Given a planar graph G, consider a graph G' obtained from G by making each vertex adjacent to a new extra vertex. Obviously, G' is planar the maximum size of an induced matching in G' equals the maximum size of an independent set in G. □

THEOREM 3.2. *The PLANAR INDUCED DIRECTED PATH problem which takes as an input a planar directed graph and two of its vertices x, y and determines whether there is a directed xy path in the graph such that no two nonconsecutive vertices are joined by an arc, is NP-complete.*

PROOF. Will appear elsewhere.

Note the striking difference between *induced dipaths* and *noninterfering dipaths* (a collection of dipaths is called noninterfering if no arc joins vertices of different paths, while backcutting arcs within each particular path are allowed). It is proved in [**DRSS**] that given two vertices in a planar digraph, one can compute in polynomial time the maximum number of noninterfering dipaths joining these vertices.

THEOREM 3.3. *The PLANAR INDUCED DIRECTED CYCLE problem which takes as an input a planar directed graph and one of its vertices x and determines whether there is an induced directed cycle passing through x, is NP-complete.*

PROOF. Follows from the method of Theorem 3.2.

Our results show that decision problems concerning induced substructures in a graph tend to be harder than corresponding decision problems concerning non-induced substructures (assuming $P \neq NP$). This is perhaps as one would expect.

References

[Co] Courcelle,B., *The monadic second order logic of graphs I: Recognizable sets of finite graphs*, Information and Comput., to appear.

[DRSS] McDiarmid,C., Reed,B., Schrijver,A., Shepherd,B., *Non-interfering dipaths in planar digraphs*, preprint.

[La] Lagergren,J., *Parallel algorithms for pathwidth and treewidth*, In: Proceedings FOCS, 1990.

[MNT] Matousek,J., Nesetril,J., Thomas,R., *On polynomial time decidability of induced minor closed classes*, Comment. Math. Univ. Carolin. 29 (1988), 703-710.

[RS] Robertson,N., Seymour,P.D., *Graph minors I–XVIII*.

DEPARTMENT OF COMPUTER SCIENCE, UNIVERSITY OF VICTORIA, VICTORIA, BRITISH COLUMBIA V8W 3P6, CANADA

DEPARTMENT OF ALGEBRA, FACULTY OF MATHEMATICS AND PHYSICS, CHARLES UNIVERSITY, SOKOLOVSKÁ 83, 186 00 PRAHA 8 PRAGUE, THE CZECH AND SLOVAK REPUBLIC

DEPARTMENT OF COMPUTER SCIENCE, UNIVERSITY OF COLOGNE, COLOGNE, GERMANY

DEPARTMENT OF COMPUTER SCIENCE, UNIVERSITY OF COLOGNE, COLOGNE, GERMANY

Induced Circuits in Graphs on Surfaces

ALEXANDER SCHRIJVER

ABSTRACT. We show that for any fixed surface S there exists a polynomial-time algorithm to test if there exists an induced circuit traversing two given vertices r and s of an undirected graph G embedded on S. (An *induced circuit* is a circuit without chords.) The general problem (not fixing S) is NP-complete. In fact, for each fixed surface S there exists a polynomial-time to find a maximum number of $r - s$ paths in G such that any two form an induced circuit.

1. Introduction

In this paper we show that the following problem is solvable in polynomial time, for any fixed compact surface S:

(1)   given: an undirected graph $G = (V, E)$ embedded on S and two vertices r and s of G;

              find: an induced circuit in G that traverses r and s.

An *induced circuit* is a circuit having no chords. The problem is NP-complete for general undirected graphs, as was shown by Bienstock [1]. In [2] the problem was shown to be solvable in polynomial time for planar graphs. In fact we show that for any fixed compact surface S the problem:

(2)   given: an undirected graph $G = (V, E)$ embedded on S and two vertices r and s of G;

              find: a maximum number of $r - s$ paths in G any two of which form an induced circuit;

is solvable in polynomial time.

Our method uses a variant of a method developed in [3] to derive, for any *fixed* k, a polynomial-time algorithm for the k disjoint paths problem in directed

---

[0]1991 Mathematics Subject Classification: Primary 05C10, 05C38, 05C85, Secondary 68Q25, 68R10
This paper is in final form and no version of it will be submitted for publication elsewhere.

planar graphs. (This problem is NP-complete for general directed graphs, even for $k = 2$.) The present method is based on cohomology over free boolean groups.

2. Free boolean groups

The *free boolean group* B_k is the group generated by $g_1, g_2, \ldots, g_k$, with relations $g_j^2 = 1$ for $j = 1, \ldots, k$. So B_k consists of all words $b_1 b_2 \ldots b_t$ where $t \geq 0$ and $b_1, \ldots, b_t \in \{g_1, \ldots, g_k\}$ such that $b_i \neq b_{i-1}$ for $i = 2, \ldots t$. The product $x \cdot y$ of two such words is obtained from the concatenation xy by deleting iteratively all occurrences of any pair $g_j g_j$. This defines a group, with unit element 1 equal to the empty word $\emptyset$.

We call $g_1, \ldots, g_k$ *generators* or *symbols*. Note that

(3) $$B_1 \subset B_2 \subset B_3 \subset \cdots.$$

The *size* $|x|$ of a word x is the number of symbols occurring in it, counting multiplicities. A word y is called a *segment* of word w if $w = xyz$ for certain words x, z. If $w = yz$ for some word z, y is called a *beginning segment* of w, denoted by $y \leq w$. This partial order gives trivially a lattice if we extend B_k with an element ∞ at infinity. Denote the meet and join by $\wedge$ and $\vee$.

We prove two useful lemmas.

LEMMA 1. *For all* $x, y, z \in B_k$ *one has:*

(4) $$x \leq y \cdot z \text{ and } z \leq y^{-1} \cdot x \iff x^{-1} \cdot y \cdot z = 1 \text{ or } y = xwz^{-1}$$
for some word w.

Proof. $\Longleftarrow$ being easy, we show $\Longrightarrow$. Let $w := x^{-1} \cdot y \cdot z$. As $x \leq y \cdot z$, $y \cdot z = xw$; and as $z \leq y^{-1} \cdot x$, $y^{-1} \cdot x = zw^{-1}$, that is, $x^{-1} \cdot y = wz^{-1}$. Hence if $w \neq 1$ then $xwz^{-1} = x \cdot w \cdot z^{-1} = y$. ∎

LEMMA 2. *Let $x, y \in B_k$. If $x \not\leq y$ then the first symbol of x^{-1} is equal to the first symbol of $x^{-1} \cdot y$.*

Proof. Let $z := x \wedge y$. So $x^{-1} \cdot y$ is the concatenation of $x^{-1} \cdot z$ and $z^{-1} \cdot y$. Since $x^{-1}z \neq 1$, the first symbol of $x^{-1} \cdot y$ is equal to the first symbol of $x^{-1} \cdot z$. Since $x^{-1}z \neq 1$ and $z \leq x$, the first symbol of $x^{-1} \cdot z$ is equal to the first symbol of x^{-1}. Hence the first symbol of x^{-1} is equal to the first symbol of $x^{-1} \cdot y$. ∎

3. The cohomology feasibility problem for free boolean groups

Let $D = (V, A)$ be a weakly connected directed graph, let $r \in V$, and let $(G, \cdot)$ be a group. Two functions $\phi, \psi : A \longrightarrow G$ are called r-*cohomologous* if there exists a function $f : V \longrightarrow G$ such that

(5)   (i) $f(r) = 1$;
        (ii) $\psi(a) = f(u)^{-1} \cdot \phi(a) \cdot f(w)$ for each arc $a = (u, w)$.

This clearly gives an equivalence relation.

Consider the following *cohomology feasibility problem (for free boolean groups)*:

(6) given: a weakly connected directed graph $D = (V, A)$, a vertex r, and a function $\phi : A \longrightarrow B_k$;

find: a function $\psi : A \longrightarrow B_k$ such that ψ is r-cohomologous to ϕ and such that $|\psi(a)| \leq 1$ for each arc a (if there is one).

We give a polynomial-time algorithm for this problem. The running time of the algorithm is bounded by a polynomial in $|A| + \sigma + k$, where σ is the maximum size of the words $\phi(a)$ (without loss of generality, $\sigma \geq 1$).

We may assume that with each arc $a = (u, w)$ also $a^{-1} := (w, u)$ is an arc of D, with $\phi(a^{-1}) = \phi(a)^{-1}$.

Note that, by the definition of r-cohomologous, equivalent to finding a ψ as in (6), is finding a function $f : V \longrightarrow B_k$ satisfying:

(7) (i) $f(r) = 1$;

(ii) for each arc $a = (u, w)$: $|f(u)^{-1} \cdot \phi(a) \cdot f(w)| \leq 1$.

We call such a function f *feasible*.

It turns out to be useful to introduce the concept of 'pre-feasible' function. A function $f : V \longrightarrow B_k$ is *pre-feasible* if

(8) (i) $f(r) = 1$;

(ii) for each arc $a = (u, w)$: if $|f(u)^{-1} \cdot \phi(a) \cdot f(w)| > 1$ then $\phi(a) = f(u)yf(w)^{-1}$ for some word y.

Pre-feasibility behaves nicely with respect to the partial order $\leq$ on the set B_k^V of all functions $f : V \longrightarrow B_k$ induced by the partial order $\leq$ on B_k as: $f \leq g \Leftrightarrow f(v) \leq g(v)$ for each $v \in V$. It is easy to see that B_k^V forms a lattice if we add an element ∞ at infinity. Let $\wedge$ and $\vee$ denote the meet and join. Then:

PROPOSITION 1. *If f_1 and f_2 are pre-feasible, then so is $f := f_1 \wedge f_2$.*

Proof. Clearly $f(r) = 1$. Suppose $|f(u)^{-1} \cdot \phi(a) \cdot f(w)| > 1$ for some arc $a = (u, w)$. We show $\phi(a) = f(u)yf(w)^{-1}$ for some y. By (4) we may assume by symmetry that $f(u) \not\leq \phi(a) \cdot f(w)$. Since $f(w) = f_1(w) \wedge f_2(w)$, there is an $i \in \{1, 2\}$ such that $f(u)^{-1} \cdot \phi(a) \cdot f_i(w)$ contains $f(u)^{-1} \cdot \phi(a) \cdot f(w)$ as a beginning segment. Without loss of generality, $i = 1$. So $|f(u)^{-1} \cdot \phi(a) \cdot f_1(w)| > 1$. As $f(u) \not\leq \phi(a) \cdot f(w)$, by Lemma 2, the first symbols of $f(u)^{-1}$ and $f(u)^{-1} \cdot \phi(a) \cdot f(w)$ are equal. Since $f(u)^{-1} \cdot \phi(a) \cdot f(w) \leq f(u)^{-1} \cdot \phi(a) \cdot f_1(w)$, it follows that the first symbols of $f(u)^{-1}$ and $f(u)^{-1} \cdot \phi(a) \cdot f_1(w)$ are equal. So $f_1(u)^{-1} \cdot \phi(a) \cdot f_1(w)$ contains $f(u)^{-1} \cdot \phi(a) \cdot f_1(w)$ as segment. Hence $|f_1(u)^{-1} \cdot \phi(a) \cdot f_1(w)| > 1$. As f_1 is pre-feasible, $\phi(a) = f_1(u)y'f_1(w)^{-1}$ for some y'. Since $f(u) \leq f_1(u)$ and $f(w) \leq f_1(w)$ this implies $\phi(a) = f(u)yf(w)^{-1}$ for some y. ∎

So for any function $f : V \longrightarrow B_k$ there exists a unique smallest pre-feasible function $\bar{f} \geq f$, provided there exists at least one pre-feasible function $g \geq f$. If no such g exists we set $\bar{f} := \infty$. In the next section we show that $\bar{f}$ can be found in polynomial time for any given f.

We first note:

PROPOSITION 2. *If $\bar{f}$ is finite then*

(9)
  (i) $f(r) = 1$;
  (ii) $|f(v)| < (\sigma + 1)|V|$ *for each vertex v;*
  (iii) $f(u) \leq \phi(a) \cdot f(w)$ *or* $f(w) \leq \phi(a)^{-1} \cdot f(u)$ *for each arc* $a = (u, w)$ *with* $|f(u)^{-1} \cdot \phi(a) \cdot f(w)| > 1$.

Proof. Let $\bar{f}$ be finite. Trivially $f(r) \leq \bar{f}(r) = 1$. Moreover, let $a_1, \ldots, a_t$ form a simple path from r to v. By induction on t one shows $|\bar{f}(v)| \leq (\sigma+1)t$. (Indeed, let $a_t = (u, v)$. If $|\bar{f}(u)^{-1} \cdot \phi(a) \cdot \bar{f}(v)| \leq 1$ then by induction $|\bar{f}(u)| \leq (\sigma+1)(t-1)$, and hence $|\bar{f}(v)| \leq \bar{f}(u)| + |\phi(a)| + 1 \leq (\sigma + 1)t$. If $|\bar{f}(u)^{-1} \cdot \phi(a) \cdot \bar{f}(v)| > 1$ then by (8) $\bar{f}(v)$ is a segment of $\phi(a)$ and hence $|\bar{f}(v)| \leq \sigma \leq (\sigma + 1)t$.) So $|f(v)| \leq |\bar{f}(v)| < (\sigma + 1)|V|$.

To see (iii), assume that $f(u) \not\leq \phi(a) \cdot f(w)$ and $f(w) \not\leq \phi(a^{-1}) \cdot f(u)$. So by Lemma 2 the first symbol of $f(u)^{-1} \cdot \phi(a) \cdot f(w)$ is equal to the first symbol of $f(u)^{-1}$. Similarly, the last symbol of $f(u)^{-1} \cdot \phi(a) \cdot f(w)$ is equal to the last symbol of $f(w)$. Since $f(u) \leq \bar{f}(u)$ and $f(w) \leq \bar{f}(w)$, it follows that $f(u)^{-1} \cdot \phi(a) \cdot f(w)$ is a segment of $\bar{f}(u)^{-1} \cdot \phi(a) \cdot \bar{f}(w)$. So $|\bar{f}(u)^{-1} \cdot \phi(a) \cdot \bar{f}(w)| > 1$. As $\bar{f}$ is pre-feasible this implies that $\phi(a) = \bar{f}(u)y\bar{f}(w)^{-1}$ for some y. Hence, since $f \leq \bar{f}$, $\phi(a) = f(u)y'f(w)^{-1}$ for some y'. So $f(u) \leq f(u)y' = \phi(a) \cdot f(w)$, contradicting our assumption. ∎

4. A subroutine finding $\bar{f}$

Let input $D = (V, A), r, \phi$ for the cohomology feasibility problem (6) be given. We may assume that for any arc $a = (u, w)$, $a^{-1} = (w, u)$ is also an arc of D, with $\phi(a^{-1}) = \phi(a)^{-1}$. Let moreover $f : V \longrightarrow B_k$ be given.

If f is pre-feasible output $\bar{f} := f$. If f violates (9) output $\bar{f} := \infty$. If none of these applies, perform the following iteration:

Iteration: Choose an arc $a = (u, w)$ satisfying $|f(u)^{-1} \cdot \phi(a) \cdot f(w)| > 1$ and $f(w) \not\leq \phi(a)^{-1} \cdot f(u)$. (Such an arc exists by (4). As (9)(iii) is not violated, we know $f(u) \leq \phi(a) \cdot f(w)$.)

Let x be obtained from $\phi(a) \cdot f(w)$ by deleting the last symbol; reset $f(u) := x$,

and iterate.

PROPOSITION 3. *At each iteration, $\sum_v |f(v)|$ strictly increases.*

Proof. Since $f(u) \leq \phi(a) \cdot f(w)$ and $|f(u)^{-1} \cdot \phi(a) \cdot f(w)| > 1$, x is strictly larger than the original $f(u)$. ∎

This directly implies:

PROPOSITION 4. *After at most $(\sigma + 1)|V|^2$ iterations the subroutine stops.*

Proof. After $(\sigma + 1)|V|^2$ iterations, by Proposition 3 there exists a vertex u such that $|f(u)| \geq (\sigma + 1)|V|$. Then (9)(ii) is violated. ∎

Moreover we have:

PROPOSITION 5. *In the iteration, resetting f does not change $\bar{f}$.*

Proof. We must show that $x \leq \bar{f}(u)$ if $\bar{f}$ is finite. If there exists y such that $\phi(a) = \bar{f}(u) y \bar{f}(w)^{-1}$ then

(10) $$f(w) \leq \bar{f}(w) \leq \bar{f}(w) y^{-1} = \phi(a)^{-1} \cdot \bar{f}(u) \leq \phi(a)^{-1} \cdot f(u)$$

(since $f(u) \leq \bar{f}(u) \leq \phi(a)$). This contradicts the choice of a in the iterations. Therefore, since $\bar{f}$ is pre-feasible, we know $|\bar{f}(u)^{-1} \cdot \phi(a) \cdot \bar{f}(w)| \leq 1$.

Since $f(w) \not\leq \phi(a^{-1}) \cdot f(u)$, by Lemma 2 the last symbol of $f(u)^{-1} \cdot \phi(a) \cdot f(w)$ is equal to the last symbol of $f(w)$. Hence (since $f(w) \leq \bar{f}(w)$) $f(u)^{-1} \cdot \phi(a) \cdot f(w) \leq f(u)^{-1} \cdot \phi(a) \cdot \bar{f}(w)$. Since $f(u) \leq \phi(a) \cdot f(w)$ it follows that $\phi(a) \cdot f(w) \leq \phi(a) \cdot \bar{f}(w)$. Let y be obtained from $\phi(a) \cdot \bar{f}(w)$ by deleting the last symbol. Then $x \leq y \leq \bar{f}(u)$, since $|\bar{f}(u)^{-1} \cdot \phi(a) \cdot \bar{f}(w)| \leq 1$. ∎

5. Algorithm for the cohomology feasibility problem

Let input $D = (V, A), r, \phi$ for the cohomology feasibility problem (6) be given. Again we may assume that for each arc $a = (u, w)$, $a^{-1} = (w, u)$ is also an arc, with $\phi(a^{-1}) = \phi(a)^{-1}$. We find a feasible function f (if there is one) as follows.

Let W be the set of pairs (v, x) with $v \in V$ and $x \in B_k$ such that there exists an arc $a = (v, w)$ with $1 \neq x \leq \phi(a)$. For every $(v, x) \in W$ let $f_{v,x}$ be the function defined by: $f_{v,x}(v) := x$ and $f_{v,x}(v') := 1$ for each $v' \neq v$. Let E be the set of pairs $\{(v, x), (v', x')\}$ from W for which $\bar{f}_{v,x} \vee \bar{f}_{v',x'}$ is finite and pre-feasible. Let E' be the set of pairs $\{(u, x), (w, z)\}$ from W for which there is an arc $a = (u, w)$ with $\phi(a) = xz^{-1}$. We search for a subset X of W such that each pair in X belongs to E and such that X intersects each pair in E'. This is a special case of the 2-satisfiability problem, and hence can be solved in polynomial time.

PROPOSITION 6. *If X exists then the function $f := \bigvee_{(v,x) \in X} \bar{f}_{v,x}$ is feasible. If X does not exist then there is no feasible function.*

Proof. First assume X exists. Since $\bar{f}_{v,x} \vee \bar{f}_{v',x'}$ is finite and pre-feasible for each two $(v, x), (v', x')$ in X, f is finite and $f(r) = 1$. Moreover, suppose $|f(u)^{-1} \cdot \phi(a) \cdot$

$f(w)| > 1$ for some arc $a = (u, w)$. By definition of f there are $(v, x), (v', x') \in X$ such that $f(u) = \bar{f}_{v,x}(u)$ and $f(w) = \bar{f}_{v',x'}(w)$ for $(v, x), (v', x') \in X$. As $\bar{f}_{v,x} \vee \bar{f}_{v',x'}$ is pre-feasible, $\phi(a) = \bar{f}_{v,x}(u) y \bar{f}_{v',x'}(w)^{-1}$ for some y. Then $|y| > 1$. Split $y = bc^{-1}$ with b and c nonempty. Then $(u, f(u)b) \in X$ or $(w, f(w)c) \in X$ since X intersects each pair in E'. If $(u, f(u)b) \in X$ then $f(u)b = f_{u,f(u)b}(u) \leq \bar{f}_{u,f(u)b}(u) \leq f(u)$, a contradiction. If $(w, f(w)c) \in X$ one obtains similarly a contradiction.

Assume conversely that there exists a feasible function f. Let X be the set of pairs $(v, x) \in X$ with the property that $x \leq f(v)$. Then X intersects each pair in E'. For suppose that for some arc $a = (u, w)$ with $\phi(a) = xz^{-1}$ and $x \neq 1 \neq z$, one has $(u, x) \notin X$ and $(w, z) \notin X$, that is, $x \not\leq f(u)$ and $z \not\leq f(w)$. This however implies $|f(u)^{-1} \cdot \phi(a) \cdot f(w)| \geq 2$, a contradiction.

Moreover, each pair in X belongs to E. For let $(v, x), (v', x') \in X$. We show that $\{(v, x), (v', x')\} \in E$, that is, $f' := \bar{f}_{v,x} \vee \bar{f}_{v',x'}$ is pre-feasible. As $\bar{f}_{v,x} \leq f$ and $\bar{f}_{v',x'} \leq f$, f' is finite and $f'(r) = 1$. Consider an arc $a = (u, w)$ with $|f'(u)^{-1} \cdot \phi(a) \cdot f'(w)| > 1$. We may assume $f'(u) = \bar{f}_{v,x}(u)$ and $f'(w) = \bar{f}_{v',x'}(w)$ (since $\bar{f}_{v,x}$ and $\bar{f}_{v',x'}$ themselves are pre-feasible). To show $\phi(a) = f'(u) y f'(w)^{-1}$ for some y, by (4) we may assume $f'(w) \not\leq \phi(a^{-1}) \cdot f'(u)$. So by Lemma 2, the last symbol of $f'(u)^{-1} \cdot \phi(a) \cdot f'(w)$ is equal to the last symbol of $f'(w)$.

Suppose now that $f'(u) \not\leq \phi(a) \cdot f'(w)$. Then by Lemma 2, the first symbol of $f'(u)^{-1} \cdot \phi(a) \cdot f'(w)$ is equal to the first symbol of $f'(u)^{-1}$. Since $f' \leq f$ this implies that $f'(u)^{-1} \cdot \phi(a) \cdot f'(w)$ is a segment of $f(u)^{-1} \cdot \phi(a) \cdot f(w)$. This contradicts the fact that $|f(u)^{-1} \cdot \phi(a) \cdot f(w)| \leq 1$.

So $f'(u) \leq \phi(a) \cdot f'(w)$. As $\bar{f}_{v',x'}(u) \leq f'(u)$ and $|f'(u)^{-1} \cdot \phi(a) \cdot f'(w)| > 1$ it follows that $|\bar{f}_{v',x'}(u)^{-1} \cdot \phi(a) \cdot f'(w)| > 1$. As $f'(w) = \bar{f}_{v',x'}(w)$ we have $|\bar{f}_{v',x'}(u)^{-1} \cdot \phi(a) \cdot \bar{f}_{v',x'}(w)| > 1$. As $\bar{f}_{v',x'}$ is pre-feasible, $\phi(a) = \bar{f}_{v',x'}(u) y \bar{f}_{v',x'}(w)^{-1}$ for some y. So $f'(u) \leq \phi(a) \cdot f'(w) = \bar{f}_{v',x'}(u)y$. Hence $\bar{f}_{v',x'}(u)y = f'(u)y'$ for some y'. It follows that $\phi(a) = f'(u)y' f'(w)^{-1}$. ∎

Thus we have:

THEOREM 1. *The cohomology feasibility problem for free boolean groups is solvable in time bounded by a polynomial in $|A| + \sigma + k$.*

6. Graphs on surfaces and homologous functions

Let $G = (V, E)$ be an undirected graph embedded in a compact surface. For each edge e of G choose arbitrarily one of the faces incident with e as the *left-hand face* of e, and the other as the *right-hand face* of e. (They might be one and the same face.) Let $\mathcal{F}$ denote the set of faces of G, and let R be one of the faces of G. We call two functions $\phi, \psi : E \longrightarrow B_k$ *R-homologous* if there exists a function $f : \mathcal{F} \longrightarrow B_k$ such that

(11)   (i) $f(R) = 1$;

                  (ii) $f(F)^{-1} \cdot \phi(e) \cdot f(F') = \psi(e)$ for each edge e, where F and F' are the left-hand and right-hand face of e respectively.

The relation to cohomologous is direct by duality. The *dual* graph $G^* = (\mathcal{F}, E^*)$ of G has as vertex set the collection $\mathcal{F}$ of faces of G, while for any edge e of G there is an edge e^* of G^* connecting the two faces incident with e. Let D^* be the directed graph obtained from G^* by orienting each edge e^* from the left-hand face of e to the right-hand face of e. Define for any function ϕ on E the function ϕ^* on E^* by $\phi^*(e^*) := \phi(e)$ for each $e \in E$. Then ϕ and ψ are R-homologous (in G), if and only if ϕ^* and ψ^* are R-cohomologous (in D^*).

7. Enumerating homology classes

Let $G = (V, E)$ be an undirected graph embedded in a surface and let $r, s \in V$, such that no loop is attached at r or s. We call a collection $\Pi = (P_1, \ldots, P_k)$ of $r - s$ walks an $r - s$ *join* (of *size* k) if:

(12)   (i) each P_i traverses r and s only as first and last vertex respectively;

            (ii) each edge is traversed at most once by the $P_1, \ldots, P_k$;

            (iii) P_i does not cross itself or any of the other P_j;

            (iv) $P_1, \ldots, P_k$ occur in this order cyclically at r.

Note that any solution of (2) can be assumed to be an $r - s$ join.

For any $r - s$ join $\Pi = (P_1, \ldots, P_k)$ let $\phi_\Pi : E \longrightarrow B_k$ be defined by:

(13) $\phi_\Pi(e) := g_i$ if walk P_i traverses e $(i = 1, \ldots, k)$;
                          := 1   if e is not traversed by any of the P_i.

Let R be one of the faces of G. Note that if ϕ is R-homologous to ϕ_Π then for each vertex $v \neq r, s$ we have

(14) $$\phi(e_1)^{\varepsilon_1} \cdot \ldots \cdot \phi(e_t)^{\varepsilon_t} = 1,$$

where $F_0, e_1, F_1, \ldots, F_{t-1}, e_t, F_t$ are the faces and edges incident with v in cyclic order (with $F_t = F_0$), and where $\varepsilon_j := +1$ if F_{j-1} is the left-hand face of e_j and F_j is the right-hand face of e_j, and $\varepsilon_j := -1$ if F_{j-1} is the right-hand face of e_j and F_j is the left-hand face of e_j. (If $F_{j-1} = F_j$ we should be more careful.) This follows from the fact that (14) holds for $\phi = \phi_\Pi$ and that (14) is invariant for R-homologous functions.

We now consider the following problem:

(15) given: a connected undirected graph cellularly embedded on a surface S, vertices r, s of G, such that $G - \{r, s\}$ is connected and r and s are not connected by an edge, a face R of G, and a natural number k;

find: functions $\phi_1, \ldots, \phi_N : E \longrightarrow B_k$ such that for each $r - s$ join Π of size k, ϕ_Π is R-homologous to at least one of $\phi_1, \ldots, \phi_N$.

(A graph is *cellularly embedded* if each face is homeomorphic with an open disk.)

THEOREM 2. *For any fixed surface S, problem (15) is solvable in time bounded by a polynomial in $|V| + |E|$.*

Proof. If e is any edge connecting two different vertices $\neq r, s$, we can contract e. Any solution of (15) for the modified graph directly yields a solution for the original graph (by (14)). So we may assume $V = \{r, s, v\}$ for some vertex v. Similarly, we may assume that G has no loops that bound an open disk.

Call two edges *parallel* if and only if they form the boundary of an open disk in S not containing R. Let p be the number of parallel classes and let f' denote the number of faces that are bounded by at least three edges. So $2p \geq 3f'$. By Euler's formula, $4 + f' \geq p + \chi(S)$, where $\chi(S)$ denotes the Euler characteristic of S. This implies $12 + 2p \geq 12 + 3f' \geq 3p + 3\chi(S)$ and hence $p \leq 12 - 3\chi(S)$. That is, for fixed S, p is bounded.

Let E' be a subset of E containing one edge from every parallel class. Note that any B_k-valued function on E is R-homologous to a B_k-valued function that has value 1 on all edges not in E'.

Let $\Pi = (P_1, \ldots, P_k)$ be an $r - s$ join such that no P_i traverses two edges e, e' consecutively that are parallel. For any 'path' e, v, e' in E' of length two, with e and e' incident with vertex v and e and e' not parallel, let $f(\Pi, e, v, e')$ be the number of times the P_i contain $\tilde{e}, v, \tilde{e'}$, for some $\tilde{e}$ parallel to e and some $\tilde{e'}$ parallel to e'. (Here e or e' is assumed to have an orientation if it is a loop.)

Now up to R-homology and up to a cyclic permutation of the indices of $P_1, \ldots, P_k$, Π is fully determined by the numbers $f(\Pi, e, v, e')$. This follows directly from the fact that the P_i do not have (self-)crossings.

So to enumerate $\phi_1, \ldots, \phi_N$ it suffices to choose for each path e, v, e' a number $g(e, v, e') \leq |E|$. Since $|E'| = p \leq 9 - 3\chi(S)$ there are at most $(|E|+1)^{(12-3\chi(S))^2}$ such choices. For each choice we can find in polynomial time an $r - s$ join Π with $f(\Pi, e, v, e') = g(e, v, e')$ for all e, v, e' if it exists. Enumerating the ϕ_Π gives the required enumeration. ∎

8. Induced circuits

THEOREM 3. *For each fixed surface S, there is a polynomial-time algorithm that gives for any graph $G = (V, E)$ embedded on S and any two vertices r, s of*

G *a maximum number of* $r-s$ *paths each two of which form an induced circuit.*

Proof. It suffices to show that for each fixed natural number k we can find in polynomial time k $r-s$ paths each two of which form an induced circuit, if they exist.

We may assume that $G - \{r,s\}$ is connected, that r and s are not connected by an edge, and that G is cellularly embedded. Choose a face R of G arbitrarily. By Theorem 2 we can find in polynomial time a list of functions $\phi_1, \ldots, \phi_N : A \longrightarrow B_k$ such that for each $r-s$ join Π, ϕ_Π is R-homologous to at least one of the ϕ_j.

Consider the (directed) dual graph $D^* = (\mathcal{F}, A^*)$ of G (see Section 6). We extend D^* to a graph $D^+ = (\mathcal{F}, A^+)$ as follows.

For every pair of vertices F, F' of D^* and every $F - F'$ path π (not necessarily directed) on the boundary of one face or of two adjacent faces of D^*, extend the graph with an arc a_π from F to F'. (Note that there are only a polynomially bounded number of such paths.) For each $\phi : A \longrightarrow B_k$ define $\phi^+ : A^+ \longrightarrow B_k$ by $\phi^+(e^*) := \phi(e)$ and

$$(16) \qquad \phi^+(a_\pi) := \phi(e_1)^{\varepsilon_1} \cdot \ldots \cdot \phi(e_t)^{\varepsilon_t}$$

for any path $\pi = (e_1^*)^{\varepsilon_1} \ldots (e_t^*)^{\varepsilon_t}$. (Here $\varepsilon_1, \ldots, \varepsilon_t \in \{+1, -1\}$.)

By Theorem 1 we can find, for each $j = 1, \ldots, N$ in polynomial time a function ϑ satisfying

$$(17) \qquad \text{(i) } \vartheta \text{ is } R\text{-cohomologous to } \phi_j^+ \text{ in } D^+, \text{ and}$$
$$\qquad \text{(ii) } |\vartheta(b)| \leq 1 \text{ for each arc } b \text{ of } D^+,$$

provided that such a ϑ exists.

If we find a function ϑ, for $i = 1, \ldots, k$ let Q_i be a shortest $r-s$ path traversing only the set of edges e of G with $\vartheta(e^*) = g_i$. If such paths $Q_1, \ldots, Q_k$ exist, and any two of them form an induced circuit, we are done (for the current value of k).

We claim that, doing this for all $\phi_1, \ldots, \phi_N$, we find paths as required, if they exist. For let $\Pi := (P_1, \ldots, P_k)$ form a collection of k $r-s$ paths any two of which form an induced circuit. Since Π is an $r-s$ join, there exists a $j \in \{1, \ldots, N\}$ such that ϕ_Π and ϕ_j are R-homologous.

We first show that there exists a function ϑ satisfying (17), viz. $\vartheta := \phi_\Pi^+$. To see this, we first show that ϕ_Π^+ is R-cohomologous to ϕ_j^+ in D^+. Indeed, ϕ_Π and ϕ_j are R-homologous in G. Hence there exists a function $f : \mathcal{F} \longrightarrow B_k$ such that $f(R) = 1$ and such that

$$(18) \qquad f(F)^{-1} \cdot \phi_\Pi(e) \cdot f(F') = \phi_j(e)$$

for each edge e, where F and F' are the left-hand and right-hand face of e respectively. This implies:

$$(19) \qquad f(F)^{-1} \cdot \phi_\Pi^+(e^*) \cdot f(F') = \phi_j^+(e^*).$$

Moreover, for every pair of vertices F_0, F_t of D^* and every $F_0 - F_t$ path $\pi = (e_1^*)^{\varepsilon_1} \ldots (e_t^*)^{\varepsilon_t}$ in D^* on the boundary of at most two faces of D^* we have (assuming $(e_i^*)^{\varepsilon_i}$ runs from F_{i-1} to F_i for $i = 1, \ldots, t$):

(20)
$$\begin{aligned}
&f(F_0)^{-1} \cdot \phi_\Pi^+(a_\pi) \cdot f(F_t) \\
&= (f(F_0)^{-1} \cdot \phi_\Pi(e_1)^{\varepsilon_1} f(F_1)) \cdot (f(F_1)^{-1} \cdot \phi_\Pi(e_2)^{\varepsilon_2} f(F_2)) \cdot \\
&\quad \ldots \cdot (f(F_{t-1})^{-1} \cdot \phi_\Pi(e_t)^{\varepsilon_t} f(F_t)) \\
&= \phi_j(e_1)^{\varepsilon_1} \cdot \phi_j(e_2)^{\varepsilon_2} \cdot \ldots \cdot \phi_j(e_t)^{\varepsilon_t} = \phi_j^+(a_\pi).
\end{aligned}$$

So ϕ_Π^+ and ϕ_j^+ are R-cohomologous.

Next we show that $|\phi_\Pi^+(b)| \le 1$ for each arc b of D^+. Indeed, for any edge e of G we have $\phi_\Pi^+(e^*) = \phi_\Pi(e) \in \{1, g_1, \ldots, g_k\}$. So $|\phi_\Pi^+(e^*)| \le 1$. Moreover, for any path $\pi = (e_1)^{\varepsilon_1} (e_2)^{\varepsilon_2} \ldots (e_t)^{\varepsilon_t}$ as above, $\phi_\Pi^+(a_\pi) = \phi_\Pi(e_1)^{\varepsilon_1} \cdot \ldots \cdot \phi_\Pi(e_t)^{\varepsilon_t}$. Since there exist two vertices v', v'' of G such that each of $e_1, \ldots, e_t$ is incident with at least one of v', v'', we know that there exists at most one $i \in \{1, \ldots, k\}$ such that P_i traverses at least one of the edges $e_1, \ldots, e_t$. Hence there is at most one generator occurring in $\phi_\Pi(e_1)^{\varepsilon_1} \cdot \ldots \cdot \phi_\Pi(e_t)^{\varepsilon_t}$. That is, $|\phi_\Pi^+(a_\pi)| \le 1$. This shows that $\vartheta := \phi_\Pi^+$ satisfies (17).

Conversely, we must show that if ϑ satisfies (17), then ϑ gives paths $Q_1, \ldots, Q_k$ as above. Indeed, since ϑ is R-cohomologous to ϕ_Π^+, for each $i = 1, \ldots, k$, the set of edges e of G with $\vartheta(e^*) = g_i$ contains an $r - s$ path (since $\zeta := \phi_\Pi^+$ has the property that the subgraph $(V, \{e \in E | \zeta(e^*) \text{ contains the symbol } g_i \text{ an odd number of times}\})$ of G has even degree at each vertex except at r and s, and since this property is maintained under R-cohomology). Choose for each i such a path Q_i. Suppose that, for some $i \ne j$, there exists an edge $e = \{v, v'\}$ with Q_i traversing v and Q_j traversing v' ($v, v' \notin \{r, s\}$). Then there exist faces F_0 and F_t of G and an $F_0 - F_t$ path $\pi = (e_1)^{\varepsilon_1} \ldots (e_t)^{\varepsilon_t}$ in D^* on the boundary of the faces v and v' of D^* such that $\vartheta(e_1^*)^{\varepsilon_1} \cdot \ldots \cdot \vartheta(e_t^*)^{\varepsilon_t}$ contains both symbol g_i and symbol g_j. Now

(21) $$\vartheta(a_\pi) = \vartheta(e_1^*)^{\varepsilon_1} \cdot \ldots \cdot \vartheta(e_t^*)^{\varepsilon_t},$$

since this equation is invariant under R-cohomology and since it holds when ϑ is replaced by ϕ_Π^+. So $\vartheta(a_\pi)$ contains both symbol g_i and g_j. This contradicts the fact that $|\vartheta(a_\pi)| \le 1$.

So there is no edge connecting internal vertices of Q_i and Q_j. Replacing each Q_i by a chordless path Q_i' in G that uses only vertices traversed by Q_i, we obtain paths as required. ∎

We refer to [4] for an extension of the methods described above.

Acknowledgement. I am grateful to Paul Seymour for very carefully reading preliminary versions of this paper and for giving several helpful suggestions.

References

1. D. Bienstock, private communication, 1989.
2. C. McDiarmid, B. Reed, A. Schrijver, and B. Shepherd, Induced circuits in planar graphs, Report BS-R9106, CWI, Amsterdam, 1991.
3. A. Schrijver, Finding k disjoint paths in directed planar graphs, Report BS-R9206, CWI, Amsterdam, 1992.
4. A. Schrijver, Disjoint paths in graphs on surfaces and combinatorial group theory, preprint, 1991.

CWI, Kruislaan 413, 1098 SJ Amsterdam, The Netherlands,
and
Department of Mathematics, University of Amsterdam, Plantage Muidergracht 24, 1018 TV Amsterdam, The Netherlands.
E-mail address: lex@cwi.nl

Tree-Representation of Directed Circuits

András Frank and Tibor Jordán

ABSTRACT. We prove that a strongly connected directed graph $G = (V, E)$ has a spanning tree T so that each fundamental circuit belonging to T is a directed circuit if and only if G has precisely $|E| - |V| + 1$ directed circuits. Another characterization of such directed graphs will also be provided in terms of forbidden minors.

1. Introduction, Preliminaries

A *join (strong join)* J of an undirected graph is a subset of edges so that $|J \cap C| \leq |C|/2$ ($|J \cap C| < |C|/2$) for every circuit C of the graph.

The investigations of joins was initiated by P. Sole and T. Zaslavsky while the problem of determining a maximum strong join is due to D. Welsh [1990]. In [Frank, 1992] a min-max theorem was provided for the maximum cardinality of a join along with a polynomial time algorithm to compute the largest join. A. Fraenkel and M. Loebl [1991] proved that the maximum strong join problem is NP-complete even if the graph is planar and bipartite. We proved in [Frank, Jordán and Szigeti, 1992] that for every graph the maximum cardinality of a strong join is at most $\lfloor (|V| - 1)/2 \rfloor$ and provided an algorithm to decide if a given bipartite graph is extreme, that is, it attains this bound.

Suppose that a bipartite graph $B = (U, V; F)$ has a perfect matching M so that for every element e of M an edge parallel to e also belongs to G. In this case clearly no element of M may belong to any strong join and the maximum strong join problem can be reformulated as follows.

1991 Mathematics Subject Classification. 05C38

This paper is in final form and no version of it will be submitted for publication elsewhere.

Define a directed graph $G = (V, E)$ so that $uv \in E$ if $uv' \in F$ where v' denotes the node in U for which $vv' \in M$. It is not difficult to prove that B is extreme if and only if G has a spanning tree T so that every fundamental circuit belonging to T is a directed circuit. (*A fundamental circuit* is one having precisely one non-tree edge). We shall call such a tree a *circuit-representing tree* or, in short, a *CR-tree*. It is also true that the set of edges in B corresponding to the edges of a CR-tree of G is a maximum strong join of B. The digraph D_2 on two nodes with two parallel edges in both directions clearly has no CR-tree.

The purpose of the present paper is to provide characterizations for digraphs having a CR-tree as well as a polynomial time algorithm to find a CR-tree if there is any.

Let $G = (V, E)$ be a directed graph. For $X \subseteq V$ let $\delta(X)$ denote the number of edges leaving X. G is called *strongly connected* if there is a directed path from u to v for every $u, v \in V$. This is equivalent to saying that $\delta(X) \geq 1$ for every $\emptyset \neq X \subset V$. We call a set X *tight* if $\delta(X) = 1$. Let T be a spanning tree of G and $e = xy$ an edge of T. Then $T - e$ has two components. Define $T(e)$ to be the node-set of the component of $T - e$ containing x. It is easy to see that T is a CR-tree if and only if $T(e)$ is tight for every edge e of T.

By an *ear-decomposition* of G we mean a sequence $\mathcal{P} := \{P_1, P_2, \ldots, P_t\}$ where P_1 is a circuit of G, each other P_i is a path in G so that each edge of G belongs to precisely one P_i ($i = 1, \ldots, t$) and precisely the end-nodes of P_i ($i = 2, \ldots, t$) belong to $P_1 \cup \ldots \cup P_{i-1}$. Each path P_i is supposed to be simple except that the two end-nodes may coincide. The number t of paths is called the *length* of the decomposition.

It is well-known that a digraph G has an ear-decomposition if and only if G is strongly connected. Moreover, for any strongly connected subgraph $H = (U, A)$ of G any ear-decomposition of H is the starting segment of an ear-decomposition of G. The length of an ear-decomposition depends only on the graph and equals $|E| - |V| + 1$. It also follows easily that every strongly connected digraph $G = (V, E)$ has at least $|E| - |V| + 1$ directed circuits.

2. Characterizations of CR-trees

Let $G = (V, E)$ be a strongly connected digraph. We call a simple directed path $P := \{v_0, e_1, v_1, e_2, \ldots, e_k, v_k\}$ *unique* if P is the only simple path from v_1 to v_k. We consider the empty set and a path $\{v_0\}$ as *trivial* unique paths.

PROPOSITION 2.1 *A non-trivial path P is unique if and only if there is a family $\{X_1, \ldots, X_k\}$ of tight sets for which $X_1 \subset X_2 \subset \ldots \subset X_k$ and e_i leaves X_i for every i, $1 \leq i \leq k$.*

Proof. Suppose first the existence of such a family. Let, indirectly, P' be another simple path from v_1 to v_k. Then there is a first edge e_i of P not belonging to P'. Since there is an edge e of P' leaving X_i, we conclude that $o(X_i) \geq 2$, contradicting the tightness of X_i.

Assume now that P is unique. For each $i, 1 \leq i \leq k$ let X_i denote the set of nodes reachable from $\{v_1, \ldots, v_{i-1}\}$ without using the edge e_i. From the definition $X_i \subseteq X_{i+1}$. We claim that $v_j \notin X_i$ for $i < j$, or equivalently, there is no path in $G - e_i$ from $\{v_1, \ldots, v_{i-1}\}$ to $\{v_i, \ldots, v_{k+1}\}$. Indeed, if such a path P' existed, choose it minimal and let s and t denote the first and last node of P', respectively. By the minimality no internal node of P' belongs to P. Hence by replacing the segment of P from s to t by P' we would obtain another simple path from v_1 to v_{k+1}, contradicting the uniqueness of P.

Since the only edge leaving X_i is e_i, each X_i is tight and the family $\{X_1, \ldots, X_k\}$ satisfies the requirements.

□□□

Note that the proof above can easily be turned into a polynomial-time algorithm that either finds two distinct paths from v_1 to v_k or constructs the family $\{X_1, \ldots, X_k\}$ in question.

Let us call an edge $e = xy \in E$ *uni-cyclic* if e is contained in exactly one directed circuit and *multi-cyclic* otherwise. We call an edge $e = xy$ *essential* if $G - e$ is not strongly connected. Otherwise e is *non-essential*. In other words, $e = xy \in E$ is uni-cyclic if there is a unique path from y to x and e is essential if $\{x, e, y\}$ is a unique path. Therefore these properties can be tested in polynomial-time.

PROPOSITION 2.2 *Every directed subpath of a CR-tree T is unique.*

Proof. Let $P := \{v_0, e_1, v_1, e_2, \ldots, e_k, v_k\}$ be a subpath of T. Recall that $T(e)$ denotes the node-set of the component of $T - e$ containing the tail of e. Since T is a CR-tree, the only edge leaving $T(e)$ is e, that is, $T(e)$ is tight for each $e \in T$. Hence the family $\{T(e_i) : i = 1, \ldots, k\}$ satisfies the properties in Proposition 2.1 and therefore P is unique.

□

THEOREM 2.3 *Let T be a spanning tree of a strongly connected digraph $G = (V, E)$. The following are equivalent.*

(a) T is a CR-tree,
(b) Every directed circuit is a fundamental circuit,
(c) Every non-tree edge is uni-cyclic.

Proof. (a→b) Let T be a CR-tree. Suppose (b) fails to hold, that is, there is a directed circuit C which is not fundamental. Then, for an edge $e = xy \in C - T$, the subpath of T from y to x is directed but not unique as $C - e$ is another path from y to x. This contradicts Proposition 2.2.

(b→c) Let C be an arbitrary circuit containing a non-tree edge e. By (b) C is the fundamental circuit belonging to e, that is, e is uni-cyclic.

(c→a) If (a) is not true, then there is a non-tree edge $e = xy$ so that its fundamental circuit is not directed. Then there exists a circuit C containing e and this C contains another non-tree edge $f = uv$. Since both e and f are uni-cyclic, both paths $C - e$ and $C - f$ are unique. By Proposition 2.1 there is a tight set X (resp., Y) so that e enters X (f enters Y) and f (e) is the only edge leaving X (Y). Therefore no edge leaves $X \cup Y$ and $X \cap Y$. Since G is strongly connected, $X \cup Y = V$ and $X \cap Y = \emptyset$, that is, $X = V - Y$. We can conclude that e is the only edge entering X and f is the only edge leaving X contradicting the fact that T is a spanning tree. □□□

3. Graphs with CR-trees

In this section we provide three characterizations for digraphs $G = (V, E)$ having CR-trees. We can assume that there is no cut-edge in G. Indeed, any cut-edge e belongs to every spanning tree and to no directed circuit. Hence G has a CR-tree precisely if G/e has a CR-tree where G/e denotes a digraph arising from G by contracting e.

A second observation is that G cannot have a CR-tree if G is not strongly connected. Indeed, let T be a CR-tree of G. Every edge of $G - T$ belongs to a directed circuit, namely to its fundamental circuit. Since there is no cut-edge, every element of T belongs to a certain fundamental circuit. Hence G is strongly connected.

Henceforth we assume that G is strongly connected.

THEOREM 3.1 *A strongly connected digraph $G = (V, E)$ has a CR-tree if and only if the set K of multi-cyclic edges forms a forest. Moreover if K is a forest, any spanning tree including K is a CR-tree.*

Proof. Suppose first that T is a CR-tree of G. By Theorem 2.3 every non-tree edge is uni-cyclic, that is, K is a subset of T, and hence K is a forest.

Conversely, suppose that K is a forest. Let T be any spanning tree including K. Now property (c) in Theorem 2.3 holds and hence T is a CR-tree.

□□□

Since we can check in polynomial time if an edge is uni-cyclic or not, Theorem 3.1 suggests an algorithm to decide if a digraph has a CR-tree. A disadvantage of the theorem is that the necessity of the condition is not very straightforward. We provide two other characterizations to overcome this drawback. We will need the following:

PROPOSITION 3.2 *If P is a unique path in G and G has a CR-tree, then G has a CR-tree including P.*

Proof. Let $P := \{v_0, e_1, v_1, e_2, \ldots, e_k, v_k\}$. By Proposition 2.1 every subpath of P is unique. By induction we may assume that there is a CR-tree T of G containing each e_i ($i = 1, \ldots, k-1$). By Theorem 2.3 each non-tree edge is uni-cyclic.

If $e_k \in T$, we are done. So suppose that $e_k \notin T$ and let C denote the fundamental circuit belonging to e_k. Since T is a CR-tree, C is directed. By Proposition 2.1 there is a tight set X containing $v_1, \ldots, v_k$ and not containing v_{k+1}. There is a (unique) edge $f \in C - P$ entering X. Because e_k is the only edge leaving X and G has only fundamental circuits, f is uni-cyclic. Hence $T' := T - f + e_k$ is a tree containing all uni-cyclic edges. By Theorem 2.3 T' is a CR-tree and includes P.

□

For a strongly connected digraph $G = (V, E)$ denote $\kappa(G) := |E| - |V| + 1$. Let $\mathcal{P} := \{P_1, P_2, \ldots, P_t\}$ be an ear-decomposition of G and let G_i ($i = 1, \ldots, t$) denote the union of the first i members of $\mathcal{P}$. By induction it follows that $\kappa(G_i) = i$ for $1 \leq i \leq t$. Since G_i is strongly connected, every P_i is a subset of a directed circuit C_i of G_i. Let $R_i := C_i - P_i$ ($i = 2, 3, \ldots, t$). Clearly, each G_i has at least $\kappa(G_i)$ directed circuits.

THEOREM 3.3 *For a strongly connected digraph $G = (V, E)$ the following are equivalent:*
(a) G has a CR-tree,

(b) G has precisely $\kappa(G)$ directed circuits,
(c) R_i is a unique path in G_{i-1} ($i = 2, 3, \ldots, t$).

Proof. The equivalence of (b) and (c) is straightforward.

($a \to b$) If G has a CR-tree T, then every directed circuit of G is a fundamental circuit by Theorem 2.3. Since there are $\kappa(G)$ non-tree edges in G, the total number of directed circuits is $\kappa(G)$.

($b \to a$) Apply induction on $t = \kappa(G)$. If $\kappa(G) = 1$, then G is a circuit and $G - e$ is a CR-tree of G for any edge e of G. Let $\kappa(G) > 1$ and assume, by induction, that G_{t-1} has a CR-tree T_{t-1} and R_t is unique in G_{t-1}. By Proposition 3.2 there is a CR-tree T_{t-1} of G_{t-1} including R_t. Then $T := T_{t-1} \cup P_t - e$ is a CR-tree of G for any edge e of P_t.

□□□

Finally, we exhibit a minor-type characterization. Let us introduce three operations of a strongly connected graph $G = (V, E)$.

(α) Contracting a multi-cyclic edge e,
(β) Deleting a non-essential edge f,
(γ) Restriction to a strongly connected induced subgraph $G_\gamma = (V', E')$.

PROPOSITION 3.4 *If G has a CR-tree, then each of the operations (α), (β), (γ) results in a strongly connected digraph having a CR-tree.*

Proof. Let G_α, G_β, G_γ denote the resulting digraphs. Clearly, each of them is strongly connected. Let T be a CR-tree of G. By Theorem 3.1 e belongs to T. Hence T/e is a CR-tree of G_α.

By Proposition 2.2 every edge of T is essential. Hence $f \notin T$ and T is a CR-tree of G_β, as well.

Finally, we show that the restriction T' of T to V' is a CR-tree. This is clearly true if T' is a tree. Suppose T' is not connected and let $X \subset V'$ be a set, $\emptyset \neq X \neq V$, so that there is no edge of T' connecting X and $V' - X$. Since G_γ is strongly connected, there is an edge e from X to $V' - X$. Let C be a directed circuit in G_γ containing e. Now C is not a fundamental circuit of G, therefore G cannot have a CR-tree by Theorem 2.3, a contradiction.

□□□

Recall that D_2 denotes the digraph on two nodes with two parallel edges in both directions.

THEOREM 3.5 *A strongly connected digraph $G = (V, E)$ has a CR-tree if and only if D_2 cannot be obtained from G by successively applying operations $(\alpha), (\beta), (\gamma)$.*

Proof. Since D_2 has no CR-tree, the preceding proposition prove the "only if" part.

Suppose now that G is a counter-example to the "if" part with a minimum number of edges. Then G has no CR-tree and cannot be reduced to D_2. Therefore

$$\text{none of } G_\alpha, G_\beta, G_\gamma \text{ can be reduced to } D_2. \tag{$*$}$$

Let $\mathcal{P} := \{P_1, P_2, \ldots, P_t\}$ be an ear-decomposition of G, as before. We use the notation of Theorem 3.3. Now $t > 1$. Let x and y denote the first and last node of P_t, respectively.

CLAIM 1 $x \neq y$ *and there are two paths Q_1, Q_2 in G_{t-1} from y to x.*

Proof. G_{t-1} arises from G by operation (γ). The minimality of G and (*) imply that G_{t-1} has a CR-tree. It follows that R_t cannot be unique in G_{t-1} for otherwise there is a CR-tree T' of G_{t-1} including R_t (by Proposition 3.2) and then $T' \cup P_t - e$ would be a CR-tree of G for any edge $e \in P_t$.

□

CLAIM 2 *Both Q_1 and Q_2 consist of one edge.*

Proof. Suppose, indirectly, that Q_1, say, has more than one edge. Let e and f be the first and last edge of Q_1, respectively. Then it is easy to check that at least one of these edges, say e, has the property that in G/e there are at least two paths from y to x. G/e arises from from G_{t-1}/e by adding P_t. By Theorem 3.3 G/e does not have a CR-tree.

On the other hand e is multi-cyclic in G since e belongs to a circuit of G_{t-1} and belongs to a circuit including P_t. By (*) and the minimality of G , $G_\alpha := G/e$ has a CR-tree, a contradiction.

□

Let e_i denote the only edge of Q_i ($i = 1, 2$) and let Q be a path in G_{t-1} from x to y. The union of P_t and Q is a circuit C. Clearly every edge of P_t and Q is multi-cyclic. First erase all nodes not in C (by operation (γ)). Apply then operation (α) to all but one edges of P_t and of Q. This way we get a digraph on two nodes with at least two parallel edges in both directions. In such a graph all edges are non-essential. Thus D_2 can be obtained by operation (β), contradicting the assumption on G.

□□□

References

[1990] A. Frank, Conservative weightings and ear-decompositions of graphs, Combinatorica, to appear.

[1991] A. Frank, T. Jordán and Z. Szigeti, On strongly conservative weightings, in preparation.

[1991] A.S. Fraenkel and M. Loebl, Complexity of circuit intersection in graphs, preprint.

[1986] L. Lovász, M. Plummer, Machting Theory, Akadémiai Kiadó Budapest and North-Holland Publishing Company.

[1991] P. Sole, Th. Zaslavsky, Covering radius, maximality and decoding of the cycle code of a graph, Discrete Mathematics, to appear.

[1990] D. Welsh, oral communication,

András Frank, Research Institute for Discrete Mathematics, Institute for Operations Research, University of Bonn, Nassestr. 2, Bonn-1, Germany D-5300. On leave from: Department of Computer Science, Eötvös University, Múzeum krt. 6-8, Budapest, Hungary, H-1088.
E-mail address: or392 at dbnuor1.bitnet

Tibor Jordán, Department of Computer Science, Eötvös University, Múzeum krt. 6-8, Budapest, Hungary, H-1088.
E-mail address: H3962jor at ella.hu

Intercyclic Digraphs

WILLIAM MCCUAIG

August 27, 1992

ABSTRACT. A digraph G is intercyclic (arc-intercyclic) if G does not have two disjoint (arc-disjoint) directed cycles. We give a complete characterization of intercyclic and arc-intercyclic digraphs. Conjectures of Gallai, Younger, Kosaraju, and Metzlar follow from this result.

1. Introduction

A digraph is *intercyclic* if it does not have two disjoint dicycles (directed cycles). A digraph is *arc-intercyclic* if it does not have two arc-disjoint dicycles. In this paper we will give a good characterization of intercyclic and arc-intercyclic digraphs.

Dirac [2] characterized the simple 3-connected graphs which do not have two disjoint cycles. Lovász [9] extended this result to all graphs. This result showed that every graph has two disjoint cycles or a set T of at most 3 vertices such that $G - T$ is acyclic. Erdös and Pósa [3] proved the existence of a function f on the natural numbers such that every graph G contains k arc-disjoint cycles or $G - F$ is acyclic for some set F of at most $f(k)$ edges. Erdös and Pósa [4] also proved the existence of a function g on the natural numbers such that every graph G contains k disjoint cycles or $G - T$ is acyclic for some set T of at most $g(k)$ vertices.

Analogous results are conjectured for dicycles in digraphs. Younger [20] conjectures that for every k, there exists a (least) natural number $f(k)$ such that every digraph G contains k arc-disjoint dicycles or $G - F$ is acyclic for some set F of at most $f(k)$ arcs. This conjecture holds when restricted to planar digraphs, as follows from a theorem of Lucchesi and Younger [10]: the maximum number

---

1991 *Mathematics Subject Classification.* Primary 05C20, 05C38, 05C75; Secondary 05C10, 05C85.

Support from NSERC is gratefully acknowledged.

This paper is in final form and no version of it will be submitted for publication elsewhere.

of arc-disjoint dicycles in a planar digraph equals the minimum number of arcs meeting all dicycles. Younger also conjectures that for every k, there exists a (least) natural number $g(k)$ such that every digraph G contains k disjoint dicycles or $G - T$ is acyclic for some set T of at most $g(k)$ vertices. The existence of $g(2)$ was originally conjectured by Gallai [6]. Younger constructed an intercyclic and an arc-intercyclic digraph which showed that $f(2)$ and $g(2)$ are at least 3, and he conjectured that equality holds. (Younger's intercyclic digraph is the digraph D_7 defined in §3.) Kosaraju [7] also conjectured that $g(2) = 3$. As well, Kosaraju proved that if any 3 dicycles of a digraph have a common vertex, then all dicycles have a common vertex. It is interesting to note that if $f(k)$ or $g(k)$ exists, then both exist and are equal. This result was pointed out by Soares [13] and will be shown in §2.

Define a *k-transversal* of a digraph G to be a set T of k vertices of G such that $G - T$ is acyclic. Define a *k-arc transversal* of a digraph G to be a set S of k arcs of G such that $G - S$ is acyclic.

The conjecture of Kosaraju and Younger can be restated to say that every intercyclic digraph has a 3-transversal. A stronger version of this conjecture was given by Metzlar [12]. She conjectured that for every intercyclic digraph G, there is a function w from $V(G)$ into the nonnegative real numbers such that $\sum_{x \in V(G)} w(x) \leq 2.5$ and for every dicycle C, $1 \leq \sum_{x \in V(C)} w(x)$. Thomassen [15] has characterized the intercyclic digraphs with 2-transversals. Later, we will show how this characterization follows from a result due to Metzlar [12].

For every k, Thomassen [14] has shown that there exists a (least) natural number $\delta(k)$ such that every digraph with minimum outdegree (or minimum indegree) at least $\delta(k)$ has k disjoint dicycles. In particular, he showed $\delta(2) = 3$. As a consequence, every intercyclic digraph has $\delta^+ \leq 2$ and $\delta^- \leq 2$.

The problem of determining if two given arcs of a digraph are on a common dicycle has been shown to be NP-complete by Fortune, Hopcroft, and Wyllie [5]. Thomassen [16] proved that this problem can be solved in polynomial time when restricted to intercyclic digraphs. He proved that two arcs of an intercyclic digraph G are on a common dicycle if and only if, for every vertex v of G, $G - v$ has a dipath from the head of one of the arcs to the tail of the other. Earlier, Kostochka [8] had proven that if v is a vertex and e is an arc of a strongly-connected intercyclic digraph, then v and e are on a common dicycle.

In this paper we will give a complete characterization of intercyclic digraphs and arc-intercyclic digraphs. In the process, we will verify the conjectures of Gallai, Younger, Kosaraju, and Metzlar about these digraphs.

In §2, we describe the notation and terminology used in this paper and state some folklore theorems. We also prove a result of Metzlar. In §3, we state the main theorem which gives a characterization of intercyclic digraphs. In §4, we use the main theorem to prove Gallai's conjecture, the conjecture of Kosaraju and Younger, and Metzlar's conjecture. The proof of the main theorem is outlined in §5. The details of the proof are given in §6, §7, and §8. In §9, we outline a polynomial algorithm which either finds two disjoint dicycles in a digraph G or shows that G is intercyclic. The characterization of arc-intercyclic digraphs is given in §10.

2. Terminology and Known Results

We will use the notation and terminology of Bondy and Murty [1]. Let e be an arc from x to y. We say that x is the *tail* of e, y is the *head* of e, e is *incident to* y, e is *incident from* x, x is *adjacent to* y, and y is *adjacent from* x. An arc of G is a *loop* if it is incident to and incident from the same vertex. Arcs are *parallel* if they have the same tail and the same head. A digraph is *strict* if it has no loops or parallel arcs.

A digraph is *acyclic* if it has no dicycles. A *source* (respectively, *sink*) is a vertex which is only incident with outgoing arcs (respectively, incoming arcs). A vertex with both incoming and outgoing arcs is called an *intermediate vertex*. A sequence $x_1, \ldots, x_n$ is a *source sequence* of an acyclic digraph G if $V(G) = \{x_1, \ldots, x_n\}$ and x_i is a source of $G - \{x_j | 1 \leq j < i\}$, $i = 1, \ldots, n$.

A directed path will be called a *dipath*. An (x, y)-*dipath* is a dipath with origin x and terminus y. The other vertices on the path are called *internal vertices*. Let A and B be subsets of $V(G)$. An (A, B)-*dipath* is a dipath with origin in A and terminus in B which has *no internal vertices in* $A \cup B$. If H and K are subdigraphs of G, then an (H, K)-*dipath* is a $(V(H), V(K))$-dipath. We say that $(p_1, \ldots, p_k)$ is an $(x_1 \to y_1, \ldots, x_k \to y_k)$-*linkage* if p_i is an (x_i, y_i)-dipath for every i in $\{1, \ldots, k\}$ and $p_1, \ldots, p_k$ are pairwise disjoint. Internally disjoint dipaths which have the same origin and different termini (or vice versa) are called *openly disjoint*. We say that $(p_1, \ldots, p_k)$ is an $(x \to y_1, \ldots, x \to y_k)$-*fan* if p_i is an (x, y_i)-dipath for every i in $\{1, \ldots, k\}$ and $p_1, \ldots, p_k$ are openly disjoint. An $(x_1 \to y, \ldots, x_k \to y)$-fan is defined similarly.

A *path* of a digraph G is a subdigraph which corresponds to a path in the underlying undirected graph of G. If e is an arc of the path $x_1, \ldots, x_n$ and e is from x_i to x_{i-1} for some i in $\{2, \ldots, n\}$, then we say that e is a *backward arc* of the path. Let $P = x_1, \ldots, x_n$ be a dipath. If $1 \leq i \leq j \leq n$, then $P[x_i, x_j]$ is defined to be the dipath $x_i, \ldots, x_j$. We define $P(x_i, x_j]$ to be $P[x_i, x_j] - x_i$, $P[x_i, x_j)$ to be $P[x_i, x_j] - x_j$, and $P(x_i, x_j)$ to be $P[x_i, x_j] - \{x_i, x_j\}$.

A digraph G is *strongly-connected* if there is an (x, y)-dipath for every ordered pair (x, y) of vertices. A *strong component* of a digraph G is a maximal strongly-connected subdigraph of G. A strong component of G is *nontrivial* if it has at least one arc.

Let G and H be digraphs. We define $G \cup H$ to be the digraph with vertex set $V(G) \cup V(H)$ and arc set $A(G) \cup A(H)$. Let A, B, and C be sets of vertices. Define $G(A, B, C)$ to be the subdigraph of G with vertex set $A \cup B \cup C$ whose arc set consists of all arcs in G from a vertex in A to a vertex in B and all arcs in G from a vertex in B to a vertex in C.

Let X be a subset of the vertex set of a digraph G. Define a digraph G' as follows. The vertex set of G' is obtained from $V(G)$ by replacing every vertex x in X by two new vertices x^+ and x^-. The arc set of G' is obtained from $A(G)$ by replacing every arc e from u to v by an arc e', where e' is incident from u^+ if $u \in X$, e' is incident from u if $u \notin X$, e' is incident to u^- if $u \in X$, and e' is incident to v if $v \notin X$. We say that u *corresponds* to u^+ and u^- if $u \in X$, u *corresponds* to itself if $u \notin X$, and e *corresponds* to e'. We say that G' is obtained from G by *vertex division* and that x is *divided* into x^+ and x^-. Let H be a subdigraph of G. Let H' be the subdigraph of G' obtained from H by dividing the vertices in $V(H) \cap X$. We say that H and H' are *corresponding subdigraphs* of G and G'.

Suppose G' is obtained from G by dividing every vertex in X. Then $G' + \{x^-x^+ | x \in X\}$ is said to be obtained from G by *splitting* every vertex in X. If $x \in X$, then x^-, x^+, and x^-x^+ are said to *correspond* to x.

Let G be a digraph with an arc f from x to y. Define a digraph G/f as follows. The vertex set of G/f is obtained from $V(G)$ by replacing x and y by a new vertex w. We say that x and y *correspond* to w and that every other vertex of G *corresponds* to itself. The arc set of G/f is obtained from $A(G) - \{f\}$ by replacing every arc e by an arc e' so that they have corresponding heads and corresponding tails. We say that e *corresponds* to e' and that G/f is obtained from G by *contracting* f.

A graph which is embedded in the plane is called a *plane graph*. Let C be a cycle of a plane graph G. If $x_1, \ldots, x_k$ are vertices on C which occur in the given cyclic order around C, then we say that $x_1, \ldots, x_k$ is a *cyclic subsequence* of C. Let p_i be a path of G with origin x_i and termini y_i such that x_i and y_i are the only vertices of p_i on C, $i = 1, 2$. If the endpoints of p_1 and p_2 are distinct and x_1, x_2, y_1, y_2 is a cyclic subsequence of C, then we say that p_1 and p_2 are *skew paths* of C. If p_1 is contained in the closure of the interior of C, then we say p_1 is an *interior path* of C.

We now give three classical results and a corollary which are needed in the proof of the main theorem. The first is due to Whitney [19].

THEOREM 2.1. *Let G be a plane graph. If G is 2-connected, the face boundaries of G are cycles.*

The next theorem follows from the Jordan curve theorem which was first proven by Veblen [17].

THEOREM 2.2. *Let G be a plane graph. A cycle of G can not have two skew interior paths.*

The next theorem gives two versions of the fundamental theorem on connectivity which is due to Menger [11] and Whitney [18].

THEOREM 2.3. *Let G be a digraph and let X and Y be subsets of $V(G)$. There exist k disjoint (X,Y)-dipaths if and only if $G-Z$ has an (X,Y)-dipath for every subset Z of $V(G)$ of size less than k. Let x and y be vertices of G. There exist k arc-disjoint (x,y)-dipaths if and only if $G-F$ has an (x,y)-dipath for every subset F of $A(G)$ of size less than k.*

COROLLARY 2.1. *Let T be the set of sinks of an acyclic digraph G such that every nonsink is adjacent to at least n vertices. If $X \subseteq V(G)$ and $n \leq |X|$ then there exist n disjoint (X,T)-dipaths.*

PROOF. Suppose $S \subseteq V(G)$ and $G-S$ has no (X,T)-dipath. We want to show that $n \leq |S|$. Let W be the set of all vertices w for which there exists an (X,w)-dipath in $G-S$. If $X \subseteq S$, then $n \leq |X| \leq |S|$. If $X-S$ is nonempty, then so is W and we may choose a sink y of $G[W]$. Since $G-S$ has no (X,T)-dipath, $y \notin T$, and so y is adjacent to at least n vertices. If $yz \in A(G)$ and $z \notin S$, then $z \in W$. But then y is not a sink of $G[W]$. Hence, $N^+(y) \subseteq S$. Therefore, $n \leq |S|$. Now the result follows from Theorem 2.3. □

We now show that the functions f and g conjectured by Younger are closely related.

THEOREM 2.4. *If $f(k)$ or $g(k)$ exists, then both exist and are equal.*

PROOF. Suppose K is obtained from a digraph G by splitting every vertex. It is easy to see that G has k disjoint dicycles if and only if K has k arc-disjoint dicycles, and that G has an r-transversal if and only if K has an r-arc transversal. Hence, if $f(k)$ exists, then $g(k)$ exists and $g(k) \leq f(k)$.

Suppose H is obtained from a digraph G as follows. For every arc e of G, H has vertices t_e and h_e and $t_e h_e$ is an arc of H. For every arc e and f of G, if e is incident to vertex x and f is incident from x, then $h_e t_f$ is an arc of H. It is easy to see that G has k arc-disjoint dicycles if and only if H has k disjoint dicycles, and that G has an r-arc transversal if and only if H has an r-transversal. Hence, if $g(k)$ exists, then $f(k)$ exists and $f(k) \leq g(k)$. □

We now define classes of acyclic digraphs which are needed to construct intercyclic digraphs. Let G be an acyclic strict digraph with sources $x_1, \ldots, x_s$ and sinks $y_1, \ldots, y_t$, where $2 \leq s$ and $2 \leq t$. Suppose every vertex which is not a source or sink has indegree and outdegree at least 2. Further, suppose there is no $(x_i \to y_\ell, x_j \to y_k)$-linkage such that $1 \leq i < j \leq s$ and $1 \leq k < \ell \leq t$. We define $\mathcal{P}(x_1, \ldots, x_s; y_1, \ldots, y_t)$ to be the class of all such digraphs G. We define $\mathcal{P}_{s,t}$ to be the class of all digraphs G in $\mathcal{P}(x_1, \ldots, x_s; y_1, \ldots, y_t)$, for some $x_1, \ldots, x_s, y_1, \ldots, y_t$. The structure of the digraphs in $\mathcal{P}_{s,t}$ is given by the following theorem due to Metzlar [12].

THEOREM 2.5. *Suppose G is in $\mathcal{P}(x_1, \ldots, x_s; y_1, \ldots, y_t)$. Let G_C be the mixed graph obtained from G by adding the edges $x_1 y_1$, $x_s y_t$, $x_i x_{i+1}$, $i = 1, \ldots, s-1$,* and $y_k y_{k+1}$, $k = 1, \ldots, t-1$. Let C be the cycle induced by the new edges. Then G_C has a planar embedding where C is the outer face boundary.

PROOF. There exists a source sequence $x_1, \ldots, x_s, a_1, \ldots, a_m, y_1, \ldots, y_t$ of G. Let $V_r = \{x_1, \ldots, x_s, a_1, \ldots, a_r\}$, $G_r = C \cup G[V_r]$, and A_r be the set of vertices in V_r which are adjacent to a vertex in $V(G) - V_r$, $r = 0, \ldots, m$.

We will prove by induction on r that G_r has a planar embedding with the following properties.

(i) Every face boundary is a cycle.
(ii) The outer face F has boundary C.
(iii) There is an inner face F_r having a boundary C_r which includes $P = x_1 y_1, y_1 y_2, \ldots, y_{s-1} y_s, y_s x_t$.
(iv) $A_r \subseteq V(C_r)$.

The result holds for $r = 0$ since $G_r = C$. Suppose we have the required embedding of G_r, where $0 \leq r < m$. By definition of source sequence, $N_G^-(a_{r+1}) \subseteq A_r$ and a_{r+1} is not adjacent to any vertices in V_r. Since $A_r \subseteq V(C_r)$, we can then obtain a planar embedding of G_{r+1} by placing a_{r+1} in F_r and adding the arcs incident to a_{r+1}. The boundary of every new face will be a cycle because there are at least two arcs incident to a_{r+1}. Since a_{r+1} is not adjacent from any vertices in $\{y_1, \ldots, y_t\}$, the embedding of G_{r+1} has an inner face F_{r+1} whose boundary C_{r+1} includes P.

It is easy to see that $A_{r+1} \subseteq V(C_{r+1})$ provided there is no cyclic subsequence x_1, u, v, w, x_s of C_r such that u, v, and w are distinct, $u a_{r+1}$ and $w a_{r+1}$ are in $A(G)$, and vb is in $A(G)$ for some b in $V(G) - V_{r+1}$. Suppose such a cyclic subsequence exists. By Corollary 2.1, there exists an $(a_{r+1} \to y_k, b \to y_\ell)$-linkage (q_u, q_v) for some k and ℓ in $\{1, \ldots, t\}$, and an $(x_i \to u, x_j \to v)$-linkage (p_u, p_v) for some i and j in $\{1, \ldots, s\}$.

We claim $i < j$. Since y_1, y_t, v, u is a cyclic subsequence of C_r, we can extend the planar embedding of G_r by adding $u y_1$ and $v y_t$ across F_r. If $j < i$, then $P_u, u y_1$ and $P_v, v y_t$ are interior skew paths of the facial cycle C and we have contradicted Theorem 2.2. Hence, $i < j$. By the definition of source sequence, $V(p_u \cup p_v) \subseteq V_r$ and $V(q_u \cup q_v) \subseteq V(G) - V_r$. Thus, $p_u, u a_{r+1}, q_u$ and p_v, vb, q_v constitute an $(x_i \to y_k, x_j \to y_\ell)$-linkage of G. Then $i < j$ implies $k < \ell$. But similarly we can show that $\ell < k$ by considering two disjoint $(\{x_1, \ldots, x_s\}, \{v, w\})$-dipaths. Therefore, $A_{r+1} \subseteq V(C_{r+1})$.

We now consider the planar embedding of G_m. It can be extended to the required embedding of G_C provided there is no cyclic subsequence x_1, y_k, y_ℓ, v, u of C_m such that $u y_\ell$ and $v y_k$ are in $A(G)$. But if such a cyclic subsequence exists, we can then obtain an $(x_i \to y_k, x_j \to y_\ell)$-linkage of G such that $i < j$ and $\ell < k$. □

3. Statement of Main Theorem

We say that a digraph is in *reduced form* if it is strict, strongly-connected, and has $2 \leq \delta_H^+$ and $2 \leq \delta_H^-$. Before stating the main theorem, we first show that the problem of determining if a digraph is intercyclic can be easily reduced to the problem of determining if a digraph in reduced form is intercyclic.

Suppose we wish to determine if a digraph G is intercyclic. First we determine the strong components of G. If G is acyclic, then G is intercyclic. If G has at least two nontrivial strong components, then each will have a dicycle, and so G will not be intercyclic. If G has a unique nontrivial strong component, then G will be intercyclic if and only if this nontrivial strong component is intercyclic. Therefore, we may assume G is strongly-connected.

There are simple reductions that can sometimes be performed on G to give a digraph H with fewer arcs such that G is intercyclic if and only if H is intercyclic. G can be reduced by removing all but one of a set of the parallel arcs. If e is the only arc incident from (or incident to) a vertex v, then G can be reduced by contracting e. (This reduction can result in new loops and parallel arcs.) We refer to these reductions as *trivial reductions*. It is easy to see that G is strongly-connected if and only if H is strongly-connected, and that G has a k-transversal if and only if H has a k-transversal.

Suppose K is obtained from G via a sequence of trivial reductions and suppose K has no trivial reductions. Then K is strongly-connected, and K is intercyclic if and only if G is intercyclic. It is easy to show that G has a 1-transversal if and only if K has only one vertex. In this case, G is clearly intercyclic. Therefore, we may assume that neither G nor K has a 1-transversal and that K has at least two vertices. If K has a loop e incident with vertex x, then $K[e]$ and any dicycle of $K - x$ are disjoint dicycles of K, and so K and G are not intercyclic. Hence, we can assume that K is loopless. Since K is strongly-connected, loopless, and has no trivial reductions, K is in reduced form. Thus, we have reduced the problem of determining if G is intercyclic to the problem of determining if the digraph K in reduced form is intercyclic. We note that if we can show that K is not intercyclic and can find two disjoint dicycles of K, then it is easy to find two disjoint dicycles of G by reversing the process of trivial reductions.

Let $\mathcal{I}$ be the class of all intercyclic digraphs in reduced form. The main theorem gives a classification of all digraphs in $\mathcal{I}$. Let $\mathcal{I}_t$ be subclass of $\mathcal{I}$ consisting of digraphs where every transversal contains *at least* 3 vertices.

We define $\mathcal{T}$ to be the class of all strict digraphs H with $2 \leq \delta_H^+$ and $2 \leq \delta_H^-$ which have distinct vertices x and y such that the digraph obtained from H by dividing x and y is in $\mathcal{P}(x^+, y^+; y^-, x^-)$.

Let D_7 be the digraph shown in Figure 1.

We define $\mathcal{K}$ to be the class of all digraphs H with $2 \leq \delta_H^+$, $2 \leq \delta_H^-$, and no 2-transversal, that can be constructed as follows. Let K_H' be in $\mathcal{P}(w_0, z_0; z_1, w_1)$ and let K_H be obtained from K_H' by possibly adding one arc in $\{w_0 z_0, z_0 w_0\}$ and

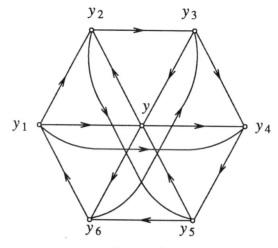

FIGURE 1.

then possibly adding one arc in $\{w_1z_1, z_1w_1\}$. Let C be a 4-dicycle x_0, x_1, x_2, x_3, x_0 which is disjoint from K_H. Let H be obtained from $C + K_H$ by adding the arcs w_1x_0, w_1x_2, z_1x_1, z_1x_3, x_0w_0, x_2w_0, x_1z_0, and x_3z_0 (see Figure 2).

We define $\mathcal{H}$ to be the class of all digraphs H with $2 \leq \delta_H^+$, $2 \leq \delta_H^-$, and no 2-transversal such that H has a set $Y = \{y_1, y_2, y_3, y_4, y_5\}$ of five vertices and arc-disjoint subdigraphs H_α, H_β, and H_γ satisfying the following conditions (see Figure 3).

(i) $y_1, \ldots, y_5$ are the only vertices of H in more than one of $V(H_\alpha)$, $V(H_\beta)$, and $V(H_\gamma)$.
(ii) $H_\alpha \in \mathcal{P}(y_4, y_3, y_1; y_5, y_2)$, $H_\beta \in \mathcal{P}(y_4, y_5; y_3, y_1, y_2)$, and $H_\gamma \in \mathcal{P}(y_1, y_2; y_3, y_4)$.
(iii) $H = H_\alpha \cup H_\beta \cup H_\gamma$.

We say that $(H_\alpha, H_\beta, H_\gamma)$ is an $\mathcal{H}$-decomposition of H with $\mathcal{H}$-separator $(y_1, y_2, y_3, y_4, y_5)$. We note that an $\mathcal{H}$-separator and $\mathcal{H}$-decomposition need not be unique.

In the next section we will show that $\mathcal{T}$, $\{D_7\}$, $\mathcal{K}$, and $\mathcal{H}$ are pairwise disjoint. The following result is the main theorem.

THEOREM 3.1. $\mathcal{I} = \mathcal{T} \cup \{D_7\} \cup \mathcal{K} \cup \mathcal{H}$.

4. Preliminary Results and Conjectures

In this section we characterize the intercyclic digraphs with 2-transversals and prove that the classes $\mathcal{T}$, $\{D_7\}$, $\mathcal{K}$, and $\mathcal{H}$ are pairwise disjoint. We also use the main theorem to prove Metzlar's conjecture and the conjecture of Kosaraju and Younger. Throughout this section we will use the notation used in §3 to define $\mathcal{T}$, D_7, $\mathcal{K}$, and $\mathcal{H}$.

Thomassen [15] characterized the intercyclic digraphs in reduced form which

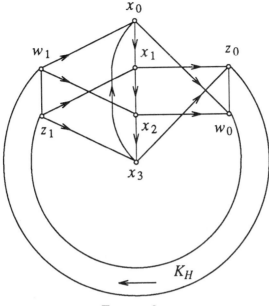

FIGURE 2.

have 2-transversals. The characterization first requires making an easy observation.

LEMMA 4.1. *G is in $\mathcal{T}$ if and only if G is an intercyclic digraph in reduced form with a 2-transversal.*

PROOF. Suppose G is an intercyclic digraph in reduced form with a 2-transversal $\{x, y\}$. Let P be the graph obtained from G by dividing x and y. Since $\{x, y\}$ is a transversal, P is an acyclic digraph with sources x^+ and y^+ and sinks x^- and y^-. All other vertices have indegree and outdegree at least 2 and P is strict because G is in reduced form. Furthermore, an $(x^+ \to x^-, y^+ \to y^-)$-linkage of P would correspond to disjoint dicycles of G. Therefore, P is in $\mathcal{P}(x^+, y^+; y^-, x^-)$. Also, G is a strict digraph with $2 \leq \delta_G^+$ and $2 \leq \delta_G^-$ because G is in reduced form. Hence, G is in $\mathcal{T}$. The converse is easy to verify. □

To complete the description of Thomassen's characterization, we just need to give the structure of the digraphs in $\mathcal{P}_{2,2}$. This is done by Theorem 2.5. Thomassen's proof of Theorem 2.5 for the case $s = t = 2$ is different from Metzler's proof.

After the following preliminary lemma, we will show that $\mathcal{T}$, $\{D_7\}$, $\mathcal{K}$, and $\mathcal{H}$ are pairwise disjoint.

LEMMA 4.2. *If $(H_\alpha, H_\beta, H_\gamma)$ is an $\mathcal{H}$-decomposition of H with $\mathcal{H}$-separator $(y_1, y_2, y_3, y_4, y_5)$, then the following statements hold.*
  (i) y_1 *is adjacent to an intermeditate vertex of H_α.*
  (ii) y_3 *is adjacent from an intermeditate vertex of H_β.*

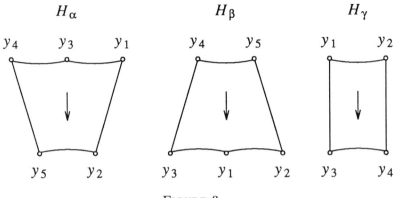

FIGURE 3.

(iii) If y_4 is not adjacent to an intermeditate vertex of H_α and y_2 is not adjacent from an intermeditate vertex of H_β, then y_4y_5 and y_5y_2 are arcs of H.

(iv) H_α and H_β each have at least 6 vertices. H_γ has at least 5 vertices.

PROOF. (i) Suppose y_1 is not adjacent to any intermediate vertex of H_α. Since $2 \leq d_H^+(y_3)$, we have $2 \leq d_{H_\alpha}^+(y_3)$. Then Corollary 2.1 implies H_α has a (y_3, y_2)-dipath p. If y_1y_5 is an arc of H_α, then (p, y_1y_5) is a $(y_3 \to y_2, y_1 \to y_5)$-linkage of H_α. But then H_α is not in $\mathcal{P}(y_4, y_3, y_1; y_5, y_2)$. Hence, $N_{H_\alpha}^+(y_1) \subseteq \{y_2\}$. It is now routine to verify that $\{y_3, y_4\}$ is a 2-transversal of H. But then H is not in $\mathcal{I}_t$.

(ii) The proof is similar to 1.

(iii) Suppose y_4 is not adjacent to an intermeditate vertex of H_α and y_2 is not adjacent from an intermeditate vertex of H_β. We can show that y_4y_2 is not an arc of H_α or H_β in the same way we proved that y_1y_5 is not an arc of H_α. Therefore, $N_{H_\alpha}^+(y_4)$ and $N_{H_\beta}^-(y_2)$ are both contained in $\{y_5\}$. If y_4y_5 or y_5y_2 is not an arc, then $\{y_1, y_3\}$ is a 2-transversal of H.

(iv) H_α and H_β have intermediate vertices by 1 and 2. and so each has at least 6 vertices. Suppose H_γ has only 4 vertices. Since $2 \leq d_H^+(y_2)$, $2 \leq d_{H_\gamma}^+(y_2)$. Hence, y_2y_3 is an arc. Similarly, y_1y_4 is an arc. Therefore, H_γ has a $(y_1 \to y_4, y_2 \to y_3)$-linkage. But then H_γ is not in $\mathcal{P}(y_1, y_2; y_3, y_4)$. □

THEOREM 4.1. *The classes $\mathcal{T}$, $\{D_7\}$, $\mathcal{K}$, and $\mathcal{H}$ are pairwise disjoint.*

PROOF. Every digraph in $\mathcal{T}$ has a 2-transversal. It is easy to verify that D_7 has no 2-transversal, and the digraphs in $\mathcal{K}$ and $\mathcal{H}$ have no 2-transversal by definition. Therefore, $\mathcal{T}$ is disjoint from the other three classes.

By their construction the digraphs in $\mathcal{K}$ each have at least eight vertices. Lemma 4.2 (iv) implies that every digraph in $\mathcal{H}$ has at least eight vertices.

Therefore, D_7 is not in $\mathcal{K}$ or $\mathcal{H}$.

Consider the set of vertices $S = \{y_1, y_2, y_5, y_3, y_4\}$ of a digraph in $\mathcal{H}$. It is easy to verify that $\{y_1, y_2, y_5\}$, $\{y_2, y_5, y_4\}$, $\{y_5, y_4, y_3\}$, $\{y_4, y_3, y_1\}$, and $\{y_3, y_1, y_2\}$ are 3-transversals contained in S. Using Corollary 2.1 and Lemma 4.2 it is routine to show that S contains at most one other 3-transversal, namely, $\{y_1, y_3, y_5\}$. Thus, every digraph in $\mathcal{H}$ has a set of vertices $S = \{v_0, v_1, v_2, v_3, v_4\}$ such that every subset of S of the the form $\{v_i, v_{i+1}, v_{i+2}\}$ is a 3-transversal, where $0 \leq i \leq 4$ and addition is modulo 4, and such that S contains at most one other 3-transversal. We now show that $\mathcal{K}$ and $\mathcal{H}$ are disjoint by showing that no digraph in $\mathcal{K}$ has such a set S of vertices. Suppose digraph G is a counterexample.

It is easy to verify that every 3-transversal of G is of one of the following forms.

(i) $\{x_i, u, v\}$, where $0 \leq i \leq 3$ and every $(\{w_0, z_0\}, \{z_1, w_1\})$-dipath of K_H uses a vertex in $\{u, v\}$.

(ii) $\{x_0, x_2, z\}$, where every (z_0, z_1)-dipath of K_H uses z.

(iii) $\{x_1, x_3, w\}$, where every (w_0, w_1)-dipath of K_H uses w.

Any two vertices in S are in a common 3-transversal, whereas no two adjacent vertices on the dicycle x_0, x_1, x_2, x_3, x_0 are in a common 3-transversal. Therefore, we can assume that $S \cap \{x_0, x_1, x_2, x_3\}$ is contained in $\{x_0, x_2\}$.

Every 3-transversal of the form $\{v_i, v_{i+1}, v_{i+2}\}$ must contain a vertex in $\{x_0, x_2\}$, and so we may assume $v_0 = x_0$ and $v_2 = x_2$. Then $\{v_1, v_2, v_3\}$ is of the form $\{x_2, u, v\}$, where every $(\{w_0, z_0\}, \{z_1, w_1\})$-dipath of K_H uses a vertex in $\{u, v\}$. It follows that $\{x_0, u, v\}$, that is, $\{v_0, v_1, v_3\}$, is also a 3-transversal. Similarly, we can show that $\{v_4, v_1, v_2\}$ is a 3-transversal. We have now shown that S contains at least seven 3-transversals. But this contradicts our assumptions about which subsets of S are 3-transversals. □

We now verify the conjectures.

THEOREM 4.2. *Every intercyclic digraph has a 3-transversal.*

PROOF. By the discussion at the start of §3, we only need to prove the theorem for digraphs in reduced form. By Theorem 3.1, it suffices to check the theorem for digraphs in $\mathcal{T}$, $\{D_7\}$, $\mathcal{K}$, and $\mathcal{H}$. For each such digraph we will give a 3-transversal which verifies the theorem.

For a digraph in $\mathcal{T}$, $\{x, y\}$ is a 2-transversal. A 3-transversal for D_7 is $\{y_1, y_3, y_5\}$. For a digraph in $\mathcal{K}$, $\{x_0, w_0, z_0\}$ is a 3-transversal. For a digraph in $\mathcal{H}$, $\{y_1, y_2, y_5\}$ is a 3-transversal. □

THEOREM 4.3. *For every intercyclic digraph G, there is a function w from $V(G)$ into the nonnegative real numbers such that $\sum_{x \in V(G)} w(x) \leq 2.5$ and for every dicycle C, $1 \leq \sum_{x \in V(C)} w(x)$.*

PROOF. As in the last theorem, it suffices to check the theorem for digraphs in $\mathcal{T}$, $\{D_7\}$, $\mathcal{K}$, and $\mathcal{H}$. For each such digraph we will give an explicit weight function w which verifies the theorem. For a digraph in $\mathcal{T}$, $w(x) = w(y) = 1$ and all other vertices have weight 0. For D_7, all vertices have weight $\frac{1}{3}$. For a digraph in $\mathcal{K}$, $w(x_0) = w(x_2) = w(w_0) = \frac{1}{2}$, $w(z_0) = 1$, and all other vertices have weight 0. For a digraph in $\mathcal{H}$, the vertices y_1, y_2, y_3, y_4, and y_5 have weight $\frac{1}{2}$ and all other vertices have weight 0. □

5. Proof Outline

In this section we will outline the proof of the main theorem. The key lemmas needed in the proof will be proven in this section. It is routine to show that all the digraphs in $\{D_7\} \cup \mathcal{T} \cup \mathcal{K} \cup \mathcal{H}$ are intercyclic digraphs in reduced form. The rest of this section and the next three sections will be devoted to proving the converse.

The proof that every intercyclic digraph in reduced form is in $\{D_7\} \cup \mathcal{T} \cup \mathcal{K} \cup \mathcal{H}$ is done by induction on the number of arcs. Let G be an intercyclic digraph in reduced form. If G has a 2-transversal, then G is in $\mathcal{T}$ by Lemma 4.1. Therefore, we may assume $G \in \mathcal{I}_t$. We will show that G has a useful reduction to a digraph H in $\mathcal{I}$ which has fewer arcs than G. Then the induction hypothesis applied to H is used to show that G is in $\{D_7\} \cup \mathcal{K} \cup \mathcal{H}$. We will next define three useful reductions: arc reductions, good contractions, and R_k-reductions.

Let $s_1 t_1$ be an arc of G. Trivially, $G - s_1 t_1$ is intercyclic, but it may not be in reduced form. Let H be the intercyclic digraph in reduced form obtained from $G - s_1 t_1$ by performing trivial reductions. If only a few trivial reductions are needed, then the reduction from G to H will be useful. If $H = G - s_1 t_1$, then we will say that G has a *type 0 reduction* at $s_1 t_1$. If $N_G^+(s_1) = \{t_1, s_2\}$ and H is obtained from $G - s_1 t_1$ by contracting $s_1 s_2$, then we say that G has a *type 1s reduction* at $s_1 t_1$. If $N_G^-(t_1) = \{s_1, t_2\}$ and H is obtained from $G - s_1 t_1$ by contracting $t_2 t_1$, then we say that G has a *type 1t reduction* at $s_1 t_1$. A *type 1 reduction* is a type 1s or type 1t reduction. If $N_G^+(s_1) = \{t_1, s_2\}$, $N_G^-(t_1) = \{s_1, t_2\}$, $s_2 \neq t_2$, and H is obtained from $G - s_1 t_1$ by contracting $s_1 s_2$ and $t_2 t_1$, then we say that G has a *type 2 reduction* at $s_1 t_1$. The four reductions just defined will be called *arc reductions*.

If x, y, and z are distinct vertices of G and xy, xz, and yz are arcs of G, then we say that xyz is a *transitive triangle* of G. Suppose xyz is a transitive triangle. If $d^+(x) = 2$, then we say that xy is *contractible*. If $d^-(z) = 2$, then we say that yz is *contractible*. If xy or yz is contractible, then we say that xyz is *reducible*. If xyz is reducible, we will sometimes put parenthesis around the contractible arc.

We now recursively define digraphs R_k for every $k \geq 2$. Let R_2 be the digraph with vertices w_1 and w_2 and the arc $w_1 w_2$. For $k \geq 3$, define R_k to be $R_{k-1} + w_k + \{w_{k-2} w_k, w_{k-1} w_k\}$. Let $\{u, v\}$ be $\{w_1, w_2\}$. We introduce this notation so that we may refer to w_1 and w_2 without specification. Denote $\{w_{k-1}, w_k\}$ by $\{u', v'\}$ so that R_k has a $(u \to u', v \to v')$-linkage $(P_{uu'}, P_{vv'})$. An R_k subdigraph of digraph G will be called *good* if it is induced, the only vertices of R_k adjacent from vertices in $V(G) - V(R_k)$ are w_1 and w_2, the only vertices of R_k adjacent to vertices in $V(G) - V(R_k)$ are w_{k-1} and w_k, and $G - V(R_k)$ has minimum indegree and outdegree at least 1.

Let R_k be a good R_k subdigraph of G. Let H be the digraph obtained from $G - V(R_k)$ by adding a new vertex ρ, adding an arc from z to ρ for every vertex z adjacent to w_1 or w_2, and adding an arc from ρ to y for every vertex y adjacent from w_{k-1} or w_k. We say that H is obtained from G by an R_k-*reduction* and that R_k is *contracted* to ρ. By the definition of good R_k subdigraph, H is an intercyclic digraph in reduced form.

In the rest of this section we will prove the central lemmas needed in the proof of the main theorem. First we show that G has an arc reduction at every arc or G has a contractible arc (Lemma 5.1). Second we show that if G has a contractible arc, then G has an R_k-reduction (Lemma 5.2). In §5 we will use the induction hypothesis and Lemma 5.1 to show that G has a contractible arc. We then use Lemma 5.2 to conclude that G has an R_k-reduction. In §6 and §7 we use the induction hypothesis and the fact that G has an R_k-reduction, to show that G is in $\{D_7\} \cup \mathcal{K} \cup \mathcal{H}$.

Note that the following lemma does not require G to be intercyclic, although we will only apply it to intercyclic digraphs.

LEMMA 5.1. *Let G be a digraph in reduced form with no 2-dicycle. Then G has an arc reduction at every arc or G has a contractible arc.*

PROOF. Let $s_1 t_1$ be an arc of G. It suffices to show that G has an arc reduction at $s_1 t_1$ or G has a contractible arc.

Suppose $d_G^+(s_1) = 2 = d_G^-(t_1)$. Let s_2 be the other vertex adjacent from s_1 and let t_2 be the other vertex adjacent to t_1. Since G is in reduced form, s_1, s_2, t_1, and t_2 are all distinct except possibly s_2 and t_2. If $s_2 = t_2$, then $s_1 s_2 t_1$ is reducible and $s_1 s_2$ and $s_2 t_1$ are contractible. Hence, we may assume s_1, s_2, t_1, and t_2 are all distinct. Since G has no 2-dicycles, $s_2 s_1$ and $t_1 t_2$ are not arcs of G.

Let H be obtained from $G - s_1 t_1$ by contracting $s_1 s_2$ to s and $t_2 t_1$ to t, and then reducing any parallel arcs. If H has a loop, then it is adjacent to s or t. But then $s_2 s_1$ or $t_1 t_2$ is an arc of G.

We now show that $2 \leq \delta_H^+$ or G has a contractible arc. Since $s_2 s_1$ and $s_2 t_1$ are not arcs of G, s_2 is adjacent to at least 2 vertices in $V(G) - \{s_1, t_1\}$. Thus, $2 \leq d_H^+(s)$. Since $t_1 s_1$ and $t_1 t_2$ are not arcs of G, t_1 is adjacent to at least 2 vertices in $V(G) - \{s_1, t_2\}$. Thus, $2 \leq d_H^+(t)$. Suppose x is in $V(H) - \{s, t\}$ and $d_H^+(x) \leq 1$. Since $2 \leq d_G^+(x)$, it is easy to see that $N_H^+(x)$ is $\{s\}$ or $\{t\}$. Hence, $N_G^+(x)$ equals $\{s_1, s_2\}$ or $\{t_1, t_2\}$. The latter is not possible because s_1 and t_2 are the only vertices adjacent to t_1. Hence, $(xs_1)s_2$ is reducible in G.

We can also show that $2 \le \delta_H^-$ or G has a contractible arc. Therefore, H is in reduced form, that is, there is an arc reduction at $s_1 t_1$ or G has a contractible arc.

The cases when $3 \le d_G^+(s_1)$ or $3 \le d_G^-(t_1)$ are easier. $\square$

Let $\mathcal{M}$ be the set of all digraphs $R_k + \{w_{k-1}w_1, w_k w_1, w_k w_2\}$, where k is odd.

LEMMA 5.2. *Let G be a digraph in $\mathcal{I}$. If G has a contractible arc, then $G \in \mathcal{M}$ or G has an R_k-reduction.*

PROOF. For every arc e of G, we define V_e be the set of vertices z such that every dicycle which includes z also includes an end of e. Choose a contractible arc e of G such that V_e is minimal. Let H be the subdigraph of G induced by the ends of e. Then H is a subdigraph R_k of G such that $e = w_i w_{i+1}$ for some i in $\{1, \ldots, k-1\}$, the only vertices of R_k incident from arcs in $A(G) - A(R_k)$ are w_1 and w_2, and the only vertices of R_k incident to arcs in $A(G) - A(R_k)$ are w_{k-1} and w_k. Suppose we choose a maximal subdigraph R_k of this type. It is easy to see that $V(R_k) \subseteq V_e$.

If $V(G) = V(R_k)$, then it is easy to show that $G \in \mathcal{M}$ because G is intercyclic. Therefore, we may assume $G - V(R_k)$ is nonempty. We will show that G has an R_k-reduction by showing that R_k is good. To do this we just need to prove that $G - V(R_k)$ has minimum indegree and minimum outdegree at least 1 and that R_k is an induced subgraph of G.

Suppose there exists w in $V(G) - V(R_k)$ such that $N^-(w) \subseteq V(R_k)$. By the choice of R_k, $N^-(w) = \{w_{k-1}, w_k\}$. Then $N^-(w) \subseteq V_e$, and so $w \in V_e$. Since $w_{k-1} w_k w$ is reducible, $f = w_k w$ is contractible. Since every dicycle using w or w_k also uses an end of e, $V_f \subseteq V_e$. But then $V_f = V_e$, because V_e was chosen to be minimal.

Suppose there exists an arc $w_{k-1} y$ of G, where $w \ne y \ne w_k$. Since G is strongly-connected there is a dicycle C through $w_{k-1} y$. Since C includes the vertex w_{k-1} in V_f, C includes w or w_k. Then it is easy to see that $A(C)$ includes an arc $z w_k$, where $z \ne w_{k-1}$, and that the (w_{k-1}, z)-dipath P of C does not include w_k or w. If $3 \le k$, then $N^-(w_k) = \{w_{k-2}, w_{k-1}\}$, and so $z = w_{k-2}$. But then $P + \{w_{k-2} w_{k-1}\}$ is a dicycle through a vertex in V_f which does not go through w or w_k. Hence, $k = 2$ and $e = w_1 w_2$. Since $3 \le d^+(w_1)$, the only way for e to be contractible is for $z(w_1 w_2)$ to be contractible. Hence, $zw_1 \in A(G)$. But now once again we have a dicycle $P + \{zw_1\}$ which includes the vertex w_1 in V_f but does not go through w or w_2. Therefore, G has no arc $w_{k-1} y$, where $w \ne y \ne w_k$. But now the subdigraph $R_k + \{w\} + \{w_{k-1} w, w_k w\}$ contradicts the maximality of R_k. Therefore, we have shown that $G - V(R_k)$ has minimum indegree and minimum outdegree at least 1.

Since $G - V(R_k)$ has minimum indegree and minimum outdegree at least 1, $G - V(R_k)$ has a dicycle C'. If R_k is not induced, then there is an arc e' from

w_{k-1} or w_k to w_1 or w_2 in $A(G) - A(R_k)$. But then $R_k + e'$ has a dicycle which is disjoint from C'. Therefore, R_k is an induced subdigraph of G. Therefore, we have shown that R_k is good. □

6. A Contractible Arc

Let G be in $\mathcal{I}_t$. Suppose every intercyclic digraph in reduced form with fewer arcs than G is in $\{D_7\}$, $\mathcal{H}$, $\mathcal{K}$, or $\mathcal{T}$. In this section we will show that G has a contractible arc. We can then conclude that G has an R_k-reduction using Lemma 5.2.

The notation used in §4 to define arc reductions will be referred to as *standard notation*. We will use standard notation in the next preliminary lemma.

LEMMA 6.1. *Suppose $x(yz)$ is reducible in H. If G has no contractible arcs, then the following statements hold.*
  (i) *If G has a type 0 reduction at $s_1 t_1$, then $z = t_1$.*
  (ii) *If G has a transitive triangle, then it has a type 0 reduction.*
  (iii) *If the arc reduction at $s_1 t_1$ is of type 1s or 2 and $d_H^-(s) = 2$, then G has a type 0 reduction. If the arc reduction at $s_1 t_1$ is of type 1t or 2 and $d_H^+(t) = 2$, then G has a type 0 reduction.*
  (iv) *If G has a type 1s reduction at $s_1 t_1$, then $A(H)$ does not contain st_1 or $t_1 s$.*
  (v) *If G has a type 1s reduction at $s_1 t_1$, then we have one of the following possibilities: G has a type 0 reduction, $z = t_1$ and $s \notin \{x, y\}$, or $y = s$ and $t_1 \notin \{x, z\}$.*
  (vi) *If G has a type 2 reduction at $s_1 t_1$, then we can not have any of the following possibilities: $z = t$ and $s \notin \{x, y\}$, $x = s$ and $t \notin \{y, z\}$, $x = s$ and $z = t$, or $s, t \notin \{x, y, z\}$.*

PROOF.   (i) Suppose G has a type 0 reduction at $s_1 t_1$. If $z \neq t_1$, then $x(yz)$ is reducible in G.
  (ii) Suppose G has a transitive triangle uvw. If $d^+(u) = 2$ or $d^-(w) = 2$, then uvw is reducible. If $3 \leq d^+(u)$ and $3 \leq d^-(w)$, then G has a type 0 reduction at uw.
  (iii) Suppose the arc reduction at $s_1 t_1$ is of type 1s or 2, and $d_H^-(s) = 2$. Let $N_H^-(s) = \{u, v\}$. If G has a type 2 reduction at $s_1 t_1$ and $u = t$, then $t_2 s_1$ and $v s_1$ are in $A(G)$ because $2 \leq d_G^-(s_1)$ and G has no 2-dicycle. Then $t_2 s_1 t_1$ is a transitive triangle of G, and so G has a type 0 reduction by 2. Thus, we may assume t is not in $N_H^-(s)$ if G has a type 2 reduction at $s_1 t_1$. Then u and v are vertices of G. Since s_1 and s_2 have indegree at least 2 in G, $u s_1$ and $v s_1$ are in $A(G)$ and we may assume $u s_2$ is in $A(G)$. Then $u s_1 s_2$ is a transitive triangle of G, and so G has a type 0 reduction by 2. The second sentence of 3 is proven similarly.
  (iv) Suppose the arc reduction at $s_1 t_1$ is of type 1s. If st_1 is an arc of H, then $s_2 t_1$ is an arc of G. But then $(s_1 s_2) t_1$ is reducible in G. If $t_1 s$ is an

arc of H, then t_1s_1 or t_1s_2 is an arc of G. If t_1s_1 is an arc of G, then G has a 2-dicycle. If t_1s_2 is an arc of G, then $(s_1t_1)s_2$ is reducible in G.

(v) Suppose the arc reduction at s_1t_1 is of type $1s$ and $s \notin \{x,y,z\}$. Then $z = t_1$, otherwise, $x(yz)$ is reducible. Suppose $s = x$. Then s_2y and s_2z are arcs of G, and so s_2yz is a transitive triangle of G. Then G has a type 0 reduction by 2. Suppose $s = z$. Then $d_H^-(z) = 2$ implies xs_1 and ys_1 are arcs of G. Thus, xys_1 is a transitive triangle of G. Suppose $s = y$. If $t_1 \in \{x,z\}$, then either st_1 or t_1s is in $A(H)$, But then we contradict 4.

(vi) Suppose the arc reduction at s_1t_1 is of type 2. Suppose $t = z$ and $s \notin \{x,y\}$. Then $d_H^-(z) = 2$ implies $N_G^-(t_2) = \{x,y\}$, and so $x(yt_2)$ is reducible in G. Suppose $s = x$ and $t \notin \{y,z\}$. Then s_2y and s_2z are in $A(G)$ and $s_2(yz)$ is reducible in G. Suppose $x = s$ and $z = t$. Since $N_G^+(s_1) = \{s_2,t_1\}$, $N_G^-(t_1) = \{s_1,t_2\}$, and sy, yt, and st are arcs of H, G has the transitive triangle s_2yt_2. Since $d_H^-(t) = 2$, we have $d_G^-(t_2) = 2$, and so $s_2(yt_2)$ is reducible in G. Suppose $s, t \notin \{x,y,z\}$. Then $x(yz)$ is reducible in G because $d_G^-(z) = 2$.

□

Let $x_1, \ldots, x_n$ be a source sequence of a digraph G in $\mathcal{P}_{s,t}$. Suppose G is embedded in the plane according to Lemma 2.5. Let G_i be the plane subdigraph $G[\{x_1, \ldots x_i\}]$, $i = 1, \ldots, n$. We refer to $G_1, \ldots, G_n$ as a *construction sequence* of G.

The graph of Figure x will be denoted by $G_{(x)}$.

THEOREM 6.1. *Suppose every digraph in $\mathcal{I}$ with fewer arcs than G is in $\mathcal{T}$, $\{D_7\}$, $\mathcal{K}$, or $\mathcal{H}$. Then G has a contractible arc.*

PROOF. We use standard notation for arc reductions, and we use the notation of §3 used to define $\mathcal{T}$, D_7, $\mathcal{K}$, and $\mathcal{H}$. Suppose G satisfies the conditions of the theorem but does not have a contractible arc. Using Lemma 5.1, we may assume G has an arc reduction at every arc. Let s_1t_1 be an arc of G and let H be the resulting digraph when G is reduced at s_1t_1. By assumption, H is in $\{D_7\}$, $\mathcal{H}$, $\mathcal{K}$, or $\mathcal{T}$. Let the arc reduction at s_1t_1 be of type i, where i is in $\{0,1,2\}$. Assume i is chosen to be as small as possible. Subject to this condition, we assume that, if possible, s_1t_1 is chosen so that H is not in $\mathcal{T}$. The notation used in defining D_7, $\mathcal{H}$, $\mathcal{K}$, and $\mathcal{T}$ will be used to describe H.

Case 1. Suppose $H = D_7$. Suppose G has a type 0 reduction at s_1t_1. If $y_2 \neq t_1$, then $y_1(yy_2)$ is reducible. If $y_4 \neq t_1$, then $y_3(yy_4)$ is reducible.

Suppose G has a type $1s$ reduction at s_1t_1. If $y_i = s$, for some i in $\{1, \ldots, 6\}$, then G has a type 0 reduction by Lemma 6.1 (iii). Hence, $y = s$. We may assume $y_1s_2 \in A(G)$. Since $sy_2 \in A(H)$, $s_2y_2 \in A(G)$. Hence, $y_1s_2y_2$ is a transitive triangle of G. But then G has a type 0 reduction by Lemma 6.1 (ii). Similarly, if G has a type $1t$ reduction at s_1t_1, we get a contradiction.

Suppose G has a type 2 reduction at s_1t_1. Then $y_i \in \{s,t\}$, for some i in $\{1,\ldots,6\}$, and so G has a type 0 reduction by Lemma 6.1 (iii).

Case 2. Suppose $H \in \mathcal{H}$. Let $y_4, y_5, b_1, \ldots, b_m$ and $y_1, y_2, c_1, \ldots, c_n$ be source sequences of H_β and H_γ, respectively. By Lemma 4.2 (iv), $4 \le m$ and $3 \le n$. If $4 = m$ or $3 = n$, then we may assume $b_2 = y_1$ or $c_2 = y_4$, respectively. By considering construction sequences of H_β, it follows that $d_H^-(b_1) = d_H^-(b_2) = 2$ and for some i in $\{4,5\}$, $y_ib_1b_2$ is a transitive triangle of H_β. Similarly, $d_H^-(c_1) = d_H^-(c_2) = 2$ and for some j in $\{1,2\}$, $y_jc_1c_2$ is a transitive triangle of H_γ.

Suppose G has a type 0 reduction at s_1t_1. Since $b_2 \ne c_2$, we have $b_2 \ne t_1$ or $c_2 \ne t_1$. But then $y_ib_1b_2$ or $y_jc_1c_2$ is reducible by Lemma 6.1 (i).

Suppose G has a type $1s$ reduction at s_1t_1. Using Lemma 6.1 (iii), $d_H^-(b_1) = 2$ implies $s \ne b_1$. Then Lemma 6.1 (v) applied to $y_ib_1b_2$ implies $t_1 = b_2$. Similarly, $t_1 = c_2$. But $b_2 \ne c_2$, and so we are done.

Suppose G has a type 2 reduction at s_1t_1. Since $\{y_4, y_5, b_1\}$ and $\{y_1, y_2, c_1\}$ are disjoint, t is not in both sets. Suppose $t \notin \{y_4, y_5, b_1\}$. Hence, either $t = b_2$ or $t \notin \{y_i, b_1, b_2\}$. By Lemma 6.1 (iii), $s \notin \{b_1, b_2\}$, and so either $s = y_i$ or $s \notin \{y_i, b_1, b_2\}$. All four possibilities for s and t lead to a contradiction by applying Lemma 6.1 (vi) to $y_ib_1b_2$. Similarly, if $t \notin \{y_1, y_2, c_1\}$, we get a contradiction.

Case 3. Suppose $H \in \mathcal{K}$. Let $t = t_1$ (respectively, $s = s_1$) if the arc reduction at s_1t_1 is of type 0 or $1s$ (respectively, type 0 or $1t$). Suppose $t \notin V(K_H)$ and $w_0, z_0, a_1, \ldots, a_n$ is a source sequence of K_H, where $2 \le n$. By considering construction sequences of K_H, it follows that aa_1a_2 is a transitive triangle, where $a \in \{w_0, z_0\}$ and $d_H^-(a_1) = d_H^-(a_2) = 2$. Since $t \notin \{a, a_1, a_2\}$, we can use Lemma 6.1 to give a contradiction for arc reductions of all types. Specifically, we use (i) for type 0, (v) for type $1s$, (ii) for type $1t$, and (iii) and (vi) for type 2. Hence, $t \in V(K_H)$. Similarly, $s \in V(K_H)$.

Let $K = G - \{x_0, x_1, x_2, x_3\}$. If K has a dicycle C, then C and x_0, x_1, x_2, x_3, x_0 are disjoint dicycles of G. Hence K is acyclic.

We now construct a superdigraph K' of K. If there is a unique vertex of K in $N_G^+(x_0) \cup N_G^+(x_2)$, then call it w_0'. If $N_G^+(x_0) \cup N_G^+(x_2)$ contains at least two vertices of K (this can occur when $w_0 = s$ or $w_0 = t$), then let w_0' be a vertex in $V(K') - V(K)$ which is adjacent to every vertex of K in $N_G^+(x_0) \cup N_G^+(x_2)$. In a similar manner we consider the vertices of K in $N_G^+(x_1) \cup N_G^+(x_3)$, $N_G^-(x_0) \cup N_G^-(x_2)$, and $N_G^-(x_1) \cup N_G^-(x_3)$ and define the vertices z_0', z_1', and w_1', respectively, of K'. It is easy to see that K' is an acyclic digraph with sources w_0' and z_0' and sinks w_1' and z_1'.

Suppose w_0' is not a vertex of K. Then K has at least two vertices in $N_G^+(x_0) \cup N_G^+(x_2)$. It follows that $w_0 = s$ or $w_0 = t$. If $w_0 = s$ and we have an arc reduction at s_1t_1 of type 0 or $1t$, then x_0 and x_2 can only be adjacent to s_1 in $V(K)$. Similarly, we can not have $w_0 = t$ and an arc reduction at s_1t_1 of type 0 or $1s$. If $w_0 = t$ and we have an arc reduction at s_1t_1 of type $1t$ or 2, then x_0 and x_2 can only be adjacent to t_2 in $V(K)$. Thus, $w_0 = s$ and the arc reduction at s_1t_1

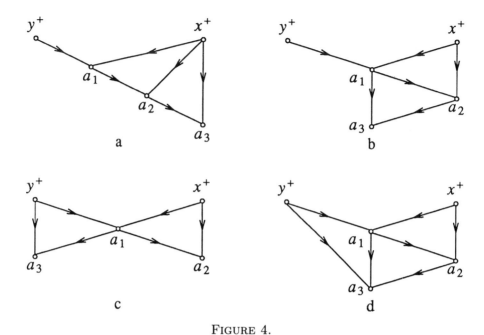

FIGURE 4.

of type $1s$ or 2. Thus, at most one of the vertices w'_0 and z'_0 is in $V(K') - V(K)$. Similarly, at most one of the vertices w'_1 and z'_1 is in $V(K') - V(K)$. Without loss of generality, we may assume the four possibilities for $V(K') - V(K)$ are $\emptyset$, $\{w'_0\}$, $\{w'_0, w'_1\}$, and $\{w'_0, z'_1\}$.

We now show that all non-source vertices of K' have indegree at least 2. This is immediate for all non-source vertices except for s_1 and s_2 when w'_0 is not a vertex of K. If w'_0 is not in $V(K)$, then we may assume $x_0 s_1$ and $x_2 s_2$ are arcs of G. If $x_0 s_2$ is an arc of G, then $x_0 s_1 s_2$ is a transitive triangle of G, and so Lemma 6.1 (ii) implies G has a type 0 arc reduction. Hence, $x_0 s_2 \notin A(G)$. Similarly, $x_2 s_1 \notin A(G)$. It follows that s_1 and s_2 have indegree at least one in K, and so both have indegree at least two in K'. Similarly, all non-sink vertices of K' have outdegree at least 2.

Suppose K' has a $(w'_0 \to w'_1, z'_0 \to z'_1)$-linkage (p, q). For all four possibilities for $V(K') - V(K)$ it is routine to construct disjoint dicycles of G using p and q.

We have shown that $\mathcal{P}(w'_0, z'_0; z'_1, w'_1)$ contains the digraph obtained from K' by removing any arc between w'_0 and z'_0 and any arc between w'_1 and z'_1. Let w'_0, $z'_0, a_1, \ldots, a_m$ be a source sequence of K' and consider construction sequences of K'. If we can find a reducible transitive triangle of K' which does not use any vertices in $V(K') - V(K)$, then we have a reducible transitive triangle of G. If $V(K') - V(K) = \emptyset$, then this is routine. Hence, we may assume $w'_0 \in V(K') - V(K)$. Then $w_0 = s$ in H, the arc reduction at $s_1 t_1$ of type $1s$ or 2, and $N^+_{K'}(w'_0) = \{s_1, s_2\}$. It follows that $6 \leq \nu(K')$ and $d^+_{K'}(w'_0) = 2$. If $V(K') - V(K)$ contains w'_1 or z'_1 we may assume a_m is in $V(K') - V(K)$.

It is easy to show that the contruction sequence of K' contains one of the digraphs in Figure 4, where $\{x^+, y^+\} = \{w_0', z_0'\}$. Since $4 \leq m$, the only vertex of $V(K') - V(K)$ in $\{x^+, y^+, a_1, a_2, a_3\}$ is w_0'. In $G_{(4a)}$, $x^+ \neq w_0'$ because $d_{K'}^+(w_0') = 2$, and so $x^+(a_1 a_2)$ is reducible in G. $G_{(4b)}$ and $G_{(4d)}$ have reducible transitive triangles $a_1(a_2 a_3)$ and $(a_1 a_2)a_3$, respectively, which avoid both x^+ and y^+. $G_{(4c)}$ has a reducible transitive triangle which avoids x^+, and one which avoids y^+. Therefore, G has a reducible transitive triangle.

Case 4. Suppose $H \in \mathcal{T}$. Let H' be obtained from H by dividing x and y, and let $x^+, y^+, a_1, \ldots, a_n$ be a source sequence of H'. Since $3 \leq \nu(H)$, we have $5 \leq \nu(H')$, and so $3 \leq n$. Without loss of generality, one of the digraphs in Figure 4 is in the construction sequence of H'.

Suppose the arc reduction at $s_1 t_1$ is of type 0. If we consider the transitive triangle $x^+ a_1 a_2$, then Lemma 6.1 (i) implies $a_2 = t_1$ in $G_{(4a)}$, $G_{(4b)}$, $G_{(4c)}$, and $G_{(4d)}$. Similarly, if we consider the transitive triangles $x^+ a_2 a_3$, $a_1 a_2 a_3$, and $y^+ a_1 a_3$ in $G_{(4a)}$, $G_{(4b)}$, and $G_{(4c)}$, respectively, then $a_3 = t_1$, a contradiction. If we consider the transitive triangle $a_1 a_2 a_3$ in $G_{(4d)}$, then $a_1 = s_1$. But then G has parallel arcs from a_1 to a_2.

Suppose the arc reduction at $s_1 t_1$ is of type 1. We may assume type 1s. By Lemma 6.1 (iii), $a_1 \neq s$. Now if we consider the transitive triangle $x^+ a_1 a_2$, then Lemma 6.1 (v) implies $a_2 = t_1$ in $G_{(4a)}$, $G_{(4b)}$, $G_{(4c)}$, and $G_{(4d)}$. But then $a_2 = t_1$ and $a_1 \neq s$ contradict Lemma 6.1 (v) when we consider the transitive triangles $x^+ a_2 a_3$, $a_1 a_2 a_3$, and $y^+ a_1 a_3$ in $G_{(4a)}$, $G_{(4b)}$, and $G_{(4c)}$, respectively. By Lemma 6.1 (iv), $a_3 \neq s$ in $G_{(4d)}$. But then $(a_1 a_2)a_3$ is reducible in G.

Suppose the arc reduction at $s_1 t_1$ is of type 2. We first consider the possibility when $a_1 \neq t$. We may assume a_2 is adjacent from x^+ and a_1. Consider the transitive triangle $x^+ a_1 a_2$. By Lemma 6.1 (v), $a_1 \neq s$ and $a_2 \neq s$. Then $x^+ = t$ or we contradict Lemma 6.1 (vi). Thus, $G_{(5a)}$ is in the construction sequence of H and $x^+ = t$. In Figure 5, we consider construction sequences of H. The arrows from $G_{(5r)}$, where $r \in \{a, b, c, d, e, h, i, j, k\}$, go to the next possible digraphs in the construction sequence of H which do not immediately give a contradiction of Lemma 6.1. For example, consider $G_{(5a)}$. If a_3 is adjacent from a_1 and a_2 only, then $a_3 \neq s$ by Lemma 6.1 (iii), and so $a_1(a_2 a_3)$ is reducible. If a_3 is adjacent from y^+ and a_1 only, then we can show that $y^+ = t$, a contradiction. Thus, $G_{(5b)}$ or $G_{(5c)}$ has to follow $G_{(5a)}$ in the construction sequence of H. Note that Lemma 6.1 (ii) implies $a_3 = s$ in $G_{(5c)}$ and $a_4 = s$ in $G_{(5h)}$. Note also that $G_{(5d)}$ and $G_{(5k)}$ are not equal to H. If $G_{(5d)} = H$, then $s = x = t$ in G. If $G_{(5k)} = H$, then $a_4 y x$ is a transitive triangle of H with $a_4 = s$ and $x = t$, and we have contradicted Lemma 6.1 (vi). In $G_{(5g)}$, $G_{(5f)}$, and $G_{(5\ell)}$, all the next possible digraphs in the construction sequence of H contradict Lemma 6.1.

We may now assume that $a_1 = t$. Similarly, we may assume that $a_{n-2} = s$. Hence, $s \neq x^+$. By Lemma 6.1 (iii), $s \neq a_2$. Since $x^+ a_1$ is an arc of H, $x^+ t_1$ or $x^+ t_2$ is an arc of G. Since $t_1 t_2$ and $s_1 t_2$ are the only arcs incident to t_1, $x^+ t_2$ is an arc of G. Since $a_1 a_2$ is an arc of H, $t_1 a_2$ or $t_2 a_2$ is an arc of G. If $t_2 a_2$ is

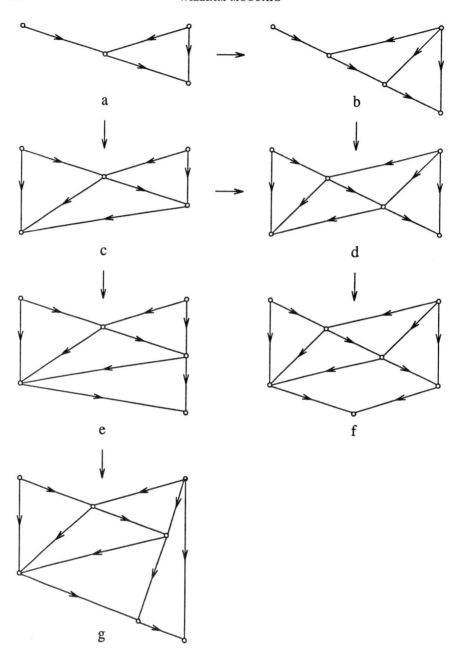

FIGURE 5.

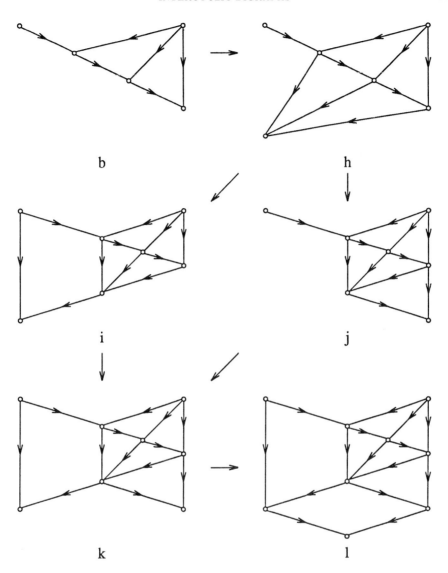

FIGURE 5. continued

an arc of G, then $x^+(t_2a_2)$ is reducible. Hence, t_1a_2 is an arc of G. Thus, t_2t_1, x^+t_2, t_1a_2, and x^+a_2 are all arcs of G.

Let uv be any arc of G. Choose another arc e incident to v. Consider the arc reduction at e. The proof up to this point has shown that G has a contractible arc or there exist vertices u' and v' such that $u'u$, $u'v'$, and vv' are arcs of G. Thus, we have the second possibility for every arc uv of G.

Let L be a subdigraph of G, where $V(L) = \{u_1, \ldots, u_n, v_1, \ldots, v_n\}$ and $A(L)$ is the union of $\{u_{i+1}u_i \mid i = 1, \ldots, n-1\}$, $\{v_iv_{i+1} \mid i = 1, \ldots, n-1\}$, and $\{u_jv_j \mid j = 1, \ldots, n\}$. We have already shown that such a subdigraph exists for $n = 2$. Suppose L is chosen so that n is maximal. Consider u_nv_n. There exists vertices u' and v' of G such that $u'u_n$, v_nv', and $u'v'$ are arcs of G.

Since G has only type 2 arc reductions, every vertex of G has indegree and outdegree 2. If $u' = u_i$, where $2 \leq i \leq n-1$, then $\{u_{i-1}, v_i, u_n\} \subseteq N_G^+(u_i)$, a contradiction. If $u' = u_1$, then $v' = v_1$. But then $u'u_n \cdots u_2u'$ and $v'v_n \cdots v_2v'$ are disjoint dicycles of G. If $u' = v_n$, then u_nv_n, v_nu_n is a 2-dicycle, and we contradict the fact that $G \in \mathcal{I}_t$. If $u' = v_i$, where $1 \leq i \leq n-1$, then $v' = v_{i+1}$. But then either $i = n - 1$ and $v' = v_n$, or $u'u_n \cdots u_iu'$ and $v'v_{i+2} \cdots v_nv'$ are disjoint dicycles of G. Therefore, $u' \notin V(L)$. Similarly, $v' \notin V(L)$. But now we get our final contradiction: $L \cup \{u', v'\} \cup \{u'u_n, v'v_n, u'v'\}$ contradicts the maximality of n. $\square$

7. R_k-reductions to $\mathcal{T}$

We have so far shown that if G is in $\mathcal{I}_t$, then G has an R_k-reduction to an intercyclic digraph H in reduced form. In this section we show that if H is in $\mathcal{T}$, then G is in $\{D_7\} \cup \mathcal{K} \cup \mathcal{H}$. Let R_k be contracted to ρ. If ρ is not in some 2-transversal S of H, then S is also a 2-transversal of G. But then G is not in $\mathcal{I}_t$. Therefore, ρ is in every 2-transversal of H. Let $K = H - \rho$ and let X be the set of all vertices x of K such that $K - x$ is acyclic. It is easy to see that K is strict and strongly-connected and that X is nonempty. Let $|X| = n$. The structures of H and K are given by the next three lemmas.

LEMMA 7.1. *Suppose $X = \{x_0, \ldots, x_{n-1}\}$ and $2 \leq n$. Then there exist arc-disjoint subdigraphs $B_0, \ldots, B_{n-1}$ of K such that the following conditions hold.*

(i) $K = \cup_{i=0}^{n-1} B_i$.

(ii) B_i *is an acyclic digraph with the unique source x_i and the unique sink x_{i+1}, for every i in $\mathcal{Z}_n$.*

(iii) x_i *is a vertex of only B_{i-1} and B_i, for every i in $\mathcal{Z}_n$.*

(iv) *For every vertex y in $V(K) - X$, there is a unique i in $\mathcal{Z}_n$ such that $y \in V(B_i)$.*

(v) *For every i in $\mathcal{Z}_n$, either B_i is the subdigraph induced by the arc x_ix_{i+1}, or B_i has two internally disjoint (x_i, x_{i+1})-dipaths.*

PROOF. Since K is strongly-connected, K has a dicycle C. Then $X \subseteq V(C)$. We may suppose the vertices in X occur on C in the cyclic order $x_0, x_1, \ldots, x_{n-1}$.

For every i in $\mathcal{Z}_n$, let Y_i be the set of vertices y in $V(K) - X$ such that all (X,y)-dipaths have origin x_i and all (y,X)-dipaths have terminus x_{i+1}. Let $V_i = Y_i \cup \{x_i, x_{i+1}\}$ and let B_i be the digraph obtained from $K[V_i]$ by removing the arc $x_{i+1}x_i$ if present.

Let y be in $V(K) - X$. Since G is strongly-connected, there exists an (X,y)-dipath and a (y,X)-dipath. Suppose there is an (X,y)-dipath p with origin x_j and a (y,X)-dipath q with terminus x_k, for some j and k in $\mathcal{Z}_n$. If $k \neq j+1$, then $p, q, C[x_k, x_j]$ is a closed directed walk of $K - x_{j+1}$. But $K - x_{j+1}$ is acyclic, and so $k = j+1$. Therefore, for every vertex y in $V(K) - X$, there is a unique i in $\mathcal{Z}_n$ such that $y \in V(B_i)$. It follows that for every i in $\mathcal{Z}_n$, every vertex in $V(K) - X$ adjacent to x_i is in Y_{i-1}, every vertex in $V(K) - X$ adjacent from x_i is in Y_i, and for every j and k in $\mathcal{Z}_n$ such that $j \neq k$, there are no arcs from a vertex in V_j to a vertex in V_k. If $x_j x_k$ is an arc, where j and k are in $\mathcal{Z}_n$ and $k \neq j+1$, then $x_j x_k, C[x_k, x_j]$ is a dicycle of $K - x_{j+1}$ and we have a contradiction. Therefore, the subdigaphs $B_0, \ldots, B_{n-1}$ are arc-disjoint, $K = \cup_{i=0}^{n-1} B_i$, and B_i has source x_i and sink x_{i+1}, for every i in $\mathcal{Z}_n$.

Let i be in $\mathcal{Z}_n$. If B_i has a dicycle, then it does not use the source x_i. But then $K - x_i$ has a dicycle. Hence, B_i is acyclic. Since every vertex z of B_i has an (x_i, z)-dipath and a (z, x_{i+1})-dipath in B_i, x_i is the only source and x_{i+1} is the only sink of B_i.

If there is a vertex z in Y_i which is on every (x_i, x_{i+1})-dipath of B_i, then z is also in X, a contradiction. Hence, either B_i is the subdigraph induced by the arc $x_i x_{i+1}$, or B_i has two internally disjoint (x_i, x_{i+1})-dipaths by Theorem 2.3. $\square$

The following lemma has a similar proof.

LEMMA 7.2. *Suppose $X = \{x\}$. Let B_0 be the digraph obtained from K by dividing x. Let $x^+ = x_0$ and $x^- = x_1$. Then B_0 is an acyclic digraph with the unique source x_0 and the unique sink x_1, and B_0 has two internally disjoint (x_0, x_1)-dipaths.*

We refer to digraphs $B_0, \ldots, B_{n-1}$ from Lemmas 7.1 and 7.2 as the *cyclic blocks* of K. If $n \geq 2$, and B_i is the subdigraph induced by the arc $x_i x_{i+1}$, then B_i will be called *trivial*. Note that if $n = 1$, then B_0 has the same form as a nontrivial cyclic block when $n \geq 2$.

Let H' be the digraph obtained from H by dividing x if $X = \{x\}$ and dividing ρ. We let $x_0 = x^+$ and $x_1 = x^-$ if $X = \{x\}$. Let B_i' be the digraph obtained from the union of B_i and $H'(\{\rho^+\}, V(B_i), \{\rho^-\})$ by *removing* the arcs $\rho^+ x_i$ and $x_{i+1}\rho^-$ if present. The following lemma describes the structure of B_i and B_i'.

LEMMA 7.3. *Suppose B_i is a nontrivial cyclic block of K. Then B_i' has no $(\rho^+ \to \rho^-, x_i \to x_{i+1})$-linkage. Also, B_i has a planar embedding M_i with an outer face boundary C_i such that the following properties hold.*

(i) *x_i and x_{i+1} are on C_i.*
(ii) *C_i is the union of two internally disjoint (x_i, x_{i+1})-dipaths P_i and Q_i.*

(iii) Let a_i be the second vertex of P_i and let b_i be the second to last vertex of Q_i. Then a_i, b_i, x_i, and x_{i+1} are all distinct, and $\rho^+ a_i$ and $b_i \rho^-$ are arcs of B_i'.

(iv) All vertices adjacent from ρ^+ in $V(B_i')$ are on $P_i[a_i, x_{i+1}]$. All vertices adjacent to ρ^- in $V(B_i')$ are on $Q_i[x_i, b_i]$.

PROOF. Suppose B_i' has a $(\rho^+ \to \rho^-, x_i \to x_{i+1})$-linkage (p, q). Lemma 7.1 implies that there is an (x_{i+1}, x_i)-dipath q' of K which uses no intermediate vertices of B_i. Then p and q, q' correspond to disjoint dicycles of H, a contradiction. By Theorem 2.5, it follows that there is a planar embedding M_i' of B_i' such that the vertices x_i, ρ^+, x_{i+1}, and ρ^- are on the outer face in the given cyclic order. Let $M_i = M_i' - \{\rho^+, \rho^-\}$. Then M_i is a planar embeding of B_i with outer face boundary C_i such that x_i and x_{i+1} are on C_i and all neighbours of ρ^+ and ρ^- in $V(B_i')$ are on C_i.

By Lemma 7.1 or 7.2, there are two disjoint (x_i, x_{i+1})-dipaths. By Corollary 2.1, for every y in $V(B_i) - \{x_i, x_{i+1}\}$, there exists an (x_i, y)-dipath and a (y, x_{i+1})-dipath such that y is the only vertex on both dipaths. Thus, the underlying graph of B_i is 2-connected. Hence, C_i is a cycle by Theorem 2.1. Then C_i is the union of 2 internally disjoint (x_i, x_{i+1})-paths P_i and Q_i. Suppose P_i or Q_i has a backward arc xy. By Corollary 2.1, there exists an (x_i, x)-dipath p and a (y, x_{i+1})-dipath q. Then p and q are skew interior paths of C_i, and so they intersect by Theorem 2.2. Let z be the first vertex of q which is on p. But now xy, $q[y, z]$, $p[z, x]$ is a dicycle in the acyclic digraph M_i. Therefore, P_i or Q_i are (x_i, x_{i+1})-dipaths.

Let a_i be a source of $B_i - x_i$. Since B_i is nontrivial, $a_i \neq x_{i+1}$. Since $2 \leq d_H^+(a_i)$, $x_i a_i$ and $\rho^+ a_i$ are arcs of B_i'. Since all vertices adjacent from ρ^+ in B_i' are on C_i, a_i is on C_i. We may assume a_i is on P_i. Suppose $x_i a_i$ is not the first arc of P_i. Then $P_i[x_i, a_i]$ is a path of length at least two because G is strict. Let w be an internal vertex of $P_i[x_i, a_i]$. By Theorem 2.2, every $(w, \{x_{i+1}, \rho^-\})$-dipath of B_i' intersects $x_i a_i$. But then a_i is on every such dipath, and so we have contradicted Corollary 2.1. Therefore, $x_i a_i$ is the first arc of P_i. Similarly, we can show that there is a sink b_i of $B_i - \{x_{i+1}\}$ such that $b_i \rho^-$ is in $A(B_i')$ and $b_i x_{i+1}$ is the last arc of P_i or Q_i.

Suppose $u\rho^-$ is in $A(B_i')$, for some $u \in V(P_i) - \{x_i\}$. But then $\rho^+ a_i$, $P_i[a_i, u]$, $u\rho^-$ and Q_i are a $(\rho^+ \to \rho^-, x_i \to x_{i+1})$-linkage of B_i'. Therefore, all vertices adjacent to ρ^- in $V(B_i')$ are in $V(Q_i) - \{x_{i+1}\}$. In particular, $b_i \in V(Q_i) - \{x_{i+1}\}$. Similarly, we can show that all vertices adjacent from ρ^+ in $V(B_i')$ are in $V(P_i) - \{x_i\}$. We can conclude that a_i is the unique source of $B_i - \{x_i\}$, b_i is the unique sink of $B_i - \{x_{i+1}\}$, and $a_i \neq b_i$. □

The structure of H and K is given by Lemmas 7.1, 7.2, and 7.3. We will use the notation of these lemmas to describe H and K. Further definitions are needed.

Let G''' be obtained from G by dividing x if $X = \{x\}$. We let $x_0 = x^+$ and $x_1 = x^-$ if $X = \{x\}$. We define B_0'' as follows. First we add four new vertices u, v, u', and v' to B_0. Then for every w in $V(B_0) - \{x_0\}$ and z in $V(B_0) - \{x_1\}$, we let $u'w$, $v'w$, zu, and zv be arcs of B_0'' if and only if $u'w$, $v'w$, zu, and zv, respectively, are arcs of G'''.

If $n \geq 2$ and B_i is a trivial cyclic block of K, then we let P_i and Q_i be $x_i x_{i+1}$. Since ρ is only adjacent to vertices of $\cup_{i=0}^{n-1} P_i$, all vertices adjacent from u' or v' in $\cup_{i=0}^{n-1} V(B_i)$ are on $\cup_{i=0}^{n-1} P_i$. Similarly, all vertices adjacent to u or v in $\cup_{i=0}^{n-1} V(B_i)$ are on $\cup_{i=0}^{n-1} Q_i$. Let I_i be the set of vertices in $\{u', v'\}$ which are adjacent to vertices in $V(P_i) - \{x_i\}$ and let $\mathcal{O}_i$ be the set of vertices in $\{u, v\}$ which are adjacent from vertices in $V(Q_i) - \{x_{i+1}\}$, $i = 0, \ldots, n-1$. We say the R_k-reduction is *type 1* if $I_0 = \{u', v'\}$, $\mathcal{O}_0 = \{u, v\}$, $I_i = \{v'\}$, and $\mathcal{O}_i = \{v\}$, $i = 1, \ldots, n-1$. We say the R_k-reduction is *type 2* if $n = 4$, $I_0 = I_2 = \{u'\}$, $\mathcal{O}_0 = \mathcal{O}_2 = \{u\}$, $I_1 = I_3 = \{v'\}$, $\mathcal{O}_1 = \mathcal{O}_3 = \{v\}$, and all the cyclic blocks of K are trivial.

The notation we have so far used to describe G, H, and K will be referred to as *standard notation*.

LEMMA 7.4. *Without loss of generality, the R_k-reduction of G to H is type 1 or 2.*

PROOF. We use standard notation. First, we will prove a restriction on $I_0, \ldots, I_{n-1}$ and $\mathcal{O}_0, \ldots, \mathcal{O}_{n-1}$. Suppose there exist i, j, k, and ℓ such that $0 \leq i \leq n-1$, $0 < j \leq k < \ell \leq n$, $u' \in I_i$, $u \in \mathcal{O}_{i+j}$, $v' \in I_{i+k}$, and $v \in \mathcal{O}_{i+\ell}$. Then there exist w in $V(P_i) - \{x_i\}$, x in $V(Q_{i+j}) - \{x_{i+j+1}\}$, y in $V(P_{i+k}) - \{x_{i+k}\}$, and z in $V(Q_{i+\ell}) - \{x_{i+\ell+1}\}$ such that $u'w$, xu, $v'y$, and zv are in $A(G)$. But then $P_{uu'}$, $u'w$, $P_i[w, x_{i+1}]$, $P_{i+1}, \ldots, P_{i+j-1}$, $Q_{i+j}[x_{i+j}, x]$, xu and $P_{vv'}$, $v'y$, $P_{i+k}[y, x_{i+k+1}]$, $P_{i+k+1}, \ldots, P_{i+\ell-1}$, $Q_{i+\ell}[x_{i+\ell}, z]$, zv are disjoint dicycles of G. Therefore, the preceeding occurance is impossible. We refer to this restriction as condition α.

It follows from Lemma 7.3 that I_i and $\mathcal{O}_i$ are nonempty, $i = 0, \ldots, n-1$. Since u' and v' are adjacent to at most one vertex in $V(R_k)$, u' and v' are in $\cup_{i=0}^{n-1} I_i$. Similarly, u and v are in $\cup_{i=0}^{n-1} \mathcal{O}_i$.

Suppose $I_m = \{u', v'\}$ and $\mathcal{O}_m = \{u, v\}$, for some m in $\{0, \ldots, n-1\}$. If $n = 1$ we are done, so assume not. Without loss of generality, we may assume $m = 0$ and $v' \in I_1$. If $u \in \mathcal{O}_1$, then condition α is contradicted, where $i = 1$, $j = k = n - 1$, and $\ell = n$. Hence, $\mathcal{O}_1 = \{v\}$. Then condition α implies $I_i = \{v'\}$, $i = 1, \ldots, n-1$. Finally, condition α implies $\mathcal{O}_i = \{v\}$, $i = 2, \ldots, n-1$. Therefore, we have a type 1 R_k-reduction. We now assume $I_i \neq \{u', v'\}$ or $\mathcal{O}_i \neq \{u, v\}$, $i = 0, \ldots, n-1$.

Suppose $v' \in I_0$ and $u \in \mathcal{O}_0$. Then we may assume $\mathcal{O}_0 = \{u\}$, otherwise, we have the previous case. Let m be the smallest integer in $\{1, \ldots, n-1\}$ such that $v \in \mathcal{O}_m$. Then $\mathcal{O}_i = \{u\}$, $i = 0, \ldots, m-1$. By condition α, $I_i = \{v'\}$, $i = m, \ldots, n-1$. Since $G \in \mathcal{I}_t$, $G - \{v', x_m\}$ has a dicycle C. Then C uses no vertices of $\cup_{i=m}^{n-1} B_i$ because $\cup_{i=m}^{n-1} I_i = \{v'\}$. Since v' and x_m are not on C, C uses u'. Hence, $\cup_{i=0}^{m-1} \mathcal{O}_i = \{u\}$ implies u is on C. Thus, $\cup_{i=0}^{m-1} B_i'' - \{x_0\}$ has a (u',u)-dipath R_u. Similarly, if we consider $G - \{u, x_m\}$, we can show that $\cup_{i=m}^{n-1} B_i'' - \{x_m\}$ has a (v',v)-dipath R_v. But then $P_{uu'}$, R_u and $P_{vv'}$, R_v are disjoint dicycles of G. Similarly, we can arrive at a contradiction if $u' \in I_0$ and $v \in \mathcal{O}_0$. We may now assume that either $I_i = \{u'\}$ and $\mathcal{O}_i = \{u\}$, or $I_i = \{v'\}$ and $\mathcal{O}_i = \{v\}$, for every i in $\{0, \ldots, n-1\}$.

Suppose $I_0 = \{u'\}$, $\mathcal{O}_0 = \{u\}$, $I_i = \{v'\}$, and $\mathcal{O}_i = \{v\}$, $i = 1, \ldots, n-1$. If we consider a dicycle of $G - \{x_0, v\}$, we can show that $B_0'' - \{x_0\}$ has a (u',u)-dipath R_u. If we consider a dicycle of $G - \{x_1, u\}$, we can show that $\cup_{i=1}^{n-1} B_i'' - \{x_1\}$ has a (v',v)-dipath R_v. But then $P_{uu'}$, R_u and $P_{vv'}$, R_v are disjoint dicycles of G. Therefore, we may assume $I_i = \{u'\}$ and $\mathcal{O}_i = \{u\}$ for at least two values of i in $\{0, \ldots, n-1\}$. Similarly, $I_i = \{v'\}$ and $\mathcal{O}_i = \{v\}$ for at least two values of i in $\{0, \ldots, n-1\}$. Hence, $4 \leq n$.

Suppose $I_0 = I_1 = \{u'\}$ and $\mathcal{O}_0 = \mathcal{O}_1 = \{u\}$. But since $I_i = \{v'\}$ and $\mathcal{O}_i = \{v\}$ for at least two values of i in $\{2, \ldots, n-1\}$, we contradict condition α. Thus, we may assume there is no i in $\{0, \ldots, n-1\}$ such that $I_i = I_{i+1}$ and $\mathcal{O}_i = \mathcal{O}_{i+1}$.

Without loss of generality, we may assume $I_0 = I_2 = \{u'\}$, $\mathcal{O}_0 = \mathcal{O}_2 = \{u\}$, $I_1 = I_3 = \{v'\}$, and $\mathcal{O}_1 = \mathcal{O}_3 = \{v\}$. If $5 \leq n$, then $I_4 = \{u'\}$ and $\mathcal{O}_4 = \{u\}$. But then condition α is contradicted. Therefore, $n = 4$ and we have a type 2 R_k-reduction. □

The following theorem follows immediately from the definitions of type 2 R_k-reduction and $\mathcal{K}$.

THEOREM 7.1. *If G has a type 2 R_k-reduction to H, then $G = D_7$ or $G \in \mathcal{K}$.*

We now finish this section by showing that if G has a type 1 R_k-reduction to H, then $G \in \mathcal{H}$. We first need two lemmas.

LEMMA 7.5. *If B_0 is nontrivial, then the following results hold.*
  (i) *For every y on $P_0[a_0, x_1)$, B_0 has a $(P_0[a_0, x_1), Q_0(x_0, b_0])$-dipath with origin y.*
  (ii) *For every z on $Q_0(x_0, b_0]$, B_0 has a $(P_0[a_0, x_1), Q_0(x_0, b_0])$-dipath with terminus z.*
  (iii) *If w is on $P_0(a_0, x_1)$, then B_0 has a $(a_0 \to b_0, w \to x_1)$-linkage.*
  (iv) *B_0' has a (ρ^+, ρ^-)-dipath.*
  (v) *Let w be on $P_0(a_0, x_1)$ and z be on $Q_0(x_0, b_0)$. If B_0' has two internally disjoint (ρ^+, ρ^-)-dipaths, then B_0 has a $(a_0 \to z, w \to b_0)$-linkage.*

PROOF. We use standard notation.
  (i) By Corollary 2.1, B_0' has a $(y \to \rho^-, y \to x_1)$-fan (p, q). All vertices adjacent to ρ^- in $V(B_0)$ are on $Q_0[x_0, b_0]$ by Lemma 7.3, and so p has a vertex on $Q_0[x_0, b_0]$. Hence, p has a subdipath p' which is a $(P_0[a_0, x_1),$

$Q_0(x_0, b_0])$-dipath of B_0. If the origin of p' is not y, then p' and some subdipath of q are interior skew paths of C_0. But then Theorem 2.2 is contradicted.

(ii) The proof similar to 1.

(iii) By 1, B_0 has a $(P_0[a_0, x_1), Q_0(x_0, b_0])$-dipath r_1 with origin a_0. Let z_1 be the terminus of r_1 and let $r = r_1, Q[z_1, b_0]$. Then $(r, P_0[w, x_1])$ is an $(a_0 \to b_0, w \to x_1)$-linkage of B_0.

(iv) By 3, B_0 has an (a_0, b_0)-dipath r. By Lemma 7.3 (iii), $\rho^+ a_0$ and $b_0 \rho^-$ are arcs of B_0'. Hence, $\rho^+ a_0, r, b_0 \rho^-$ is a (ρ^+, ρ^-)-dipath of B_0'.

(v) Let w be on $P_0(a_0, x_1)$ and z be on $Q_0(x_0, b_0)$. Suppose B_0' has two internally disjoint (y^+, y^-)-dipaths. Suppose there is a vertex c which is on every $(\{a_0, w\}, \{z, b_0\})$-dipath of B_0. By 1, B_0 has a $(P_0[a_0, x_1), Q_0(x_0, b_0])$-dipath p_1 with origin w. Let z_1 be the terminus of p_1. Then $p_1, Q_0[z_1, b_0]$ is a $(\{a_0, w\}, \{z, b_0\})$-dipath which avoids a_0. Hence, $c \neq a_0$. By 1, B_0 has a $(P_0[a_0, x_1), Q_0(x_0, b_0])$-dipath p_2 with origin a_0. Let z_2 be the terminus of p_2. Then $p_2, Q_0[z_2, b_0]$ is a $(\{a_0, w\}, \{z, b_0\})$-dipath which is disjoint from $P_0(a_0, x_1)$. Therefore, c is not on P_0. Similarly, c is not on Q_0. Let q_1 be one of the (y^+, y^-)-dipaths of B_0' which avoids c. By Lemma 7.3 (iv), all vertices adjacent from ρ^+ in B_i' are on $P_i[a_i, x_{i+1}]$ and all vertices adjacent to ρ^- in B_i' are on $Q_i[x_i, b_i]$, and so q_1 has a subdipath q_2 which is a $(P_0[a_0, x_1), Q_0(x_0, b_0])$-dipath. Let q_2 have origin w_2 and terminus z_2. But then $P_0[a_0, w_2], q_2, Q_0[z_2, b_0]$ is a $(\{a_0, w\}, \{z, b_0\})$-dipath of B_0 which avoids c. Therefore, Theorem 2.3 implies that B_0 has two disjoint $(\{a_0, w\}, \{z, b_0\})$-dipaths. Finally, Theorem 2.2 shows that these two paths give a $(a_0 \to z, w \to b_0)$-linkage.

□

LEMMA 7.6. *Suppose $G \in \mathcal{I}_t$ and $Y = \{z_1, z_2, z_3, z_4, z_5\}$ is contained in $V(G)$. Let K be obtained from G by dividing the vertices in Y. If K is acyclic and has no (z_i^+, z_j^-)-dipath for every (i, j) in $\{(1,1), (2,2), (3,3), (4,4), (5,5), (2,1), (2,5), (3,1), (3,4), (5,4)\}$, then $G \in \mathcal{H}$.*

PROOF. Let $Y^+ = \{z_1^+, \ldots, z_5^+\}$ and $Y^- = \{z_1^-, \ldots, z_5^-\}$. For every x in $V(K)$, define $Y^+(x)$ to be the set of all z^+ in Y^+ such that there exists a (z^+, x)-dipath of K, and define $Y^-(x)$ to be the set of all z^- in Y^- such that there exists a (x, z^-)-dipath of K. Let $I = V(K) - (Y^+ \cup Y^-)$, $I_\alpha = \{x \in I \mid z_3^+ \in Y^+(x)\}$, $I_\beta = \{x \in I \mid z_1^- \in Y^-(x)\}$, and $I_\gamma = \{x \in I \mid Y^+(x) = \{z_1^+, z_2^+\}$ and $Y^-(x) = \{z_3^-, z_4^-\}\}$. Let $K_\alpha = K[I_\alpha] \cup K(\{z_4^+, z_3^+, z_1^+\}, I_\alpha, \{z_5^-, z_2^-\})$, $K_\beta = K[I_\beta] \cup K(\{z_4^+, z_5^+\}, I_\beta, \{z_3^-, z_1^-, z_2^-\})$, and $K_\gamma = K[I_\gamma] \cup K(\{z_1^+, z_2^+\}, I_\gamma, \{z_3^-, z_4^-\})$. Let G_α, G_β, and G_γ be the corresponding subdigraphs of G. We will show that $(G_\alpha, G_\beta, G_\gamma)$ is an $\mathcal{G}$-decomposition of G with $\mathcal{H}$-separator $(z_1, z_2, z_3, z_4, z_5)$.

Since K is acyclic, Corollary 2.1 implies that K has two openly disjoint (z_3^+, Y^-)-dipaths. Since K has no $(z_3^+, \{z_1^-, z_3^-, z_4^-\})$-dipath, we must have a $(z_3^+ \to z_2^-, z_3^+ \to z_5^-)$-fan (p_{32}, p_{35}). Similarly, K has a $(z_4^+ \to z_1^-, z_5^+ \to z_1^-)$-fan

(p_{41}, p_{51}), a $(z_2^+ \to z_3^-, z_2^+ \to z_4^-)$-fan (p_{23}, p_{24}), and a $(z_1^+ \to z_4^-, z_2^+ \to z_4^-)$-fan (q_{14}, q_{24}). Since K has no (z_3^+, z_1^-)-dipath, $p_{32} \cup p_{35}$ and $p_{41} \cup p_{51}$ are disjoint.

Suppose $u \in I$ and $z_2^+ \in Y^+(u)$. Then $Y^-(u) = \{z_3^-, z_4^-\}$. Since $z_4^- \in Y^-(u)$, $Y^+(u) = \{z_1^+, z_2^+\}$. Thus, $u \in I_\gamma$. Similarly, if $u \in I$ and $z_4^- \in Y^-(u)$, then $u \in I_\gamma$. Thus, if u in I is adjacent with a vertex in I_γ, then $z_2^+ \in Y^+(u)$ or $z_4^- \in Y^-(u)$, and so $u \in I_\gamma$. It follows that vertices in I_γ are not adjacent with vertices in $I - I_\gamma$, and that I_γ and $I_\alpha \cup I_\beta$ are disjoint. Hence, p_{23}, p_{24}, q_{14}, and q_{24} are all dipaths of K_γ, and K_γ is disjoint from p_{32}, p_{35}, p_{41}, and p_{51}. If I_α and I_β are not disjoint, then K has a (z_3^+, z_1^-)-dipath. Thus, I_α, I_β, and I_γ are pairwise disjoint.

Suppose K_γ has a $(z_1^+ \to z_4^-, z_2^+ \to z_3^-)$-linkage (r_{14}, r_{23}). Then p_{41}, r_{14} and p_{32}, r_{23} are disjoint dicycles of K. Hence, $K_\gamma \in P(z_1^+, z_2^+; z_3^-, z_4^-)$. Suppose there exists a vertex x on all $(\{z_1^+, z_2^+\}, \{z_3^-, z_4^-\})$-dipaths of K_γ. But then the existence of p_{23} and p_{24} implies $x = z_2^+$, while the existence of q_{14} and q_{24} implies $x = z_4^-$. Thus, K_γ has a $(z_1^+ \to z_3^-, z_2^+ \to z_4^-)$-linkage (r_{13}, r_{24}).

Suppose $x \in I - (I_\alpha \cup I_\beta \cup I_\gamma)$. Since $z_2^+ \in Y^+(x)$ or $z_4^- \in Y^-(x)$ implies $x \in I_\gamma$, $z_3^+ \in Y^+(x)$ implies $x \in I_\alpha$, and $z_1^- \in Y^-(x)$ implies $x \in I_\beta$, we have $Y^+(x) \subseteq \{z_1^+, z_4^+, z_5^+\}$ and $Y^-(x) \subseteq \{z_2^-, z_3^-, z_5^-\}$. Since K has no (z_5^+, z_5^-)-dipath, we may assume by symmetry that $Y^-(x) = \{z_2^-, z_3^-\}$.

There exists a $(\{z_4^+, z_5^-\}, x)$-dipath q and an $(x \to z_2^-, x \to z_3^-)$-fan (q_2, q_3) by Corollary 2.1. Let z be the last vertex of q on p_{41} or p_{51}.

Suppose z is on p_{51}. Let $q_{53} = p_{51}[z_5^+, z], q[z, x], q_3$. Since p_{41} and p_{51} are openly disjoint, p_{41} and $p_{51}[z_5, z]$ are disjoint. By the choice of z, p_{41} and $q[z, x]$ are disjoint. Since $z_1^- \notin Y^-(x)$, p_{41} does not intersect q_3. Thus, p_{41} and q_{53} are disjoint. We have already shown that p_{35}, p_{41}, and q_{14} are pairwise disjoint. Also, q_{53} and $p_{35} \cup q_{14}$ are disjoint, otherwise, K would have a restricted path. But now p_{35}, q_{53} and p_{41}, q_{14} correspond to disjoint dicycles of G.

Suppose z is on p_{41}. Let $q_{42} = p_{41}[z_4^+, z], q[z, x], q_2$. As in the previous case we can show that q_{42} and q_{51} are disjoint. Since $z_5^- \notin Y^-(x)$, p_{35} does not intersect q_2. Since K has no (z_3^+, z_3^-)-dipath, p_{35} does not intersect $p_{41}[z_4^+, z], q[z, x]$. Thus, p_{35} and q_{42} are disjoint. Furthermore, r_{13} and r_{24} are disjoint by construction. It then follows that $r_{13}, r_{24}, p_{35}, p_{51}$, and q_{42} are pairwise disjoint, otherwise, K would have a restricted dipath. But now p_{35}, p_{51}, r_{13} and q_{42}, r_{24} correspond to disjoint dicycles of G. Therefore, $I = I_\alpha \cup I_\beta \cup I_\gamma$.

Suppose $x \in I_\alpha$. Then $z_3^+ \in Y^+(x)$, and $Y^-(x) = \{z_2^-, z_5^-\}$. Since $z_5^- \in Y^-(x)$, $Y^+(x) \subseteq \{z_1^+, z_3^+, z_4^+\}$. Thus, $A(K_\alpha)$ contains all arcs joining a vertex in I_α and a vertex in $Y^+ \cup Y^-$. Similarly, $A(K_\beta)$ contains all arcs joining a vertex in I_β and a vertex in $Y^+ \cup Y^-$.

By the constructions of K_α, K_β, and K_γ, every arc of $K(Y^+, Y^-)$ belongs to a unique K_α, K_β, or K_γ, with the possible exception of $z_4^+ z_2^-$. But if $z_4^+ z_2^- \in A(K)$, then $z_4^+ z_2^-, r_{24}$ and p_{35}, p_{51}, r_{13} correspond to disjoint dicycles of G.

If a vertex in I_α and a vertex in I_β are adjacent, then K has a (z_3^+, z_1^-)-dipath or a (z_5^+, z_5^-)-dipath.

Therefore, we have shown $K = K_\alpha \cup K_\beta \cup K_\gamma$, and only the vertices in $Y^+ \cup Y^-$ are in more than one of $V(K_\alpha)$, $V(K_\beta)$, and $V(K_\gamma)$.

If K_α has a $(z_4^+ \to z_2^-, z_3^+ \to z_5^-)$-linkage (r_{42}, r_{35}), then r_{42}, r_{24} and r_{35}, p_{51}, r_{13} correspond to disjoint dicycles of G. If K_α has a $(z_4^+ \to z_2^-, z_1^+ \to z_5^-)$-linkage (r_{42}, r_{15}), then r_{42}, r_{24} and r_{15}, p_{51} correspond to disjoint dicycles of G. If K_α has a $(z_3^+ \to z_2^-, z_1^+ \to z_5^-)$-linkage (r_{32}, r_{15}), then r_{32}, p_{23} and r_{15}, p_{51} correspond to disjoint dicycles of G. Therefore, $K_\alpha \in P(z_4^+, z_3^+, z_1^+; z_5^-, z_2^-)$. Similarly, $K_\beta \in P(z_4^+, z_5^+; z_3^-, z_1^-, z_2^-)$.

It now follows that $(G_\alpha, G_\beta, G_\gamma)$ is an $\mathcal{H}$-decomposition of G with $\mathcal{H}$-separator $(z_1, z_2, z_3, z_4, z_5)$. □

THEOREM 7.2. *If G has a type 1 R_k-reduction to H, then $G \in \mathcal{H}$.*

PROOF. We use standard notation. There exists a (ρ^+, ρ^-)-dipath of B_0' by Lemma 7.5 (iv). Suppose there is a vertex z_1 in $V(B_0)$ such that $B_0' - z_1$ has no (ρ^+, ρ^-)-dipath. If there exists a vertex z_0 such that $\cup_{i=0}^{n-1} B_i'' - \{x_0, z_0\}$ has no $(\{u', v'\}, \{u, v\})$-dipath, then $\{x_0, z_0\}$ corresponds to a 2-transversal of G. Hence, $\cup_{i=0}^{n-1} B_i'' - x_0$ has 2 disjoint $(\{u', v'\}, \{u, v\})$-dipaths p_1 and q_1. If (p_1, q_1) is a $(u' \to u, v' \to v)$-linkage, then $P_{uu'}, p_1$ and $P_{vv'}, q_1$ are disjoint dicycles of G. Therefore, we may assume (p_1, q_1) is a $(v' \to u, u' \to v)$-linkage. Since x_0 is not on p_1 and u is only in $\mathcal{O}_0$, p_1 is a dipath of B_0''. Hence, z_1 is on p_1, and so z_1 is not on q_1. Since z_1 is not on q_1 and u' is only in $\mathcal{I}_0$, x_1 is on q_1. Similarly, we can show that $G - x_1$ has a $(v' \to u, u' \to v)$-linkage (p_2, q_2) such that p_2 is a dipath of B_0'', z_1 is on p_2, and x_0 is on q_2. If $q_1[u', x_1]$ intersects $p_2(z_1, v]$, then we have a (u', v)-dipath in B_0'' which avoids z_1. Hence, $q_1[u', x_1]$ and $p_2[z_1, v]$ are disjoint. Similarly, $q_2[x_0, u]$ and $p_1[v', z_1]$ are disjoint. Let $P_{x_1 x_0}$ be an (x_1, x_0)-dipath of $\cup_{i=0}^{n-1} B_i$. But then $P_{uu'}, q_1[u', x_1], P_{x_1 x_0}, q_2[x_0, u]$ and $P_{vv'}, p_1[v', z_1], p_2[z_1, v]$ are disjoint dicycles of G. Therefore, there exist two internally disjoint (ρ^+, ρ^-)-dipaths in B_0' by Theorem 2.3.

Since there are two disjoint (ρ^+, ρ^-)-dipaths in B_0', $N^+(u) \cup N^+(v)$ contains at least two vertices on $P_0(x_0, x_1)$. The existence of p_1 and p_2 implies that both u and v are adjacent to a vertex on $P_0(x_0, x_1)$. By Lemma 7.3 (iii), $\rho a_0 \in A(H)$, and so $a_0 \in N^+(u) \cup N^+(v)$. These three observations imply that there exists w on $P_0(a_0, x_1)$ such that $u'a_0, v'w \in A(G)$ or $u'w, v'a_0 \in A(G)$. Similarly, there exists z on $Q_0(x_0, b_0)$ such that $zu, b_0v \in A(G)$ or $zv, b_0u \in A(G)$. Subject to the previous conditions, suppose w and z are chosen so that $P_0[a_0, w]$ and $Q_0[z, b_0]$ have maximum length.

Since B_0' has two internally disjoint (ρ^+, ρ^-)-dipaths, Lemma 7.5 (v) implies B_0 has an $(a_0 \to z, w \to b_0)$-linkage (p_3, q_3). If $v'w, u'a_0, zu$, and b_0v are arcs of G, then $P_{vv'}, v'w, q_3, b_0v$ and $P_{uu'}, u'a_0, p_3, zu$ are disjoint dicycles of G. We get a similar contradiction if $v'a_0, u'w, zv$, and b_0u are arcs of G. Hence, $v'w, u'a_0, zv, b_0u \in A(G)$ or $v'a_0, u'w, zu, b_0v \in A(G)$.

Suppose $2 \leq n$. Since the R_k-reduction is type 2, v is adjacent from some vertex z' on Q_1. Suppose $v'w, u'a_0, zv, b_0u \in A(G)$. By Lemma 7.5 (iii), B_0 has an $(a_0 \to b_0, w \to x_1)$-linkage (p_4, q_4). But then $P_{uu'}$, $u'a_0$, p_4, b_0u and $P_{vv'}$, $v'w$, q_4, $Q_1[x_1, z']$, $z'v$ are disjoint dicycles of G. Therefore, $v'a_0, u'w, zu, b_0v \in A(G)$. Without loss of generality, we may assume these four arcs are in $A(G)$ when $n = 1$.

Suppose B_0 has three disjoint $(\{x_0, a_0, w\}, \{z, b_0, x_1\})$-dipaths. Then we have an $(x_0 \to z, a_0 \to b_0, w \to x_1)$-linkage (p_5, q_5, r_5), otherwise, $P_0 \cup Q_0$ has two interior skew paths and Theorem 2.2 is contradicted. But then $P_{uu'}$, $u'w$, r_5, $P_{x_1 x_0}$, p_5, zu and $P_{vv'}$, $v'a_0$, q_5, b_0v are disjoint dicycles of G. Therefore, Theorem 2.3 implies the existence of two vertices s and t which intersect all $(\{x_0, a_0, w\}, \{z, b_0, x_1\})$-dipaths of B_0. Since $P_0[w, x_1]$ and $Q_0[x_0, z]$ are disjoint $(\{x_0, a_0, w\}, \{z, b_0, x_1\})$-dipaths we may assume s is on $P_0[w, x_1]$ and t is on $Q_0[x_0, z]$. If $s = x_1$, then q_3 is a (w, b_0)-dipath avoiding s and t. Hence, s is on $P_0[w, x_1)$. Similarly, t is on $Q_0(x_0, z]$.

Suppose $u'x_1$ is in $A(G)$. Then Lemma 7.5 (ii) implies B_0 has a $(P_0[a_0, x_1], Q_0(x_0, b_0])$-dipath r_1 with terminus b_0. Let w_1 be the origin of r_1. Then $P_{uu'}$, $u'x_1$, $P_{x_1 x_0}$, $Q_0[x_0, z]$, zu and $P_{vv'}$, $v'a_0$, $P_0[a_0, w_1]$, r_1, b_0v are disjoint dicycles of G. Therefore, $u'x_1 \notin A(G)$. Similarly, $x_0 u \notin A(G)$.

In this paragraph and the next we show the existence of various dipaths in B_0 which will be useful in the rest of the proof. Suppose w' is a vertex on $P_0[a_0, s)$. We will show that $B_0 - s$ has a (w', t)-dipath. By Lemma 7.5 (i), B_0 has a $(P_0[a_0, x_1], Q_0(x_0, b_0])$-dipath r_2 with origin w'. Let z_2 be the terminus of r_2. If z_2 is on $Q_0(t, b_0]$, then $P_0[b_0, w']$, $r_2[w', z_2]$, $Q_0[z_2, b_0]$ is an (a_0, b_0)-dipath of $B_0 - \{s, t\}$. Hence, z_2 is on $Q_0(x_0, t]$. Therefore, r_2, $Q_0[z_2, t]$ is a (w', t)-dipath of $B_0 - s$. Similarly, $B_0 - t$ has an (s, z')-dipath, for every z' on $Q_0(t, b_0)$.

Suppose d is a vertex on $P_0(s, x_1)$. We will show that $B_0 - V(Q_0[x_0, b_0))$ has a (d, b_0)-dipath. If w_1 is on $P_0[a_0, s)$, then $P_0[a_0, w_1]$, r_1 is an (a_0, b_0)-dipath of $B_0 - \{s, t\}$. Hence, w_1 is on $P_0[s, x_1]$. If w_1 is on $P_0[d, x_1]$, then $P_0[d, w_1]$, r_1 is the desired path. Hence, we may assume w_1 is on $P_0[s, d)$. By Lemma 7.5 (i), B_0' has a $(d, Q_0(x_0, b_0])$-dipath r_3. Then r_3 intersects r_1, otherwise, $P_0 \cup Q_0$ has two interior skew paths. Let w_3 be the first vertex of r_3 on r_1. Then $r_7 = r_3[d, w_3]$, $r_1[w_3, b_0]$ is the desired path.

Suppose there is a vertex w' on $P_0[a_0, s)$ and a vertex d on $P_0(s, x_1)$ such that $u'w'$ and $v'd$ are in $A(G)$. By the previous two paragraphs, $B_0 - s$ has a (w', t)-dipath p_6 and $B_0 - V(Q_0[x_0, b_0))$ has a a (d, b_0)-dipath r_7. Furthermore, p_6 and r_7 are disjoint, otherwise, we get an (a_0, b_0)-dipath of $B_0 - \{s, t\}$. But then $P_{uu'}$, $u'w'$, p_6, $Q_0[t, z]$, zu and $P_{vv'}$, $v'd$, r_7, b_0v are disjoint dicycles of G. Therefore, v' is not adjacent to any vertex on $P_0(s, x_1)$ or u' is not adjacent to any vertex on $P_0[a_0, s)$. In the latter case we must have $w = s$ and $N_G^+(u') = \{v', s\}$. Thus, v' is not adjacent to any vertex on $P_0(s, x_1)$ or $N_G^+(u') = \{v', s\}$. Similarly, v is not adjacent to any vertex on $Q_0(x_0, t)$ or $N_G^-(u) = \{v, t\}$.

Suppose v' is not adjacent to any vertex on $P_0(s, x_1)$ and v is not adjacent from any vertex on $Q_0(x_0, t)$. Since $G \in \mathcal{I}_t$, $G - \{s, t\}$ has a dicycle C. Since w and z were chosen so that $P_0[a_0, w]$ and $Q_0[z, b_0]$ have maximum length, u' is not adjacent to any vertex on $P_0(s, x_1)$ and u is not adjacent from any vertex on $Q_0(x_0, t)$. We have also shown that $u'x_1$ and x_0u are not arcs of G. It now follows that $v'x_1$ and x_0v are the only possible $(\{x_0, u', v'\}, \{x_1, u, v\})$-dipaths of $B_0'' - \{s, t\}$, otherwise, $B_0 - \{s, t\}$ has an $(\{x_0, a_0, w\}, \{z, b_0, x_1\})$-dipath. Hence, C uses no vertices in $V(B_0) - \{x_0, x_1\}$. Then C uses u' or v'. Since v' and the vertices on $P_0[a_0, s)$ are the only possible vertices adjacent from u', C uses v'. Similarly, C uses v. Hence, we may assume $P_{vv'}$ is a subdigraph of C. If B_0'' has a (u', u)-dipath r_4, then C and $P_{uu'}$, r_4 are disjoint dicycles of G. Thus, B_0'' has no $(u'u)$-dipath. If u' is adjacent to a vertex w' on $P_0(x_0, s)$, then we have shown that B_0 has a (w', t)-dipath r_2. But then $u'w'$, r_2, $Q_0[z_2, z]$, zu is a (u', u)-dipath of B_0''. Hence, $w = s$ and $N_G^+(u') = \{s, v'\}$. Similarly, $z = t$ and $N_G^-(u) = \{v, t\}$.

In the previous paragraph we have shown that if B_0'' has a (u', u)-dipath, then we may assume v' is adjacent to some vertex d on $P_0(s, x_0)$. We have shown that this implies $N_G^+(u') = \{s, v'\}$. We will deal with this possibility in cases 1 and 2. Also, we have shown that if B_0'' has no (u', u)-dipath, then $N_G^+(u') = \{s, v'\}$ and $N_G^-(u) = \{v, t\}$. We will deal with this possibility in case 3.

Case 1. Suppose B_0'' has a (u', u)-dipath, v' is adjacent to some vertex d on $P_0(s, x_0)$, and $N_G^+(u') = \{s, v'\}$. Suppose there exists a vertex z' on $Q_0[t, b_0)$ such that $z'u$ is in $A(G)$ and B_0 has an (s, z')-dipath r_5.

If there exist two disjoint $(\{s, d\}, \{z', b_0\})$-dipaths in B_0, then we must have an $(s \to z', d \to b_0)$-linkage (p_7, q_7) in B_0. But then $P_{uu'}$, $u's$, p_7, $z'u$ and $P_{vv'}$, $v'd$, q_7, b_0v are disjoint dicycles of G. Therefore, there exists a vertex c which intersects every $(\{s, d\}, \{z', b_0\})$-dipath of B_0.

We now show that c is not on P_0 or Q_0. If $b_0 = c$, then $r_5[c, z']$, $Q_0[z', b_0]$ is a closed directed walk of B_0. Hence, $b_0 \neq c$. By Lemma 7.5 (i), B_0 has a $(P_0[a_0, x_1], Q_0(x_0, b_0])$-dipath r_6 with origin s. Let z_6 be the terminus. Then c is on r_6, $Q_0[z_6, b_0]$, and so c is not on $P_0(s, x_1]$. We have shown that $B_0 - V(Q_0(x_0, b_0))$ has a (d, b_0)-dipath r_7. Then c is on r_7, and so c is not on $P_0[x_0, d)$ or $Q_0[x_0, b_0)$.

It follows from the previous paragraph that c is on every $(P_0[s, x_1], Q_0(x_0, x_1])$-dipath. Hence, every $(P_0[a_0, x_1], Q_0[t, b_0])$-dipath uses t or c.

We now show that B_0 has a $(P_0[a_0, x_1], Q_0(x_0, b_0])$-dipath r with origin s and terminus z'. Since c is not on $P_0 \cup Q_0$, c is an internal vertex of r_6. The (s, z')-dipath r_5 has a subdipath r_5' which is a $(P_0[a_0, x_1], Q_0(x_0, b_0])$-dipath. Then the terminus z_8 of r_5' is on $Q_0(x_0, z']$ and c is an internal vertex of r_5'. Let r_8 be $r_6[s, c]$, $r_5[c, z_8]$. By Lemma 7.5 (ii), B_0 has a $(P_0[a_0, x_1], Q_0(x_0, b_0])$-dipath r_9 with terminus z'.

Let z_9 be the last vertex of r_9 on $r_8 \cup P_0$. If z_9 is on $P_0[s, x_1)$, then c is on $r_9(z_9, t)$. But c is on r_8, and so we contradict the choice of z_9. If z_9 is on $P_0[a_0, s)$, then $r_9(z_9, t)$ and r_8 are interior skew paths of $P_0 \cup Q_0$. Hence, z_9 is on r_8. Then we can let r be $r_8[s, z_9]$, $r_9[z_9, t]$.

If v is adjacent from a vertex z_3 on $Q_0[x_0, t)$, then $P_{vv'}$, $v'd$, $P_0[d, x_1]$, $P_{x_1 x_0}$, $Q_0[x_0, z_3]$, $z_3 v$ and $P_{uu'}$, $u's$, r, $z'u$ are disjoint dicycles of G. Hence, $V(Q_0[x_0, t))$ $\cap N_G^-(v) = \emptyset$. If $2 \leq n$, then $v \in \mathcal{O}_1$ because we have a type 1 R_k-reduction, and so we can get a similar contradiction. Therefore, $n = 1$ and x_0 corresponds to x_1 in G.

We now show that G is in $\mathcal{H}$ by applying Lemma 7.6 with $(z_1, z_2, z_3, z_4, z_5)$ equal to (v', x_0, s, t, c) or (v', s, x_0, c, t). Let $Y = \{v, x_0, s, t\ c\}$ and G_s be the graph obtained from G by dividing the vertices in Y. Let C be any dicycle of G. We first show that C uses two vertices in Y. Suppose C uses vertices of R_k. Since R_k is acyclic, C must also use vertices of $G - V(R_k)$. Since s is the only vertex of $G - V(R_k)$ adjacent from u', C uses v' or s. If C does not use x_0, then C has a $(P_0[a_0, x_1], Q_0[t, b_0))$-dipath because all vertices of $G - V(R_k)$ adjacent to u or v are on $Q[t, b_0]$. Then C uses t or c. Suppose C does not use any vertices of R_k. Then C corresponds to an (x_0, x_1)-dipath of B_0 because $n = 1$. Such a dipath must also use s or t. Therefore, G_s is acyclic and has no (z^+, z^-)-dipath for every z in Y.

Finally, we show that G_s does not have any of the other restricted dipaths. Suppose G_s has an (x_0^+, c^-)-dipath. Then $G - \{v', s, t\}$ has an (x_0, c)-dipath, and so $B_0 - \{s, t\}$ has an (x_0, c)-dipath p_8. But then p_8, $r_7[c, b_0]$ is an (x_0, b_0)-dipath of B_0 which avoids s and t. Thus, G_s has no (x_0^+, c^-)-dipath. Suppose G_s has a $(x_0^+, (v')^-)$-dipath. Then u or v is adjacent from some vertex z_8 on $Q_0(t, b_0)$ such that $B_0 - \{s, t\}$ has an (x_0, z_8)-dipath q_8. But then $q_8, Q_0[z_8, b_0]$ is an (x_0, b_0)-dipath of B_0 which avoids s and t. Thus, G_s has no $(x_0^+, (v')^-)$-dipath. Since c is on every $(P_0[s, x_1], Q_0(x_0, b_0])$-dipath of B_0, every $(s, \{v', t\})$-dipath of G uses c or x_0 Hence, G_s has no $(s^+, (v')^-)$-dipath and no (s^+, t^-)-dipath. Finally, if G_s has a (c^+, t^-)-dipath and a (t^+, c^-)-dipath, then B_0 has a (c, t)-dipath and a (t, c)-dipath. But then B_0 has a dicycle. Hence, G_s has no (c^+, t^-)-dipath or no (t^+, c^-)-dipath.

Case 2. Suppose B_0'' has a (u', u)-dipath, v' is adjacent to some vertex d on $P_0(s, x_0)$, and $N_G^+(u') = \{s, v'\}$. Suppose b_0 is the only vertex z' on $Q_0[t, b_0)$ such that $z'u$ is in $A(G)$ and B_0 has an (s, z')-dipath r_5. We have shown that there is an (s, z')-dipath, for every z' on $Q_0(t, b_0)$. Therefore, $z = t$ and t and b_0 are the only vertices on Q_0 adjacent to u.

Suppose v is adjacent from a vertex z_4 on $Q_0(t, b_0)$. Let $p_9 = P_0[s, d], r_7$. We have shown that B_0 has an (a_0, t)-dipath q_9 which does not use s. Since $B_0 - \{s, t\}$ does not have an (a_0, b_0)-dipath, p_9 and q_9 do not intersect. But then $P_{uu'}$, $u's$, p_9, $b_0 u$ and $P_{vv'}$, $v' a_0$, q_9, $Q_0[t, z_4]$, $z_4 v$ are disjoint dicycles of G. Therefore, v is not adjacent from any vertex on $Q_0(t, b_0)$. Suppose v is adjacent from a vertex z_5 on $Q_0[x_0, t]$. By Lemma 7.5 (i), B_0 has an (s, b_0)-dipath r_{10} which is disjoint from $P_0(s, x_1)$. Since B_0 has no (s, z)-dipath, r_{10} is disjoint from $Q_0[x_0, t]$. But then $P_{uu'}$, $u's$, r_{10}, $b_0 u$ and $P_{vv'}$, $v'd$, $P_0[d, x_1]$, $P_{x_1 x_0}$, $Q_0[x_0, z_5]$, $z_5 v$ are disjoint dicycles of G. Therefore, v is not adjacent from any vertex on $Q_0(x_0, t]$. If $2 \leq n$, then $v \in \mathcal{O}_1$ because we have a type 1 R_k-reduction, and so we can get a similar contradiction. Therefore, $n = 1$ and x_0 corresponds to x_1 in G.

We have shown that $N_G^-(v) = \{u, b_0\}$. We also know that t and b_0 are the only vertices on Q_0 adjacent to u, and that B_0 has no (s, t)-dipath. Hence, every $(P_0[a_0, x_1], Q_0(x_0, b_0])$-dipath uses t or b_0. As in case 1, we can now show that G is in $\mathcal{H}$ by applying Lemma 7.6 with $(z_1, z_2, z_3, z_4, z_5)$ equal to (v', x_0, s, t, b_0) or (v', s, x_0, c, t).

Case 3. Suppose B_0'' has no (u', u)-dipath, $N_G^+(u') = \{s, v'\}$, and $N_G^-(u) = \{v, t\}$. If $v = v'$, then $\{v, x_0\}$ is a 2-transversal of G because B_0'' has no (u', u)-dipath. But then G is no longer in $\mathcal{I}_t$. Thus, $v \neq v'$.

We now show G is in $\mathcal{H}$ by using Lemma 7.6 with $(z_1, z_2, z_3, z_4, z_5) = (t, v, s, v', x_0)$. Let $Y = \{t, v, s, v'\ x_0\}$ and G_s be the graph obtained from G by dividing the vertices in Y. Let C be any dicycle of G. We first show that C uses two vertices in Y. Suppose C uses vertices of R_k. Since R_k is acyclic, C must also use vertices of $G - V(R_k)$. Since s is the only vertex of $G - V(R_k)$ adjacent from u', C uses v' or s. Similarly, C uses v or t. If C does not use any vertices of R_k, then C has a subdipath corresponding to an (x_0, x_1)-dipath of B_0. Hence, C uses x_0, and s or t. Therefore, G_s is acyclic and has no (z^+, z^-)-dipath for every z in Y. Finally, we show that G_s does not have any of the other restricted dipaths. Since s is the only vertex of $G - V(R_k)$ adjacent from u', every $(v, \{t, x_0\})$-dipath of G uses s or v'. Hence, G_s has no (v^+, t^-)-dipath or (v^+, x_0^-)-dipath. Similarly, G_s has no $(s^+, (v')^-)$-dipath or $(x_0^+, (v')^-)$-dipath because t is the only vertex of $G - V(R_k)$ adjacent to u. If G_s has an (s^+, t^-)-dipath, then B_0 has an (s,t)-dipath r_{11}. But then $u's$, r_{11}, tu is a (u', u) path of B_0''. □

8. R_k-reductions to $\{D_7\} \cup \mathcal{K} \cup \mathcal{H}$

We have shown that if G is in $\mathcal{I}_t$, then G has an R_k-reduction to an intercyclic digraph H in reduced form. In the last section we proved that if H is in $\mathcal{T}$, then G is in $\{D_7\} \cup \mathcal{K} \cup \mathcal{H}$. In this section we prove that if H is in $\{D_7\} \cup \mathcal{K} \cup \mathcal{H}$, then G is in $\{D_7\} \cup \mathcal{K} \cup \mathcal{H}$. Throughout this section we use the notation of §3 used in defining D_7, $\mathcal{K}$, and $\mathcal{H}$.

THEOREM 8.1. *If G has an R_k-reduction to D_7, then $G \in \mathcal{K}$.*

PROOF. Let R_k be contracted to ρ. Suppose $\rho = y_1$. We may assume $y_6 u$ and $y_4 v$ are in $A(G)$. If $u'y$ and $v'y_2$ are in $A(G)$, then $P_{uu'}$, $u'y$, yy_6, $y_6 u$ and $P_{vv'}$, $v'y_2$, $y_2 y_3$, $y_3 y_4$, $y_4 v$ are disjoint dicycles of G. If $u'y_2$ and $v'y$ are in $A(G)$, then $P_{uu'}$, $u'y_2$, $y_2 y_5$, $y_5 y_6$, $y_6 u$ and $P_{vv'}$, $v'y$, yy_4, $y_4 v$ are disjoint dicycles of G. Hence, $\rho \neq y_1$. Similarly, $\rho \neq y_i$, $i = 2, \ldots, 6$. Therefore, $\rho = y$. By symmetry we may assume $y_1 u$, $y_3 u$, and $y_5 v$ are in $A(G)$.

Suppose $v'y_2 \in A(G)$. If $u'y_4 \in A(G)$, then $P_{uu'}$, $u'y_4$, $y_4 y_1$, $y_1 u$ and $P_{vv'}$, $v'y_2$, $y_2 y_5$, $y_5 v$ are disjoint dicycles of G. If $u'y_6 \in A(G)$, then $P_{uu'}$, $u'y_6$, $y_6 y_3$, $y_3 u$ and $P_{vv'}$, $v'y_2$, $y_2 y_5$, $y_5 v$ are disjoint dicycles of G. Hence, $u'y_4, u'y_6 \notin A(G)$, and so $u'y_2$, $v'y_4$, and $v'y_6$ are in $A(G)$. But then $P_{uu'}$, $u'y_2$, $y_2 y_3$, $y_3 u$ and

$P_{vv'}, v'y_4, y_4y_5, y_5v$ are disjoint dicycles of G. Therefore, $v'y_2 \notin A(G)$, and so $u'y_2 \in A(G)$.

If $v'y_4 \in A(G)$, then $P_{uu'}, u'y_2, y_2y_3, y_3u$ and $P_{vv'}, v'y_4, y_4y_5, y_5v$ are disjoint dicycles of G. Hence, $v'y_4 \notin A(G)$, and so $u'y_4 \in A(G)$. Since $v'y_2, v'y_4 \notin A(G)$, $v'y_6 \in A(G)$. By symmetry, $y_1v, y_3v \notin A(G)$. Each of y_5u and $u'y_6$ could be in $A(G)$.

Now $G \in \mathcal{K}$, where $K = G[V(R_k) \cup \{y_5, y_6\}]$, $x_0 = y_4$, $x_1 = y_1$, $x_2 = y_2$, $x_3 = y_3$, $w_0 = y_5$, $z_0 = u$, $w_1 = u'$, and $z_1 = y_6$. □

THEOREM 8.2. *If G has an R_k-reduction to H in $\mathcal{K}$, then $G \in \mathcal{K}$.*

PROOF. Let R_k be contracted to ρ. K'_H has a $(w_0 \to z_1, z_0 \to w_1)$-linkage (p, q) by Corollary 2.1. If $2 \leq d_{K'_H}(w_0)$, then there is a (w_0, w_1)-dipath r in K'_H by Corollary 2.1. If $d_{K'_H}(w_0) = 1$, then $w_0z_0 \in A(H)$, and so w_0z_0, q is a (w_0, w_1)-dipath r in K_H.

Suppose $\rho = x_0$. We may assume x_3u and w_1v are in $A(G)$. If $u'x_1, v'w_0 \in A(G)$, then $P_{uu'}, u'x_1, x_1x_2, x_2x_3, x_3u$ and $P_{vv'}, v'w_0, r, w_1v$ are disjoint dicycles of G. If $u'w_0, v'x_1 \in A(G)$, then $P_{uu'}, u'w_0, p, z_1x_3, x_3u$ and $P_{vv'}, v'x_1, x_1z_0, q, w_1v$ are disjoint dicycles of G. Hence, $\rho \neq x_0$. Similarly, $\rho \neq x_i$, $i = 1, 2, 3$.

Suppose ρ is an intermediate vertex of K'_H. Let G' be obtained from K'_H by expansion at ρ. If G' has a $(w_0 \to w_1, z_0 \to z_1)$-linkage (p_1, q_1), then p_1, w_1x_0, x_0w_0 and q_1, z_1x_1, x_1z_0 are disjoint dicycles of G. Hence, $G' \in \mathcal{P}(w_0, z_0; z_1, w_1)$. It is now easy to see that $G \in \mathcal{K}$, where $K'_G = G'$.

Suppose $\rho = w_0$. Suppose x_0 and x_2 are adjacent to distinct vertices in $\{u, v\}$. We may assume x_0u and x_2v are in $A(G)$. Choose distinct y_1 and y_2 in $V(K_H) - \{w_0\}$ such that $u'y_1, v'y_2 \in A(G)$. There exist two disjoint $(\{y_1, y_2\}, \{w_1, z_1\})$-dipaths p_2 and q_2 in K'_H by Corollary 2.1. If (p_2, q_2) is a $(y_1 \to w_1, y_2 \to z_1)$-linkage, then $P_{uu'}, u'y_1, p_2, w_1x_0, x_0u$ and $P_{vv'}, v'y_2, q_2, z_1x_1, x_1x_2, x_2v$ are disjoint dicycles of G. If (p_2, q_2) is a $(y_1 \to z_1, y_2 \to w_1)$-linkage, then $P_{uu'}, u'y_1, p_2, z_1x_3, x_3x_0, x_0u$ and $P_{vv'}, v'y_2, q_2, w_1x_2, x_2v$ are disjoint dicycles of G. Therefore, only one of u of v is adjacent from x_0 or x_2. We may assume that $x_0u, x_2u, z_0v \in A(G)$ and $x_0v, x_2v \notin A(G)$.

Let G_1 be obtained from K_H by expansion at ρ. Let G'_1 be obtained from G_1 by removing uz_0, z_0u, w_1z_1, and z_1w_1 if present. Since $z_0v \in A(G)$, $z_0w_0 \in A(H)$. Hence, $w_0z_0 \notin A(H)$, and so $u'z_0, v'z_0 \notin A(G)$. Therefore, G'_1 is an acyclic digraph with sources u and z_0 and sinks w_1 and z_1. If G'_1 has a $(u \to w_1, z_0 \to z_1)$-linkage (p_3, q_3), then p_3, w_1x_0, x_0u and q_3, z_1x_1, x_1z_0 are disjoint dicycles of G. Hence, $G'_1 \in \mathcal{P}(u, z_0; z_1, w_1)$. It is now easy to see that $G \in \mathcal{K}$, where $K'_G = G'$.

Similarly, if $\rho \in \{w_1, z_0, z_1\}$, then $G \in \mathcal{K}$. □

In order to prove that G is in $\{D_7\} \cup \mathcal{K} \cup \mathcal{H}$ when H is in $\mathcal{H}$, we need to prove two lemmas.

LEMMA 8.1. *Suppose $H \in \mathcal{H}$ and $(H_\alpha, H_\beta, H_\gamma)$ is an $\mathcal{H}$-decomposition of H with $\mathcal{H}$-separator $(y_1, y_2, y_3, y_4, y_5)$. If z is an intermediate vertex of H_α such that $H_\alpha - z$ has no $(\{y_1, y_3\}, y_5)$-dipath, then H has an $\mathcal{H}$-decomposition with $\mathcal{H}$-separator (y_1, y_2, y_3, y_4, z).*

PROOF. The result follows easily from Lemma 7.6 with $\{z_1, z_2, z_3, z_4, z_5\} = (y_1, y_2, y_3, y_4, z)$. □

LEMMA 8.2. *Suppose H is in $\mathcal{H}$ and has $\mathcal{H}$-decomposition $(H_\alpha, H_\beta, H_\gamma)$ with $\mathcal{H}$-separator $(y_1, y_2, y_3, y_4, y_5)$. Then the following statements hold.*
  (i) *y_1 is adjacent to an intermeditate vertex of H_γ.*
  (ii) *H_α has two openly disjoint $(y_3, \{y_2, y_5\})$-dipaths. H_β has two openly disjoint $(\{y_4, y_5\}, y_1)$-dipaths.*
  (iii) *If a is an intermediate vertex of H_α which is adjacent from y_1, then H_α has a $(y_3 \to y_5, a \to y_2)$-linkage.*
  (iv) *If a is an intermediate vertex of H_γ which is adjacent from y_1, then H_γ has a $(a \to y_3, y_2 \to y_4)$-linkage.*
  (v) *If a is an intermediate vertex of H_α which is adjacent to y_2 and there exist two openly disjoint $(\{y_3, y_1\}, y_5)$-dipaths in H_α, then H_α has a $(y_3 \to y_5, y_1 \to a)$-linkage.*
  (vi) *If $1 \leq d^+_{H_\alpha}(y_4)$, H_α has two openly disjoint $(\{y_3, y_1\}, y_5)$-dipaths, and a is an intermediate vertex of H_α which is adjacent to y_2, then H_α has a $(y_4 \to y_5, y_3 \to a)$-linkage.*

PROOF. (i) By Lemma 4.2 (iv), H_γ has an intermeditate vertex x. By Corollary 2.1, H_γ has a (y_1, x)-dipath. Hence, y_1 is adjacent to an intermeditate vertex of H_γ.

(ii) Since $2 \leq d^+_H(y_3)$, $2 \leq d^+_{H_\alpha}(y_3)$. Then H_α has two openly disjoint $(y_3, \{y_2, y_5\})$-dipaths by Corollary 2.1. Similarly, we can prove H_β has two openly disjoint $(\{y_4, y_5\}, y_1)$-dipaths.

(iii) Suppose a is an intermediate vertex of H_α which is adjacent from y_1. Suppose there exists a vertex c which is on every $(\{y_3, a\}, \{y_2, y_5\})$-dipath of H_α. By 2, H_α has two openly disjoint $(y_3, \{y_2, y_5\})$-dipaths, and so $c = y_3$. By Corollary 2.1, H_α has an $(a, \{y_2, y_5\})$-dipath, and so $c \neq y_3$. Therefore, Theorem 2.3 implies that H_α has two disjoint $(\{y_3, a\}\{y_2, y_5\})$-dipaths. Since H_α is in $\mathcal{P}(y_4, y_3, y_1; y_5, y_2)$, these dipaths give a $(y_3 \to y_5, a \to y_2)$-linkage.

(iv) The proof is very similar to 3.

(v) The proof is very similar to 3.

(vi) Suppose $1 \leq d^+_{H_\alpha}(y_4)$ and H_α has two openly disjoint $(\{y_1, y_3\}, y_5)$-dipaths, p_1 and p_3, where p_i has origin y_i, $i = 1, 3$. Let a be an intermeditate vertex of H_α which is adjacent to y_2. Suppose there exists a vertex c_1 which is on every $(\{y_3, y_4\}, \{y_5, a\})$-dipath of H_α. Since $1 \leq d^+_{H_\alpha}(y_4)$, Corollary 2.1 implies H_α has an two disjoint $(\{y_3, y_4\},$

$\{y_2, y_5\}$)-dipaths. Since H_α is in $\mathcal{P}(y_4, y_3, y_1; y_5, y_2)$, these dipaths give a $(y_4 \to y_5, y_3 \to y_2)$-linkage (q_4, q_3). If p_1 and q_3 are disjoint, then we contradict the fact that H_α is in $\mathcal{P}(y_4, y_3, y_1; y_5, y_2)$. Let x_1 be the first vertex of q_3 on p_1. Then c_1 is on $q_3[y_3, x_1]$, $p_1[x_1, y_5]$. Since c_1 is on q_4 and p_3, c_1 is not on q_3 or $p_1[y_1, y_5]$. Thus, $c_1 = y_5$. By Corollary 2.1, H_α has a $(\{y_3, y_4\}, a)$-dipath, and so $c_1 \neq y_5$. Therefore, Theorem 2.3 implies that H_α has two disjoint $(\{y_3, y_4\}, \{a, y_5\})$-dipaths. Since H_α is in $\mathcal{P}(y_4, y_3, y_1; y_5, y_2)$, these dipaths give a $(y_4 \to y_5, y_3 \to a)$-linkage. □

THEOREM 8.3. *If G has an R_k-reduction to H in $\mathcal{H}$, then $G \in \mathcal{H}$.*

PROOF. Let $(H_\alpha, H_\beta, H_\gamma)$ be an $\mathcal{H}$-decomposition of H with $\mathcal{H}$-separator $(y_1, y_2, y_3, y_4, y_5)$ and let $Y = \{y_1, y_2, y_3, y_4, y_5\}$. Let R_k be contracted to ρ in H.

Case 1. Suppose $\rho \notin Y$. Then ρ is an intermediate vertex of H_i, for some i in $\{\alpha, \beta, \gamma\}$. We can then show that $G \in \mathcal{H}$ by using Lemma 7.6 with $(z_1, z_2, z_3, z_4, z_5) = (y_1, y_2, y_3, y_4, y_5)$.

Case 2. Suppose $\rho = y_1$. Since $N_H^-(y_1) \subseteq V(H_\beta)$, $N_G^-(u) \cup N_G^-(v) \subseteq V(H_\beta - y_1)$. Hence, there are distinct vertices w_1 and z_1 in $V(H_\beta - y_1)$ such that $w_1 u$ and $z_1 v$ are in $A(G)$. By Corollary 2.1, there are two disjoint $(\{y_4, y_5\}, \{w_1, z_1\})$-dipaths in H_β. Without loss of generality, H_β has a $(y_5 \to w_1, y_4 \to z_1)$-linkage (p_1, q_1).

Suppose there exists w_2 in $V(H_\gamma - y_1)$ and z_2 in $V(H_\alpha - y_1)$ such that $u' w_2$ and $v' z_2$ are in $A(G)$. By Lemma 8.2 (iii), (iv) there is a $(y_3 \to y_5, z_2 \to y_2)$-linkage (p_2, q_2) in H_α and a $(w_2 \to y_3, y_2 \to y_4)$-linkage (p_3, q_3) in H_γ. But then $P_{uu'}, u'w_2, p_3, p_2, p_1, w_1 u$ and $P_{vv'}, v'z_2, q_2, q_3, q_1, z_1 v$ are disjoint dicycles of G. Suppose there exists w_3 in $V(H_\alpha) - \{y_1, y_2\}$ and z_3 in $V(H_\gamma) - \{y_1, y_3\}$ such that $u' w_3$ and $v' z_3$ are in $A(G)$. By Corollary 2.1, there exists a (w_3, y_5)-dipath p_4 in H_α and a (z_3, y_4)-dipath q_4 in H_γ. But then $P_{uu'}, u'w_3, p_4, p_1, w_1 u$ and $P_{vv'}, v'z_3, q_4, q_1, z_1 v$ are disjoint dicycles of G.

By Lemmas 4.2 (i) and 8.2 (i), y_1 is adjacent to intermediate vertices of H_α and H_γ. It then follows from the previous paragraph that either $N_G^+(v') = \{y_3, u'\}$ or $N_G^+(u') = \{y_2, v'\}$. If $N_G^+(v') = \{y_3, u'\}$, then we can show that $G \in \mathcal{H}$ by using Lemma 7.6 with $(z_1, z_2, z_3, z_4, z_5) = (u', y_2, y_3, y_4, y_5)$. Similarly, if $N_G^+(u') = \{y_2, v'\}$, then we can show $G \in \mathcal{H}$.

Case 3. Suppose $\rho = y_5$. By Lemma 4.2 (iii), $1 \leq d_{H_\alpha}^+(y_4)$ or $1 \leq d_{H_\beta}^-(y_2)$. Without loss of generality, we may assume $1 \leq d_{H_\alpha}^+(y_4)$.

Suppose $d_{H_\beta}^-(y_2) = 0$. Then $(H_\alpha, H_\gamma, H_\beta)$ is an $\mathcal{H}$-decomposition of H with $\mathcal{H}$-separator $(y_4, y_5, y_3, y_1, y_2)$. Now Lemma 8.1 with $(z_1, z_2, z_3, z_4, z_5) = (y_4, y_5, y_3, y_1, y_2)$ allows us to choose y_2 so that H_α has two openly disjoint $(\{y_3, y_4\}, y_2)$-dipaths. Therefore, if $d_{H_\beta}^-(y_2) = 0$, then H_α has two openly disjoint $(\{y_3, y_4\}, y_2)$-dipaths.

If H_α does not have two openly disjoint $(\{y_1, y_3\}, y_5)$-dipaths, then Lemma 8.1 implies there is an intermediate vertex z of H_α such that H has an $\mathcal{H}$-decomposition with $\mathcal{H}$-separator (y_1, y_2, y_3, y_4, z). Then we have case 1. Therefore, we may assume H_α has two openly disjoint $(\{y_1, y_3\}, y_5)$-dipaths.

If v is not adjacent from any intermediate vertex of H_α, then we can show that $G \in \mathcal{H}$ by using Lemma 7.6 with $(z_1, z_2, z_3, z_4, z_5) = (y_1, y_2, y_3, y_4, u)$. Therefore, we may assume v is adjacent from an intermediate vertex of H_α. Similarly, the same holds for u.

By Corollary 2.1, H_α has a $(\{y_1, y_3\}, x)$-dipath, for every intermediate vertex x. This fact and the conclusions of the previous two paragraphs can be used to find intermediate vertices w_4 and z_4 of H_α such that $w_4 u$ and $z_4 v$ are in $A(G)$ and there are two disjoint $(\{y_1, y_3\}, \{w_4, z_4\})$-dipaths p_5 and q_5 of H_α. We may assume (p_5, q_5) is a $(y_1 \to w_4, y_3 \to z_4)$-linkage. Similarly, we can find intermediate vertices w_5 and z_5 of H_β such that $u'w_5$ and $v'z_5$ are in $A(G)$ and there are two disjoint $(\{w_5, z_5\}, \{y_1, y_3\})$-dipaths p_6 and q_6 of H_β. If (p_6, q_6) is a $(w_5 \to y_1, z_5 \to y_3)$-linkage then $P_{uu'}, u'w_5, p_6, p_5, w_4 u$ and $P_{vv'}, v'z_5, q_6, q_5, z_4 v$ are disjoint dicycles of G. Therefore, we may assume (p_6, q_6) is a $(w_5 \to y_3, z_5 \to y_1)$-linkage.

Case 3.a. Suppose $d^-_{H_\beta}(y_2) = 0$. Therefore, H_α has two openly disjoint $(\{y_3, y_4\}, y_2)$-dipaths.

Suppose there exists a vertex x on all $(\{y_3, y_4\}, \{w_4, z_4\})$-dipaths in H_α. Then $x \in V(q_5)$, and so $x \notin V(p_5)$. Since there are two openly disjoint $(\{y_3, y_4\}, y_2)$-dipaths, there exists such a dipath r_4 such that $x \notin V(r_4)$. Then r_4 and p_5 intersect, otherwise, $(r_4, [p_5, w_4 y_5])$ is either a $(y_4 \to y_2, y_1 \to y_5)$-linkage or a $(y_3 \to y_2, y_1 \to y_5)$-linkage of H_α. But then $r_4 \cup p_5$ contains a $(\{y_3, y_4\}, w_4)$-dipath which does not include x. Therefore, H_α has two disjoint $(\{y_3, y_4\}, \{w_4, z_4\})$-dipaths p_7 and q_7 by Theorem 2.3.

Suppose (p_7, q_7) is a $(y_4 \to w_4, y_3 \to z_4)$-linkage. By Corollary 2.1, H_α has a (w_4, y_2)-dipath p_8 and a (z_4, y_2)-dipath q_8. If p_8 does not intersect q_7, then $([p_7, p_8], [q_7, z_4 y_5])$ is a $(y_4 \to y_2, y_3 \to y_5)$-linkage of H_α. Thus, p_8 and q_7 intersect, and so there is a (w_4, z_4)-dipath in H_α. If q_8 does not intersect p_5, then $([q_5, q_8], [p_5, w_4 y_2])$ is a $(y_3 \to y_2, y_1 \to y_5)$-linkage of H_α. Thus, q_8 and p_5 intersect, and so there is a (z_4, w_4)-dipath in H_α. But H_α is acyclic, and so it cannot have a (w_4, z_4)-dipath and a (z_4, w_4)-dipath. Therefore, (p_7, q_7) is a $(y_3 \to w_4, y_4 \to z_4)$-linkage.

By Lemmas 8.2 (i) and Corollary 2.1, H_γ has a (y_1, y_4)-dipath r_2. But then $P_{uu'}, u'w_5, p_6, p_7, w_4 u$ and $P_{vv'}, v'z_5, q_6, r_2, q_7, z_4 v$ are disjoint dicycles of G.

Case 3.b. Suppose $1 \leq d^-_{H_\beta}(y_2)$. By assumption, $1 \leq d^+_{H_\alpha}(y_4)$.

Suppose there exists a vertex x on all $(\{y_3, y_4\}, y_5)$-dipaths of H_α such that $x \neq y_5$. Then $x \in V(q_5)$, and so $x \notin V(p_5)$. Since $1 \leq d^+_{H_\alpha}(y_4)$, there exists two disjoint $(\{y_3, y_4\}, \{y_2, y_5\})$-dipaths p_9 and q_9 in H_α by Corollary 2.1. Since H_α is in $\mathcal{P}(y_4, y_3, y_1; y_5, y_2)$, we may assume (p_9, q_9) is a $(y_3 \to y_2, y_4 \to y_5)$-linkage. Then $x \in V(q_9)$, and so $x \notin V(p_9)$. Then p_5 and p_9 intersect, otherwise,

$([p_5, w_4y_5], p_9)$ is a $(y_1 \to y_5, y_3 \to y_2)$-linkage of H_α. But then $p_5 \cup p_9$ contains a $(\{y_3, y_4\}, y_5)$-dipath which does not include x. Therefore, H_α has two openly disjoint $(\{y_3, y_4\}, y_5)$-dipaths by Theorem 2.3.

Since all vertices adjacent to y_5 are in $V(H_\alpha)$, both u and v are adjacent from vertices in $V(H_\alpha) - \{y_5\}$. By Corollary 2.1, H_α has a $(\{y_1, y_3\}, x)$-dipath, for every intermediate vertex x. These two facts and the conclusion of the previous paragraph can be used to find vertices w_6 and z_6 in $V(H_\alpha) - \{y_5\}$ such that w_6u and z_6v are in $A(G)$ and there are two disjoint $(\{y_3, y_4\}, \{w_6, z_6\})$-dipaths p_{10} and q_{10} of H_α.

If (p_{10}, q_{10}) is a $(y_3 \to w_6, y_4 \to z_6)$-linkage, then $P_{uu'}, u'w_5, p_6, p_{10}, w_6u$ and $P_{vv'}, v'z_5, q_6, r_2, q_{10}, z_6v$ are disjoint dicycles of G. Therefore, we may assume (p_{10}, q_{10}) is a $(y_4 \to w_6, y_3 \to z_6)$-linkage. Similarly, we can find vertices w_7 and z_7 in $V(H_\beta) - \{y_5\}$ such that uw_7 and vz_7 are in $A(G)$ and there exists a $(w_7 \to y_2, z_7 \to y_1)$-linkage (p_{11}, q_{11}). By i and iv of Lemma 7.2, H_γ has a $(y_2 \to y_4, y_1 \to y_3)$-linkage (p_{12}, q_{12}). But then $P_{uu'}, u'w_7, p_{11}, p_{12}, p_{10}, w_6u$ and $P_{vv'}, v'z_7, q_{11}, q_{12}, q_{10}, z_6v$ are disjoint dicycles of G.

Case 4. Suppose $\rho = y_2$. If $d^+_{H_\alpha}(y_4) = 0$, then $(H_\gamma, H_\beta, H_\alpha)$ is an $\mathcal{H}$-decomposition of H with $\mathcal{H}$-separator $(y_1, y_3, y_2, y_5, y_4)$. Now we have case 3. Therefore, we may assume that $1 \leq d^+_{H_\alpha}(y_4)$.

Since $N^+_H(y_2) \subseteq V(H_\gamma)$, there are distinct vertices w_9 and z_9 in $V(H_\gamma - y_2)$ such that $u'w_9$ and $v'z_9$ are in $A(G)$. Without loss of generality, Corollary 2.1 implies that H_γ has a $(w_9 \to y_3, z_9 \to y_4)$-linkage (p_{14}, q_{14}).

Let w_8 be in $V(H_\alpha) - \{y_1, y_2\}$ and z_8 be in $V(H_\beta) - \{y_2\}$. Suppose w_8u and z_8v are in $A(G)$. Since $1 \leq d^+_{H_\alpha}(y_4)$ and H_α has two openly disjoint $(\{y_3, y_1\}, y_5)$-dipaths, Lemma 7.2vi implies the existence of a $(y_3 \to w_8, y_4 \to y_5)$-linkage (p_{13}, q_{13}) in H_α. Since $z_8 \neq y_4$, there is a (y_5, z_8)-dipath r_7 in H_β by Corollary 2.1. But then $P_{uu'}, u'w_9, p_{14}, p_{13}, w_8u$ and $P_{vv'}, v'z_9, q_{14}, q_{13}, r_7, z_8v$ are disjoint dicycles of G. Suppose w_8v and z_8u are in $A(G)$. Since there exist two openly disjoint $(\{y_3, y_1\}, y_5)$-dipaths in H_α, Lemma 7.2v implies the existence of a $(y_3 \to y_5, y_1 \to w_8)$-linkage (p_{15}, q_{15}) in H_α. By Corollary 2.1, there are two disjoint $(\{y_4, y_5\}, \{y_1, z_8\})$-dipaths p_{16} and q_{16} in H_β. If (p_{16}, q_{16}) is a $(y_5 \to y_1, y_4 \to z_8)$-linkage, then $(p_{16}, [q_{16}, z_8y_2])$ is a $(y_5 \to y_1, y_4 \to y_2)$-linkage of H_β. Therefore, (p_{16}, q_{16}) is a $(y_5 \to z_8, y_4 \to y_1)$-linkage. But then $P_{uu'}, u'w_9, p_{14}, p_{15}, p_{16}, z_8u$ and $P_{vv'}, v'z_9, q_{14}, q_{16}, q_{15}, w_8v$ are disjoint dicycles of G.

The preceding paragraph implies that $(V(H_\alpha) \cup V(H_\beta)) - \{y_1, y_2\}$ does not intersect both $N^-_G(u)$ and $N^-_G(v)$. We also know that y_2 (and hence u or v) is adjacent from vertices in both $V(H_\alpha) - \{y_1, y_2\}$ and $V(H_\beta) - \{y_2\}$. It follows that $N^-_G(v) = \{u, y_1\}$ and for every w in $N^-_H(y_2) - \{y_1\}$, $wu \in A(G)$. We can then show that $G \in \mathcal{H}$ by using Lemma 6.6 with $(z_1, z_2, z_3, z_4, z_5) = (y_1, u, y_3, y_4, y_5)$. $\square$

9. A Polynomial Algorithm

We now present a polynomial algorithm which either finds 2 disjoint dicycles in a digraph G or shows that G is intercyclic. In §3, we showed that this problem can easily be reduced to digraphs in reduced form. Hence, we may assume G is in reduced form.

The algorithm requires the following subroutine. When applied to a digraph G in reduced form, the subroutine will have four possible outcomes. It will either find two disjoint dicycles of G, show that G is in $\mathcal{M}$, find an R_k-reduction to a digraph H in reduced form, or find an arc reduction to a digraph H in reduced form.

First, we determine if G has a contractible arc. (It is easy to show that every intercyclic digraph in reduced form has a contractible arc using the main theorem. Hence, if G has no contractible arc, then we know immediately that G is not intercyclic. But if this is the case, we still have to find two disjoint dicycles.)

Suppose G has no contractible arc and no 2-dicycle. Then Lemma 4.1 shows that there is an arc reduction at every arc. Let H be obtained from G by an arc reduction.

Suppose G has no contractible arc and G has a 2-dicycle xy, yx. If $G - \{x, y\}$ is acyclic, then it has a source z. But then $x(yz)$ is contractible. Hence, $G - \{x, y\}$ has a dicycle. Such a dicycle C is very easy to find. Then C and xy, yx are disjoint dicycles of G.

Suppose G has a contractible arc. Then the proof of Lemma 4.2 can be used to find an R_k-reduction of G to a digraph H, find two disjoint dicycles of G, or show that G is in $\mathcal{M}$.

We now require a second subroutine. Suppose H is obtained from G using the first subroutine. If H is intercyclic, then suppose we know the structure of H. If H is not intercyclic, then suppose we have found two disjoint dicycles of H. When applied to H, the second subroutine will either show that G is intercyclic and give its structure, or it will find two disjoint dicycles of G.

If we have found two disjoint dicycles of H, then it is easy to find two disjoint dicycles of G.

Suppose H is intercyclic and we know the structure of H. If H is obtained from G using an R_k-reduction, then the results and proofs of §6 and §7 can be used to either show that G is intercyclic and give its structure, or to find two disjoint dicycles of G. If H is obtained from G using an arc-reduction, then the results and proofs of §5 can be used to find two disjoint dicycles of G.

The two subroutines can now be used to construct the algorithm. First we find a sequence of digraphs $G = G_0, \ldots, G_n$ in reduced form such that G_i is obtained from G_{i-1} using the first subroutine, $i = 1, \ldots, n$, and G_n is in $\mathcal{M}$ or we have found two disjoint dicycles in G_n. Then the second subroutine is sequentially applied to $G_n, \ldots, G_0$ to either show that G is intercyclic and give its structure, or to find two disjoint dicycles of G.

10. Arc-intercyclic Digraphs

A digraph is *arc-intercyclic* if it does not have two arc-disjoint dicycles. We say that a digraph is in *arc-reduced form* if it is strongly-connected and loopless, does not have parallel arcs incident with a vertex of indegree or outdegree 1, and does not have two adjacent vertices which both have indegree 1 or both have outdegree 1. Arc-reduced form will play the same role for arc-intercyclic digraphs as reduced form does for intercyclic digraphs.

Before stating the main theorem, we first show that the problem of determining if a digraph is arc-intercyclic can easily be reduced to the problem of determining if a digraph in arc-reduced form is arc-intercyclic.

Suppose we wish to determine if a digraph G is arc-intercyclic. As with intercyclic digraphs, we may assume G is strongly-connected and has a dicycle. There are simple reductions that can sometimes be performed on G to give a digraph H with fewer arcs such that G is arc-intercyclic if and only if H is arc-intercyclic. If ux and xv are the only arcs incident with vertex x, then G can be reduced by removing x and adding a new arc from u to v. If parallel arcs are incident to (respectively, incident from) a vertex of outdegree 1 (respectively, indegree 1), then G can be reduced by removing one of the parallel arcs. If vertices u and v have outdegree 1 and G has an arc e from u to v, then G can be reduced by contracting e. (This reduction can result in new parallel arcs.) We get a similar reduction if there exist adjacent vertices with indegree 1. We refer to these reductions as *trivial reductions*. It is easy to see that G is strongly-connected if and only if H is strongly-connected, and that G has a k-arc transversal if and only if H has a k-arc transversal.

Suppose K is obtained from G via a sequence of trivial reductions and suppose K has no trivial reductions. Then K is strongly-connected, and K is arc-intercyclic if and only if G is arc-intercyclic. It is easy to show that G has a 1-arc transversal if and only if K has only one arc. In this case, G is clearly arc-intercyclic. Therefore, we may assume that neither G nor K has a 1-arc transversal and that K has at least two arcs. If K has a loop e, then $K[e]$ and any dicycle of $K - e$ are arc-disjoint dicycles of K, and so K and G are arc-intercyclic. Hence, we can assume that K is loopless. Since K is strongly-connected, loopless, and has no trivial reductions, K is in arc-reduced form. Thus, we have reduced the problem of determining if G is arc-intercyclic to the problem of determining if the digraph K in arc-reduced form is arc-intercyclic. We note that if we can show that K is not arc-intercyclic and can find two arc-disjoint dicycles of K, then it is easy to find two arc-disjoint dicycles of G by reversing the process of trivial reductions.

In the next theorem we show that there is a natural bijection between vertex-intercyclic digraphs in reduced form and arc-intercyclic digraphs in reduced form.

The characterization of vertex-intercyclic digraphs will then give a characterization of arc-intercyclic digraphs.

A *perfect matching* of a digraph is a subset M of its arc set such that M contains no loops and every vertex is incident with exactly one arc in M.

THEOREM 10.1. *G is an arc-intercyclic digraph in arc-reduced form if and only if G can be obtained from a vertex-intercyclic digraph H in reduced form by splitting all the vertices of H.*

PROOF. Suppose G is obtained from a vertex-intercyclic digraph H in reduced form by splitting all the vertices of H. For every x in $V(H)$, let x^- and x^+ be the vertices of G corresponding to x. Let C_1 and C_2 be dicycles of G. If a dicycle of G uses x^- or x^+, then it must use x^-x^+. Hence, C_1 and C_2 correspond to dicycles of H. Then these dicycles of H have a common vertex v. Hence, C_1 and C_2 both use v^-v^+. Therefore, G is arc-intercyclic.

We now show that G is in reduced form. Since H is strongly-connected, so is G. The method of construction of G precludes loops. Every vertex of G has indegree or outdegree at least 2 because every vertex of H has indegree and outdegree at least 2. Since H has no parallel arcs, neither does G. Suppose G has vertices u and v with outdegree 1 and there is an arc e from u to v. Since u and v have outdegree 1, they have indegree at least 2, and so $u = y^-$ and $v = z^-$ for some y and z in $V(H)$. But y^- is only adjacent to y^+. Thus, G does not have adjacent vertices of outdegree 1. Similarly, G does not have adjacent vertices of indegree 1. Therefore, G is in reduced form.

Conversely, let G be an arc-intercyclic digraph in reduced form. First we prove that every vertex of G has indegree 1 or outdegree 1. Suppose G has a vertex x with indegree and outdegree at least 2. Let G' be obtained from G by dividing x. If G' has 2 arc-disjoint (x^+, x^-)-dipaths, then they correspond to arc-disjoint dicycles of G. Thus, Theorem 2.3 implies that G' has an arc e such that $G' - e$ has no (x^+, x^-)-dipath. Let U (respectively, V) be the set of all vertices u such that there exists an (x^+, u)-dipath (respectively, (u, x^-)-dipath) in $G' - e$. Then U and V are disjoint, and so $G'[U]$ and $G'[V]$ cannot both have dicycles. We may assume $G'[U]$ is acyclic. Let z be a sink of $G'[U]$. It is easy to see that if u is a vertex in U and u is adjacent to w in $G' - e$, then w is in U. Hence, e is the only outgoing arc of z in G'. Since x^+ has outdegree at least 2 in G', $x^+ \neq z$, and so we may choose a sink y of $G'[U] - z$. Then y is only adjacent to z in G'. But now we can contradict the fact that G is in reduced form. If there is only one arc from y to z, then G can be reduced by contracting that arc. If there are at least 2 arcs from y to z, then G can be reduced by removing one of the arcs.

Let X^- (respectively, X^+) be the set of vertices with outdegree 1 (respectively, indegree 1). Since G is in reduced form and no vertex of G has both indegree and outdegree at least 2, $V(G)$ is the disjoint union of X^+ and X^-. Let M be the set of arcs of G from a vertex in X^- to a vertex in X^+. If $G[X^+]$ or $G[X^-]$

has an arc e, then G can be reduced by contracting e. It follows with a bit of thought that M is a perfect matching between X^+ and X^-. For every x^- in X^-, let e_x be the arc in M incident from x^- and let x^+ be the end of e_x in X^+.

Let H be obtained from G be contracting all the arcs in M. It now suffices to show that H is a vertex-intercyclic digraph in reduced form. If H has vertex-disjoint dicycles, then they correspond to vertex-disjoint (and, hence, arc-disjoint) dicycles of G. Thus, H is vertex-intercyclic. Since G is strongly-connected, so is H. For every e_x in M, let x be the vertex of H corresponding to x^-, x^+, and e_x. Since M is a perfect matching of G, every vertex of H is of this form. For every vertex x of H, $2 \leq d_G^-(x^-) = d_H^-(x)$ and $2 \leq d_G^+(x^+) = d_H^+(x)$. If H has parallel arcs from x to y, then G has parallel arcs from x^+ to y^-. But then G can be reduced by removing one of the arcs from x^+ to y^-. Suppose H has a loop at vertex x. Then $H - x$ is acyclic. Let w be a sink of $H - x$. Since w can only be adjacent to x and since H has no parallel arcs, $d_H^+(w) \leq 1$. Hence, H is loopless. Therefore, H is in reduced form. Therefore, G is obtained from a vertex-intercyclic digraph in reduced form by splitting all its vertices. □

11. Acknowledgements

The author thanks Alice Metzlar and Daniel Younger for interesting discussions about intercyclic digraphs. He also thanks the referees for carefully reading this paper and for helpful suggestions.

REFERENCES

1. J. A. Bondy and U. S. R. Murty, *Graph theory with applications*, North-Holland, New York, 1981.
2. G. A. Dirac, *Some results concerning the structure of graphs*, Canad. Math. Bull. **6** (1963), 183–210.
3. P. Erdös and L. Pósa, *On the maximal number of disjoint circuits in a graph*, Publ. Math. Debrecen **9** (1962), 3–12.
4. _____, *On independent circuits contained in a graph*, Canad. J. Math. **17** (1965), 347–352.
5. S. Fortune, J. Hopcroft, and J. Wyllie, *The directed subgraph homeomorphism problem*, J. Theoret. Comput. Sci. **10** (1980), 111–121.
6. T. Gallai, *Problem 6*, Theory of Graphs, Proc. Colloq. Tihany 1966 (New York), Academic Press, 1968, p. 362.
7. S. R. Kosaraju, *On independent circuits of a digraph*, J. Graph Theory **1** (1977), 379–382.
8. A. V. Kostochka, *A problem on directed graphs*, Acta Cybernet. **6** (1983), 89–91, (in Russian with English summary).
9. L. Lovász, *On graphs not containing independent circuits*, Mat. Lapok **16** (1965), 289–299, (in Hungarian).
10. C. L. Lucchesi and D. H. Younger, *A minimax theorem for directed graphs*, J. London Math. Soc. **17** (1978), 369–374.
11. K. Menger, *Zur allgemeinen kurventheorie*, Fund. Math. **10** (1927), 96–115.
12. A. Metzlar, *Minimum transversal of cycles in intercyclic digraphs*, Ph.D. thesis, Department of Combinatorics and Optimization, University of Waterloo, Waterloo, Ontario, Canada, September 1989.
13. J. Soares, August 1991, private communication.
14. C. Thomassen, *Disjoint cycles in digraphs*, Combinatorica **3** (1983), 393–396.

15. _____, *The 2-linkage problem for acyclic digraphs*, Discrete Math. **55** (1985), 73–87.
16. _____, *On digraphs with no two disjoint directed cycles*, Combinatorica **7** (1987), no. 1, 145–150.
17. O. Veblen, *Theory of plane curves in non-metrical analysis situs*, Trans. Amer. Math. Soc. **6** (1905), 83–98.
18. H. Whitney, *Congruent graphs and the connectivity of graphs*, Amer. J. Math. **54** (1932), 150–168.
19. _____, *Non-separable and planar graphs*, Trans. Amer. Math. Soc. **34** (1932), 339–362.
20. D. H. Younger, *Graphs with interlinked directed circuits*, Proceedings of the Midwest Symposium on Circuit Theory 2, 1973, pp. XVI 2.1 – XVI 2.7.

DEPARTMENT OF MATHEMATICS AND STATISTICS, P.O. BOX 3055, UNIVERSITY OF VICTORIA, VICTORIA, BRITISH COLUMBIA, CANADA, V8W 3P4

Current address: Department of Mathematics, Physical Sciences Division, Scarborough Campus, University of Toronto, 1265 Military Trail, Scarborough, Ontario, CANADA M1C 1A4

E-mail address: mccuaig@math.toronto.edu

Eulerian Trails Through a Set of Terminals in Specific, Unique and All Orders

JØRGEN BANG-JENSEN AND SVATOPLUK POLJAK

Abstract

Let G be an eulerian digraph and X be a set of terminals (specified vertices). We consider the question of whether or not G admits an eulerian trail visiting the terminals (i) in a specific (pre-given) order, (ii) in unique order, and (iii) in any required order.

The problem (ii) was solved earlier by Ibaraki and Poljak. Here we show that an instance (G, X) which is infeasible for problem (iii) and is minimal with respect to edge contractions must have restricted degrees of vertices.

We also formulate some topological obstructions for problem (i), which arise from certain embeddings of G on a surface. The studied problems are closely related with linking problems in digraphs.

1 Eulerian trail problems

A digraph $G = (V, E)$ consists of a set V of vertices and a set E of directed edges. For technical reasons we allow multiple parallel edges in the same direction, but loops are excluded. A digraph G is said to be *eulerian* if its underlying undirected graph is connected and $d^+(v) = d^-(v)$ for every vertex v, where $d^-(v)$ and $d^+(v)$ denote the in- and out-degree of a vertex v. A digraph G is *strongly connected* if for every ordered pair of distinct vertices x, y there exists a directed path in G with endvertices x, y.

1991 Mathematics Subject Classification. Primary 05C45.
The second author was supported by the A. von Humboldt Foundation.
This paper is a preliminary version and the detailed version will be submitted elsewhere.

A sequence $T = (v_0, e_1, v_1, e_2, v_2, \ldots, v_{t-1}, e_t, v_t)$ such that e_i is a directed edge from vertex v_{i-1} to v_i, $i = 1, \ldots, k$, and all e_i's are distinct, is called a *trail*. If $v_0 = v_k$, a trail is called *closed*. An *eulerian trail* is a trail containing all edges of G.

It is well known that every eulerian digraph admits an eulerian trail. We study a more specific problem - existence of trails going through some specified vertices in a pre-specified order. Let $x_1, \ldots, x_k$ be a k-tuple of (not necessarily distinct) vertices which will be called *terminals*. We say that a trail $T = (v_0, e_1, v_1, e_2, v_2, \ldots, v_{t-1}, e_t, v_t)$ *visits the terminals in the order* $x_1, \ldots, x_k$ if $x_1 = v_{i_1}, x_2 = v_{i_2}, \ldots, x_k = v_{i_k}$ for some $0 \leq i_1 \leq \ldots \leq i_k \leq t$. (We do not exclude some additional occurences of terminals in a trail. In general, a trail may visit given terminals in several different orders.) Since a trail visiting the terminals $x_1, x_2, \ldots, x_k$ in the given order can be extended to a closed trail, there exists a trail visiting the terminals in any order which is a cyclic rotation of $x_1, x_2, \ldots, x_k$. Based on the following lemma, we could restrict ourselves only to eulerian trails. However, it is convenient to work also with non-eulerian trails.

LEMMA 1.1 *Let G be an eulerian digraph. Assume that there is a trail visiting some terminals in the order $x_1, \ldots, x_k$. Then there exists an eulerian trail visiting the terminals in the same order.* □

We will consider the following three problems on eulerian trails.

Specific Trail Problem (ST-problem).

INSTANCE: An eulerian digraph G and an ordered k-tuple of terminals $x_1, \ldots, x_k$.

QUESTION: Does there exist a trail visiting the terminals in the order $x_1, \ldots, x_k$?

Unique Trail Problem (UT-problem).

INSTANCE: An eulerian digraph G and a k-tuple of terminals $x_1, \ldots, x_k$.

QUESTION: Do all eulerian trails visit the terminals in the same cyclical order?

All Trail Problem (AT-problem).

INSTANCE: An eulerian digraph G and a k-tuple of terminals $x_1, \ldots, x_k$.

QUESTION: Does there exist, for every permutation π of $\{1, \ldots, k\}$ a trail visiting the terminals in the order $x_{\pi(1)}, \ldots, x_{\pi(k)}$?

Let us remark that if two or more terminals in the UT-problem are located in the same vertex, then there are always several cyclical orders for visiting the terminals, i.e. a necessary (but not sufficient) condition for the existence of unique order is that the terminals are located in distinct vertices. We will denote by k-ST, k-UT and k-AT the corresponding problems when the number of terminals is restricted to k. The ST-problem seems to be the most important among the three problems, since it is equivalent to the weak linking problem (see Lemma 3.1). However, the remaining two problems occur naturally in the study of the ST-problem. Some of the above problems have already been studied in [6], and the following results were obtained.

- The UT-problem is polynomially solvable.

- The 3-ST-problem is polynomially solvable.

- The ST-problem is NP-complete.

Further, it has been conjectured in [6] that the k-ST-problem is polynomial for any fixed k. It is easy to see that for $k = 3$, the problems 3-ST, 3-UT, and 3-AT are equivalent. The reason is that for $k = 3$ there are only two distinct cyclical orders of terminals, (x_1, x_2, x_3) and (x_1, x_3, x_2). Moreover, we may assume that one eulerian trail T of G is already given (since it may be constructed by a polynomial time algorithm). The trail T visits the terminals in one of the possible orders, say (x_1, x_2, x_3). Hence the question to decide is whether there is a trail visiting the terminals in the other order. Our next goal is to find a solution for the 4-ST-problem, and in this note we present some partial results on it.

Let us remark that a related problem of constrained eulerian trails was formulated by W. R. Pulleyblank and J. A. Tomlin [7] in connection with a helicopter schedulling problem. An instance of their problem is a triple (G, v, A) where G is an undirected eulerian graph, v a base node, and A a set of directed arcs (which are not part of G). The task is to find an eulerian trail, starting and ending at v, such that for each precedence arc from A, the first visit to the tail precedes the last visit to the head.

The paper is organized as follows. In Section 2 we recall a solution of the UT-problem of [6] in a slightly different formulation. In Section 3 we prove a result on vertex degrees of minimal infeasible AT-instances. In the last section, we formulate some topological obstructions for the 4-ST-problem, as a generalization of 3-ST-problem.

2 The Unique Trail Problem.

In this section we recall the solution of the UT-problem, since it suggests a possible approach to the remaining two problems which will be followed in the consecutive sections.

We introduce the operation of contraction of (G, X), which will be also used later in other sections. Let G/e denote the graph obtained from G by contracting an edge e. We admit contraction of an edge e even if both ends of e are terminals. The contraction does not change the set X of terminals, because we allow more terminals to be located in the same vertex. Clearly, if (G, X) admits several cyclic orders of visiting terminals, then $(G/e, X)$ admits several cyclic orders as well, but the converse need not be true. (In particular, $(G/e, X)$ admits several cyclic orders when both ends of e are terminals.) We say that (G, X) is *UT-minimal*, if (G, X) admits a unique cyclical order of visiting terminals by an eulerian trail, but $(G/e, X)$ admits several orders whenever an edge is contracted.

THEOREM 2.1 *[6] Let (G, X) be a UT-minimal instance. Then*

(i) $d^+(x) = d^-(x) = 1$ for every terminal x, and $d^+(u) = d^-(u) = 2$ for every non-terminal u,

(ii) G admits a planar representation such that every face is a directed cycle, and

(iii) all terminals lie on one common face.

□

Observe that the condition (ii) is equivalent with the property that the four edges incident to a non-terminal vertex u are oriented alternatively out of and in to the vertex u (in the planar representation).

The condition of minimality can be replaced by a more technical notion of irreducibility. Let us say that an instance (G, X) is *2-irreducible* if there is no subset $S, |S| > 1$, of vertices such that one of the following holds.

(i) $|\delta^+(S)| = |\delta^-(S)| \leq 2$, $<S>$ is connected and $S \cap X = \emptyset$, or

(ii) $|\delta^+(S)| = |\delta^-(S)| = 1$, and $|S \cap X| = 1$.

Here $\delta^-(S)$ (and $\delta^+(S)$) denote the set of edges from $V \setminus S$ to S (and from S to $V \setminus S$), and $<S>$ the subgraph of G induced by S. For a subset S of V we let G/S denote the digraph obtained from G by contracting S into one vertex s in such a way that all edges between S and $V - S$ in G now go between s and $V - S$ and possible loops are deleted. Note that G/S is eulerian whenever G is eulerian. It is not difficult to see the following

LEMMA 2.2 *Let (G, X) be an instance of the UT-problem which admits a unique order, and S be a subset satisfying one of (i) and (ii). Then $(G/S, X)$ admits a unique order as well.* □

It is also easy to see that G/S can be realized by a series of edge contractions, and hence every minimal UT-instance is 2-irreducible. Thus, the following theorem is a generalization of the previous one.

THEOREM 2.3 *[6] Let (G, X) be a UT-instance which is 2-irreducible and admits eulerian trail with unique order of terminals. Then (i), (ii) and (iii) of Theorem 2.1 hold.* □

The polynomial time algorithm for the UT-problem is a consequence of Theorem 2.3. The algorithm proposed in [6] consists of the following steps. (1) Reduce an instance (G, X) to an 2-irreducible one. (This can be done by an application of the network flow algorithm.) (2) Check the degree conditions. (3) Using a planarity test, decide whether G has a planar drawing, and if yes, then test the remaining conditions of Theorem 2.3. (The planar drawing is unique for 2-irreducible UT-instances which admit unique cyclical order to visit terminals, see [6].)

REMARK. The notion of *2-irreducibility* formulated here is weaker than the notion of *irreducibility* used in [6] where it was required, in addition, that (G, X) does not contain any non-terminal vertex of in- and out-degree one. However, using the general definition of *irreducibility* given just before Theorem 3.3 we can see that this additional condition would be automatically satisfied by any AT-infeasible and irreducible instance.

3 Degree Conditions on Minimal Instances.

The purpose of this section is present an upper bound on degrees of an AT-minimal instance (as in section 2 minimality is with respect to edge contractions). We also need to recall a connection between the eulerian trail problems and the weak linking problem.

The weak k-linking problem consists of deciding whether or not k given pairs of terminals $(s_1, t_1), \ldots, (s_k, t_k)$ can be connected by a collection of edge disjoint paths in a digraph G. Note that it is an easy consequence of Edmonds' branching theorem [2] that a digraph G is k-edge-linked, i.e. has a weak k-linking for any choice of the terminals $(s_1, t_1), \ldots, (s_k, t_k)$, if and only if G is k-edge-connected. (We recall that a digraph is k-edge-connected if $G \setminus E'$ is strongly connected whenever a set E' of at most $k-1$ edges is deleted.) Still the weak k-linking problem is known to be NP-complete already for $k = 2$ by a result of [3]. The problem is known to be polynomially solvable in some special cases.

- G is acyclic and k fixed (see [3]).

- G is a tournament, or more generaly, a semi-complete digraph, and $k = 2$ (see [1]),

- $G + H$, where H is the demand digraph, is acyclic and planar (see [4]),

- $G + H$ is eulerian, and $k = 3$ (see [6]).

Let us recall that the *demand digraph* is the digraph consisting of the demand edges $t_i s_i, i = 1, \ldots, k$. We study the weak linking problem when $G + H$ is an eulerian digraph where G is the supply graph and H the demand graph. This restricted version is called the *Eulerian weak k-linking problem*.

The problem is easy for $k = 2$ (see [4]). A solution for $k = 3$ has been given in [6], since we have

LEMMA 3.1 *The k-ST-problem is equivalent to the Eulerian weak k-linking problem.*

PROOF. The $k - ST$ problem is a special case of the Eulerian k-weak linking problem by the following reduction. Let $(G, x_1, \ldots, x_k)$ be an instance of the $k - ST$-problem with $s_i = x_i$ and $t_i = x_{i+1}, i = 1, \ldots, k$.

Conversely, given an instance $G, (s_1, t_1), \ldots, (s_k, t_k)$ where $G + H$ is eulerian, we construct an instance of the $k - ST$-problem as follows. Let $\tilde{G}$ be a graph obtained from G by adding new vertices $x_1, \ldots, x_k$, and directed edges $x_i s_i, t_i x_{i+1}, i = 1, \ldots, k$. Clearly, $\tilde{G}$ is an eulerian digraph, and it admits a closed trail visiting the terminals in the order $x_1, \ldots, x_k$

if and only if G admits a weak k-linking for the prescribed pairs (s_i, t_i) $i = 1, \ldots, k$, of terminals. □

Let G be a digraph and $S = (s_1, \ldots, s_k)$ and $T = (t_1, \ldots, t_k)$ two collections of terminals (the terminals need not be distinct). Let us say that an instance (G, S, T) is *linked* if, for every permutation π of $\{1, \ldots, k\}$, there exists a weak linking connecting terminals $(s_1, t_{\pi(1)}), \ldots, (s_k, t_{\pi(k)})$.

We will call an instance (G, S, T) *eulerian* if the digraph $G + H$ is eulerian, where H is a demand digraph consisting of the edges $t_{\pi(i)} s_i$, $i = 1, \ldots, k$ for arbitrary permutation π.

Equivalently, (G, S, T) is *eulerian* if G is (weakly) connected and

$$|\delta^-(u)| + |\{i : s_i = u\}| = |\delta^+(u)| + |\{i : t_i = u\}|$$

for every vertex u of G. For a subset U of vertices, let $d(U) = |\delta^-(U)| + |\{i : s_i \in U\}|$. In particular, $d(u) = |\delta^-(u)| + |\{i : s_i = u\}|$. Clearly, $d(U) = |\delta^+(U)| + |\{i : t_i \in U\}|$ for an eulerian instance (G, S, T).

We introduce a more general notion of reducibility as a generalization of 2-reducibility, which was used in the solution of the UT-problem. Let us say that a subset U is $d(U)$-*reducible*, if $(<U>, A, B)$ is linked where $<U>$ denotes the subgraph of G induced by U, $A = (a_1, \ldots, a_{d(U)})$ and $B = (b_1, \ldots, b_{d(U)})$ are collections of "new" terminals where every a_i denotes either a head of an edge from $\delta^-(U)$ or an "old" terminal s_i which is located in U, and similarly, every b_i denote either a tail of an edge from $\delta^+(U)$ or an "old" terminal t_i which is located in U.

LEMMA 3.2 *Let (G, S, T) be an eulerian instance, and U a $d(U)$-reducible subset. Then (G, S, T) is linked if and only if the contraction $(G/U, S, T)$ is linked.*

PROOF. Clearly, if (G, S, T) is linked, then any contraction is linked as well. Conversely, we show that (G, S, T) is linked provided $(G/U, S, T)$ is linked. Let $\bar{u}$ denote the vertex of G/U obtained by contracting the set U. For sake of clarity, assume that there are no terminals in U.

Let π be a given permutation, and let $P'_1, \ldots, P'_k$ be a weak linking in G/U connecting the prescribed pairs of terminals. We may assume that every path P'_i goes at most once through the vertex $\bar{u}$, and let $a'_i \bar{u}$ and $\bar{u} b'_i$ be the edges on path P'_i incident to $\bar{u}$. Then, there are edges $a'_i a_i$ and $b_i b'_i$ entering and leaving U in G. By our assumption, $<U>$ contains a linking $Q_1, \ldots, Q_k$ connecting terminals a_i, b_i. Now, we can insert Q_i into P'_i in order to obtain a linking in G. □

Let us say that an eulerian instance (G, S, T) is *irreducible*, if it does not contain a r-reducible subset for any $r \leq k - 1$, where $k = |S| = |T|$.

THEOREM 3.3 *Let $(G, S, T), |S| = |T| = k$, be an eulerian instance which is irreducible and not linked. Then $d(u) \leq k - 1$ for every vertex u of G.*

PROOF. We will use the induction on k. Clearly, the statement is true for $k = 1, 2$, since every instance of this size is linked. Let $k \geq 3$, and assume that $d(u) \geq k$ for some vertex u. We distinguish two cases.

Case (i) Assume that there exists a collection $P_1, \ldots, P_k$ of k edge disjoint paths from $s_1, \ldots, s_k$ to u such that P_i starts at s_i. Then, using the assumption that the instance (G, S, T) is eulerian, there exist also paths $P'_1, \ldots, P'_k$ from u to $t_1, \ldots, t_k$, such that $P_1, \ldots, P_k, P'_1, \ldots, P'_k$ are mutually edge disjoint. The existence of the paths $P'_1, \ldots, P'_k$ is ensured by the following claim, where $\tilde{G} = G + H$.

CLAIM 3.4 *Let $\tilde{G}$ be an eulerian digraph. Assume that $P_1, \ldots, P_k$ are edge disjoint paths from $x_1, \ldots, x_k$ to u. Then there exist a collection $P'_1, \ldots, P'_k$ of edge disjoint paths from u to $x_1, \ldots, x_k$ in $G \setminus (E(P_1) \cup \ldots \cup E(P_k))$.*

PROOF. Assume that $P'_1, \ldots, P'_k$ do not exist. By Menger's Theorem, there exists a set W not containing u such that the number of edges entering W in $G \setminus (E(P_1) \cup \ldots \cup E(P_k))$ is less then the number of x_i's in W. This contradicts with the fact that $\tilde{G}$ is eulerian and contains the paths $P_1, \ldots, P_k$. This proves the claim.

Now, it is easy to show that (G, S, T) is linked. Let π be a given permutation. Then $(P_i, P'_{\pi(i)}), i = 1, \ldots, k$, is an $(s_i, t_{\pi(i)})$-linking.

Case (ii) Assume that the paths $P_1, \ldots, P_k$ do not exist. Then, by Menger's Theorem, there exists a set W not containing u and such that the number of terminals s_i in W exceeds $|\delta^+(W)|$. Set $p = |\{i : s_i \in W\}|$, and let $U = V \setminus W$ be the complement of W. Then the number of s_i's in U is $k-p$. Since $|\delta^+(W)| = |\delta^-(U)| < p$, and hence $d(U) \leq k-1$, we may use the assumption on the irreducibility of U. It is easy to see that the irreducibility of (G, S, T) implies that $(<U>, A, B)$ is not linked, where A and B were introduced in the definition of irreducibility. We also have that $|U| > 1$, since U contains u, and $d(u) \geq k$ while $d(U) < k$.

CLAIM 3.5 *The instance $(<U>, A, B)$ is irreducible.*

The proof of the claim follows from our assumption that (G, S, T) is irreducible, and the fact that, for every $u \in U$, $d(u)$ is the same in $<U>$ and G, since we add a new terminal for any edge entering or leaving U.

Thus, we conclude that $(<U>, A, B)$ is irreducible and not linked. Hence, $d_U(v) \leq k - 2$ for every vertex v of $<U>$, by the induction hypothesis, where d_U denotes the d with respect to $(<U>, A, B)$. However, $d_U(v) = d(v)$, and hence $d(u) \leq k - 2$, a contradiction. □

Let (G, X) be an instance of an AT-problem. Let us say that (G, X) is *AT-minimal*, if (G, X) does not admit an eulerian trail visiting the terminals for every given order, but $(G/e, X)$ does whenever an edge e is contracted.

COROLLARY 3.6 *Let (G, X) be AT-minimal. Then $d^+(u) \leq k - 1$ for every non-terminal u, and $d^+(x) \leq k - 2$ for every terminal x.*

PROOF. Set $S = T = X$ and apply Theorem 3.3 to (G, S, T). Since (G, X) is AT-minimal, (G, S, T) is $(k-1)$-irreducible, because any set-contraction can be realized as a sequence of edge contractions. Clearly, we have $d(u) = d^+(u)$ for every non-terminal vertex, and $d(x) \geq d^+(x)+1$ for every terminal. (In fact, $d^+(x)$ is decreased by the number of terminals located at the vertex.) □

4 Topological obstructions

The purpose of this section is to formulate some necessary conditions for the existence of trails in specific order. Two necessary combinatorial conditions, namely the *directed cut criterion* and the *covering criterion*, have been formulated by A. Frank in [4] for the problem of weak k-linking (the formulation of the problem was recalled in Section 3). Let G and H denote the supply and the demand graph, respectively. A directed cycle in $G + H$ is called *good* if it contains exactly one demand edge.

DIRECTED CUT CRITERION. $d_G^+(S) \geq d_H^-(S)$ for every $S \subset V$.

COVERING CRITERION. There is no subset of edges of less than k edges covering all good cycles of $G + H$.

It is easy to see that the covering criterion is stronger than the former directed cut criterion. However, these criteria seem to be too weak to certificate the infeasibility in most cases. We propose to investigate a different type of conditions, based on the existence of certain topological embeddings. This approach has been succesful to provide the full solution of the UT-problem, the special case of which is the 3-ST-problem. Let us recall the topological criterion for this special case.

PROPOSITION 4.1 *Let G be an eulerian digraph and x_1, x_2, x_3 a triple of terminals, such that $d^+(x_i) = 1, i = 1, 2, 3$, and $d^+(u) = 2$ for every non-terminal u. Assume that G has a planar drawing such that*

(i) the terminals are located on one common face which is a directed cycle going through the terminals in the order x_3, x_2, x_1, and

(ii) for every non-terminal vertex u, the four edges incident to it are oriented alternatively out and in.

Then G does not admit a trail visiting the terminals in the order x_1, x_2, x_3.

PROOF. Let F be the face contaning the terminals. For a contradiction, assume that there is a trail T visiting the terminals in the required order x_1, x_2, x_3. Let P_1 and P_2 denote the segment of T from x_1 to x_2, and x_2 to x_3, respectively. Let $x_1 u_1$ be the unique edge leaving x_1, and $u_2 x_2$ be the unique edge entering x_2. Clearly, $x_1 u_1$ and $u_2 x_2$ are the first and the last edge of paths P_1. Let $P_1[u_1, u_2]$ be the segment of P_1 between these two vertices, and let $F[u_1, u_2]$ the directed path of the face F from u_1 to u_2. Then $P_1[u_1, u_2] \cup F[u_1, u_2]$ is a closed curve which separates the plane into two regions, one containing x_2 and the other one containing x_3. Hence P_2 and P_1 must have a common vertex, say z. But the edges incident to z, which all belong to P_1 and P_2, are not oriented alternatively out and in, which contradicts to the part (ii) of the lemma. □

It has been shown in section 2 that every infeasible instance (G, x_1, x_2, x_3) of the 3-ST-problem can be embedded in the plane to meet the conditions of Proposition 4.1. It is natural to extend the criterion of Proposition 4.1 to the k-ST-problem with $k > 3$. In particular, we have

PROPOSITION 4.2 *Let G be an eulerian digraph and x_1, x_2, x_3, x_4 be terminals such that $d^+(x_i) = 1, i = 1, 2, 3, 4$ and $d^+(u) = 2$ for every non-terminal u. Assume that G has a drawing on a surface S such that*

(i) For every non-terminal vertex u, the four edges incident to it are oriented alternatively out and in;

and one of the following (ii) or (iii) is satisfied

(ii) S is the plane, the terminals x_1 and x_3 are are located on a common face F_1, the terminals x_2 and x_4 are are located on a common face F_2 (distinct from F_1), and the faces F_1 and F_2 have the same orientation;

(iii) S is the projective plane, the four terminals are located on one common face which is a directed cycle going through the terminals in the order x_4, x_3, x_2, x_1.

Then G does not admit a trail visiting the terminals in the order x_1, x_2, x_3, x_4. □

PROPOSITION 4.3 *Let G be an eulerian digraph and x_1, x_2, x_3, x_4, x_5 be terminals terminals such that $d^+(x_i) = 1, i = 1, \ldots, 5$ and $d^+(u) = 2$ for every non-terminal u. Assume that G has a drawing on the torus such that*

(i) For every non-terminal vertex u, the four edges incident to it are oriented alternatively out and in; and

(ii) the five terminals are located on one common face which is a directed cycle going through the terminals in the order x_5, x_4, x_3, x_2, x_1.

Then G does not admit a trail visiting the terminals in the order x_1, x_2, x_3, x_4, x_5. □

Let $(G, x_1, x_2, \ldots, x_k)$ be an instance of the k-ST-problem. Obviously, a trail T visiting the terminals in the order $x_1, \ldots, x_k$ cannot exist unless there exists, for every $i = 1, \ldots, k$ a trail T_i visiting the $k - 1$ terminals $x_1, \ldots, x_{i-1}, x_{i+1}, \ldots, x_k$ in that order. However, it has been shown in [5] that for every k there exists an infeasible instance $(G, x_1, x_2, \ldots, x_k)$ for which all trails $T_i, i = 1, \ldots, k$, exist. This shows that the k-ST-problem cannot be solved by an inspection of $(k - 1)$-subcases.

Finally, let us summarize the partial results on the 4-ST-problem. Let (G, x_1, x_2, x_3, x_4) be a given instance of the problem. First of all, we may check whether all instances (G, x_2, x_3, x_4), (G, x_1, x_3, x_4), (G, x_1, x_2, x_4), and (G, x_1, x_1, x_3) of 3-ST-problem are feasible. If not, (G, x_1, x_2, x_3, x_4) is not feasible. As a next step, we contract every subset U which is 3-reducible. This can be done efficiently, because we may use the polynomial time algorithm of [6] for the 3-ST-problem as a subroutine. Hence assume that (G, x_1, x_2, x_3, x_4) is 3-irreducible. Now, we can apply corollary 3.6. If there exists either a non-terminal vertex u with indegree at least 4, or a terminal vertex x_i with indegree at least 3, then the instance is feasible. If the in-degrees of non-terminals and terminals are 2 and 1, respectively, we can apply the criterion of proposition 4.2, and test the infeasibility by embedding to the plane or the projective plane. However, it is easy to find instances of 4-ST-problem which cannot be decided by this procedure.

References

[1] J. Bang-Jensen, Edge-disjoint in- and out-branchings in tournaments and related path problems. Journal of Combinatorial Theory Ser. B **51** (1991), 1–23.

[2] J. Edmonds, Edge-disjoint branchings, in *Combinatorial Algorithms*, Academic Press New York, 1973, pp. 91–96.

[3] S. Fortune, J. Hopcroft and J. Wyllie, The directed subgraph homeomorphism problem, Theoretical Computer Science **10** (1990), 111–121.

[4] A. Frank, Connectivity in Graphs, in *Handbook of Combinatorics*, North-Holland, to appear.

[5] P. Hemelik, Master Thesis, Charles University, 1991.

[6] T. Ibaraki and S. Poljak, Weak three-linking in Eulerian digraphs, Siam Journal on Discrete Mathematics **4** (1991), 84–98.

[7] W. R. Pulleyblank, personal communication.

JØRGEN BANG-JENSEN
DEPARTMENT OF MATHEMATICS AND COMPUTER SCIENCE, ODENSE UNIVERSITY, CAMPUSVEJ 55, DK-5230 ODENSE M, DENMARK
E-mail address: jbj@imada.ou.dk

SVATOPLUK POLJAK
DEPARTMENT OF APPLIED MATHEMATICS, FACULTY OF MATHEMATICS AND PHYSICS, CHARLES UNIVERSITY, MALOSTRANSKE NAM. 25, 118 00 PRAHA 1, CZECHOSLOVAKIA

Current address: Institut für Diskrete Mathematik, Universität Bonn, Nassestrasse 2, D-5300 Bonn 1, Germany

E-mail address: or429@dbnuor1.bitnet

2-Reducible Cycles Containing Two Specified Edges in $(2k+1)$-Edge-Connected Graphs

HARUKO OKAMURA

ABSTRACT. Let G be a k-edge-connected graph and let f_1 and f_2 be edges. We call a cycle C (not necessarily simple) *2-reducible* if $G - E(C)$ is $(k-2)$-edge-connected. The author [5] has shown that if k is even, then G has a 2-reducible cycle containing f_1 and f_2. When k is odd, this is not always true. We here characterize the graphs G having no 2-reducible cycle containing f_1 and f_2, when the distance between f_1 and f_2 is 1 or there is a $(k+1)$-cut containing f_1 and f_2.

1. Introduction

We consider finite undirected multigraphs without loops. Let G be a graph and let $V(G)$ and $E(G)$ be the set of vertices and edges of G, respectively. We allow repetition of vertices (but not edges) in a path or cycle. k is a natural number. When k is fixed and G is k-edge-connected, we call a cycle (or path) C *2-reducible* if $G - E(C)$ is $(k-2)$-edge-connected.

Let G be k-edge-connected ($k \geq 2$) and let f_1 and f_2 be edges. It is known that G has a 2-reducible cycle containing f_1 and f_2, if f_1 and f_2 are incident (Okamura [4] and Mader [3]) or if k is even (Okamura [5]). On the other hand, when k is odd, we can construct G having vertices x and y at distance 3 such that each cycle containing x and y is not 2-reducible (Huck and Okamura [1]). We here characterize the graphs G having no 2-reducible cycle containing f_1 and f_2 in two further cases in theorems 1 and 2.

Let $X, Y, \{x, y\} \subseteq V(G)$ and $X \cap Y = \emptyset$. We often denote $\{x\}$ by x. We denote by $\partial(X, Y; G)$ the set of edges with one end in X and the other in Y, and define $\partial(X; G) := \partial(X, V(G) - X; G)$, $e(X, Y; G) := |\partial(X, Y; G)|$ and $e(X; G) := |\partial(X; G)|$. We denote by $\lambda(X, Y; G)$ the maximum number of edge-disjoint paths between X and Y. We set $\lambda(X; G) := \min_{x, y \in X} \lambda(x, y; G)$ and $\lambda(G) := \lambda(V(G); G)$. In such expressions we often omit G. For $X \subseteq V(G)$, G/X denotes the graph obtained from G by identifying all the vertices in X and deleting

---

1991 Mathematics Subject Classification. Primary 05C40, 05C38, 05C75.
This paper is in final form and no version of it will be submitted for publication elsewhere.

any resulting loops. In G/X, X denotes the corresponding new vertex and for $Y \subset V(G)$ with $Y \cap X \neq \emptyset$, Y denotes $(Y - X) \cup \{X\}$.

It is useful to relax slightly the hypothesis that G is k-edge-connected, and to permit up to two so-called "dummy" vertices which may not be k-edge-connected to the other vertices. More precisely, we say that $S \subseteq V(G)$ is *dummy*, if (1.1) below holds.

(1.1) $S = \emptyset$, $\{b\}$, or $\{b, b'\}$, $e(b') = k - 1$, $e(b, b') \leq e(b)/2$, and $e(b) \leq k - 1$ is even.

The theorems become simpler if $\lambda(G) \geq k$ (when the dummy S is empty), but the forms with dummies will be useful in applications.

Our first main result is as follows.

Theorem 1. *Suppose that $k \geq 5$ is odd, $V(G) = T \cup W \cup S$ (disjoint union), $\lambda(V(G) - S) \geq k$, $T = \{u_1, u_2, v_1, v_2\}$, $|T| = 4$, S is dummy, $f_i \in \partial(u_i, v_i)$ ($i = 1, 2$) and $f_0 \in \partial(v_1, v_2)$. Then (1), (2) and (3) below are equivalent.*

(1) *For each cycle C containing f_1 and f_2, $\lambda(V(G) - S; G - E(C)) \leq k - 3$.*

(2) *There are $X, Y \subseteq V(G)$ such that $X \cap T = \{v_1, u_2\}$, $Y \cap T = \{v_1, v_2\}$ and $e(X) = e(Y) = k + 1$.*

(3) $V(G) = X_1 \cup X_2 \cup Y_1 \cup Y_2$ *(disjoint union), for $i = 1, 2$, $e(X_i) = e(Y_i) = k$, $u_i \in X_i$, $v_i \in Y_i$, $e(X_i, Y_i) = 1$ and,*

$$e(X_1, Y_2) = e(X_2, Y_1) = e(Y_1, Y_2) = e(X_1, X_2) = (k - 1)/2.$$

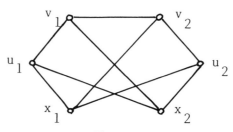

Figure 1.

Theorem 1 does not hold for $k = 3$; for example, if $k = 3$, then the graph in figure 1 satisfies (1), but does not satisfy (2) and (3). Incidentally, for $k \geq 2$ even and $S = \emptyset$, an analogue to theorem 1 is already known; it is shown in [6] that (1.2) and (1.3) below are equivalent.

(1.2) G has no 2-reducible cycle containing f_1, f_0 and f_2.

(1.3) There is an $X \subseteq V(G)$ such that $X \cap T = \{v_1, u_2\}$ and $e(X) \leq k + 1$.

Our second main result is the following.

Theorem 2. *Suppose that:*

(i) $k \geq 5$ is odd, $V(G) = W \cup S$ (disjoint union), S is dummy and $\lambda(V(G) - S) \geq k$,

(ii) $V(G) = X \cup Y$, $X \cap Y = \emptyset$ and $e(X) = k + 1$,

(iii) $\{u_1, u_2\} \subseteq X \cap W$, $\{v_1, v_2\} \subseteq Y \cap W$ and $f_i \in \partial(u_i, v_i)$ ($i = 1, 2$).

Then (1) and (2) below are equivalent.

(1) For each cycle C containing f_1 and f_2, $\lambda(V(G) - S; G - E(C)) \leq k - 3$.

(2) There exist X_1, X_2, Y_1, Y_2 so that $X = X_1 \cup X_2$, $Y = Y_1 \cup Y_2$ (disjoint union), and for $i = 1, 2$, $e(X_i) = e(Y_i) = k$, $u_i \in X_i$, $v_i \in Y_i$, $e(X_i, Y_i) = 1$ and

$$e(X_1, X_2) = e(Y_1, Y_2) = e(X_1, Y_2) = e(X_2, Y_1) = (k-1)/2.$$

Theorem 2 also does not hold for $k = 3$; again, the graph in figure 1 is a counterexample, with $X = \{u_1, u_2, x_1, x_2\}$ and $Y = \{v_1, v_2\}$.

To prove theorems 1 and 2, we need some lemmas, the following theorems 3, 4 and 5.

Theorem 3. If $k \geq 5$ is odd, $V(G) = T \cup W \cup S$ (disjoint union), $\lambda(V(G) - S) \geq k$, $T = \{u, v_1, v_2\}$ (where possibly $v_1 = v_2$), S is dummy, $f_i \in \partial(u, v_i)$ ($i = 1, 2$) and $e(u) = k + 1$, then one of the following holds.

(1) There is a cycle C containing f_1 and f_2 such that $\lambda(V(G) - S; G - E(C)) \geq k - 2$ and $\lambda(a, u; G - E(C)) = k - 1$ for some $a \in V(G) - u$.

(2) $V(G) = \{u\} \cup X_1 \cup X_2$ (disjoint union), where $v_i \in X_i$ and $e(X_i) = k$ for $i = 1, 2$.

(3) $V(G) = \{u, b\} \cup X_1 \cup X_2$ (disjoint union), where for $i = 1, 2$, $v_i \in X_i$, $e(X_i) = k$, $4 \leq e(b) \leq k - 3$, $e(X_i, u) \leq (k-1)/2$, $e(X_i \cup \{b\}) \geq k + 2$, $e(b, u) \geq 2$ and $e(X_1, X_2) \geq 3$.

Again, theorem 3 does not hold when $k = 3$; the graph in figure 2 is a counterexample.

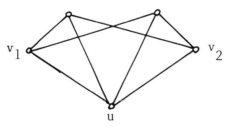

Figure 2

Theorem 4. Suppose that:

(i) $k \geq 5$ is odd, $V(G) = W \cup S \cup S'$ (disjoint union), $\lambda(V(G) - S - S') \geq k$, S is dummy and each vertex in S' has even degree,

(ii) $V(G) = X \cup Y$, $X \cap Y = \emptyset$, $e(X) = k + 1$, $X - W = S$ and $Y - W = S'$,

(iii) $\{u_1, u_2\} \subseteq X \cap W$, $\{v_1, v_2\} \subseteq Y \cap W$ and $f_i \in \partial(u_i, v_i)$ $(i = 1, 2)$,

(iv) $F \subseteq E(G/X)$, $F \cap \partial(X; G/X) = \{f_1, f_2\}$ and

$$\lambda(V(G) - S - S'; G/X - F) \geq k - 2.$$

Then one of the following holds.

(1) G/Y has a cycle C containing f_1 and f_2 such that

$$\lambda(V(G) - S - S'; G - F - E(C)) \geq k - 2,$$

(2) $X = X_1 \cup X_2$, $X_1 \cap X_2 = \emptyset$, where for $i = 1, 2$, $u_i \in X_i$ and $e(X_i) = k$.

In theorem 4, we usually take F to be the edge set of a path or cycle in G/X containing f_1 and f_2.

Theorem 5. *If $k \geq 3$ is odd, $V(G) = T \cup W \cup S$ (disjoint union), $\lambda(V(G) - S) \geq k$, $T = \{u_1, u_2, u_3\}$, $|T| = 3$, $f_i \in \partial(u_i, u_{i+1})$ $(i = 1, 2)$, and each vertex in S has even degree, then there are distinct edges g_i $(1 \leq i \leq (k-1)/2)$ in $\partial(u_1) - f_1$, such that for $1 \leq i \leq (k-1)/2$, G has a cycle C_i containing f_1, f_2 and g_i with $\lambda(V(G) - S; G - E(C_i)) \geq k - 2$.*

We set $\alpha := (k-1)/2$. Let X, $\{x, y\} \subseteq V(G)$. We set $\bar{X} := V(G) - X$, and $N(x; G) := \{a \in V(G) - x \mid e(a, x) > 0\}$. We write $P = P[x, y]$ to denote that P is a path between x and y, and for $a, b \in V(P)$, $P(a, b)$ denotes the subpath between a and b. We sometimes give a cycle by the edge set. If $|X| \geq 2$, $|\bar{X}| \geq 2$, and $e(X) = k$, then we call X and $\partial(X)$ a *k-set* and a *k-cut* respectively. If X is a *k*-set and $e(Z) \geq k + 1$ for each $Z \subseteq X$ with $Z \neq X$ and $|Z| \geq 2$, then we call X a *minimal k-set*. For $a, b \in N(x)$ with $a \neq b$, and for $f \in \partial(x, a)$ and $g \in \partial(x, b)$, $G_x^{a,b}$ and $G^{f,g}$ denote the graph $(V(G), (E(G) \cup \{h\}) - \{f, g\})$, and is called a *lifting of G at x*, where h is a new edge between a and b. We call $G^{f,g}$ *admissible* if for each $y \neq z \in V(G) - x$, $\lambda(y, z; G^{f,g}) = \lambda(y, z; G)$. For $K \subseteq V(G) \cup E(G)$, we define

$$\mathcal{C}(G, K, X) = \mathcal{C}(G, K, X, k-2)$$

$$:= \{C \mid C \text{ is a cycle in } G \text{ meeting } K \text{ such that } \lambda(X; G - E(C)) \geq k - 2\}$$

and we define $\mathcal{C}(G, K) = \mathcal{C}(G, K, k-2) := \mathcal{C}(G, K, X, k-2)$ where $X = \{x \in V(G) \mid e(x) \geq k\}$.

2. Preliminaries

We prepare some lemmas. Lemma 1 is obvious, and lemma 2 follows by simple counting.

Lemma 1. *If $\{x, y\} \subseteq X \subseteq V(G)$, $z \in \bar{X}$, $e(X) = k$, and $e(z, X; G) = k$, then $\lambda(x, y; G/\bar{X}) = \lambda(x, y; G)$.*

Lemma 2. *If $X, Y \subseteq V(G)$, then*

$$e(X-Y) + e(Y-X) = e(X) + e(Y) - 2e(X \cap Y, \overline{X \cup Y}),$$

$$e(X \cap Y) + e(X \cup Y) = e(X) + e(Y) - 2e(X-Y, Y-X).$$

Lemma 3. (Mader [2]). *If $x \in V(G)$, $e(x) \neq 3$, $|N(x)| \geq 2$, and x is not a cut-vertex, then there is an admissible lifting of G at x.*

Lemma 4. (Mader [3]). *If $\lambda(G) \geq 2$, $u \in V(G)$, and $\{f_1, f_2\} \subseteq \partial(u)$, then there is a cycle C containing f_1, f_2 such that for each $x \neq y \in V(G)$, $\lambda(x, y; G - E(C)) \geq \lambda(x, y; G) - 2$.*

Lemma 5. *Suppose that*

(i) *$k \geq 3$ is odd, $V(G) = X \cup Y$, $X \cap Y = \emptyset$ and $e(X) = k+1$,*

(ii) *$W \subseteq V(G)$, $W \cap X \neq \emptyset \neq W \cap Y$, $\lambda(W; G/X) \geq k$, $\lambda(W; G/Y) \geq k$ and each vertex in W has even degree, and*

(iii) *either (a) or (b) below holds:*

(a) *for some $x \in X$, $\lambda(x, Y) = k+1$,*

(b) *there is no $Z \subseteq X$ such that $e(Z) = k$ and $e(Z, Y) = (k+1)/2$.*

Then $\lambda(W; G) \geq k$.

Proof. First we prove that (a) implies (b). If (b) does not hold, then for some $Z_1 \subseteq X$, $e(Z_1) = k$ and $e(Z_1, Y) = (k+1)/2$. Let $Z_2 = X - Z_1$; then $e(Z_2) = e(Z_2, Y) + e(Z_2, Z_1) = k$. For each $x \in X$, $e(x, Y) \leq k$ and so (a) does not hold. Thus (a) implies (b). Next we assume that (i), (ii) and (b) hold. It is easy to see $\lambda(W) \geq k-1$. Suppose that $\lambda(W) = k-1$. Then for some $Z \subseteq V(G)$, $Z \cap W \neq \emptyset \neq \overline{Z} \cap W$ and $e(Z) = k-1$. We may assume that $Z \cap X \cap W \neq \emptyset$, for otherwise $\overline{Z} \cap X \cap W \neq \emptyset$ and we may take $\overline{Z}$ instead of Z. Now $e(Z \cap X) \geq k$. If $e(Z \cap X)$ is even, then $e(Z \cap X) \geq k+1$ and by lemma 2, $e(\overline{Z} \cup X) = e(Y - Z) \leq k - 1$ and so $Y - Z \subseteq \overline{W}$. Thus $(Y \cap Z) \cap W \neq \emptyset$ and $e(Y \cap Z) \geq k$. But $e(Y \cap Z) = e(Z - X) \geq k+1$, since $e(Y \cap Z) = e(Z) - e(Z \cap X) \equiv 0 \pmod{2}$. By lemma 2, $e(X - Z) \leq k - 1$ and $X - Z \subseteq \overline{W}$, and so $\overline{Z} \cap W = \emptyset$, a contradiction. Therefore $e(Z \cap X)$ is odd. Now $e(X-Z)$, $e(Z-X)$ and $e(Z \cup X)$ are odd. By lemma 2 $e(Z \cap X) = e(Z - X) = k$, and so $e(Z \cap X; Y) = e(Z - X; Y) = (k+1)/2$, which contradicts (b). Thus our assumption that $\lambda(W) = k-1$ was false, and the result follows. ∎

Lemma 6. *Suppose that $k \geq 3$ is odd, $V(G) = W \cup S' \cup S$ (disjoint union), S is dummy, $\lambda(W) \geq k$, $\lambda(V(G) - b) \geq k - 1$, each vertex in $S' \cup S$ has even degree, $\{x_1, x_2, x_3\} \subseteq N(b)$, $\lambda(W; G_b^{x_1, x_2}) \geq k$ and $\lambda(W; G_b^{x_i, x_3}) < k$ ($i = 1, 2$). Then there are disjoint $X_1, X_2, X_3 \subseteq V(G) - b$ such that $x_i \in X_i$, $e(X_i) = k$ ($1 \leq i \leq 3$) and $e(X_i, X_3) = (k-1)/2$ ($i = 1, 2$).*

Proof. For $i = 1, 2$, there exists $Y_i \subseteq V(G) - b$ such that $\{x_i, x_3\} \subseteq Y_i$, $Y_i \cap W \neq \emptyset \neq \overline{Y_i} \cap W$ and $e(Y_i) = k$ or $k+1$. Since $\lambda(W; G_b^{x_1, x_2}) \geq k$, we have $x_1 \notin Y_2, x_2 \notin Y_1$, and if $(Y_1 \cup Y_2) \cap W \neq \emptyset$, then $e(Y_1 \cup Y_2) \geq k+2$. Now $2k - 2 \leq e(Y_1 - Y_2) + e(Y_2 - Y_1) \leq 2k$ by lemma 2. Consequently, if $e(Y_1 - Y_2) = e(Y_2 - Y_1) = k$, then $e(Y_1 \cap Y_2, \overline{Y_1 \cup Y_2}) = 1$, $e(Y_i) = k+1$ ($i = 1, 2$)

and $e(Y_1 \cap Y_2)$ and $e(Y_1 \cup Y_2)$ are odd. By lemma 2, $e(Y_1 \cap Y_2) = k$, $e(Y_1 \cup Y_2) = k + 2$ and the result holds. Thus we may assume that $e(Y_1 - Y_2) = k - 1$. Since $Y_1 - Y_2 \subseteq S' \cup S$, it follows that $Y_1 \cap Y_2 \cap W = Y_1 \cap W \neq \emptyset$ and $(\overline{Y_1 \cup Y_2}) \cap W = \overline{Y_2} \cap W \neq \emptyset$, and so $e(Y_1 \cap Y_2) \geq k$ and $e(Y_1 \cup Y_2) \geq k + 2$. By lemma 2, $e(Y_i) = k + 1$ ($i = 1, 2$) and $e(Y_1 \cap Y_2) = k$. Then $e(Y_1 - Y_2)$ is odd, a contradiction. ∎

3. Proof of Theorem 3

In this section (1), (2) and (3) denotes (1), (2) and (3) in theorem 3 respectively. We will prove that if (1) does not hold, then (2*) below holds, by using induction on $|E(G)|$. A minimal counterexample G is considered. Let $\alpha := (k-1)/2$.

(2*) $V(G) = A \cup X_1 \cup X_2$ (*disjoint union*), for $i = 1, 2$, $v_i \in X_i$ and $e(X_i) = k$, and either $A = \{u\}$, or $A = \{u, b\}$ and $e(b) \leq k - 3$.

We denote by $I(G, u) = I(G, u, f_1, f_2)$ the set of (C, a) satisfying (1), and we denote by $L(G, u)$ the set of (A, X_1, X_2) satisfying (2*).

(3.1) *If* (2*) *holds, then* (2) *or* (3) *holds.*

Proof. If $A = \{u\}$, then (2) follows. Assume $A = \{u, b\}$. If $e(X_2, u) = \alpha + 1$ or $e(X_1 \cup \{b\}) = k$, then $(u, X_1 \cup \{b\}, X_2) \in L(G, u)$, and so $e(X_i, u) \leq \alpha$ and $e(X_i \cup \{b\}) \geq k + 2$ (since it is odd) for $i = 1, 2$, and $e(b) \geq 4$. By lemma 2 $e(X_1) + e(X_2) = e(X_1 \cup \{b\}) + e(X_2 \cup \{b\}) - 2e(b, u)$, and thus $e(b, u) \geq 2$. But

$$e(A) = e(b) + e(u) - 2e(b, u) \leq (k-3) + (k+1) - 4 = 2k - 6.$$

Hence $e(X_i, A) \leq k - 3$ ($i = 1, 2$) and $e(X_1, X_2) \geq 3$. ∎

(3.2) *G has no k-set, and if $|S| = 2$, then $e(S) \geq k + 1$.*

Proof. Let $X \subseteq V(G) - u$ be a k-set. By induction either $I(G/X, u) \neq \emptyset$ or $L(G/X, u) \neq \emptyset$. Assume first that there is a $(C_1, a) \in I(G/X, u)$. If $\{v_1, v_2\} \subseteq X$, then by lemma 4 there is a $C_2 \in C(G/\overline{X}, \{f_1, f_2\})$ and $(C_2, a) \in I(G, u)$. If $\{v_1, v_2\} \not\subseteq X$, and C_1 meets X, then we can extend C_1 to a cycle in G by lemma 4 and $I(G, u) \neq \emptyset$. If $L(G/X, u) \neq \emptyset$, then clearly $L(G, u) \neq \emptyset$. Thus there is no such X. If $|S| = 2$ and $e(S) \leq k - 1$, then by (1.1), $e(S) = k - 1$ and $\lambda(b', \overline{S}) = k - 1$. Then either $I(G/S, u) \neq \emptyset$ or there is $(u, X_1, X_2) \in L(G/S, u)$, and so (1) or (2*) holds in G, a contradiction. ∎

(3.3) *If $x \in (T \cup W) - u$, then $e(x) = k$.*

Proof. If $e(x) \geq k + 1$, then by (3.2) $\lambda(x, u) = k + 1$ and by lemma 4, (1) follows, a contradiction. ∎

(3.4) $v_1 \neq v_2$ *and* $e(v_1, v_2) = 0$.

Proof. If $v_1 = v_2$, then by (3.3) there is a vertex $x \neq v_1$ of odd degree, and so $\lambda(x, u; G - \{f_1, f_2\}) = k - 1$ by (3.2). Thus $v_1 \neq v_2$. If there exists $g \in \partial(v_1, v_2)$, then let $C := \{f_1, f_2, g\}$. By (3.2) $G - E(C)$ has no $(k-2)$-cut, and so if there is an $x \in V(G) - T$ with $e(x) \geq k - 1$, then $\lambda(x, u; G - E(C)) = k - 1$ and $I(G, u) \neq \emptyset$. Thus either $V(G) = T$, or $V(G) = T \cup \{b\}$ and $e(b) \leq k - 3$. Hence (2*) follows, a contradiction. ∎

(3.5) *If* $u \in X \subseteq V(G) - \{v_1, v_2\}$ *and* $|X| \geq 2$, *then* $e(X) \geq k + 2$.

Proof. By (3.2), $e(X) \geq k + 1$. Assume $e(X) = k + 1$; then $\lambda(u, \bar{X}) = k + 1$. If there exists $(C, a) \in I(G/X, X)$, then $\lambda(u, a; G - E(C)) = k - 1$ and by lemma 5, $\lambda(T \cup W; G - E(C)) = k - 2$, and so $(C, a) \in I(G, u)$. Thus there exists $(A, Y_1, Y_2) \in L(G/X, X)$ and $e(Y_1, Y_2) > 0$. By (3.2), $Y_i = \{v_i\}$ ($i = 1, 2$), which contradicts (3.4). ∎

(3.6) *If* $b \in S$, *then* $V(G) \neq T \cup \{b\}$ *and either* $e(b, v_1) = 0$ *or* $e(b, v_2) = 0$.

Proof. If $V(G) = T \cup \{b\}$, then by (3.4) $e(b, v_i) = \alpha$ and $e(u, v_i) = \alpha + 1$ for $i = 1, 2$, and so $\{v_1, b\}$ is a k-set, contrary to (3.2). Thus there exists $x \in W \cup S - b$ with $e(x) \geq k - 1$. Suppose that $e(b, v_1)$, $e(b, v_2) \neq 0$; let $g_i \in \partial(b, v_i)$ for $i = 1, 2$, and let $C := \{f_1, f_2, g_1, g_2\}$. By (3.5) $\lambda(T \cup W; G - E(C)) \geq k - 2$. Since $(C, x) \notin I(G, u)$, there is a $(k + 2)$-set X with $\{u, b\} \subseteq X \subseteq V(G) - \{v_1, v_2, x\}$. Choose (X, x) such that $|X|$ is minimum. Since $k + 2$ is odd, X has a vertex y of odd degree, and so there is a $(k + 2)$-set Y with $\{u, b\} \subseteq Y \subseteq V(G) - \{v_1, v_2, y\}$. By lemma 2, $e(X - Y) + e(Y - X) \leq 2k - 4$, and so $Y - X = \emptyset$, contrary to the minimality of $|X|$. Hence not both $e(b, v_1)$, $e(b, v_2)$ are non-zero. ∎

(3.7) $S = \emptyset$.

Proof. Let $b \in S$ and $N(b) = \{x_1, \ldots, x_n\}$. Then $n \geq 3$ by (3.2). By lemma 3, for some $i \neq j$, $G_b^{x_i, x_j}$ is admissible, say for $(i, j) = (1, 2)$. By (3.5) $e(b, u) < e(b)/2$, and so we can choose x_1, x_2 from $V(G) - u$. If $I(G_b^{x_1, x_2}, u) \neq \emptyset$, then $I(G, u) \neq \emptyset$, a contradiction; and so there is an $(A, X_1, X_2) \in L(G_b^{x_1, x_2}, u)$. We claim:

(3.7.1) (a) *or* (b) *below holds:*

(a) $A = \{u, b\}$, *and for some* $(i, j) \in \{(1, 2), (2, 1)\}$, $X_i = \{v_i\}$, $\{x_1, x_2, v_j\} \subseteq X_j$, $e(b, v_i; G) > 0$ *and* $e(X_j; G) = k + 2$.

(b) $A = \{u\}$, *for some* $(i, j) \in \{(1, 2), (2, 1)\}$, $b \in X_i$ *and* $\{x_1, x_2\} \subseteq X_j$, *and* $e(X_1; G) = e(X_2; G) = k + 2$.

For if $A = \{u\}$, then we may let $b \in X_1$. By (3.2), $\{x_1, x_2\} \subseteq X_2$ and (b) follows. If $A = \{u, b\}$, then we may let $x_1 \in X_2$. If $x_2 \in X_1$, then $e(X_i; G) = k$ ($i = 1, 2$) and by (3.2) $X_i = \{v_i\}$ ($i = 1, 2$). Thus $V(G) = T \cup \{b\}$, contrary to (3.6). Hence $x_2 \in X_2$, $X_1 = \{v_1\}$, and $e(X_2; G) = k + 2$. Hence $e(b, X_2) \leq e(b)/2$, for otherwise $(u, \{v_1\}, X_2 \cup \{b\}) \in L(G, u)$. If $e(b, v_1) = 0$, then $e(b, u) \geq e(b)/2$, contrary to (3.5). This proves (3.7.1).

Assume that (a) or (b) holds in (3.7.1) with $(i, j) = (1, 2)$. We can choose $x_3 \in X_1 \cap N(b)$. For if not, then (b) holds and $X_1 \cap N(b) = \emptyset$; but then $e(X_1 - b) < k$, since $e(b) \geq 4$, a contradiction. If (b) holds, then we choose $\{x_1, x_2, X_1, X_2\}$ such that $|X_1|$ is minimum. We claim that $\lambda(T \cup W; G_b^{x_i, x_3}) \geq k$ for some $i \in \{1, 2\}$, say for $i = 1$. For otherwise by lemma 6 and (3.2), $e(x_i) = k$ ($1 \leq i \leq 3$) and $e(x_i, x_3) = \alpha$ ($i = 1, 2$), and so $N(x_3) = \{b, x_1, x_2\}$, $x_3 \neq v_1$, $|X_1| \geq 2$ and $e(X_1 - x_3) \leq k + 2 - 3$, a contradiction. We may also assume

(3.7.2) $x_1 \neq b'$ or $\lambda(T \cup W; G_b^{x_2, x_3}) < k$.

Let $(B, Y_1, Y_2) \in L(G_b^{x_1, x_3}, u)$.

Case 1. (a) holds for both (A, X_1, X_2) and (B, Y_1, Y_2). Now $x_3 = v_1$, $\{v_1, x_1\} \subseteq Y_1$, $Y_2 = \{v_2\}$ and $e(b, v_i) > 0$ $(i = 1, 2)$, contrary to (3.6).

Case 2. (a) holds for (A, X_1, X_2) and (b) holds for (B, Y_1, Y_2). Now $x_3 = v_1$, $\{x_1, x_2, v_2\} \subseteq X_2$, $e(X_2) = e(Y_1) = e(Y_2) = k + 2$, $\{v_1, x_1\} \subseteq Y_1$ and $\{b, v_2\} \subseteq Y_2$. Thus $Y_1 - X_2 = \{v_1\}$ and by lemma 2,

$$e(X_2 - Y_1) \leq e(Y_1) + e(X_2) - e(Y_1 - X_2) - 2 \leq k + 2.$$

If $e(X_2 - Y_1) = k$, then by (3.2) $X_2 - Y_1 = \{v_2\}$, and so $Y_2 = \{b, v_2\}$. By (3.6), $e(b, v_2) = 0$ and $e(Y_2) \geq k + 4$, a contradiction. Thus $e(X_2 - Y_1) = k + 2$. Since $Y_2 = \{b\} \cup (X_2 - Y_1)$, we have $e(Y_2) = e(b) + e(X_2 - Y_1) - 2e(b, X_2 - Y_1)$, and so $e(b, X_2 - Y_1) = e(b)/2$. Hence $e(b, X_2) \geq e(b)/2 + 1$ and $e(X_2 \cup \{b\}) \leq e(X_2) - 2 = k$, contrary to (3.2).

Case 3. (b) holds for both (A, X_1, X_2) and (B, Y_1, Y_2). Let $b \in Y_r$ and $\{x_1, x_3\} \subseteq Y_s$ for $(r, s) = (1, 2)$ or (2.1). Let $Z_1 := X_1 \cap Y_r$, $Z_2 := X_1 \cap Y_s$, $Z_3 := X_2 \cap Y_r$ and $Z_4 := X_2 \cap Y_s$. If $(r, s) = (2, 1)$, then $v_2 \in Z_3$. If $(r, s) = (1, 2)$ and $Z_3 = \emptyset$, then $Y_1 \subseteq X_1 - x_3$, contrary to the minimality of $|X_1|$. Thus $Z_3 \neq \emptyset$. By lemma 2 we have

(3.7.3) For $(i, j) = (1, 4), (4, 1)$,

$$e(Z_2) + e(Z_3) = e(Z_2 \cup Z_i) + e(Z_3 \cup Z_i) - 2e(Z_i, Z_j \cup \{u\})$$

$$= 2k + 4 - 2e(Z_i, Z_j \cup \{u\}).$$

Thus for $(i, j) = (1, 4), (4, 1)$, $2e(Z_i, Z_j \cup \{u\}) \leq 2k + 4 - (2k - 1) = 5$, and so $e(Z_1, Z_4) + e(Z_p, u) \leq 2$ $(p = 1, 4)$. Since $e(Z_1, Z_4) \geq e(b, x_1) \geq 1$, we have $e(u, Z_i) \leq 1$ $(i = 1, 4)$. By lemma 2

(3.7.4) $\quad e(Z_4) + e(Z_1 \cup \{u\}) \leq e(Z_2 \cup Z_4) + e(Z_2 \cup Z_1 \cup \{u\})$

$$= e(Z_2 \cup Z_4) + e(Z_3 \cup Z_4) = 2k + 4.$$

If $(r, s) = (1, 2)$, then $\{b, v_1\} \subseteq Z_1$ and $v_2 \in Z_4$, and so $e(Z_1) \geq k + 1$ and $e(Z_4) \geq k$. Hence $e(Z_1 \cup \{u\}) \geq e(Z_1) + e(u) - 2e(Z_1, u) \geq 2k \geq k + 5$, contrary to (3.7.4). Thus $(r, s) = (2, 1)$, $v_1 \in Z_2$ and $v_2 \in Z_3$. Then $e(Z_1 \cup \{u\}) \geq e(Z_1) + e(u) - 2 \geq e(b) + k - 1 \geq k + 3$, and so $e(Z_4) \leq k + 1$. Hence $x_2 \notin Z_4$ (since $G_b^{x_1, x_2}$ is admissible) and $x_2 \in Z_3$. By (3.6), $v_1 \neq x_3$ or $v_2 \neq x_2$, and so $e(Z_2) + e(Z_3) \geq 2k + 1$. By (3.7.3), $e(Z_2) + e(Z_3)$ is even and $\leq 2k + 2$, and thus $e(Z_2) + e(Z_3) = 2k + 2$, $e(Z_1, Z_4) = 1$ and $e(u, Z_1) = e(u, Z_4) = 0$. By (3.7.4),

$$2k + 4 \geq e(Z_4) + e(Z_1 \cup \{u\}) = e(Z_4) + e(Z_1) + e(u)$$

and so $e(Z_1) + e(Z_4) \leq k + 3$. Thus $S = \{b, b'\}$ where $Z_1 = \{b\}$, $Z_4 = \{b'\}$, $e(b) = 4$ and $e(b') = k - 1$. By (3.7.2), for some $Z_0 \subseteq V(G) - b$, $\{x_2, x_3\} \subseteq Z_0$, $\overline{Z}_0 \cap (T \cup W) \neq \emptyset$ and $e(Z_0) = k + 1$. Now $Z_i - Z_0 \neq \emptyset$, $i = 2$ or 3 (for otherwise $\overline{Z}_0 = \{b, u\}$ or $\{b, b', u\}$, and $e(Z_0) \geq k + 5$), say for $i = 2$. If for $(i, j) = (2, 3)$ or (3,

2), $e(Z_i) = k$ and $e(Z_j) = k + 2$, then by (3.2), $e(u, Z_i) \leq \alpha$ and $e(u, Z_j) \leq \alpha + 1$, which contradicts $e(u) = k + 1$. Hence $e(Z_2) = e(Z_3) = k + 1$ and $e(u, Z_i) = \alpha + 1$ ($i = 2, 3$). By lemma 2 $e(Z_2 - Z_0) = e(Z_0 - Z_2) = k$ and $e(Z_0 \cap Z_2, \overline{Z_0 \cup Z_2}) = e(x_3, b) = 1$, and so $Z_0 - Z_2 = \{x_2\}$. Thus $u \notin Z_0$, $Z_2 - Z_0 = \{v_1\}$ and $e(u, v_1) = \alpha + 1$, contrary to (3.2). ∎

(3.8) $W \neq \emptyset$, and if $x_1 \neq x_2 \in W$, then $e(x_1, x_2) = 0$.

Proof. If $W = \emptyset$, then by (3.7), $V(G) = T$ and (2*) holds. Thus $W \neq \emptyset$. Assume $x_1, x_2 \in W$ and $g \in \partial(x_1, x_2)$. In $G - g$, $\{x_1, x_2\}$ is considered as dummy. Assume there exists $(C_1, a) \in I(G - g, u)$. If $\{x_1, x_2\} \subseteq V(C_1)$ and $e(\{x_1, x_2\}; G - E(C_1)) \leq k - 3$, then we replace the path $P[x_1, x_2]$ in C_1 with $u \notin V(P)$ by g in G and produce a cycle C in G; and if not, then let $C := C_1$. Then $(C, a) \in I(G, u)$, a contradiction. Thus $I(G - g, u) = \emptyset$, and so there exists $(u, X_1, X_2) \in L(G - g, u)$. If $\{x_1, x_2\} \subseteq X_1$, then $(u, X_1, X_2) \in L(G, u)$, and so we may let $x_i \in X_i$ and $e(X_i; G) = k + 1$ ($i = 1, 2$). We choose $\{x_1, x_2, X_1, X_2\}$ such that $|X_1|$ is minimum.

Case 1. $|X_1| = 2$.

Now $e(v_1, x_1) = \alpha$ and by (3.2), $e(v_1, u) \leq \alpha$, and thus there is an $h_1 \in \partial(v_1, X_2)$. If for $G_1 := G/(X_1 \cup \{u\})$, there exists $(C, a) \in I(G_1, X_1 \cup \{u\}, h_1, f_2)$, then $\lambda(a, u; G - E(C) - f_1) = k - 1$ and (1) holds in G; and if not, then by induction and (3.2), $X_2 = \{v_2, x_2\}$ and $h_1 \in \partial(v_1, x_2)$. For $h_2 \in \partial(x_2, v_2)$, let $C := \{f_1, f_2, h_1, h_2\}$. Now $e(u, x_2) = k - e(x_2, \{v_2, x_1, v_1\}) \leq \alpha - 1$, and thus $e(\{u, x_2\}) \geq k + 4$, $\lambda(x_1, u; G - E(C)) = k - 1$ and (1) follows.

Case 2. $|X_1| \geq 3$.

If $N(x_1) \cap X_1 = \{v_1\}$, then $e(X_1 - x_1) = k$, contrary to (3.2). Thus for some $x_3 \in X_1 - \{v_1, x_1\}$, there are $h \in \partial(x_1, x_3)$ and $(u, Y_1, Y_2) \in L(G - h, u)$. Now $e(Y_i; G) = k + 1$ ($i = 1, 2$) and $Y_1 - X_1 \neq \emptyset$, since $|X_1|$ is minimum. By lemma 2 $e(X_1 - Y_1) = e(Y_1 - X_1) = k$. Thus $e(X_1 \cap Y_1)$ and $e(\overline{X_1 \cup Y_1})$ are odd. But $|X_1 \cap Y_1| \geq 2$ and by (3.2), $e(X_1 \cap Y_1) \geq k + 2$ and $e(\overline{X_1 \cup Y_1}) \geq k + 2$, contrary to lemma 2. This proves (3.8). ∎

By (3.3) $|W|$ is even, by (3.7) and (3.8) we may let $W = \{x_1, x_2\}$, and $N(x_i) = T$ ($i = 1, 2$). If $e(x_1, u) < \alpha$, then for $g_i \in \partial(x_1, v_i)$ ($i = 1, 2$), $\lambda(x_2, u; G - \{f_1, f_2, g_1, g_2\}) = k - 1$ and (1) holds. Thus $e(x_i, u) = \alpha$ ($i = 1, 2$) and $e(u, v_i) = 1$ ($i = 1, 2$). Then $e(v_i, x_j) = \alpha$ ($i, j = 1, 2$), and so $e(x_1) = 3\alpha \geq k + 1$, a contradiction. This completes the proof of theorem 3. ∎

4. Proof of theorems 4 and 2

Let $\alpha := (k - 1)/2$. First we prove theorem 4. If there is a cycle $C \in \mathcal{C}(G/Y, \{f_1, f_2\}, W)$ such that $\lambda(a, Y; G/Y - E(C_2)) = k - 1$, for some $a \in X$, then by lemma 5, $\lambda(W; G - F - E(C)) \geq k - 2$; and if not, then by using theorem 3 in G/Y, either (2) in theorem 4 holds, or $X = \{b\} \cup X_1 \cup X_2$ (disjoint union), for $i = 1, 2$, $u_i \in X_i$, $e(X_i) = k$, $e(X_i, Y) \leq \alpha$, $e(X_i \cup \{b\}) \geq k + 2$ and $e(X_1, X_2) \geq 3$. Let $g \in \partial(X_1, X_2)$. Now

$$\lambda(W; ((G/Y)/X_1) X_2 - \{f_1, f_2, g\}) \geq k - 2$$

and by lemma 4 we can choose $C \in \mathcal{C}(G/Y, \{f_1, f_2, g\}, W)$. Let $G_1 := (G/X_1)/X_2$ and $G_2 := G_1 - F - E(C)$. In G_2, $X = \{b, X_1, X_2\}$ and there is no $Z \subseteq X$ such that $e(Z; G_2) = k - 2$ and $e(Z, Y; G_2) = \alpha$. Thus by lemma 5 with $k - 2$ instead of k, $\lambda(W; G_2) \geq k - 2$. It is easy to see that $\lambda(W; G - F - E(C)) \geq k - 2$ and hence theorem 4 is proved. ∎

Next we prove theorem 2. By lemma 4 there are $C_1 \in \mathcal{C}(G/X, \{f_1, f_2\})$ and $C_2 \in \mathcal{C}(G/Y, \{f_1, f_2\})$. Let $C := C_1 \cup C_2$ (we construct the cycle C by combining C_1 and C_2 in G). Assume that C_2 (respectively, C_1) cannot be chosen such that $\lambda(W; G - E(C)) \geq k - 2$. Then by theorem 4, $X = X_1 \cup X_2$, $X_1 \cap X_2 = \emptyset$, for $i = 1, 2$, $u_i \in X_i$ and $e(X_i) = k$ (respectively, $Y = Y_1 \cup Y_2$, $Y_1 \cap Y_2 = \emptyset$, for $i = 1, 2$, $v_i \in Y_i$ and $e(Y_i) = k$). It easily follows that $e(X_1, X_2) = e(Y_1, Y_2) = \alpha$. Let G_1 be the graph obtained from G by contracting X_i to u_i and Y_i to v_i ($i = 1, 2$). Then $V(G_1) = \{X_1, X_2, Y_1, Y_2\}$ and $\lambda(G_1 - E(C)) \leq k - 3$. Thus $e(\{X_1, Y_2\}; G_1) = k + 1$, and (2) in theorem 2 follows. ∎

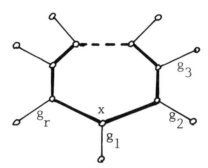

Figure 3

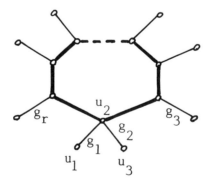

Figure 4

5. Proof of theorem 5

We may assume that G is 2-connected, and we may assume

(5.1) $e(u_2) = k$ or $k+1$, and $e(x) = k$ for each $x \in W \cup \{u_3\}$.

Proof. Let $x \in V(G)$ and $\partial(x) = \{g_1, ..., g_r\}$. If $x \neq u_2$ (respectively, $x = u_2$, $g_1 = f_1$, and $g_2 = f_2$) and $r \geq k+1$ (respectively, $r \geq k+2$), then we replace x and $\partial(x)$ by the graph in Figure 3 (respectively, Figure 4), in which heavy edges represent $\alpha = (k-1)/2$ parallel edges, producing a new graph G'. If the result holds in G', then it also holds in G. ∎ Let $F(G) = F(G, T \cup W) := \{g \in \partial(u_1) - f_1 \mid \mathcal{C}(G, \{f_1, f_2, g\}, T \cup W) \neq \emptyset\}$. We prove $|F(G)| \geq \alpha$ by induction on $|E(G)|$.

(5.2) *If X is a k-set with $|X \cap T| \leq 1$, then $X \cap T = \{u_1\}$.*

Proof. Assume $X \cap T \neq \{u_1\}$. Then $|F(G/X)| \geq \alpha$. For $g \in F(G/X)$ and $C_1 \in \mathcal{C}(G/X, \{f_1, f_2, g\})$, if $\partial(X) \cap E(C_1) = \{h_1, h_2\}$, then by lemma 4 there is a $C_2 \in \mathcal{C}(G/\overline{X}, \{h_1, h_2\})$ and a cycle $C_1 \cup C_2$ in G belonging to $\mathcal{C}(G, \{f_1, f_2, g\})$. Thus $|F(G)| \geq \alpha$. ∎

(5.3) *If $X \subseteq V(G)$, $X \cap T = \{u_2\}$ and $|X| \geq 2$, then $e(X) \geq k+2$.*

Proof. Otherwise by (5.2), $e(X) = k+1$. Let $g \in F(G/X)$, and $C \in \mathcal{C}(G/X, \{f_1, f_2, g\})$. If $e(u_2) = k+1$, then $\lambda(u_2, \overline{X}; G - E(C)) = k-1$; and if $e(u_2) = k$, then $X - u_2$ has a vertex x of odd degree and $\lambda(x, \overline{X}; G - E(C)) = k-1$ by (5.2). Thus $\lambda(T \cup W; G - E(C)) = k-2$ by lemma 5. ∎

(5.4) $S = \emptyset$.

Proof. Assume $x \in S$ and $e(X) \geq 4$. We may let $|N(x)| \geq 2$. In an admissible lifting G_x (see lemma 3), the result holds. ∎

(5.5) *If $x_1, x_2 \in W$ and $h \in \partial(x_1, x_2)$, then h is contained in a k-cut.*

Proof. Suppose not; then $|F(G - h, V(G) - \{x_1, x_2\})| \geq \alpha$. But for $g \in F(G - h, V(G) - \{x_1, x_2\})$ and $C_1 \in \mathcal{C}(G - h, \{f_1, f_2, g\})$, we can extend C_1 to C in $\mathcal{C}(G, \{f_1, f_2, g\})$ (see the proof of (3.8)). ∎

(5.6) $e(u_1) \leq k+1$.

Proof. If $e(u_1) \geq k+2$, then by lemma 3, there is an admissible lifting G_1 at u_1. Then $\lambda(V(G_1) - u_1; G_1) \geq k$ and $e(u_1; G_1) \geq k$, and thus $\lambda(G_1) \geq k$ and the result holds in G_1. ∎

If G has a k-set, then by (5.2) let X be a minimal k-set with $X \cap T = \{u_1\}$. By (5.5) for each $x \subset X - u_1$, $N(x) \subseteq \{u_1\} \cup \overline{X}$. By (5.6), $e(\{u_1\} \cup \overline{X}) \leq 2k-1$, and so $|X| = 2$, say $X = \{u_1, x\}$. Then $e(u_1, x) = \alpha + 1$. If $g \in F(G/X)$ for some $g \in \partial(x, \overline{X})$, then $\partial(u_1, x) \subseteq F(G)$, and if not, then $F(G/X) \subseteq F(G)$. Thus $|F(G)| \geq \alpha$ and G has no k-set. By (5.5), for each $x \in W$, $N(x) = T$. For each $g \in \partial(u_1) - \partial(u_1, u_2)$, G has a cycle C containing $\{f_1, f_2, g\}$ of length 3 or 4. By (5.3), $\lambda(G - E(C)) \geq k-2$. Thus $\partial(u_1) - \partial(u_1, u_2) \subseteq F(G)$ and $e(u_1, u_2) \leq \alpha + 1$ (for otherwise $e(\{u_1, u_2\}) \leq k-1$). This completes the proof of theorem 5. ∎

6. Proof of theorem 1

Let $\alpha := (k-1)/2$. We shall need the following.

Lemma 7. Suppose that

(i) $k \geq 5$ is odd, $\lambda(G) = k$, G is k-regular, and G has no k-cut

(ii) $V(G) = \{v_1, v_2\} \cup W$ (disjoint union), $W \neq \emptyset$, $f_0 \in \partial(v_1, v_2)$ and for each $x, y \in W$, $e(x, y) = 0$ or α

(iii) for each $x, y \in W$ and $g \in \partial(x, y)$, $G - g$ has paths $P_1[v_1, v_2]$ and $P_2[v_1, v_2]$ such that $V(P_1) \cap V(P_2) = \{v_1, v_2\}$, $x \in V(P_1)$, $y \in V(P_2)$, and $\lambda(V(G) - \{x, y\}; G - g - E(P_1 \cup P_2)) = k - 2$; and $\lambda(G - E(C_i)) = k - 3$ for some $i \in \{1, 2\}$ where C_1, C_2 are the cycles $C_1 := P_1(v_1, x_2) \cup P_2(y, v_2) \cup \{f_0, g\}$ and $C_2 := P_2(v_1, y) \cup P_1(x, v_2) \cup \{f_0, g\}$.

Then for some $n \geq 2$, $W = \{x_1, ..., x_n\}$, $e(x_i, x_{i+1}) = \alpha$ $(1 \leq i \leq n-1)$, for $(i, j) = (1, 2)$ or $(2, 1)$, $e(x_1, v_i) = 1$ and $e(x_1, v_j) = \alpha$, and for $(r, s) = (1, 2)$ or $(2, 1)$, $e(x_n, v_r) = 1$ and $e(x_n, v_s) = \alpha$.

Proof. We denote by $\Omega(x, y)$ the set of (P_1, P_2, C_1, C_2) given in (iii). Let $W = \{x_1, ..., x_n\}$; then $n \geq 2$. For $1 \leq i \leq n$, $|N(x_i)| = 3$ and $e(x_i, \{v_1, v_2\}) > 0$. Since $G - \{v_1, v_2\}$ is connected, we may let $e(x_i, x_{i+1}) = \alpha$ $(1 \leq i \leq n-1)$. Let $(P_1, P_2, C_1, C_2) \in \Omega(x_1, x_2)$.

Case 1. $n \geq 3$ and $e(x_1, x_n) = \alpha$.

Suppose that two adjacent members of W are both adjacent to v_1, say $e(x_i, v_1) = 1$ $(i = 1, 2)$. By the symmetry we may let $\lambda(G - E(C_1)) = k - 3$. For some $p \geq 3$, $V(C_1) = \{v_1, x_1, x_2, ..., x_p, v_2\}$ and $e(Z; G - E(C_1)) = k - 3$ for some $Z \subseteq V(G) - V(P_1(x_1, v_2))$ with $v_1, x_2 \in Z$. If $|\partial(Z) \cap E(C_1)| \geq 6$, then $e(Z; G - E(C_1)) \geq 3(\alpha - 1) \geq k - 2$, a contradiction, and so $|\partial(Z) \cap E(C_1)| = 4$ and $e(Z; G) = k + 1$. Thus $e(Z \cup \{x_1\}) = k$, a contradiction. Thus $x_1, x_2, ..., x_n, x_1$ are adjacent alternately to v_1 and to v_2, and n is even and we may let $e(x_i, v_1) = 1$ for odd $1 \leq i \leq n$ and $e(x_i, v_2) = 1$ for even $1 \leq i \leq n$. Now $|E(C_1)| = 4$ and $\lambda(G - E(C_1)) = k - 2$. Thus $\lambda(G - E(C_2)) = k - 3$. For some $3 \leq p < q \leq n$, $V(C_2) = \{v_1, v_2, x_q, ..., x_n, x_1, ..., x_p\}$. For some $\{v_1, x_1\} \subseteq Z \subseteq V(G) - \{v_2, x_2\}$, $e(Z; G - E(C_2)) = k - 3$. Thus $x_n \in Z$, $e(x_n, v_2) = 1$, and so $e(Z; G - E(C_2)) \geq k - 2$.

Case 2. $n = 2$ or $e(x_1, x_n) = 0$.

Now $V(P_1) = \{v_1, x_1, v_2\}$. By the symmetry we may let $e(x_2, v_1) > 0$. If $e(G - E(C_2)) = k - 3$, then $e(x_1, v_1) = \alpha$ and $e(x_1, v_2) = 1$. Thus $e(G - E(C_1)) = k - 3$. Hence, $e(Z; G - E(C_1)) = k - 3$ for some Z with $\{x_1, v_2\} \subseteq Z \subseteq V(G) - \{v_1, x_2\}$. Thus $|\partial(Z) \cap E(C_1)| = 4$ and $e(Z; G) = k + 1$. Since $e(Z - x_1) \geq k$, we have $e(x_1, v_2) = \alpha$. ∎

Now we prove theorem 1. In what follows, (1), (2) and (3) denote (1), (2) and (3) in theorem 1 respectively. Assume (2) holds, and let $X_1 := X \cup Y$, $X_2 := X - Y$, $Y_1 := X \cap Y$, and $Y_2 := Y - X$. Then by lemma 2, $e(X_i) = e(Y_i) = k$ $(i = 1, 2)$ and (3) holds. It is easy to see that (3) implies (1) and (2).

Thus it suffices to prove that if (6.1) below holds, then
$C(G, \{f_1, f_2\}, T \cup W, k-2) \neq \emptyset$.

(6.1) *One of the following holds:*

(i) *for each* $X \subseteq V(G)$ *with* $X \cap T = \{v_1, v_2\}$, $e(X) \neq k+1$.

(ii) *for each* $X \in V(G)$ *with* $X \cap T = \{v_1, u_2\}$, $e(X) \neq k+1$.

If v_1 is a cut-vertex and a block B contains f_1 but not f_0 and f_2, then by lemma 4 there are $C_1 \in C(B, \{f_1\})$ and $C_2 \in C(G/V(B), \{f_0, f_2\})$; but then $C_1 \cup C_2 \in C(G, \{f_0, f_1, f_2\})$ as required. Thus if G is not 2-connected, then one block B contains $\{f_1, f_0, f_2\}$ and by induction the result holds in B, and so it also holds in G. Therefore we may assume G is 2-connected and we may assume (see (5.1))

(6.2) $e(v_i) = k$ or $k+1$ $(i = 1, 2)$ and $e(x) = k$ for each $x \in \{u_1, u_2\} \cup W$.

Let $C^*(G) = C^*(G, \{f_1, f_2\}) :=$

$\{C \in C(G, \{f_1, f_2\}) |\ C$ has no repeated vertices$\}$.

We will prove that if (6.1) holds, then $C^*(G) \neq \emptyset$, by induction on $|E(G)|$. By using induction in G/S, it is easy to see

(6.3) *If* $|S| = 2$, *then* $e(S) \geq k+1$.

(6.4) *If* X *is a* k-*set, then* $X \cap T = \{u_1, v_1\}$ *or* $\{u_2, v_2\}$.

Proof. Assume that $|X \cap T| \leq 2$ and $X \cap T \neq \{u_i, v_i\}$ $(i = 1, 2)$. By induction or lemma 4, there is a $C_1 \in C^*(G/X)$. If $|X \cap T| \leq 1$, then we can construct a cycle $C \in C^*(G)$ from C_1 (by using lemma 4 in $G/\overline{X}$ if $E(C_1) \cap \partial(X) \neq \emptyset$). If $|X \cap T| = 2$, then for $C_2 \in C^*(G/\overline{X})$, $C_1 \cup C_2 \in C^*(G)$. ∎

(6.5) *If* $X \subseteq V(G)$ *and* $X \cap T = \{v_1, u_2\}$ *or* $\{v_1, v_2\}$, *then* $e(X) \geq k+2$.

Proof. Otherwise by (6.4) $e(X) = k+1$. Let $X \cap T = \{v_1, u_2\}$ (the proof is similar for the case $X \cap T = \{v_1, v_2\}$). By theorem 2 and (6.4) $V(G) = T$ and there are $g_1 \in \partial(v_1, u_2)$ and $g_2 \in \partial(u_1, v_2)$. By (6.1) we deduce that $\lambda(T; G - \{f_1, g_1, f_2, g_2\}) = k-2$. ∎

By (6.5), we deduce:

(6.6) $e(u_1, u_2) = 0$.

(6.7) $S = \emptyset$.

Proof. If $b \in S$ and $e(b) \geq 4$, let $N(b) = \{x_1, ..., x_n\}$. Then $n \geq 3$ by (6.4). By lemma 3 $G_b^{x_i, x_j}$ is admissible for some $1 \leq i \neq j \leq n$, say for $(i, j) = (1, 2)$. By (6.3) we can choose x_1, x_2 from $T \cup W$. For each $1 \leq i \neq j \leq n$, let

$M_{i,j} := \{X \subseteq V(G) - b |\ \{x_i, x_j\} \subseteq X, e(X) = k+3,$ and $X \cap T = \{v_1, u_2\}$ or $\{u_1, v_2\}\}$.

If $M_{1,2} = \emptyset$, then there is a $C_1 \in C^*(G_b^{x_1, x_2})$. Let C_2 be the corresponding cycle in G. If C_2 is simple, then $C_2 \in C^*(G)$. Assume C_2 is not simple. If $f_0 \in E(C_2)$, then a simple subcycle C_3 in C_2 with $f_0 \in E(C_3)$ belongs to $C^*(G)$. If $f_0 \notin E(C_2)$, then there are two edge-disjoint paths $P_1[v_1, v_2]$ and $P_2[v_1, v_2]$ in C_2, $f_i \in E(P_i)$ ($i = 1, 2$) and $b \in V(P_1) \cap V(P_2)$. Then

$$P_1(v_1, b) \cup P_2(b, v_2) \cup \{f_0\} \in C^*(G).$$

Thus $M_{1,2} \neq \emptyset$. Choose $X \in M_{1,2}$ such that $|X|$ is maximum. We may let $X \cap T = \{v_1, u_2\}$ and $x_3 \notin X$. By (6.5) $e(X \cup \{b\}) \geq k + 2$, and so $e(b, N(b) - X) \geq e(b)/2$. Since $e(\{b, x_3\}) > k$, it follows that $|N(b) - X| \geq 2$, say $x_4 \in N(b) - X$.

(6.7.1) $\lambda(T \cup W; G_b^{x_i, x_j}) \geq k$ for some i, j with $i = 1$ or 2 and $j = 3$ or 4.

Otherwise by lemma 6, for some disjoint k-sets $X_1, X_2, X_3 \subseteq V(G) - b$, $x_i \in X_i$ ($1 \leq i \leq 3$) and $e(X_i, X_3) = \alpha$ ($i = 1, 2$), and for some disjoint k-sets $Y_1, Y_2, Y_3 \subseteq V(G) - b$, $x_i \in Y_i$ ($i = 1, 2$), $x_4 \in Y_3$ and $e(Y_i, Y_3) = \alpha$ ($i = 1, 2$). Assume that for some $r \in \{1, 2\}$, $|X_r| \geq 2$. If $|X_r \cap T| = \{u_1, v_1\}$, then by lemma 2 $e(X - X_r) + e(X_r - X) \leq 2k - 1$, a contradiction. Thus by (6.4) we may let $|X_1| = 1$ and $X_2 \cap T = \{u_2, v_2\}$. Now $\{u_1, b\} \subseteq X \cup X_2$ and by lemma 2, $e(X \cap X_2) = k$ and $e(X \cup X_2) = k + 1$. Then $X \cap X_2 = \{u_2 = x_2\}$ and $e(X - X_2)$ is odd, and so by lemma 2, $e(X - X_2) = k$ and $e(X_2 - X) = k + 1$. Thus $X - X_2 = \{v_1 = x_1\}$ and $X_3 \subseteq \overline{X \cup X_2}$, since X_1, X_2, X_3 are disjoint. Now $e(v_1, \overline{X}) \geq \alpha + 3$ and $e(X - v_1) < k$, a contradiction. Hence $|X_i| = |Y_i| = 1$ ($i = 1, 2$). If $|X_3| = 1$, then $e(x_3, X) = 2\alpha$ and $e(X \cup \{x_3\}) \leq k$, contrary to (6.4). Thus $|X_3| \geq 2$. If $X_3 \cap T = \{u_2, v_2\}$, then $e(X - X_3, X_3 - X) \geq e(v_1, v_2) \geq 1$, and so $e(X \cap X_3) = k$, $e(X \cup X_3) = k + 1$, $X \cap X_3 = \{u_2\}$ and $e(\{x_1, x_2\}, X_3 - X) = 0$. Thus $e(x_i, u_2) = \alpha$ ($i = 1, 2$) and $e(X_3 - u_2) < k$, a contradiction. Hence $X_3 \cap T = \{u_1, v_1\}$. By lemma 2 $e(X_3 - X) = k$ and $X_3 - X = \{x_3 = u_1\}$. Similarly $x_4 = u_1$, a contradiction. Now (6.7.1) is proved.

Assume $\lambda(T \cup W; G_b^{x_3, x_1}) \geq k$, and let $Y \in M_{1,3}$. If $Y \cap T = \{v_1, u_2\}$, then $X - Y \neq \emptyset$, since $|X|$ is maximum. By lemma 2

$$e(X - Y) + e(Y - X) \leq 2k + 6 - 8 = 2k - 2,$$

a contradiction. If $Y \cap T = \{u_1, v_2\}$, then $e(X - Y) + e(Y - X) \leq 2k + 4$. By (6.5) $e(X - Y) = e(Y - X) = k + 2$, and so $e(X \cap Y)$ and $e(X \cup Y)$ are odd. Thus $e(X \cap Y) \geq k$ and $e(\overline{X \cup Y}) \geq k + 2$, contrary to lemma 2. This proves (6.7). ∎

There are $V_1, V_2 \subseteq V(G)$ such that $V_i \cap T = \{v_i\}$ and $e(V_i) \leq k + 1$ ($i = 1, 2$) ($\{v_1\}$ and $\{v_2\}$ satisfy this condition). We choose V_i such that $|V_i|$ is maximum ($i = 1, 2$). Then

(6.8) $V_1 \cap V_2 = \emptyset$, there is a path $P^*[u_1, u_2]$ in $G - (V_1 \cup V_2)$, and if for $i = 1, 2$, $Z \subseteq V(G)$, $Z \cap T = \{v_i\}$ and $e(Z) \leq k + 1$, then $Z \subseteq V_i$.

Proof. If $V_1 \cap V_2 \neq \emptyset$, then by lemma 2 $e(V_1 \cup V_2) = k$, contrary to (6.4). Assume that $V_1 \cup V_2$ separates u_1 from u_2. Now

$$e(V_1 \cup V_2) = e(V_1) + e(V_2) - 2e(V_1, V_2) \leq 2k.$$

By (6.4), $V(G) - (V_1 \cup V_2) = \{u_1, u_2\}$, $e(V_i) = k + 1$ ($i = 1, 2$) and $e(V_1, V_2) = 1$.

By (6.4), $e(u_i, V_i) = \alpha + 1$ ($i = 1, 2$) and $e(u_1, V_2) = e(u_2, V_1) = \alpha$. For $(i, j) = (1, 2), (2, 1)$, let $G_i := G/(V_j \cup \{u_j\})$. We will prove that for $(i, j) = (1, 2), (2, 1)$, there exists $C_i \in \mathcal{C}(G_i, \{h_i, f_i\})$ for some $h_i \in \partial(u_j, V_i)$. For let $h \in \partial(u_2, V_1)$. We may assume that $\mathcal{C}(G_1, \{h, f_1\}) = \emptyset$, and so by theorem 2, $V_1 = \{v_1, x_1\}$ and $e(x_1, V_2 \cup \{u_2\}) = 1$, and there exists $g \in \partial(u_2, v_1)$ with $\mathcal{C}(G_1, \{g, f_1\}) \neq \emptyset$ as required. Let h_1, h_2, C_1, C_2 be as above. We can choose C_1 and C_2 such that $h_2 \in E(C_1)$ and $h_1 \in E(C_2)$. Then $C_1 \cup C_2 \in \mathcal{C}(G, \{f_1, f_2\})$. If $Z - V_i \neq \emptyset$, then $V_i - Z \neq \emptyset$ and by lemma 2 $e(Z - V_i) + e(V_i - Z) \leq 2k - 2$, contrary to (6.7). Thus $Z \subseteq V_i$, as required. ∎

In what follows, let $P^*[u_1, u_2]$ be a path in $G - (V_1 \cup V_2)$.

(6.9) (i) *or* (ii) *below does not hold:*

  (i) *There is a k-set X with $X \cap T = \{u_1, v_1\}$.*

  (ii) *There is a $(k + 2)$-set Y with $Y \cap T = \{v_1, u_2\}$.*

Proof. Assume that both (i) and (ii) hold. By lemma 2 $e(X - Y) = e(Y - X) = k$. Thus $X - Y = \{u_1\}$, $Y - X = \{u_2\}$ and $e(X \cap Y)$ and $e(X \cup Y)$ are even. Hence $e(X \cap Y) = e(X \cup Y) = k + 1$. By (6.5) $(X \cap Y) \cup (\overline{X \cup Y})$ separates u_1 from u_2, which contradicts (6.8). ∎

(6.10) *If $x_1, x_2 \in W$, $g \in \partial(x_1, x_2)$ and $\lambda(V(G) - \{x_1, x_2\}; G - g) = k$, then $e(x_1, x_2) = \alpha$ and $G - g$ satisfies (6.1).*

Proof. Assume $e(x_1, x_2) < \alpha$. If there exists $C \in \mathcal{C}^*(G - g, \{f_1, f_2\})$, then $\lambda(G - E(C)) = k - 2$, since $e(\{x_1, x_2\}; G - E(C)) \geq k - 2$. Thus this is not the case, and $G - g$ does not satisfy (6.1). For some $X, Y \subseteq V(G)$, $g \in \partial(X) \cap \partial(Y)$, $X \cap T = \{v_1, u_2\}$, $Y \cap T = \{v_1, v_2\}$ and $e(X) = e(Y) = k + 2$. We may let $x_1 \in X$ and $x_2 \in \bar{X}$.

Case 1. $x_1 \in Y$.

Let $Z_1 := X \cap Y$ and $Z_2 := \overline{X \cup Y}$. By lemma 2, $e(Z_1, Z_2) = 2$ and $e(X - Y) = (Y - X) = k$ and so $X - Y = \{u_2\}$, $Y - X = \{v_2\}$, $e(u_2, v_2) = 1$ and $e(Z_1) = e(Z_2) = k + 1$. Thus $e(u_2, Z_i) = e(v_2, Z_i) = \alpha$ ($i = 1, 2$). If there exists $C \in \mathcal{C}(G/Z_1, \{f_1, f_0, f_2\})$, then $\lambda(G - E(C)) = k - 2$, by lemma 5 and since $\lambda(x_1, \bar{Z}_1; G - E(C)) = k - 1$. Thus $\mathcal{C}(G/Z_1, \{f_1, f_0, f_2\}) = \emptyset$. Let $h_1 \in \partial(u_2, Z_2)$; then $\lambda((G/Z_1)/Z_2 - \{f_1, f_0, f_2, h_1\}) = k - 2$. Thus by theorem 4, $Z_2 = \{u_1, x_2\}$ and $\alpha = e(u_1, x_2) = e(u_2, x_2) = e(u_1, v_2)$. Let $h_2 \in \partial(u_1, v_2)$ and $h_3 \in \partial(u_2, Z_1)$; then $\lambda(G/Z_1 - \{f_1, h_2, f_2, h_3\}) = k - 2$, and so by theorem 4, $Z_1 = \{v_1, x_1\}$ and $\alpha = e(v_1, x_1) = e(u_2, x_1) = e(v_2, v_1)$. Then $e(\{u_1, v_2, v_1\}) = k$, contrary to (6.4).

Case 2. $x_1 \notin Y$.

Let $Z_1 := Y - X$ and $Z_2 := X - Y$. By lemma 2, $X \cap Y = \{v_1\}$, $\overline{X \cup Y} = \{u_1\}$, $e(v_1) = k$, $e(u_1, v_1) = 1$, $e(Z_1) = e(Z_2) = k + 1$ and G is the same type of graph as in case 1. ∎

(6.11) *If $x_1, x_2 \in W$, $g \in \partial(x_1, x_2)$ and $\lambda(V(G) - \{x_1, x_2\}; G - g) \geq k$, then*

(i) $G - g$ has paths $P_1[v_1, v_2]$ and $P_2[v_1, v_2]$ such that $f_i \in E(P_i)$ ($i = 1, 2$), $V(P_1) \cap V(P_2) = \{v_1, v_2\}$, $x_r \in V(P_1)$ and $x_s \in V(P_2)$ for $(r, s) = (1, 2)$ or $(2, 1)$, and $\lambda(V(G) - \{x_1, x_2\}; G - g - E(P_1 \cup P_2)) = k - 2$; and

(ii) $e(Z; G - E(C)) = k - 3$ where $C := P_1(v_1, x_r) \cup P_2(x_s, v_2) \cup \{f_0, g\}$ in G, for some $Z \subseteq V(G) - V(P_1(v_2, x_r))$ with $V(P_2(v_1, x_s)) \subseteq Z$.

Proof. By (6.10) $G - g$ satisfies (6.1). Let $C_1 \in \mathcal{C}^*(G - g)$, and let $P_1[v_1, v_2]$ and $P_2[v_1, v_2]$ be two paths in C_1. If $e(\{x_1, x_2\}; G - E(C_1)) \geq k - 2$, then $C_1 \in \mathcal{C}^*(G)$, and so for $(r, s) = (1, 2)$ or $(2, 1)$, $x_r \in V(P_1)$, $x_s \in V(P_2)$, and we may let $f_i \in E(P_i)$ ($i = 1, 2$). For C given in (ii) $\lambda(G - E(C)) = k - 3$, and thus (ii) easily follows. ∎

We denote by $\Omega(x_1, x_2)$ the set of (P_1, P_2, C, Z) given in (6.11).

(6.12) *If X_1, X_2 are k-sets with $X_i \cap T = \{u_i, v_i\}$ ($i = 1, 2$), then $V(G) = X_1 \cup X_2$.*

Proof. Otherwise for some disjoint k-sets Y_1, Y_2 with $X_i \subseteq Y_i$ ($i = 1, 2$) and $V(G) \neq Y_1 \cup Y_2$, $G_1 := (G/Y_1)/Y_2$ has no k-set (Y_i might equal X_i). Let $W_1 := W - Y_1 - Y_2$. Then $W_1 \neq \emptyset$ and $V(G_1) = \{Y_1, Y_2\} \cup W_1$. By (6.10) and (6.11) G_1 satisfies the condition of lemma 7. Thus $W_1 = \{x_1, ..., x_n\}$, $e(x_i, x_{i+1}) = \alpha$ ($1 \leq i \leq n - 1$) and for $(i, j) = (1, 2)$ or (2.1), $e(x_1, Y_i) = 1$ and $e(x_1, Y_j) = \alpha$, say for $(i, j) = (1, 2)$. Let $(P_1, P_2, C, Z) \in \Omega(x_1, x_2)$. We can choose Z with $Y_1 \subseteq Z \subseteq V(G) - Y_2$, because otherwise $\lambda(G_1 - E(C)) \geq k - 2$ and $\lambda(G - E(C)) \geq k - 2$. If $x_2 \in V(P_1)$, then $x_1 \in Z \subseteq V(G) - x_2$ and

$$e(Z; G - E(C)) \geq e(Z - x_1; G - E(C)) \geq k - 2,$$

a contradiction. Thus $x_1 \in V(P_1)$, $\mathcal{C}(G/\overline{Y}_1, \{f_0, f_1, h\}) \neq \emptyset$ for each $h \in \partial(x_1, Y_1)$ and $\mathcal{C}(G/\overline{Y}_2, \{f_0, f_2, g\}) = \emptyset$ for each $g \in \partial(x_1, Y_2)$.

Case 1. $e(x_n, Y_1) = 1$ and $e(x_n, Y_2) = \alpha$.

Similarly $\mathcal{C}(G/\overline{Y}_2, \{f_0, f_2, g\}) = \emptyset$ for each $g \in \partial(x_n, Y_2)$, which contradicts theorem 5.

Case 2. $e(x_n, Y_1) = \alpha$ and $e(x_n, Y_2) = 1$.

By theorem 5, $\mathcal{C}(G/\overline{Y}_2, \{f_0, f_2, g\}) \neq \emptyset$ for each $g \in \partial(Y_2) - \partial(Y_2, x_1) - f_0$, and $\mathcal{C}(G/\overline{Y}_1, \{f_0, f_1, g\}) \neq \emptyset$ for each $g \in \partial(Y_1) - \partial(Y_1, x_n) - f_0$. Thus $e(Y_1, Y_2) = 1$ and $n \geq 4$. For some $2 \leq i, j \leq n - 1$ with $|i - j| = 1$, there are $h_1 \in \partial(x_i, Y_1)$ and $h_2 \in \partial(x_j, Y_2)$. It is easy to see that for $g \in \partial(x_i, x_j)$, $\lambda(G_1 - \{f_0, h_1, g, h_2\}) = k - 2$. Thus $\mathcal{C}(G, \{f_0, f_1, h_1, g, h_2, f_2\}) \neq \emptyset$, a contradiction. ∎

(6.13) *G has no k-cut.*

Proof. Otherwise by (6.4) and (6.12) there are minimal k-sets X_1, X_2 such that $X_i \cap T = \{u_i, v_i\}$ ($i = 1, 2$) and $V(G) = X_1 \cup X_2$. If $|E(P^*) \cap \partial(X_1)| \geq 3$, then starting from u_2 along P^*, let y be the first vertex in X_1. Now $V_1 \cup X_2$ does not separate u_1 from y, since $e(V_1 \cup X_2) \leq (k + 1) + k - 2 = 2k - 1$. Thus we can choose P^* such that $|E(P^*) \cap \partial(X_1)| = 1$. If $|V(P^*) \cap X_i| \leq 2$ ($i = 1, 2$), then $\lambda(G - E(P^*) - \{f_1, f_0, f_2\}) = k - 2$. Thus let $\{u_1, y_1, y_2\} \subseteq V(P^*) \cap X_1$ and

$e(u_1, y_1) > 0$. Let D be a component containing y_1 and y_2 in $G - (T \cup X_2)$, let $V(D) = \{x_1, ..., x_n\}$ ($n \geq 2$) and by (6.10), we may let $e(x_i, x_{i+1}) = \alpha$ ($1 \leq i \leq n - 1$). Let $(P_1, P_2, C, Z) \in \Omega(x_1, x_2)$. We can choose Z in X_1.

Case 1. $n \geq 3$ and $e(x_1, x_n) = \alpha$.

We may let $x_1 = y_1$ and $e(u_1, x_1) = 1$. Then $x_i \in V(P_i)$ ($i = 1, 2$), $x_2 \in Z$ and $x_n \notin Z$. Since $e(Z; G - E(C)) = k - 3$, we have $e(x_2, X_2) = e(x_n, v_1) = 0$, and so $e(x_2, v_1) = 1$. Similarly $e(x_n, v_1) = 1$, a contradiction.

Case 2. $n = 2$ or $e(x_1, x_n) = 0$.

We will prove that (a) or (b) below holds.

(a) $e(x_1, u_1) = 1$ and $e(x_1, X_2) = \alpha$.

(b) $e(x_1, v_1) = \alpha$ and $e(x_1, X_2) = 1$.

Assume first $x_1 \in V(P_2)$. Then $V(P_2) \cap X_1 = \{v_1, x_1\} \subseteq Z$ and $x_2 \notin Z$. Now $e(Z - x_1; G - E(C)) \geq k - 2$, and so $e(x_1, X_2) = 1$ and $e(x_1, \{u_1, v_1\}) = \alpha$. If there exists $g \in \partial(x_1, u_1)$, then let $C_1 := \{f_0, f_1, g\} \cup P_2(x_1, v_2)$. It follows that for some $Z_1 \subseteq X_1$, $e(Z_1; G - E(C_1)) \leq k - 3$, and $\{v_1, x_1\} \subseteq Z_1 \subseteq X_1 - \{u_1, x_2\}$, since $e(Z_1; G - E(C_1)) \geq e(Z_1; G - E(C))$. But $e(Z_1) = k + 1$ and $e(x_1, Z_1) \geq \alpha + 2$, and so $e(Z_1 - x_1) \leq (k + 1) - 3$, a contradiction. Thus $e(x_1, u_1) = 0$ and (b) holds. Assume next that $x_1 \in V(P_1)$. Then $\{v_1, x_2\} \subseteq Z \subseteq X_1 - x_1$, and it is easy to see that $|\partial(Z) \cap E(C)| = 4$ and $e(Z) = k + 1$. If $u_1 \in Z$, then $e(x_1, X_2) = \alpha$, $e(x_1, u_1) = 1$, and (a) holds, since $e(Z \cup \{x_1\}; G - E(C)) \geq k - 2$. Thus $u_1 \notin Z$, and so $Z \subseteq V_1$ by (6.8). If $x_n \in Z$, then $\{x_2, x_3, ..., x_n\} \subseteq Z \subseteq V_1$ and $V(P^*) \cap X_1 = \{u_1, x_1\}$, a contradiction. Thus $x_n \notin Z$. If $x_3 \notin Z$, then $e(Z - x_2) \leq e(Z) - 3 = k - 2$, a contradiction. Thus $x_3 \in Z$ and $e(x_i, v_1) = 1$ ($i = 2, 3$). Then $\Omega(x_2, x_3) = \emptyset$, contrary to (6.11).

Similarly for x_n instead of x_1, (a) or (b) holds. If for both x_1 and x_n, (a) holds, then by theorem 5 for some $g \in \partial(\{x_1, x_n\}, X_2)$, there is a $C_1 \in C(G/X_1, \{f_0, f_2, g\})$, say $g \in \partial(x_1, X_2)$. Then $\lambda(G/X_2 - \{f_0, f_1, h, g\}) = k - 2$ for $h \in \partial(u_1, x_1)$, and $C(G, \{f_0, f_1, f_2\}) \neq \emptyset$, a contradiction. Thus for x_1 or x_n, (b) holds, say for x_1.

Case 2.1. $e(x_n, v_1) = \alpha$ and $e(x_n, X_2) = 1$.

By (6.4) $e(\{v_1, x_1, ..., x_n\}) \geq k + 1$ and $e(\{u_1\} \cup x_2) \leq 2k$. Thus $\{u_1\} \cup X_2$ is not a separating set, and so $X_1 = \{u_1, v_1, x_1, ..., x_n\}$. Since $e(u_1, v_1) = 1$ and $e(u_1, X_2) \leq \alpha$, we have $e(u_1, \{x_2, x_3, ..., x_{n-1}\}) \geq \alpha \geq 2$. Thus $e(u_1, x_p) = 1$ for some p with $2 \leq p \leq n - 2$. Let $(P'_1, P'_2, C', Z') \in \Omega(x_p, x_{p+1})$. Then $x_p \in V(P'_1)$, $e(x_{p+1}, X_2) = 1$ and $V(C') \cap X_1 = \{v_1, u_1, x_p, x_{p+1}\}$. Since $V(P^*) \cap \{x_1, ..., x_n\} \neq \emptyset$, we have $e(X_1 - u_1) \geq k + 2$, and it follows that $e(G - E(C')) = k - 2$.

Case 2.2 $e(x_n, u_1) = 1$ and $e(x_n, X_2) = \alpha$.

For $h_1 \in \partial(u_1, x_n)$ and $h_2 \in \partial(x_n, X_2)$, $\lambda(G/X_2 - \{f_0, f_1, h_1, h_2\}) = k - 2$. Thus for each $h \in \partial(x_n, X_2)$, $C(G/X_1, \{h, f_0, f_2\}) = \emptyset$, and so by theorem 5 for each $g \in \partial(X_2) - \partial(x_n, X_2) - f_0$, $C(G/X_1, \{g, f_0, f_2\}) \neq \emptyset$. If $n = 2$, then

$e(X_1 - \{x_1, x_2\}) = k$, contrary to (6.12). Thus $n \geq 3$. If $e(x_2, X_2) = 1$, then $x_2 \in V(P_1)$ and for $g \in \partial(x_2, X_2)$, $C(G, \{f_0, f_1, g, f_2\}) \neq \emptyset$. Hence $e(x_2, u_1) = 1$. By (6.11) $e(x_3, u_1) = 0$ and $n \geq 4$. Let $(P_1', P_2', C', Z') \in \Omega(x_2, x_3)$. Then $V(P_1') \cap X_1 = \{v_1, u_1, x_2, x_1\}$, and so $\lambda(G/X_2 - E(P_1') - f_0) = k - 2$ and $C(G, \{f_0, f_1, f_2\}) \neq \emptyset$. ∎

By (6.6) there exists $y \in V(P^*) - T$. Let D be the component of $G - T$ containing y, let $V(D) = \{x_1, ..., x_n\}$, and let $e(x_i, x_{i+1}) = \alpha$ ($1 \leq i \leq n - 1$). If $n = 1$, then by (6.5) $\lambda(G - E(P^*) - \{f_0, f_1, f_2\}) = k - 2$, a contradiction. Thus $n \geq 2$. Let $(P_1, P_2, C, Z) \in \Omega(x_1, x_2)$.

Case 1. $n \geq 3$ and $e(x_1, x_n) = \alpha$.

We may let $e(x_1, u_1) = 1$. Then $x_i \in V(P_i)$ ($i = 1, 2$). If $e(x_2, u_2) = 1$, then $e(Z; G - E(C)) = k - 2$. Thus $e(x_2, v_1) = 1$. Similarly $e(x_n, v_1) = 1$. Then $e(Z; G - E(C)) = k - 2$, since $\{v_1, x_2\} \subseteq Z \subseteq V(G) - \{x_1, x_n\}$.

Case 2. $n = 2$ or $e(x_1, x_n) = 0$.

We will prove that for $(i, j) = (1, 2)$ or $(2, 1)$, $e(x_1, u_i) = 1$ and $e(x_1, v_j) = \alpha$. If $x_1 \in V(P_2)$, then $V(P_2) = \{v_1, x_1, u_2, v_2\}$ and it follows that $e(x_1, v_1) = \alpha$ and $e(x_1, u_2) = 1$ in the same way as case 2 in (6.13). Thus $x_1 \in V(P_1)$, and $e(x_1, u_1), e(x_1, v_2) > 0$ and by the symmetry we have $e(x_1, v_2) = \alpha$ and $e(x_1, u_1) = 1$.

Case 2.1. $e(x_i, u_1) = 1$ and $e(x_i, v_2) = \alpha$ ($i = 1, n$).

Now $x_1 \in V(P_1)$, and so $x_2 \in V(P_2)$, $n \geq 3$ and $e(x_2, \{v_1, u_2\}) = 1$. Let $P[v_1, v_2]$ be a path with $x_2 \in V(P) \subseteq T \cup \{x_1, x_2\}$. If $e(x_2, v_1) = 1$, then $f_1, f_2 \notin E(P)$, and if $e(x_2, u_2) = 1$, then $\{f_1, f_2\} \subseteq E(P)$. Thus $\Omega(x_2, x_3) = \emptyset$, contrary to (6.11).

Case 2.2. $e(x_1, u_1) = e(x_n, u_2) = 1$ and $e(x_1, v_2) = e(x_n, v_1) = \alpha$. Suppose that $n = 2$; then $V(P^*) = \{u_1, x_1, x_2, u_2\}$. let $C := P^* \cup \{f_1, f_0, f_2\}$ and $X \subseteq V(G)$. If $|\partial(X) \cap E(C)| = 6$ and $v_1 \in X$, then $e(X) \geq e(X - x_1) + k \geq 2k$. Assume $|\partial(X) \cap E(C)| = 4$ and $|X \cap T| \leq 2$. If $X \cap T = \{u_1\}$, then $X \cap V(C) = \{u_1, x_2\}$ and $e(X) \geq e(X - x_2) + k$. If $X \cap T = \{u_1, v_1\}$, then $X \cap V(C) = \{u_1, v_1, x_2\}$ and $e(X) \geq e(X - x_2) + 1 \geq k + 2$ by (6.13). Since $x_1, x_2 \notin V_1 \cup V_2$ and by (6.5), we have $\lambda(G - E(C)) \geq k - 2$, a contradiction. Thus $n \geq 3$ and $e(x_2, \{v_1, u_2\}) = 1$. Now we can deduce $\Omega(x_2, x_3) = \emptyset$ in the same way as case 2.1. ∎

REFERENCES

1. A. Huck and H. Okamura, "Counterexamples to a conjecture of Mader about cycles through specified vertices in n-edge-connected graphs", *Graphs and Combin.*, in press.
2. W. Mader, "A reduction method for edge-connectivity in graphs", *Ann. Discrete Math.* 3 (1978), 145-164.
3. W. Mader, "Paths in graphs, reducing the edge-connectivity only by two", *Graphs and Combin.* 1 (1985), 81-89.

4. H. Okamura, "Paths and edge-connectivity in graphs", *J. Combin. Theory, Ser. B*, 37 (1984), 151-172.
5. H. Okamura, "Paths in k-edge-connected graphs", *J. Combin. Theory, Ser. B*, 45 (1988), 345-355.
6. H. Okamura, "Cycles containing three consecutive edges in $2k$-edge-connected graphs", *Topics in Combinatorics and Graph Theory* (eds. R. Bodendiek and R. Henn), Physica-Verlag Heidelberg (1990), 549-553.

Faculty of Engineering, Osaka City University, Sugimoto, Osaka 558, Japan.

Edge – disjoint cycles in n – edge – connected graphs

ANDREAS HUCK

ABSTRACT Okamura proved the following: If $n \geq 4$ is an even integer, G is an n – edge – connected graph and f_1, g_1 are edges of G, then there exists a cycle C_1 in G containing f_1 and g_1 such that $G - E(C_1)$ is $(n-2)$ – edge – connected. Clearly, if $n \geq 6$ and if f_2 and g_2 are edges of G not contained in C_1, there is another cycle C_2 in $G - E(C_1)$ containing f_2 and g_2 such that $G - E(C_1) - E(C_2)$ is $(n-4)$ – edge – connected. We prove that C_2 can be chosen in such a way that in addition $G - E(C_2)$ is $(n-2)$ – edge – connected.

1. Introduction

In this paper we consider graphs which are finite, undirected, without loops and in which multiple edges are possible. For integers $u \leq v$ we set $[u:v] := \{u, u+1, ..., v\}$. In proofs the sign $\#$ always indicates that we have obtained a contradiction.

Let $G = (V, E)$ be a graph. $V(G) = V$ and $E(G) = E$ denote the set of the vertices of G and the set of the edges of G respectively. If $X, Y \subseteq V$, then $[X, Y]_G$ denotes the set of all edges of G connecting a vertex of X with a vertex of Y. Moreover define $\overline{X} := V - X$, $\delta(G; X, Y) := |[X, Y]_G|$ and $\delta(G; X) := \delta(G; X, \overline{X})$ (the *degree* of X). A vertex $x \notin X$ is called a *neighbour* of X if there is an edge connecting x with a vertex of X. $N(G; X)$ denotes the set of all neighbours of X. Moreover $E(G; X)$ denotes the set of all edges connecting two vertices of X and for $A \subseteq E$, $V(G; A)$ is the set of all vertices incident to an edge of A. Finally define $\mu(G) := |V| + |E|$ (the *size* of G).

Paths in G are allowed to pass through a vertex more than once but using an edge more than once is forbidden. If P is a path from a vertex x to a vertex y, then P is also called an x, y – *path*. Moreover, if $X, Y \subseteq V$ with $x \in X$ and $y \in Y$, then P is called an X, Y – *path*. P is called a *cycle* if $x = y$ and $|V(P)| \geq 2$. Now assume that P passes through each vertex of G at most once.

1991 mathematics subject classification 05C38, 05C40
This paper is a final version and it will not be published elsewhere

Then P is called a *simple path* or *simple cycle* if $x \neq y$ or $x = y$ respectively. If P is a simple x, y – path and if $X \subseteq V$ with $X \cap V(P) = \{x, y\}$, then P is called X – *admissible*.

Two paths are called *disjoint* if they have no edge in common. For $X, Y \subseteq V$ with $X \cap Y = \emptyset$, $\lambda(G; X, Y)$ denotes the maximal number of pairwise disjoint X, Y – paths in G. If $X \cap Y \neq \emptyset$, we define $\lambda(G; X, Y) := \infty$. Moreover if $|X| \geq 2$, we let $\lambda(G; X)$ be the maximal number n such that for each $x, y \in X$ there are at least n pairwise disjoint x, y – paths in G. If $|X| \leq 1$, define $\lambda(G; X) := \infty$. Finally define $\lambda(G) := \lambda(G; V)$. If $\lambda(G) \geq n$, then G is called n – *edge* – *connected*.

When using the notation defined above and if $X = \{x\}$ or $Y = \{y\}$, then we also write x or y instead of X or Y respectively. Moreover if no misunderstanding is possible, we also write $\delta(X)$ instead of $\delta(G; X)$, $\lambda(x, y)$ instead of $\lambda(G; x, y)$ etc.

If $X \subseteq V$ is not empty, G/X denotes the graph obtained from G by contracting X to a single vertex x. When doing so, we always identify X with x so that $V(G/X) = (V - X) \cup \{X\}$. Moreover for each $y \in V - X$, $[X, y]_{G/X} = [X, y]_G$. Define $G/\emptyset := G$. If $X_1, ..., X_m \subseteq V$ are pairwise disjoint, we define inductively $G/X_1, ..., X_m := (G/X_1, ..., X_{m-1})/X_m$, and for $Y \subseteq V$ define $Y/X_1, ..., X_m := (Y - X_1 - ... - X_m) \cup \{X_i; i \in [1 : m], X_i \cap Y \neq \emptyset\}$.

If $X \subseteq V$ with $X \neq \emptyset \neq \overline{X}$, then $(X, \overline{X})$ is called a *cut*. We call $(X, \overline{X})$ an n – *cut*, $(\leq n)$ – *cut*, $(< n)$ – *cut* etc. if $\delta(G; X, \overline{X}) = n$, $\delta(G; X, \overline{X}) \leq n$, $\delta(G; X, \overline{X}) < n$ etc. respectively. If $D \subseteq V$ with $X \cap D \neq \emptyset \neq \overline{X} \cap D$, then we say: $(X, \overline{X})$ *divides* D.

In [2] Okamura proved the following: If $n \geq 4$ is an even integer, G is an n – edge – connected graph and f_1, g_1 are edges of G, then there exists a cycle C_1 in G containing f_1 and g_1 such that $G - E(C_1)$ is $(n - 2)$ – edge – connected. Clearly, if $n \geq 6$ and if f_2 and g_2 are two edges of G not contained in C_1, there is another cycle C_2 in $G - E(C_1)$ containing f_2 and g_2 such that $G - E(C_1) - E(C_2)$ is $(n - 4)$ – edge – connected. In this paper we prove the following more general result:

THEOREM 1: *Let $n \geq 6$ be even and let $G = (V, E)$ be a graph with $\lambda(G) \geq n$. Moreover let C_1 be a cycle in G with $\lambda(G - E(C_1)) \geq n - 2$ and $f, g \in E - E(C_1)$. Then there exists a cycle C_2 in $G - E(C_1)$ containing f and g such that $\lambda(G - E(C_1) - E(C_2)) \geq n - 4$ and $\lambda(G - E(C_2)) \geq n - 2$.*

By using this theorem and Proposition 3 in [1] the following corollary easily follows:

COROLLARY 1: *Let $n \geq 6$ be even and let $G = (V, E)$ be a graph with $\lambda(G) \geq n$. Moreover let f_1, g_1, f_2, g_2 be pairwise distinct edges of G. Then for $i = 1, 2$ there exists a cycle C_i in G containing f_i and g_i such that C_1 and C_2 are disjoint, $\lambda(G - E(C_i)) \geq n - 2$ for $i = 1, 2$ and $\lambda(G - E(C_1) - E(C_2)) \geq n - 4$.*

2. Preliminaries

LEMMA 1: *Let $G = (V, E)$ be a graph, $D \subseteq V$ and $x \in V - D$. Moreover let $\lambda(G; D) \geq n$ and $\lambda(G; x, D) \geq n$. Then $\lambda(G; D + x) \geq n$.*

Proof: Straightforward by Menger's Theorem. □

If $G = (V, E)$ is a graph, then $D \subseteq V$ is called *nice* if each $x \in V - D$ has even degree.

LEMMA 2: *Let $n \geq 0$ be even, $G = (V, E)$ be a graph, $D \subseteq V$ be nice and $a \in D$ with odd degree. Moreover let $\lambda(G; D) \geq n$. Then there exists $b \in D - a$ such that $\lambda(G; a, b) \geq n + 1$.*

Proof: By induction on $\mu(G)$.
Case 1: G has a D – dividing n – cut $(X, \overline{X})$ with $|X|, |\overline{X}| \geq 2$.
Let $a \in X$. Since $\lambda(G/\overline{X}; D/\overline{X}) \geq n$ and $\delta(G/\overline{X}; \overline{X}) = n$, there exists $b \in X \cap D$ with $\lambda(G/\overline{X}; a, b) \geq n+1$ by the induction hypothesis. $\lambda(G; a, b) \geq n+1$ easily follows.
Case 2: If otherwise.
By reasons of parity, there exists $b \in D$ with odd degree. If $\lambda(G; a, b) \leq n$, then Case 1 would happen. □

LEMMA 3: *Let $m, n, \alpha \geq 0$ be even with $m \leq n$, $G = (V, E)$ be a graph and $D \subseteq V$ be nice. Moreover let $(X, \overline{X})$ be a D – dividing $(\leq n + \alpha + 1)$ – cut of G such that $\lambda(G/\overline{X}; D/\overline{X}) \geq m$ and $\lambda(G/X; D/X) \geq n$. Then $\lambda(G; D) \geq m - \alpha$.*

Proof: Choose $a \in X \cap D$ and $b \in \overline{X} \cap D$ such that if $\delta(X, \overline{X}) = n + \alpha + 1$, then $\lambda(G; a, \overline{X}) \geq m + 1$ and $\lambda(G; b, X) \geq n + 1$ (Lemma 2 !). Then for each $z \in X \cap D$ and for each $z \in \overline{X} \cap D$ it easily follows that $\lambda(G; z, b) \geq m - \alpha$ and $\lambda(G; z, a) \geq m - \alpha$ respectively. If $x, y, z \in V$ with $\lambda(G; x, y) \geq k$ and $\lambda(G; y, z) \geq k$, then by Menger's Theorem, $\lambda(G; x, z) \geq k$. Using this fact it is easy to conclude $\lambda(G; x, y) \geq m - \alpha$ for each $x, y \in D$. □

LEMMA 4: *Let $m, n, \alpha \geq 0$ be even with $m \leq n$, $G = (V, E)$ be a graph and $U \subseteq D \subseteq V$ such that D is nice. Moreover let $(X, \overline{X})$ be a U – dividing $(\leq n + 1)$ – cut such that $\lambda(G/\overline{X}; D/\overline{X}) \geq m$, $\lambda(G/X; D/X) \geq n$ and $\lambda(G/(U + \overline{X}); D/(U + \overline{X})) \geq m + \alpha$, $\lambda(G/(U + X); D/(U + X)) \geq n + \alpha$. Then $\lambda(G; D) \geq m$ and $\lambda(G/U; D/U) \geq m + \alpha$.*

Proof: By Lemma 3, $\lambda(G; D) \geq m$. Now assume that $\lambda(G/U; D/U) \leq m + \alpha - 1$. Then there exists a D – dividing $(\leq m + \alpha - 1)$ – cut $(Y, \overline{Y})$ of G with $U \subseteq Y$. Let $A_1 := X \cap Y$, $A_2 := X \cap \overline{Y}$, $A_3 := \overline{X} \cap Y$ and $A_4 := \overline{X} \cap \overline{Y}$. Then $A_1 \cap U \neq \emptyset \neq A_3 \cap U$. For $i \in [1:4]$ let $\delta_i := \delta(G; A_i)$, and for distinct $i, j \in [1:4]$ let $\delta_{ij} := \delta(G; A_i, A_j)$. Hence $\delta_1 \geq m$ and $\delta_3 \geq n$ by $\lambda(G/\overline{X}; D/\overline{X}) \geq m$ and $\lambda(G/X; D/X) \geq n$ respectively. Moreover $\delta(X, \overline{X}) = \delta_{13} + \delta_{14} + \delta_{23} + \delta_{24}$,

$\delta(Y, \overline{Y}) = \delta_{12} + \delta_{14} + \delta_{32} + \delta_{34}$ and $(\delta_1 + \delta_2 + \delta_3 + \delta_4)/2 = \sum_{i<j} \delta_{ij}$.

Case 1: $A_2 \cap D \neq \emptyset \neq A_4 \cap D$.

Then $\delta_2 \geq m + \alpha$ and $\delta_4 \geq n + \alpha$ by $\lambda(G/(U + \overline{X}); D/(U + \overline{X})) \geq m + \alpha$ and $\lambda(G/(U + X); D/(U + X)) \geq n + \alpha$ respectively. Moreover $\delta(X, \overline{X})$, $\delta_1 + \delta_2$ and $\delta_3 + \delta_4$ have the same parity. Now we obtain

$$\begin{aligned} m + \alpha - 1 &\geq \delta(Y, \overline{Y}) \\ &\geq \delta_{12} + \delta_{34} \\ &= (\delta_1 + \delta_2 + \delta_3 + \delta_4)/2 - \delta(X, \overline{X}) \\ &\geq \begin{cases} (2m + 2n + 2\alpha)/2 - n & \text{if } \delta(X, \overline{X}) = n \\ (2m + 2n + 2\alpha + 2)/2 - n - 1 & \text{if otherwise} \end{cases} \\ &= m + \alpha \ \#. \end{aligned}$$

Case 2: $A_2 \cap D = \emptyset$.

Then $A_4 \cap D \neq \emptyset$ and hence $\delta_4 \geq n + \alpha$. Moreover δ_2 is even because D is nice. $\delta_{21} \geq \delta_{23} + \delta_{24} + m - n$ follows:

⌐ Otherwise $\delta_{21} \leq \delta_{23} + \delta_{24} + m - n - 2$ and hence $m \leq \delta_1 = \delta_{12} + \delta_{13} + \delta_{14} \leq \delta_{23} + \delta_{24} + m - n - 2 + \delta_{13} + \delta_{14} = \delta(X, \overline{X}) + m - n - 2 \leq m - 1 \ \#$. ⌐

Now we obtain $n + \alpha \leq \delta_4 = \delta_{34} + \delta_{14} + \delta_{24} \leq \delta_{34} + \delta_{14} + \delta_{21} - m + n \leq \delta(Y, \overline{Y}) - m + n \leq n + \alpha - 1 \ \#$.

Case 3: $A_4 \cap D = \emptyset$.

Then $A_2 \cap D \neq \emptyset$ and hence $\delta_2 \geq m + \alpha$. Moreover δ_4 is even because D is nice. $\delta_{43} \geq \delta_{41} + \delta_{42}$ follows:

⌐ Otherwise $\delta_{43} \leq \delta_{41} + \delta_{42} - 2$ and hence $n \leq \delta_3 = \delta_{31} + \delta_{32} + \delta_{34} \leq \delta_{31} + \delta_{32} + \delta_{41} + \delta_{42} - 2 = \delta(X, \overline{X}) - 2 \leq n - 1 \ \#$. ⌐

Now we obtain $m + \alpha \leq \delta_2 = \delta_{12} + \delta_{32} + \delta_{42} \leq \delta_{12} + \delta_{32} + \delta_{43} \leq \delta(Y, \overline{Y}) \leq m + \alpha - 1 \ \#$. □

We write $PRM_{L3}(m, n, \alpha, G, D, X, \overline{X})$ if m, n, α, G, D, X and $\overline{X}$ satisfy the premises of Lemma 3. $PRM_{L4}(m, n, \alpha, G, U, D, X, \overline{X})$ indicates that m, n, α, G, U, D, X and $\overline{X}$ satisfy the premises of Lemma 4.

Let $G = (V, E)$ be a graph and $m \geq 3$ be odd. Moreover let $x \in V$ and $\delta := \delta(x) \geq 3$. Assume that the graph G^* is obtained from G by replacing x in G by a graph B according to the following figure:

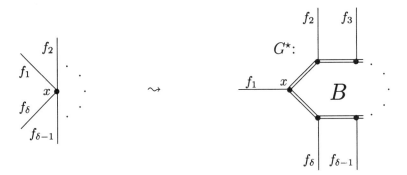

Here $a \rule{1cm}{0.4pt} b$ always indicates that there are exactly $(m-1)/2$ edges between a and b. Moreover let $G^\star/V(B) = G$ by identifying x and $V(B)$. B is called a *limitation*, and the transition from G to $G^\star$ is called m - *limiting relative to* x. Note that in $G^\star$ the degree of every vertex in $V(B)$ is m.

Assume that $G^\star$ is obtained from G by limitings relative to some vertices $x_1, \ldots, x_\ell$ of G. If $P^\star$ is an a,b - path in $G^\star$ for $a,b \in V$, then we obtain an a,b - path P in G from $P^\star$ in an obvious way. We call P *the path induced by* $P^\star$.

LEMMA 5 [1, LEMMA 3]: *Let $G = (V, E)$ be a graph, $m \geq 3$ be odd, $n \in [0 : m-1]$, $x \in V$ with $\delta(x) \geq 3$ and $D \subseteq V$ with $\lambda(G; D) \geq n$. Moreover let $G^\star$ be obtained from G by m - limiting relative to x. Set $D^\star := D$, if $x \notin D$, and $D^\star := D + V(B)$ otherwise where B is the limitation by which x is replaced. Then $\lambda(G^\star; D^\star) \geq n$.*

PROPOSITION 1 [1, PROPOSITION 3]: *Let $n \geq 2$ be an even integer and $G = (V, E)$ be a graph with $\lambda(G) \geq n$. Moreover let $A \subseteq E$ with $|A| < n/2$ and $f, g \in E - A$. Then there exists a cycle C in $G - A$ such that $f, g \in E(C)$ and $\lambda(G - E(C)) \geq n - 2$.*

Let $G = (V, E)$ be a graph, let P be an x, y - path and let h be an edge not contained in $G - E(P)$. If $x = y$, then we define $G_h^P := G - E(P)$. Now let $x \neq y$. Then G_h^P denotes the graph obtained from $G - E(P)$ by adding h between x and y. Moreover define $G^P := G_f^P$ where f is the first edge passed through by P. Now assume that $f, g \in E$, $s \in V(f) \cap V(g)$ and $V(f) \neq V(g)$, and let h be an edge not contained in $G - f - g$. Then $G_h^{f,g}$ denotes the graph obtained from $G - f - g$ by adding h between x and y where $x \in V(f) - s$ and $y \in V(g) - s$. We call $G_h^{f,g}$ a *splitting* and we define $G^{f,g} := G_f^{f,g}$. If $P^\star$ is a path in $G_h^{f,g}$, then we obtain a path P in G from $P^\star$ in an obvious way. P is called *the path induced by* $P^\star$.

PROPOSITION 2: *Let $n \geq 2$ be an even integer and $G = (V, E)$ be a graph.*

Moreover let $W \subseteq V$ with $|W| \geq 2$, $c \in W$, $A \subseteq E$ with $|A| < n/2$ and $f \in [c] - A$. Assume that $\lambda(G) \geq n - 2$ and $\lambda(G/W) \geq n$. Then there exists a c, W - path P in $G - A$ such that $f \in E(P)$ and $\lambda(G^P) \geq n - 2$.

Proof: Let $G^\star$ be the graph obtained from G by $(n-1)$ - limiting relative to every vertex $x \in W$ with $\delta(G; x) \geq n$ and by $(n+1)$ - limiting relative to every vertex $x \in V - W$ with $\delta(G; x) \geq n+2$. When doing so, we let $f \in [c]_{G^\star}$. Let $W^\star$ be the set of all vertices of $G^\star$ which belong to W or to a limitation by which a vertex of W is replaced in $G^\star$. Then by Lemma 5, $\lambda(G^\star) \geq n - 2$ and $\lambda(G^\star/W^\star) \geq n$. Moreover Lemma 7 in [1] states that Proposition 2 is true if for each $x \in W$, $\delta(G; x) \leq n - 1$ and for each $x \in V - W$, $\delta(G; x) \leq n + 1$. Thus there is a $c, W^\star$ - path $P^\star$ in $G^\star - A$ with $f \in E(P^\star)$ and $\lambda((G^\star)^{P^\star}) \geq n - 2$. The path in G induced by $P^\star$ satifies the conclusion of Proposition 2. □

The proofs of the next two lemmas will be given in the following chapters.

LEMMA 6: Let $n \geq 6$ be even, $G = (V, E)$ be a graph, $U, S \subseteq V$ with $U \cap S = \emptyset$, $c \in V - S$ and $f, g \in [c]$ such that

(i) $\delta(x) \leq n + 1$ for each $x \in V$

(ii) $\delta(x) \leq n - 1$ for each $x \in U$

(iii) $\delta(s)$ is even for each $s \in S$

(iv) $|S| \leq 1$

(v) $\lambda(G; V - S) \geq n - 2$

(vi) $\lambda(G/U; (V - S)/U) \geq n$

Then there exists a cycle C in G such that $f, g \in E(C)$, $\lambda(G - E(C); V - S) \geq n - 4$ and $\lambda((G - E(C))/U; (V - S)/U) \geq n - 2$.

LEMMA 7: Let $n \geq 6$ be even, $G = (V, E)$ be a graph, $U, W, S \subseteq V$ with $U \cap S = \emptyset$, $c \in W - S$ and $f \in [c]$ such that

(i) $\delta(x) \leq n + 1$ for each $x \in V$

(ii) $\delta(x) \leq n - 1$ for each $x \in U + W$

(iii) $\delta(x) \leq n - 3$ for each $x \in U \cap W$

(iv) $\delta(s)$ is even for each $s \in S$

(v) $|S| \leq 1$, $|W| \geq 2$

(vi) $\lambda(G; V - S) \geq n - 4$

(vii) $\lambda(G/U; (V - S)/U) \geq n - 2$

(viii) $\lambda(G/W; (V - S)/W) \geq n - 2$

(ix) $\lambda((G/U, W); ((V - S)/U, W)) \geq n$ if $U \cap W = \emptyset$, and $\lambda(G/(U + W); (V - S)/(U + W)) \geq n$ if $U \cap W \neq \emptyset$

Then there exists a c, W - path P in G such that $f \in E(P)$, $\lambda(G^P; V-S) \geq n-4$ *and* $\lambda((G^P)/U; (V - S)/U) \geq n - 2$.

Note that in Lemma 6 and Lemma 7, U may be the empty set.

3. Proof of Lemma 6

Let $n \geq 6$ be a fixed even number. We write $PRM_{L6}(G, U, S, c, f, g)$ if G, U, S, c, f and g satisfy the premises of Lemma 6. In this case we let $CON_{L6}(G, U, S, c, f, g)$ be the set of all cycles satisfying the conclusion of Lemma 6. Note that by (i) and (ii), each cycle in $CON_{L6}(G, U, S, c, f, g)$ contains a simple subcycle in $CON_{L6}(G, U, S, c, f, g)$.

The proof of Lemma 6 is by induction on $2\mu(G) - |S|$. Let $p \geq 0$ be an integer and assume the following:

ASSUMPTION 1: *If* $PRM_{L6}(G, U, S, c, f, g)$ *and* $2\mu(G) - |S| < p$, *then* $CON_{L6}(G, U, S, c, f, g) \neq \emptyset$.

Now let $PRM_{L6}(G, U, S, c, f, g)$, $2\mu(G) - |S| = p$ and $CON := CON_{L6}(G, U, S, c, f, g) = \emptyset$. We have to find a contradiction. By (iii) and (v), $\delta(c) \geq 2$. Therefore let w.l.o.g. $f \neq g$.

CLAIM 1: *For each* $x \in V - c$, $\{f, g\} \not\subseteq [x]$.

Proof: Otherwise the cycle C with $E(C) = \{f, g\}$ is contained in CON by (v) and (vi) #. □

CLAIM 2: *For each* $s \in S$, $\delta(s) \geq 2$.

Proof: If not, $\emptyset \neq CON_{L6}(G-s, U, \emptyset, c, f, g) \subseteq CON$ by Assumption 1 #. □

CLAIM 3: G *has no* $(\leq n - 1) - cut$ $(X, \overline{X})$ *with* $|X|, |\overline{X}| \geq 2$.

Proof: Let $(X, \overline{X})$ be such a cut with $c \in X$. Then $(X, \overline{X})$ divides $V - S$ by (iv). Set $G^\star := G/\overline{X}$, $U^\star := U/\overline{X}$ and $S^\star := X \cap S$. Moreover let $G' := G/X$, $U' := U/X$ and $S' := \overline{X} \cap S$. By (vi), $\overline{X} \in U^\star$ and $X \in U'$. Moreover by Assumption 1 there is a simple $C^\star \in CON_{L6}(G^\star, U^\star, S^\star, c, f, g)$. If $\overline{X} \notin V(C^\star)$, then by $PRM_{L4}(n - 4, n - 2, 2, G - E(C^\star), U, V - S, X, \overline{X})$ we obtain $C^\star \in CON$ #. Therefore $\overline{X} \in V(C^\star)$. Let $\{h_1, h_2\} := E(C^\star) \cap [X, \overline{X}]_G$. Then by Assumption 1 there is a simple $C' \in CON_{L6}(G', U', S', X, h_1, h_2)$. $C^\star$ and C' yield a cycle C in G. By $PRM_{L4}(n - 4, n - 4, 2, G - E(C), U, V - S, X, \overline{X})$,

$C \in CON$ #. □

CLAIM 4: G has no $(\leq n+1)$ - cut $(X, \overline{X})$ with $U \subseteq X$ and $|X|, |\overline{X}| \geq 2$.

Proof: Let $(X, \overline{X})$ be such a cut. Then $(X, \overline{X})$ divides $V - S$ by (iv). Set $G^\star := G/\overline{X}$, $S^\star := S \cap X$, $G' := G/X$ and $S' := S \cap \overline{X}$.
Case 1: $c \in X$.

By Assumption 1 there is a simple $C^\star \in CON_{L6}(G^\star, U, S^\star, c, f, g)$. If $\overline{X} \notin V(C^\star)$, then by $PRM_{L3}(n - 4, n, 0, G - E(C^\star), V - S, X, \overline{X})$ and $PRM_{L3}(n - 2, n, 0, (G - E(C^\star))/U, (V - S)/U, X/U, \overline{X})$ we obtain $C^\star \in CON$ #. Therefore $\overline{X} \in V(C^\star)$. Let $\{h_1, h_2\} := E(C^\star) \cap [X, \overline{X}]_G$. Then by Assumption 1 there is a simple $C' \in CON_{L6}(G', \emptyset, S', X, h_1, h_2)$. $C^\star$ and C' yield a simple cycle C in G. By $PRM_{L3}(n - 4, n - 2, 0, G - E(C), V - S, X, \overline{X})$ and $PRM_{L3}(n - 2, n - 2, 0, (G - E(C))/U, V - S, X/U, \overline{X})$ we obtain $C \in CON$ #.
Case 2: $c \in \overline{X}$.

By Assumption 1 there is a simple $C' \in CON_{L6}(G', \emptyset, S', c, f, g)$. If $X \notin V(C')$, then by $PRM_{L3}(n - 2, n - 2, 2, G - E(C'), V - S, \overline{X}, X)$ and $PRM_{L3}(n - 2, n, 0, (G - E(C'))/U, (V - S)/U, \overline{X}, X/U)$ we obtain $C' \in CON$ #. Therefore $X \in V(C')$. Let $\{h_1, h_2\} := E(C') \cap [X, \overline{X}]_G$. Then by Assumption 1 there is a simple $C^\star \in CON_{L6}(G^\star, U, S^\star, \overline{X}, h_1, h_2)$. $C^\star$ and C' yield a simple cycle C in G. By $PRM_{L3}(n - 4, n - 2, 0, G - E(C), V - S, X, \overline{X})$ and $PRM_{L3}(n - 2, n - 2, 0, (G - E(C))/U, (V - S)/U, X/U, \overline{X})$ we obtain $C \in CON$ #. □

CLAIM 5: $S = \emptyset$.

Proof: Let $s \in S$. Then $|N(s)| \geq 2$:

⌐ Assume that $N(s) \subseteq \{x\}$. Then $\delta(\{s, x\}) \leq \delta(x) - 2$ by Claim 2. If $x \in U$ or $x \in V - U$, then by (ii) and (v) or by (i) and (vi) respectively we obtain $V = \{s, c\}$ and thus $f, g \in [s]$ contradicting Claim 1. ⌐

Let $h, i \in [s]$ with $V(h) \neq V(i)$. Moreover, if $\{f, g\} \cap \{h, i\} \neq \emptyset$, let $h \in \{f, g\}$. Set $G^\star := G^{h,i}$. Then $PRM_{L6}(G^\star, U, S, c, f, g)$ ((v) and (vi) follow by Claim 3 and Claim 4 respectively). Therefore by Assumption 1 there exists $C^\star \in CON_{L6}(G^\star, U, S, c, f, g)$ which induces a cycle of CON #. □

CLAIM 6: For each $x \in V - U - c$, $\delta(x) = n + 1$.

Proof: Otherwise by Assumption 1 there is a simple $C \in CON_{L6}(G, U, \{x\}, c, f, g)$. By $\delta(G - E(C); x) \geq n - 2$ and Lemma 1, $C \in CON$ #. □

CLAIM 7: $E(V - U - c) = \emptyset$.

Proof: Let $h \in E(V - U - c)$. Then $PRM_{L6}(G - h, U, S, c, f, g)$ ((v) and (vi) follow by Claim 3 and Claim 4 respectively and by Claim 6). Therefore by Assumption 1, $\emptyset \neq CON_{L6}(G - h, U, S, c, f, g) \subseteq CON$ #. □

Case 1: $c \in U$.

By (v) and Proposition 1 there exists a cycle C in G such that $f, g \in E(C)$ and $\lambda(G - E(C)) \geq n - 4$. Moreover by (ii) we can assume that C is simple. By Claim 7, $N(G; x) \subseteq U$ for each $x \in V - U$. Hence for each $x \in V - U$, $\delta(G - E(C); x, U) \geq n - 2$. Therefore $C \in CON$ #.

Case 2: $c \in V - U$.

$|N(G - f - g; c)| \geq 2$ holds:

⌐ Assume that $N(G - f - g; c) \subseteq \{x\}$. Then $\delta(G; c, x) \geq n - 2$ by (vi). If $x \in U$, then $\delta(G; \{c, x\}) \leq 3 < n - 2$ by (ii), and if $x \in V - U$, then $\delta(G; \{c, x\}) \leq 5 < n$ by (i). Therefore by (v) and (vi) we obtain $V = \{c, x\}$ and thus $f, g \in [x]_G$ contradicting Claim 1. ⌐

Let $h, i \in [c] - f - g$ with $V(h) \neq V(i)$, and set $G^\star := G^{h,i}$. Then by $\delta(G^\star; c) \geq n - 2$ and Claim 3, $\lambda(G^\star) \geq n - 2$. Therefore by Proposition 1 there is a cycle C in $G^\star - h$ such that $f, g \in E(C)$ and $\lambda(G^\star - E(C)) \geq n - 4$. By $\delta(G^\star; c) \leq n - 1$ we can assume that C is simple. By $h \notin E(C)$, C is also a simple cycle in G with $\lambda(G - E(C)) \geq n - 4$. By $CON = \emptyset$, $\lambda((G - E(C))/U) \leq n - 3$. Let $(X, \overline{X})$ be an $(\leq n - 3)$ – cut of $G - E(C)$ with $U \subseteq X$. By $\delta(G - E(C); x) \geq n - 2$ for each $x \in V - U$ we obtain $|\overline{X}| \geq 2$. Also $|X| \geq 2$ (otherwise since C is simple, $\delta(G; X) \leq n - 1$ contradicting (vi)). Moreover $c \in \overline{X}$ (otherwise by Claim 7, $\delta(G - E(C); X, \overline{X}) \geq |\overline{X}|(n - 2) \geq n - 2$ #). By Claim 4 and (i), for each $x \in \overline{X} - c$, $\delta(G; x, c) \leq n/2$ and thus $\delta(G; x, U) \geq n/2 + 1$ by Claim 6 and Claim 7. Therefore for each $x \in \overline{X} - c$, $\delta(G - E(C); x, X) \geq n/2 - 1$. Now $|\overline{X}| = 2$ and thus $|E(C) \cap [X, \overline{X}]_G| \leq 4$. Therefore $\delta(G; X, \overline{X}) \leq \delta(G - E(C); X, \overline{X}) + 4 \leq n + 1$ contradicting Claim 4. Now the proof of Lemma 6 is complete. ∎

4. Proof of Lemma 7

Let $n \geq 6$ be a fixed even number. We write $PRM_{L7}(G, U, W, S, c, f)$ if G, U, W, S, c and f satisfy the premises of Lemma 7. In this case we let $CON_{L7}(G, U, W, S, c, f)$ be the set of all paths satisfying the conclusion of Lemma 7. Let $P \in CON_{L7}(G, U, W, S, c, f)$. Then $CON_{L7}(G, U, W, S, c, f)$ contains a W – admissible subpath of P:

⌐ Assume that $|V - S| \geq 2$. Let $e \in W$ such that P is a c, e – path. By (ii) and (iii), $[c]_G \cap E(P) = \{f\}$ and hence $e \neq c$. Let P' be a simple c, e – subpath of P. Clearly, $P' \in CON_{L7}(G, U, W, S, c, f)$. Moreover P' contains a W – admissible subpath Q. Choose $d \in W$ such that Q is a c, d – path. Moreover let R be the simple d, e – subpath of P'. Then G_g^Q is obtained from $G_h^{P'}$ by replacing h by the c, e – path $R + g$. Hence $Q \in CON_{L7}(G, U, W, S, c, f)$.

Now let $|V - S| \leq 1$. Then by (v), $V = W = \{c, s\}$ for $s \in S$. Therefore $f \in E(W)$ and the path Q with $E(Q) = \{f\}$ is a W – admissible subpath of P contained in $CON_{L7}(G, U, W, S, c, f)$. ⌟

Let $p \geq 0$ be an integer. We assume the following:

ASSUMPTION 2: *If $PRM_{L7}(G, U, W, S, c, f)$ and $2\mu(G) - |S| < p$, then $CON_{L7}(G, U, W, S, c, f) \neq \emptyset$.*

Now let $PRM_{L7}(G, U, W, S, c, f)$, $2\mu(G) - |S| = p$ and $CON := CON_{L7}(G, U, W, S, c, f) = \emptyset$. We have to find a contradiction. Let $d \in V(f) - c$.

CLAIM 1: $d \notin W$.

Proof: Otherwise the path P with $E(P) = \{f\}$ is contained in CON by (vi) and (vii) #. □

CLAIM 2: *For each $s \in S$, $\delta(s) \geq 2$.*

Proof: Assume that $\delta(s) = 0$. Then $|W - s| \geq 2$:

⌈ Otherwise $W = \{c, s\}$ and hence $\delta(W) \leq n - 1$ by (ii). Moreover by Claim 1, $V - W \neq \emptyset$. Thus by (ix), $U \cap W = \{c\}$. Therefore $\delta(W) \leq n - 3$ by (iii) contradicting (viii). ⌟

Therefore by Assumption 2, $\emptyset \neq CON_{L7}(G - s, U, W - s, \emptyset, c, f) \subseteq CON$ #. □

CLAIM 3: $|V| \geq 4$.

Proof: By Claim 1 and (v), $|V| \geq 3$. Now assume that $|V| = 3$. Then by Claim 1 and (v) we may let $V = \{c, d, e\}$ and $W = \{c, e\}$.
Case 1: $e \in N(d)$.
Let $g \in [e, d]$ and let P the simple c, e – path with $E(P) = \{f, g\}$. Then by (vi), (viii) and $d \notin W$ we obtain $\delta(G^P; z) \geq n - 4$ for each $z \in V - S$ and thus $\lambda(G^P; V - S) \geq n - 4$ since $|V| = 3$. If $d \in U$, then $U \cap W \neq \emptyset$ by (ii) and (ix). Hence $(G^P)/U = G/U$ and thus $P \in CON$ by (vii) #. Therefore $d \in V - U - W$ and thus by (ix), $\delta(G^P; d) \geq n - 2$ if $d \notin S$. Now by $|V/U| \leq 3$ and (vii), $\lambda((G^P)/U; (V - S)/U) \geq n - 2$ and thus $P \in CON$ #.
Case 2: $N(d) = \{c\}$.
Then $\delta(c) = \delta(d) + \delta(e)$ and $\delta(W) = \delta(d)$. Hence by Claim 2 and (vi), $\delta(c) \geq n - 2$ and thus $c \notin U$ by (iii). If $d \in U$, then by (ii) and (ix) we obtain $e \in U \cap W$ and thus $S = \emptyset$. Therefore by (vi), $\delta(c) \geq 2(n - 4) \geq n + 2$ contradicting (i). Hence $d \in V - U - W$. Moreover $d \in S$ (otherwise $\delta(d) \geq n$ by (ix) and thus $\delta(c) \geq n + 2$ by Claim 2 contradicting (i)). Now by (ii) and Claim 2, $\delta(\{c, d\}) \leq n - 3$ contradicting (vii). □

CLAIM 4: G has no $(\leq n-3)$ - cut $(X,\overline{X})$ with $|X|,|\overline{X}| \geq 2$.

Proof: Let $(X,\overline{X})$ be such a cut with $c \in X$. Then $(X,\overline{X})$ divides $V-S$ by (v). Set $G^\star := G/\overline{X}$, $U^\star := U/\overline{X}$, $W^\star := W/\overline{X}$ and $S^\star := X \cap S$. Moreover let $G' := G/X$, $U' := U/X$, $W^\star := W/X$ and $S' := \overline{X} \cap S$. By (vii) and (viii), $\overline{X} \in U^\star \cap W^\star$ and $X \in U' \cap W'$. Moreover by Assumption 2 there is a $W^\star$ - admissible $P^\star \in CON_{L7}(G^\star, U^\star, W^\star, S^\star, c, f)$. If $\overline{X} \notin V(P^\star)$, then by $PRM_{L4}(n-4, n-4, 2, G^{P^\star}, U, V-S, X, \overline{X})$ we obtain $P^\star \in CON$ #. Therefore $\overline{X} \in V(P^\star)$. Let $h \in E(P^\star) \cap [X,\overline{X}]_G$. Then by Assumption 2 there is a W' - admissible $P' \in CON_{L7}(G', U', W', S', X, h)$. $P^\star$ and P' yield a W - admissible path P in G. By $PRM_{L4}(n-4, n-4, 2, G^P, U, V-S, X, \overline{X})$, $P \in CON$ #. □

CLAIM 5: G has no $(\leq n-1)$ - cut $(X,\overline{X})$ with $U \subseteq X$, and $|X|,|\overline{X}| \geq 2$.

Proof: Let $(X,\overline{X})$ be such a cut. Then $(X,\overline{X})$ divides $V-S$ by (v). Set $G^\star := G/\overline{X}$, $W^\star := W/\overline{X}$ and $S^\star := S \cap X$. Moreover let $G' := G/X$, $W' := W/X$ and $S' := S \cap \overline{X}$. Then by (ix), $\overline{X} \in W^\star$ and $X \in W'$.

Case 1: $c \in X$.

By Assumption 2 there is a $W^\star$ - admissible $P^\star \in CON_{L7}(G^\star, U, W^\star, S^\star, c, f)$. If $\overline{X} \notin V(P^\star)$, then by $PRM_{L3}(n-4, n-2, 0, G^{P^\star}, V-S, X, \overline{X})$ and $PRM_{L3}(n-2, n-2, 0, (G^{P^\star})/U, (V-S)/U, X/U, \overline{X})$ we obtain $P^\star \in CON$ #. Therefore $\overline{X} \in V(P^\star)$. Let $h \in E(P^\star) \cap [X,\overline{X}]_G$. Then by Assumption 2 there is a W' - admissible $P' \in CON_{L7}(G', \emptyset, W', S', X, h)$. $P^\star$ and P' yield a W - admissible path P in G. By $PRM_{L3}(n-4, n-2, 0, G^P, V-S, X, \overline{X})$ and $PRM_{L3}(n-2, n-2, 0, (G^P)/U, (V-S)/U, X/U, \overline{X})$ we obtain $P \in CON$ #.

Case 2: $c \in \overline{X}$.

By Assumption 2 there is a W' - admissible $P' \in CON_{L7}(G', \emptyset, W', S', c, f)$. If $X \notin V(P')$, then by $PRM_{L3}(n-4, n-2, 0, G^{P'}, V-S, X, \overline{X})$ and $PRM_{L3}(n-2, n-2, 0, (G^{P'})/U, (V-S)/U, X/U, \overline{X})$ we obtain $P' \in CON$ #. Therefore $X \in V(P')$. Let $h \in E(P') \cap [X,\overline{X}]_G$. Then by Assumption 2 there is a $W^\star$ - admissible $P^\star \in CON_{L7}(G^\star, U, W^\star, S^\star, \overline{X}, h)$. $P^\star$ and P' yield a W - admissible path P in G. By $PRM_{L3}(n-4, n-2, 0, G^P, V-S, X, \overline{X})$ and $PRM_{L3}(n-2, n-2, 0, (G^P)/U, X/U, \overline{X})$ we obtain $P \in CON$ #. □

CLAIM 6: G has no $(\leq n-1)$ - cut $(X,\overline{X})$ with $W \subseteq X$ and $|\overline{X}| \geq 2$.

Proof: Let $(X,\overline{X})$ be such a cut. Then $(X,\overline{X})$ divides $V-S$ by (v). Set $G^\star := G/\overline{X}$, $U^\star := W/\overline{X}$ and $S^\star := S \cap X$. Moreover let $G' := G/X$, $U' := W/X$ and $S' := S \cap \overline{X}$. Then by (ix), $\overline{X} \in U^\star$ and $X \in U'$. By Assumption 2 there is a W - admissible $P^\star \in CON_{L7}(G^\star, U^\star, W, S^\star, c, f)$. If $\overline{X} \notin V(P^\star)$, then by $PRM_{L4}(n-4, n-2, 2, G^{P^\star}, U, V-S, X, \overline{X})$ we obtain $P^\star \in CON$ #. Therefore $\overline{X} \in V(P^\star)$. Let $\{h_1, h_2\} := E(P^\star) \cap [X,\overline{X}]_G$. Then by Lemma 6 there is a simple $C' \in CON_{L6}(G', U', S', X, h_1, h_2)$. $P^\star$ and C' yield a W - admissible path P in G. By $PRM_{L4}(n-4, n-4, 2, G^P, U, V-S, X, \overline{X})$,

$P \in CON \#$. □

CLAIM 7: G has no $(\leq n+1) - cut$ $(X, \overline{X})$ with $U + W \subseteq X$ and $|\overline{X}| \geq 2$.

Proof: Let $(X, \overline{X})$ be such a cut. Then $(X, \overline{X})$ divides $V - S$ by (v). Set $G^\star := G/\overline{X}$, $S^\star := S \cap X$, $G' := G/X$ and $S' := S \cap \overline{X}$. By Assumption 2 there is a W – admissible $P^\star \in CON_{L7}(G^\star, U, W, S^\star, c, f)$. If $\overline{X} \notin V(P^\star)$, then by $PRM_{L3}(n-4, n, 0, G^{P^\star}, V - S, X, \overline{X})$ and $PRM_{L3}(n-2, n, 0, (G^{P^\star})/U, (V-S)/U, X/U, \overline{X})$ we obtain $P^\star \in CON \#$. Therefore $\overline{X} \in V(P^\star)$. Let $\{h_1, h_2\} := E(P^\star) \cap [X, \overline{X}]_G$. Then by Lemma 6 there is a simple $C' \in CON_{L6}(G', \emptyset, S', X, h_1, h_2)$. $P^\star$ and C' yield a W – admissible path P in G. By $PRM_{L3}(n-4, n-2, 0, G^P, V - S, X, \overline{X})$ and $PRM_{L3}(n-2, n-2, 0, (G^P)/U, (V-S)/U, X/U, \overline{X})$ we obtain $P \in CON \#$. □

CLAIM 8: G has no $(\leq n+1) - cut$ $(X, \overline{X})$ with $U \subseteq X$, $W \subseteq \overline{X}$ and $|X| \geq 2$.

Proof: Let $(X, \overline{X})$ be such a cut. Then $(X, \overline{X})$ divides $V - S$ by (v). Set $G^\star := G/\overline{X}$, $S^\star := S \cap X$, $G' := G/X$ and $S' := S \cap \overline{X}$. By Assumption 2 there is a W – admissible $P' \in CON_{L7}(G', \emptyset, W, S', c, f)$. If $X \notin V(P')$, then by $PRM_{L3}(n-2, n-2, 2, G^{P'}, V - S, \overline{X}, X)$ and $PRM_{L3}(n-2, n, 0, (G^{P'})/U, (V-S)/U, \overline{X}, X/U)$ we obtain $P' \in CON \#$. Therefore $X \in V(P')$. Let $\{h_1, h_2\} := E(P') \cap [X, \overline{X}]_G$. Then by Lemma 6 there is a simple $C^\star \in CON_{L6}(G^\star, U, S^\star, \overline{X}, h_1, h_2)$. $C^\star$ and P' yield a W – admissible path P in G. By $PRM_{L3}(n-4, n-2, 0, G^P, V - S, X, \overline{X})$ and $PRM_{L3}(n-2, n-2, 0, (G^P)/U, (V-S)/U, X/U, \overline{X})$ we obtain $P \in CON \#$. □

CLAIM 9: $S = \emptyset$.

Proof: Let $s \in S$. By Claim 2, $N(s) \neq \emptyset$. Moreover $|N(s)| = 1$:

⌐ Assume that not. Choose $g, h \in [s]$ with $V(g) \neq V(h)$. Moreover if $f \in \{g, h\}$, let $g = f$. Let $G^\star := G^{g,h}$. Then $PRM_{L7}(G^\star, U, W, S, c, f)$ ((vi), (vii), (viii) and (ix) follow straightforward by Claim 4, Claim 5, Claim 6 and Claim 7,8 respectively). Therefore by Assumption 2 there is $P^\star \in CON_{L7}(G^\star, U, W, S, c, f)$ which induces a path in $CON \#$. ⌐

Let $a \in N(s)$. Then $\delta(\{a, s\}) \leq \delta(a) - \delta(s) \leq \delta(a) - 2$ by (iv). $s \in W$ follows:

⌐ Otherwise $s \in V - U - W$. If $a \in U \cap W$, $a \in W - U$, $a \in U - W$ or $a \in V - U - W$, then by Claim 3 and by (iii), (ii), (ii) or (i) we easily obtain a contradiction to (vi), (vii), (viii) or (ix) respectively. ⌐

Moreover $a \in U$ (otherwise by (i), $\delta(\{a, s\}) \leq n - 1$ contradicting Claim 3 and Claim 5) and hence by (iii), (vi) and Claim 3, $a \notin W$. $\delta(s) = 2$ follows by (ii), (vi) and Claim 3, and Assumption 2 yields a $(W+a)$ – admissible $P^* \in CON_{L7}(G-s, U, W-s+a, \emptyset, c, f)$. Let $x \in (V(P^*) \cap (W-s+a)) - c$. Then $x = a$ (otherwise $P^* \in CON$ #). Let P be the c, s – path obtained from P^* by adding s and an edge of $[a, s]_G$. Then G^P is obtained from $(G-s)_h^{P^*}$ by replacing h by a simple c, a – path which uses the vertices c, s and a. Now $P \in CON$ follows #. □

CLAIM 10: *For each $x \in W - U - c$, $\delta(x) = n-1$, and for each $x \in V - U - W$, $\delta(x) = n+1$.*

Proof: Otherwise by Assumption 2 there is a W – admissible $P \in CON_{L7}(G, U, W, \{x\}, c, f)$. Then $\delta(G^P; x) \geq n-2$. Therefore by Lemma 1, $P \in CON$ #. □

CLAIM 11: $E(V - U - c) = \emptyset$.

Proof: Let $h \in E(V - U - c)$. Then $PRM_{L7}(G-h, U, W, S, c, f)$ ((vi), (vii), (viii) and (ix) follow by Claim 4, Claim 5, Claim 6 and Claim 7,8 respectively and by Claim 10). Therefore by Assumption 2, $\emptyset \neq CON_{L7}(G-h, U, W, S, c, f) \subseteq CON$ #. □

Case 1: $c \in U$.

By (vi), (viii) and Proposition 2 there exists a c, W – path P in G such that $f \in E(P)$ and $\lambda(G^P) \geq n-4$. Moreover by (iii) we can assume that P is W – admissible. By Claim 11, $N(G; x) \subseteq U$ for each $x \in V - U$. Hence by construction, for each $x \in V - U$ we obtain $\delta(G^P; x, U) \geq n-2$. Therefore $P \in CON$ #.

Case 2: $c \in W - U$.

$|N(G - f; c)| \geq 2$ holds:

⌐ Assume that $N(G - f; c) \subseteq \{x\}$. Then $\delta(G; c, x) \geq n-3$ by (vii). If $x \in U + W$, then by (ii) we obtain $\delta(G; \{c, x\}) \leq 3 \leq n-3$ contradicting Claim 3 and Claim 4. If $x \in V - U - W$, then by (i), $\delta(G; \{c, x\}) \leq 5 \leq n-1$ contradicting Claim 3 and Claim 5. ⌐

Choose $g, h \in [c] - f$ with $V(g) \neq V(h)$, and set $G^* := G^{g,h}$. Then by $\delta(G^*; c) \geq n-4$, Claim 4 and Claim 6 we obtain $\lambda(G^*) \geq n-4$ and $\lambda(G^*/W) \geq n-2$. Therefore by Proposition 2 there is a c, W – path P in $G^* - g$ such that $f \in E(P)$ and $\lambda((G^*)^P) \geq n-4$. By $\delta(G^*; c) \leq n-3$ we can assume that P is W – admissible, and by $h \notin E(P)$, P is also a W – admissible path in G with $\lambda(G^P) \geq n-4$. Therefore $\lambda((G^P)/U) \leq n-3$ by $CON = \emptyset$. Hence there exists an $(\leq n-3)$ – cut $(X, \overline{X})$ of G^P with $U \subseteq X$. By $\delta(G^p; x) \geq n-2$ for each $x \in V - U$ we obtain $|\overline{X}| \geq 2$. Also $|X| \geq 2$ (otherwise since P is W – admissible, we obtain $\delta(G; X) \leq \delta(G^P; X) + 2 \leq n-1$ if $X \cap W = \emptyset$ and $\delta(G; X) = \delta(G^P; X) \leq n-3$ if $X \cap W \neq \emptyset$ contradicting (ix) and (vii) respectively). Moreover by (vii) and Claim 11, for each $x \in \overline{X} - c$ we obtain

$\delta(G^P; x, U + c) \geq n - 2$ even if $x \in V(P) \cap W$. Therefore $c \in \overline{X}$ (otherwise $\delta(G^P; X, \overline{X}) \geq |\overline{X}|(n-2) \geq n-2$ ⌗). By Claim 5, for each $x \in (\overline{X} \cap W) - c$ we obtain $\delta(G; c, x) \leq n/2 - 1$ and thus $\delta(G; x, U) \geq n/2$ by Claim 10 and Claim 11. Moreover by Claim 5, for each $x \in \overline{X} - W$ we obtain $\delta(G; c, x) \leq n/2$ and thus $\delta(G; x, U) \geq n/2 + 1$ by Claim 10 and Claim 11. Therefore for each $x \in \overline{X} - c$, $\delta(G^P; x, X) \geq n/2 - 1$. Now $|\overline{X}| = 2$ follows. Let $x \in \overline{X} - c$. If $x \in W$, then $|E(P) \cap [X, \overline{X}]_G| \leq 2$ and thus $\delta(G; X, \overline{X}) \leq \delta(G^P; X, \overline{X}) + 2 \leq n - 1$ which contradicts Claim 5. Therefore $x \in \overline{X} - W$. Then the edge by which P is replaced in G^P connects c with a vertex of X. Moreover $|E(P) \cap [X, \overline{X}]_G| \leq 3$. Thus once more we obtain $\delta(G; X, \overline{X}) \leq \delta(G^P; X, \overline{X}) + 2 \leq n - 1$ contradicting Claim 5. This completes the proof of Lemma 7. ∎

5. Proof of Theorem 1

Theorem 1 follows by the following proposition:

PROPOSITION 3: *Let $n \geq 6$ be even, $G = (V, E)$ be a graph and $U \subseteq V$ with $\lambda(G) \geq n - 2$ and $\lambda(G/U) \geq n$. Moreover let $f, g \in E$. Then there exists a cycle C in G through f and g such that $\lambda(G - E(C)) \geq n - 4$ and $\lambda((G - E(C))/U) \geq n - 2$.*

⌐ Assume that Proposition 3 is true. Let n, $G = (V, E)$, C_1, f and g satisfy the premises of Theorem 1. Then $\lambda((G - E(C_1))/V(C_1)) \geq n$ so that by Proposition 3 with $U = V(C_1)$, there is a cycle C_2 in $G - E(C_1)$ with $\lambda(G - E(C_1) - E(C_2)) \geq n - 4$ and $\lambda((G - E(C_1) - E(C_2))/V(C_1)) \geq n - 2$. Assume that $\lambda(G - E(C_2)) \leq n - 3$. Then there exists an $(\leq n - 3)$ - cut $(X, \overline{X})$ of $G - E(C_2)$. By construction, $(X, \overline{X})$ divides $V(C_1)$. Therefore $|E(C_1) \cap [X, \overline{X}]_G| \geq 2$ and hence $\lambda(G - E(C_1) - E(C_2); X, \overline{X}) \leq n - 5$ contradicting $\lambda(G - E(C_1) - E(C_2)) \geq n - 4$. ⌐

Now we prove Proposition 3. Let n, $G = (V, E)$, U, f and g satisfy the premises of Proposition 3. W.l.o.g. we can assume that $\delta(x) \leq n + 1$ for each $x \in V$ and $\delta(x) \leq n - 1$ for each $x \in U$ (if not, consider a graph which arises from G by $(n - 1)$ - and $(n + 1)$ - limiting relative to each vertex $x \in U$ with $\delta(x) \geq n$ and to each vertex $x \in V - U$ with $\delta(x) \geq n + 2$ respectively). If $h, i \in E$, then we call a cycle C in G h, i - *regular* if $h, i \in E(C)$, $\lambda(G - E(C)) \geq n - 4$ and $\lambda((G - E(C))/U) \geq n - 2$. Let $M := \{h \in E;$ there exists an f, h - regular cycle in $G\}$. If $c \in V(f)$, then $PRM_{L6}(G, U, \emptyset, c, f, f)$ and thus $f \in M$ by Lemma 6. Assume that $g \notin M$. Then it is easy to find an $h \in E - M$ with $V(h) \cap V(M) \neq \emptyset$. By construction there exists $i \in M$ with $V(i) \cap V(h) \neq \emptyset$ and an f, i - regular cycle D in G. D is simple by the degree - conditions in G. Moreover by $E(D) \subseteq M$, $h \notin E(D)$. Let $c \in V(i) \cap V(h)$. Then by Lemma 7 there exists a $V(D)$ - admissible $P_1 \in CON_{L7}(G - E(D), U, V(D), \emptyset, c, h)$. Let $\{c, d\} := V(P_1) \cap V(D)$. Moreover let P_2 be the simple c, d - path in D containing f and let P_3 be the other simple c,d – path in D. P_1 and P_2 yield a cycle C in G and we obtain $(G-E(C))^{P_3} = G^\star$ for $G^\star = (G-E(D))^{P_1}$. Now by $\lambda(G^\star) \geq n-4$ and $\lambda(G^\star/U) \geq n-2$, C is f,h – regular and thus $h \in M$ #. ∎

6. Additional notes

Clearly by Theorem 1 and Proposition 1, Corollary 1 follows. Also by Proposition 1 and by an easy induction on ℓ we obtain the following statement:

Let $n \geq 2$ be even and $\ell < n/4+1$. Moreover let $G = (V,E)$ be a graph with $\lambda(G) \geq n$, and let $f_1,...,f_\ell, g_1,...,g_\ell$ be pairwise distinct edges of G. Then for each $i \in [1:\ell]$ there exists a cycle C_i containing f_i and g_i such that $C_1,...,C_\ell$ are pairwise disjoint and that for each $i \in [1:\ell]$, $\lambda(G-E(C_1)-...-E(C_i)) \geq n-2i$.

Therefore the following conjecture arises:

CONJECTURE 1: Let $n \geq 2$ be even and $\ell < n/4+1$. Moreover let $G = (V,E)$ be a graph with $\lambda(G) \geq n$, and let $f_1,...,f_\ell, g_1,...,g_\ell$ be pairwise distinct edges of G. Then for each $i \in [1:\ell]$ there exists a cycle C_i containing f_i and g_i such that $C_1,...,C_\ell$ are pairwise disjoint and that for each $I \subseteq [1:\ell]$, $\lambda(G - \bigcup_{i \in I} E(C_i)) \geq n - 2|I|$.

In this conjecture $\ell < n/4+1$ is necessary: Let $\ell \geq n/4+1$ and let $G = (V,E)$ be a graph with exactly three vertices a,b,c, $\delta(a,b) = \delta(a,c) = n/2$ and $\delta(b,c) = 2\ell - 2$. Let $\{f_1,...,f_{\ell-1}, g_1,...,g_{\ell-1}\} := [b,c]$, $f_\ell \in [a,b]$ and $g_\ell \in [a,c]$. Then it is easy to see that $\lambda(G) \geq n$ and that there are no cycles as described in Conjecture 1.

REFERENCES

[1] A. Huck, "Edge – disjoint paths and cycles in n – edge – connected graphs", submitted to *J. of Graph Theory*

[2] H. Okamura, "Paths in k – edge – connected graphs", *J. of Combinatorial Theory, Series B*, 45 (1988), 345 – 355

Author's address: Andreas Huck, Institut für Mathematik, Universität Hannover, Welfengarten 1, W - 3000 Hannover 1, Germany

Finding Disjoint Trees in Planar Graphs in Linear Time

B.A. REED

N. ROBERTSON

A. SCHRIJVER

P.D. SEYMOUR

ABSTRACT. We show that for each fixed k there exists a linear-time algorithm for the problem: *given:* an undirected plane graph $G = (V, E)$ and subsets $X_1, \ldots, X_p$ of V with $|X_1 \cup \cdots \cup X_p| \leq k$; *find:* pairwise vertex-disjoint trees $T_1, \ldots, T_p$ in G such that T_i covers X_i ($i = 1, \ldots, p$).

1. Introduction

Consider the following *disjoint trees problem*:

given: an undirected graph $G = (V, E)$ and subsets $X_1, \ldots, X_p$ of V;

find: pairwise vertex-disjoint trees $T_1, \ldots, T_p$ in G such that T_i covers X_i ($i = 1, \ldots, p$).

(We say that tree T_i *covers* X_i if each vertex in X_i is a vertex of T_i.)

Robertson and Seymour [5] gave an algorithm for this problem that runs, for each fixed k, in time $O(|V|^3)$ for inputs satisfying $|X_1 \cup \cdots \cup X_p| \leq k$. (Recently, Reed gave an improved version with running time $O(|V|^2 \log |V|)$.) In this paper we show that if we moreover restrict the input graphs to planar graphs there exists a *linear-time* algorithm:

THEOREM. *There exists an algorithm for the disjoint trees problem for planar graphs that runs, for each fixed k, in time $O(|V|)$ for inputs satisfying*

1991 Mathematics Subject Classification: Primary 05C10, 05C38, 05C85; secondary 68Q25, 68R10

This paper is in final form and no version of it will be submitted for publication elsewhere.

$|X_1 \cup \cdots \cup X_p| \leq k.$

If we do not fix an upper bound k on $|X_1 \cup \cdots \cup X_p|$, the disjoint trees problem is NP-hard (D.E. Knuth, see [1]), even when we restrict ourselves to planar graphs and each X_i is a pair of vertices (Lynch [2]).

Our result extends a result of Suzuki, Akama, and Nishizeki [7] stating that the disjoint trees problem is solvable in linear time for planar graphs for each fixed upper bound k on $|X_1 \cup \cdots \cup X_p|$, when

(1)   there exist two faces f_1 and f_2 such that each vertex in $X_1 \cup \cdots \cup X_p$ is incident with at least one of f_1 and f_2.

(In fact, they showed more strongly that the problem (for nonfixed k) is solvable in time $O(k|V|)$. Indeed, recently Ripphausen, Wagner, and Weihe [4] showed that it is solvable in time $O(|V|)$.)

Equivalent to a linear-time algorithm for the disjoint trees problem (for fixed k) is one for the following "realization problem". Let $G = (V, E)$ be a graph and let $X \subseteq V$. For any $E' \subseteq E$ let $\Pi(E')$ be the partition $\{K \cap X | K$ is a component of the graph (V, E') with $K \cap X \neq \emptyset\}$ of X. We say that E' *realizes* Π if $\Pi = \Pi(E')$. We call a partition of X *realizable in* G if it is realized by at least one subset E' of E. Now the *realization problem* is:

given: a graph $G = (V, E)$ and a subset X of V;

find: subsets $E_1, \ldots, E_N$ of E such that each realizable partition of X is realized by at least one of $E_1, \ldots, E_N$.

We give an algorithm for the realization problem for planar graphs that runs, for each fixed k, in time $O(|V|)$ for inputs satisfying $|X| \leq k$. In [3] we extend this result to graphs embedded on any fixed compact surface.

2. Realizable partitions

We will use the following lemma of Robertson and Seymour [6], saying that any vertex that is "far away" from X and also is not on any "short" curve separating X, is irrelevant for the realization problem and can be left out from the graph.

Let $G = (V, E)$ be a plane graph (that is, a graph embedded in the plane $\mathcal{R}^2$). For any curve C on $\mathcal{R}^2$, the *length* length(C) of C is the number of times C meets G (counting multiplicities). We say that a curve C *separates* a subset X of $\mathcal{R}^2$ if X is contained in none of the components of $\mathcal{R}^2 \setminus C$. (So C separates X if C intersects X.)

LEMMA. *There exists a computable function* $g : \mathcal{N} \longrightarrow \mathcal{N}$ *with the following property. Let* $G = (V, E)$ *be a plane graph, let* $X \subseteq V$ *and let* $v \in V$ *be such that each closed curve* C *traversing* v *and separating* X *satisfies* length$(C) \geq g(|X|)$; *then each partition of* X *realizable in* G *is also realizable in* $G - v$.

[$G - v$ is the graph obtained from G by deleting v and all edges incident with v.]

Moreover, we use the following easy proposition, enabling us to reduce the realization problem to smaller problems.

PROPOSITION 1. *Let $G = (V,E)$ be an undirected graph and let $X \subseteq V$. Moreover, let $V_1, \ldots, V_n, Y$ be subsets of V such that*

(2)   (i) *each edge of G is contained in at least one of $V_1, \ldots, V_n$;*

         (ii) $X \subseteq Y$ *and* $V_i \cap V_j \subseteq Y$ *for each $i, j \in \{1, \ldots, t\}$ with $i \neq j$.*

Let $E_{i,1}, \ldots, E_{i,N_i}$ form a solution for the realization problem with input $\langle V_i \rangle, V_i \cap Y$ ($i = 1, \ldots, n$). Then the sets $E_{1,j_1} \cup \cdots \cup E_{n,j_n}$, where j_i ranges over $1, \ldots, N_i$ (for $i = 1, \ldots, n$), form a solution for the realization problem with input G, X.

[$\langle W \rangle$ denotes the subgraph of G induced by W.]

3. Proof of the theorem

We show that, for each fixed k, there exists a linear-time algorithm for the realization problem for plane graphs $G = (V,E)$ and subsets X of V with $|X| \leq k$. We may assume that G is connected.

For any subset W of V let $\delta(W)$ be the set of vertices in W that are adjacent to at least one vertex in $V \setminus W$. Let $W^o := W \setminus \delta(W)$.

Let H be the graph with vertex set V, where two vertices v, v' are adjacent if and only if there exists a face of G that is incident with both v and v'. For any subset W of V, let $\kappa(W)$ denote the number of components of the subgraph of H induced by W. Note that $\kappa(W)$ can be computed in linear time.

We say that W is *linked* if $\kappa(W) = 1$. Observe that if $W \neq \emptyset$ then

(3) W is linked if and only if G does not contain a circuit C splitting W.

Here we say that C *splits* W if C does not intersect W and $\emptyset \neq W \cap \text{int} C \neq W$, where $\text{int} C$ denotes the (open) area of $\mathcal{R}^2$ enclosed by C.

We apply induction on $\kappa(X)$. If $\kappa(X) \leq 2$, the problem can be reduced to one satisfying (1). Indeed, if $\kappa(X) = 2$ we can find in linear time a collection F of faces of G such that the subspace $X \cup \bigcup_{f \in F} f$ of $\mathcal{R}^2$ has two connected components and such that $|F| \leq |X|$. Choose two faces $f, f' \in F$ and a vertex $v \in X$ incident with both f and f'. "Open" the graph at v, by splitting v into two new vertices, joining f and f' to form one new face. After this is repeated $|F| - 3$ times, the faces in F are replaced by two faces f_1 and f_2 and the vertices in X are split (or not) to a set X' of $|X| + |F| - 2$ vertices, such that each vertex in X' is incident with f_1 or f_2. By the result of Suzuki, Akama, and Nishizeki [7] we can solve the realization problem for the new graph and X' in linear time. This directly gives a solution for the realization problem for the original realization problem. We proceed similarly if $\kappa(X) = 1$.

If $\kappa(X) > 2$ we proceed as follows. Let $X_1, \ldots, X_t$ be the components of the subgraph of H induced by X. (So $t = \kappa(X) \leq k$.) We may assume that $\delta(X_i) = X_i$ for each $i = 1, \ldots, t$ (by attaching to each vertex in X_i a new vertex of valency 1). Let p be a nonnegative integer. A *p-neighbourhood* is a collection $W_1, \ldots, W_t$ of pairwise disjoint linked subsets of V with the following properties:

(4)   (i) for $i = 1, \ldots, t$, $W_i \supseteq X_i$, and if $W_i \neq X_i$ then $|\delta(W_i)| = p$

        (ii) for all distinct $i, j \in \{1, \ldots, t\}$, there are p vertex-disjoint paths in G between W_i and W_j.

We note:

PROPOSITION 2. *Let $W_1, \ldots, W_t$ be a p-neighbourhood. Let $i, j \in \{1, \ldots, t\}$ be distinct, and let T be a set of vertices intersecting each path from W_i to W_j such that $|T| = p$. Then T is linked.*

Proof. Suppose not. Let C be a circuit in G splitting T. Let U_i and U_j be the sets of vertices that can be reached from W_i and W_j, respectively, without intersecting T. So $U_i \cap U_j = \emptyset$. Then $U_i \cap C = \emptyset$ or $U_j \cap C = \emptyset$, since otherwise all vertices in C belong both to U_i and U_j. We may assume that $U_j \cap C = \emptyset$. Hence we may assume moreover that U_i is contained in int$C$ (as U_i is linked). Then each path from W_i to W_j intersects $T \cap \text{int}C$, contradicting the facts that there exist p disjoint such paths and that $|T \cap \text{int}C| < |T| = p$. ∎

In particular, $\delta(W_i)$ is linked for all i. (If $W_i = X_i$ then $\delta(W_i) = \delta(X_i) = \overset{\bullet}{X}_i$.)

Call a p-neighbourhood $W_1, \ldots, W_t$ *maximal* if for each $i = 1, \ldots, t$ and for each linked U satisfying $W_i \subset U \subseteq V \setminus \bigcup_{j \neq i} W_j$ one has $|\delta(U)| > p$.

First we describe an algorithm which, given a p-neighbourhood $W_1, \ldots, W_t$, finds a maximal p-neighbourhood:

1. Choose $i \in \{1, \ldots, t\}$. Determine an inclusionwise maximal set U satisfying $W_i \subseteq U \subseteq V \setminus \bigcup_{j \neq i} W_j$ and $|\delta(U)| = p$. Replace W_i by U. If no such U exists, we leave W_i unchanged.

2. Repeat for all $i \in \{1, \ldots, t\}$ in turn. This gives a maximal p-neighbourhood.

Note that by Proposition 2, $\delta(U)$ in Step 1 is linked, and hence U is linked. Note moreover that Step 1 can be performed in time $O(p|V|)$ with the Ford-Fulkerson augmenting path method (one augmenting path can be found in time $O(|V|)$). See also [4].

Second we give an algorithm which, given a maximal p-neighbourhood, finds either a $p+1$-neighbourhood or a reduction for the realization problem:

1. If there exist $i \neq j$ and a vertex v such that both $W_i \cup \{v\}$ and $W_j \cup \{v\}$ are linked, apply Proposition 1 to $V_1 := W_i \cup \{v\}, V_2 := W_j \cup \{v\}, V_3 := V \setminus (W_i^o \cup W_j^o)$ and $Y := X \cup \delta(W_i) \cup \delta(W_j) \cup \{v\}$.

Otherwise, for each $i = 1, \ldots, t$ with $|\delta(W_i)| = p$, choose a vertex $v_i \in V \setminus W_i$ such that $W_i \cup \{v_i\}$ is linked, and let $U_i := W_i \cup \{v_i\}$; for all other i let $U_i := W_i$.

2. If there exist $i \neq j$ such that there do not exist $p + 1$ disjoint paths connecting U_i and U_j, find a subset U of V such that $U_i \subseteq U, U_j \subseteq U' := V \setminus U^o$ and $|\delta(U)| = p$. Apply Proposition 1 to $V_1 := W_1, \ldots, V_t := W_t, V_{t+1} := (U \setminus (W_1^o \cup \cdots \cup W_t^o)) \cup \delta(U), V_{t+2} := (U' \setminus (W_1^o \cup \cdots \cup W_t^o)) \cup \delta(U)$ and $Y := X \cup \delta(W_1) \cup \cdots \cup \delta(W_t) \cup \delta(U)$.

3. Otherwise, $U_1, \ldots, U_t$ form a $p+1$-neighbourhood.

PROPOSITION 3. *In Step 1, if there exist i and j as stated, then $\kappa(V_h \cap Y) < t$ for $h = 1, 2, 3$.*

Proof. Without loss of generality, $i = 1$ and $j = 2$. We have $\kappa(V_1 \cap Y) = \kappa(X_1 \cup \delta(W_1) \cup \{v\}) \leq 2 < t$, since both X_1 and $\delta(W_1) \cup \{v\}$ are linked. Similarly, $\kappa(V_2 \cap Y) \leq 2 < t$.

Finally, $\kappa(V_3 \cap Y) < t$, since $V_3 \cap Y = X_3 \cup \cdots \cup X_t \cup \delta(W_1) \cup \delta(W_2) \cup \{v\}$, where $X_3, \ldots, X_t$ and $\delta(W_1) \cup \delta(W_2) \cup \{v\}$ are linked (as $\delta(W_1) \cup \{v\}$ and $\delta(W_2) \cup \{v\}$ are linked). ∎

PROPOSITION 4. *Let $A, B \subseteq V$ such that $\delta(A)$ and $\delta(B)$ are linked, and such that $B \not\subseteq A^o$ and $A^o \cup B^o \neq VG$. Then $\delta(A) \cup (A \cap \delta(B))$ is linked.*

Proof. Suppose $\delta(A) \cup (A \cap \delta(B))$ is not linked. Let C be a circuit in G splitting $\delta(A) \cup (A \cap \delta(B))$. Since $\delta(A)$ is linked, we may assume that $\delta(A) \subset \text{int}C$. Since C splits $\delta(A) \cup (A \cap \delta(B))$, we know that there are vertices in $A \cap \delta(B)$ that are in the exterior of C.

Since G is connected, there exists a path in G from a vertex in A in the exterior of C to a vertex of C disjoint from $\delta(A)$, and hence C intersects A. Therefore, $VC \subseteq A$. Hence every vertex of G in the exterior of C belongs to A. As $\delta(B)$ is linked and as $\delta(B)$ does not intersect C (because $A \cap \delta(B)$ does not intersect C), we have that $\delta(B)$ is contained in the exterior of C. As $B \not\subseteq A^o$ this implies that each vertex in int$C$ is contained in B. So $A^o \cup B^o = VG$, contradicting the assumption. ∎

PROPOSITION 5. *In Step 2, if there exist i and j as stated, then $\kappa(V_h \cap Y) < t$ for $h = 1, \ldots, t + 2$.*

Proof. Without loss of generality, $i = 1$ and $j = 2$. By the maximality of W_1 we know that U intersects at least one of $W_2, W_3 \ldots, W_t$. So U intersects at least two of $W_1, \ldots, W_t$. Similarly, U' intersects at least two of $W_1, \ldots, W_t$.

For each $h = 1, \ldots, t$ we have $\kappa(V_h \cap Y) \leq 2 < t$, since $V_h \cap Y = X_h \cup \delta(W_h) \cup (W_h \cap \delta(U))$ and since $\delta(W_h) \cup (W_h \cap \delta(U))$ is linked by Proposition 4. (Note that $U \not\subseteq W_h^o$ since U intersects at least two of $W_1, \ldots, W_t$, and that $U^o \cup W_h^o \neq VG$ since U' intersects at least two of $W_1, \ldots, W_t$.)

Next we show $\kappa(V_{t+1} \cap Y) < t$. Note that $V_{t+1} \cap Y = \delta(U) \cup (U \cap (\delta(W_1) \cup \cdots \cup \delta(W_t)))$. Since U' intersects at least two of $W_1, \ldots, W_t$, it suffices to show that if U' intersects W_h then $\delta(U) \cup (U \cap \delta(W_h))$ is linked.

Suppose U' intersects W_h and $\delta(U) \cup (U \cap \delta(W_h))$ is not linked. As $\delta(U)$ and $\delta(W_h)$ are linked (by Proposition 2), Proposition 4 implies that $W_h \subseteq U^o$ or $W_h^o \cup U^o = VG$. However, $W_h \subseteq U^o$ contradicts the fact that W_h intersects U'. Moreover, $W_h^o \cup U^o = VG$ contradicts the fact that there is another $W_{h'}$ intersecting U'.

This shows $\kappa(V_{t+1} \cap Y) < t$. Similarly, $\kappa(V_{t+2} \cap Y) < t$. ∎

Finally we give the algorithm which finds a reduction:

Starting with the 0-neighbourhood $X_1, \ldots, X_t$, for $p = 0, 1, \ldots, 2g(k) - 1$ apply the above algorithms to find a reduction or a $2g(k)$-neighbourhood.

If we find a $2g(k)$-neighbourhood $W_1, \ldots, W_t$, then for all distinct $i, j \in \{1, \ldots, t\}$, find a shortest path $P_{i,j}$ in H between W_i and W_j. Among all $P_{i,j}$ choose one, $P := P_{1,2}$ say, of minimum length.

If length$(P) > 2g(k)$, delete from G all vertices of P except the first $g(k)$ and the last $g(k)$. If length$(P) \leq 2g(k)$ leave G unchanged. Call the new graph G'.

Let R be the set of vertices in P that are not deleted. Apply Proposition 1 to G' and $V_1 := W_1, V_2 := W_2, V_3 := V \setminus (W_1^o \cup W_2^o)$ and $Y := X \cup \delta(W_1) \cup \delta(W_2) \cup R$.

PROPOSITION 6. *In G', $\kappa(V_h \cap Y) < t$ for $h = 1, 2, 3$.*

Proof. $\kappa(V_1 \cap Y) = \kappa(X_1 \cup \delta(W_1)) \leq 2 < t$. Similarly, $\kappa(V_2 \cap Y) < t$. Finally, $\kappa(V_3 \cap Y) = \kappa(X_3 \cup \cdots \cup X_t \cup \delta(W_1) \cup \delta(W_2) \cup R) < t$ since $\delta(W_1) \cup \delta(W_2) \cup R$ is linked. ∎

PROPOSITION 7. *Deleting the vertices does not affect realizability.*

Proof. Let Q be the set of vertices deleted. We must show that for any vertex $v \in Q$, any closed curve C traversing v and separating X has at least $g(k)$ intersections with $G - (Q \setminus \{v\})$ (since it means by the lemma that we can delete v, even after having deleted all other vertices in Q). In other words, any closed curve in $\mathcal{R}^2$ intersecting Q and separating X should have at least $g(k) - 1$ intersections with $G - Q$.

Let C be a closed curve intersecting Q and separating X, having a minimum number p of intersections with $G - Q$. We may assume that C intersects G only in vertices of G. Suppose $p \leq g(k) - 2$. It is not difficult to see that, by the minimality of p, there exist $x, y \in Q$ on C (possibly $x = y$) such that, if we denote by K and K' the two (closed) $x - y$ parts of C, then one of these parts, K say, intersects G only in Q, while K' intersects Q only in the end points x and y of K'. We may assume that K is part of P. Hence as P is a shortest path, length$(K) \leq$ length$(K') = p + 2$. So length$(C) =$ length$(K) +$ length$(K') - 2 \leq 2p + 2 \leq 2g(k) - 2$.

Hence C does not intersect any face incident with any point in any W_i, since otherwise C would contain a curve of length at most $g(k) - 1$ connecting Q and W_i, contradicting the minimality of P. As C separates X, there exist $i \neq j$ such that W_i and W_j are in different components of $\mathcal{R}^2 \setminus C$. This contradicts the facts that there exist $2g(k)$ pairwise disjoint paths from W_i to W_j and that length$(C) < 2g(k)$. ∎

Acknowledgement. We are grateful to a referee for giving several helpful suggestions improving the presentation of our results.

References

1. R.M. Karp, On the computational complexity of combinatorial problems, Networks 5 (1975) 45–68.
2. J.F. Lynch, The equivalence of theorem proving and the interconnection problem, (ACM) SIGDA Newsletter 5 (1975) 3:31–36.
3. B.A. Reed, N. Robertson, A. Schrijver, and P.D. Seymour, Finding disjoint trees in graphs on surfaces in linear time, preprint, 1992.
4. H. Ripphausen, D. Wagner, and K. Weihe, The vertex-disjoint Menger problem in planar graphs, preprint, 1992.
5. N. Robertson and P.D. Seymour, Graph Minors XIII. The disjoint paths problem, 1986, submitted.
6. N. Robertson and P.D. Seymour, Graph Minors XXII. Irrelevant vertices in linkage problems, preprint, 1992.
7. H. Suzuki, T. Akama, and T. Nishiseki, An algorithm for finding a forest in a planar graph — case in which a net may have terminals on the two specified face boundaries (in Japanese), Denshi Joho Tsushin Gakkai Ronbunshi 71-A (1988) 1897–1905 (English translation: Electron. Comm. Japan Part III Fund. Electron. Sci. 72 (1989) 10:68–79).

DEPARTMENT OF COMPUTER SCIENCE, UNIVERSITÉ DU QUÉBEC À MONTRÉAL, MONTRÉAL, QUÉBEC, CANADA
E-mail addresses: breed@watserv1.uwaterloo.ca, breed@mipsmath.uqam.ca

DEPARTMENT OF MATHEMATICS, OHIO STATE UNIVERSITY, COLUMBUS, OHIO 43210, U.S.A.
E-mail address: robertso@function.mps.ohio-state.edu

CWI, KRUISLAAN 413, 1098 SJ AMSTERDAM, THE NETHERLANDS,
AND
DEPARTMENT OF MATHEMATICS, UNIVERSITY OF AMSTERDAM, PLANTAGE MUIDERGRACHT 24, 1018 TV AMSTERDAM, THE NETHERLANDS
E-mail address: lex@cwi.nl

BELLCORE, 445 SOUTH ST., MORRISTOWN, NEW JERSEY 07962, U.S.A.
E-mail address: pds@breeze.bellcore.com

Surface Triangulations Without Short Noncontractible Cycles

TERESA M. PRZYTYCKA and JÓZEF H. PRZYTYCKI

ABSTRACT. We discuss three methods of constructing surface triangulations that do not have short noncontractible cycles (equivalently, that have high representativity). The three methods are: the covering spaces technique, a combinatorial method, and a method that applies hyperbolic geometry. Using the first method we show that for any genus g and $n \geq c_1 g \log \log g$ there exists a triangulation of a genus g surface with an n vertex graph such that the representativity is at least $c_1' \sqrt{n/g} \sqrt{\log \log g}$ (where c_1, c_1' are constants). Using the second method we show that for any genus g and $n > c_2 g \log g$ there exists a triangulation of a genus g surface with an n vertex graph such that the representativity is at least $c_2' \sqrt{n/g} \sqrt{\log g}$ (where c_2, c_2' are constants). Finally, the third method allows us to develop an argument which leads to the conjecture that, for any g and n sufficiently large, a surface of genus g can be triangulated with representativity at least $c_3 \sqrt{n/g} \log g$ (where c_3 is a constant).

1 Introduction

Properties of graph embeddings have been recently investigated in a number of papers [1,20,19]. Robertson and Seymour [18] introduced the following concept of representativity of a graph embedding. Let G be a graph embedded in a surface Σ. Denote by $\pi(\Sigma)$ the family of all noncontractible closed paths on Σ.

1991 Mathematics Subject Classification. Primary 05C15

This paper is in final form and no version of it will be submitted for publication elsewhere.

The *representativity* (or the *face width*), $\rho(\Sigma, G)$, of an embedded graph G is equal to the minimum over $\pi(\Sigma)$ of the number of intersections of a given path with the graph G. Such a minimal closed path can be chosen in such a way that it cuts G only in vertices, and thus representativity can be equivalently defined as the length of a shortest noncontractible *facial walk* (that is, a walk of type $v_1, f_1, v_2, f_2, \ldots, v_k, f_k, v_1$ where, for any i, v_i is a vertex and f_i is a face of the graph). Therefore the representativity of a surface triangulation is equal to the length of the shortest noncontractible cycle (or equivalently to the *edge width* of the triangulation). Recently much attention has been given to the investigation of how combinatorial and topological properties of an embedding depend on the representativity of the embedding. It has been informally stated [1] that representativity measures how well a given embedding approximates the surface. Robertson and Vitray [19] consider as a major effect of high representativity the fact that it makes the embedding "highly locally planar" and that "the locally Euclidean property of the surface is mirrored by the locally planar property of the embedded graph".

In this paper, we address the problem of triangulating a surface such that the representativity of the embedding is maximized. Let $f(\Sigma, n)$ be the maximum representativity that can be achieved by triangulating the surface Σ with an n vertex graph. Joan Hutchinson [13] showed that if Σ is an orientable surface without boundary then $f(\Sigma, n) = O(\sqrt{n/g} \log g)$ where g is the genus of the surface[1]. Hutchinson conjectured that $f(\Sigma, n) = O(\sqrt{n/g})$. This conjecture has been disproved by the authors [16] and replaced by the conjecture that $f(\Sigma, n) = \Theta(\sqrt{n/g} \log g)$ (*i.e.* a surface of genus g can be triangulated with representativity at least $c_3 \sqrt{n/g} \log g$ where c_3 is a constant). Thus, we conjecture that the upper bound given by Hutchinson is tight up to a constant.

Our graph theoretical terminology follows [2]. Thus a *cycle* in a graph does not have self-intersections while a *closed walk* can repeat both edges and vertices. Unless otherwise specified, we use *surface* to describe a compact, connected, orientable 2- manifold. Informally, this describes a sphere with g handles (g is the genus of the surface) and d boundary components. We use $\Sigma_{g,d}$ to denote a genus g surface with d boundary components and we adopt the notation that $\Sigma_g = \Sigma_{g,0}$.

To be consistent with standard topological terminology we assume that a (closed) path on a surface may have self-intersections while a *simple (closed)*

---

[1] In the paper we use $\log n$ to denote $\log_2 n$. We use O, Θ, and Ω notation in its standard meaning [12].

path is not allowed to intersect itself.

A graph is said to be embedded in an orientable surface $\Sigma_{g,d}$ if it can be drawn on the surface in such a way that no two edges cross and the boundary of the surface (if any) is a part of the graph. If the graph G is embedded in a surface Σ the complement of G relative to Σ is a collection of open sets called open faces. If all of the open faces are open discs, we say that the embedding is a *2-cell embedding*. In this paper we will consider only 2-cell embeddings.

An embedding is called a *triangulation* if every face is bounded by three edges. A cycle C on a surface Σ is called *noncontractible* if neither of the components of $\Sigma - C$ is homeomorphic to an open disc.

The paper is organized as follows. In Section 2, we discuss the covering space technique of constructing triangulations of high representativity. The section contains a proof of a result not published elsewhere, and thus it contains more technical details than the other sections. In Section 2, using the covering space technique, we show that, for any genus g and $n \geq c_1 g \log \log g$, $f(\Sigma_g, n) = \Omega(\sqrt{n/g}\sqrt{\log \log g})$ (where c_1 is a constant). A main step of the construction is to obtain triangulations of surfaces with boundary such that the representativity of the triangulation is $\Omega(\sqrt{n/g} \log \log g)$. Then, based on these triangulations, we construct high representativity triangulations of closed surfaces. The second step involves technicalities that are not important for understanding the covering space method, and we postpone the details of this step to the appendix. To make Section 2 accessible for a reader not familiar with algebraic topology we include various comments and definitions that may be skipped by a reader familiar with the topic.

In Section 3, we review the combinatorial technique (presented with more details in [17]) which allows us to triangulate a genus g surface with representativity $\Omega(\sqrt{n/g}\sqrt{\log g})$. However we need to assume that $n \geq c_2 g \log g$ where c_2 is a constant.

Finally, in Section 4, we relate the problem of constructing a high representativity triangulation with the problem of computing the length of the shortest closed geodesic in a hyperbolic structure. We introduce the notion of an approximation of a surface with a geometrical structure, with the help of a triangulation; and discuss a construction of triangulations that approximate a given hyperbolic structure. This section provides evidence for the conjecture that any genus g surface can be triangulated with representativity $\Theta(\sqrt{n/g}\log g)$ for n big enough. As in Section 2, in this section we also give definitions and informal descriptions that are needed to make the section accessible for a reader not familiar with hyperbolic geometry.

In this paper we concentrate on orientable surfaces. However most of the results presented can be extended to nonorientable surfaces.

We believe the theorem of M.Hall, which we use in the second section of the paper, will find more applications in covering graph theory. We thank Professor J.McCool for pointing out this theorem to us. We also thank Professors P.Buser and S.Wolpert for helpful discussions.

2 Constructing High Representativity Triangulations Using Covering Space Technique

The covering space technique was used in the construction disproving Hutchinson's conjecture [13,16]. Here we improve the main result of [16]. We increase the representativity of the constructed triangulations from $\sqrt{n/g}\log^* g$ to $\sqrt{n/g}\sqrt{\log\log g}$ using a theorem of M.Hall [9]. [2]

In the first subsection, we provide basic topological facts and definitions. In the second subsection, we show a construction of a covering space with the property that all simple closed paths that are in preimages of homotopically non-zero closed paths of a base surface are "long" (formal definitions are provided in Subsection 1). Finally, in the last subsection, we triangulate the surface constructed in the second subsection.

The constants achieved are not the best possible and have been chosen to keep technical computations simple.

2.1 Basic Definitions

In this section, we describe the concept of a covering. To minimize technicalities, we restrict this presentation to the case of surfaces. For a more general treatment of the topic see [14,11]. In this paper, we only use coverings involving surfaces and graphs.

Definition 2.1 *Let $\widetilde{X}$ and X be two surfaces. A continuous mapping $p : \widetilde{X} \mapsto X$ is said to be a covering map if each point $x \in X$ has an open neighborhood U_x such that $p^{-1}(U_x)$ is the disjoint sum of open subsets of $\widetilde{X}$ each of which is mapped homeomorphically onto U_x by p. The surface $\widetilde{X}$ is called the covering surface, and the surface X is called the base surface.*

---

[2] Let $\log^k n$ denote the function log composed k times. Then $\log^* n = k$ iff $\log^k n \leq 1$ and $\log^{k-1} n > 1$.

Definition 2.2 *The covering $p : \widetilde{X} \mapsto X$ is called a k-fold covering if $|p^{-1}(x)| = k$ for any $x \in X$.*

It is sometimes convenient to imagine that in a k-fold covering, for each U_x from Definition 2.1, there exist k copies of U_x in the covering space $\widetilde{X}$ each of them mapped onto U_x by p.

Having chosen a point $x \in X$ we can consider the set of all closed paths from x to x. The point x is called a *base point* for those paths.

Definition 2.3 *Two closed paths with a base point x are equivalent relative to the base point x if they are homotopic relative to the base point x (one can be transformed continuously to the other in such a way that endpoints are not moved).*

In Definition 2.3, we introduced an equivalence relation on closed paths with a base point. In the definition below we do not fix a base point.

Definition 2.4 *Two closed paths are equivalent if they are homotopic (one can be transformed continuously to the other).*

Theorem 2.5 *The set of equivalence classes of closed paths based at $x \in X$ forms a group.*

The group formed by the set of equivalence classes of closed paths based at $x \in X$ is denoted by $\pi_1(X, x)$ and called the *fundamental group* or the *first homotopy group* of X with the base point x.

We use a special kind of covering called a regular covering. This covering has a number of properties which will be useful in the construction. However to define regular covering we need a few more facts:

Theorem 2.6 *If $p : \widetilde{X} \mapsto X$ is a covering with $\tilde{x}_0 \in \widetilde{X}$, $x_0 \in X$ such that $p(\tilde{x}_0) = x_0$ then the induced homomorphism $p_* : \pi_1(\widetilde{X}, \tilde{x}_0) \mapsto \pi_1(X, x_0)$ is a monomorphism.*

Note that $p_*(\pi_1(\widetilde{X}, \tilde{x}_0))$ is a subgroup of $\pi_1(X, x_0)$. We define regular covering as follows:

Definition 2.7 *A covering $p : \widetilde{X} \mapsto X$ is said to be regular if the group $p_*(\pi_1(\widetilde{X}, \tilde{x}_0))$ is a normal subgroup of $\pi_1(X, x_0)$ (that is, for each $g \in p_*(\pi_1(\widetilde{X}, \tilde{x}_0))$ and $h \in \pi_1(X, x_0), hgh^{-1} \in p_*(\pi_1(\widetilde{X}, \tilde{x}_0))$).*

A covering (resp., a regular covering) $p : \widetilde{X} \mapsto X$, defines a subgroup (resp., a normal subgroup) of $\pi_1(X, x_0)$. Conversely, a subgroup (resp., a normal subgroup) of $\pi_1(X, x_0)$ defines a covering (resp., a regular covering). The multiplicity (the folding number) of the covering defined in this way is equal to the index of the subgroup (see [14,11]).

The first *homology group* (denoted by $H_1(X)$) is obtained from the first homotopy group $\pi_1(X, x_0)$ by abelianization [3]. (Thus in the definition of the first homology group there is no need for a base point.) We use h_a to denote the abelianization homomorphism (here $h_a : \pi_1(X, x_0) \mapsto H_1(X)$).

The following standard technique for construction of a regular covering will be used later in the paper. Let G be a group and h a homomorphism $h : \pi_1(X, x_0) \mapsto G$ that is "onto". Then the kernel, $ker(h) \subset \pi_1(X, x_0)$, defines a regular cover of X. Since the index of $ker(h)$ in $\pi_1(X, x_0)$ is equal to the order, $r(G)$, of the group G, the multiplicity of the covering is equal to $r(G)$.

We divide all simple closed paths on a surface into two classes: a simple closed paths γ on surface X is called a *separating path* if $X - \gamma$ is disconnected, and otherwise it is called a *nonseparating path*. More generally we divide all closed paths into *homologically non-trivial* paths (this generalizes non-separating closed simple paths) and *homologically trivial* closed paths (this generalizes separating closed simple paths).

Let γ be a path on the surface X which begins at x and ends at y. Let $\tilde{x} \in p^{-1}(x)$. Then γ uniquely defines the path from $\tilde{x}$ to some $\tilde{y}$ where $\tilde{y} \in p^{-1}(y)$. Assume that γ is a closed path with a base point x. Let γ^k denote the closed path composed of the sequence of k paths γ. Consider the smallest number r such that the path in the covering space which starts at $\tilde{x}$ and corresponds to γ^r is closed. We call the number r the *developing number* of γ with respect to the given covering of γ with base points $x, \tilde{x}$. The closed path which corresponds to γ^r in the covering space is called a *generalized lift* of γ in the given covering and with the given base point. [4] Developing number has the following properties:

P1: In a regular covering the developing number does not depend on the choice of base points $x \in X$ and $\tilde{x} \in p^{-1}(x)$.

P2: In a regular covering, generalized lifts of equivalent closed paths have equal developing numbers.

Theorem 2.8 *In a k-fold covering $p : \widetilde{\Sigma} \mapsto \Sigma_{g,d}$ if $\chi(\Sigma_{g,d})$ is Euler characteristic of $\Sigma_{g,d}$ then the covering surface $\widetilde{\Sigma}$ has the Euler characteristic $\chi(\widetilde{\Sigma}) = k\chi(\Sigma_{g,d}) = k(2 - 2g - d)$.*

We say that the word. g, over the alphabet $\{a_1^1, a_1^{-1}, a_2^1, a_2^{-1}, \ldots, a_k^1, a_k^{-1}\}$ is *reduced* (resp., *cyclically reduced*) if no two consecutive (resp., cyclically consecutive) letters in g are inverses of each other. Let G be a free group with

---

[3] Abelianization of a group is obtained by adding to the group presentation, for any elements a, b of the group, the relation $ab = ba$.

[4] The concept of generalized lift is in a close relation to the standard topological notion of a *lift*.

a free generating set $\{a_1, a_2, \ldots, a_k\}$ and let $g \in G$. The *length*, $|g|$, of g is the number of symbols in g where g is represented by a reduced word.

2.2 Construction of the Covering Space

Our goal is to show how to construct a triangulation without short noncontractible cycles. In our construction we start with $\Sigma_{1,1}$ which is the simplest surface with boundary and non-zero genus. Then we construct any surface Σ_g ($g > 1$) by constructing covering surfaces of $\Sigma_{1,1}$ with the property that all generalized lifts of any "short" nontrivial closed path on $\Sigma_{1,1}$ are "long", gluing them together along boundary components, and capping off the remaining boundary components with discs.

In this section, we show how to construct a regular covering of $\Sigma_{1,1}$ (the punctured torus) in which all generalized lifts of any "short" closed path are "long". We start with introducing a combinatorial measure of the length of a closed path, called *complexity*.

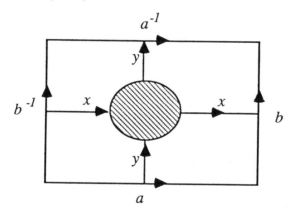

Figure 2.1 The presentation of $\Sigma_{1,1}$

Consider the presentation of $\Sigma_{1,1}$ as a square, with opposite edges identified (Figure 2.1). Let x and y be the oriented intervals dual to the fundamental cycles a and b as shown in Figure 2.1.

Let $Z = \{x, y, x^{-1}, y^{-1}\}$, $B = \{x, y\}$ and let $z \in B$. With any closed oriented path γ we associate a word over Z in the following way. Start from any point on the path and travel along the path according to the orientation. The word associated with the path corresponds to the order in which the given path cuts x and y (without loss of generality it is enough to consider only paths in general position[5]). If a path cuts z as in Figure 2.2a then the corresponding symbol in the word is z, and otherwise (Figure 2.2 b) this symbol is z^{-1}. In particular, the closed path corresponding to the word x intersects the interval x in one point.

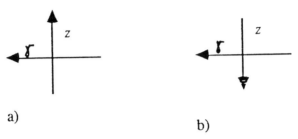

a)   b)

Figure 2.2

Thus the homotopy class of an oriented closed curve can be described by an element of the free group $\pi_1(\Sigma_{1,1})$ generated by B.

Definition 2.9 *The complexity of a closed path γ on $\Sigma_{1,1}$ is defined to be the length of a cyclically reduced word over Z that describes the homotopy class of γ.*

Informally, if we measure the length of a path by the number of crossings of the given path with x and y then the complexity of a closed path γ is the length of the shortest path over all paths that are equivalent to γ and are in general position with respect to intervals x and y.

Definition 2.10 *Let X be a covering space of $\Sigma_{1,1}$. Then the complexity of a closed path on X is equal to the complexity of its projection on the base surface.*

The following properties of a regular covering are crucial for our construction:

P3: The complexity of a generalized lift of γ with developing number r is equal to r times the complexity of γ. In particular the complexity of a generalized lift of γ is always greater than or equal to the complexity of γ. This follows from the simple fact that if w is a cyclically reduced word then w^r is also cyclically reduced and $|w^r| = r|w|$.

P4: Two paths whose projections on $\Sigma_{1,1}$ are homotopic have equal complexities. In particular, the complexities of two homotopic paths are equal.

For any constant l, we are going to construct a covering space of $\Sigma_{1,1}$ such that for any closed path C on $\Sigma_{1,1}$ the complexity of any generalized lift of C

---

[5] We say that a path is in general position with respect to a set of paths if it does not cut a common point of these paths and whenever it cuts a path from the given set of paths it is perpendicular to the path it crosses.

is at least l. By property P4, it suffices to work only with representatives of homotopy classes. The following lemma gives an upper bound on the number of different homotopy classes of complexity less than l.

Lemma 2.11 *There are at most $2 \cdot 3^{l-1}$ nonhomotopic paths on $\Sigma_{1,1}$ of complexity less than l.*

Proof: The number of different homotopy classes of closed paths of complexity s is equal to the number of cyclically reduced words of the group $\pi_1(\Sigma_{1,1})$ of length s. Thus there are at most $4 \cdot 3^{s-1}$ different homotopy classes of closed paths of complexity s. Thus there are at most

$$1 + 4\sum_{s=0}^{l-2} 3^s = 1 + 4\frac{3^{l-1} - 1}{2} = 2(3^{l-1}) - 1 < 2 \cdot 3^{l-1}$$

different homotopy classes of complexity less than l. □

2.2.1 The Basic Idea of the Construction

Let l be a constant. We construct a covering in which all generalized lifts of any homotopically nontrivial closed path on $\Sigma_{1,1}$ have complexity at least l. To do this, for any homotopically nontrivial closed path, we construct a regular covering of $\Sigma_{1,1}$ such that all generalized lifts of a given path (thus also of all closed paths in the same homotopy class) have complexity at least l. We use a different technique for homologically trivial and homologically nontrivial closed paths. Finally, we construct a regular covering that is a common covering for all coverings constructed before. In this regular covering of $\Sigma_{1,1}$ all generalized lifts of all homotopically nontrivial closed paths $\Sigma_{1,1}$ have complexity at least l.

2.2.2 Lifting a Homologically Nontrivial Closed Path

In this subsection we construct a regular covering of $\Sigma_{1,1}$ with the property that every generalized lift of a fixed closed, non-zero-homologous path has complexity at least l, where l is a constant.

We start with the following lemma.

Lemma 2.12 *Let Σ be a surface with boundary and let $\{b_1, \ldots, b_k\}$ be a base of $H_1(\Sigma)$. Let γ be a fixed closed non-zero-homologous path on Σ such that $h_a(\gamma) = ib_1$, where h_a is the abelianization homomorphism and i is a non-zero integer. Then for any constant s there exists a regular s-fold covering $p : \widetilde{\Sigma} \mapsto \Sigma$ such that the developing number of γ is equal to $\frac{s}{gcd(s,i)}$.*

Proof: We use the construction described in Subsection 2.1. Define the homomorphism
$$h_\gamma : H_1(\Sigma) \mapsto Z_s$$
by $h_\gamma(b_1) = 1$ and $h_\gamma(b_j) = 0$ for $j > 1$.

The composition $h_\gamma h_a$ is "onto" so the kernel $ker(h_\gamma h_a)$ defines an s-fold regular covering of Σ. Let $\tilde\gamma$ be a generalized lift of γ. Since $r = \frac{s}{\gcd(s,i)}$ is the smallest number such that $h_\gamma h_a(\gamma^r) = 0$, the developing number of γ equals to r. □

We will use the following two corollaries to this lemma.

Corollary 2.13 *For a non-zero-homologous closed path γ on $\Sigma_{1,1}$ there exists a regular s-fold covering Σ_γ such that any generalized lift of γ has complexity at least s.*

Proof: Choose $B = \{x, y\}$ to be the basis of $H_1(\Sigma_{1,1})$ (recall the definition of B from the beginning of this section). Since γ is non-zero-homologous we can write $h_a(\gamma) = i_1 x + i_2 y$ where $i_1 \neq 0$ or $i_2 \neq 0$. Assume that $i_1 \neq 0$. Then
$$complexity(\gamma) \geq |i_1|.$$

Let $\tilde\gamma$ be a generalized lift guaranteed by Lemma 2.12. Thus the developing number of $\tilde\gamma$ is $\frac{s}{\gcd(s,i_1)}$. Therefore
$$complexity(\tilde\gamma) \geq \frac{s}{\gcd(s,i_1)}|i_1| \geq s.$$

□

Corollary 2.14 *Let Σ be a surface with boundary and s be a constant. For any closed path γ on Σ such that $h_a(\gamma)$ is a member of a basis of $H_1(\Sigma)$, there is a regular s-fold covering $p : \widetilde\Sigma \mapsto \Sigma$ such that the developing number of $\tilde\gamma$ is s.*

Proof: Immediately from Lemma 2.12 by assigning $i = 1$. □

The covering announced at the beginning of this subsection is guaranteed by Corollary 2.13 applied with $s = l$.

2.2.3 Lifting a Homologically Trivial Closed Path

Let γ be a homotopically nontrivial closed path that is homologous to zero and let l be a constant. In this subsection we construct a regular covering of $\Sigma_{1,1}$ such that any generalized lift of γ has complexity at least l. The construction proceeds in three steps. First, using a corollary of a theorem of Hall, we construct a (not necessarily regular) covering of $\Sigma_{1,1}$ in which at least one generalized lift of γ is homologically nontrivial. Then, to this particular lift of γ, we apply the construction from the previous subsection. We obtain a (not necessarily regular) covering of $\Sigma_{1,1}$ in which at least one generalized lift of γ has complexity at least l. Finally, we construct a regular covering of $\Sigma_{1,1}$ with the required property.

The following theorem, which is a special case of a theorem of M.Hall [9,3, 10,21], will be used in the proof of the main theorem of this subsection. Since we need to refer to some facts from the proof of the theorem, for the completeness of the presentation, we give a proof of this special case.

Theorem 2.15 (Hall) *Let G be a finitely generated free group and w a nontrivial element of G (i.e. $w \neq 1$). Then there exists a subgroup H of G of finite index such that w is an element of a free generating set of H.*

Proof: Let $A = \{a_1, a_2, \ldots, a_k\}$ be a free generating set of G. Let Γ be the wedge of oriented circles with vertex x_0 and edges labeled with elements from A (see Figure 2.3). Thus $\pi_1(\Gamma, x_0) = G$.

We can assume, without loss of generality, that w is represented by a cyclically reduced word. Otherwise we can write $w = gw'g^{-1}$ where w' is cyclically reduced and $g \in G$. If H' is a subgroup of G that has the properties of the theorem with respect to w' then $H = gH'g^{-1}$ also has the properties of the theorem with respect to w and the index of H equals the index of H'.

Let W be the subgroup of G generated by w, let $p : \widetilde{\Gamma} \mapsto \Gamma$ be the covering defined by W and let $\tilde{x}_0$ be a fixed point in $p^{-1}(x_0)$. Thus $p_*(\pi_1(\widetilde{\Gamma}, \tilde{x}_0)) = W$ and therefore the fundamental group of $\widetilde{\Gamma}$ is isomorphic to W.

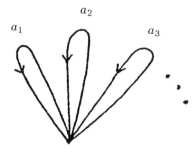

Figure 2.3 The wedge of oriented circles

Let γ be the closed path in Γ representing w and let $\tilde{\gamma}$ be the generalized lift of γ starting at $\tilde{x}_0$. Note that $\tilde{\gamma}$ generates $\pi_1(\widetilde{\Gamma}, \tilde{x}_0)$. Since every embedding of a graph into a graph induces a monomorphism on fundamental groups, it follows that the embedding $i : \tilde{\gamma} \hookrightarrow \tilde{\Gamma}$ induces the isomorphism $i_* : \pi_1(\tilde{\gamma}, \tilde{x}_0) \mapsto \pi_1(\tilde{\Gamma}, \tilde{x}_0)$. Therefore $\tilde{\gamma}$ is a cycle with $|w|$ edges (for otherwise W would not be generated by one cyclically reduced element).

Let $p/_{\tilde{\gamma}}$ be the restriction of p to $\tilde{\gamma}$. We extend the mapping, $p/_{\tilde{\gamma}}$, to a covering $\bar{p} : \bar{\Gamma} \mapsto \Gamma$ where $\bar{\Gamma}$ is obtained from $\tilde{\gamma}$ by adding edges (but not vertices) in the following way. The labels and orientation of edges in γ induces, via p, labels and orientation of edges in $\tilde{\gamma}$. Then we add to $\tilde{\gamma}$ labeled directed edges in such a way that each vertex has exactly k incoming and exactly k outgoing edges each labeled with a different element from A. A simple counting argument shows that this is always possible. The projection $\bar{p}$ is yielded by the labeling of edges in $\bar{\Gamma}$. By construction, $\bar{p} : \bar{\Gamma} \mapsto \Gamma$ is a covering. The multiplicity of this covering is equal to the number of vertices in $\bar{\Gamma}$, that is the number of edges in $\tilde{\gamma}$ (which, as noted before, is equal to $|w|$). Let $\tilde{w}$ be the element of $\pi_1(\bar{\Gamma}, \tilde{x}_0)$ representing $\tilde{\gamma}$. Since, for any graph G and a cycle C of G with $x_0 \in C$ there exists a free generating set, X, of $\pi_1(G, x_0)$ containing C, it follows that $\tilde{w}$ is an element of a free generating set of $\pi_1(\bar{\Gamma}, \tilde{x}_0)$.

Let $H = \bar{p}_*(\pi_1(\bar{\Gamma}, \tilde{x}_0))$. Thus H is a subgroup of G. By definition $p_*(\tilde{w}) = w$. Thus w is an element of a free generating set of H (compare Figure 2.4). Furthermore we have the following property:

P5: The index of H in G is equal to $|w|$.
□

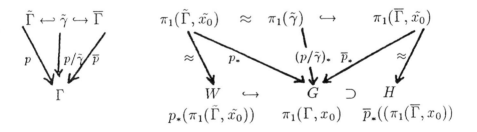

Figure 2.4

Now we are in the position to prove the main theorem of this subsection.

Theorem 2.16 *Let γ be a closed homotopically nontrivial and homologically trivial path in $\Sigma_{1,1}$ of complexity s. Then there exists an s-fold covering $\widetilde{\Sigma}$ of $\Sigma_{1,1}$ such that a generalized lift of γ is an element of a basis of $H_1(\widetilde{\Sigma})$.*

Proof: Let $G = \pi_1(\Sigma_{1,1}, x_0)$ and let w be a cyclically reduced word representing γ in the free group G. From Theorem 2.15 it follows that there exists a subgroup H of G such that $w \in H$ and w belongs to a free generating set of H. Furthermore, by property P5 given in the proof of Theorem 2.15, the index of H in G is $|w|$. It is well known that the abelianization homomorphism h_a sends elements of a free generating set of H to a base of $h_a(H)$. Thus H defines an s-fold covering, $\widetilde{\Sigma}$, of $\Sigma_{1,1}$ such that the generalized lift of γ starting from $\tilde{x}_0$ (where $\tilde{x}_0$ is a fixed point from $p^{-1}(x)$) represents an element of a basis of $H_1(\widetilde{\Sigma})$ and therefore is homologically nonzero. □

Now we are ready to show the construction of the regular covering for a closed path γ that is zero-homologous.

Theorem 2.17 *For any homotopically nontrivial zero-homologous closed path γ on $\Sigma_{1,1}$ which has complexity less than s there exists a regular covering $p_\gamma : \widetilde{\Sigma} \mapsto \Sigma_{1,1}$ of multiplicity at most s^{2s^2} such that any generalized lift of γ has complexity at least s.*

Proof: Let $p : \bar{\Sigma} \mapsto \Sigma_{1,1}$ be the s-fold covering guaranteed by Theorem 2.16 and let $p' : \widetilde{\Sigma} \mapsto \bar{\Sigma}$ be the regular s-fold covering guaranteed by Corollary 2.14. The composition $pp' : \widetilde{\Sigma} \mapsto \Sigma$ defines an s^2-fold covering of $\Sigma_{1,1}$. Using standard tools from covering spaces theory [14,11] (see also Theorem 4.A in [16]), we construct a covering p_γ that is a regular covering of both $\Sigma_{1,1}$ and $\widetilde{\Sigma}$. The multiplicity of this covering is at most s^{2s^2}. Since p_γ is a regular covering of $\Sigma_{1,1}$, by property P4 of regular coverings all generalized lifts of all paths on $\Sigma_{1,1}$ that are homotopic to γ have equal complexity. Since p_γ is a regular covering of $\widetilde{\Sigma}$, by property P3 the complexity of a generalized lift of γ is at least s. □

2.2.4 Construction of a Common Covering

In this section we prove the theorem that guarantees the existence of the covering announced in Section 2.2.1.

Theorem 2.18 *Let $l > 1$ be a constant and $\gamma_1, \ldots, \gamma_k$ be a family of closed, homotopically nontrivial and pairwise nonhomotopic paths on $\Sigma_{1,1}$ of complexity less than l. Then there exists an m-fold regular covering $p : \widetilde{\Sigma} \mapsto \Sigma_{1,1}$, where*

$$m \leq l^{4l^2 3^{l-1} - 4l^2} < \frac{2^{8^l}}{l^{4l^2}}$$

(and for $l \geq 8$, $m < \frac{2^{8^{l-1}}}{l^{4l^2}}$) such that each generalized lift of γ_i ($i = 1, ..., k$) has complexity at least l.

Proof: By Lemma 2.11, $k < 2 \cdot 3^{l-1} - 1$ (we exclude the homotopically trivial closed path). For each γ_i, $i \leq k$, construct a regular covering $p_i : \Sigma_i \mapsto \Sigma_{1,1}$ such that each generalized lift of γ_i has complexity at least l. If γ_i is nonhomologous to zero use Corollary 2.13, and otherwise use the construction from Theorem 2.17. By Corollary 2.13 and Theorem 2.17, p_i is at most an l^{2l^2}-fold covering. Using standard tools from covering spaces theory [14,11] (see also Theorem 4.A in [16]), we construct a regular m-fold covering $p : \widetilde{\Sigma} \mapsto \Sigma_{1,1}$ that is also a covering of each of Σ_i where $m \leq (l^{2l^2})^{2 \cdot 3^{l-1} - 2}$. Thus $m \leq l^{4l^2 3^{l-1} - 4l^2} < \frac{2^{8^l}}{l^{4l^2}}$. Furthermore, if $l \geq 8$ then $l^{4l^2 3^{l-1} - 4l^2} < \frac{2^{8^{l-1}}}{l^{4l^2}}$. □

Corollary 2.19 For any $l > 4$ the m-fold covering constructed in Theorem 2.18 has the following parameters:

1. If d is the number of the boundary components of the covering space then $l \leq d \leq m/l$.
2. Let r be the developing number of the boundary, $\delta\Sigma_{1,1}$, of $\Sigma_{1,1}$. Then $\frac{m}{l} \geq r = \frac{m}{d} \geq l$.
3. If g is the genus of the covering space then $\frac{m-l+2}{2} \geq g > \frac{1}{2}m(1 - \frac{1}{l})$.

Proof: Let $B = \{x, y\}$ be the base of $\pi_1(\Sigma_{1,1})$ chosen at the beginning of Section 2.2. Consider the l-fold covering, $p_x : \Sigma_x \mapsto \Sigma_{1,1}$, constructed as in Corollary 2.13 for the closed path corresponding to x. Let $\delta\Sigma_{1,1}$ be the boundary of $\Sigma_{1,1}$. Because $\delta\Sigma_{1,1}$ is a homologically trivial, its developing number in this covering is equal to one. Thus, the covering space Σ_x has l boundary components. Since our m-fold covering of $\Sigma_{1,1}$ is also a covering of Σ_x it follows that $l \leq d$. On the other hand, $\delta\Sigma_{1,1}$ is a homologically trivial closed path with complexity 4. Therefore, by Theorem 2.18, we apply to this path the construction from Theorem 2.15 prior to the construction of the common covering. This implies that the developing number of the boundary component is at least l. Since the complexity of $\delta\Sigma_{1,1}$ is four, the complexity of a generalized lift of $\delta\Sigma_{1,1}$ is at least $4l$. Furthermore, because $\Sigma_{1,1}$ has only one boundary component and the covering is regular, $m = rd$. Thus $d \leq \frac{m}{l}$. These prove properties 1 and 2. Since the Euler characteristic of $\Sigma_{1,1}$ is -1, by Theorem 2.8 the Euler characteristic of the covering space is $-m$ and therefore $m = d + 2g - 2$. This implies property 3. □

2.3 Construction of Triangulations

Let T be a triangulation of $\Sigma_{1,1}$ in a general position to x, y such that any edge of T cuts $x \cup y$ at most once. We call such a triangulation a *base triangulation*. Let $n(T)$ denote the number of vertices in T. Let $\widetilde{\Sigma}$ be a covering space of $\Sigma_{1,1}$ and p the covering mapping. The triangulation T defines a triangulation $\widetilde{T}$ of the covering space such that vertices of $\widetilde{T}$ are preimages of vertices of T and edges of $\widetilde{T}$ are defined by preimages of edges of T. The triangulation $\widetilde{T}$ is called a *covering triangulation* of the base triangulation T. Note that the length of a closed path in $\widetilde{T}$ is greater than or equal to its complexity. Therefore we can use the ideas from the previous section to construct, for any base triangulation T, a covering triangulation $\widetilde{T}$ in which all noncontractible paths have length greater than any given constant. This gives a high representativity triangulation of a surface with boundary.

Theorem 2.20 *For a given $l > 4$, let $\widetilde{\Sigma}$ be the covering space of $\Sigma_{1,1}$ constructed as in Theorem 2.18. Let $T_{1,1}$ be the triangulation of $\Sigma_{1,1}$ as presented in Figure 2.5 and let $\widetilde{T}$ be the triangulation of $\widetilde{\Sigma}$ that is the covering triangulation of $T_{1,1}$. Then the representativity of $\widetilde{T}$ is at least l and $l > \frac{1}{3\sqrt{10}}\sqrt{\frac{n(\widetilde{T})}{g}}\log\log g$ where g is the genus of $\widetilde{\Sigma}$. Furthermore $n(\widetilde{T}) < 10g$.*

Proof: Let m be the multiplicity of the covering $p : \widetilde{\Sigma} \mapsto \Sigma_{1,1}$ as given by Theorem 2.18. Then $n(\widetilde{T}) = 4m$ and, by Corollary 2.19, $g > \frac{1}{2}m(1 - \frac{1}{l})$. Therefore $\frac{n(\widetilde{T})}{g} < 8\frac{1}{1-\frac{1}{l}}$. Thus, for $l > 4$, $\frac{n(\widetilde{T})}{g} < 10$. Furthermore $\log\log g < 3l$. Therefore $l > \frac{1}{3\sqrt{10}}\sqrt{\frac{n(\widetilde{T})}{g}}\log\log g$ □

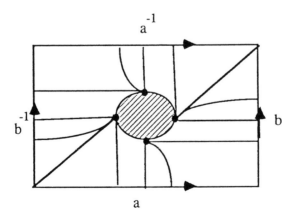

Figure 2.5 The triangulation $T_{1,1}$ of $\Sigma_{1,1}$

To get the result for a closed surface (that is, a surface without boundary) we need to cup off boundary components with discs or glue pairs of boundary components using an annular collar. Discs and/or annuli should be triangulated in such a way that the identification does not introduce noncontractible cycles of length less than l. The details of the construction are presented in the Appendix. The construction presented in the Appendix leads us to the following theorem:

Theorem 2.21 *For any $g > 1$ there exists an integer N_g such that for any $n \geq N_g$*

$$f(\Sigma_g, n) \geq \frac{1}{12}\sqrt{\frac{n}{g}}\sqrt{\log \log g}.$$

Furthermore if $g \geq 2^{8^{10}}$ then $N_g \leq \frac{5}{3} g \log \log g$ and otherwise $N_g \leq 7g$.

3 The Combinatorial Technique

In this section we sketch a combinatorial technique that allows us to construct triangulations of high representativity. For a full presentation we refer the reader to [17]. We use a result of Erdös and Sachs [7,8] and a method similar to that used by Buser [4]. To describe the result of Erdös and Sachs we need to introduce a few more definitions.

Let G be a graph with at least one cycle. The *girth* of G is equal to the length of the shortest cycle in G. The number of edges incident with a vertex is called the *valency* of the vertex (a loop is counted twice). A graph all of whose vertices have valency 3 is called a *cubic graph*.

The following theorem is a special case of theorem of Erdös and Sachs [7,8].

Theorem 3.1 ([7,8]) *For any $l \geq 2$ and for any $2^l - 1 > N \geq 2^{l-1} - 1$ there exists a cubic graph of $2N$ vertices with girth at least l.*

Below we sketch the idea of the construction implied by this theorem.

Theorem 3.2 ([17]) *For any $g \geq 2$ and any $n \geq 4.5 g \log g$*

$$f(\Sigma_g, n) \geq \tfrac{1}{6}\sqrt{\tfrac{n}{g}}\sqrt{\log g}.$$

Proof: (sketch) Let $l \geq 2$. Take the cubic graph, G_0, of $2N$ vertices and girth at least l that is guaranteed by Theorem 3.1. Since G_0 is cubic it has $3N$ edges. Thus its cyclomatic number is $N + 1$. Consider the boundary of the regular neighborhood of G_0 embedded in the 3-dimensional space R^3 (see Figure. 3.1).

This boundary forms an orientable surface Σ_g of genus $g = N + 1$.

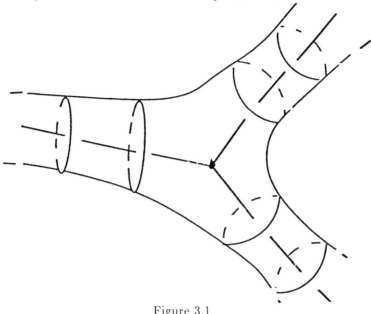

Figure 3.1

For each edge, e, of the graph we choose a simple closed path $C(e)$ corresponding to e as illustrated in Figure 3.2.

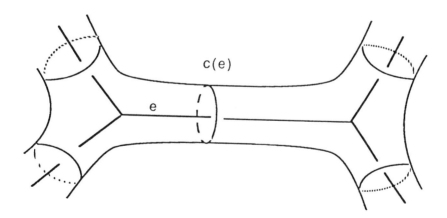

Figure 3.2

Observe that if, for any edge e from G_0, we cut the surface along cycle $C(e)$ then the surface breaks into $2N$ *pairs of pants* (where by a pair of pants we understand a sphere with three holes) such that each pair of pants corresponds to one vertex (see Figure 3.3).

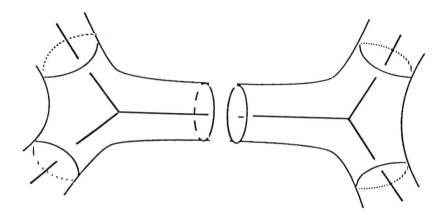

Figure 3.3. Breaking the surface into pairs of pants

Let $PP(v)$ denote the pair of pants corresponding to the vertex v. We can embed G_0 on Σ_g by embedding every vertex, v, on $PP(v)$. Furthermore if $C(e)$ is a common cycle of two pairs of pants then we join the vertices corresponding to these two pairs of pants by an edge (also denoted by e) cutting the cycle $C(e)$ in one point as in Figure 3.4.

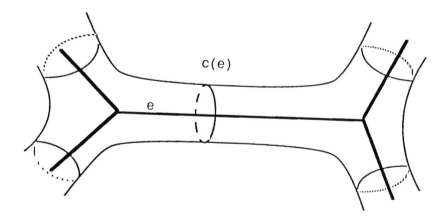

Figure 3.4

All cycles of this embedding are noncontractible. Now we construct a new embedded graph G_1 by adding, for each edge e of G_0, a vertex, $v(e)$, subdividing

e. We embed $v(e)$ on the common point of the cycle $C(e)$ and the edge e. Let l' be equal to l if l is even or $l+1$ otherwise. From G_1 we construct a new graph G_2 together with its 2-cell embedding. For each edge e we construct a cycle, $c^0(e), c^1(e), ..., c^{l'-1}(e), c^0(e)$ where $c^0(e) = v(e), c^i(e) \in C(e)$, (see Figure 3.5).

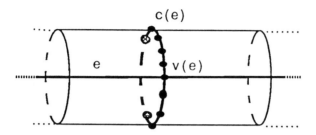

Figure 3.5. C(e)

The number of vertices of G_2 is $3Nl'+2N$. Since each closed noncontractible path in G_2 either contains some $C(e)$ or a cycle in G_1, all noncontractible cycles of G_2 are of length at least l. Now each pair of pants looks like presented on Figure 3.6.

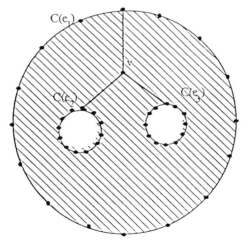

Figure 3.6.

The last step of our construction is to triangulate each pair of pants. First, for each pair of pants $PP(v)$, we add a vertex v' and edges $(c^{l'/2}(e_1), v')$,

$(c^{l'/2}(e_2), v')$, $(c^{l'/2}(e_3), v')$. Next, we subtriangulate each of the three resulting faces as in Figure 3.7(a).

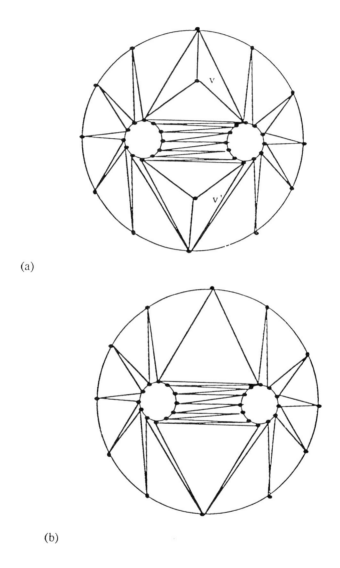

(a)

(b)

Figure 3.7. Triangulation of a pair of pants

Observe that the role of vertices v and v' is symmetric. Note that any cycle going through one of the vertices v and v' can be replaced by a shorter cycle which does not go through such a vertex. Thus we can remove all such vertices from the triangulation. (The remaining embedded graph is also a triangulation).

The resulting triangulation of a pair of pants is presented in Figure 3.7 (b). We denote this triangulation by G_3.

All noncontractible cycles in this triangulation have length at least l (see [17] for a formal proof).

Thus, for a fixed genus $g = N + 1$, we can construct a triangulation all of whose non-contractible cycles have length at least l. The number of vertices of the triangulation is equal to $n_g = 3Nl'$, where $2^l > N + 1 \geq 2^{l-1}$. By routine computation (see [17]) $n_g \leq 4.5g \log g$ and $l \geq \frac{1}{\sqrt{3.6}}\sqrt{\frac{n_g}{g}}\sqrt{\log g}$. Furthermore, $n_g \geq \frac{3}{2}(2g - 2)$. Therefore we can use Lemma 5.8 with $\varepsilon = \frac{2}{5}$ (i.e. $\frac{1-\varepsilon}{\varepsilon} = \frac{3}{2}$ and $\frac{1-\varepsilon}{\sqrt{4-\varepsilon}} = \frac{1}{\sqrt{10}}$), to get

$$f(\Sigma_g, n) \geq \frac{1}{6}\sqrt{\frac{n}{g}}\sqrt{\log g}.$$

□

4 The Hyperbolic Geometry Technique

A graph that is embedded in a surface can be considered as a structure that approximates this surface. In this context the representativity of a graph is often viewed as a parameter of the approximation.

In this section we formalize the concept of approximating an orientable surface with the help of a triangulation. We relate the problem of the approximation of a surface to a geometrical structure of the surface (i.e. a spherical, a Euclidean, or a hyperbolic structure depending on the Euler characteristic of a surface). We discuss whether the representativity of a triangulation can be used to measure how good the approximation is.

Informally, when approximating a surface with the help of a triangulation, we would like to spread the vertices of the triangulation evenly on the surface. If we adopt this intuition, an approximation should depend on the metrical properties of a surface. The metric of a surface is determined by a geometrical structure of the surface. We will give an informal description of the notion of a geometrical structure later in this section. It is important to note that a surface has many geometrical structures. Therefore a triangulation which is a "good approximation" of one of the structures may not be a "good approximation" of another one.

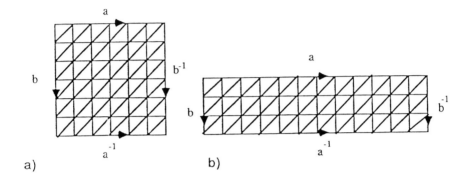

Figure 4.1

Consider first two different triangulations of the torus, each of them using the same number of vertices $n = k^2$, where k is an even number (Figure 4.1). The representativity of the first triangulation is $\sqrt{n}$, and the representativity of the second triangulation is $\frac{\sqrt{n}}{2}$. One may ask which of our two triangulations approximates the torus better. Informally, these two triangulations approximate two different Euclidean structures of the torus. (The metrics on the tori are induced by the Euclidean metric on the plane taking into account the identifications of edges of the rectangles (see Figure 4.1)). One cannot claim that one of the triangulations presented in Figure 4.1 approximates the torus better than the other. However, the first triangulation approximates a geometrical structure that has a bigger *injectivity radius* (that is, the maximum radius of a ball which can be embedded in the surface with an arbitrary point of the surface as its center). Thus paraphrasing the terminology of Robertson and Vitray, the first triangulation is more "locally planar" than the other.

A surface of genus greater than one possesses a hyperbolic structure. Below, we introduce informally the notion of the hyperbolic structure. It will be modeled on the hyperbolic plane, H^2, which we define first. For a formal discussion of the topic we refer the reader to [22].

We use the Poincaré model of H^2. Let $Int(D^2) = \{z \in C : |z| < 1\}$ where C is the set of complex numbers. We introduce a metric on $Int(D^2)$ by deforming the Euclidean metric in every point, z, by the function $h(z) = \frac{2}{1-|z|^2}$. We call the new metric the *hyperbolic metric*. It is important to observe that the deformation does not depend on a direction but only on the point. Thus the Poincaré model is *conformal*, that is, hyperbolic angles between curves are equal to the Euclidean ones. The boundary of $Int(D^2)$ is called the circle at infinity (for the distance of any point on the circle from the center is infinity). In the Poincaré model, geodesics (that is, curves of minimal distance) are curves determined by circles perpendicular to the circle at infinity (including the straight lines through the point 0). Inversions with respect to geodesics preserve D^2 and are isometries of H^2. Furthermore, these inversions generate all isometries of H^2.

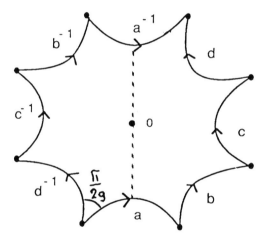

Figure 4.2

A hyperbolic surface (that is, a surface with hyperbolic structure) is obtained by gluing together pieces of H^2 using isometries [6] (that is, we do the gluing in such a way that the metrics on the common part of any two pieces are identical). Below we show a practical construction of a hyperbolic structure on a surface of genus $g > 1$ (see Figure 4.2). Consider a regular $4g$-gon in the hyperbolic plane with vertex angle $\pi/2g$ and identification of the opposite sides (where the sides of the polygon are drawn along geodesics). This identification gives a

---

[6] In the language of differential geometry, a hyperbolic surface is defined as a Riemannian surface with constant negative (equal to -1) curvature.

hyperbolic structure on a surface of genus g.[7] To see this, note that identified edges have the same hyperbolic length, all vertices of the polygon are identified to one vertex, and the sum of all angles along this vertex is equal to 2π. (Note that the last property is not true when we work with the Euclidean regular 4g-gon).

Since the area of a triangle with angles α, β, γ is equal to $\pi - (\alpha + \beta + \gamma)$ we can easily check that the area of the constructed hyperbolic structure is equal to $2\pi(2g - 2)$. This fact is true for any hyperbolic structure on a surface of genus $g > 1$ and is known as the Gauss-Bonnet theorem.

Theorem 4.1 (Gauss-Bonnet) *The area of a hyperbolic surface of genus $g > 1$ is equal to $-2\pi\chi(\Sigma_g) = 2\pi(2g - 2)$.*

Consider a surface Σ together with a metric $\hat{d}$. Now we introduce the definition of an approximation of a pair $(\Sigma, \hat{d})$ with the help of a triangulation.

Let T be a triangulation of Σ and let u, v be two vertices of T. Let γ be a curve on Σ joining u and v (if $u = v$ then γ is a closed curve). We define $\hat{d}_\gamma(u, v)$ (respectively, $d_\gamma(u, v)$) to be the length of the shortest curve in $(\Sigma, \hat{d})$ (respectively, the shortest path in T) joining u and v that is homotopic to γ relative to u and v.

Definition 4.2 *We say that a triangulation T of a surface Σ is a β-approximation of Σ (where $0 < \beta \leq 1$) with respect to the metric $\hat{d}$ if and only if for every two vertices u, v of T, and any path γ on Σ joining u and v, the following is true:*

$$\beta\sqrt{\frac{n}{g}} \leq \frac{d_\gamma(u,v)}{\hat{d}_\gamma(u,v)} \leq \frac{1}{\beta}\sqrt{\frac{n}{g}}.$$

We treat the factor $\sqrt{\frac{n}{g}}$ as a "scaling" factor (we will discuss this in more detail later). In what follows we assume that $\hat{d}$ is a hyperbolic metric.

Conjecture 4.3 *There exists a constant β ($0 < \beta \leq 1$) such that, for any hyperbolic surface Σ and $n \gg g$ there exists an n vertex triangulation that is a β-approximation of Σ. Conversely, for any triangulation T of a surface Σ_g of genus greater than one, there exists a hyperbolic structure that is β-approximated by T.*

Observe that if a triangulation T is a β-approximation of a surface then the subdivision T_i of T (recall Figure 2.1) is a β'-approximation of the same surface where, for $n(T) \gg g$, β' is approximately equal to β. To see this observe that

---

[7] A genus g surface has more than one geometrical structure. The geometrical structures of a surface of genus $g > 1$ are precisely the hyperbolic structures forming a Teichmüller's space (which topologically is homeomorphic to the open $6g - 6$ dimensional ball).

in a single subdivision step the distance between any two vertices increases 2 times while the number of vertices grows approximately 4 times (recall formula (1) from the proof of Lemma 2.28).

We discuss now arguments towards Conjecture 4.3. We split the discussion into two parts.

Part I. First we would like to argue that, for any hyperbolic structure, there is a triangulation which approximates this structure up to some constant factor. Assume that we spread n vertices evenly on the surface [8]. Then $\frac{n}{2\pi(2g-2)}$ gives the "density" of the vertices of the triangulation.

We require that our "evenly spread" triangulation has the following properties:

There exist constants $\delta > 0$ and $1 \geq \epsilon > 0$ such that

T1: All angles of the triangulation are greater than δ,

T2: For any two edges of the triangulation the ratio of the length of the shorter edge to the length of the longer edge is in the interval $[\epsilon, \frac{1}{\epsilon}]$.

Let L (respectively, l) be the length of the longest (respectively, the shortest) edge of the triangulation. Then the area of any face is $O(L^2)$. On the other hand, by property T1, the area of any face is $\Omega(l^2)$. Therefore the ratio of the area of the smallest face to the area of the largest belongs to the interval $[c, \frac{1}{c}]$ where c is a constant depending on δ and ϵ. Thus, by the Gauss-Bonnet theorem, the area of each such triangle is $\Theta(\frac{2\pi(2g-2)}{f})$ where f is the number of triangles. Therefore the area of a triangle is $\Theta(\frac{g}{n})$ and the hyperbolic length of an edge is $\Theta(\sqrt{\frac{g}{n}})$.

I a). Now we prove the inequality:

$$\beta'\sqrt{\frac{n}{g}} \leq \frac{d_\gamma(u,v)}{\hat{d}_\gamma(u,v)}.$$

Take two points u, v of the triangulation and a curve γ joining them. Let $\tilde{d}_\gamma(u,v)$ be the hyperbolic length of the path in the triangulation that realizes $d_\gamma(u,v)$. Then

$$\hat{d}_\gamma(u,v) \leq \tilde{d}_\gamma(u,v) \leq d_\gamma(u,v) L \leq c' d_\gamma(u,v) \sqrt{\frac{g}{n}}$$

where c' is a constant.

Thus assigning $\beta' = \frac{1}{c}$ we obtain the required inequality.

---

[8] Imagine, for example, n electrons placed on our hyperbolic surface. We can expect them to spread evenly on the surface

I b). To prove the inequality

$$\frac{d_\gamma(u,v)}{\hat{d}_\gamma(u,v)} \leq \frac{1}{\beta''}\sqrt{\frac{n}{g}}$$

we project the path that realizes $\hat{d}_\gamma(u,v)$ on the edges of the triangulation. By property T1, the hyperbolic length of this projection is $O(\hat{d}_\gamma(u,v))$. Since the hyperbolic length of an edge is $\Theta(\sqrt{\frac{g}{n}})$ this implies the above inequality.

Finally we can chose $\beta = min(\beta', \beta'')$.

It remains to prove that there exist "evenly spread" triangulations that have properties T1 and T2 defined above. An important step towards construction of such triangulations is the following theorem:

Theorem 4.4 (Colin de Verdière, Marin [6]) *On a given hyperbolic surface there exists a family $\mathcal{T}_n$ of geodesic triangulations such that*

1. *The upper limit of diameters of a triangle tends to 0 as $n \mapsto \infty$.*
2. *$\exists N, \forall n \geq N$ the angles of the triangulation $\mathcal{T}_n$ are in the interval $[\frac{2}{7}\pi, \frac{5}{14}\pi]$.*

Part II. We would like to show that given a triangulation of a genus g surface, one can construct a hyperbolic structure which is approximated by this triangulation. The following theorem could be an important step towards such construction.

Theorem 4.5 (Buser [5]) *Given a triangulation of a closed surface there exists a hyperbolic structure in which all edges of the triangulation are drawn along geodesics.*

If Conjecture 4.3 is true then to obtain a triangulation of maximal representativity we need to find a triangulation that approximates the hyperbolic surface with maximal shortest geodesic. The length of the shortest geodesic is equal to twice the injectivity radius. Unfortunately, for $g > 1$, it is not known either what the value of the maximum injectivity radius is or which hyperbolic structure achieves this maximum. However if Conjecture 4.3 is true then a lower bound for the length of the shortest nontrivial geodesic could be used to to get a lower bound for the length of the shortest noncontractible cycle and vice versa.

Let $\varphi(\Sigma_g)$ denote the maximum over all hyperbolic structures of Σ_g of the length of the shortest geodesic. Using Theorem 3.1 Buser [4] proved that $\varphi(\Sigma_g) = \Omega(\sqrt{\ln g})$. Using a simple area argument, it is not difficult to prove that $\varphi(\Sigma_g) \leq 2\ln 4g$ [4]. Thus the best currently known upper and lower bounds for the length of the shortest noncontractible cycle and the length of the shortest geodesic agree.

Conjecture 4.6 *There exists a constant α such that for any $g > 1$ there exists a hyperbolic surface of genus g whose shortest geodesic is at least $\alpha \log g$ where α is a constant.*

To motivate this conjecture consider the geometric structure presented in Figure 4.2. In this structure the edges and the diameters (from the midpoint of an edge to the midpoint of the opposite edge) have length $2\text{arc cosh ctg}(\pi/4g)$. For big g this value can be approximated by $\ln g$. If one could prove that these edges and diameters are the shortest geodesics of this structure the Conjecture 4.6 would be proven.

Note, that Conjecture 4.3, part Ia, and Conjecture 4.6 imply the following conjecture

Conjecture 4.7
$$f(\Sigma_g, n) = \Theta(\sqrt{\frac{n}{g}} \log g).$$

5 Appendix:

In the Appendix, we show how the construction of Section 2.2 can be used to obtain triangulations of closed surfaces with representativity claimed in Theorem 2.21.

To get a triangulation of a closed surface (that is, a surface without boundary) from one or more surfaces with boundary we need to cap off boundary components with discs or glue pairs of boundary components using an annular collar. Discs and/or annuli should be triangulated in such a way that the identification does not introduce noncontractible cycles of length less than l. The following simple lemmas describe triangulations of discs and annuli that have the required property.

Consider a triangulation of an annulus with two boundary components $\delta_1 A, \delta_2 A$. Let v, u be two vertices of this triangulation that belong to the same boundary component. We use $\delta(v, u)$ to denote the length of shortest paths between u and v that entirely belong to the boundary component.

Lemma 5.1 *Consider a triangulation of an annulus as presented in Figure 5.1 a. The boundary components are cycles of length k and the depth of the triangulation is j. This triangulation has the following properties;*

1. *The number of vertices of the triangulation is $(k+1)j$.*
2. *Each noncontractible cycle has length at least k.*
3. *Any path joining different boundary component has length at least j.*

4. Any path joining two vertices, u and v, from the same boundary component has length at least $\delta(u,v)$.

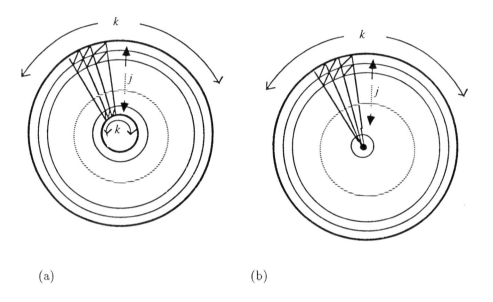

Figure 5.1 The triangulation of an annulus and a disc

Lemma 5.2 *Consider the triangulation of a disc obtained from the triangulation of an annulus from Lemma 5.1 by contracting edges of the boundary component $\delta_2 A$ and deleting multiedges (see Figure 5.1 b). Then the length of any path joining to vertices u, v on the disc boundary is at least $\min\{2j, \delta(u,v)\}$.*

Lemma 5.3 *Consider a triangulation of an annulus obtained from the triangulation from Lemma 5.1 by contracting consecutive edges of boundary component $\delta_2 A$ so that l ($l > 2$) edges of $\delta_2 A$ remain and deleting multiedges. Then*

1. *Each noncontractible cycle has length at least l.*

2. *Any path joining two vertices, u and v, from the boundary component $\delta_2 A$ has length at least $\delta(u,v)$.*

3. *Any path joining two vertices, u and v, from the boundary component $\delta_1 A$ has length at least $\min\{2j, \delta(u,v)\}$*

4. *Any path joining two vertices from different boundary components has length at least j.*

Let T_0 be a triangulation of a surface $\Sigma_{g,d}$. Let T_1 be the triangulation obtained from T_0 by subdividing each face of T_0 as in Figure 5.2 b.

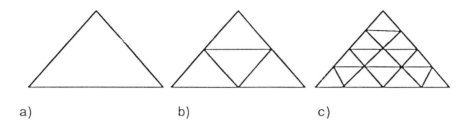

Figure 5.2

Thus the triangulation T_1 is a subdivision of the triangulation T_0. Let T_i be the triangulation obtained from T_0 by i iterations of the subdivision described above (that is, for $i > 1, T_i = (T_{i-1})_1$). A sequence $T_0, T_1, \ldots$ of subdividing triangulations satisfies the properties stated in the following simple lemma:

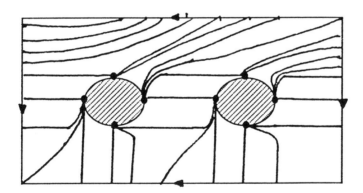

Figure 5.3 The triangulation $T_{1,2}$ of $\Sigma_{1,2}$

Lemma 5.4 *Let f_i be the number of faces of the triangulation T_i. Then, for any $i > 1$*

1. $f_i = 4f_{i-1}$,
2. $\rho(\Sigma_{g,d}, T_i) = 2\rho(\Sigma_{g,d}, T_{i-1})$.

Lemma 5.5 *Let $T_{1,2}$ be the triangulation of $\Sigma_{1,2}$ with $n_0 = 8$ vertices, $f_0 = 12$ faces, and $e_0 = 22$ edges as presented in Figure 5.3.[9] Let $(T_{1,2})_i$ be the*

---

[9] Note that $T_{1,2}$ is a 2-fold covering of the triangulation of $\Sigma_{1,1}$ from Figure 2.5.

triangulation of $\Sigma_{1,2}$ obtained from $T_{1,2}$ by i subdivision steps, and let n_i be the number of vertices of the triangulation $(T_{1,2})_i$. Then

$$n_i = 6 \cdot 4^i + 2^{i+2} - 2.$$

Now we are in the position to give a lower bound for the function $f(g,n)$ that is, for the maximum representativity of an n-vertex triangulation of a closed surface of genus g. To illustrate our approach we start with the following example.

Example 5.6

Let $l > 4$ and let g_l be the genus of the surface $\tilde{\Sigma} = \Sigma_{g_l,d}$ constructed as in Theorem 2.18 ($\tilde{\Sigma}$ is the m-fold covering of $\Sigma_{1,1}$). Let $T_{g_l,d}$ be the triangulation of $\Sigma_{g_l,d}$ that is the covering triangulation of $T_{1,1}$. The number of vertices of this triangulation is equal to $4m$. Let T_{g_l} be the triangulation of Σ_{g_l} obtained from $T_{g_l,d}$ by cupping off boundary components of $\Sigma_{g_l,d}$ by discs triangulated as in Lemma 5.2 with $j = \lfloor \frac{l}{2} \rfloor$ and $k = \frac{4m}{d}$. By Lemma 5.2 the representativity of T_{g_l} is at least l. The number of vertices, $n(T_{g_l})$, is bounded by

$$n(T_{g_l}) < n(T_{g_l,d}) + d(\frac{4m}{d} \cdot \frac{l}{2}) - d(\frac{4m}{d}) \leq 2lm.$$

Since, by Lemma 2.18, $\log\log g_l < 3l$ and by Corollary 2.19, $g_l > \frac{1}{2}m(1 - \frac{1}{l}) > \frac{3}{8}m$ we obtain

$$\frac{1}{4}\sqrt{\frac{n(T_{g_l})}{g_l}}\sqrt{\log\log g_l} < l.$$

Now, we generalize the construction from Example 5.6 to an arbitrary g.

Lemma 5.7 *For any $g > 1$, there exists an integer N_g such that*

$$f(\Sigma_g, N_g) \geq \frac{1}{4}\sqrt{N_g/g}\sqrt{\log\log g}.$$

Furthermore if $g \geq 2^{8^{10}}$ then $N_g < \frac{5}{3}g\log\log g$, and otherwise $N_g \leq 7g$.

Proof: For any $l > 4$ define $\Sigma_{g_l,d}$ and $T_{g_l,d}$ as in the Example 5.6.

Consider any $g > 1$. Let l be an integer such that

$$2^{8^{l-1}} \leq g < 2^{8^l}.$$

We consider separately the cases of $l < 11$ and $l \geq 11$.

CASE I: $g \geq 2^{8^{10}}$ (i.e. $l \geq 11$).

Let $g = ig_l + k$ where $k < g_l$. Since $g_l < 2^{8^{l-1}}/l^{4l^2}$ we have $i \geq l^{4l^2}$. Construct Σ_g and T_g as follows. Consider i copies of $\Sigma_{g_l,d}$ and k copies of $\Sigma_{1,2}$. Each copy of $\Sigma_{g_l,d}$ is triangulated with T_{g_l} as in Example 5.6. Each copy of $\Sigma_{1,2}$ is triangulated with $(T_{1,2})_{k_l}$, where $(T_{1,2})_i$ is defined as in Lemma 5.5 and $k_l = \lceil \log \frac{l}{3} \rceil$ (that is $l \leq 3 \cdot 2^{k_l} < 2l$). Note that $(T_{1,2})_{k_l}$ and $T_{g_l,d}$ have representativity at least l. Furthermore, for $l > 4$, each boundary component of $T_{g_l,d}$ has at least $4l$ vertices and each boundary component of $(T_{1,2})_{k_l}$ has less than $2l$ vertices.

Now we glue together i copies $T_{g_l,d}$ and k copies of $(T_{1,2})_{k_l}$ as in Figure 5.4 using collars triangulated as in Lemma 5.5.

triangulated collar

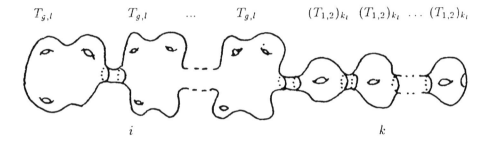

Figure 5.4

Then we cap off remaining boundary components with discs triangulated as in Lemma 5.2. The resulting surface has genus g and is triangulated with representativity at least l. After approximating the number of vertices used in this triangulation (see a more detailed calculation below) we obtain the inequality stated in the lemma.

CASE II: $g < 2^{8^{10}}$.

If $g \geq 4$ then we construct Σ_g and its triangulation T_g, by cyclically gluing $g-1$ copies of $T_{1,2}$ as in Figure 5.5. The representativity of $T_{1,2}$ is three. If the gluing of $g-1$ copies of $T_{1,2}$ is done carefully then we do not introduce cycles of length two. Therefore $\rho(\Sigma_g, T_g) = 3$. The number of vertices is equal to $4(g-1)$. Therefore $\frac{n}{g} < 4$ and so $n < 4g$. Finally

$$\sqrt{\frac{n}{g}}\sqrt{\log\log g} \leq 2\sqrt{\log\log g} < 2\sqrt{3\cdot 10} < 12 = 4\rho(\Sigma_g, T_g).$$

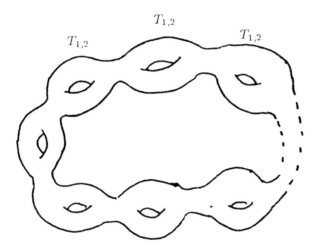

Figure 5.5

For $1 < g < 4$. Σ_g and its triangulation T_g is obtained by gluing g copies of $T_{1,2}$ (as before one has to be careful not to introduce noncontractible cycles of length two) and cupping off two remaining boundary components by discs triangulated as in Lemma 5.2 with $j = 1$. The representativity of the triangulation is 3. The number of vertices is equal to $4g + 4 + 2 = 4g + 6$. Therefore

$$\frac{n}{g} < 4 + \frac{6}{g} \leq 7.$$

Finally

$$\sqrt{\frac{n}{g}}\sqrt{\log\log g} \leq \sqrt{7} < 12.$$

Now we give a more detailed computation for case I.

CASE I : Details of the computation.

First we approximate the number, n, of vertices of T_g.

By Lemma 5.5, the total number of vertices on copies of $(T_{1,2})_{k_l}$ (not counting annuli connecting them) is $k(6\cdot 4^{\lceil\log\frac{l}{3}\rceil} + 2^{\lceil\log\frac{l}{3}\rceil+2} - 2) \leq k(6\cdot 4(\frac{l}{3})^2 + \frac{8}{3}l - 2) \leq 6kl^2$.

- The total number of vertices on annuli connecting copies of $(T_{1,2})_{k_l}$ plus the vertices on the disc cupping off the remaining boundary component on the last copy of $(T_{1,2})_{k_l}$ is at most kl^2.
- By an argument similar to the one used in Example 5.6 the number of remaining vertices of T_g is at most $i \cdot 2lm$.

Thus
$$n \leq 2ilm + 7kl^2.$$

Therefore
$$\frac{n}{g} \leq \frac{2ilm + 7kl^2}{ig_l + k}.$$

But $g < 2^{8^l}$ and $ig_{l'} + k > ig_l > i\frac{1}{2}m(1-\frac{1}{l})$ and therefore
$$\frac{n}{g} \leq \frac{2ilm + 7kl^2}{i\frac{1}{2}m(1-\frac{1}{l})}.$$

Since $i \geq l^{4l^2}$, $k < g_l$, and $l \geq 11$ we obtain
$$\frac{n}{g} \leq \frac{4l}{1-\frac{1}{l}} + \frac{1}{l} < \frac{4l}{1-\frac{1}{9}} + \frac{1}{l} = \frac{9}{2}l + \frac{1}{l} < 5(l-1).$$

Therefore
$$n < 5(l-1)g \leq \frac{5}{3} g \log \log g.$$

Finally,
$$\sqrt{\frac{n}{g}} \sqrt{\log \log g} \leq \sqrt{5l} \sqrt{\log \log g} < \sqrt{5l}\sqrt{3}\sqrt{l} < 4l.$$

□

In the last lemma, we constructed, for any $g > 1$, a triangulation of Σ_g such that the representativity of this triangulation is at least $c\sqrt{\frac{n}{g}}\sqrt{\log \log g}$ for some constant c. Now we will extend this result to any sufficiently large n.

Lemma 5.8 *Assume that for a genus g there exists an integer $N_g \geq \frac{1-\varepsilon}{\varepsilon}(2g-2)$ such that*
$$f(\Sigma_g, N_g) \geq c\sqrt{\frac{N_g}{g}} \phi(g)$$

where $\phi(g)$ is a function of g and c is a constant. Then for any $n \geq N_g$
$$f(\Sigma_g, n) \geq c\frac{1-\varepsilon}{\sqrt{4-\varepsilon}} \sqrt{\frac{n}{g}} \phi(g).$$

Proof: Let $n_0 = N_g$ and let T_0 be a triangulation of Σ_g that has representativity at least $c\sqrt{\frac{N_g}{g}}\phi(g)$. Subtriangulate T_0 iteratively according to the scheme presented in Figure 5.2. Let n_i, e_i, f_i be respectively the number of vertices, edges and faces after the ith iteration. By Euler's formula we have for every i:

$$n_i - e_i + f_i = 2 - 2g.$$

Since we are dealing with a triangulation it follows that $2e_i = 3f_i$. Thus $e_i = 3(n_i + 2g - 2)$.

Furthermore $n_i = n_{i-1} + e_{i-1}$, and therefore

$$n_i = n_{i-1} + 3(n_{i-1} + 2g - 2) = 4n_{i-1} + 3(2g - 2) \tag{1}$$

which implies

$$n_i = 4^i n_0 + 3(2g - 2)\sum_{j=0}^{i-1} 4^j = 4^i n_0 + (4^i - 1)(2g - 2).$$

Thus

$$4^i n_0 = n_i - (4^i - 1)(2g - 2). \tag{2}$$

Let T_i be the triangulation resulting from T_0 after the ith subdivision step. By Lemma 5.4 we have

$$\rho(\Sigma_g, T_i) = 2^i \rho(\Sigma_g, T_0).$$

However, by the assumption of the lemma

$$\rho(\Sigma_g, T_0) \geq c\sqrt{\frac{n_0}{g}}\phi(g),$$

and thus for any ε

$$\rho(\Sigma_g, T_i) \geq 2^i c\sqrt{\frac{n_0}{g}}\phi(g) = c\frac{\phi(g)}{\sqrt{g}}\sqrt{4^i n_0(1-\varepsilon) + 4^i \varepsilon n_0}.$$

Therefore, by (2),

$$\rho(\Sigma_g, T_i) \geq c\phi(g)\frac{1}{\sqrt{g}}\sqrt{(1-\varepsilon)(n_i - (4^i - 1)(2g - 2)) + 4^i \varepsilon n_0} =$$

$$c\phi(g)\frac{1}{\sqrt{g}}\sqrt{(1-\varepsilon)n_i - (4^i - 1)(2g - 2)(1-\varepsilon) + 4^i \varepsilon n_0} =$$

$$c\phi(g)\frac{1}{\sqrt{g}}\sqrt{(1-\varepsilon)n_i + 4^i \varepsilon n_0 - 4^i(2g - 2)(1-\varepsilon) + (2g - 2)(1-\varepsilon)} >$$

$$c\phi(g)\frac{1}{\sqrt{g}}\sqrt{(1-\varepsilon)n_i + 4^i\varepsilon n_0 - 4^i(2g-2)(1-\varepsilon)} =$$

$$c\phi(g)\frac{1}{\sqrt{g}}\sqrt{(1-\varepsilon)n_i + 4^i\varepsilon(n_0 - \frac{1-\varepsilon}{\varepsilon}(2g-2))}$$

Thus, by the assumption of the lemma we have

$$\rho(\Sigma_g, T_i) \geq c\phi(g)\frac{1}{\sqrt{g}}\sqrt{(1-\varepsilon)n_i}\ .$$

This implies that,

$$f(\Sigma_g, n_i) \geq c\phi(g)\sqrt{\frac{n_i}{g}}\sqrt{(1-\varepsilon)}. \qquad (3)$$

For any n there exists i such that $n_{i+1} > n \geq n_i$. By (1) and by the assumption of the lemma it follows that

$$n_{i+1} \leq 4n_i + \frac{3\varepsilon}{1-\varepsilon}n_0.$$

But, for any i, we have $n_i \geq n_0$, and so

$$n_{i+1} \leq 4n_i + \frac{3\varepsilon}{1-\varepsilon}n_i = n_i(4 + \frac{3\varepsilon}{1-\varepsilon}).$$

Thus

$$n_i \geq n_{i+1}\frac{1-\varepsilon}{4-\varepsilon}.$$

Observe that if we subtriangulate a triangle using one additional vertex as in Figure 5.6 then we do not decrease the representativity of the triangulation.

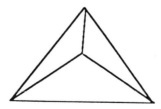

Figure 5.6 Adding a vertex to a triangulation

Because $n \geq n_i$ it follows that

$$f(\Sigma_g, n) \geq f(\Sigma_g, n_i).$$

Thus

$$f(\Sigma_g, n) \geq c\sqrt{1-\varepsilon}\sqrt{\frac{n_i}{g}}\phi(g) \geq c\sqrt{1-\varepsilon}\sqrt{\frac{1-\varepsilon}{4-\varepsilon}}\sqrt{\frac{n_{i+1}}{g}}\phi(g) >$$

$$c\frac{1-\varepsilon}{\sqrt{4-\varepsilon}}\sqrt{\frac{n}{g}}\phi(g).$$

□

Now we will use the above lemma to prove Theorem 2.21 of Section 2.

Theorem 5.9 (Theorem 2.21) *For any $g > 1$ there exists an integer N_g such that for any $n \geq N_g$*

$$f(\Sigma_g, n) \geq \frac{1}{12}\sqrt{\frac{n}{g}}\sqrt{\log\log g}.$$

Furthermore if $g \geq 2^{8^{10}}$ then $N_g \leq \frac{5}{3}g\log\log g$ and otherwise $N_g \leq 7g$.

Proof: Let $g < 2^{8^{10}}$. By Case II of the proof of Lemma 5.7, for $g \geq 4$ we can set $N_g = 4(g-1)$ and for $2 \leq g \leq 3$ we can set $N_g = 4g + 6$. So in both cases $N_g \geq 2(2g-2)$.

Let $g \geq 2^{8^{10}}$. Thus we deal with CASE I of the proof of Lemma 5.7. Therefore we have

$$n(T_g) \geq i \cdot n(T_{g_l}) > i \cdot n(T_{g_l,d}).$$

By Corollary 2.19, $n(T_{g_l,d}) > 8g_l$, and since $(i+1)g_l > g$, we have further

$$n(T_g) > 8ig_l = 8\frac{i}{i+1}(i+1)g_l > 8\frac{i}{i+1}g > 4g > 2(2g-2).$$

Therefore we can use Lemma 5.8 for $\varepsilon = \frac{1}{3}$ (i.e. $\frac{1-\varepsilon}{\varepsilon} = 2$). Because $\frac{1-\varepsilon}{\sqrt{4-\varepsilon}} = \frac{2}{\sqrt{33}} > \frac{1}{3}$ we obtain

$$f(\Sigma_g, n) \geq \frac{1}{12}\sqrt{\frac{n}{g}}\sqrt{\log\log g}.$$

as required. □

References

[1] D.Archdeacon, Densely embedded graphs, *J. Combn. Theory.* Ser. B, 54 (1992), 13-36.

[2] J.A.Bondy, U.S.R.Murty, *Graph Theory With Applications,* The Macmillan Press (1976).

[3] R.G. Burns, A note on free groups, *Proc. Amer. Math. Soc.* 23 (1969) 14-17.

[4] P.Buser, Riemannsche Flächen mit grosser Kragenweite, *Comment. Math., Helvetici,* 53 (1978) 395-407.

[5] P.Buser, Private communication, 1988.

[6] Y.Colin de Verdière, A.Marin, Triangulations presque équilatérales des surfaces, *J. Differential Geometry* 32, (1990) 199-207.

[7] P.Erdös, H.Sachs, Regulare Graphen gegebener Taillenweite mit minimaler Knotenzahl, Wiss. Z. Univ. Halle-Wittenberg Math.-Nat. 12 (1963), 251-257

[8] B.Bollobás, *Extremal Graph Theory with Emphasis on Probabilistic Methods,* Regional conference series in mathematics, 62 (1986), AMS.

[9] M.Hall, Coset representation in free groups, *Trans. Amer. Math.* Soc. 49 (1949) 422-432.

[10] J.Hempel, *3-Manifolds,* Ann. of Math. Studies 86, Princeton University Press, 1976.

[11] C.Kośniowski, *A First Course in Algebraic Topology,* Cambridge University Press, 1980.

[12] D.Knuth, *The Art of Computer Programming, Vol.1. Fundamental Algorithms,* Addison-Wesley, Reading MA, 1986.

[13] J.Hutchinson, On short non-contractible cycles in embedded graphs, *SIAM J. Disc. Math,* (1988) 185-192.

[14] J.Massey, *Algebraic Topology; An Introduction,* Harcourt, Brace and World, New York, 1967.

[15] J.McCool, Private communication, 1987.

[16] T.M.Przytycka, J.H.Przytycki, On lower bound for short noncontractible cycles in embedded graphs, *SIAM J. Disc. Math,* (1990) 281-293.

[17] T.M.Przytycka, J.H.Przytycki, A simple construction of high representativity triangulations, manuscript, to be submitted.

[18] N.Robertson, P.D.Seymour, Graph minors. VII. Disjoint paths on a surface, *J.Comb. Theory*, Ser B, 48 (1990), 212-254.

[19] N.Robertson, R.Vitray, Representativity of surface embeddings. *Algorithms and Combinatorics, Volume 9, Paths, Flows and VLSI-Layout.* Springer-Verlag, Berlin, Heidelberg, Volume Editors: B. Korte, L. Lovász, H. J. Promel, and A. Schrijver, (1990), 293 - 328.

[20] C.Thomassen, Embedding of graphs with no short non-contractible cycles, *J. Comb. Theory*, B, 48 (1990), 155-177.

[21] J.R.Stallings, Finite graphs and free groups, *Contemp. Math.* 44, 1985, 79-84.

[22] W. Thurston, *The Geometry and Topology of 3-Manifolds,* Princeton University Press, to appear.

Teresa M. Przytycka
Department of Computer Science, University of California, Riverside, CA 92521
On leave from
Instytut Informatyki, Uniwersytet Warszawski, Warsaw, Poland.

Józef H. Przytycki
Department of Mathematics, University of California, Riverside, CA 92521
On leave from
Department of Mathematics, Uniwersytet Warszawski, Warsaw, Poland.

Representativity and Flexibility on the Projective Plane

R. P. VITRAY

ABSTRACT. Let G be a graph embeddable on a surface, S. The *flexibility* of G (with respect to S) is the number of distinct (up to homeomorphism) labeled embeddings of G into S. The flexibility of the topologically minimal 3-representative projective planar embeddings has been determined along with global projective twists which explain the flexibilities. This result, along with Whitney twists, can be used to analyze the number of embeddings of any 3-representative projective planar graph. As a consequence it can be shown that every 3-connected, 4-representative projective planar embedding is *rigid* (i.e. has only one labeled embedding). Thus an alternative method is supplied for obtaining embedding results shown independently by Negami. It is conjectured that a 3-connected, k+3-representative embedding on a surface with k crosscaps is rigid.

1. Introduction

In 1932 and 1933 Hassler Whitney published a series of monumental graph theoretic papers in which he defined a combinatorial dual of a graph and showed the existence of such a dual was equivalent to the existence of a representation of the graph on the plane with a finite set of points joined by curves intersecting only at those points. This led to two related results. First, all embeddings, up to isomorphism, of a 2-connected graph in the plane can be obtained by a sequence of twists of blocks of the graph attached at two vertices; and, second, any 3-connected graph has, up to isomorphism, at most one embedding in the plane. It remains an open area of research to extend Whitney's results to nonplanar surfaces. A natural first step is to analyze the projective plane where already the situation is more complicated. Here, graphs which are 3-connected often have nonisomorphic drawings which are not in general related by Whitney twists.

In their landmark work on graph minors, Robertson and Seymour have developed a tool for attacking this problem in the concept of representativity. For any fixed surface of positive genus, the embeddings of representativity at least n form an upper ideal under the quasi-ordering of minor inclusion. Robertson and Seymour have shown that for any such

---

1991 Mathematics Subject Classification. Primary 05C10, 57M15.
This paper is in final form and no version of it will be submitted for publication elsewhere.

set the collection of nonisomorphic minor minimal members is finite. There is no limit on the size of such a finite collection but in those instances where the minor minimal embeddings can be found interesting results can be obtained pertaining to the entire set.

On the projective plane, hereafter denoted PP, the complete set of nonisomorphic minor minimal 3-representative embeddings is known [1]. An analysis of these embeddings yields a theory explaining the flexibility of projective planar embeddings.

2. Definitions

A *surface* is a compact two-dimensional manifold without boundary. For convenience, Σ will be used exclusively to denote nonspherical surfaces. An O-arc is a subset of Σ homeomorphic to the unit circle and an *essential* O-arc is one that is not contractible to a point. If Ψ is an embedding of a graph G in a surface Σ then G is the *underlying graph of* Σ. In an abuse of language, graph theoretic terms will also be used in reference to an embedding (so, for example, the point in Σ which is the image of a vertex of G will be referred to as a vertex of Ψ). If v is a vertex of Ψ then the *wheel neighborhood of v* is the union of the closures of the faces of Ψ which have v in their boundary. The *representativity* of Ψ, denoted $\rho(\Psi)$, is given by:

(1) $\qquad \rho(\Psi) = \text{MIN}(\{|C \cap \Psi| : C \text{ is an essential O-arc in } \Sigma\})$.

Let Δ be a triangular face of an embedding, Ψ, in Σ. If we draw three internally disjoint paths from a point, v, in Δ to the three vertices on the boundary of Δ, delete the edges in the boundary of Δ, and assign v to be a vertex, we obtain a new embedding, Ψ', in Σ. Ψ' is said to have been obtained from Ψ by *popping a triangle*. The inverse operation, *punching a triad*, can be used to obtain Ψ from Ψ' and the two operations together are referred to as *Y-Δ operations on Σ* (or simply Y-Δ operations if Σ is understood). Note that the Y-Δ operations defined combinatorially for a graph G do not necessarily carry over to an embedding Ψ of G. In particular if a triangle of G does not bound a face of Ψ then it cannot be popped. It is elementary to show that the representativity of an embedding is invariant under Y-Δ operations ([1]).

For any two embeddings, Ψ_1 and Ψ_2, of a graph G in Σ, a *isomorphism from Ψ_1 to Ψ_2* is a homeomorphism from Σ to Σ which maps the image of G under Ψ_1 to the image of G under Ψ_2. If such a homeomorphism exists, Ψ_1 and Ψ_2 are said to be *isomorphic*. If Ψ_1 and Ψ_2 are both embeddings in Σ then Ψ_1 is a *subembedding* of Ψ_2 if Ψ_1 is isomorphic to an embedding obtained from Ψ_2 by a sequence of edge and vertex deletions. Also, Ψ_1 is a *minor* of Ψ_2, denoted $\Psi_1 \leq_m \Psi_2$, if Ψ_1 is isomorphic to an embedding obtained from a subembedding of Ψ_2 by a sequence of contractions of nonloop edges where the contractions correspond to the quotient map of Σ which identifies the points of the edge. The set of nonisomorphic minor minimal embeddings on Σ of representativity at least n will be denoted $M_n(\Sigma)$. It follows from the work on Robertson and Seymour that $M_n(\Sigma)$ is finite for any choice of n and Σ (although this result is not needed here). For many questions, including the one of flexibility, it is appropriate to restrict the edge contractions to edges which are incident with divalent vertices. Under this restriction, Ψ_1 is an *embedding minor* of Ψ_2, denoted $\Psi_1 \leq_e \Psi_2$. Under embedding inclusion, there are still a finite set of nonisomorphic minimal embeddings of representativity at least n, denoted $E_n(\Sigma)$, and they can be obtained from $M_n(\Sigma)$ by expanding vertices of valence greater than 2 in all possible ways, where expansion is the inverse operation to contraction.

A *labeling* of a graph G is a one to one mapping from $\{1, 2,...., |V(G)|\}$ to the vertices of G. Two labelings of G are equivalent if there is an automorphism of G which induces the identity map on the labels. Suppose Ψ_1 and Ψ_2 are both embeddings in Σ of a graph G with labeling f. The two embeddings are *f-isomorphic* if and only if there exists a isomorphism from Ψ_1 to Ψ_2 which preserves the labeling of the vertices. A graph G is *n-flexible* on Σ (or simply n-flexible if Σ is understood) if n is the maximum number of embeddings of G which are not f-isomorphic for some fixed labeling f (it is clear that n does not depend on the particular choice of f). A *rigid* graph is one which is 1-flexible. Any flexibility of a graph embedding which can be explained via the results of Whitney will be referred to as a *Whitney twist*.

3. Projective Flexibility

All of the members of $M_3(PP)$ can be obtained from K_6 by Y-Δ operations (see [1]) and all of the members of $E_3(PP)$ can be obtained from the members of $M_3(PP)$ by a finite sequence of vertex expansions. It is natural therefore to begin our investigation of the flexibility of these embeddings by determining the flexibility of K_6 on PP.

PROPOSITION 1. *The graph K_6 is 12-flexible on the projective plane.*

Proof: It is well known and easy to show using Euler's formula that any embedding of K_6 in the projective plane is isomorphic to Γ_0, the embedding shown in Figure 1 (where the antipodal points of the outer circle are identified).

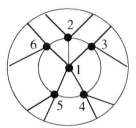

Figure 1

Further note that since K_6 is a complete graph all of its labelings are equivalent. It follows that any two labeled embeddings of K_6 in PP which are related by a rotation and/ or reflection of PP are f-isomorphic; so, an upper bound for the number of f-isomorphism equivalence classes can be found by examining the sequence of labelings around the vertex labeled 1. This is equivalent to solving the well known combinatorial problem of counting the number of ways of seating 5 people at a round table where two seatings are different if and only if there is some pair sitting next to each other in one of the seatings who are not sitting next to each other in the other seating. The solution to this problem is easily seen to be 12 (see nearly any elementary discrete mathematics text). Therefore, K_6 is at most 12-flexible on the projective plane. Finally, by inspection, none of the twelve embeddings generated in the above described manner are f-isomorphic.|

Fortunately, there are only fifteen embeddings in E3(PP) (see [1]). It is thus possible to examine each of them as in proposition 1. This has been done and yields the following theorem.

THEOREM 2. *The flexibility of the elements of E3(PP) are as shown in figure 2, where the flexibility of each embedding is given in parentheses.*

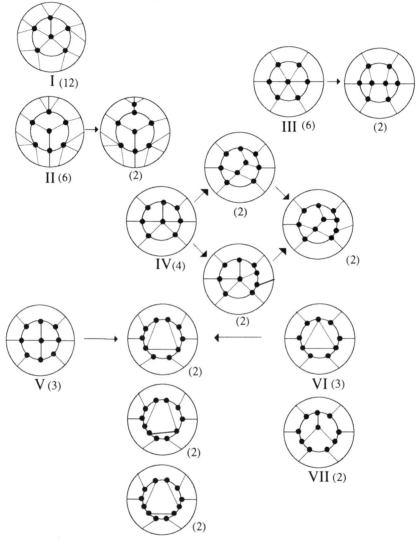

Figure 2

The seven embeddings denoted by roman numerals are the members of M3(PP). Note that I is an embedding of K_6 and that VII is an embedding of the Petersen graph. Also the graph embedded in V is isomorphic to the graph embedded in VI. Thus the underlying graphs of these embeddings are not the seven graphs which can be obtained from K_6 by combinatorial Y-Δ operations (the missing graph cannot be embedded on the projective plane).

If Γ is a projective embedding and $\rho(\Gamma) \geq 3$ then Γ is *q-twistable* if there exist vertices 1, 2, 3, 4, 5 and 6 which are related as in (a) of figure 3,

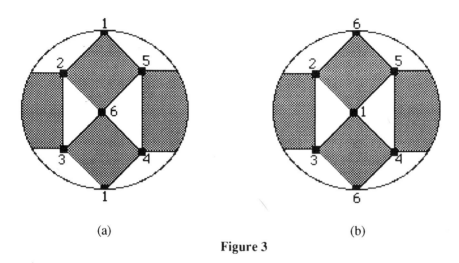

Figure 3

where the shaded regions represent any planar embedding (the "q" stands for quadrangle). If Γ is a q-twistable projective embedding of a graph G with labeling f then there is a projective embedding, Γq, of G with the same labeling which is a subset of the structure in (b) of Figure 3. The drawing Γq is obtained from Γ by detaching Γ at the vertices labeled 1 and 6 and flipping the two quadrangular regions containing these vertices in their boundaries. Since the vertices with distinct labels are distinct, it is easy to see Γq will not, in general, be f-isomorphic to Γ. The set of vertices labelled 1, 2, 3, for example, are on a common region of Γ but are not on a common region of Γq, (note $\rho(\Gamma) \geq 3$ implies that the regions must be distinct). The drawing Γq is called a *q-twist* of Γ.

The embedding Γ is t-twistable if there exist vertices 1, 2, 3, 4, 5, 6 and 7,

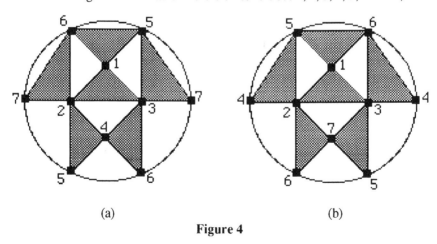

Figure 4

which are related as in the structure in (a) of Figure 4, where, again, the shaded regions represent any planar embedding (the "t" in t-twistable stands for triangle). If Γ is a t-twistable projective embedding of a graph G with labeling f then there is a projective embedding, Γt, of G with the same labeling which is a subset of the structure in (b) of Figure 4. The drawing Γ_t is obtained from Γ by fixing the shaded region with vertices of attachment 1, 2, 3; flipping the shaded region with vertices of attachment 1, 5, 6; sliding the regions with vertices of attachment 2, 6, 7 and 3, 5, 7 down to the bottom of the drawing; and, sliding the regions with vertices of attachment 3, 4, 6 and 2, 4, 5 up to the sides of the drawing. The drawing Γ_t is called a *projective t-twist* of Γ.

In the case that Γ is an element of $E_3(PP)$, inspecting the various embeddings of the drawings in figure 2, leads to the following theorem (a similar result is derived by Negami via a different method [2]).

THEOREM 3. *If Ψ and Ψ' are elements of $E_3(PP)$ with the same underlying graph and equivalent labelings then Ψ' can be obtained from Ψ by a finite number of projective q-twists and t-twists.*

More generally, suppose Γ is an embedding on the projective plane of a graph G with a labeling f such that $\rho(\Gamma) \geq 3$. There exists an embedding from among the embeddings shown in Figure 2 such that $\Psi \leq_e \Gamma$ ([3]). Thus each embedding of G in PP corresponds to one of the embeddings of Ψ enumerated in Theorem 2 (although some of the possible embeddings of Ψ may not correspond to embeddings of G). If C is a connected component of Γ which is distinct from Ψ then C must be contained in one of the faces of Ψ, each of which is homeomorphic to the plane. It follows that C is a planar embedding which may be embedded in any of the faces of Ψ, and the manner of embedding for any particular face is entirely determined by Whitney twists. Any block of Γ distinct from Ψ but contained in the same connected component of Γ as Ψ is also contained in a face of Ψ with a single vertex of attachment, v. Such a block may be embedded in any face of Ψ which is incident with v in a manner which is also determined by Whitney twists. Hence, we need only explain the flexibility of the block of Γ which contains Ψ to explain any flexibility not explained by Whitney, i.e. we may as well assume that G is 2-connected.

Let β be a bridge of Ψ in Γ. By inspection of the embeddings in figure 2, there is a vertex, v, of Ψ such that β is a subset of $W(v)$, the wheel neighborhood of v. Also, $W(v)$ is homeomorphic to a closed disk in $\mathbf{R}^2$, so, if Γ and Γ' induce f-isomorphic embeddings of Ψ then their embeddings of β can only differ by Whitney twists and theorem 3 has the following corollary.

COROLLARY 3.1. *Let Γ and Γ' be projective embeddings of representativity at least 3 with the same underlying graph and equivalent labelings then Γ' can be obtained from Γ by a finite number of projective q-twists, t-twists, and Whitney twists.*

Note that if Γ is a 3-representative projective embedding we may choose Ψ, a minor of Γ from among the embeddings of figure 2, with minimum flexibility. Evidently, the number of different embeddings which can be obtained from Γ by q-twists or t-twists is equal to the number of different embeddings of Ψ. Furthermore, 3-connected graphs do not have Whitney twists. It follows that a 3-connected, 3-representative projective embedding has flexibility 1, 2, 3, 4, 6 or 12.

Observe that an embedding which is q-twistable or t-twistable cannot be 4-representative. Thus theorem 3 has the following as a second corollary (a slight strengthening of another result obtained independently by Negami [2]).

COROLLARY 3.2. *If Γ is a projective embedding of a 3-connected graph with $\rho(\Gamma) \geq 4$ then Γ is rigid.*

4. Related Results and Questions

It has been shown that if Γ is an embedding of a graph G on a surface Σ_g with g handles and $\rho(\Gamma) \geq 2g + 3$ then Γ is a genus embedding of G and any other embedding of G on Σ_g can be obtained from Γ by a sequence of Whitney twists ([3]). In particular, if G is 3-connected then Γ is rigid. On the other hand, Archdeacon has shown that for any constant B there exists embeddings Ψ_1 and Ψ_2 embedded on different surfaces each with the same underlying graph and each with representativity at least B. It is not known whether $2g + 3$ is the best possible lower bound. The analogous lower bound for Σ_k a surface with k crosscaps would be $k + 3$, but this bound has not been shown to work in general (although it has been shown on some surfaces, note corollary 3.2 above).

Randby has shown that minor minimal n-representative projective embeddings for any fixed n are always related by Y-Δ operations. Using a refinement of representativity, Schriver has been able to show an analogous result holds for all orientable surfaces. The question remains open for nonorientable surfaces other than the projective plane.

For flexibility, the best possible result would be to find a finite collection of "twists" which would determine the flexibility of embeddings on any surface. Alternatively, one might show that no such collection exists.

REFERENCES

1. R. Vitray, *2 and 3 Representative Projective Planar Embeddings*, Journal of Combinatorial Theory, Series B, Volume 54, 1992.

2. S. Negami, *Re-embeddings of Projective-Planar Graphs*, Journal of Combinatorial Theory, Series B, Volume 44, 1988.

3. N. Robertson and R. Vitray, Representativity of Surface Embeddings, in *Paths, Flows, and VLSI-Layout* (B. Korte, L. Lovasz, H.J. Promel, and A. Schrijver, eds.), Springer, Berlin, 1990, pp. 293-328.

On Non-null Separating Circuits In Embedded Graphs

XIAOYA ZHA AND YUE ZHAO

ABSTRACT. Suppose G is a graph embedded in a surface Σ of orientable genus $g \geq 2$ or non-orientable genus $k \geq 2$. We prove that representativity $\rho \geq 6$ for orientable surfaces and $\rho \geq 5$ for non-orientable surfaces are sufficient conditions for the existence of a circuit C in G which separates Σ into two nontrivial parts. The proof yields a polynomially bounded algorithm to find such separating circuits. Examples where such circuits do not exist are given when $\rho = 2$.

1. Introduction

Under what conditions do there exist non-null-homotopic separating circuits in a graph embedded in a surface of orientable genus $g \geq 2$ or non-orientable genus (number of cross-caps) $k \geq 2$? The interest here is not just in finding structural conditions implying the existence of a certain type of circuit in an embedded graph, but also in the potential applications of such a circuit to induction on the genus of a surface. Suppose there exists a non-null-homotopic circuit in an embedded graph which separates the surface into two parts each containing some of the graph. We cut the surface along this separating circuit and cap off the boundary circuits in the two connected parts by disks. Each disk may include any minor of the other part which embeds in it; often this minor will consist of one vertex joined by edges to vertices of the separating circuit. The resulting graphs are minors of the original graph and are embedded in surfaces with lower genus. Therefore, for classes of embedded graphs closed under this type of surface minor, we can perform this surgery and apply induction on the genus.

1991 *Mathematics Subject Classification.* Primary 05C10.
Supported in part by NSF Grant number DMS-8903132.
This paper is in final form and no version of it will be submitted for publication elsewhere.

Define an *embedding* Ψ of a graph G on a surface Σ to be a triple (Σ, G, V) such that G is a closed subset of the surface Σ, and V is a finite subset of G such that the connected components of $G \backslash V$ are a finite number of open 1-cells. As usual, an element of V is called a *vertex*, the closure of each 1-cell of $G \backslash V$ is called an *edge*, and a connected component of $\Sigma \backslash G$ is called a *face* of Ψ. We denote by $V(\Psi)$, $E(\Psi)$, $F(\Psi)$ the sets of vertices, edges and faces of Ψ, respectively. Evidently, the pair (G, V) defines a finite topological graph in the usual sense. All graphs considered here are topological and we will use standard graph theoretical terminology [6] regarding them. In particular, a *path* is a homeomorphic image of the interval $[0,1]$ contained in G whose endpoints are vertices of G and a *circuit* is a homeomorphic image of the unit circle contained in G. A *subgraph* of (G, V) is a graph (G', V') with $G' \subseteq G$ and $V' = G' \cap V$ while a *topological subgraph* is a graph (G', V') with $G' \subseteq G$ and $V' \subseteq V$. More generally, a *minor* of (G, V) is a graph (G''', V'') obtained from a subgraph (G', V') by contracting (sequentially) some edges to vertices. Note that when contracting an edge of the graph in an embedding (Σ, G, V), in order to preserve the surface embedding structure, we adopt the convention that the edge must always have distinct endpoints. To simplify notation we will usually refer to G as the *graph* of an embedding (Σ, G, V), rather than the pair (G, V).

Let Σ be a closed surface which is not the 2-sphere. A closed curve Γ in Σ is *essential* if it is not null-homotopic. If Ψ is an embedding of a graph G in Σ, then the *representativity* of Ψ is defined to be $\rho(\Psi) = \min\{|\Gamma \cap G| : \Gamma$ is an essential closed curve in $\Sigma\}$. Since representativity is a parameter which describes the degree to which the embedded graph G is a discrete "approximation" of the surface Σ, we use representativity to give some sufficient conditions for the existence of essential separating circuits in G. A partial answer to the opening question of this paper derives from the following observation. The set of embeddings on a fixed surface of graphs which contain no non-null-homotopic separating circuits forms a proper minor-closed subclass $\mathcal{L}$ of embeddings. Consequently, the representativity of the graphs in $\mathcal{L}$ is bounded, according to the results in Section 9 of [3], and so if the representativity is high enough, then G contains an essential separating circuit. We conjecture that $\rho(\Psi) \geq 3$ is sufficient to assure the existence of an essential separating circuit. Robertson and Thomas [4] have shown that if the surface is the Klein bottle and $\rho(\Psi) \geq 3$, then G has an essential separating circuit (in their paper, they use nested separating circuits to find the orientable genus of graphs embedded in the Klein bottle, and this can be seen as an application of essential separating circuits). Richter and Vitray [5] recently proved that if $\rho(\Psi) \geq 11$, then G has essential separating circuits.

In this paper we will prove that $\rho(\Psi) \geq 6$ for orientable surfaces (with at least 2 handles) and $\rho(\Psi) \geq 5$ for non-orientable surfaces (with at least 2 cross-caps) is sufficient for the existence of such circuits in embedded graphs. In Section 2 we prove some lemmas which are crucial to our main result. Section 3 contains the main theorem. In Section 4, a polynomial-time algorithm is given for finding a separating circuit under these sufficient conditions. In Section 5, we present examples with $\rho(\Psi) = 2$ which have no essential separating circuits in their embedded graphs.

2. Some Lemmas

Let Ψ be an embedding of G in Σ, where Σ is not simply connected. A *3-component* of a graph G can be represented by a maximal 3-connected topological subgraph of G, where two of these are considered equivalent if they differ only in the choice of pairwise internally disjoint paths in G for their edges. Given an embedding $\Psi = (\Sigma, G, V)$ there may be equivalent representations G_1, G_2 of a 3-component of G which are homeomorphically distinct as subembeddings of Ψ. A 3-component of G is *essential* if it contains an essential circuit. Robertson and Vitray have the following result, where the essential 3-component is unique up to a homeomorphism of Σ fixing its vertex-set pointwise (see [6] for details).

Proposition 2.1 ([6], page 308). *If $\rho(\Psi) \geq 3$ then G has an unique essential 3-component, say H. Moreover, $\rho(\Psi_H) = \rho(\Psi)$, where Ψ_H is the restriction of the embedding Ψ to H.*

Since Ψ_H is a minor of Ψ, the existence of essential separating circuits in H would imply the existence of essential separating circuits in G. As we are studying what representativity bound implies the existence of essential separating circuits, note that if $\rho(\Psi) \geq 3$ then the assumption that G is 3-connected can be made without loss of generality.

Let G be a 3-connected graph embedded in Σ with $\rho(\Psi) \geq 3$. For each vertex v, the symmetric difference of the boundaries of the faces incident with v is a circuit C passing through all the neighbors of v and bounding a closed disk D_v containing v. This closed disk is the *wheel neighborhood* of v ([6], page 298). The existence, for all $v \in V(G)$, of a wheel neighborhood D_v with at least three edges incident to v, is necessary and sufficient for $\rho(\Psi) \geq 3$ and G to be 3-connected. This criterion generalizes as follows [2]. Let k be a positive integer. Then necessary and sufficient conditions for $\rho(\Psi) \geq 2k + 1$ are that, for any $v \in V(G)$, there exist k disjoint nested circuits bounding disks containing v, and necessary and sufficient conditions for $\rho(\Psi) \geq 2k$ are that, for any face f of Ψ, there exist k disjoint nested circuits bounding disks containing f.

Suppose Ψ is an embedding and f is a face of Ψ. Denote the boundary of f by ∂f and the closure of f by $\overline{f}(= f \cup \partial f)$. Note that if $\rho(\Psi) \geq 2$ and G is nonseparable, then ∂f is a circuit. Let f, g be two faces of Ψ. If $\partial f \cap \partial g \neq \emptyset$, we say f and g are *attached*. If f and g are attached, then $\partial f \cap \partial g$ is the union of connected components, called the *attachments* of f and g. Clearly, if $\rho(\Psi) \geq 2$ and G is nonseparable, then each attachment is either a path or a vertex of G.

A closed curve Γ in Σ is Ψ-*minimal* if Γ is simple, essential, $\Gamma \cap G \subseteq V(G)$ and $|\Gamma \cap G| = \rho(\Psi)$. It is easy to see that a Ψ-minimal curve Γ always exists in Σ when $\rho(\Psi)$ is defined. Let Ψ be an embedding of a 3-connected graph G with

$\rho(\Psi) \geq 3$. Then any vertex has a wheel neighborhood and any two faces have at most one attachment, which is either an edge or a vertex of G. The consecutive faces through which Γ passes will have one attachment and non-consecutive faces will have disjoint boundaries. Hence we immediately have the following lemma.

Lemma 2.2. *Let G be a 3-connected graph and Ψ be an embedding of G in a surface Σ with $\rho(\Psi) \geq 3$. If Γ is a Ψ-minimal curve in Σ, then G has a circuit C which is (freely) homotopic to Γ.*

Actually as both Γ and C are simple curves they are in fact ambient isotopic, as can be seen by pushing Γ into the face boundaries along which it passes to form C.

Let G be a 3-connected graph and Ψ be an embedding of G in Σ with $\rho(\Psi) = k \geq 3$. Let Γ be a Ψ-minimal curve in Σ and $f_1, f_2, ..., f_k$ be the k faces in sequence through which Γ passes (called the *face ring* for Γ), and $v_1, v_2, ..., v_k$ be the vertices such that $v_i \in \partial f_{i-1} \cap \partial f_i$ for $i = 1, 2, ..., k$ (with $f_0 = f_k$) and $\Gamma \cap G = \{v_1, v_2, ..., v_k\}$. We describe the face ring $f_1, ..., f_k$ and Γ as shown in Figure 1 with the top line and the bottom line identified as indicated by the arrows. We can formally assign two sides to the section of Γ, denoted by $\Gamma(v_1, ..., v_k)$, between v_1 (through $v_2, ..., v_{k-1}$) and v_k. For $i = 1, 2, ..., k$, denote by $D^l_{v_i}$ ($D^r_{v_i}$) the left (right) half disk into which Γ divides the wheel neighborhood D_{v_i}, including $\overline{f}_{i-1}$ and $\overline{f}_i$, where i is read modulo k. Note that if $\partial f_{i-1} \cap \partial f_i$ is an edge, then one of $D^l_{v_i}$ and $D^r_{v_i}$ is exactly D_{v_i}.

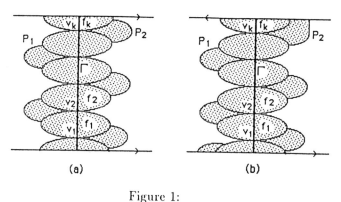

Figure 1:

Let D be a closed disk in Ψ. Then the interior of D is denoted by D°. The following lemma is easy to prove.

Lemma 2.3. *Let G be a 3-connected graph and Ψ be an embedding of G in Σ with $\rho(\Psi) \geq 4$. Let $\Gamma, f_i, D^l_{v_i}, D^r_{v_i}, i = 1, 2, ..., k$ be defined as above. Then*
(1) $D^r_{v_i} \cap D^l_{v_{i+1}}$, $D^l_{v_i} \cap D^r_{v_{i+1}}$ and $D^r_{v_i} \cap D^r_{v_{i+1}}$ are closed disks;
(2) $(D^r_{v_i})^\circ \cup (D^l_{v_{i+1}})^\circ$, $(D^l_{v_i})^\circ \cup (D^r_{v_{i+1}})^\circ$ and $(D^r_{v_i})^\circ \cup (D^r_{v_{i+1}})^\circ$ are simply connected.

Proof: As $\rho(\Psi) \geq 4$, each face f_i has a closed disk neighborhood D_i, which is bounded by a circuit C_i in G such that ∂f_i and C_i are nested and $\partial f_i \cap C_i = \emptyset$. Clearly, $D_{v_i} \cup D_{v_{i+1}}$ is contained in D_i. It is easy to see that $D^r_{v_i} \cap D^l_{v_{i+1}} = D^l_{v_i} \cap D^r_{v_{i+1}} = \overline{f}_i$, which is a closed disk. If f_i is the only face which has both v_i and v_{i+1} on its boundary, then $D^r_{v_i} \cap D^r_{v_{i+1}} = \overline{f}_i$. If there is another face f which also has v_i and v_{i+1} on its boundary and is on the right side of Γ (see Figure 2), then f is the only face besides f_i which has this property and $\partial f \cap \partial f_i$ is an edge, since G is 3-connected. Therefore $D^r_{v_i} \cap D^r_{v_{i+1}} = \overline{f}_i \cup \overline{f}$ is a closed disk. Hence (1) is true.

The truth of (2) follows immediately from (1). □

Remark: Since $D^r_{v_i}$ and $D^l_{v_{i+1}}(D^r_{v_{i+1}})$ may meet in more than one component (see Figure 2), the boundaries of $(D^r_{v_i})° \cup (D^l_{v_{i+1}})°$ and $(D^r_{v_i})° \cup (D^r_{v_{i+1}})°$ may be not simple. However, the closed walk in the graph which is traced out in these boundaries following the circular order of the open disks $(D^r_{v_i})° \cup (D^l_{v_{i+1}})°$ or $(D^r_{v_i})° \cup (D^r_{v_{i+1}})°$ is well-defined. Here we do not distinguish the walk and the point set of the subgraph it traverses.

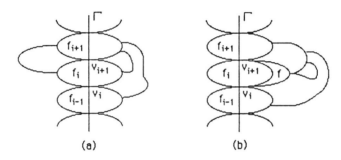

Figure 2:

Let Σ be a surface. A closed curve in Σ is 2-*sided* if it is orientation-preserving and 1-*sided* if it is orientation-reversing. Our method involves the following important lemma:

Lemma 2.4. *Let G be a 3-connected graph, and Ψ be an embedding of G in Σ with $\rho(\Psi) \geq 5$ (or with $\rho(\Psi) \geq 3$ when G is trivalent). Let Γ be a Ψ-minimal curve in Σ. It follows that:*

(1) if Γ is 1-sided, then there exists a Möbius band M in Σ containing Γ which is bounded by a circuit C in G;

(2) if Γ is 2-sided, then there exists an annulus A in Σ containing Γ which is bounded by two circuits C_1, C_2 in G.

Proof: Assume Ψ is an embedding of a 3-connected graph G in a surface Σ with $\rho(\Psi) \geq 3$ and Γ is a Ψ-minimal curve in Σ. Let $f_1, f_2, ..., f_k$ be the face ring through which Γ passes and $v_1, v_2, ..., v_k$ be the vertices such that $v_i \in \partial f_{i-1} \cap \partial f_i$ for $i = 1, 2, ..., k$ (with $f_0 = f_k$) and $\Gamma \cap G = \{v_1, v_2, ..., v_k\}$. We want to expand $N = \overline{f}_1 \cup \overline{f}_2 \cup ... \cup \overline{f}_k$ to an annulus (or a Möbius band). Since consecutive face boundaries meet at a vertex or an edge and non-consecutive faces have disjoint boundaries, if $\partial f_{i-1} \cap \partial f_i$ is an edge, for $i = 1, 2, ..., k$, then N is as required already. This is true when G is trivalent, and thus justifies the claim in parentheses.

Now we assume G is not trivalent and $\rho(\Psi) \geq 5$. Let $H = (\bigcup_{i\ even} (D_{v_i}^l)^\circ) \cup (\bigcup_{i\ odd} (D_{v_i}^r)^\circ)$. By Lemma 2.3 and its following remark, H is bounded by two walks P_1 and P_2 in G as shown in Figure 1. In Figure 1(a), P_1 and P_2 are two closed walks, while in Figure 1(b), P_1 and P_2 form one closed walk after identifying their endvertices. Note that P_1 and P_2 need not be simple. When k is even, H is formed by alternate right and left open half disks of $(D_{v_i})^\circ$. When k is odd, the two consecutive open half disks $(D_{v_k}^r)^\circ$ and $(D_{v_1}^r)^\circ$ are both right open half disks, while the other open half disks are chosen alternately.

Claim 1: *If Γ is 2-sided then $P_1 \cap P_2 = \emptyset$ and if Γ is 1-sided then P_1 and P_2 meet only at their endvertices.*

Suppose the claim is false. Then there is a common vertex w of P_1 and P_2 which is an internal vertex of P_1 or P_2. Then $w \in P_1 \cap \partial D_{v_i}$ and $w \in P_2 \cap \partial D_{v_j}$, for some $i, j \leq k$. Note that w may not be one of the vertices v_s, for $s \in \{1, 2, ..., k\}$, because for odd s, v_s is not on P_2 and for even s, v_s is not on P_1. Let Γ_1 be a simple curve on Σ which consists of a segment Γ_1' of Γ between v_i and v_j meeting G in the fewest number of vertices and a curve Γ_1'' from v_i to v_j meeting G only at v_i, w and v_j and passing through P_1 and P_2 at w. Note that Γ_1 intersects Γ in Γ_1' only and crosses Γ through this intersection. Therefore Γ_1 is homologically non-null, that is, Γ_1 is a nonseparating simple curve in Σ. Since $min\{|j-i|, k-|j-i|\} \leq \lfloor \frac{k}{2} \rfloor$ and $|\Gamma_1 \cap G| \leq min\{|j-i|, k-|j-i|\} + 2$, it follows that $|\Gamma_1 \cap G| < k$ when $k \geq 5$. This implies $\rho(\Psi) < k$, which is a contradiction. Hence Claim 1 is true.

Claim 2: *For $i = 1, 2$, if P_i is not simple then we can find a simple walk $P_i' \subset P_i$ which is homotopic to P_i.*

Suppose P_1 is not simple, and let w be a repeated vertex of P_1. Since P_1 traverses edges in the wheel neighborhoods of $v_1, v_2, ..., v_k$, we may assume that w repeats on D_{v_i}, D_{v_j} with $1 \leq i < j \leq k$. As $D_{v_i} \cap D_{v_j} \neq \emptyset$ implies $i - j \leq 2$, we can find a simple closed curve Γ_2 in Σ which consists of a segment Γ_2' of Γ between v_i and v_j meeting G in the fewest number of vertices and a curve Γ_2'' from v_i to v_j meeting G only at v_i, w and v_j. Now $i - j \leq 2$ implies that $|\Gamma_2 \cap G| < k$ if $k \geq 5$. Since $\rho(\Psi) = k$, we know that Γ_2 is null-homotopic and so bounds a disk. Thus we can delete the part of P_1 bounded by Γ_2 and obtain $P_1^\#$ which is contained in P_1 and is homotopic to P_1. If $P_1^\#$ is still not simple, we repeat this procedure until we obtain a simple walk P_1' homotopic to P_1. Hence Claim 2 is also true.

By Claim 1 and Claim 2, if Γ is 1-sided, then $P_1' \cup P_2'$ is a circuit, and if Γ is 2-sided, then P_1' and P_2' are two circuits. When Γ is 1-sided, let M be the closed region bounded by $P_1' \cup P_2'$ and containing H, and when Γ is 2-sided, let A be the closed region bounded by P_1' and P_2' and containing H. By the construction of P_1' and P_2', M is a Möbius band and A is an annulus. Thus Lemma 2.4 is true. □

A θ-*configuration* Θ is formed by three internally disjoint simple curves Δ_1, Δ_2 and Δ_3 with common distinct ends. Note that Θ contains three simple closed curves Γ_1, Γ_2 and Γ_3, where $\Gamma_1 = \Delta_2 \cup \Delta_3$, $\Gamma_2 = \Delta_3 \cup \Delta_1$, $\Gamma_3 = \Delta_1 \cup \Delta_2$. The following is obvious.

Lemma 2.5. *Let Θ be a θ-configuration in a surface Σ and Γ_1, Γ_2, Γ_3 be its three simple closed curves. Then*

(1) *if two are null-homotopic then the third is also null-homotopic;*

(2) *either all three are 2-sided or one is 2-sided and two are 1-sided.*

A closed curve Γ in Σ is Ψ-*minimal 1-sided* if Γ is simple, 1-sided, $\Gamma \cap G \subseteq V(G)$ and $|\Gamma \cap G| \leq |\Gamma' \cap G|$ for all 1-sided closed curves Γ' in Σ. This is well-defined for all non-orientable embeddings. The following is a strengthening of Lemma 2.4 for non-orientable surfaces.

Lemma 2.6. *Let G be a 3-connected graph, and Ψ be an embedding of G in a surface Σ with $\rho(\Psi) \geq 5$ (or with $\rho(\Psi) \geq 3$ when G is trivalent). Let Γ be a Ψ-minimal 1-sided curve in Σ. Then there exists a Möbius band M containing Γ which is bounded by a circuit C in G.*

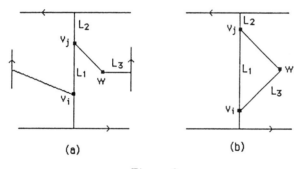

Figure 3:

Proof: If Γ is also a Ψ-minimal curve, then this follows from Lemma 2.4. Therefore we assume $|\Gamma \cap G| = k > \rho(\Psi)$.

Let $f_1, f_2, ..., f_k$ be a k-face ring of Γ, and $\Gamma \cap G = \{v_1, ..., v_k\}$ be as in the proof of Lemma 2.4. Form N, P_1 and P_2 as before. If G is trivalent and $\rho(\Psi) \geq 3$, then N is the required Möbius band.

Now assume G is not trivalent. If P_1 and P_2 meet at an internal vertex or P_s is not simple for $s=1$ or 2, then there exists a repeated vertex w. As in the proof of Lemma 2.4, this w is on the rims of D_{v_i} and D_{v_j}, where $1 \leq i < j \leq k$. We obtain a θ-configuration as shown in Figure 3, with v_i and v_j as its two ends, such that $\Delta_1 \cup \Delta_2 = \Gamma$ and Δ_3 is a simple curve on Σ joining x_i and x_j and meeting G only at x_i, w, x_j. Figure 3(a) represents the case where P_1 and P_2 meet at an internal vertex, and Figure 3(b) represents the case where P_s is not simple, $s=1$ or 2. Let $\Gamma_1 = \Delta_2 \cup \Delta_3$ and $\Gamma_2 = \Delta_3 \cup \Delta_1$. We know $\Gamma = \Delta_1 \cup \Delta_2$

is 1-sided. If $j - i \leq 2$, then $|\Gamma_2 \cap G| \leq 4$. Since $\rho(\Psi) \geq 5$, Γ_2 is null-homotopic and bounds a disk. This can only happen when P_1 (or P_2) is not simple, as Γ_2 is essential in the case of Figure 3(a). We delete the part of P_1 bounded by Γ_2 and obtain $P_1^\#$, which is contained in P_1 and is homotopic to P_1, as we did in the proof of Lemma 2.4.

If $3 \leq j - i \leq k - 3$, then we have $5 \leq |\Gamma_2 \cap G| \leq k - 1$ and similarly $|\Gamma_1 \cap G| < k$. But then Γ_1 and Γ_2 must be 2-sided as Γ is Ψ-minimal 1-sided. This contradicts Lemma 2.5(2).

If $j - i = k - 2$, or $j - i = k - 1$, then $|\Gamma_1 \cap G| \leq 4$, and we are back to the first case.

Therefore, P_1 and P_2 meet only at endvertices, and for $i = 1, 2$ we can find a simple walk $P_i' \subseteq P_i$ which is homotopic to P_i. Thus $P_1' \cup P_2'$ bounds a Möbius band including Γ. This completes our proof. □

3. The Main Theorem

In this section, we will prove the main result of this paper, the following theorem.

Theorem 3.1. *Let G be a connected graph and Ψ be an embedding of G in a surface Σ of orientable genus $g \geq 2$ or non-orientable genus $k \geq 2$. Suppose either*

(1) Σ has a separating Ψ-minimal curve and $\rho(\Psi) \geq 3$, or

(2) Σ is a non-orientable surface and $\rho(\Psi) \geq 5$ (or $\rho(\Psi) \geq 3$ when G is trivalent), or

(3) Σ is an orientable surface and $\rho(\Psi) \geq 6$ (or $\rho(\Psi) \geq 5$ when G is trivalent).

Then there exists an essential separating circuit in G.

Proof: Theorem 3.1 is true in case (1) by Lemma 2.2.

Suppose Σ is a non-orientable surface and $\rho(\Psi) \geq 5$ (or $\rho(\Psi) \geq 3$ if G is trivalent). Let Γ be a Ψ-minimal 1-sided curve in Σ. By Lemma 2.6, Γ is contained in a Möbius band which is bounded by a circuit C in G. Clearly, C is an essential separating circuit. Therefore Theorem 3.1 holds in case (2).

Now we assume Σ is an orientable surface and $\rho(\Psi) \geq 6$ (or $\rho(\Psi) \geq 5$ when G is trivalent). Let Γ be a Ψ-minimal simple curve in Σ. By Theorem 3.1(1), we may assume Γ is nonseparating. By Lemma 2.4(2), there exists an annulus A containing Γ which is bounded by two circuits C_1 and C_2 in G. By cutting Σ along C_1, we obtain a new surface, which has two boundary circuits C_1 and C_1', where C_1' is a copy of C_1 cut away from A. Capping C_1 and C_1' off by two disks f_1 and f_1', respectively, gives a closed surface Σ' and an embedding Ψ' of a graph G' on Σ'. The Euler characteristic of Σ' has been increased by 2. Now find a shortest face chain $g = \{g_1, g_2, ..., g_l\}$ in Ψ' joining a vertex of C_2 and a vertex of C_1'. Without loss of generality, we may assume that the face g_1 is attached to a face g_0 in the right half disk $D_{v_1}^r$, in the notation used in the proof of Lemma

2.4(2) (if g_1 is attached to v_i for some even i, then the proof is actually easier). As $C_1' \cap C_2 = \emptyset$ it follows that $l \geq 1$. Let Γ_1 be a simple curve in Σ' from x_0 to x_l with $\Gamma_1 \cap G' = \{x_0, x_1, ..., x_l\}$ as shown in Figure 4, where the x_i's are vertices in G' such that $x_0 \in \partial g_1 \cap C_2$, $x_l \in \partial g_l \cap C_1'$ and $x_i \in \partial g_i \cap \partial g_{i+1}$ for $i = 1, 2, ..., l - 1$.

Claim: *There exists a closed disk B bounded by a circuit in G' which contains Γ_1 and is such that $\partial B \cap C_2$ and $\partial B \cap C_1'$ are two paths.*

Assign two sides to Γ_1 (which is possible since Γ_1 is a curve homeomorphic to the interval [0,1] in Σ'), say its *top* and *bottom*. Consider all the faces incident with C_2 and x_1. Clearly $C_2 \cap D_{x_1} \neq \emptyset$ since g_1 is a face joining C_2 and x_1. Without loss of generality, choose g_1, Γ_1 (fixing $g_2, ..., g_{l-1}$) so that there is no other face g_1' which joins C_2 and x_1 and is above Γ_1. Choose g_l similarly.

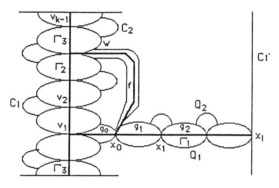

Figure 4:

Now take the symmetric difference of the boundaries of all faces on the top of Γ_1 which are incident with x_i, for $i = 0, 1, ..., l$. The faces $g_1, ..., g_l$ are included here, but the face f_1' is not included, and neither are the faces incident with x_0 in the annular region between C_1 and C_2. This forms two walks Q_1 and Q_2 from x_0 to x_l, with Q_1 below or touching Γ_1 and Q_2 above Γ_1. Let R denote the open region containing $g_1, ..., g_l$ and bounded by Q_1, Q_2. We want to show:

(a) $Q_1 \cap Q_2 = \{x_0, x_l\}$, and Q_1 and Q_2 are homotopic to Γ_1.

(b) Q_1 is simple, and if Q_2 is not simple, then we can find a path $Q_2' \subset Q_2$ which is homotopic to Q_2.

(c) If $\overline{R}$ is not a closed disk or $\partial R \cap C_2$ (or $\partial R \cap C_1'$) is not a path, then there is an open disk R' containing R such that $\overline{R'}$ is a closed disk such that $\partial R' \cap C_2$ and $\partial R' \cap C_1'$ are paths in G.

Proof of (a): The proof is similar to the part of the proof of Lemma 2.4 where we proved that P_1 and P_2 do not meet each other when the surface is orientable or meet only at endvertices when the surface is non-orientable. If $w \in V(\Psi') \cap Q_1 \cap Q_2$ exists with w internal to Q_1 or Q_2, then for some $i, j \leq l$ we can find a closed simple curve Γ_1' on Σ', containing the segment of Γ_1 between x_i and x_j, which crosses Γ_1 and passes through x_i, w, x_j consecutively. This

w is on ∂D_{x_i} and ∂D_{x_j}. Extend Γ_1 to C_1 (through g_0 to v_1 if $x_0 \neq v_1$) and following Γ from v_1 to x_l (crossing the first face joining Γ to x_l if $x_l \neq v_i$ for some i) to form a closed simple curve Γ_1^* in Σ. Then Γ_1^* is essential in Σ, as it crosses C_1 only once. Since Γ_1' crosses Γ_1^* only once in Σ, it is also essential in Σ. Therefore $|\Gamma_1' \cap G'| = |\Gamma_1' \cap G| \geq \rho(\Psi) \geq 5$. This implies $|j - i| \geq 3$. Hence we can find a shorter face chain, which passes through x_i, w and x_j, to join C_2 and C_1'. This contradicts the minimality of $g = (g_1, ..., g_l)$. Therefore $Q_1 \cap Q_2 = \{x_0, x_l\}$.

It is clear that Q_1 is homotopic to Γ_1. Note that any vertex of G' on the circuits C_1 and C_1' has two nested half circuits which together with the corresponding sections of C_1 or C_1' bound closed disks containing this vertex. Any vertex on C_2 has a wheel neighborhood and a half circuit (or a circuit) together with a section (or vertex, respectively) of C_1 bounding a disk containing this wheel neighborhood. Any other vertex has at least two nested circuits in G' bounding disks containing that vertex. These statements imply that Q_2 is also homotopic to Γ_1. Thus (a) is true.

Proof of (b): As Q_1 is contained in $\partial g_1 \cup \partial g_2 \cup ... \cup \partial g_l$ it is clear from the definition of shortest face chains that Q_1 is simple when $\rho(\Psi) \geq 3$. If Q_2 is not simple, then there is a repeated vertex w and i and j ($i < j$) such that w is on the boundaries of D_{x_i} and D_{x_j}. By the assumption that g is a shortest face chain joining C_2 and C_1', it follows that $j - i \leq 2$. Therefore we can find a simple closed curve Γ_1'' in Σ' which passes through x_i, w, x_j consecutively and meets G' in at most four vertices. Since Γ_1'' does not cross C_1 and C_1', if Γ_1'' is essential in Σ' then it is also essential in Σ. Thus $|\Gamma_1'' \cap G'| = |\Gamma_1'' \cap G| \leq 4$ implies that Γ_1'' is null-homotopic in Σ'. As Γ_1'' bounds a disk which contains the closed section of Q_2 between the two repetitions of w, we can delete this section and obtain a walk $Q_2' \subset Q_2$. Clearly Q_2' is homotopic to Q_2. This procedure can be repeated until we obtain a path Q_2'' which is homotopic to Q_2 and joins C_2 and C_1'. Thus (b) is also true.

Proof of (c): This needs more argument. By (b), we may assume that Q_2 is simple. Since R is bounded by Q_1 and Q_2 and both Q_1 and Q_2 are homotopic to Γ_1, R is null-homotopic in Σ'. As $Q_1 \cup Q_2$ is a circuit, it follows that $\overline{R}$ is a closed disk. Suppose $\partial R \cap C_2$ has more than one component. Let w be a vertex in one of the components of $\partial R \cap C_2$ which does not contain x_0. Since g_1 is the only face in R joining C_2 and x_1, there must exist a face f which is above Γ_1 and is incident with x_0 and w. Then w is either on the rim of the wheel neighborhood of v_i for some odd i or w is one of the v_i for some even i. Now there exist two simple closed curves Γ_2 and Γ_3 in Σ' such that $\Gamma_2 \cap G' = \{w, x_0, v_1, ..., v_i\}$ and $\Gamma_3 \cap G' = \{x_0, w, v_i, v_{i+1}, ..., v_k, v_1\}$. To insure that either $|\Gamma_2 \cap G'| < k$ or $|\Gamma_3 \cap G'| < k$ (so that at least one of Γ_2 and Γ_3 bounds a disk), we need

$$\lfloor \frac{k-2}{2} + 4 \rfloor < k,$$

which implies $k \geq 7$. If G is trivalent, then the annulus A is formed by the union of the $\overline{f}_i$ for $i = 1, ..., k$, where $(f_1, ..., f_k)$ is the face ring through which the Ψ-minimal 2-sided curve Γ passes, and C_2 is the union of sections of ∂f_i, for $i = 1, ..., k$. Therefore the condition that $k \geq 5$ is sufficient to imply either $|\Gamma_2 \cap G'| < k$ or $|\Gamma_3 \cap G'| < k$. Hence, for a trivalent graph, we only need $k \geq 5$.

If Γ_2 bounds a disk in Σ', then we expand R to R_1 by adding in all the faces together with the parts of their boundaries contained in Γ_2 but not contained in the annular region between C_1 and C_2. Repeat this procedure until no such Γ_2's occur. Let the resulting open disk be R'. If $\partial R' \cap C_2$ has more than one component, and Γ_2 is the closed simple curve on Σ' as defined above, then Γ_2 is essential in Σ'. This implies that $|\Gamma_2 \cap G'| = |\Gamma_2 \cap G| \geq k$, and therefore $|\Gamma_3 \cap G'| = |\Gamma_3 \cap G| < k$. So Γ_3 is null-homotopic in Σ, and hence null-homotopic in Σ', as Γ_3 does not cross C_1 and C_1' in Σ'. Since C_2 is also null-homotopic in Σ', it follows by Lemma 2.5(1) that Γ_2 is null-homotopic in Σ', as C_2, Γ_2 and Γ_3 form a Θ-configuration. This is a contradiction. Thus (c) holds if $\rho(\Psi) \geq 7$. If $\rho(\Psi) = 6$, since in Lemma 2.4, H is constructed by taking the union of the half disks $(D_{v_i}^l)^\circ$ and $(D_{v_i}^r)^\circ$ alternately, either $|\Gamma_2 \cap G'| < 6$ or $|\Gamma_3 \cap G'| < 6$. By the argument above, we can show that $\partial R' \cap C_2$ is also a path if $\rho(\Psi) = 6$. Similarly we can prove that $\partial R' \cap C_1'$ is a path. Clearly, R' is homotopic to Γ_1, and therefore it is null-homotopic in Σ'. If G is trivalent, then a similar argument will show that $\partial R' \cap C_2$ and $\partial R' \cap C_1'$ are paths when $\rho(\Psi) \geq 5$. Thus (c) is true. By (a), (b) and (c), the above claim is true with $B = \overline{R'}$.

The union of the annulus A and the band B gives a torus with a disk removed. As Σ is an orientable surface of genus ≥ 2, the complement of $A \cup B$ in Σ is a non-trivial surface with a disk removed. This completes our proof. □

Remark: If the surface is non-orientable with genus at least 2, then the union of the annulus A and the disk B gives a torus with a disk removed or a Klein bottle with a disk removed (the later case occurs when the disk B is connected to C_1' in a twisted way). This is not the case we are considering here. However we do need this in Section 4, where we give a polynomial-time algorithm to find essential separating circuits, since we do not have a polynomial-time algorithm to find a Ψ-minimal 1-sided curve. If $k \geq 3$ then the complement of $A \cup B$ in Σ is a non-trivial surface with a disk removed. The case $k = 2$ follows from the result in [4].

4. Finding an Essential Separating Circuit in Polynomial-Time

The proof of our main theorem yields a polynomially bounded algorithm, as will be explained in this section.

Theorem 4.1. *Let G be a nonseparable graph embedded in a surface Σ with $\rho(\Psi) \geq 6$ (or with $\rho(\Psi) \geq 5$ when G is trivalent). Then there is a polynomial-time algorithm to find an essential separating circuit in G.*

Remark 1: We will start the algorithm with a Ψ-minimal curve Γ by applying the polynomial-time algorithm given in [6] for computing $\rho(\Psi)$ from input Ψ. The input Ψ may be given by the graph G and the facial walks of the embedding. The facial walks are circuits under the hypotheses of Theorem 4.1.

Remark 2: When $G(\Psi)$ is nonseparable and $\rho(\Psi) \geq 2$, the *radial graph* $R(\Psi)$ of Ψ is a graph embedded on Σ obtained by placing a vertex in each face f of Ψ and connecting each such vertex by edges to all the vertices of G on the boundary of f. The choice of $R(\Psi)$ is unique up to a homeomorphism of Ψ fixing G. Call the added vertices the *facial* vertices of $R(\Psi)$. The vertex set of $R(\Psi)$ is the union of the set $F'(\Psi)$ of facial vertices and $V(\Psi)$. The edge set of $R(\Psi)$ consists of all the edges chosen joining facial vertices and vertices in $V(\Psi)$. Note that $R(\Psi)$ is a bipartite graph embedded on Σ, and a Ψ-minimal curve in Σ, which meets $V(\Psi)$ and $F(\Psi)$ alternately, corresponds to a circuit in $R(\Psi)$.

Proof of Theorem 4.1:

Step 1. Find a Ψ-minimal curve Γ in Σ. For $k = \rho(\Psi)$ denote the vertices of G through which Γ passes by $v_1, ..., v_k$ and the faces through which Γ passes by $f_1, ..., f_k$ in cyclic order and with $v_i \in \partial f_{i-1} \cap \partial f_i$, for $i = 1, ..., k$, with indices reduced modulo k. This uses polynomial time (see [6], page 312).

Step 2. Test if Γ is a separating curve. Since $\rho(\Psi) \geq 5$, Γ is separating if and only if $\{v_1, ..., v_k\}$ forms a non-trivial vertex cutset of G. This is just a connectivity test.

Step 3. If Γ is separating, then the symmetric difference of $\partial f_1, ..., \partial f_k$ gives two essential separating circuits C_1, C_2 which are homotopic to Γ. If Γ is non-separating, assign left and right sides to the segment of Γ from v_1 to v_k. Starting from v_1, find the symmetric difference C of the alternate left and right half disks of $v_1, ..., v_k$. Then, as $\rho(\Psi) \geq 6$, C is either a closed walk or the disjoint union of two closed walks. If a component of C is not simple, remove a closed sub-walk W to obtain $C' \subset C$. The subwalk W can be obtained in the following way: Let w be a repeated vertex on C. Then w is on the rims of the wheel neighborhoods of two vertices v_i and v_j, for $1 \leq i < j \leq k$. We know that $j - i \leq 2$. The walk W is between two appearances of w, and its edges are on the rims of $D_{v_i}, D_{v_{i+1}}$ (or $D_{v_{i+2}}$ if $j - i = 2$). It is homotopically trivial as shown in the proof of Lemma 2.4. Repeat this process to find C'' with two simple components and homotopic to C. Rename C'' as C. If the graph is trivalent, then take the symmetric difference C of $\partial f_1, ..., f_k$, which is a circuit if Γ is 1-sided and two disjoint circuits if Γ is 2-sided.

Step 4. If C is a circuit, then C is an essential separating circuit which separates a cross-cap from the rest of Σ. This is the required separating circuit.

Step 5. If $C = C_1 \cup C_2$ where C_1 and C_2 are disjoint circuits in G, then by Lemma 2.4, we know that C_1 and C_2 bound an annulus. Cut Σ along C_1, splitting C_1 into two circuits C_1 and C_1'. Cap off C_1 and C_1' by two disks f_1 and f_1' to obtain an embedding Ψ'.

Step 6. Find a shortest face chain from C_2 to C_1'. Process this in the radial graph $R(\Psi')$ by finding a shortest path joining the facial vertices for f_1 and f_1'.

Step 7. Expand the shortest face chain to a null-homotopic band B by the method used in the proof of the main theorem. Then the symmetric difference of the boundaries of the annulus and the null-homotopic band gives a separating circuit which separates a torus or a Klein bottle (with the interior of a closed disk removed) from the rest of Σ.

Since each step uses polynomial time, the algorithm is polynomially bounded.

5. Examples of Embeddings Without Essential Separating Circuits

In this section we present some examples of embeddings with representativity 2 which do not have essential separating circuits. For nonseparable graphs, an embedding Ψ with $\rho(\Psi) \geq 2$ is an embedding in which the boundary of each face is a circuit in the graph (such an embedding is called a *closed 2-cell embedding* [6] and a *strong embedding* in [1]). A *cycle double cover* of a connected graph G is a family of cycles (a cycle is the edge-disjoint union of circuits) of G such that each edge of G appears in exactly two of these cycles.

Let G be a nonseparable graph and $S = \{c_1, ..., c_l\}$ be a cycle double cover of G. Suppose each $c_i \in S$ is a circuit in G. Then c_i contains at most one pair of edges incident to a fixed vertex v. Let E_v be the set of edges which are incident with $v \in V(G)$. For each $v \in V(G)$, define a graph G_v with vertex set E_v and such that two vertices e and e' of G_v are adjacent if and only if e and e' are contained as edges in some circuit $c_i \in S$. Note that G_v is a 2-regular graph and when connected it is itself a circuit. Since G_v is a circuit if and only if the edge rotation at v guarantees a disk neighborhood of v in the 2-complex formed by pasting disjoint open disks onto the circuits $c_1, ..., c_l$, the following proposition is obvious.

Proposition 5.1. *A nonseparable graph G has a surface embedding Ψ with $\rho(\Psi) \geq 2$ if and only if G has a cycle double cover $S = \{c_1, ..., c_l\}$ such that c_i is a circuit in G for $1 \leq i \leq l$, and the induced graph G_v is connected (i.e., is a circuit) for all $v \in V(G)$.*

Figure 5: An orientable embedding of $2C_{2n+2}$ of genus n

Examples 5.2. Let $2C_m$, for $m \geq 3$, denote the m-circuit with each edge duplicated in parallel. Four circuits c_1, c_2, c_3 and c_4 are defined by the four broken curves in Figure 5 and Figure 6. It is easy to see that the following embedding induced by the cycle double cover $S = \{c_1, c_2, c_3, c_4\}$ of $2C_m$ is a ($\rho \geq 2$) embedding. The example in Figure 5 is an orientable embedding and the examples in Figure 6 are non-orientable embeddings. To check the non-existence of separating circuits, we observe that there are only four faces, and if there is any essential separating circuit, there must be two faces in each part, and the separating circuit cannot be formed by two parallel edges (a 2-circuit). A case by case check (three cases) shows the non-existence of essential separating circuits. A calculation of the Euler characteristic gives the genus.

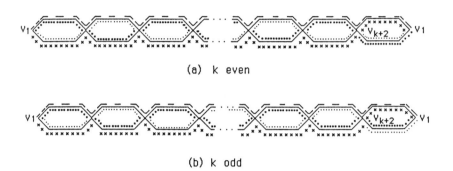

(a) k even

(b) k odd

Figure 6: A non-orientable embedding of $2C_{k+2}$ of genus k

Acknowledgment

The authors are grateful to Professor Bojan Mohar and Professor Bruce Richter. The upper bound in the sufficient condition of the main theorem was improved by their suggestions. The authors also would like to thank Professor Neil Robertson for his valuable advice, support, discussions and patience in helping us to revise this paper.

References

1. F. Jaeger, "A survey of the cycle double cover conjecture", *Annals of Discrete Mathematics* 27 (1985), 1-12.
2. N. Robertson, "Graph embedding problems", Lecture at the Seattle Graph Minors Conference.
3. N. Robertson, P. D. Seymour, "Graph minors. VII. Disjoint paths on a surface", *J. Combinatorial Theory, Ser. B*, 45 (1988), 212-254.
4. N. Robertson, R. Thomas, "On the orientable genus of graphs embedded in the Klein bottle", *J. Graph Theory* 15 (1991), 407-419.
5. R. B. Richter, R. Vitray, "On the existence of essential cycles in embedded graphs", preprint.
6. N. Robertson, R. Vitray, " Representativity of surface embeddings", *Algorithms and Combinatorics, Vol. 9, Paths, Flows, and VLSI-Layout* (1990), Springer-Verlag, 293-328.

DEPARTMENT OF MATHEMATICS, THE OHIO STATE UNIVERSITY, COLUMBUS, OHIO 43210

E-mail address: zha@function.mps.ohio-state.edu, zhao@function.mps.ohio-state.edu

Projective-Planar Graphs with Even Duals II

SEIYA NEGAMI

ABSTRACT. A face of a 2-cell embedding of a graph is said to be *even* if its boundary walk has even length. A graph G is said to have an *even dual* on a closed surface F^2 if there is an embedding of G on F^2 with only even faces. It will be shown that a connected graph G has an even dual on the projective plane if and only if G has a 2-fold planar bipartite covering and contains no two disjoint odd cycles. Moreover, the forbidden structures for graphs with projective-planar even duals will be determined.

1. Introduction

If a graph G is 2-cell embedded in a closed surface F^2 so that each face is bounded by a closed walk of even length (such a face is called an *even face*), then its dual G^* on F^2 is *even*, that is, each vertex of G^* has even degree. So we say that G has an even dual on F^2 if there is such an embedding of G on F^2. For example, a planar connected graph has an even dual on the sphere if and only if it is bipartite.

Recently, the author [5] discussed the relationship between even duals and planar coverings of projective-planar graphs and showed the following theorem:

THEOREM 1. *Let G be a nonplanar connected graph. Then G has an even dual on the projective plane if and only if either G is a projective-planar bipartite graph or its canonical bipartite covering $B(G)$ is planar.*

1991 Mathematics Subject Classification. Primary 05C10.
This paper is in final form and no version of it will be submitted for publication elsewhere.

A graph $\tilde{G}$ is called an *n-fold covering* of G if there is an n-to-1 surjection $p : V(\tilde{G}) \to V(G)$, called the *projection*, which induces a bijection $p|_{N(\tilde{v})} : N(\tilde{v}) \to N(v)$ between the neighborhoods of corresponding vertices $\tilde{v} \in p^{-1}(v)$ and $v \in V(G)$. In particular, $B(G)$ in the theorem is the covering of G obtained in the following way and is called the *canonical bipartite covering* of G. When G is bipartite, set $B(G) = G$. Otherwise, prepare two copies u_1 and u_2 of each vertex $u \in V(G)$ and join u_1 to v_2 and u_2 to v_1 by edges if there is an edge $uv \in E(G)$. The resulting graph is $B(G)$. This has been defined and discussed in [3].

By the general arguments in [5], the two alternatives in the theorem both imply that G has a 2-fold planar bipartite covering. Indeed, it is easy to see that any graph with an even dual on the projective plane satisfies this condition. The converse implication does not hold however without the nonplanarity of G. But we shall establish the following refinement of Theorem 1, eliminating the non-planarity hypothesis.

THEOREM 2. *Let G be a connected graph. Then G has an even dual on the projective plane if and only if G has a 2-fold planar bipartite covering and does not contain two disjoint odd cycles.*

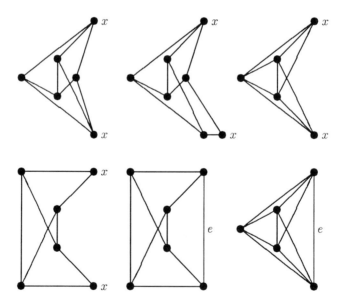

Figure 1. Obstructions for projective-planar even duals

Moreover, we shall determine the forbidden structures, called *obstructions*, for those graphs which have even duals on the projective plane. Each obstruction is obtained from one of the six graphs in Figure 1 in the following way. Subdivide the first four in Figure 1 to be bipartite so that there is a path of odd length joining two x's and identify the x's to a single vertex. Subdivide the last two so that any odd cycle, if any, passes through the edge e, which also may be subdivided.

THEOREM 3. *Let G be a connected graph which is not bipartite. Then G has an even dual on the projective plane if and only if G contains no two disjoint odd cycles and no obstruction.*

To prove these theorems, we shall develop a theory of pairs of graphs and their free involutions with "equivariant minors" in Section 2, and establish a characterization of those graphs with free involutions that can be embedded on the sphere so that their involutions extend to involutions on the sphere in Section 3. The proofs of Theorems 2 and 3 will be given, in Section 4, as an application of this theory.

2. Free involutions on graphs

Let G be a graph which may have multiple edges but no loop and let $\tau : G \to G$ be a free involution on G, that is, an automorphism $\tau \in Aut(G)$ of period 2 which has no fixed vertex and leaves no edge invariant. The pair (G, τ) is called an *involutive pair* of G and is supposed to satisfy these assumptions throughout this paper, unless stated otherwise. In particular, the pair (G, τ) is said to be a *connected involutive pair* if G is connected. Then the quotient of G by the action of τ is a graph, denoted by G/τ, and G covers G/τ doubly. The natural covering $p : G \to G/\tau$ is the same one as is defined combinatorially in the introduction if both G and G/τ are simple, and G/τ has loops if τ leaves multiple edges invariant pairwise.

Now let F^2 be a closed surface and $h : F^2 \to F^2$ an involution on F^2, that is, a homeomorphism of period 2 which may have fixed points, and call the piar (F^2, h) an *involutive pair* of F^2. An involutive pair (G, τ) is said to be *embeddable* in an involutive pair (F^2, h) of F^2 if there is an embedding $f : G \to F^2$ such that $hf = f\tau$. Roughly speaking, we embed G on F^2 so that the action on F^2 generated by h realizes the symmetry of G under τ, and we say that τ *extends* to h. If $g : F^2 \to F^2$ is a homeomorphism, then (F^2, ghg^{-1}) also is an involutive pair of F^2 and is said to be *conjugate* to (F^2, h). It is clear that an involutive pair (G, τ) is embeddable in (F^2, h) if and only if it is embeddable in any involutive pair conjugate to (F^2, h).

As is well-known, any graph G decomposes uniquely into subgraphs $B_1, \ldots, B_n$, called *blocks* of G, so that each of them is either a 2-connected subgraph in G or isomorphic to K_2 and is maximal with this property and that B_i and B_j have at most one vertex, called a *cut vertex*, in common if $i \neq j$. Since any automorphism preserves this block decomposition, $\tau(B_i)$ coincides with one of these blocks for any involution $\tau : G \to G$, say $B_j = \tau(B_i)$. If B_i met B_j only at a cut vertex $v \in V(B_i) \cap V(B_j)$, then v would be fixed by τ, contrary to our assumption of τ being free. Thus, either $\tau(B_i) = B_i$ or $\tau(B_i) \cap B_i = \emptyset$. We say that B_i is *invariant* in the former case and is *equivariant* under τ in both cases.

LEMMA 4. *Any connected involutive pair (G, τ) has a unique invariant block, which contains an invariant cycle.*

Proof. Let P be a path in G with end points v and $\tau(v)$. If P is a shortest such path for all $v \in V(G)$, then $P \cup \tau(P)$ is an invariant cycle. Then the block containing this invariant cycle is invariant and any other block is not invariant since otherwise a cut vertex separating them would be fixed. ∎

Let (G, τ) be an involutive pair and let S be a subgraph or a set of vertices or edges of G. Then S is said to be *equivariant* under τ if either $\tau(S) = S$ or $\tau(S)$ is disjoint from S. When S is a vertex cut of G, each component of $G - S$ is called a *fragment* of S. For a technical reason, we require an *equivariant fragment* F of S to be disjoint from $\tau(S)$ in addition to the above condition. If F is an equivariant fragment, then τ leaves $H = G - F \cup \tau(F)$ invariant and hence $(H, \tau|_H)$ is an involutive pair.

For example, suppose that G has an invariant edge cut $\{uv, \tau(uv)\}$ whose removal decomposes G into two invariant subgraphs H_1 and H_2 with $u \in V(H_1)$ and $v \in V(H_2)$. Then both $S_1 = \{u, \tau(u)\}$ and $S_2 = \{u, \tau(v)\}$ are equivariant cuts of G if H_1 has more than two vertices. In this case, if $F = H_1 - u$ is connected, then F is a fragment of S_2 but is not equivariant since $\tau(u) \in F \cap \tau(S)$. On the other hand, $H_1 - S_1$ is an equivariant fragment of S_1 if it is connected.

LEMMA 5. *Let (G, τ) be a connected involutive pair with $|V(G)| \geq 5$. If G is not 3-connected, then there exists an equivariant cut of one or two vertices which has an equivariant fragment.*

Proof. If G is not 2-connected, take one of the end blocks, say B, which contains a unique cut vertex v. Then $\{v\}$ is an equivariant cut with an equivariant fragment $B - v$. In this case, they are not invariant; otherwise, v would be fixed by τ.

If G is 2-connected, take a 2-vertex cut $S = \{u, v\}$ one of whose fragments, say F, is minimal under inclusion. That is, we suppose that F does not include any other fragment of any other 2-vertex cut. Because τ is free and has period 2, S has to be equivariant (i.e. $\tau(S) = S$ or $\tau(S) \cap S = \emptyset$). If $\tau(S) = S$, then both F and $\tau(F)$ are components of $G - S$ and hence either $F = \tau(F)$ or $F \cap \tau(F) = \emptyset$, that is, F is an equivariant fragment of S.

Now suppose that $\tau(S) \cap S = \emptyset$. If $\tau(S) \subset F$, then $\tau(S)$ would cut off a fragment included in F, contrary to the minimality of F. If F contains only one of $\tau(u)$ and $\tau(v)$, say $\tau(u)$, then F consists of only $\tau(u)$ since otherwise $\{u, \tau(u)\}$ or $\{v, \tau(u)\}$ would cut off a fragment smaller than F. Since G is 2-connected, $\tau(u)$ has to be adjacent to both u and v in this case, but the edge $u\tau(u)$ would be fixed by τ, contrary to τ being free. Thus, $\tau(S) \cap F = \emptyset$. This implies that $\tau(F) \cap F = \emptyset$ and hence S and F are each equivariant. ∎

LEMMA 6. *Let (G, τ) be a connected involutive pair and let $S = \{u, \tau(u)\}$ be an invariant set of two vertices of G. Then G does not decompose into two* connected subgraphs B and $\tau(B)$ which meet at S if and only if G contains an invariant cycle disjoint from S.

Proof. The sufficiency is clear since an invariant cycle can be contained in neither B nor $\tau(B)$. Suppose that every invariant cycle contains S. By Lemma 4, G has an invariant cycle C, which has to contain u and $\tau(u)$. If B is a path in C with end points u and $\tau(u)$, then $B \cup \tau(B) = C$ and $B \cap \tau(B) = S$. Assume that B is a connected subgraph of G with $B \cap \tau(B) = S$ which maximizes $|E(B \cup \tau(B))|$. It remains to show that $B \cup \tau(B) = G$.

Suppose that $B \cup \tau(B) \neq G$. Then there is an edge xy of G with $x \in V(B)$ which belongs to neither B nor $\tau(B)$ since G is connected. If $y \notin V(\tau(B))$, then there would be a connected subgraph B' with $xy \in E(B')$ for which $B' \cap \tau(B') = S$ and $B \subset B'$, contrary to the maximality of B. Hence $y \in V(\tau(B))$. In this case, $\tau(x) \in V(\tau(B))$ and $\tau(y) \in V(B)$. Since B is connected, there is a path P in B which joins $\tau(y)$ to x. If P passed through neither u nor $\tau(u)$, then $Pxy\tau(P)\tau(x)\tau(y)$ would form an invariant cycle disjoint from u and $\tau(u)$. Thus, any path between $\tau(y)$ and x in B has to pass through one of u and $\tau(u)$, and hence B decomposes into two connected subgraphs B_1 and B_2 such that $x \in V(B_1)$, $\tau(y) \in V(B_2)$ and $B_1 \cap B_2 \subset S$. In this case, the connected subgraph $B_1 \cup \tau(B_2) + xy$ contradicts the maximality of B. Therefore, $B \cup \tau(B) = G$. ∎

Let (G, τ) be an involutive pair and uv an edge of G. We consider deletion and contraction of both uv and $\tau(uv)$ at a time, called *equivariant deletion* and *equivariant contraction* of edges. It is clear that $(G - \{uv, \tau(uv)\}, \tau|_{G-\{uv,\tau(uv)\}})$ is an involutive pair, but so is $(G/\{uv, \tau(uv)\}, \tau')$ only when $\tau(u) \neq v$; otherwise, the involution τ' induced by τ would have a fixed vertex $u = v$, which we forbid here.

An involutive pair (H, σ) is said to be an *equivariant minor* or simply a *minor* of (G, τ) if (G, τ) can be deformed into (H, σ) by a finite sequence of equivariant deletions and contractions of edges. To keep our assumption of involutive pairs, we forbid contraction of edges lying on a cycle of length 2. If we contract edges on an invariant cycle of length 2, then the resulting involution is not free, as mentioned above, and the involution σ on H would have a fixed vertex. Thus, this is not the case if (H, σ) satisfies our assumption. On the other hand, if the cycle C of length 2 is not invariant, contraction of edges on C and on $\tau(C)$ yield two loops in the resulting graph. However, these loops will be removed later to make a loopless graph at the final stage. So first delete edges corresponding to the loops and next contract the others.

LEMMA 7. *Let (G, τ) be an involutive pair and (H, σ) one of its minors. If (G, τ) is embeddable in (F^2, h), then so is (H, σ).*

Proof. Delete and contract edges of G on F^2 symmetrically. The surface F^2 will not be pinched as long as we keep our deformation rule; because each cycle has length at least 2 at any stage. ∎

The equivariant minor relation makes naturally the set of involutive pairs a partially ordered set, so that $(H, \sigma) < (G, \tau)$ for an involutive pair (G, τ) and an equivariant minor (H, σ) of it. By Lemma 7, the embeddability of involutive pairs is inherited from the upper to the lower. So we might be able to establish a Kuratowski type theorem, making a complete list of forbidden minors, that is, the list of minimal elements among those that are not embeddable in a given involutive pair of a surface. In fact, we shall do it for involutive pairs of the sphere in the next section.

The following two lemmas are prepared in order to reduce the embeddability of involutive pairs with low connectivity to those with 3-connected graphs.

LEMMA 8. *Let (G, τ) and (F^2, h) be involutive pairs of a graph G and of a closed surface F^2 and suppose that G decomposes into three subgraphs B, H, $\tau(B)$ so that:*

1. *B and H have a unique vertex v in common,*

2. *H is invariant under τ,*

3. *B is disjoint from $\tau(B)$ and*

4. *B is planar.*

Then if $(H, \tau|_H)$ is embeddable in (F^2, h), so is (G, τ).

Proof. Embed H on F^2 so that $\tau|_H$ extends to h and let A be one of faces of H incident to v. Then we can add a planar embedding of B in such a small neighborhood of v within A that $h(B)$ is disjoint from B. Now τ extends to h. ∎

LEMMA 9. *Let (G, τ) and (F^2, h) be involutive pairs of a graph G and of a closed surface F^2 and suppose that G decomposes into three subgraphs B, H, $\tau(B)$ so that:*

1. *B and H have precisely two vertices u and v in common,*

2. *H is invariant under τ,*

3. *$B \cap \tau(B) = \{u, v\}$ or $= \emptyset$ and*

4. *$B + uv$ is planar.*

Let H' be the graph obtained from H by adding edges uv and $\tau(u)\tau(v)$ and let τ' be the free involution on H' naturally induced by τ. Then if (H', τ') is embeddable in (F^2, h), so is (G, τ).

Proof. Embed H' on F^2 so that τ' extends to h. Since $B + uv$ is planar, we can embed B on the plane so that both u and v are incident to the infinite face. Then we can add two copies of this planar embedding of B along uv and $\tau(u)\tau(v)$ so that τ extends to h. ∎

By Lemma 5, if G is not 3-connected, then G has a decomposition compatible with τ. The only case which Lemmas 8 and 9 do not cover is that G decomposes into two invariant connected subgraphs H and B which have precisely two vertices u and v in common. Since B contains an invariant cycle, $(H + \{uv, vu\}, \tau')$ is an equivariant minor of (G, τ), where $\tau'|_H = \tau|_H$ and $\tau'(uv) = vu$. However, when $H + \{uv, vu\}$ is embedded on a surface F^2 so that $uv \cup vu$ does not bound a face, then we might not be able to add B to such an embedding of H even if $B + uv$ is planar.

3. Spherical pairs

Involutions on the sphere are classified into three types, namely the antipodal map, the half rotation and the reflexion. That is, any involution is conjugate to one of these standard involutions, which maps each point $(x, y, z) \in S^2$ to $(-x, -y, -z)$, $(-x, -y, z)$, $(x, y, -z)$, respectively, if S^2 is regarded as the unit sphere in $\mathbf{R}^3$. The antipodal map has no fixed point and the half rotation fixes only the north and south poles $(0, 0, \pm 1)$, while the reflexion fixes the equator $\{(x, y, 0) : x^2 + y^2 = 1\}$ pointwise.

An involutive pair (G, τ) is said to be *spherical* if G can be embedded on the sphere so that τ extends to an involution on the sphere. A spherical pair (G, τ) and its embedding on the sphere are said to be *antipodal* or *rotative* if τ extends to the antipodal map or the half rotation, respectively. Actually, any spherical pair (G, τ) is either antipodal or rotative if G is connected. If G were embedded on S^2 so that τ extends to the reflexion, then the equator would separate G since τ is free.

As is pointed out in [1], the uniqueness of duals of 3-connected planar graphs, proved by Whitney [8], implies the following fact:

THEOREM 10. (Whitney) *A 3-connected planar graph is uniquely and faithfully embeddable in the sphere.*

That is, a 3-connected planar graph G has a unique embedding on the sphere, up to homeomorphisms and automorphisms, and all the automorphisms of G extend to homeomorphisms so that the symmetry of G under $Aut(G)$ is realized on the sphere.

However, we are concerned only with the extendability of free involutions, rather than that of all automorphisms, and for us the assumption that G is 3-connected is unnecessarily strong. Relaxing this assumption, the author has already shown a sufficient condition for a free involution to extend a homeomorphism (Lemma 5 and Theorem 8 in [5]), which can be rephrased to the following in our terminology:

LEMMA 11. *Let (G, τ) be a connected involutive pair and suppose that G/τ is either 2-connected or nonplanar. Then the involutive pair (G, τ) is spherical if and only if G is planar.*

Moreover, we would like a characterization of spherical pairs without reference to their quotients. This will be obtained later as Theorem 15.

The planarity of G by itself does not guarantee the sphericity of an involutive pair (G, τ) in general. For example, the two involutive pairs (M_1, τ_1) and (M_2, τ_2)

as given in Figure 2 are not spherical. Both τ_1 and τ_2 leave each cycle $x_i y_i$ of length 2 invariant and interchange two tridents and two triangles, respectively, which join the three invariant cycles. If M_i could be embedded on the sphere so that τ_i extends, then the three invariant cycles would be placed on the sphere in parallel around the axis, but then their joints, tridents or triangles, could not be embedded in the sphere.

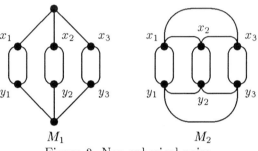

Figure 2. Non-spherical pairs

If an involutive pair (G, τ) is antipodal, then an embedding of G/τ in the projective plane can be obtained as the quotient of its antipodal embedding. Conversely, given a projective-planar embedding of a graph G', we get an antipodal pair (G, τ) such that G covers G', taking the 2-fold covering of the projective plane with the sphere. So we need to discuss when a spherical pair is antipodal.

For example, the two involutive pairs given in Figure 3 are antipodal but not rotative. Their graphs are the cube and the octahedron, so they are called the *antipodal cube pair* and the *antipodal octahedron pair*. They are drawn on paper to be invariant under the rotations around their center through π, which give their free involutions. If we arrange them on the sphere stereographically, their involutions can be seen as just the antipodal map. Since the cube and the octahedron are 3-connected, such spherical embeddings are unique and their involutions extend to only the antipodal map.

 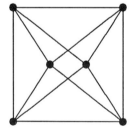

Figure 3. The antipodal cube pair and octahedron pair

LEMMA 12. *A connected spherical pair (G, τ) is antipodal and rotative if and only if G splits into two subgraphs H and $\tau(H)$ which meet in two vertices v and $\tau(v)$.*

Proof. By Lemma 4, we may assume that G is 2-connected. Embed G on the sphere so that τ extends to the half rotation h. Then each face of G is bounded by a cycle. Let C_1 and C_2 be the boundary cycles of the two faces each of which contains one of the two fixed points of h. It is clear that C_1 and C_2 have to have at least two vertices v and $\tau(v)$ in common in any antipodal embedding of (G, τ). In the first rotative embedding, the whole of G is contained within the annular region between C_1 and C_2, which implies that G splits into two subgraphs H and $\tau(H)$ at v and $\tau(v)$.

Conversely if G splits like that, we can embed G in two ways so that τ extends to the antipodal map and the half rotation, turning H with $\tau(H)$ fixed. ∎

LEMMA 13. *A connected spherical pair (G, τ) is rotative but not antipodal if and only if G contains two disjoint invariant cycles.*

Proof. The sufficiency is clear. To show the necessity, embed G on the sphere S^2 so that τ extends to the half rotation h. Similarly to the proof of Lemma 12, G can be assumed to be placed within the annular region between two invariant cycles C_1 and C_2, which bound a face with a fixed points of h at its center, respectively. If C_1 and C_2 have a vertex v and hence $\tau(v)$ in common, then G would have a decomposition as in Lemma 12 and would be antipodal. Thus, C_1 and C_2 are disjoint from each other. ∎

LEMMA 14. *A connected spherical pair (G, τ) is antipodal but not rotative if and only if either the antipodal cube pair or octahedron pair is an equivariant minor of (G, τ).*

Proof. Since the antipodal cube and octahedron pairs are not rotative, the sufficiency is clear by Lemma 7.

To prove the necessity, we may suppose that (G, τ) is antipodal but not rotative and is minimal among those with respect to the equivariant minor relation. If G is not 3-connected, then contradictions follow immediately from Lemmas 8 and 9 except the case when G decomposes into two invariant subgraphs H and B which meet at precisely two vertices. However, if the exceptional case cannot be reduced to one of the others, then each of H and B contains an invariant cycle with these cycles disjoint from each other. By Lemma 13, (G, τ) would be rotative, a contradiction. Thus, G is 3-connected.

Embed G on the sphere S^2 so that τ extends to the antipodal map. By Lemma 4, G has an invariant cycle C, which may be considered as being placed along the equator of S^2. Suppose that C is a shortest invariant cycle. Then there is no chord of C, that is, no edge $xy \in E(G) - E(C)$ with $x, y \in V(C)$. Thus, there is a vertex v not on C and three paths in G which join v to three distinct vertices u_1, u_2 and u_3 on C. If there is an edge $e \in E(G)$ both of whose ends do not belong to C, then equivariant contraction of e and $\tau(e)$ yields an antipodal embedding of $G/\{e, \tau(e)\}$. The resulting minor of (G, τ) is antipodal and rotative by the minimality of (G, τ) and hence there is a vertex $x \in V(C)$ such that $\{x, \tau(x)\}$ separates $G/\{e, \tau(e)\}$ by Lemma 12. But it would be a 2-vertex cut of G, contrary to G being 3-connected. Thus, v is adjacent to u_1, u_2, u_3. Let Y be the subgraph of G induced by $\{vu_1, vu_2, vu_3\}$.

Similarly, equivariant deletion of vu_2 and $\tau(vu_2)$ yields a minor of (G, τ) which is antipodal and rotative and there is a 2-vertex cut $\{x, \tau(x)\}$ of $G - \{vu_2, \tau(vu_2)\}$ with $x \in V(C)$. Let Q be the path on C joining x and $\tau(x)$ through u_2. Then $Q - \{x, \tau(x)\}$ does not contain any neighbor of v except u_2 and $\{u_1, u_3\} \subset \tau(Q)$. If $\{u_1, u_3\} = \{x, \tau(x)\}$, then there is a fourth vertex u_4 in Q which is adjacent to v since otherwise $\{x, \tau(x)\}$ would be a cut of G. In this case, $C \cup Y \cup \tau(Y) + \{vu_4, \tau(vu_4)\}$ shrinks to the octahedron. If $\tau(u_2) \notin \{u_1, u_3\}$ and $\{u_1, u_3\} \neq \{x, \tau(x)\}$, then $C \cup Y \cup \tau(Y)$ is an invariant subdivision of the cube. If either u_1 or u_3, say u_1, coincides with $\tau(u_2)$, consider equivariant deletion of vu_3 and $\tau(vu_3)$. Then $\{u_1, u_2\}$ is a cut of $G - \{vu_3, \tau(vu_3)\}$ and a fourth neighbor of v can be found in the path on C joining u_1 and u_2 not through u_3, similarly to the first case. Thus, (G, τ) has the antipodal octahedron pair as a minor in this case. ∎

The following two theorems are our goals in this section:

THEOREM 15. *A connected involutive pair (G, τ) is spherical if and only if G is planar and neither (M_1, τ_1) nor (M_2, τ_2), given in Figure 2, is an equivariant minor of (G, τ).*

Proof. Since (M_1, τ_1) and (M_2, τ_2) are not spherical, the necessity is clear. To show the sufficiency, suppose that (G, τ) is not spherical with G planar and is minimal among those, that is, any proper minor of (G, τ) is spherical. Then G has to be 2-connected by Lemma 8.

First assume that G is not 3-connected. Then G has an equivariant 2-cut $S = \{u, v\}$ with an equivariant fragment F by Lemma 5. Put $H = G - F \cup \tau(F)$ and $B = \langle S \cup F \rangle$, where $\langle X \rangle$ denotes the subgraph induced by X, that is, one obtained from X by adding all the edges both of whose end points belong to X. Hence B can be obtained from F by adding S and all the edges joining S to F and meets H at S. Let H' be H with two edges uv and $\tau(u)\tau(v)$ added. Then τ induces a free involution τ' on H' with $\tau'(uv) = \tau(u)\tau(v)$ and the involutive pair (H', τ') is an equivariant minor of (G, τ) and hence is spherical.

If $B \neq \tau(B)$, then G would decompose into H, B and $\tau(B)$ as just in Lemma 9 and would be spherical. Thus, $B = \tau(B)$ and hence uv and $\tau(u)\tau(v)$ are a pair of multiple edges and form an invariant cycle of length 2. (We write simply vu for $\tau(u)\tau(v)$ here.) Moreover, we can assume that B does not decompose into two subgraph B' and $\tau(B')$ which meet at S. Then B contains an invariant cycle C which is disjoint from S by Lemma 6. Since $B + \{uv, vu\}$ has two disjoint invariant cycles, namely C and $uv \cup vu$, the involutive pair of $B + \{uv, vu\}$ is rotative by Lemma 13, and hence B can be embedded on the plane so that $\tau|_B$ extends and both u and v are incident to the infinite face.

Embed H' on the sphere S^2 so that τ' extends to some involution on S^2. If the invariant cycle $uv \cup vu$ bounded a face, we could add B to the face and would get a spherical embedding of (G, τ). Thus, H decomposes into two subgraphs H_1 and H_2 which lie in the two 2-cell regions bounded by $uv \cup vu$ separately. If τ' extends to the antipodal map, then $\tau'(H_1) = H_2$ and we can re-embed H_1 and H_2 in one of the regions together so that τ' extends to a half rotation, which implies the same contradiction as above.

Therefore, τ' extends to a half rotation and both H_1 and H_2 are invariant under τ'. If H_i decomposed into two subgraphs H_i' and $\tau'(H_i')$ which meet at S, then we could replace H_i' and $\tau'(H_i')$ along uv and vu, respectively, so that $uv \cup vu$ bounds a face afterward. Thus, H_i does not decompose and contains an invariant cycle C_i which is disjoint from S by Lemma 6. Considering the quotients of B, H_1 and H_2, we find two disjoint tridents Y and $\tau(Y)$ which join u and v, respectively, to C, C_1 and C_2. Then $C \cup C_1 \cup C_2 \cup Y \cup \tau(Y)$ forms a subdivision of M_1, invariant under τ.

Now suppose that G is 3-connected. Replace each pair or set of multiple edges with a single edge to get a simple 3-connected graph G' and consider an involution τ' on G' which has no fixed vertex but may have invariant edges. Embed G' on the sphere S^2. By Whitney's theorem, the spherical embedding of G' is unique and τ' extends to an involution h on S^2.

If h were conjugate to the antipodal map or a half rotation, then an antipodal or rotative embedding of (G, τ) would be obtained by adding multiple edges to G', contrary to (G, τ) not being spherical. (In this case, if an edge e of G' is invariant under τ' and contains a fixed point x of h, then e corresponds to multiple edges of G which form an invariant cycle, and x has to be an isolated point at the middle of e. Put multiple edges around x symmetrically.)

Therefore, h is the reflexion and fixes the equator of S^2 pointwise. Since τ' has no fixed vertex, the equator crosses only edges of G', which form an edge cut and which correspond to invariant cycles of length 2 in G. Since G' is 3-connected, the edge cut has to contain at least three edges. Let $x_1 y_1$, $x_2 y_2$ and $x_3 y_3$ be such three edges and suppose that x_i and y_i belong separately to the upper and lower halves of G'. The three vertices x_1, x_2 and x_3 are joined to one another by paths in the upper half. The union of those paths can be assumed to be either a trident Y or a cycle Δ. Then we have an invariant subdivision of either M_1 or M_2 in G which consists of three pairs of multiple edges $x_i y_i$'s and $Y \cup \tau(Y)$ or $\Delta \cup \tau(\Delta)$, respectively. ∎

THEOREM 16. *A connected involutive pair (G, τ) is antipodal if and only if the following three conditions hold:*

1. *G contains no two disjoint invariant cycles.*

2. *G contains neither a subdivision K of K_5 nor of $K_{3,3}$ with $K \cap \tau(K) = \emptyset$.*

3. None of the involutive pairs given in Figure 4 is an equivariant minor of (G, τ).

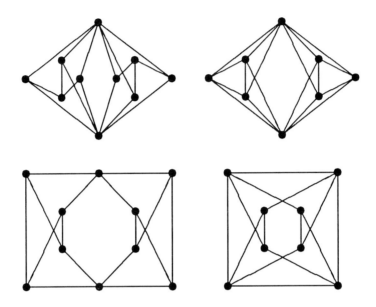

Figure 4. Forbidden minors for antipodal pairs

The involution of each involutive pair in Figure 4 is given as the rotation around the center of each graph through π. The fourth can be excluded from the list since it contains two disjoint invariant cycles.

Proof. By Lemma 13, if G contains two disjoint invariant cycles, then (G, τ) is not antipodal. Since all of the graphs in Figure 4 are not planar, (G, τ) cannot be spherical if it has one of them as a minor. Thus, the necessity follows.

To show the sufficiency, suppose that (G, τ) is not antipodal and is minimal among those with respect to the equivariant minor relation. If G is not 3-connected, there is an equivariant cut S with an equivariant fragment F by Lemma 5. Let $H = G - F \cup \tau(F)$ and $B = \langle S \cup F \rangle$.

If B is not invariant, then we have an antipodal minor obtained by shrinking B and $\tau(B)$ to edges and $B + uv$ (or B if S is a singleton) has to be nonplanar by Lemmas 8 and 9. In this case, $B + uv$ contains a subdivision K of K_5 or $K_{3,3}$ and either $K \subset B$ and $K \cap \tau(K) = \emptyset$ or (G, τ) has one of the first three in Figure 4 as a minor. If B is invariant and if we cannot reduce this case to the previous, then each of B and H contains an invariant cycle. These two cycles are disjoint from S by Lemma 6 and hence from each other.

Now suppose that G is 3-connected and let G'' be the simple graph obtained from G by replacing each pair or set of multiple edges with a single edge. Then

G'' also is 3-connected and has an edge $uv \in E(G'')$ such that G''/uv is 3-connected, as is well-known (see [7]). If uv corresponds to an invariant cycle $uv \cup vu$ of multiple edges, then $G - \{u,v\}$ is invariant under τ and is connected since G is 3-connected, and hence it contains an invariant cycle by Lemma 6, which is disjoint from $uv \cup vu$. Thus, we may assume that uv and $\tau(uv)$ are disjoint from each other and can be contracted, keeping our assumption on involutive pairs.

Let (G', τ') be the involutive pair obtained from (G, τ) by contracting uv and $\tau(uv)$ and let w be the vertex of G' to which uv shrinks. Then (G', τ') is antipodal by the minimality of (G, τ). If G' is not 3-connected, then there is an equivariant cut S' with an equivariant fragment F' and S' has to contain w since G is 3-connected. If $S' \cap \tau'(S') = \emptyset$, then $\tau'(S')$ could be regarded as a cut of G/uv which cuts off $\tau'(F')$, contrary to G/uv being 3-connected. Thus, $\tau'(S') = S' = \{w, \tau'(w)\}$.

If F' is invariant, then F' and $G - F'$ contain invariant cycles, which are disjoint from each other. If F' is not invariant and if $G' - F' \cup \tau(F') \cup S'$ has an invariant component R, then each of $\langle F' \cup \tau(F') \cup S' \rangle$ and R contains an invariant cycle and those cycles are disjoint from each other. These cases are however contrary to (G', τ') being antipodal. Thus, all the fragments of S' are equivariant but not invariant. Then $S = \{u, v, \tau(u), \tau(v)\}$ is a 4-vertex cut of G and has the same fragments as has S'.

If S has only two fragments, that is, if $G = \langle F' \cup \tau(F') \cup S \rangle$, then we consider the equivariant minor $(G - \{uv, \tau(uv)\}, \tau|_{G-\{uv,\tau(uv)\}})$ of (G, τ). By the minimality of (G, τ), this minor is antipodal and $G - \{uv, \tau(uv)\}$ can be embedded on the sphere S^2 so that its involution extends to the antipodal map $h : S^2 \to S^2$. Since F' and $\tau(F')$ are connected, there is a simple closed curve Γ on S^2 with $h(\Gamma) = \Gamma$ which contains S and separates F' and $\tau(F')$. Then u, v, $\tau(u)$ and $\tau(v)$ have to lie along Γ in this order, so we can add uv and $\tau(uv)$ to this embedding and get an antipodal embedding of (G, τ) on S^2, contrary to the hypothesis of (G, τ). Thus, S has at least four fragments.

Shrink each fragment of S to a vertex. The resulting graph H admits a free involution σ induced by τ and consists of S and several independent vertices with only edges between S and them. Let $(x, \sigma(x))$ and $(y, \sigma(y))$ be any two pairs of vertices of H not belonging to S. They are adjacent to at least three vertices in S since G is 3-connected. If x is adjacent to u and $\tau(u)$ and if y is adjacent to v and $\tau(v)$, then $xu\sigma(x)\tau(u)$ and $yv\sigma(y)\tau(v)$ are disjoint invariant cycles, which are pulled back to those in G. If we cannot find such invariant cycles in H, then both x and y have degree 3 and are adjacent to only u, v and $\tau(u)$. In this case, $\{u, \tau(u)\}$ would be a 2-vertex cut of G, contrary to G being 3-connected.

Therefore, we can assume that G' is 3-connected and can be embedded on the sphere S^2 so that τ' extends to the antipodal map h. Let $St(w)$ be the union of the closures of faces incident to w. Since G' is 3-connected, $St(w)$ is a closed 2-cell on S^2 and is bounded by a cycle C. If $\tau'(w)$ were in C, then w and $\tau'(w)$ could be joined by a simple curve Γ passing through a face, and $\Gamma \cup h(\Gamma)$ would be a simple closed curve on S^2 which separates G' at $\{w, \tau'(w)\}$, contrary to G'

being 3-connected. Thus, $St(w) \cap h(St(w)) \subset C \cap \tau'(C)$.

Similarly to Thomassen in [6], we can find a subdivision K of either K_5 or $K_{3,3}$ whose vertex set is $V(C) \cup \{u, v\}$ and which contracts to a "wheel" with w at the center. The center w is adjacent to three (or four) vertices on C when K is a subdivision of K_5 (or $K_{3,3}$, respectively). We call these neighbors of w on C *feet* of w here. By the above arguments, $K \cap \tau(K) \subset C \cap \tau(C)$. We shall deform $K \cup \tau(K)$ up to equivariant minors, keeping this condition, so that $C = \tau(C)$ afterward, as follows.

Contract or delete all the edges inside the annular region between C and $\tau(C)$ on S^2. Then C and $\tau(C)$ meet in several vertices $w_1, \ldots, w_k$ which cut C and $\tau(C)$ into several arcs. If the arc A between w_i and w_{i+1} on C contains no foot of w inside, then $K \cup \tau(K) - A \cup \tau(A)$ contains a subgraph with the same structure as $K \cup \tau(K)$. If A contains a foot of w and if w_i is not a foot of w, then the graph obtained from $K \cup \tau(K)$ by contracting the path between the foot and w_i contains another $K \cup \tau(K)$.

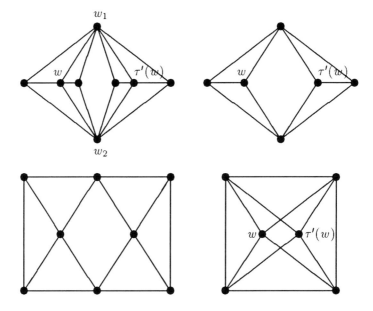

Figure 5. Forbidden minors with uv and $\tau(uv)$ contracted

After these deformations are carried out as far as possible, we have either $C = \tau(C)$ as we want or $C \cap \tau(C) = \{w_1, w_2\}$. In the latter case, $K \cup \tau(K)$ is isomorphic to the first one in Figure 4, where K and $\tau(K)$ correspond to the left and right halves which meet in $\{w_1, w_2\}$ and contain C and $\tau(C)$, respectively. So we can suppose that G can be deformed into $K \cup \tau(K)$ with $C = \tau(C)$ by deleting and contracting edges equivariantly. Moreover, $K \cup \tau(K)$ can be contracted to the second or the fourth in Figure 4, according to which K is, K_5 or $K_{3,3}$. Compare those with ones in the same position of Figure 5, which are $K \cup \tau(K)/\{uv, \tau(uv)\}$ and whose peripheral cycles are $C = \tau(C)$. For each graph in Figure 5, w and $\tau'(w)$ split to be horizontal edges uv and $\tau(uv)$ in Figure 4, respectively. ∎

4. Proof of Theorems

Let $\tilde{G}$ be a 2-fold covering of a graph G with projection $p : \tilde{G} \to G$. Then it admits a free involution $\tau : \tilde{G} \to \tilde{G}$ such that $\tau(u) = v$ for each pair of distinct vertices u and v with $p(u) = p(v)$. Such a covering $\tilde{G}$ is said to be *antipodal* if the involutive pair $(\tilde{G}, \tau)$ is antipodal.

LEMMA 17. *A connected graph G has an even dual on the projective plane if and only if there is a bipartite antipodal covering of G.*

Proof. Let $q : S^2 \to P^2$ be the 2-fold covering of the projective plane P^2 by the sphere S^2. Then its covering transformation $h : S^2 \to S^2$ is just the antipodal map. If G is embedded in P^2 with an even dual, then $\tilde{G} = q^{-1}(G)$ is a 2-fold planar covering of G with a free involution $\tau = h|_{\tilde{G}}$ and the involutive pair $(\tilde{G}, \tau)$ is antipodal. Since each face of G is lifted homeomorphically to a face of $\tilde{G}$, all the faces of $\tilde{G}$ are even and hence $\tilde{G}$ is bipartite.

Conversely suppose that $(\tilde{G}, \tau)$ is an antipodal pair with $\tilde{G}$ planar bipartite graph which covers G doubly. Then its antipodal embedding has only even faces, which are equivariant but not invariant under h; if a face were invariant, then it would contain a fixed point of h. Thus, the quotient of this antipodal embedding, $\tilde{G}/\tau = q(\tilde{G})$, is an embedding of G in P^2 with an even dual. ∎

Combining this criterion in the above lemma with our results in the previous section, we can prove easily Theorems 2 and 3 as follows.

Proof of Theorem 2. When G is embedded in the projective plane P^2 with an even dual, any odd cycle of G, if any, has to be essential, that is, does not bound any 2-cell region; if it did, the 2-cell region would contain an odd face. Since the projective plane cannot contain two disjoint essential closed curves, G does not have two disjoint odd cycles. Thus, with Lemma 17, the conditions in Theorem 2 are necessary for G to have an even dual on the projective plane.

Now suppose that G has a 2-fold planar bipartite covering $\tilde{G}$ with projection $p : \tilde{G} \to G$ and contains no two disjoint odd cycles. Negami has already shown in [2, 4] that a connected graph can be embedded in the projective plane if and only if it has a 2-fold planar covering. So G can be embeddable in the projective plane and has an even dual if G is bipartite.

Hence assume that G is not bipartite and let $\tau : \tilde{G} \to \tilde{G}$ be the free involution on $\tilde{G}$ with $\tilde{G}/\tau = G$. By the properties of $B(G)$ shown in [3], $\tilde{G}$ is equivalent to $B(G)$ in this case and a cycle C in G can be lifted isomorphically to a cycle in $\tilde{G}$ if and only if the length of C is even. In other words, an equivariant cycle C in $\tilde{G}$ is invariant under τ if and only if $p(C)$ is an odd cycle.

If the involutive pair $(\tilde{G}, \tau)$ is not spherical, then $\tilde{G}$ contains three invariant cycles C_1, C_2, C_3, which shrink to invariant cycles of multiple edges in M_1 or

M_2 by Theorem 15. Then $p(C_1)$, $p(C_2)$ and $p(C_3)$ have to be odd cycles in G and they are disjoint from one another. Thus, $(\tilde{G}, \tau)$ is spherical and is either antipodal or rotative. If it were not antipodal, $\tilde{G}$ would contain two disjoint invariant cycles C_1 and C_2 by Lemma 13, and hence $p(C_1)$ and $p(C_2)$ would be two disjoint odd cycles in G. Therefore, $(\tilde{G}, \tau)$ is antipodal. ∎

Proof of Theorem 3. Let G be a connected graph which is not bipartite. Then $B(G) \ne G$ and every 2-fold bipartite covering of G is equivalent to $B(G)$, as is shown in [3]. Thus, G has an even dual on the projective plane if and only if $B(G)$ is antipodal.

By Theorem 16, we have already had a characterization for non-antipodal $B(G)$ in terms of forbidden equivariant minors. Split several vertices of those minors so that the results contain no two invariant cycles and classify forbidden invariant subgraphs for antipodal pairs. Then we have six forbidden subgraphs obtained as 2-fold coverings of ones in Figure 1 in the following way. Make two copies of each of the first four in Figure 1 and join them at x's so that they admit free involutions and that two x's form a cut in each. For each of the last two in Figure 1, make two copies of that with the edge e deleted and join them with two edges so that these edges form a 2-edge cut.

Therefore, $B(G)$ is not antipodal if and only if one of the followings holds:

1. $B(G)$ contains two disjoint invariant cycles.
2. $B(G)$ contains a subdivision K of either K_5 or $K_{3,3}$ with $K \cap \tau(K) = \emptyset$.
3. $B(G)$ contains an invariant subdivision of one of the six as mentioned above.

In the first case, those invariant cycles project to two disjoint odd cycles in G. In the second case, G contains a subdivision $p(K)$ of K_5 or $K_{3,3}$ which is isomorphic to K and is bipartite. This can be regarded as one of the last two in Figure 1 with no odd cycle through the edge e. In the third case, G contains one of the six in Figure 1.

If two distinct vertices x and y in $B(G)$ project to the same vertex in G, then there is a path of odd length which joins x and y since they belong to the two partite sets separately. Thus, the six in Figure 1 should be subdivided as mentioned before Theorem 3. Conversely, if G contains one of such forbidden subgraphs, then $B(G)$ contains a forbidden invariant subgraphs which covers it since any odd cycle cannot be lifted to a cycle in $B(G)$, and hence $B(G)$ is not antipodal. ∎

By the general observation in [5], a connected graph G, not bipartite, has an even dual on a closed surface F^2 if and only if $B(G)$ can be embedded on a 2-fold covering of F^2 so that the free involution on $B(G)$ extends to the covering transformation. For example, any 2-fold covering of the torus is homeomorphic to a torus. So we conjecture that a nonplanar graph G has an even dual on the torus if and only if either G is bipartite or $B(G)$ is toroidal. Note that a toroidal graph may have three or more odd cycles which are disjoint from one another even if it has an even dual on the torus.

REFERENCES

1. S. Negami, Uniqueness and faithfulness of embedding of toroidal graphs, Discrete Math. **44** (1983), 161–180.
2. S. Negami, Enumeration of projective-planar embeddings of graphs, Discrete Math. **62** (1986), 299–306.
3. S. Negami, The virtual k-factorability of graphs, J. Graph Theory **11** (1987), 359–365.
4. S. Negami, The spherical genus and virtually planar graphs, Discrete Math. **70** (1988), 159–168.
5. S. Negami, Projective-planar graphs with even duals, to appear in J. Graph Theory.
6. C. Thomassen, Kuratowski's theorem, J. Graph Theory **5** (1981), 225–241.
7. W.T. Tutte, A theory of 3-connected graphs, Indag. Math. **23** (1961), 441–455.
8. H. Whitney, Congruent graphs and the connectivity of graphs, Amer. J. Math. **54** (1932), 150–168.

DEPARTMENT OF MATHEMATICS, FACULTY OF EDUCATION, YOKOHAMA NATIONAL UNIVERSITY, 156 TOKIWADAI, HODOGAYA-KU, YOKOHAMA 240, JAPAN
E-mail address: negami@ms.ed.ynu.ac.jp

2-Factors, Connectivity and Graph Minors

NATHANIEL DEAN AND KATSUHIRO OTA

February 1, 1992

ABSTRACT. One of the most celebrated theorems of Tutte states that every 4-connected planar graph is hamiltonian and, as a corollary, has a 2-factor. We extend this latter result by showing that (i) every 4-connected graph embeddable in the torus or the Klein bottle has a 2-factor and (ii) every edge of a 4-connected planar or projective planar graph is contained in a 2-factor. We also consider a best possible function t of any surface Σ such that every $t(\Sigma)$-connected graph G embeddable in Σ has a 2-factor. Our approach does not rely on the usual topological methods, but is more combinatorial.

1. Introduction

One of the most celebrated theorems of Tutte states that *every 4-connected planar graph is hamiltonian*. Although every known proof of this result is tedious and topological, the corollary that such graphs have a 2-factor can be proved by methods which are more combinatorial. In fact, we show that every 4-connected graph embeddable in the torus or the Klein bottle has a 2-factor. This result and an even stronger result for projective planar graphs will be proved with the aide of a somewhat technical result (Lemma 2.1) on the number of edges in certain members of minor closed families. Our approach demonstrates that methods other than (and perhaps more elegant than) the topological approach can be used to attack problems of this type (for example, see Tutte[10], Thomassen[8], and Thomas and Yu[9]).

The graphs considered here are finite and contain no loops or multiple edges. Let G be a graph, and let f be an integer-valued function defined on $V(G)$. An f-**factor** is a spanning subgraph H of G such that $d_H(v) = f(v)$ for each vertex v. Let S and T be disjoint subsets of $V(G)$. A component C of $G - S - T$ is

---

1991 *Mathematics Subject Classification.* Primary 05C10, 05C70.

Key words and phrases. factors, connectivity, graph minors, surfaces.

This paper is in final form and no version of it will be submitted for publication elsewhere.

called **odd** if $\sum_{x \in V(C)} f(x) + |E(V(C), T)|$ is odd. Otherwise, C is **even**. Let $h(f, S, T)$ denote the number of odd components of $G - S - T$, and define

(1) $$\delta(f, S, T) = \sum_{t \in T}(d(t) - f(t)) + \sum_{s \in S} f(s) - |E(S,T)| - h(f, S, T).$$

When f is a constant k over $V(G)$ we use the notation $\delta(k, S, T)$ instead of $\delta(f, S, T)$ and $h(k, S, T)$ instead of $h(f, S, T)$, and an f-factor is then called a k-factor. This paper concentrates on 2-factors (that is, a collection of disjoint cycles that cover every vertex of the graph).

The following theorem of Tutte is fundamental to this paper as it is to almost every discussion of factors in graphs. It is used in Section 2 after proving a technical lemma (i.e., the Bipartite Minors Lemma) to conclude that *every 4-connected toroidal or Klein bottle graph contains a 2-factor.*

THEOREM A (TUTTE[10]). *Let G be a graph and $f : V(G) \to \mathbb{Z}$. Then G has an f-factor if and only if $\delta(f, S, T) \geq 0$ for all disjoint subsets S, T of $V(G)$.*

Stronger results for the plane and the projective plane can be proved using a more recent Tutte-like theorem of Liu[6]. By combining Liu's theorem with the Bipartite Minors Lemma we find that *every edge of a 4-connected planar or projective planar graph is contained in a 2-factor.*

THEOREM B (LIU[6]). *Let G be a graph and $f : V(G) \to \mathbb{Z}$. If $\delta(f, S, T) \geq 2$ for all disjoint subsets S, T of $V(G)$ with $S \cup T \neq \emptyset$, then every edge of G is contained in an f-factor.*

In Section 3 we consider a best possible function t of any surface Σ such that every $t(\Sigma)$-connected graph G embeddable in Σ has a 2-factor. We refer the reader to Beineke and White [1] for a quick introduction to topological graph theory. With a few exceptions, most of our notation and terminology is the same as for Bondy and Murty [2].

2. The Bipartite Minors Lemma

The section presents a basic lemma which will be used together with the theorems of Tutte and Liu to quickly establish some results on the existence of 2-factors in certain minor closed families of graphs (for example, graphs embeddable in a surface of non-negative Euler characteristic). The lemma is proved by contradiction using a series of claims, and it depends very strongly on the edge density of the bipartite members of the family.

A graph H is said to be a **minor** of a graph G if H can be obtained from a subgraph of G by contracting edges. Since we are only concerned with simple graphs, loops and multiple edges are deleted after the contraction. For each integer r let $\mathcal{B}_r$ denote the largest family of graphs with the following properties:

(i) If $G \in \mathcal{B}_r$ and H is a minor of G, then $H \in \mathcal{B}_r$.

(ii) Every bipartite member B of $\mathcal{B}_r$ with $|V(B)| \geq 4$ satisfies $|E(B)| \leq 2|V(B)| - r$.

Notice that $\mathcal{B}_r \supseteq \mathcal{B}_{r+1}$. Further, $\mathcal{B}_0$ contains all toroidal and Klein bottle graphs, $\mathcal{B}_2$ contains all projective planar graphs, and $\mathcal{B}_4$ contains all planar graphs. We use the following definition to express a bound on $\delta(2, S, T)$.

$$\rho(r) = \begin{cases} r & \text{if } r \leq 1 \\ 2 & \text{otherwise} \end{cases}$$

LEMMA 2.1 (BIPARTITE MINORS LEMMA). *Let G be a 4-connected member of $\mathcal{B}_r$. Then $\delta(2, S, T) \geq \rho(r)$ for every pair of disjoint subsets S, T of $V(G)$ with $S \cup T \neq \emptyset$.*

Assume there is a 4-connected member G of $\mathcal{B}_r$ containing two disjoint subsets S, T of $V(G)$ such that $S \cup T \neq \emptyset$ and $\delta(2, S, T) < \rho(r)$. Since f equals 2 at every vertex, Equation 1 can be written as

$$\delta(2, S, T) = 2|S| + \sum_{t \in T} d(t) - |E(S, T)| - 2|T| - h(2, S, T) \tag{2}$$

where a component C of $G - S - T$ is called odd if $|E(V(C), T)|$ is odd. Let G_0 be the graph obtained from G by removing all edges in S and T, removing all even components, and contracting each odd component to a single vertex. Then $G_0 \in \mathcal{B}_r$. We define the following subsets of $V(G_0)$:

$$U = V(G_0) - S - T$$
$$H_1 = \{u \in U : |N_{G_0}(u) \cap S| \geq 3\}$$
$$H_2 = \{u \in U : |N_{G_0}(u) \cap S| = 2 = |N_{G_0}(u) \cap T|\}$$
$$H_3 = \{u \in U : |N_{G_0}(u) \cap S| \leq 2 \text{ and } |N_{G_0}(u) \cap T| \geq 3\}$$

Clearly $h(2, S, T) = |U|$.

Claim 1. $|S \cup T| \geq 4$.

PROOF. If not, then $|S \cup T| \leq 3$ and $G - S - T$ has only one component (even or odd). Hence, $h(2, S, T) \leq 1$ and so $2 \geq \rho(r) > \delta(2, S, T) \geq 2|S| + 4|T| - |E(S, T)| - 2|T| - 1 = 2|S \cup T| - |E(S, T)| - 1 \geq |S \cup T| \geq 1$. Hence, $|S \cup T| = 1 = \delta(2, S, T)$. If $T = \emptyset$, then $\delta(2, S, T) = 2|S| = 2$, a contradiction. If $S = \emptyset$, then $\delta(2, S, T) = d(T) - 2 - h(2, S, T) \geq 5 - 2 - 1 = 2$ (contradiction) if $d(T) \geq 5$. If $d(T) = 4$, then $h(2, S, T) = 0$ and so $\delta(2, S, T) = 2$ (contradiction). □

It follows that $h(2, S, T) = |H_1| + |H_2| + |H_3|$. For each vertex $u \in U$, let $C(u)$ denote the corresponding odd component.

Claim 2. For every $u \in H_2$ there is a vertex $t(u) \in T$ such that $|E(C(u), t(u))| \geq 2$.

PROOF. Since $C(u)$ is an odd component and $|E(u, T)| = 2$, $|E(C(u), T)| \geq 3$. □

For each vertex $u \in H_2$, let $N_{G_0}(u) \cap S = \{s_1, s_2\}$. If $s_1 t(u) \notin E(G_0)$ or $s_2 t(u) \notin E(G_0)$, say $s_i t(u) \notin E(G_0)$, we construct a graph G_1 from G_0 by deleting u and adding the edge $s_i t(u)$. In general, for $j \geq 1$, if there is a vertex $u \in H_2$ such that $s_1 t(u) \notin E(G_j)$ or $s_2 t(u) \notin E(G_j)$ where $N_{G_j}(u) \cap S = \{s_1, s_2\}$ (equivalently, $N_{G_0}(u) \cap S = \{s_1, s_2\}$), then we perform the above operations to produce the graph G_{j+1}. After finitely many steps we cannot proceed and we are left with the graph G_m.

Claim 3. $G_m \in \mathcal{B}_r$.

PROOF. The graph G_m can be obtained by edge contractions and edge deletions. □

Let H_2'' be the remaining vertices of H_2, and let $H_2' = H_2 - H_2''$. Then for each vertex $u \in H_2''$ with $N_{G_m}(u) \cap S = \{s_1, s_2\}$, we have both $s_1 t(u) \in E(G_m)$ and $s_2 t(u) \in E(G_m)$.

For each $x \in T$ we define $b(x) = |\{u \in H_2'' : t(u) = x\}|$. The next claim follows from the definitions.

Claim 4. $|H_2''| = \sum_{x \in T} b(x)$.

Claim 5. For every vertex $x \in T$, $d_G(x) \geq 4 + b(x)$.

PROOF. When $b(x) = 0$, the claim is trivial. If $b(x) \geq 2$, then $d_G(x) \geq 2 + 2b(x) \geq 4 + b(x)$. Thus, we may assume that $b(x) = 1$ and, hence, there is a vertex $u \in H_2''$ such that $t(u) = x$. Let $\{s_1, s_2\} = N_{G_m}(u) \cap S$ and $\{x, y\} = N_{G_m}(u) \cap T$. Since both $s_1 x \in E(G_m)$ and $s_2 x \in E(G_m)$, if $s_1 x \notin E(G_0)$ or $s_2 x \notin E(G_0)$, there must be a vertex $u' \in H_2'$ such that $t(u') = x$. Let these vertices be denoted by $u_1', \ldots, u_k'$. Of course, $d_G(x) \geq |E(C(u), x)| + |E(C(u_1'), x)| + 1 \geq 5$ if $k = 1$ and $d_G(x) \geq |E(C(u), x)| + |E(C(u_1'), x)| + |E(C(u_2'), x)| \geq 6$ if $k \geq 2$. Hence, we may assume that $s_1 x \in E(G_0)$ and $s_2 x \in E(G_0)$. Since G is 4-connected, if $G - \{x, y, s_1, s_2\}$ contains a component other than $C(u)$, then $d_G(x) \geq 5$. If $C(u)$ is the only component, then $S = \{s_1, s_2\}$ and $T = \{x, y\}$, and hence

$$\delta(2, S, T) = 2|S| + \sum_{x \in T} d_G(x) - |E(S, T)| - 2|T| - h(2, S, T)$$
$$\geq 2 \cdot 2 + 2 \cdot 4 - 4 - 2 \cdot 2 - 1$$
$$\geq 3,$$

a contradiction. □

Claim 6. $|E(S, T)| \leq 2|S| + 2|T| - |H_1| - |H_2'| - |H_3| - r$.

PROOF. Consider the bipartite subgraph B of G_m with partite sets $S \cup H_3$ and $T \cup H_1$. Because of Claim 1 and the definition of $\mathcal{B}_r$, we have $|E(B)| \leq 2|V(B)| - r$. On the other hand, $|E(B)| \geq |E(S,T)| + 3|H_1| + 3|H_3| + |H_2'|$ and $|V(B)| = |H_1| + |H_3| + |S| + |T|$. Hence, $|E(S,T)| + 3|H_1| + 3|H_3| + |H_2'| \leq 2|H_1| + 2|H_3| + 2|S| + 2|T| - r$. Thus, the claim follows. □

By Claims 4, 5 and 6,

$$\begin{aligned}\delta(2,S,T) &= 2|S| + \sum_{x \in T} d_G(x) - |E(S,T)| - 2|T| - h(2,S,T) \\ &\geq 2|S| + \sum_{x \in T}(4 + b(x)) - (2|S| + 2|T| - |H_1| - |H_2'| - |H_3| - r) \\ &\quad - 2|T| - (|H_1| + |H_2| + |H_3|) \\ &= 2|S| + 4|T| + |H_2''| - 2|S| - 2|T| + |H_1| + |H_2'| + |H_3| + r \\ &\quad - 2|T| - |H_1| - |H_2| - |H_3| \\ &= r,\end{aligned}$$

the final contradiction.

THEOREM 2.1. *Every 4-connected graph embeddable in the torus or the Klein bottle has a 2-factor.*

PROOF. Every toroidal or Klein bottle graph is a member of $\mathcal{B}_0$. Combine Tutte's f-factor theorem with the lemma. $\square$

THEOREM 2.2. *Every edge of a 4-connected planar or projective planar graph is contained in a 2-factor.*

PROOF. Every planar and projective planar graph is a member of $\mathcal{B}_2$. Combine Liu's theorem with the lemma. $\square$

We expect that further results along these lines can be proved using other extensions of Tutte's f-factor theorem.

3. Other Surfaces and Higher Connectivity

The results of the previous section have not yet been extended to graphs of higher connectivity nor to smaller values of the parameter r (and thus to more complicated surfaces). This section only serves to provide some directions for a generalization of the ideas presented thus far. We use $\chi(\Sigma)$ to denote **the Euler characteristic of a surface** Σ. Hence, $\chi(\Sigma) = 2 - 2h$ if Σ is homeomorphic to the orientable surface S_h (i.e., a sphere with h handles) and $\chi(\Sigma) = 2 - k$ if Σ is homeomorphic to the nonorientable surface N_k (i.e., a sphere with k crosscaps).

Let $t(\Sigma)$ denote the smallest integer k such that every k-connected graph embeddable in Σ has a 2-factor. We may also define a similar function $c(\Sigma)$ where now we insist that the graph be hamiltonian. The existence of t follows from the existence of c which was proved by Duke[4], but we state it in a slightly more general form than what appears in [4].

THEOREM C (DUKE[4]). $c(\Sigma) \leq 3 + \sqrt{9 - 3\chi}$, *if* $\Sigma \neq S_0$.

This inequality implies that $t(N_1) \leq 5$ and $t(S_1), T(N_2) \leq 6$. However, from Theorem 2.1 we see that $t(N_1) = t(S_1) = t(N_2) = 4$.

OPEN PROBLEM 3.1. Determine the function $t(\Sigma)$.

4. Further Remarks

It should be pointed out that Theorem 2.1 follows from an old and still unsettled conjecture of Grunbaum[5] and Nash-Williams[7] which states that *every 4-connected toroidal graph is hamiltonian*. The first author of this paper has made a conjecture that implies Theorem 2.2.

CONJECTURE 4.1 (DEAN[3]). Every 4-connected graph embeddable in the projective plane is hamiltonian-connected.

The planar case of Theorem 2.2 follows from the result of Thomassen[8] that *every 4-connected planar graph is hamiltonian-connected*.

REFERENCES

1. L. W. Beineke and A. T. White, Topological graph theory. *Selected Topics in Graph Theory*, Academic Press, New York (1978) 15-49.
2. J. A. Bondy and U. S. R. Murty, *Graph Theory with Applications*, American Elsevier, New York (1976).
3. N. Dean, On bicritical graphs, talk presented at the 21st Southeastern International Conf. on Combinatorics, Graph Theory and Computing (Feb. 1990), Boca Raton, Florida.
4. R. A. Duke, On the genus and connectivity of hamiltonian graphs, *Discrete Math.* **2** (1972) 199-206.
5. B. Grünbaum, Polytopes, Graphs, and Complexes, *Bull. Am. Math. Soc.* **76** (1970) 1131-1201.
6. G. Liu, On f-covered graphs, *Congressus Numerantium* **61** (1988) 81-86.
7. C. St. J. A. Nash-Williams, Unexplored and semi-explored territories in graph theory, *New Directions in Graph Theory*. Academic, New York (1973) 149-186.
8. C. Thomassen, A theorem on paths in planar graphs, *J. Graph Theory* **7** (1983) 169-176.
9. R. Thomas and X. Yu, manuscript (1991).
10. W. T. Tutte, The factors of graphs, *Canad. J. Math.* **4** (1952) 314-328.

BELLCORE, 445 SOUTH STREET, MORRISTOWN, NJ 07960-1910, USA
E-mail address: nate@bellcore.com

KEIO UNIVERSITY, 3-14-1 KOHOKUKU, YOKOHAMA, 223 JAPAN
E-mail address: ohta@math.keio.ac.jp

A Conjecture in Topological Graph Theory

JOHN PHILIP HUNEKE

The purpose of these comments is not to present new results but rather to publicize a conjecture which, if valid, would be a beautiful characterization of projective planar graphs, and if invalid should yield a large planar graph counterexample.

One rationale for characterizing projective planar graphs is that the graph genus problem in NP-complete [9]. But for a graph embedded on the projective plane, its representativity can be computed in polynomial time [8] and hence also its genus by the following:

THEOREM 1[3]. *The genus of a graph embedded in the projective plane is the integer part of half its representativity.*

Kuratowski [5] characterized planar graphs as those which do not contain a subdivision of K_5 or $K_{3,3}$. Analogously, Archdeacon [1] characterized projective planar graphs as those which do not contain a subdivision of one of the 103 graphs of Glover, Huneke and Wang [4]. Although this list of 103 contains only 35 which are minor-minimal and the list can be inductively generated from five graphs by splitting a vertex and deleting edges, this characterization is unwieldy. Negami's "1-2-∞ Conjecture" [6] states that any connected graph which has a finite n-fold topological cover by a planar graph has a 2-fold (topological) cover

---

1991 *Mathematics Subject Classification.* Primary 05C10
This paper is in final form and no version of it will be submitted for publication elsewhere.

by a planar graph. A connected graph embeds on the projective plane if and only if it has a 2-fold planar cover. Hence we have:

CONJECTURE (NEGAMI). *A graph embeds on the projective plane if and only if it has an n-fold cover by a planar graph.*

Since the sphere is a 2-fold cover of the projective plane, one direction of the conjecture is immediate. To see the converse it would suffice to show that the 100 connected graphs in [4] do not have an n-fold planar cover. Archdeacon and Richter [2] as well as Negami [7] describe a reduction from 100 graphs to considering only two specific graphs, $K_{4,4} - K_2$, named E_2 in [4], and $K_{1,2,2,2} = K_7 - 3K_2$, named A_2 in [4], where $3K_2$ denotes three disjoint edges, a 3-matching.

The following Theorem is given here as an example to be mimicked.

THEOREM 2. *The graph $K_{3,5}$ has no n-fold planar cover.*

As a sketch of a proof, observe that a graph which n-fold covers $K_{3,5}$ must be bipartite, and for a planar embedding of the graph each valence 5 vertex is in a region boundary with more than 4 edges. To see this is not possible, subdivide each of these larger regions, R, with a new vertex, v_R, and new edges $[v_R, u]$ for u a vertex with valence 5 on the boundary of R. This yields a new planar bipartite graph with all vertices in one bipartition having valence at least 6, and the others having valence at least 3. Such a graph would extend to a planar graph with every valence at least 6, which would contradict the Euler formula for a graph embedded in the plane. ∎

Attempts to prove that there is no planar cover for the two graphs above might be analogous to this sketch of a proof. On the other hand, attempts to find a planar graph counter-example to Negami's conjecture might also be pursued by carefully analyzing local constraints of planar graphs which cover either of the two. Settling this conjecture either way, whether it is valid or not, would be a pretty result.

References

1. D. S. Archdeacon, A Kuratowski Theorem for the Projective Plane, Ph.D. Dissertation, The Ohio State University (1980).
2. D. S. Archdeacon and R. B. Richter, On the parity of planar covers, J. Graph Theory, 14 (1990) 199-204.
3. J. R. Fiedler, J. P. Huneke, R. B. Richter and N. Robertson, Computing the orientable genus of projective graphs, resubmitted (1991).
4. H. H. Glover, J. P. Huneke, and C. S. Wang, 103 graphs which are irreducible for the projective plane, J. Combinatorial Theory, B, 27 (1979) 332-370.
5. K. Kuratowski, Sur le problème des courbes gauches en topologie, Fund. Math. 15 (1930) 271-283.
6. S. Negami, The spherical genus and virtually planar graphs, Discrete Math. 70 (1988) 159-168.

7. S. Negami, Graphs which have no finite planar covering, Bull. Inst. Math. Academia Sinca 16 (1988) 377-384.
8. N. Robertson and R. P. Vitray, Representativity of surface embeddings. Algorithms and Combinatorics, Vol. 9. Paths, Flows and VLSI-Layout, Editors: Korte, Lovàsz, Schrijver and Prömel, Springer-Verlag, Berlin Heidelberg (1990) 293-328.
9. C. Thomassen, The graph genus problem is NP-complete, J. of Algorithms, 10 (1989) 568-576.

DEPARTMENT OF MATHEMATICS, OHIO STATE UNIVERSITY, 231 WEST 18TH AVENUE, COLUMBUS, OHIO 43210

E-mail address: huneke@function.mps.ohio-state.edu

On The Closed 2-Cell Embedding Conjecture

XIAOYA ZHA

ABSTRACT. We introduce a vertex version of the face chain method which can be used to obtain some closed 2-cell embeddings from old ones. We also introduce two new operations following the face chain method. After giving some general discussion of face chains in surfaces, especially the non-orientable surfaces, we outline a proof of the theorem which shows that all 2-connected graphs embeddable in N_5 (the sphere with 5 crosscaps), including double toroidal graphs, have closed 2-cell embeddings in some surfaces. As a corollary, such graphs have cycle double covers.

1. Introduction

In this paper, all graphs are finite, loopless and 2-connected. A *circuit* is a nontrivial simple closed walk and a *cycle* is a nontrivial closed trail. A *surface*, denoted by Σ, is a compact 2-manifold without boundary. The definitions of *graph embeddings* in surfaces and their *rotation schemes* (*rotation projections*) follow those used in [2] (see Section 3.2). We will identify the graph and the point-set of its embedded image as a harmless convenience when discussing graph embeddings. In an embedding the subgraph induced by a circuit is a simple closed curve, and the subgraph induced by a cycle is expressible as an edge-disjoint union of such simple closed curves.

A *closed 2-cell embedding* (called a *strong embedding* in [5], [6] and a *circular embedding* in [8]) of a graph G in some surface Σ is an embedding such that the closure of every face is homeomorphic to a closed disk. In a closed 2-cell embedding the boundary of every face is a circuit in the graph. The *closed 2-cell embedding conjecture* states that every 2-connected graph G has a closed 2-cell embedding in some surface. It is well-known that a plane embedding of any 2-connected planar graph is a closed 2-cell embedding. Negami ([7], Lemma 2.1), and independently Robertson and Vitray ([9], page 303) showed that every

---

1991 *Mathematics Subject Classification*. Primary 05C10.
Supported in part by NSF Grant number DMS-8903132.
This paper is in final form and no version of it will be submitted for publication elsewhere.

2-connected projective planar graph has a closed 2-cell embedding either in the sphere or in the projective plane. Richter, Seymour and Siran [8] proved that every 3-connected planar graph has a closed 2-cell embedding in some surface other than the sphere; also they characterize those planar graphs which have this property.

A *cycle double cover* of a graph G is a family of cycles $C = (c_1, ..., c_n)$ such that every edge in $E(G)$ is contained in exactly two of the C_i's. The *cycle double cover conjecture* states that any 2-edge-connected graph has a cycle double cover. Clearly, the existence of a closed 2-cell embedding implies the existence of a cycle double cover (where all the cycles are circuits) of a graph simply by taking all face boundaries as a set of cycles. Since closed 2-cell embeddings of the 2-connected components of a graph generate a cycle double cover of that graph, the closed 2-cell embedding conjecture implies the cycle double cover conjecture and so provides a topological approach to that problem.

An *edge-strong embedding* (called an *unitary embedding* in [4]) is an embedding such that every edge is on the boundary of two different faces. Note that the boundary of a face in an edge-strong embedding may be not a circuit in the graph. An edge-strong embedding is weaker than a closed 2-cell embedding but still implies the cycle double cover of that graph. A graph G is N_k *embeddable* if G can be embedded in the surface N_k, the non-orientable surface of k cross-caps. Huneke, Richter and Younger [4] proved that any 2-edge connected graph which is N_3 embeddable has an edge-strong embedding in some surface.

In this paper, we will give a general discussion of the face chain method and outline a proof of the theorem which shows that all 2-connected N_5 embeddable graphs (including double toroidal graphs) have closed 2-cell embeddings in some surface. In Section 2, we will generalize the edge version of some operations in the face chain method to their vertex versions. In Section 3, we will introduce two new operations involving face chains with which we may derive new closed 2-cell embeddings. In Section 4 we will study these operations in general surfaces. In Section 5 we will outline the proof of the existence of closed 2-cell embeddings of all 2-connected N_5 embeddable graphs. The detailed proof of this part is quite long and will appear in separate papers. As a corollary, such graphs have cycle double covers.

2. Vertex versions of some operations

In the previous approaches to the cycle double cover conjecture the circuit chain method is widely used (see [1], [10]). Suppose G is a graph which has a cycle double cover. The circuit chain method shows that for two non-adjacent vertices x, y in G joined by a certain type of sequence of circuits, one can obtain a cycle double cover of the graph $G \cup \{xy\}$ from the original cycle double cover of G, where xy is a new edge added to G. There is a surface embedding version of this method which can be found in [4], [6]. The surface embedding version of the circuit chain method is to find a face chain joining two non-adjacent vertices

x, y and to use this to construct an embedding of $G \cup \{xy\}$. Suppose Ψ is a closed 2-cell embedding in some surface. Since the sequence of face boundaries in a face chain form a circuit chain, a face chain is a special kind of circuit chain. But on the other hand, a face chain involves the surface embedding and hence one can hope to apply some topological arguments.

The face chains used in [4] and [6] require that any two consecutive faces must have at least one edge in common. Therefore in the resulting embedding one may avoid repeated edges, but repeated vertices are harder to avoid since nonadjacent faces in the sequence may have common vertices on their boundaries. Hence embeddings formed by using such face chains are edge-strong embeddings. To improve the face chains used in [4], [6] and elsewhere, consecutive faces in a face chain used here have common vertices but not necessarily common edges. We introduce some notation first.

Let Ψ be an embedding of a 2-connected graph G in Σ. When the faces of Ψ are 2-cells the embedding is called an *open 2-cell embedding*. Open 2-cell embeddings exist for any connected graph; for example any embedding of minimum orientable genus. Denote by ∂f the boundary of a face and by $\overline{f}(= f \cup \partial f)$ the closure of f. Let Ψ be an open 2-cell embedding of G and f be a face of Ψ. The boundary ∂f is a subgraph of G. Following the circular order of the open disk f this subgraph is traced out by a closed walk in G, unique up to rotations and reversal of direction, called the *facial walk* of f. An edge e is said to be a *monofacial edge* of Ψ if it appears in a facial walk twice. A monofacial edge e is *consistent* if the edge e is traversed twice in the same direction in the facial walk, otherwise, it is *inconsistent* (all the monofacial edges in an orientable embedding are inconsistent). Similarly, a vertex v is said to be a *multiple vertex* of a face f if it traversed more than once in the facial walk of f. As usual, the two appearances of the endvertex of a facial walk are counted as one. If a multiple vertex appears in a facial walk only twice, we call it a *double vertex*. Define *consistent* and *inconsistent double vertices* in the same way. Consistent and inconsistent double vertices are well-defined because each vertex has a small closed disk neighborhood which meets the incident edges in initial segments only. These segments are given a circular order by their intersections with the disk neighborhood boundary. Each time a facial walk traverses a vertex it uses consecutive segments in the order. Two traversals cannot cross in the order and therefore will be either consistent or inconsistent. Call monofacial edges and double vertices of a face f the *double-attachments* of f.

Let f, g be two faces. If $\partial f \cap \partial g \neq \emptyset$, we say f and g are *attached*. Clearly, if Ψ is a closed 2-cell embedding, then $\partial f \cap \partial g$ is the union of connected components, where each component is either a common edge, possibly subdivided, or a common vertex. Denote by $\|\partial f \cap \partial g\|$ the number of connected components of $\partial f \cap \partial g$. If $\|\partial f \cap \partial g\| = k$, we say the faces f and g have k attachments.

Let f and g be two attached faces of Ψ. Assign local orientations to the facial walks of f and g. The directions of the two sides of each attachment of

f and g will be either the same or opposite. For a chosen local orientation, call the attachment *consistent* if the directions are the same and *inconsistent* if the directions are opposite. Let a and b be two attachments of f and g. We say these two attachments are of the *same type* if they are both consistent or inconsistent, and otherwise they are of *different types*.

Let x and y be two vertices. A *face chain* $C = (f_1, f_2, ..., f_n)$ which *joins* x and y is a sequence of faces of Ψ such that $\partial f_i \cap \partial f_{i+1} \neq \varnothing$ for $i = 1, 2, ..., n-1$ and $x \in \partial f_1, y \in \partial f_n$. A face chain is *simple* if $\partial f_i \cap \partial f_j \neq \varnothing$ implies $|i - j| = 1$. Clearly any face chain joining x and y contains a simple sub-face chain which also joins x and y.

We now introduce the vertex versions of three operations which generalize their edge versions. The edge version of the first operation can be found in [3]. The edge versions of the second and third operations can be found in [4] and [6].

OPERATION 2.1. Let Ψ be an open 2-cell embedding of G in Σ and $e = xy$ be a consistent monofacial edge with facial walk $xeyP_1xeyP_2$. By viewing the local embedding as a rotation projection and putting an $'\times'$ on e, we obtain an embedding Ψ' in a surface Σ'. In Figure 2.1, the $'\times'$ on an edge means this edge is twisted, that is, the facial walk will cross the edge from one side to the other side. If an edge has an $'\times'$ on it originally, after adding another $'\times'$, it becomes an ordinary edge. The facial walk $xeyP_1xeyP_2$ in Ψ is divided into two facial walks $xeyP_1$ and $xeyP_2$ in Ψ', with all other facial walks unchanged. The edge e is no longer a monofacial edge in Ψ'. Similarly, if v is a consistent double vertex of face f (so that v appears in this facial walk twice), and $ab...cd...a$ is the rotation of all edges at v with $a, b, c, d \in \partial f$, we change the rotation projection as shown in Figure 2.1(b). Again the original facial walk of f is broken into two facial walks with all other facial walks unchanged. We obtain a new embedding of G with fewer double vertices.

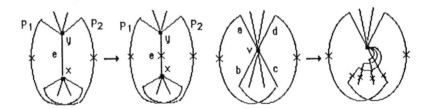

Figure 2.1 :

Remark: Suppose a face f of Ψ has two or more consistent double-attachments. If we apply Operation 2.1 on one of these consistent double-attachments, all other consistent double-attachments are either no longer double-attachments or remain consistent double-attachments of the new embedding. Therefore, if an open 2-cell embedding Ψ of G has no inconsistent double-attachment, by repeatedly applying Operation 2.1, we will end up with a closed 2-cell embedding. Thus if an embedding has no inconsistent double-attachment and no other multiple vertices, then this embedding can be transformed using Operation 2.1 into a closed 2-cell embedding.

The next operation is similar to the one above except that the edge (or vertex) is not a monofacial edge (or a double vertex).

OPERATION 2.2. Let Ψ be an open 2-cell embedding of a graph G and f_1, f_2 be two faces of Ψ such that $\partial f_1 \cap \partial f_2 \neq \emptyset$. Choose an attachment from $\partial f_1 \cap \partial f_2$. If this attachment is an edge, put an $'\times'$ on this edge. If this attachment is a vertex v, and $ab...cd...a$ is the rotation of all edges at v with $a, b \in \partial f_1$ and $c, d \in \partial f_2$, change the rotation projection as indicated in Figure 2.1(b). Call such an attachment a *passing attachment* from f_1 to f_2.

OPERATION 2.3. Let Ψ be a closed 2-cell embedding of G in Σ and x, y be two non-adjacent vertices of G. Let $C = (f_1, f_2, ..., f_n)$ be a simple face chain joining x and y. For $i = 1, ..., n - 1$, choose a passing attachment from f_i to f_{i+1}, and apply Operation 2.2 on these attachments. Change the rotation projection by adding a new edge xy as shown in Figure 2.2(a) to obtain an embedding Ψ' of $G^+ = G \cup \{xy\}$ in Σ'. The faces $f_1, ..., f_n$ of Ψ have turned into two faces g_1 and g_2 of Ψ', while the other faces of Ψ are unchanged. The $'\bigcirc'$ on the edge xy means an $'\times'$ if the number of edges with $'\times'$'s on the facial walk of g_1 (or of g_2, whose number of $'\times'$'s has the same parity as for g_1) is odd, or nothing otherwise.

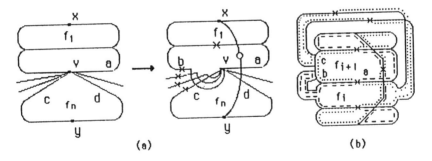

Figure 2.2 :

Remark: Note that Operation 2.3 may create double-attachments in the new facial walks as shown in Figure 2.2(b). In this example f_i and f_{i+1} have three attachments a, b and c. If we use a as the passing attachment, then b is part of the common boundary of two new faces, but c becomes an inconsistent double-attachment of the dotted face. In general, if Ψ is a closed 2-cell embedding and Operation 2.3 does not create inconsistent double-attachments, then either Ψ' is a closed 2-cell embedding already, or it can be transformed using Operation 2.1 into a closed 2-cell embedding. Moreover, if x and y are not on the boundary of the same face in Ψ, then the embedding obtained by contracting the edge xy in Ψ' is also a closed 2-cell embedding.

3. Two new operations

We introduce two new operations in this section which can be used to construct new closed 2-cell embeddings based on given closed 2-cell embeddings.

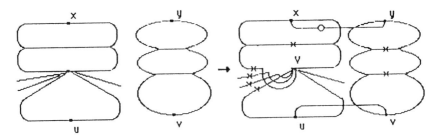

Figure 3.1 :

OPERATION 3.1. Let Ψ be a closed 2-cell embedding with $x, y, u, v \in V(G)$ and $xy, uv \notin E(G)$. Let $G^+ = G \cup \{xy, uv\}$, where xy and uv are two new edges. Suppose C_1 is a simple face chain joining x and u, C_2 is a simple face chain joining y and v, and no face of C_1 is attached to a face of C_2. For every two consecutive faces in C_1 and C_2, choose a passing attachment and apply Operation 2.2. Change the rotation projection by adding two edges xy and uv as shown in Figure 3.1. The '○' on the edge xy is placed by the same rule as in Operation 2.3. Similar remarks to those for Operation 2.3 hold here.

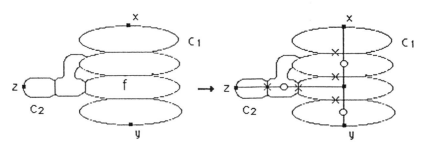

Figure 3.2 :

OPERATION 3.2. Let Ψ be a closed 2-cell embedding of G in a surface Σ and $x, y, z \in V(G)$. Without loss of generality, we may assume no two of x, y, z are on the boundary of the same face. Let C_1 be a simple face chain joining x and y, and C_2 be a simple face chain joining z and a vertex of C_1 and such that only the last face of C_2 is attached to any face of C_1 (if z is on the boundary of a face in C_1 then C_2 is degenerate). Let u be a new vertex, ux, uy, uz be three new edges, and $G^+ = G \cup \{u, ux, uy, uz\}$. Let f_1, f_2 be faces of C_1, C_2, respectively, such that $\partial f_1 \cap \partial f_2 \neq \emptyset$, and let e be an attachment edge or vertex) between f_1 and f_2. For every two consecutive faces in C_1 and C_2, choose a passing attachment and apply Operation 2.2. Change the rotation projection by adding the new vertex u and three new edges ux, uy, uz, as shown in Figure 3.2. The '○' on the edge ux means a $'\times'$ if the number of the edges with $'\times'$ in the segment of the new facial walk between x and f is odd, or nothing otherwise. The $'\bigcirc'$'s on the edges uy and uz are defined similarly.

4. Good face chains and new embeddings

In this section we give a general discussion of the face chain method. As mentioned in the remark after Operation 2.3, applying Operation 2.3 may create inconsistent monofacial edges or multiple vertices in the resulting embedding. Our purpose is to find so-called good face chains and apply some of the operations in Section 2 and Section 3 to derive new closed 2-cell embeddings.

Let Σ be a surface and Ψ be a closed 2-cell embedding of G in Σ. Let $C = (f_1, ..., f_n)$ be a simple face chain joining x and y, with $x, y \in V(G)$ and $xy \notin E(G)$. The face chain C is *good* if the resulting embedding obtained by applying Operation 2.3 on C has no inconsistent double-attachment; otherwise it is *bad*. Let C be a face chain joining x and y. In general the attachments of f_{i-1} and f_i and the attachments of f_i and f_{i+1} may be in no particular order on the face boundary of f_i. If ∂f_i is a disjoint union of two paths p_i and p_{i+1} such that $\partial f_{i-1} \cap \partial f_i \subseteq p_i$ and $\partial f_i \cap \partial f_{i+1} \subseteq p_{i+1}$, then we say $\partial f_{i-1} \cap \partial f_i$ and $\partial f_i \cap \partial f_{i+1}$ are *separated* on ∂f_i, otherwise they are *alternated* on ∂f_i.

Let C be a simple face chain in a closed 2-cell embedding. Suppose there exist three consecutive faces $f_{i-1}, f_i,$ and f_{i+1} in C such that $||\partial f_{i-1} \cap \partial f_i|| = ||\partial f_i \cap \partial f_{i+1}|| = 2$, $\partial f_{i-1} \cap \partial f_i$ and $\partial f_i \cap \partial f_{i+1}$ are alternated on ∂f_i, with the attachments of f_{i-2} and f_{i+2} placed as in Figure 4.1, and such that two attachments of $\partial f_{i-1} \cap \partial f_i$ (as well as $\partial f_i \cap \partial f_{i+1}$) are of different types. Then there does not exist a suitable choice of passing attachments from $\partial f_{i-1} \cap \partial f_i$ and $\partial f_i \cap \partial f_{i+1}$ for us to apply Operation 2.3 and obtain an embedding without an inconsistent double-attachment. Hence such a face chain is bad.

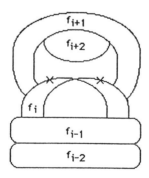

Figure 4.1 :

In a given face chain, two consecutive faces can be attached in a very complicated way which often creates inconsistent double-attachment when applying Operation 2.3. However, the attachments of two faces in an embedding by an orientable face chain is not so important. We have the following theorem.

THEOREM 4.1. *Let Ψ be a closed 2-cell embedding and C be a simple face chain in Ψ. Denote by $U(C)$ the union of the closures of all faces in C. If $U(C) \subset \Sigma$ does not contain orientation reversing simple curves, then C is good.*

Proof: After applying Operation 2.3, all the faces in C, together with the new edge xy, have turned into two faces g_1 and g_2 of a new embedding, while all the other faces remain unchanged. Our only concern is whether the face boundaries of g_1 and g_2 are circuits or not. If not, suppose that g_1 has double-attachments. Since the face chain is simple, these double-attachments must be components of the common boundaries of two consecutive faces, say f_i and f_{i+1}. By the construction, the part of the facial walk of g_1 from f_i to f_{i+1} contains one '×'. Since $U(C)$ is orientable, all the attachments of f_i and f_{i+1} are inconsistent. Therefore after adding an '×' to its facial walk, these double-attachments in g_1 are consistent. Hence C is good. □

The next two corollaries follow immediately.

COROLLARY 4.2. *Let Ψ be a closed 2-cell embedding in an orientable surface. Then any simple face chain C in Ψ joining non-adjacent vertices x and y is good.*

COROLLARY 4.3. *Let G be a 2-connected graph with $x, y \in V(G)$ and $xy \notin E(G)$. Let $G^+ = G \cup \{xy\}$ be a new graph obtained by adding a new edge xy to G. If there exists a closed 2-cell embedding Ψ of G in some orientable surface, then there exists a closed 2-cell embedding Ψ' of G^+ in some non-orientable surface.*

Remark: If x and y are not on the boundary of the same face in Ψ, then the embedding obtained by contracting the edge xy in Ψ' is also a closed 2-cell embedding.

The behavior of a simple face chain in a non-orientable embedding is much different. Generally, it is hard for a simple face chain in a non-orientable embedding to be good. However, under certain circumstances, we are still able to find good simple face chains. In a non-orientable surface, two faces can be attached in a very complicated way; now we introduce a model to describe the arrangement of all attachments between two faces.

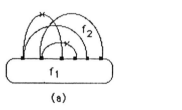 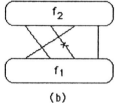

(a)   (b)

Figure 4.2 :

Let f_1 and f_2 be two faces with n attachments. Choose an attachment as the first attachment, and give a local orientation to the facial walks of f_1 and f_2 so that this first attachment is inconsistent (it is not necessary that a first attachment be inconsistent). Draw two circuits representing the facial walks of f_1 and f_2, mark n points on each circuit to represent the n attachments (again, these attachments may be either edges or vertices). For each attachment, draw a segment if it is inconsistent or a segment with an $'\times'$ if it is consistent, connecting the corresponding points on these two circuits. Figure 4.2 gives an example following these instructions. Figure 4.2(a) is a local embedding of two faces given by rotation projection, while Figure 4.2(b) is the model to show in a more clear way how these two faces are attached.

Let $C = (f_1, ..., f_n)$ be a simple face chain of an embedding Ψ which joins two vertices x and y. We use the above attachment instructions to discuss how to implement Operation 2.3. Let f_{i-1}, f_i, f_{i+1} be three consecutive faces in the face chain C. Assume we have already chosen all the passing attachments from f_1 to f_{i-1}. For the face pair f_{i-1} and f_i, call the passing attachment between f_{i-2} and f_{i-1} the *entry* and the passing attachment between f_i and f_{i+1} the *exit*. The question is how to find a suitable passing attachment between f_{i-1} and f_i to avoid inconsistent double-attachment in new facial walks. We state the following facts for better understanding.

(4.1) We know that Operation 2.2 will add an $'\times'$ in the facial walk. Under the attachment description, if the passing attachment is inconsistent, then after Operation 2.2 it becomes a segment with an $'\times'$, and therefore the new facial walks should cross the passing attachment. If the passing attachment is consistent, namely, with an $'\times'$ on it, then after Operation 2.2, the two $'\times'$'s will cancel out, and so the passing attachment is a segment without an $'\times'$, and therefore the new facial walks will not cross the attachment.

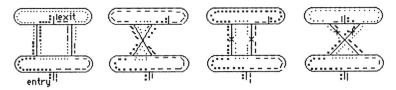

Figure 4.3 :

(4.2) We may choose two passing attachments between two consecutive faces. Simply by observation, these two passing attachments must be of the same type; both are segments with $'\times'$'s or segments without $'\times'$'s. The four possible two passing attachments are shown in Figure 4.3. The dashed and two dotted curves represent new facial walks.

(4.3) The resulting facial walks may be not simple; i.e., they may have double-attachments. If they are consistent, we can reduce them easily by Operation 2.1. If they are inconsistent, then the double-attachments are caused by consistent attachments if the passing attachment is inconsistent, or by inconsistent attachments if the passing attachment is consistent.

Now we are ready to prove the following lemma.

LEMMA 4.4. *Let Ψ be a closed 2-cell embedding of G in some surface. Let*

$x, y \in V(G)$, $xy \notin E(G)$ and $C = (f_1, ..., f_n)$ be a simple face chain joining x and y. For a given i, if $\|\partial f_i \cap \partial f_{i+1}\| \leq 4$ then, except for one case, by choosing suitable passing attachments, Operation 2.3 will not create inconsistent double-attachments in the resulting embedding.

Proof: Assume we have located the entry and all the points representing these attachments. Each attachment segment l has two ends in the facial walk of f_i and f_{i+1}, respectively. Call these the *bottom end* and *top end*. By our convention, we may assume the first attachment segment is the one without an $'\times'$.

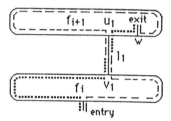

Figure 4.4 :

Case 1: The exit w is on the right of the top end u_1 of the first segment l_1.

Use l_1 as the passing attachment. It is clear that the resulting facial walks are simple at this point. The local rotation projection is shown in Figure 4.4. Two new facial walks are represented by the dotted curve and dashed curve.

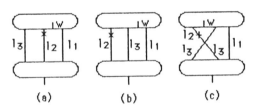

Figure 4.5 :

Case 2: The exit w is on the left of the top end u_1 of the first segment l_1.

If we use l_1 as the passing attachment and the resulting facial walk has an inconsistent double-attachment, then by the remarks in (4.3), there must be an attachment segment l_2 with an $'\times'$ on it whose top end u_2 is on the left of w and bottom end v_2 is on the left of v_1. Now we use l_2 as the passing attachment. If the resulting facial walks have an inconsistent double-attachment, there must be an ordinary attachment segment l_3 in one of the three cases shown in Figure 4.5. We have now three attachment segments, and one of them is an attachment segment with an $'\times'$. In the cases of Figure 4.5(a) and 4.5(b) use two ordinary attachment segments as two passing attachments; in the case of Figure 4.5(c) use the third segment l_3 as the passing attachment. By the same argument, if the result fails to be a closed 2-cell embedding, then there must be an attachment segment with an $'\times'$ appearing as in one of the thirteen sub-cases shown in Figure

4.6. Among these sub-cases, the first four are illustrated in Figure 4.5(a), the second four in Figure 4.5(b) and the last five in Figure 4.5(c). The choices of the passing attachments in some cases are not unique, however, except for the last one, we can find a single (or a pair of) attachment segment(s) with an $'\times'$ (or $'\times'$s) as the passing attachment(s) to apply Operations 2.2 and 2.3. In Figure 4.6, we put a $'\square'$ to identify the passing attachment. It is easy to check that the resulting embedding is simple at this portion. Thus the lemma is true. □

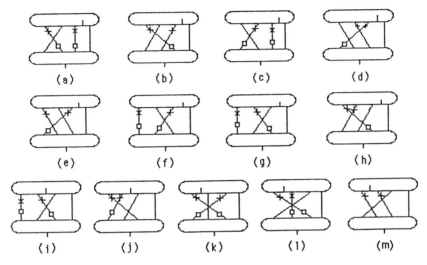

Figure 4.6 :

In a face chain $C = (f_1, ..., f_n)$, if the attachments between faces f_{i-1}, f_i and f_i, f_{i+1} are separated on ∂f_i, for $i = 2, ..., n-1$, then it is easy to prove inductively the following theorem.

THEOREM 4.5. *Let $C = (f_1, ..., f_n)$ be a simple face chain joining x and y with $\|\partial f_i \cap f_{i+1}\| \leq 4$ and not satisfying the the case of Figure 4.6(m). If for each i, $\partial f_{i-1} \cap \partial f_i$ and $\partial f_i \cap \partial f_{i+1}$ are separated on ∂f_i, then there exists a closed 2-cell embedding of $G \cup \{xy\}$ in some non-orientable surface.*

The significance of Theorem 4.5 is that for an embedding of a 3-connected graph, if the surface has small orientable or non-orientable genus, then two faces cannot have a large number of attachments.

5. Outline of the proof that N_5 embeddable graphs have closed 2-cell embeddings

By the techniques developed in earlier sections, we are able to prove that all 2-connected N_5 embeddable graphs have closed 2-cell embeddings in some surfaces. Since the proof is quite long, we only outline the proof here. The detailed proof is in [11] and [12].

We first make some reductions. We know that in order to show the existence of a cycle double cover for a 2-connected graph, one may always assume the graph is cyclically 4-connected (see [5]). We will show now that this is also true for the existence of a closed 2-cell embedding of any 2-connected graph. Let G be a 2-connected graph. Suppose G has a non-trivial 2-vertex separation, say $G = G_1 \cup G_2$, $G_1 \cap G_2 = \{x, y\} \subset V(G)$. Separate G at x and y, and add a virtual edge e joining x and y to G_1 and G_2, respectively, to form G'_1 and G'_2. If we have closed 2-cell embeddings for G'_1 and G'_2 in surfaces Σ_1 and Σ_2, respectively, then remove the edge e from Σ_1 and Σ_2 by deleting the interior of small disks around e, and identifying the two boundaries (the disk-sum of Σ_1 and Σ_2 at e). The resulting embedding is also a closed 2-cell embedding. Therefore we may assume the graph G is topologically 3-connected (it may contain some divalent vertices, which have no effect on finding the closed 2-cell embedding). Now suppose G has a 3-vertex separation, say $G = G_1 \cup G_2, G_1 \cap G_2 = \{x, y, z\} \subset V(G)$, where G_1 and G_2 each contain circuits. Separate G at x, y, z and add a new vertex u and three new edges ux, uy, uz to G_1 and G_2 to form G'_1 and G'_2, respectively. If we have closed 2-cell embeddings of G'_1 and G'_2 in surfaces Σ_1 and Σ_2, respectively, we can obtain an embedding of G by removing the interior of small closed disks about the new vertex and edges in Σ_1, Σ_2 and taking the disk-sum of Σ_1 and Σ_2 over these disks. When taking the disk-sum, the order of x, y, z on the boundary of the disk in Σ_1 must match the order of x, y, z on the boundary of the disk in Σ_2. If the orders are not same, we can easily make them same by taking the 'mirror' embedding in Σ_2. The resulting embedding is also a closed 2-cell embedding of G. Therefore we may always assume the graph G is topologically cyclically 4-connected.

Remark: Since we are only able to deal with N_5 embeddable graphs, we have to be careful in intermediate steps not to increase the genera of the resulting graphs. Clearly, the 2-separation reduction will not increase the genera of the resulting graphs. But the 3-separation reduction may increase the genera of the resulting graphs when G_1 or G_2 is a triangle. Therefore when we perform the 3-separation reduction we should avoid those which will increase the genera of the resulting graphs.

Now let Ψ be an embedding of G in Σ. If it is not a closed 2-cell embedding, by a result of Robertson and Vitray ([9], page 297), there exists an essential simple curve Γ in Σ which meets the graph only at a vertex. If the embedding is of highest Euler characteristic, by Operation 2.1, we know Γ must be orientation preserving. Cut Σ along Γ, delete the pendant edge (if any), then cap off the boundaries with two disks. We obtain a new embedding Ψ' of a new graph G' embedded in a new surface Σ' with its Euler characteristic increased by 2. Call this *cut and paste* surgery. If the embedding Ψ' is a closed 2-cell embedding, then we construct a closed 2-cell embedding of G by applying Operation 2.3, which replaces the deleted edge (if there is no deleted edge, the operation will add a new edge and we can contract this new edge later to obtain the original graph), based on a face chain in Ψ'. If Ψ' is not a closed 2-cell embedding, we perform the cut and paste surgery and obtain an embedding Ψ'' of a graph G'' in a new surface Σ'' with the Euler characteristic increased by 4. Then we construct a closed 2-cell embedding of G based on Ψ'' by using two face chains and adding two edges (we may contract them later if necessary to obtain the original graph). This can be done by applying Operations 2.3, 3.1 and 3.2.

We start from toroidal graphs. Theorem 3.1 and the condition that G is topologically 3-connected will easily give closed 2-cell embeddings for all toroidal graphs.

If Σ is the double torus, we perform cut and paste surgery. If the resulting embedding is a closed 2-cell embedding, then Theorem 3.1 will give a closed 2-cell embedding of G. If the resulting embedding is not a closed 2-cell embedding, we perform the cut and paste surgery again. The resulting surface Σ'' is the sphere. We then construct a closed 2-cell embedding of G by Operations 2.3, 3.1 and 3.2.

If the surface is N_2 (the Klein bottle), after the cut and paste surgery we see that G is a toroidal graph and we are back to the previous case.

If the surface is N_3, after the cut and paste surgery Σ' is the projective plane, and it is not hard to obtain a closed 2-cell embedding of G from the projective embedding of G' by some properties of the embedding Ψ'.

If the surface is N_4 and N_5, then we need some topological properties of the simple curves in N_2 and N_3. In [12] we classify all non-homotopic simple curves in N_2 and N_3. By discussing the dual circuit through any pair of attachments between two attached faces, we can bound the attachment number of two faces in any given embedding. Therefore we can find a face chain with a bounded number of attachments between any two consecutive faces. If the case shown in Figure 4.1 happens, we try to find another face chain to replace the previous face chain. A careful argument will show the existence of closed 2-cell embeddings for all 2-connected N_5 embeddable graphs.

Acknowledgment

I am grateful to Professor Neil Robertson for his valuable advice, support and stimulating discussions through the course of this work.

References

1. B. R. Alspach, L. Goddyn, C. Q. Zhang, "Graphs with the circuit cover property", (preprint).
2. J. Gross, T. W. Tucker, *Topological Graph Theory*, John Wiley & Sons, 1987.
3. G. Haggard, "Edmonds' characterization of disk embeddings", *Proceedings of the Eighth Southeastern Conference on Combinatorics, Graph Theory and Computing* (1977), Utilitas Mathematica, Winnipeg, 291-302.
4. J. P. Huneke, R. B. Richter, D. H. Younger, "Embeddings and double covers", (preprint).
5. F. Jaeger, "A survey of the cycle double cover conjecture", *Annals of Discrete Mathematics* **27** (1985), 1-12.

6. C. Little, R. Ringeisen, "On the strong graph embedding conjecture", *Proceedings of the Ninth Southeastern Conference on Combinatorics, Graph Theory and Computing* (1978), Utilitas Mathematica, Winnipeg, 479-487.
7. S. Negami, "Re-embedding of projective-planar graphs", *J. Combin. Theory, Ser. B* **44** (1988), 276-299.
8. R. B. Richter, P. D. Seymour, J. Siran, "Circular embeddings of planar graphs in non-spherical surfaces", preprint.
9. N. Robertson, R. Vitray, "Representativity of surface embeddings", *Algorithms and Combinatorics, Vol. 9, Paths, Flows, and VLSI-Layout* (1990), Springer-Verlag, 293-328.
10. P. D. Seymour, "Sums of circuits", *Graph Theory and Related Topics* (1979), Academic Press, New York, Berlin, 341-355.
11. X. Zha, "Closed 2-cell embeddings of double toroidal graphs", preprint.
12. X. Zha, "Closed 2-cell embeddings of 5 cross-cap embeddable graphs", preprint.

DEPARTMENT OF MATHEMATICS, THE OHIO STATE UNIVERSITY, COLUMBUS, OHIO 43210
E-mail address: zha@function.mps.ohio-state.edu

Cycle Cover Theorems and Their Applications

CUN-QUAN ZHANG

ABSTRACT In this paper, we survey cycle cover theorems and their applications: a compatible cycle decomposition theorem, an even cycle decomposition theorem and a strong embedding theorem for graphs with no K_5-minor. Some related problems, such as the cycle double cover conjecture, the equivalence of the Chinese postman problem and the shortest cycle cover problem, and the 4-flow conjecture, are also surveyed. A complete proof of the strong embedding theorem is also included in this paper.

We follow the terminology and notation of [3]. Note that 'cycles' in this paper are 2-regular connected subgraphs. Since the cycle cover problem is discussed in this paper, all graphs considered here are 2-edge-connected.

1. Cycle Cover Theorems

A weight $w: E(G) \to \{0,...,n\}$ is called *eulerian* if the total weight of each edge-cut is even. A *(1,2)-eulerian weight* of G is an eulerian weight $w: E(G) \to \{1,2\}$. A graph G together with a weight w is denoted by (G,w). If (G,w) has a family F of cycles such that each edge e of G is contained in precisely $w(e)$ cycles of F, then the family F is called a *cycle w-cover* and G is *cycle w-coverable*. An eulerian weight w is *admissible* if for every cut T and every edge e of T, $w(e) \le \frac{w(T)}{2}$. The weight w being admissible is a necessary condition for (G,w) to be cycle w-coverable, but it is not sufficient. For example, that the Petersen graph with a (1,2)-eulerian weight w (fig. 1) has no cycle w-cover can be found in many papers (see [11], [17], [2], etc).

The following theorem was first proved by Seymour and recently generalized by Alspach, Goddyn and Zhang (note, Theorem 3 is proved in a submitted paper).

AMS Subject Classification   05C38, 05C45, 05C70, 05C15
Partially supported by National Science Foundation under the grant DMS-9104824
This paper is a preliminary version and the detailed version will be published elsewhere

Theorem 1   (Seymour [17])
Every planar graph is cycle w-coverable for any admissible eulerian weight w.

Theorem 2   (Alspach and Zhang [2])
Every cubic graph containing no Petersen-minor is cycle w-coverable for any (1,2)-eulerian weight w.

Theorem 3   (Alspach, Goddyn and Zhang [1])
Every graph containing no Petersen-minor is cycle w-coverable for any admissible eulerian weight w.

Theorem 4   (Zhang [23])
Every graph admitting a nowhere-zero 4-flow is cycle w-coverable for any (1,2)-eulerian weight w.

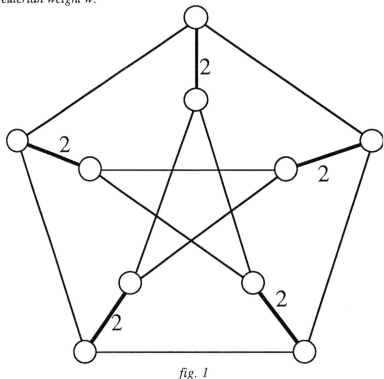

fig. 1

Refer to [22] or [14] for the definition of integer flow. Note that a graph admitting a nowhere-zero 4-flow may not be cycle w-coverable for some admissible eulerian weight w because of the following example (fig. 2). The graph is a cubic bipartite graph and, by Jaeger's 3-flow Theorem (See [15] or [14]), therefore admits a nowhere-zero 3-flow. However, it does not have a cycle w-cover for the following eulerian weight w.

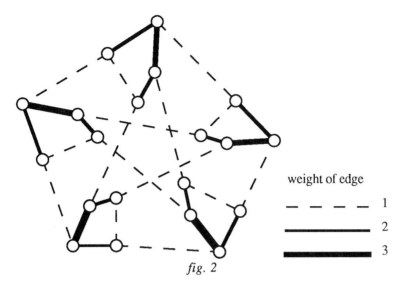

fig. 2

In the following sections, we are going to discuss some applications of these cycle cover theorems.

2. The Cycle Double Cover Conjecture

The following conjecture is a well-known conjecture. Its motivation and relation to other problems can be found in the surveys [13] and [14].

Cycle double cover conjecture (Szekeres [20], Seymour [17], etc.)
Every 2-edge-connected graph G has a family F of cycles such that each edge of G is contained in two cycles of F.

Let w_2 be an eulerian weight of G with $w_2(e) = 2$ for every edge e of G. Being cycle w_2-coverable is equivalent to being cycle double coverable. Thus the following result is an immediate corollary of Theorem 3.

Corollary 5
A 2-connected graph G is cycle double coverable if G contains no Petersen-minor.

3. The Chinese Postman Problem and The Shortest Cycle Cover Problem

A closed trail covering all edges of G is called a *postman tour* of G. The *Chinese postman problem* (abbreviated to *CPP*) is to find a shortest postman tour of G (see [5]).

The *shortest cycle cover problem* (abbreviated to *SCC*) is to find a family F of cycles covering every edge of G such that the total length of F is as small as possible.

CPP and SCC look like very similar to each other. But CPP is solvable by a polynomial algorithm (Edmonds & Johnson [5]), while SCC might be an NP-complete problem (conjectured by Itai, Lipto, Papadimitriou and Rodeh [10]). For the Petersen graph, the solution of CPP is of length 20, while the solution of SCC is of length 21. If the length of the solution of CPP of a graph G equals the length of the solution of SCC, we say CPP and SCC are *equivalent* for G (denoted by $|CPP| = |SCC|$).

A solution T of CPP in a graph G is called a *Chinese postman tour* of G. Define an eulerian weight w_T such that if the postman passes through an edge μ times then $w_T(e) = \mu$. It is obvious that w_T is a (1,2)-eulerian weight of G with the least total weight. Actually, T is a Chinese postman tour if and only if w_T is a smallest (1,2)-eulerian weight of G. Let F be a shortest cycle cover of G. Define an eulerian weight w_F such that if an edge e is contained in μ cycles of F then $w_F(e)=\mu$. Note that w_F is an admissible eulerian weight of G, but it may not be smallest. And a smallest (1,2)-eulerian weight w_T of G may not have a cycle w_T-cover because a single edge is NOT considered as a cycle in the SCC problem. Consequently, $|CPP| \le |SCC|$ for any 2-edge-connected graph, and $|CPP| = |SCC|$ if and only if G is cycle w-coverable for some smallest (1,2)-eulerian weight w. That motivated the following.

Theorem 6   (Guan and Fleischner [9])
The Chinese postman problem and the shortest cycle cover problem are equivalent for every 2-connected planar graph G.

By Theorem 3 and 4, we have the following generalization of Theorem 6.

Corollary 7   (Alspach, Goddyn, Zhang [2], [1] and Jackson, Zhang [12], [23])
The Chinese postman problem and the shortest cycle cover problem are equivalent for a graph G if either G contains no Petersen-minor or G admits a nowhere-zero 4-flow.

4. Compatible Cycle Decompositions of Eulerian Graphs

Let $G=(V,E)$ be a graph. Let v be a vertex of G and let $P(v)$ be a partition of the set of all edges incident with v. An element of $P(v)$ is called a *forbidden part at v*. Let

$$P = \bigcup_{v \in V(G)} P(v)$$

which is called *a set of forbidden parts* of G. A graph G together with a set of forbidden parts P is denoted by (G,P).

A cycle decomposition C of $E(G)$ is *compatible* with a set P of forbidden parts if $|E(C) \cap P| \leq 1$ for every $C \in C$ and every $P \in P$.

A cut T of G is called a *bad cut* of (G,P) if there is a forbidden part P of P such that $2|P \cap T| > |T|$. It is obvious that a necessary condition for (G,P) to have a compatible cycle decomposition is that (G,P) has no bad cut.

Let G be a 2-connected graph. Construct an eulerian graph G'' by replacing each edge e of G by a pair of parallel edges $\{e', e''\}$. For a vertex v, let $P(v)$ be the set of all pairs of parallel edges $\{e', e''\}$ incident with v. It is obvious that G has a CDC (cycle double cover) if and only if G'' has a cycle decomposition compatible with $P = \bigcup_{v \in V(G)} P(v)$. The following theorem was proved by Fleischner and Frank (it is a generalization of a prior result in [6]).

Theorem 8 (Fleischner and Frank, [8])
Let G be a planar eulerian graph and P be a set of forbidden parts of G with no bad cut. Then (G,P) has a compatible cycle decomposition.

As we mentioned above, having no bad cut is a necessary condition for (G,P) to have a compatible cycle decomposition. However it is not a sufficient condition because of the example (K_5, P^*), where K_5 is the complete graph with 5 vertices $\{v_0, v_1,...,v_4\}$ and
  $P^* = \{$2-path $v_i v_j v_k$: either $k=j+1$ and $i=j-1$, or $k=j+2$ and $i=j-2$, mod 5$\}$.
By Kuratowski's Theorem (see [3]), K_5 and $K_{3,3}$ are the only two forbidden minors for planar graphs. However the graph $K_{3,3}$ is not an exception for the problem of compatible cycle decomposition. A natural question is whether a graph containing no K_5-minor has a compatible cycle decomposition for any set of forbidden parts with no bad cut. By applying Theorem 3, we answered the question in the following theorem (note, Theorem 9 is proved in a submitted paper)

Theorem 9 (Zhang [24])
Let G be an eulerian graph containing no K_5-minor and let P be a set of forbidden parts of G with no bad cut. Then (G,P) has a compatible cycle decomposition.

Let $L = e_0 e_1 ... e_{r-1}$ be an eulerian tour of an eulerian graph G. The following set
  $P_L = \{e_i e_{i+1}: i=0,...,r-1, \bmod r\}$
is called a set of forbidden parts induced by L. The following well-known conjecture was due to Sabidussi.

Conjecture 10 (Sabidussi, see [7])
Let G be an eulerian graph with minimum degree at least 4 and L be an eulerian tour of G. Then (G,P_L) has a compatible cycle decomposition.

By Theorem 9, this conjecture holds for all graphs containing no K_5-minor.

5. Even Cycle Decompositions of Eulerian Graphs

Let G be an eulerian graph. A cycle decomposition F of G is called *even* if each cycle of F is of even length. A necessary condition for an eulerian graph to have an even cycle decomposition is that each block of G must have an even number of edges. However this necessary condition is not sufficient. For the complete graph K_5, every even cycle is of length four and $|E(K_5)| = 10$ is not a multiple of four. The following theorem was proved by Seymour.

Theorem 11   (Seymour [18], or see [8])

Let G be a planar eulerian graph containing no odd block. Then G has an even cycle decomposition.

Some ideas introduced in [8] and Theorem 9 are applied in the proof of the following generalization of Theorem 11 (note, Theorem 12 is proved in a submitted paper).

Theorem 12   (Zhang [25])
Let G be an eulerian graph containing no K_5-minor and containing no odd block. Then G has an even cycle decomposition.

The even cycle decomposition problem seems more flexible than the problems about cycle w-covers and compatible cycle decompositions. The author believes that K_5 is the only counterexample to the even cycle decomposition problem if the connectivity of an eulerian graph is sufficiently high.

Conjecture 13
Let G ($G \neq K_5$) be a 3-connected eulerian graph containing an even number of edges. Then G has an even cycle decomposition.

Note, the connectivity cannot be reduced. Some eulerian graphs, which are 2-edge-connected or 2-connected and have no even cycle decomposition, are illustrated in the following figures. (fig. 3)

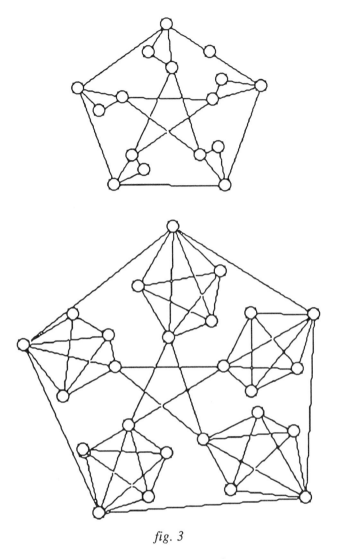

fig. 3

6. Strong Embedding

Let G be a graph and S be a compact connected 2-manifold without boundary. The graph G is said to be *embedded* in the surface S if it can be drawn in S so that edges intersect only at their common vertices. If G is embedded in a surface S, then we regard G as a topological subspace of S and each component of $S \setminus G$ is called a *face* of the embedding. An embedding of G in S is a *strong-embedding* if every face is homeomorphic to the open disk and the boundary of each face is a cycle of G. It is obvious that if G has a strong embedding in some surface, then the set of boundaries of faces is a cycle double cover of G. Conversely, if F is a cycle double cover of a graph G, then consider each cycle of F as the boundary of a disk; then joining of these disks at their edges yields a surface S. Therefore the graph G is strongly embedded on the surface S. But the surface S may not be a 2-manifold, because a neighborhood at some vertex of G may not be homeomorphic to a open disk.

However, for a cubic graph G, the cycle double cover problem and the strong embedding problem are equivalent. The following result is a direct corollary of Theorem 2.

Corollary 14 (Alspach and Zhang [2])
Every 2-connected cubic graph containing no Petersen-minor has a strong embedding on some 2-manifold surface.

With the aid of some previous results, Corollary 14 can be generalized for graphs with no K_5-minor.

Theorem 15[1]
Every 2-connected graph containing no K_5-minor has a strong embedding on some 2-manifold surface.

Let $G=(V,E)$ be a 2-connected graph containing no K_5-minor. Since a graph with a Petersen-minor also has a K_5-minor, by Theorem 3 G has a cycle double cover F. Let v be a vertex of G. Let $E(v)$ be the set of all edges incident with v. Construct a simple graph F_v for each vertex v of G as follows. The vertex set of F_v is $E(v)$ and two vertices e' and e'' of F_v are adjacent in F_v if and only if the edges e' and e'' of G are contained in some cycle C of F. Obviously, each component of F_v is either a cycle or a single edge. (Note that if two edges e' and e'' incident with v are contained in two cycles of F, the degrees of e' and e'' are both one). Let $c(F_v)$ be the number of components of the graph F_v. The following proposition is obvious.

Proposition
A graph G has a strong embedding if and only if G has a cycle double cover F such that $c(F_v)=1$ for each vertex v of G.

Proof of Theorem 15

Let F be a cycle double cover of G such that $\sum_{v \in V(G)} c(F_v)$ is as small as possible. Assume that $c(F_z) > 1$ for some vertex z of G. Let $E_1, E_2,...,E_r$ be the components of the graph F_z, and F_μ be the set of cycles of F containing some edges of E_μ for $\mu=1,...,r$. Construct a graph H such that $V(H)=F$ and $C', C'' \in F$ are

---

[1] The proof of this theorem was completed in "The Graph Minor" - AMS summer research workshop, Seattle, 1991. The author wishes to thank the co-chairs of the workshop for their invitation and financial support. The complete proof of the theorem is included in this paper.

adjacent in H if and only if $[V(C') \cap V(C'')]\setminus\{z\} \neq \emptyset$. Since $G\setminus\{z\}$ is connected, so is H. Let Q be a shortest path of H joining a pair from distinct $F_{\mu'}$ and $F_{\mu''}$. Let $Q = C_1...C_s$ and $C_1 \in F_{\mu'}$, $C_s \in F_{\mu''}$. By the choice of Q,

$$\left(\bigcup_{i=2}^{s-1} V(C_i)\right) \cap \{z\} = \emptyset \text{ and } \left(\bigcup_{i=2}^{s-1} E(C_i)\right) \cap E(z) = \emptyset.$$

Let G' be the subgraph of G induced by the set of all edges in $\bigcup_{i=1}^{s} E(C_i)$. Obviously, the degree of each vertex of G' is 2, 3 or 4, and G' is 2-connected.

Construct G'' from G' by replacing every edge of G' contained in two cycles of Q by two parallel edges. G'' is therefore eulerian with $d_{G''}(v) = 2$ or 4 for each vertex v of G''. Define a set P of forbidden parts on G'' as follows:

    i)   if $d_{G''}(v) = 2$, let each forbidden part incident with v consist of only one edge;

    ii)   if $d_{G''}(v) = 4$ and $d_{G'}(v) = 2$, let a pair of parallel edges in G'' be e' and e'', and another pair of parallel edges be e^* and e^{**}. Let $\{e', e''\}$, $\{e^*, e^{**}\}$ be the forbidden parts incident with v;

    iii)   if $d_{G''}(v) = 4$ and $d_{G'}(v) = 3$, let the two parallel edges in G'' be e' and e'', and another two edges incident with v be e^* and e^{**}. Let $\{e', e''\}$, $\{e^*\}$ and $\{e^{**}\}$ be the forbidden parts incident with v;

    iv)   if $d_{G''}(v) = d_{G'}(v) = 4$, let e'_1 and e'_2 be two edges incident with v and contained in a cycle C_i of Q and e''_1 and e''_2 be two edges incident with v and contained in a cycle C_{i+1} of Q. Case 1, the edges e'_1, e'_2, e''_1 and e''_2 are in the same component of F_v. Let the component of F_v containing e'_1, e'_2, e''_1 and e''_2 be a cycle $e'_1 e'_2 ... e''_1 e''_2 ...... e'_1$. Then let $\{e'_2, e''_1\}$ and $\{e'_2, e''_1\}$ be the forbidden parts incident with v. Case 2, the edges $\{e'_1, e'_2\}$, $\{e''_1, e''_2\}$ are in two distinct components of F_v. Then let $\{e'_1, e'_2\}$ and $\{e''_1, e''_2\}$ be the forbidden parts incident with v.

Since G' and G'' are 2-connected and each forbidden part contains at most two edges, (G'', P) does not have a bad cut. Since G contains no K_5-minor, so does G''. By Theorem 9, G'' has a compatible cycle decomposition F^*. Thus G has a cycle double cover $F' = (F \setminus Q) \cup F^*$. Obviously,

$$\sum_{v \in V(G)} c(F'_v) < \sum_{v \in V(G)} c(F_v).$$

This is a contradiction and completes the proof of the theorem.

7. Conjectures

7.1 A weak 4-flow conjecture

The following conjecture proposed by Tutte is a well-known refinement of the 4-color problem.

4-flow Conjecture (Tutte [21], or see [14], [22])
Every 2-edge-connected graph containing no Petersen-minor admits a nowhere-zero 4-flow.

Though the 4-color problem has been verified by computer searching, this conjecture is still open. The best known result, due to Jaeger [13], is that every 4-edge-connected graph admits a nowhere-zero 4-flow. And it is also known that the 4-flow conjecture is equivalent to the following problem.

4-flow Conjecture (the even-subgraph-cover version)
Every 2-edge-connected graph containing no Petersen-minor has a double cover F consisting of at most 4 even subgraphs.

An even graph H is a graph in which the degree of every vertex is even. Obviously, an even graph is a union of several edge-disjoint cycles.

The author would like to propose the following problem, which is similar to the weak 3-flow conjecture ([14]).

Conjecture 16 (Weak 4-flow conjecture)
There is an integer k such that every 2-edge-connected graph containing no Petersen-minor has a cycle double cover consisting at most k even subgraphs.

Note that a similar conjecture (the 5-cycle double cover conjecture) proposed by Celmins ([4]) is closely related to the 5-flow conjecture and is much stronger than the cycle double cover conjecture. But it is not related to the 4-flow conjecture which concerns graphs with no Petersen-minor. One might hope that Theorem 3 would be helpful in proving Conjecture 16.

7.2 Repeated edges in a shortest cycle cover

A cycle cover F of a 2-edge-connected graph G is a family of cycles such that each edge of G is contained in some cycle of F. Let t_F be the least integer such that each edge of G is contained in at most t_F cycles of F. Denote
$$t_{CC}(G) = \min \{t_F \colon \text{for every cycle cover } F \text{ of } G\}$$
and
$$t_{SCC}(G) = \min \{t_F \colon \text{for every shortest cycle cover } F \text{ of } G\}.$$

The cycle double cover conjecture is equivalent to the problem that $t_{CC}(G) \le 2$ for every 2-edge-connected graph. And the 6-flow theorem (Seymour, [19]) and the 8-flow theorem (Jaeger, [15]) imply that $t_{CC}(G) \le 3$ for every 2-edge-connected graph. By Theorem 3 and 4, every 2-edge-connected graph G containing no Petersen-minor or admitting a nowhere-zero 4-flow has the property that $t_{SCC}(G) \leq 2$. However, $t_{SCC}(P_{10}) = 2$ for the Petersen graph itself. Thus, it seems that the appearance of a Petersen-minor may not raise t_{SCC}. The author wishes to propose the following conjecture.

Conjecture 17
  For any 3-connected graph G, $t_{SCC}(G) \leq 2$.

Note that the connectivity in the conjecture cannot be reduced since the 2-connected graph in fig. 4 has $t_{SCC} = 3$.

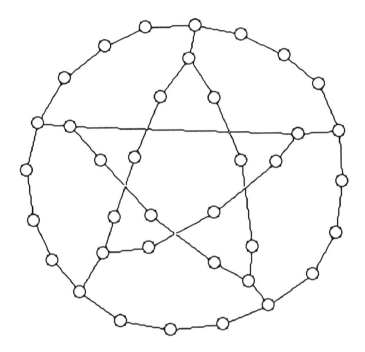

fig. 4

Conjecture 18 (a weak version of Conjecture 17)
  There is an integer k such that for every 3-connected graph G, $t_{SCC}(G) \leq k$.

REMARKS.
  According to the Robertson-Seymour Theory, finding a given graph minor is polynomially solvable. Recently, McGuinness and Kézdy ([16]) found a practical algorithm for finding a K_5-minor, which takes time $O(n^2)$ for a graph of order n.

  Recently, N. Robertson (personal communication) suggested an alternative proof of Theorem 15. His proof is based on Wagner's Theorem.

REFERENCES

1. B. Alspach, L. Goddyn and C.Q. Zhang, Graphs with the circuit cover property, (submitted)
2. B. Alspach and C. Q. Zhang, Cycle coverings of cubic multigraphs, *Discrete Mathematics* (to appear)
3. J.A. Bondy and U.S.R. Murty, *Graph Theory with Applications*, Macmillan, London and Elsevier, New York.
4. U. A. Celmins, On cubic graphs that do not have an edge 3-coloring, Ph. D. Thesis, Waterloo. 1984
5. J. Edmonds and E.L. Johnson, Matchings, Euler tours and the Chinese postman, *Mathematical Programming*, 5, (1973) 88-124.
6. H. Fleischner, Eulersche Linien und Kreisuberdeckungen die vorgebene Durchgange inden Kanten vermeiden, *J. Combinatorial Theory*, B, 29, (1980) 145-167.
7. H. Fleischner, Eulerian graphs, in *Selected Topics in Graph Theory 2*, (L.W. Beineke and R.J. Wilson, eds), Academic Press, New York, 1983, 17-53.
8. H. Fleischner and A. Frank, On cycle decomposition of eulerian graph, *J. Combinatorial Theory*, B, 50, (1990) 245-253.
9. M. Guan and H. Fleischner, On the minimum weighted cycle covering problem for planar graphs, *Ars Combinatoria*, 20, (1985) 61-68.
10. A. Itai, R.J. Lipto, C.H. Papadimitriou and M. Rodeh, Covering graphs with simple circuits, *SIAM J. Computing* 10 (1981) 746-750
11. A. Itai and M. Rodeh, Covering a graph by circuits, Automata, *Languages and Programming, Lecture Notes in Computer Science 62*, Springer-Verlag, Berlin, (1978) 289-299.
12. B. Jackson, Shortest circuit covers and postman tours in graphs with a nowhere zero 4-flow, *SIAM J.Discrete Math.* (to appear)
13. F. Jaeger, A survey of the cycle double cover conjecture, *Annals of Discrete Mathematics*, 27, (1985) 1-12.
14. F. Jaeger, Nowhere-zero flow problems, in *Selected Topics in Graph Theory*, Vol. 3, (L.W. Beineke and R.J. Wilson, eds), Academic Press, New York, (1988) 71-95.
15. F. Jaeger, Flows and generalized coloring theorems in graphs, *J. Combinatorial Theory*, B 26 (1979), 205-216.
16. P. J. McGuinness, and A. E. Kézdy, Sequential and parallel algorithms to find a K_5 minor, preprint.
17. P. D. Seymour, Sums of circuits, in *Graph Theory and Related Topics* (J.A. Bondy and U.S.R. Murty, eds), Academic Press, New York, 1978, 341-355.
18. P. D. Seymour, Even circuits in planar graphs, *J. Combinatorial Theory*, B, 31, (1981) 178-192.
19. P. D. Seymour, Nowhere-zero 6-flows, *J. Combinatorial Theory*, B 30 (1981) 130-135.
20. G. Szekeres, Polyhedral decompositions of cubic graphs, *J. Austral. Math. Soc.*, 8, (1973) 367-387.
21. W. T. Tutte, On the embedding of linear graphs in surfaces, *Proc. London Math. Soc.*, Ser. 2, 51, (1949) 474-489.
22. D. H. Younger, Integer flows, J. Graph Theory, 7, (1983) 349-357.
23. C. Q. Zhang, Minimum cycle coverings and integer flows, *J. Graph Theory*, 14 (1990) 537-546.
24. C. Q. Zhang, On compatible cycle decompositions of eulerian graphs, (submitted)
25. C. Q. Zhang, On even cycle decompositions of eulerian graphs, (submitted)

DEPARTMENT OF MATHEMATICS
WEST VIRGINIA UNIVERSITY
MORGANTOWN
WV 26506

Email address: CQZHANG@WVNVM.WVNET.EDU

Cones, Lattices and Hilbert Bases of Circuits and Perfect Matchings

LUIS A. GODDYN

ABSTRACT. There have been a number of results and conjectures regarding the cone, the lattice and the integer cone generated by the (real-valued characteristic functions of) circuits in a binary matroid. In all three cases, one easily formulates necessary conditions for a weight vector to belong to the set in question. Families of matroids for which such necessary conditions are sufficient have been determined by Seymour; Lovász and Seress; Alspach, Fu, Goddyn and Zhang, respectively. However, circuits of matroids are far from being well understood. Perhaps the most daunting (and important) problem of this type is to determine whether the circuits of a matroid form a *Hilbert basis*. That is, for which matroids does the integer cone coincide with those vectors which belong to both the cone and the lattice? Additionally, all of the above questions have been asked with regard to perfect matchings in graphs.

We present a survey of this topic for circuits in matroids, and also for perfect matchings in graphs. There are some striking similarities, especially with regard to the role that Petersen's graph plays in both of these subjects. A possible explanation is that much of the theory of perfect matchings is captured by the circuits of certain 1-element extensions of graphic matroids called *grafts*. For example, a possible extension to the class of grafts of the following result would imply the Four-color theorem: *The circuits of a graph form a Hilbert basis if and only if the graph has no Petersen minor.*

1. Introduction and Notations

A fruitful setting for studying a combinatorially defined collection of subsets of a ground set E is to consider the corresponding collection of real-valued characteristic functions. This observation underlies much of the theory of polyhedral combinatorics and integer programming. Our aim is to compare the collection

---

1991 *Mathematics Subject Classification.* Primary 05C70; Secondary 05C38, 05B35.

The author was supported by the National Sciences and Engineering Research Council operating grant #A-4699 and SFU's President's Research Grant #A-2326.

This paper is in final form and no version of it will be submitted for publication elsewhere.

of circuits in a matroid with the collection of perfect matchings in a graph by considering properties of their characteristic functions.

For $S \subseteq E$ we denote by χ^S the $\{0,1\}$-characteristic vector of S in $\mathbb{Q}^E$. For any collection $\mathcal{S}$ of such subsets of E, we define the *linear hull, cone, lattice* and *integer cone* of $\mathcal{S}$ as follows.

$$Lin.Hull(\mathcal{S}) := \{\sum_{S \in \mathcal{S}} \alpha_S \chi^S : \alpha_S \in \mathbb{Q}\}$$

$$Cone(\mathcal{S}) := \{\sum_{S \in \mathcal{S}} \alpha_S \chi^S : \alpha_S \in \mathbb{Q}_{\geq 0}\}$$

$$Lat(\mathcal{S}) := \{\sum_{S \in \mathcal{S}} \alpha_S \chi^S : \alpha_S \in \mathbb{Z}\}$$

$$Int.Cone(\mathcal{S}) := \{\sum_{S \in \mathcal{S}} \alpha_S \chi^S : \alpha_S \in \mathbb{Z}_{\geq 0}\}$$

We have the following four containments.

$$Int.Cone(\mathcal{S}) \begin{array}{c} \subset Cone(\mathcal{S}) \subset \\ \subset Lat(\mathcal{S}) \subset \end{array} Lin.Hull(\mathcal{S}).$$

Throughout this paper $\mathcal{S}$ shall be either the collection $\mathcal{C} = \mathcal{C}(M)$ of circuits in a matroid $M = (E, \mathcal{C})$ on the ground set E, or the collection of perfect matchings (1-factors) $\mathcal{M} = \mathcal{M}(G)$ in a graph $G = (V, E)$.

It can be argued that the integer cone is the most interesting of the four sets defined above. A vector p belongs to $Int.Cone(\mathcal{S})$ if and only if there is a list of subsets in $\mathcal{S}$ such that each $e \in E$ belongs to precisely $p(e)$ members of the the list. Such a list is often called a *cover* of the weighted set (E, p). If p is the constant unit vector $\mathbf{1}$, then a cover of (E, p) is a *decomposition* of E into subsets from $\mathcal{S}$. A cover of $(E, 2)$ is often called a *double cover* of E.

When $\mathcal{S} = \mathcal{C}$, we are in the area of *circuit covers* and *circuit decompositions*, where numerous papers [1, 2, 4, 5, 11, 12, 13, 15, 18, 20, 22, 25, 28, 29, 30, 43, 51, 52, 55, 57] have been written, especially for graphic matroids. Many of these papers are concerned with circuit covers which have additional conditions on parameters such as the number of circuits in the cover, or the total length of the circuits. Here, we are concerned only with the existence of circuit covers of fixed vectors p.

Where $\mathcal{S} = \mathcal{M}$, we are studying *perfect matching covers* of graphs. The case $p = \mathbf{1}$ is concerned with *1-factorizations* of graphs, where we have the classical *Four-color Theorem* and the stronger *4-flow conjecture* of Tutte [53]. For $p = 2$ we have *perfect matching double covers* with unsolved conjectures of Fulkerson [16] and Seymour [47].

Although combinatorially less interesting, the cone and the lattice of $\mathcal{S}$ are generally easier to determine than the integer cone. For example, the both the cone and the lattice generated by (characteristic vectors of) perfect matchings in a graph have been well characterized [9, 34], whereas it is *NP-hard* to determine whether **1** belongs to the integer cone of perfect matchings of a cubic graph [21]. The study of the cone and the lattice is further motivated by the formula

(1) $$Int.Cone\,(S) \subseteq Cone\,(S) \cap Lat\,(S)$$

which provides necessary conditions for a vector p to belong to the integer cone of S.

Understandably, it is of special interest to know when equality holds in (1).

DEFINITION 1.1. *A set of vectors S for which equality holds in (1) is called a* Hilbert basis.

This concept is closely related to *total dual integrality*, and has been studied by various authors [17, 6, 37, 38]. In our setting, the *Hilbert basis problem* is to determine classes of matroids and graphs for which C and M form Hilbert bases. This problem will be addressed in Sections 3 and 6, respectively.

It must be emphasised that the cone and the lattice of S are worthy of independent study. For example, the characterizations of both the cone [9] and the lattice [34] of perfect matchings are landmarks in graph theory. Both the cone and the lattice of circuits in a graph have simpler descriptions [43] than those of perfect matchings. However, they easily become intractable for more general classes of matroids. For example, determining whether a vector belongs to the cone of circuits in a cographic matroid is NP-complete [31].

Those who work with either circuits or perfect matchings agree that Petersen's graph plays an anomalous role. (This is particularly evident when considering the Hilbert basis problem.) This observation suggests that these two areas may be related. In fact, connections between circuits and perfect matchings in graphs are already well established. For example, the Chinese Postman problem [10] is closely related to both matchings and Euler tours. As another example, the Four-color theorem is equivalent to the statement that any bridgeless planar graph is the union of two subgraphs, each of which is the edge-disjoint union of circuits. One can not say, however, that such connections satisfactorily explain the predominating role of Petersen's graph.

In Section 7, we describe another connection between circuits and perfect matchings, which is expressed via certain 1-element extensions of graphic matroids called *grafts*. It is through the integer cone of circuits in grafts that we see a possible explanation of the role of Petersen's graph in the theory of circuits and perfect matchings.

We shall assume basic familiarity with graphs and matroids as in [3] and [54]. Thus a *bridge* or *coloop* of a matroid $M = (E, C)$ is any element contained in no circuit. Two non-bridge elements are *in series* or *coparallel* if no circuit contains exactly one of them. Recall that "coparallel" is an equivalence relation on E. A *bond* or *cocircuit* is a minimal subset of E intersecting all bases of M. The *dual*,

M^* of M has ground set E and its circuits are the bonds of M. If $G = (V, E)$ is a graph then $M(G)$ denotes the matroid on $E(G)$ whose circuits are the polygons in G. For binary matroids (including graphs) a *cycle* is any (element-) disjoint union of circuits in M, including the empty cycle. Thus a cycle in a graph is the edge set of any subgraph whose vertices all have even degree. A matroid is *binary* if it is isomorphic to a set of vectors with linear dependence over $\mathrm{GF}(2)$ or, equivalently, if it has no U_4^2-minor (that is, no minor isomorphic to U_4^2, the *uniform* matroid of rank 2 on 4 elements). The cycles of a binary matroid under the symmetric difference operator Δ form a vector space of dimension $|E| - rank(M)$ in $\mathrm{GF}(2)^E$ called the *cycle space* of M. Dually, *parallel* elements, *loops* and *cocycles* in M are are defined to be coparallel elements, coloops and cycles in M^*, respectably. A cocycle is sometimes called a *cut*.

Where convenient, we identify a subset of E with its characteristic vector, a graph G with its matroid $M(G)$, and a subset of edges of a graph with the subgraph it induces.

2. The Cone and Lattice of Circuits

Although there is no known non-trivial characterization of $Cone\,(C)$, $Lat\,(C)$, and $Int.Cone\,(C)$ for general matroids, the linear hull of circuits has an easy description. Any vector in $Lin.Hull\,(C)$ must clearly be zero on bridges and constant on coparallel classes. In fact, these two conditions characterize the linear hull.

PROPOSITION 2.1. *For any matroid* $M = (E, C)$, $Lin.Hull\,(C) = \{p \in \mathbb{Q}^E : p(e) = 0 \text{ for any bridge } e, \text{ and } p(f) = p(g) \text{ for } f, g \text{ coparallel }\}$.

PROOF. Let $[e]$ denote the set of elements which are coparallel with e. It is enough to show that, for any element e in a bridgeless matroid, $\chi^{[e]} \in Lin.Hull\,(C)$. We use the following observation of Seymour, [**44**, (3.2)].

OBSERVATION 2.2. *If M is bridgeless then* $\mathbf{1} \in Cone\,(C)$.

(Here $\mathbf{1} = \chi^E$ is the vector of ones.) As M is bridgeless, so is $M\backslash[e]$, hence $\chi^E, \chi^{E\backslash[e]} \in Cone\,(C)$. Subtracting, we have $\chi^{[e]} \in Lin.Hull\,(C)$. □

We now examine the cone of circuits. Since no circuit in a matroid meets a bond in exactly one element, any weight vector $p \in Cone\,(C)$ must be *balanced*. That is, no element in M has more than half the total weight of any bond containing it. Seymour [**44**] has characterized those matroids for which the cone of circuits is precisely the set of non-negative balanced vectors.

THEOREM 2.3. *For any matroid* $M = (E, C)$, $Cone\,(C) \subseteq \{p \in \mathbb{Q}_{\geq 0}^E : p(e) \leq p(B\backslash e) \text{ for all } e \in B, \text{ for all bonds } B\}$, *with equality if and only if M has no minor isomorphic to any of* U_4^2, $M^*(K_5)$, F_7^*, *or* R_{10}. □

(As usual, $p(S)$ denotes $\sum_{e \in S} p(e)$ for any $S \subseteq E$.) A matroid for which equality holds in Theorem 2.3 is said to have the *Sums of Circuits Property*. In particular, all graphs have the Sums of Circuits Property [43]. It follows from Theorem 2.3 (and is not difficult to show directly) that the Sums of Circuits Property is preserved under taking minors. To see that each of the four obstructing minors (see [44] or [48] for their definitions) do not have this property consider the following four weight vectors $p \in \mathbb{Q}_{\geq 0}^E$.

U_4^2:: $p(e_0) = 2$ for some fixed $e_0 \in E$ and $p(e) = 1$ for the remaining 3 elements.

$M^*(K_5)$:: $p(e) = 1$ for all edges e in some fixed subgraph of K_5 isomorphic to $K_{2,3}$ and $p(e) = 2$ for the remaining 4 edges.

F_7^*:: $p(e) = 1$ for all elements e in some fixed 4-circuit and $p(e) = 2$ for the remaining 3 elements.

R_{10}:: $p(e) = 3$ for all elements e in some fixed 3-subset of elements not contained in any 4-circuit, and $p(e) = 1$ for the remaining 7 elements.

One easily checks that each of these vectors is balanced. To show that they are not in $Cone\,(C)$ we use *Farkas' Lemma*. That is, we describe a weight vector $s \in \mathbb{Q}^E$ which has positive inner product with p, but for which each circuit in the matroid has non-positive weight. For U_4^2 we take $s(e_0) = 2$ and $s(e) = -1$ for the remaining 3 elements. In each of the other three examples we take $s(e) = -1$ if $p(e) = 1$ and $s(e) = 1$ otherwise.

It appears unlikely that there is a good description of $Cone\,(C)$, even for cographic matroids, as it is known [31] that the membership problem for the cone of bonds in K_n is *NP-hard*. Some work [8] has been done toward finding facets of this cone.

Likewise, the lattice of circuits appears to be difficult to characterize for general matroids. Indeed it is not easy to imagine non-trivial necessary conditions for a vector to belong to $Lat\,(C)$. The situation is better, although not yet settled, for binary matroids.

In a binary matroid M, any circuit intersects any bond in an even number of elements. Thus for a weight vector to belong to $Lat\,(C)$, it is necessary that p be *eulerian*, that is, p must be integer-valued and each bond in M must have even total weight. This condition turns out to be sufficient for an important class of binary matroids.

PROPOSITION 2.4. *For any binary matroid $M = (E, C)$, $Lat(C) \subseteq Lin.Hull(C)$ $\cap \{p \in \mathbb{Z}^E : p(B)$ is even for all bonds $B\}$, with equality if M has no F_7^*-minor.*

PROOF. Assume that M has no F_7^*-minor and that it contains neither bridges nor two elements in series. Let p be an integer weight vector such that each bond has even weight. The set F of edges having odd weight belongs to the cycle space of M since F is orthogonal to every bond (over GF(2)). It suffices to show that $2\chi^{\{e\}} \in Lat\,(C)$ for any $e \in E$, since this would imply that the even-valued vector

$p + \chi^F$ — and hence p — belongs to $Lat(C)$. To this end, we need only find two circuits C_1, C_2 such that $C_1 \cap C_2 = \{e\}$, whereby $2\chi^{\{e\}} = \chi^{C_1} + \chi^{C_2} - \chi^{C_1 \Delta C_2}$. The existence of C_1, C_2 in bridgeless binary matroids with no F_7^* minors was established by Cunningham [7]. This also follows from a theorem of Seymour [45] which states that all binary matroids with no F_7^*-minor have the *Integer Max-Flow Min-Cut Property*. □

A matroid for which equality holds in Proposition 2.4 is said to have the *Lattice of Circuits Property*. In particular, all graphic and cographic matroids, and indeed all regular matroids have the Lattice of Circuits Property, as do all matroids with the Sums of Circuits Property. Unlike the Sums of Circuits Property, the class of matroids with the Lattice of Circuits Property is not is not closed under taking minors (although it is closed under element-contraction). For example, although F_7^* does not have the Lattice of Circuits Property, exactly one of the two 1-element extensions of F_7^* does.

Recently, Lovász and Seress [35] have characterized the class of binary matroids with the Lattice of Circuits Property. We shall state their result without proof. We need a definition. In [44] Seymour defines, for $k = 1, 2, 3$, the *k-sum* of two binary matroids M_1, M_2 to be the matroid on $E(M_1) \Delta E(M_2)$ whose circuits are all subsets of the form $C_1 \Delta C_2$ where $C_i \in C(M_i)$, $i = 1, 2$. In particular, $k = 1$ if $E(M_1) \cap E(M_2) = \emptyset$; $k = 2$ if $E(M_1) \cap E(M_2)$ consists of a single element which is not a loop in each M_i; and $k = 3$ if $E(M_1) \cap E(M_2)$ is a circuit of cardinality 3 in each M_i.

DEFINITION 2.5. *Any matroid which can be obtained from copies of the Fano plane F_7 via 1-, 2- and 3-sums shall be called a* Fano-cycle.

THEOREM 2.6. [35] *A binary matroid M has the Lattice of Circuits Property if and only if the dual matroid M^* contains no Fano-cycle as a submatroid.* □

EXAMPLE 2.7. *The affine geometry $AG(2,3)$ is a Fano-cycle of cardinality 8, since it is the 3-sum of two Fano planes. As $AG(2,3)$ is self dual, Theorem 2.6 asserts that $AG(2,3)^* \cong AG(2,3)$ does not have the Lattice of Circuits Property. Indeed this follows directly from the fact all circuits in $AG(2,3)$ have cardinality four.*

In general, the lattice of circuits in a binary matroid can be arbitrarily "sparse".

EXAMPLE 2.8. *The binary projective geometry of dimension m, $PG(2, m)$, is the binary matroid represented by the $2^{m+1} - 1$ non-zero binary $(m + 1)$-tuples. For example, $PG(2,2) \cong F_7^*$. For $0 \leq k \leq m$, a k-flat is any submatroid of $PG(2, m)$ which is isomorphic to $PG(2, k)$. The cocircuits of $PG(2,m)$ are precisely the complements of its $(m - 1)$-flats, and thus have cardinality 2^m. It follows that any vector in lattice of cocircuits of $PG(2, m)$ (that is, $Lat(C(PG(2,m)^*))$) has total weight divisible by 2^m. In fact, $p \in Lat(C(PG(2,m)^*))$*

if and only if $p(S)$ is divisible by 2^k, for every k-flat S of $PG(2,m)$, $0 \leq k \leq m$ [35].

3. Hilbert Bases of Circuits

The *circuit cover problem* is to determine whether a given weight vector belongs to the integer cone of circuits of a given matroid. We are interested in finding classes of matroids for which this the circuit cover problem can be solved. Having seen that the cone and the lattice of circuits are often characterizable, we are naturally led to the Hilbert basis problem for circuits (recall Definition 1.1). That is, we would like to determine those matroids M for which (M,p) has a circuit cover for all $p \in Cone\,(C) \cap Lat\,(C)$.

The circuits of a matroid do not always form a Hilbert basis.

EXAMPLE 3.1. *Let P_{10} denote Petersen's graph and let p_{10} denote the weight vector which takes the value 2 on some fixed 1-factor of P_{10}, and 1 on the complementary 2-factor. One easily checks that p_{10} is balanced and eulerian, and hence belongs to $Cone\,(C) \cap Lat\,(C)$, by Theorems 2.3 and 2.4. However, $p_{10} \notin Int.Cone\,(C)$ since $p_{10} - \chi^C \notin Cone\,(C)$, for all $C \in \mathcal{C}$.*

Every matroid for which we know that C does not form a Hilbert basis contains Petersen's graph as a minor. On this flimsy evidence one might propose the following.

CONJECTURE 3.2. *If a matroid contains no P_{10}-minor then C forms a Hilbert basis.*

As we shall see, progress has been made toward this conjecture, but mostly for graphs and other binary matroids. We direct the reader's attention to the strikingly similar Conjecture 6.6 regarding perfect matchings in graphs.

A basic problem with dealing with Conjecture 3.2 is that we do not know $Cone\,(C)$ for general matroids. Thus it makes sense restrict our attention to matroids for which this cone has a nice description, namely, the matroids with the Sums of Circuits Property. Recall from Theorems 2.3 and 2.4 that, for such matroids, the cone (lattice) of circuits is precisely the set of balanced (eulerian) weight vectors. In 1979, Seymour verified Conjecture 3.2 for planar graphs by showing that every balanced, eulerian edge-weighted planar graph has a circuit cover. In 1981, Seymour [44] characterized the matroids with the Sums of Circuits Property and, in the same paper, proposed that Conjecture 3.2 holds for such matroids. Recently, Alspach, Goddyn and Zhang [1], shed Seymour's planarity restriction, and verified Conjecture 3.2 for the class of graphic matroids. Using this result, and Seymour's matroid decomposition theorems, Fu and Goddyn [15] have since shown that the conjecture holds for all matroids with the Sums of Circuits Property. We state this result in an alternate form.

We say that a matroid has the *Circuit Cover Property* if the integer cone of circuits is precisely the set of balanced and eulerian weight vectors. Thus, any matroid with the Sums of Circuits Property has the Circuit Cover Property exactly when its circuits form a Hilbert basis.

THEOREM 3.3. *A matroid has the Circuit Cover Property if and only if it has no minor isomorphic to any of U_4^2, $M^*(K_5)$, F_7^*, R_{10}, $M(P_{10})$.* □

Perhaps the most relevant aspect of Theorem 3.3 is that planarity restrictions on graphs have been dropped. The literature abounds with graph properties which are known to hold for planar graphs, and which are conjectured to hold for wider classes of graphs. Such problems include classical "nuts" such as the *circuit double cover conjecture* and Tutte's *Nowhere-zero flow* conjectures. Thus it is of interest whenever a planarity restriction can be dropped (or relaxed) from the hypothesis of a known theorem. For example, Theorem 3.3 has already been used to extend results involving Even Circuit Decompositions [57] and Compatible Circuit Decompositions of Eulerian graphs [56] from the class of planar graphs to the class of graphs with no K_5-minor. We refer the interested reader to [1] for more applications Theorem 3.3.

Little is known about the integer cone of circuits in non-binary matroids. As observed by Sebő [38], the circuits of uniform matroids U_n^k do indeed form a Hilbert basis. This follows from the fact that the circuits of U_n^k are precisely the bases of U_n^{k+1} (when $k < n$), and from the following consequence of Edmonds' matroid intersection theorem.

THEOREM 3.4. *The bases of any matroid form a Hilbert basis.* □

4. Range-Restricted Circuit Covers

Perhaps we are asking too much of matroids when we require their circuits to form Hilbert bases. One way to weaken the Hilbert-basis property is to ask whether some restricted subset of $Cone\,(C) \cap Lat\,(C)$ is contained within $Int.Cone\,(C)$. Our intention is to determine the point at which non-Hilbert matroids such a $M(P_{10})$ cease to behave anomalously. In this way we obtain a more sensitive test of "how bad" such anomalous matroids really are.

A notorious problem of this type is the *circuit double cover conjecture*.

CONJECTURE 4.1. [50, 43] *For any bridgeless graph, $2 \in Int.Cone\,(C)$.*

Of course, the "bridgeless" condition is only there to assure the membership of 2 in the cone of circuits. We refer the interested reader to [25, 18, 19, 52, 30, 4, 5]. The circuit double cover conjecture is perhaps the most interesting of the *uniform* circuit cover problems, where the weight vector p is required to be constant on E. For integers greater than 2, the uniform circuit cover problem is has been completely solved for a large class of binary matroids, which includes all graphs.

PROPOSITION 4.2. *For any binary matroid with no F_7^*-minor and for any integer $r \neq 2$, $r \in Int.Cone\,(C)$ if and only if $r \in Cone\,(C) \cap Lat\,(C)$.*

PROOF. The "only if" direction is trivial.

Conversely, suppose that r is odd. For r to be in the lattice of circuits of M, it is necessary that all cocircuits in M have even cardinality. That is, M is eulerian, and thus $E(M)$ is the disjoint union of circuits. That is, $1 \in Int.Cone(C)$ and so $r \in Int.Cone(C)$. Note that, for odd r, the statement of the theorem holds for all binary matroids.

For even r, we note that $r \in Cone(C)$ if and only if the matroid M is bridgeless. Using matroid decompositions, Jamshy and Tarsi [28] showed that $r \in Int.Cone(C)$ for any bridgeless binary matroid with no F_7^*-minor if and only if the same holds for any bridgeless graph. It suffices to prove that $4, 6 \in Int.Cone(C)$ for any bridgeless graph, since every larger even integer is in the integer cone generated by $\{4, 6\}$. Indeed, Jaeger [24] proved the case $r = 4$ and Fan [12] proved the case $r = 6$. Both of these results are consequences of Seymour's 6-flow theorem for bridgeless graphs [46]. □

Incidently, Jamshy and Tarsi [28] also show that if the circuit double cover conjecture is true, then $2 \in Int.Cone(C)$ for any bridgeless binary matroid with no F_7^*-minor.

Little is known about uniform circuit covers of general matroids. The obvious necessary conditions are that the matroid be bridgeless and that the vector in question belong to the lattice of circuits. The following example of M. Laurent [33] shows that these conditions do not suffice.

EXAMPLE 4.3. *Let M denote the matroid whose circuits are the even edge cuts of K_{12} (Using the terminology of Section 7, M is the matroid obtained by deleting the element τ from the dual of the graft $(K_{12})_V$). We have that $2 \in Lat(C)$, but $2 \notin Int.Cone(C)$.*

It would be interesting to characterize those matroids for which membership in $Lat(C)$ suffices for the uniform circuit cover problem.

We generalize to non-uniform circuit covers by allowing the range of the weight vector to take two or more fixed values. Our general aim is to classify those ranges (subsets of positive integers) for which we have a nice characterization such as in Proposition 4.2. For sake of brevity, we shall confine most of our discussion to graphic matroids.

DEFINITION 4.4. *Let R be a set of positive integers. We say that R is a* good range *(for the class of graphs) if for any graph $G = (V, E)$ and any weight vector $p \in R^E$, $p \in Int.Cone(C)$ if and only if $p \in Cone(C) \cap Lat(C)$. A range that is not good is said to be* bad.

In view of Theorems 2.3 and 2.4, a range R is good for the class of graphs if every balanced, eulerian weighted graph (G, p) with $p \in R^E$ has a circuit cover. If R is good then so is any subset of R. Example 3.1 shows that $\{1, 2\}$ is a bad range for graphs. More bad ranges can be obtained by modifying this example.

PROPOSITION 4.5. *If $\{1,k\} \subseteq R$, for some $k \geq 2$, then R is bad for the class of graphs.*

PROOF. I thank P. Seymour for the following construction. Let F denote a fixed 1-factor in Petersen's graph P_{10}. For any $k \geq 2$, let $P_{10}^{(k)}$ denote the graph obtained from P_{10} by replacing each edge in F with $k-1$ parallel edges. We consider the weight vector $p_{10}^{(k)} \in \{1,k\}^E$ which takes the value k on exactly one of the $k-1$ edges in each of the five parallel classes defined above, and which takes the value 1 elsewhere. One easily checks that $p_{10}^{(k)}$ is balanced. As any circuit cover of $(P_{10}^{(k)}, p_{10}^{(k)})$ would have to use $k-2$ digons from each parallel class, it follows from Example 3.1 (which is the case $k=2$) that $p_{10}^{(k)} \notin Int.Cone\,(C(P_{10}^{(k)}))$. □

If some range containing 2 is good, then in particular, the circuit double cover conjecture is true. This gives us a way of strengthening the circuit double cover conjecture. For example, Seymour [**44**, (16.6)] proposed the following.

CONJECTURE 4.6. *The set of positive even integers is good for the class of graphs.*

On the other hand, extending a range does not always affect it goodness.

PROPOSITION 4.7. *Let R be a range of even integers and let $r \in R$ be such that $r \geq \max R/2$. Then for any odd integer $k > r$, $R \cup \{k\}$ is good (for the class of graphs) if and only if R is good.*

PROOF. The "only if" part is trivial. Conversely, suppose that R is good and let $p' \in R'^E \cap Cone\,(C) \cap Lat\,(C)$, where $R' := R \cup \{k\}$. As p' is eulerian, the set F of elements having weight k form a cycle. Clearly, $p := p' - (k-r)\chi^F$ belongs to R^E and is eulerian. We claim that p is balanced. Suppose not. Then for some cocircuit B and some $e \in B$ we have $p(e) > p(B\backslash\{e\})$ whereas $p'(e) \leq p'(B\backslash\{e\})$. As $|F \cap B|$ is even, this implies $|F \cap B\backslash\{e\}| \geq 2$ and $e \notin F$. From this we have $\max R \geq p(e) > p(B\backslash\{e\}) \geq 2r$, contradicting the hypothesis and proving the claim. Thus, $p \in Int.Cone\,(C)$. Since $\chi^F \in Int.Cone\,(C)$, we have $p' \in Int.Cone\,(C)$. □

In particular, $\{2,3\}$ is good if and only if the circuit double conjecture is true. We recall that the range $\{1,2\}$ is bad. For larger consecutive pairs we have the following.

PROPOSITION 4.8. *For the class of graphs, the range $\{k, k+1\}$ is good for all $k \geq 3$.*

PROOF. When k is even, this follows from Propositions 4.2 and 4.7. It suffices to prove the cases $k=3$ and $k=5$, since $4 \in Int.Cone\,(C)$ for any bridgeless graph. For these two cases we use refined versions of the results of Jaeger and Fan mentioned in the proof of Proposition 4.2. Jaeger [**24**] actually proved that any bridgeless graph contains 7 cycles $C_1, \ldots, C_7$ such that every edge is contained in exactly 4 of them. If $p \in \{3,4\}^E \cap Cone\,(C(G)) \cap Lat\,(C(G))$ then G is bridgeless and the set F of edges having weight 3 form a cycle. We consider the 7 cycles of the form $F \Delta C_i$. Since $7 = 3 + 4$ one easily sees that any edge in F is contained in exactly 3 of these cycles and that any edge in $E \backslash F$ is contained in exactly 4 of them. As each cycle decomposes into circuits, we have $p \in Int.Cone\,(C)$.

The proof for $k = 5$ is exactly analogous, using the fact $11 = 5 + 6$ and Fan's observation [12] that any bridgeless graph contains 11 cycles such that every edge is contained in exactly 6 of them (actually, Fan shows that only 10 cycles are needed, but we may take the empty cycle to be the eleventh). □

We remark that the this proof works equally well for any regular matroid which has a nowhere-zero 6-flow.

Little more is known about good and bad ranges. Indeed, it is frightfully easy to pose difficult conjectures. One which is most likely to be true, but for which I know of no proof is the following.

CONJECTURE 4.9. *For some $k \geq 2$, the range $\{k, k+2\}$ is good for the class of graphs.*

This conjecture has a very different flavor between odd and even values of k. At the other extreme, the boldest conjecture of this type that one can possibly make is the converse of Proposition 4.5.

CONJECTURE 4.10. *A range R is good (for the class of graphs) if and only if $\{1, k\} \not\subseteq R$, for all $k \geq 2$.*

The following table summarizes results regarding good and bad ranges for the class of graphic matroids.

Range	Status	Comments
$\{2\}$	Good?	Conjecture 4.1
$\{k\}, k \neq 2$	Good	Proposition 4.2
$\{1, k\}, k \geq 2$	Bad	Proposition 4.5
$\{2, 3\}$	Good?	Equivalent to Conjecture 4.1
$\{k, k+1\}, k \geq 3$	Good	Proposition 4.8
$\{k, k+2\}, k \geq 2$	Good?	Conjecture 4.9
$\{2k : k \in \mathbb{Z}_{\geq 0}\}$	Good??	Conjecture 4.6
$\mathbb{Z}_{\geq 0} \backslash \{1\}$	Good????	Conjecture 4.10

Perhaps some study into good and bad ranges of cographic matroids is warranted. Here we suspect that all ranges are good for this class of matroids.

CONJECTURE 4.11. *The bonds of any graph form a Hilbert basis.*

As $M(P_{10})$ is not cographic, this conjecture would follow immediately from Conjecture 3.2. However, this has only yet been verified for the class of cographic matroids with no $M^*(K_5)$-minor [15]. Again, the major problem is our lack of knowledge about the cone of cuts in graphs [8]. In contrast to the graphic matroids, it is easy to show that any range of cardinality 1 is good for the class of cographic matroids (see [28]). On the other hand, one cannot use flow theory to prove statements such as Proposition 4.8 for the class of cographic matroids, since the chromatic number (which is the dual flow number) of graphs is not bounded.

We conclude this section by pointing out that I know of no reason why Conjecture 4.10 cannot be extended to the class of all matroids. Further, it is possible to formulate a common generalization to the bold Conjectures 3.2 and 4.10, although it is probably imprudent to speculate further on the matter. Still, it would be very interesting to find any example of a matroid with a weight vector in $Cone\,(C) \cap Lat\,(C) \backslash Int.Cone\,(C)$ which is not based on Petersen's graph (as in Proposition 4.5).

5. Cone and Lattice of Perfect Matchings

Let M denote the set of perfect matchings (as subsets of edges) in a graph $G = (V, E)$. As with circuits in graphs, each of $Lin.Hull\,(M)$, $Cone\,(M)$ and $Lat\,(M)$ has been well characterized, and there exist polynomial-time membership tests for these three subsets of R^E. These results are more complicated than the corresponding ones for circuits, and we shall only state them roughly. We refer the reader to [9, 34] for further details.

One begins by "preprocessing" the fixed graph G. First, those edges of G which are contained in no perfect matchings are deleted. Then we perform a *brick decomposition* on the resulting graph as follows. A *tight cut* is a edge cut which intersects each perfect matching in exactly one edge. For example, any *trivial* edge cut (that is, an edge cut in which one of its two *shores* contains a single vertex) is tight. Any non-trivial tight cut yields two proper minors of G obtained by contracting each of the shores of the cut. In a brick decomposition, non-trivial tight cuts are recursively found in each of these minors. A similar reduction is performed whenever one of the minors has a vertex-cut of cardinality less than three. Any multiple edge occurring in a minor is replaced with a single edge. (If G is edge-weighted then this new edge is assigned the total weight of the parallel class it replaces.) The result of a brick decomposition of G is a list of simple 3-connected non-bipartite minors which contain no non-trivial tight cuts. Each member of this list is called a *brick* of G. It turns out that this list of bricks is independent (up to re-ordering and isomorphism) of the particular tight-cut decomposition chosen for G. Lovász [34] points out that a list of bricks for G can be obtained in polynomial time.

THEOREM 5.1. *For any graph G containing an even number of vertices, $Lin.Hull\,(M) = \{p \in \mathbb{Q}^E : \exists r \in \mathbb{Q},\ p(B) = r,\ \text{for all trivial cuts and tight cuts } B \text{ encountered during a brick decomposition of } G\}$.* □

The cone of perfect matchings follows from Edmonds' well known characterization [9] of the convex hull. An *odd cut* is an edge cut such that both of its shores contain an odd number of vertices.

THEOREM 5.2. *For any graph G containing an even number of vertices, $Cone\,(M) = \{p \in \mathbb{Q}^E_{\geq 0} : \exists r \in \mathbb{Q},\ p(B) = r,\ \text{for all trivial cuts}\ B,\ \text{and}\ p(B') \geq r$ for all odd cuts $B'\}$.* □

The lattice of perfect matchings was characterized by Lovász [34]. Here, bricks of G which are isomorphic to Petersen's graph P_{10} play a central role. We recall that any brick resulting from a brick decomposition of a weighted graph (G,p) naturally inherits a weight function, which we shall also denote by p.

THEOREM 5.3. *For any graph G containing an even number of vertices, $Lat\,(M) = Lin.Hull\,(M) \cap \{p \in \mathbb{Z}^E : p(C_5)$ is even, for every circuit C_5 of length five contained in any brick of G isomorphic to $P_{10}\}$.* □

In particular, the lattice of perfect matchings is just the set of integer vectors contained in the linear hull, provided that G has no P_{10}-minors. This fact was observed for cubic graphs by Seymour [47]. The necessity of the condition on $p(C_5)$ in Theorem 5.3 follows from the observation that each of the 6 perfect matchings of P_{10} intersect C_5 in an even number of edges.

In summary, given any weighted graph (G,p), one can determine in polynomial time whether p belongs to the cone, the lattice or the linear hull of perfect matchings in G.

6. Perfect Matching Covers

Some well-known results and conjectures address the *Perfect Matching Cover Problem*, the problem of determining whether a particular integer vector belongs to $Int.Cone\,(M)$. We recall the necessary condition that the vector in question belongs to $Cone\,(M) \cap Lat\,(M)$, and that $M(G)$ is said to form a Hilbert basis if this condition is also sufficient.

For uniform vectors k with $k > 0$, $k \in Cone\,(M)$ if and only if, for some $r \geq 1$, G is an r-regular graph with an even number of vertices such that all odd cuts have size at least r. Following Seymour [47], we call such graphs r-*graphs*. For example, a cubic graph is a 3-graph if and only if it is bridgeless. We note that, for any r-graph, $2 \in Lat\,(M)$. Furthermore, if an r-graph has no P_{10}-brick then $1 \in Lat\,(M)$. We also note that $1 \in Int.Cone\,(M)$ if and only if the graph has a 1-factorization.

Unlike circuit covers, Perfect Matching Cover Problems can often be reduced to problems regarding uniform weight vectors by adding parallel edges to graphs. For example, we have the following.

OBSERVATION 6.1. *Let $\mathcal{G}$ be any family of graphs containing no P_{10}-minors, and which closed under duplicating edges. Then $M(G)$ forms a Hilbert basis for every $G \in \mathcal{G}$ if and only if $1 \in Int.Cone\,(M(H))$ for every r-graph $H \in \mathcal{G}$.*

PROOF. The "only if" direction follows immediately from the fact that $1 \in Cone(M(H)) \cap Lat(M(H))$ for any r-graph $H \in \mathcal{G}$. For the converse, let $p \in Cone(M(G)) \cap Lat(M(G))$ where $G \in \mathcal{G}$. In (G,p), every trivial bond has weight r for some $r \in \mathbb{Z}_{\geq 0}$. Let H be the r-regular graph obtained from G by replacing each edge e by $p(e)$ parallel edges. As $G \in \mathcal{G}$, so is H. Since $p \in Cone(M(G))$, $1 \in Cone(M(H))$ so H is an r-graph. By hypothesis, $1 \in Int.Cone(M(H))$. As any perfect matching in H corresponds to one in G, we have $p \in Int.Cone(M(G))$. □

Much of the work that has been done regarding perfect matching covers of r-graphs deals with the case $r = 3$. Indeed, Seymour [47, (3.5)] has proposed that this is really the only interesting case.

CONJECTURE 6.2. *If $r \geq 4$ then any r-graph has a perfect matching whose deletion yields an $(r-1)$-graph.*

This conjecture is not yet known to be true for any $r \geq 4$.

Using the above terminology, we list some known results and conjectures regarding perfect matchings

THEOREM 6.3. *(Four-color theorem) For any planar 3-graph, $1 \in Int.Cone(M)$.*
□

I do not know the origin of the following natural generalization, though it is implied by Conjecture (7.3) in [49].

CONJECTURE 6.4. *For any planar r-graph with $r \geq 0$, $1 \in Int.Cone(M)$.*

The case $r = 4$ of this conjecture has been has been investigated by Jaeger and others (see [26, 27]), and is known to imply the Four-color Theorem. By Observation 6.1, Conjecture 6.4 is equivalent to the assertion that the perfect matchings of any planar graph form a Hilbert basis.

Another well-known strengthening of the Four-color Theorem is still open [53].

CONJECTURE 6.5. *(Tutte's 4-flow conjecture for cubic graphs) For any 3-graph which has no P_{10}-minor, $1 \in Int.Cone(M)$.*

By replacing "3-graph" by "r-graph" in Tutte's conjecture, Lovász [35] proposed a very strong conjecture which would imply Conjectures 6.4 and 6.5 and the Four-color Theorem.

CONJECTURE 6.6. *If a graph contains no P_{10}-minor then its perfect matchings form a Hilbert basis.*

We note that this conjecture would hold true provided both Conjecture 6.5 and Conjecture 6.2 were true.

Little is known about whether $M(G)$ forms a Hilbert basis when G contains a P_{10}-minor. It is perhaps surprising that the perfect matchings of P_{10} form a Hilbert basis; this fact follows from the observation that the six perfect matchings in P_{10} are linearly independent in $\mathbb{Q}^E$. However, M is not always a Hilbert basis.

EXAMPLE 6.7. *Let $P_{10} + e$ denote the (unique) graph obtained from P_{10} by joining any two non-adjacent vertices with a new edge e. Let p be the weight function which takes the value 0 on e and takes the value 1 elsewhere. As $P_{10} + e$ is a brick different from P_{10}, it follows that $p \in Cone\,(M) \cap Lat\,(M)$. However, $p \notin Int.Cone\,(M)$, since this would imply that P_{10} has a 1-factorization. Thus $M(P_{10} + e)$ is not a Hilbert basis.*

Clearly, M is not a Hilbert basis for any graph containing $P_{10}+e$ as a subgraph.

Seymour [47] proposed the following analog of the Circuit Double Cover Conjecture.

CONJECTURE 6.8. *(Perfect Matching Double Cover Conjecture) For any r-graph, $2 \in Int.Cone\,(M)$.*

The special case $r = 3$ of Conjecture 6.8 was first proposed by Fulkerson [16] and is still open. Incidently, Fulkerson's conjecture is equivalent to a strengthening of Jaeger's observation as referred to in the proof of Theorem 4.8.

CONJECTURE 6.9. *Any bridgeless graph contains exactly 6 cycles such that any edge is contained in 4 of them.*

The equivalence of these two conjectures becomes evident for cubic graphs when one considers that the complement of a perfect matching is a cycle. By "blowing up" vertices, one can see that Conjecture 6.9 holds for all graphs provided it holds for cubic graphs.

Unlike the case with circuit covers, the Perfect Matching Cover Problem has not been solved for larger uniform vectors k, $k > 2$. By the fact $1 \in Cone\,(M)$ for any r-graph we have that, for any r-graph G, there exists $k \geq 1$ such that $k \in Int.Cone\,(M)$. However, it is not known whether k can be picked independently of G. This gives the following weak Fulkerson-type conjecture.

CONJECTURE 6.10. *There exists $k \geq 2$ such that, for any r-graph G, $k \in Int.Cone\,(M(G))$.*

An even weaker conjecture was proposed by B. Jackson [23].

CONJECTURE 6.11. *There exists $k \geq 2$ such that any r-graph contains $k + 1$ perfect matchings with empty intersection.*

A form of Jaeger's 8-flow theorem states that any bridgeless cubic graph G is the union of 3 of its cycles. In fact, one can modify Jaeger's proof to ensure that at least one of the three cycles is a 2-factor of G. If one can show that all three cycles can be chosen to be 2-factors then, by taking complements, Conjecture 6.11 will have been proven for $r = 3$ and $k = 2$. Jackson [23] asked the following question. Can one show that at least two of the three cycles are 2-factors of G? Surely, this very special consequence of Fulkerson's conjecture must be true.

7. Circuits, Perfect Matchings and Grafts

The vague similarities between circuits and perfect matchings might be explained by considering certain 1-element extensions of graphic matroids.

DEFINITION 7.1. *Let $A = A(G)$ denote the vertex-edge $\{0,1\}$-valued incidence matrix of a connected graph G. Thus the columns of A represent the graphic binary matroid $M(G)$ of rank $|V(G)| - 1$ with linear independence over GF(2). Let $T \subseteq V$ and let τ denote the $\{0,1\}$-valued column vector which is the characteristic vector of T. Then $[A\ \tau]$ represents a binary matroid of rank $|V(G)| - 1$ on the ground set $E \cup \{\tau\}$, which we denote by G_T. Following Seymour [48], we call the matroid G_T a graft.*

Grafts are precisely the binary 1-element extensions of graphic matroids. A graft G_T is interesting only when $|T|$ is even, since τ is otherwise a coloop in G_T. If $|T| = 0$ then τ is a loop in G_T. If $|T| = 2$ then $G_T \cong G + e$ where e is a new edge joining the vertices in T. For larger subsets T, grafts can be non-graphic and even non-regular. Seymour [48, p. 339] shows how the matroids $F_7, F_7^*, M^*(K_5), M^*(K_{3,3})$ and R_{10} are all grafts G_T, where G has at most 7 vertices. For $T \subseteq V$, a *T-join* is any subset $S \subseteq E(G)$ such that $T = \{v \in V : v$ is incident with an odd number of non-loop edges in $S\}$ (in some papers, T-joins are also required to be acyclic). A *T-cut* is an edge-cut in G which contains an odd number of vertices of T on each shore. T-joins and T-cuts are closely related to matchings and have been studied by various authors [32, 14, 36, 37, 39, 40, 41, 42, 47, 10]. One easily checks that, when $|T|$ is even, the cycles of a graft G_T are precisely the cycles of G, together with sets of the form $\{\tau\} \cup J$ where J is any T-join in G. The cocycles of G_T are precisely the cuts of G which are not T-cuts, together with the sets of the form $\{\tau\} \cup B$ where B is any T-cut in G.

If $T = V$ and G has a perfect matching, then then the circuits of G_T which contain τ and which have minimum cardinality are precisely the subsets of the form $\{\tau\} \cup F$ where $F \in M(G)$. In this way, we obtain a connection between $C(G_T)$ and $M(G)$. In particular, the Uniform Perfect Matching Cover Problem for graphs may be posed as a Circuit Cover Problem for grafts.

EXAMPLE 7.2. *Let $k \geq 1$ and $r \geq 3$. Let G be any r-regular graph, and set $T = V$. Consider the weight function p on the graft G_T where $p(\tau) = rk$, and*

$p(e) = k$ for all $e \in E(G)$. If p is a non-negative linear combination of circuits in $C(G_T)$ then all circuits having positive coefficients must be of the form $\tau \cup F$, $F \in M(G)$. This gives us the following facts.
  (i) $p \in Int.Cone\,(C(G_T))$ if and only if $k \in Int.Cone\,(M(G))$.
  (ii) $p \in Cone\,(C(G_T))$ if and only if $k \in Cone\,(M(G))$.
  (iii) $p \in Lat\,(C(G_T))$ if $k \in Lat\,(M(G))$.
In particular, $p \in Cone\,(C(G_T))$ if and only if G is an r-graph, and $p \in Lat\,(C(G_T))$ if either k is even or G has no P_{10}-brick.

As it is NP-hard to decide whether $1 \in Int.Cone\,(M)$ for 3-graphs [21], Example 7.2 implies that determining whether a vector is in the integer cone of circuits is NP-hard for the class of grafts. However, the complexity of the latter problem remains unknown when "grafts" is replaced by "graphs".

Example 7.2 also serves to connect two of the main conjectures presented earlier in this paper. We have seen that Conjecture 3.2 holds for the class of graphs. We shall see that if this conjecture were to hold true for the class of grafts, then the Four-color Theorem would follow, as well as many of the open problems discussed in Section 6. We begin with a curious property of Petersen's graph. In general, if a graft G_T contains a graphic matroid $M(H)$ as a minor, then one cannot deduce that G contains $M(H\backslash e)$ as a minor, for some $e \in E(H)$. For example, let G be the polygon of length four, and let T be a pair of non-adjacent vertices in G. Then $G_T/\tau \cong H$ where H is the graph consisting of two digons joined at a vertex. However, one easily sees that $H\backslash e$ is not a minor of G for any $e \in V(H)$.

If H is Petersen's graph, however we have a different story. Note that $P_{10}\backslash e$ is independent of e up to isomorphism.

LEMMA 7.3. *If a graft G_T contains $M(P_{10})$ as a minor, then G contains $P_{10}\backslash e$ as a minor.*

PROOF. Suppose that $G_T/S\backslash R \cong P_{10}$ where S, R are disjoint subsets of $E(G) \cup \{\tau\}$. If $\tau \notin S \cup R$ then, as in [46, (10.2)], $G_T\backslash R/S \cong (G\backslash R/S)_{T'}$ for some $T' \subseteq V(G\backslash R/S)$. Deleting any element from $(G\backslash R/S)_{T'} = P_{10}$ yields $P_{10}\backslash e$ so, in particular, $P_{10}\backslash e \cong (G\backslash R/S)_{T'}\backslash \tau = G\backslash R/S$ and we are done. If $\tau \in R$ then $P_{10} \cong G_T\backslash R/S = G\backslash(R - \{\tau\})/S$ is a minor of G, and again we are done. Thus we assume that $\tau \in S$. Here we have $G_T\backslash R/(S - \{\tau\}) \cong G'_{T'}$ where $G' = G\backslash R/(S - \{\tau\})$ and T' is some subset of $V(G')$.

It remains to show that G' contains $P_{10}\backslash e$ as a minor given that $G'_{T'}/\tau \cong P_{10}$. Suppose that G' contains a bridge $f \in E(G)$. Then, in $G'_{T'}$, either f is a bridge or f is coparallel with τ. In the first case, f is also a bridge of $G'_{T'}/\tau \cong P_{10}$, a contradiction. In the second case we have $P_{10} \cong G'_{T'}/\tau \cong G'_{T'}/f \cong (G'/f)_{T''}$ for some $T'' \subseteq V(G'/f)$. Deleting any element from $(G'/f)_{T''}$ yields $P_{10}\backslash e$ so, in particular, $P_{10}\backslash e \cong (G'/f)_{T''}\backslash \tau = G'/f$. Thus $P_{10}\backslash e$ is a minor of G', provided G' contains a bridge.

Thus we assume that G' is a 2-edge-connected graph with 15 edges. We claim that $G' = P_{10}$ and hence that G has a P_{10}-minor. It is well known that Petersen's graph is the only 2-edge connected graph on at most 15 edges which is not the union of two of its cycles (this property is equivalent to having a 4-*nowhere-zero flow*). Suppose that $G' \not\cong P_{10}$. Then G' is the the union of two of its cycles, say $E(G') = C_1 \cup C_2$. Since both the extension and contraction operations preserve cycles in a matroid, both C_1 and C_2 are cycles in $G'_{T'}/\tau$, and their union is all of $G'_{T'}/\tau \cong P_{10}$. This contradiction establishes our claim and completes the proof. □

THEOREM 7.4. *If Conjecture 3.2 holds for grafts then Conjecture 6.6 holds for graphs which have no minor isomorphic to $P_{10}\backslash e$.*

PROOF. Suppose that $C(G_T)$ forms a Hilbert basis for any graft G_T having no P_{10}-minor. By Observation 6.1, it suffices to show that $1 \in Int.Cone\,(M(G))$ for any r-graph G which has no minor isomorphic to $P_{10}\backslash e$. Let (G_T, p) be the weighted graft obtained from G as in Example 7.2 with $k = 1$. By *1.* of the example, we need to show that $p \in Int.Cone\,(C(G_T))$. By *2.* and *3.*, $p \in Cone\,(C(G_T)) \cap Lat\,(C(G_T))$, so it suffices to show that $C(G_T)$ forms a Hilbert basis. This follows from the hypothesis since, by Lemma 7.3, G_T does not contain a P_{10}-minor. □

Theorem 7.4 demonstrates both the relevance and the ominous difficulty of Conjecture 3.2. Were it to hold for grafts, the Four-color Theorem and the stronger Conjecture 6.4 would be immediate corollaries. It would be nice if the forbidden-minor restriction in the conclusion of Theorem 7.4 could be be dropped. This would make Tutte's 4-flow Conjecture (6.5) a consequence of Conjecture 3.2. To drop this restriction only requires an argument for those graphs G which have a $P_{10}\backslash e$-minor, but no P_{10}-minor. Although we are tantalizingly close to such a result, a new idea may be needed, since there exists a 3-graph which contains no P_{10}-minor although the graft G_T (with $T = V(G)$) does.

It remains to address the problem of characterizing the cone, the lattice and the integer cone of circuits in grafts. It seems unlikely that the lattice and the cone have simple descriptions, as grafts have neither the Lattice of Circuits nor the Sums of Circuits property (recall Theorems 2.3, 2.4). Indeed, $F_7^* \cong G_T$ where $G = K_{3,2}$, $T = V(G) - v$, and v is a vertex of degree 3. The smallest 3-graph G for which G_T contains a F_7^*-minor (where $T = V(G)$) is the triangular prism (the complement of a circuit of length 6). It is interesting that this graph arizes as an anomaly in matching theory, particularly with regard to ear-decompositions (see, for example [**34**, (3.2), (3.3)]). The complexity of the cone and the lattice of circuits in grafts is also attested by the effort that was required to characterize the special cases $Cone\,(M(G))$ and $Lat\,(M(G))$ (Theorems 5.2 and 5.3).

On the other hand this success in matching theory, and our increasing understanding of T-joins [**39, 40**] is encouraging. The circuits of grafts are not impossibly complicated. For example, the lattice is fairly tame in that grafts cannot contain the dual of the projective geometry $PG(2,m)$ as a minor, for any $m > 2$ (see Example 2.8). The cone of circuits is especially worthy of further investigation. Indeed, it is far more important to know the cone than the lattice when investigating whether circuits form a Hilbert basis. It is reasonable to guess that this class of matroids will predominate much of the future research on circuits in matroids.

Acknowledgement. I wish to thank Bill Jackson, András Sebő, and Paul Seymour for their valuable suggestions and stimulating discussions.

Added in Proof. Conjecture 3.2 is false. Let f be an element of the cographic matroid $M^*(K_6)$ and let $p(f) = 4$ and $p(e) = 2$, $e \in E - \{f\}$. Then $p \in Cone\,(C) \cap Lat\,(C)$, but $p \notin Int.Cone\,(C)$. This matroid is not a graft. I thank M. Laurent for these counterexamples.

REFERENCES

1. B. R. ALSPACH, L. A. GODDYN AND C-Q ZHANG, Graphs with the circuit cover property, *Trans. Amer. Math. Soc.*, submitted.
2. J. C. BERMOND, B. JACKSON AND F. JAEGER, Shortest coverings of graphs with cycles, *J. Combin. Theory Ser. B* **35** (1983), 297-308.
3. J. A. BONDY AND U. S. R. MURTY, "Graph Theory with Applications" North Holland, New York, 1980.
4. J. A. BONDY, Small cycle double covers of graphs, *in* "Cycles and Rays", G. Hahn, G. Sabidussi and R. Woodrow, eds., *Nato ASI Series C*, vol. 301, Klummer Academic Publishers, Dordrecht/Boston/London, 1990, pp. 21-40.
5. P. CATLIN, Double cycle covers and the Petersen graph, *J. Graph Theory* **13** (1989), 465-483.
6. W. COOK, J. FONLOUPT AND A. SCHRIJVER, An integer analogue of Carathéodory's Theorem, *J. Combin. Theory Ser. B* **40** (1986), 63-70.
7. W. H. CUNNINGHAM, Chords and disjoint paths in matroids, *Discrete Mathematics* **19** (1977), 7-15.
8. M. DEZA AND M. LAURENT, New results on facets of the cut cone, em in "Integer Programming and Combinatorial Optimization (IPCO) Proceedings", R. Kannan, W. R. Pulleyblank, Eds., University of Waterloo Press, Waterloo, 1990, pp.171-184.
9. J. EDMONDS, Maximum matching and a polyhedron with (0,1) vertices, *J. Res. Nat. Bur. Standards B* **69** (1965), 125-130.
10. J. EDMONDS AND E. L. JOHNSON, Matching, Euler-Tours, and the Chinese Postman, *Math. Programming* **5** (1973), 88-124.
11. G. FAN, Covering weighed graphs by even subgraphs, *J. Combin. Theory Ser. B* **49** (1990), 137-141.
12. G. FAN, Integer flows and cycle covers, *J. Combin. Theory Ser. B*, to appear.
13. H. FLEISCHNER AND A. FRANK, On circuit decompositions of planar Eulerian graphs, *J. Combin. Theory Ser. B* **50** (1990), 245-253.
14. A. FRANK, Conservative weightings and ear-decompositions of graphs. *in* "Integer Programming and Combinatorial Optimization (IPCO) Proceedings", R. Kannan, W. R. Pulleyblank, Eds., University of Waterloo Press, Waterloo, 1990, pp.217-230.
15. X. FU AND L. A. GODDYN, Matroids with the circuit cover property, in preparation.

16. D. R. FULKERSON, Blocking and antiblocking pairs of polyhedra, *Math. Programming* **1** (1971), 168-194.
17. F. R. GILES AND W. R. PULLEYBLANK, Total dual integrality and integer polyhedra, *Linear Algebra and Its Applications* **25** (1979), 191-196.
18. L. A. GODDYN, Cycle double covers of graphs with Hamilton paths, *J. Combin. Theory Ser. B* **46** (1989), 253-254.
19. L. A. GODDYN, "Cycle Covers of Graphs", Ph. D. Thesis, Dept. of Combinatorics and Optimization, University of Waterloo, Waterloo, Canada, 1988.
20. G. HAGGARD, Loops in duals, *Amer. Math. Monthly* **87** (1980), 654-656.
21. I. HOLYER, The NP-completeness of edge-coloring, *SIAM J. Comput.* **10** (1981), 718-720.
22. B. JACKSON, Shortest circuit covers and postman tours in graphs with a nowhere zero 4-flow, *SIAM J. Discrete Math.*, submitted.
23. B. JACKSON, personal communication.
24. F. JAEGER, Flows and generalized coloring theorems in graphs, *J. Combin. Theory Ser. B* **26** (1979), 205-216.
25. F. JAEGER, A survey of the cycle double cover conjecture, *in* "Cycles in Graphs", B. R. Alspach and C. D. Godsil, Eds., Annals Disc. Math., Vol. 27, pp. 1-12, North-Holland, Amsterdam/New York/Oxford, 1985.
26. F. JAEGER AND H. SHANK, On the edge-coloring problem for a class of 4-regular maps. *J. Graph Theory*, **5** (1981), 269-275.
27. F. JAEGER AND G. KOESTER, Vertex signatures and edge-4-colorings of 4-regular plane graphs. *J. Graph Theory*, **14** (1990), 399-403.
28. U. JAMSHY AND M. TARSI, Cycle covering of binary matroids, *J. Combin. Theory Ser. B* **46** (1989), 154-161.
29. U. JAMSHY, A. RASPAUD AND M. TARSI, Short circuit covers for regular matroids with a nowhere zero 5-flow, *J. Combin. Theory Ser. B* **42** (1987), 354-357.
30. U. JAMSHY AND M. TARSI, Short cycle covers and the cycle double cover conjecture, *J. Combin. Theory Ser. B*, submitted.
31. R. KARZANOV, Metrics and undirected cuts, *Math. Programming* **32** (1985), 183-198.
32. E. KORACH, "Packing of T-cuts and Other Aspects of Dual Integrality", Ph. D. Thesis, Dept. of Combinatorics and Optimization, University of Waterloo, Waterloo, Canada, 1982.
33. M. LAURENT, Personal communication.
34. L. LOVÁSZ, Matching structure and the matching lattice, *J. Combin. Theory Ser. B* **43** (1987), 187-222.
35. L. LOVÁSZ, Á. SERESS, Personal communication,
36. L. LOVÁSZ, 2-matchings and 2-covers of hypergraphs, *Acta Math. Acad. Sci. Hung.* **26** (1975), 433-444.
37. A. SCHRIJVER, On total dual integrality, *Linear Algebra and Its Applications* **38** (1981), 27-32.
38. A. SEBÖ, Hilbert bases, Carathéodory's theorem and combinatorial optimization, *in* "Integer Programming and Combinatorial Optimization (IPCO) Proceedings", R. Kannan, W. R. Pulleyblank, Eds., University of Waterloo Press, Waterloo, 1990, pp.431-456.
39. A. SEBÖ, Undirected distances and the postman-structure of graphs. *J. Combin. Theory Ser. B* **49** (1990), 10-39.
40. A. SEBÖ, Finding the t-join structure of graphs. *Math. Programming* **36** (1986), 123-134.
41. A. SEBÖ, The Schrijver system of odd join polyhedra, *Combinatorica* **8** (1986), 103-116.
42. A. SEBÖ, A quick proof of Seymour's theorem on t-joins, *Discrete Mathematics* **64** (1987), 101-103.
43. P. D. SEYMOUR, Sums of circuits, *in* "Graph Theory and Related Topics", J. A. Bondy and U. S. R. Murty, Eds., pp. 341-355, Academic Press, New York/Berlin, 1979.
44. P. D. SEYMOUR, Matroids and multicommodity flows, *Europ. J. Combinatorics* **2** (1981), 257-290.
45. P. D. SEYMOUR, Matroids with the max-flow min-cut property. *J. Combin. Theory Ser. B* **23** (1977), 189-222.

46. P. D. SEYMOUR, Nowhere-zero 6-flows, *J. Combin. Theory Ser. B* **30** (1981), 130-135.
47. P. D. SEYMOUR, On multi-colourings of cubic graphs, and conjectures of Fulkerson and Tutte. *Proc. London Math. Soc.* (3) **38** (1979), 423-460.
48. P. D. SEYMOUR, Decomposition of regular matroids. *J. Combin. Theory Ser. B* **28** (1980), 305-359.
49. P. D. SEYMOUR, On Tutte's extension of the Four-Colour Problem *J. Combin. Theory Ser. B* **31** (1981), 82-94.
50. G. SZEKERES, Polyhedral decompositions of cubic graphs, *J. Austral. Math. Soc.* **8** (1973), 367-387.
51. M. TARSI, Nowhere zero flows and circuit covering in regular matroids, *J. Combin. Theory Ser. B* **39** (1985), 346-352.
52. M. TARSI, Semi-duality and the cycle double cover conjecture, *J. Combin. Theory Ser. B* **41** (1986), 332-340.
53. W. T. TUTTE, On the algebraic theory of graph colorings, *J. Combin. Theory* **1** (1966), 15-50.
54. D. WELSH, "Matroid Theory", Academic Press, San Francisco, 1976.
55. C-Q ZHANG, Minimum coverings and integer flows, *J. Graph Theory*, to appear.
56. C-Q ZHANG, On compatible cycle decompositions of eulerian graphs, preprint.
57. C-Q ZHANG, On even cycle decompositions of eulerian graphs, preprint.

DEPARTMENT OF MATHEMATICS AND STATISTICS, SIMON FRASER UNIVERSITY, BURNABY, BRITISH COLUMBIA, CANADA V5A 1S6

E-mail address: goddyn@cs.sfu.ca

Regular Maps From Voltage Assignments

PAVOL GVOZDJAK
JOZEF ŠIRÁŇ

ABSTRACT. The theory of voltage graphs enables to lift a given map to new, so-called derived maps which cover the original one. This paper introduces a method of lifting map automorphisms of a map to automorphisms of the derived map. As an application, constructions of infinite classes of orientable as well as nonorientable regular maps are presented.

1. Introduction

The theory of (ordinary) voltage graphs introduced by Gross [7] provides a powerful means for constructing maps, i.e., graph embeddings. There is an extensive literature dealing with voltage graphs; for an excellent survey the reader is referred to the book by Gross and Tucker [8]. The existing papers concentrate mostly on lifting graphs (and maps) to obtain new and interesting embeddings. The aim of this paper is to study the lifting of map automorphisms (rather than lifting maps) by means of (ordinary) voltage assignments in both Abelian as well as non-Abelian groups.

In a more general framework, the problem of lifting map automorphisms can be viewed as a specification of the question of lifting a continuous mapping to a covering space. As regards covering spaces without branch points, a complete answer to the latter, in terms of fundamental groups, is well known in algebraic topology (see e.g. Massey [12], or Djokovic [4] for graphs only). A similar, but homology-type criterion for lifting map automorphisms in unbranched coverings can be found in Surowski [15]. However, voltage graph constructions often lead to branched coverings of surfaces, where the above results are not applicable. This case was considered by S.Wilson [19] who studied lifts of automorphisms in branched coverings of maps derived from special current (=dual voltage) assignments in cyclic groups.

1991 Mathematics Subject Classification. Primary 05C10
This paper is in final form and no version of it will be submitted for publication elsewhere.

The main result of our paper is a simple necessary and sufficient condition for lifting map automorphisms in branched coverings of maps arising from ordinary voltage assignments in an arbitrary group. As an application of our technique of lifting map automorphisms, introduced in Section 4, we present constructions of new regular maps (i.e., maps with the largest possible number of symmetries) from the known ones.

In order to make the paper self-contained, in Section 2 and 3 we briefly review a combinatorial approach to maps which consists in representing them by means of three involutory permutations acting on the set of doubly directed edges of the underlying graph (cf. [11] for a similar approach). This has the additional advantage of a unified treatment of both orientable as well as nonorientable maps. As regards voltage graphs and lifting techniques, a preliminary acquaintance with them is assumed, e.g. in the extent of the book [8].

2. Maps and their algebraic description

A graph is a finite 1-dimensional cell complex. Thus, in graphs we allow both multiple- and self-adjacencies. Let G be a graph and Σ a closed surface (orientable or not). An *embedding* of G in Σ is a continuous injection $j : G \to \Sigma$. The embedding j is said to be *cellular* if every component of the set $\Sigma - j(G)$ is homeomorphic to an open disc; such components are *faces*. A cellular embedding j induces a cell decomposition of Σ; this decomposition will be called a *map* (on Σ). We do not distinguish between the graphs G and $j(G)$ and we refer to G as the *underlying graph* of the map. Note that the underlying graph of a map is necessarily connected.

There are several ways that maps can be described combinatorially (e.g. [10,14,18]). Here we adopt the approach which originated in [11]. It has the advantage that it is suitable for both orientable as well as nonorientable maps.

Let M be a map on a surface Σ, with underlying graph G. Loosely speaking we can imagine M as a "drawing" of G on Σ. Bearing this in mind, we first endow each edge of G by two kinds of orientations: *longitudinal orientation* (which results in a directed edge, or arc, in the usual sense) and a *transversal orientation* (indicated by an arrow on Σ crossing perpendicularly the original edge). An edge with a longitudinal and a transversal orientation is called a *flag*; this terminology complies with that used in combinatorial geometry [5]. It follows that each edge of G gives rise to four possible flags; one of them is depicted in Fig.1.

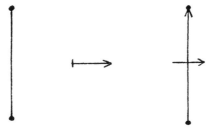

Fig. 1. A flag arising from an edge.

Let $F(G)$ be the set of all flags of G. It is clear that $|F(G)| = 4|E(G)|$ where $|E(G)|$ denotes the edge set of G. Our next step is to introduce three involutory permutations L,T and R on $F(G)$ which are determined by the map M in a very natural way.

Let L and T be permutations on $F(G)$ which reverse the longitudinal and transversal orientations of flags, respectively. More precisely, if x is a flag then Lx is the flag obtained from x by reversing its longitudinal orientation and letting the transversal orientation unchanged; T is defined analogously. Observe that the group $<L,T>$ generated by L and T is isomorphic to the group $Z_2 \times Z_2$. Moreover, every orbit of $<L,T>$ consists of exactly four flags (namely the four flags x, Lx, Tx, LTx arising from one edge).

In order to define the third involutory permutation R of $F(G)$ associated with the map M, consider two consecutive edges e and f appearing on the boundary of a face J of M. Let x and y be flags obtained from e and f, respectively in the way indicated in Fig.2:

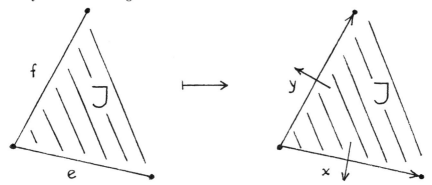

Fig. 2.

Then we put $R(x) = y$ and $R(y) = x$. It is easy to see that by means of this rule the permutation R is well defined (just observe that a flag x uniquely determines a "corner" on Σ at the initial vertex of x, as depicted in Fig.2).

It can be shown that the permutations L, T and R of $F(G)$ characterize the map M. To see this let us first list some more properties of L, T and R. Obviously R, L and T are fixed-point free involutions. For each vertex u of G there are two disjoint orbits of the permutation RT, consisting of flags with initial vertex u. Similarly, to every face of M there corresponds a pair of orbits of the permutation RL. Finally, note that the group $<L,T,R>$ acts transitively on $F(G)$.

Now let X be a nonempty set such that $|X| = 4m$. Let L, T and R be fixed-point free permutations on X satisfying the following four conditions:

(1) $R^2 = L^2 = T^2 = (LT)^2 = id$;

(2) the group $<L,T,R>$ acts transitively on X;

(3) every orbit of the group $<L,T>$ has cardinality 4;

(4) for every $x \in X$ the orbits $O_{RT}(x)$ and $O_{RT}(Tx)$ of the permutation RT are disjoint; the similar holds for $O_{RL}(x)$ and $O_{RL}(Lx)$

Then the map $M = M(L,T,R)$ determined by L, T and R can be reconstructed as follows. The vertex set of the underlying graph G is $V(G) =$

$\{O_{RT}(x) \cup O_{RT}(Tx); x \in X\}$. The edge set $E(G)$ is simply the set of orbits of the group $<L,T>$. The faces of M are determined by the unions $O_{RL}(x) \cup O_{RL}(Lx), x \in X$. The incidence among vertices, edges and faces thus described is given by a nonempty intersection of the corresponding orbits (or pairs of orbits).

A more detailed analysis shows that the orbits $O_{RT}(x)$ and $O_{RT}(Tx)$ have not only the same length but can be written in the form

$$O_{RT}(x) = (x, RTx, (RT)^2 x, ..., (RT)^{k-1}x, (RT)^k x = x), \text{ and}$$

$$O_{RT}(Tx) = (Tx, T(RT)^{-1}x, T(RT)^{-2}x, ..., T(RT)^{-k}x = Tx).$$

This can be interpreted as assigning two possible local orientations to the vertex $O_{RT}(x) \cup O_{RT}(Tx)$ on the surface determined by M. The similar can be done with faces: the orbits $O_{RL}(x)$ and $O_{RL}(Lx)$ represent the two possible orientations of the face $O_{RL}(x) \cup O_{RL}(Lx)$. Using these facts we can easily determine whether the map M is orientable or not: It is sufficient to consider the action of the group $<RL, RT>$ on X. A routine calculation shows that this group has at most two orbits on X. If $<RL, RT>$ acts transitively on X then the surface determined by M is nonorientable. In the case when $<RL, RT>$ has two orbits on X then the surface of M admits two opposite global orientations, i.e., M is orientable.

Note also that the permutations L^*, T^*, R^* on the same set X defined by $L^* = T, T^* = L$ and $R^* = R$ determine the *dual map* M^* to the original map M.

Let us conclude with a remark concerning orientable maps $M = M(L, T, R)$. Since $<RL, RT> = <RT, LT>$, as a consequence of the above analysis we obtain the fact that M is fully characterized by means of the permutations $P = RT$ and $Q = LT$. Moreover, since a global orientation of the supporting surface can be chosen it is not necessary to consider transversal orientations of edges of the underlying graph G. Thus, the permutations P and Q can be interpreted as acting on the set of arcs (=longitudinally directed edges) of G. The permutation P (often called *rotation*) cyclically permutes, for each vertex u, the arcs emanating from u consistently with the chosen global orientation. The permutation Q is simply the arc reversing involution. The pair (P, Q) is known as the Heffter-Edmonds embedding scheme for orientable maps.

3. Voltage assignments and derived maps

In this section we briefly describe one of the main tools for constructing new maps from old ones. It is the technique of lifting maps by means of voltage assignments. An extensive exposition of this method (in a somewhat different language) can be found in [8].

Let $M(L, T, R)$ be a map and let G be its underlying graph. Consider a finite group Γ. A mapping $\alpha : F(G) \to \Gamma$ assigning group elements to flags is called a *voltage assignment* if $\alpha Tx = \alpha x$ and $\alpha Lx = (\alpha x)^{-1}$ for every flag $x \in F(G)$. Thus, the only nontrivial condition imposed on a voltage assignment is that longitudinally opposite flags receive mutually inverse elements of the group. The pair (G, α) is said to be a *voltage graph*.

The voltage assignment α can be naturally extended to oriented walks in G. As expected, by an *oriented walk* in G we understand a sequence $W = x_1 x_2 \ldots x_n$ of flags of G such that, for each $i \leq n-1$, the terminal vertex of the flag x_i coincides with the initial vertex of x_{i+1} (i.e., only longitudinal orientations of edges have to be consistent). For an oriented walk $W = x_1 x_2 \ldots x_n$ we now set $\alpha W = \prod_{i=1}^{n} \alpha x_i$. Observe that if $LW = Lx_n Lx_{n-1} \ldots Lx_1$ is the opposite walk to W then $\alpha LW = (\alpha W)^{-1}$. The walk W is *closed* if the initial vertex of x_1 and terminal vertex of x_n are identical. We also admit a degenerated closed walk W_0 of zero length (consisting just of a single vertex) and put $\alpha W_0 = 1_\Gamma$, the unit element of Γ. Closed walks are of special importance in the theory of voltage graphs. It is easy to see that, for a given vertex $u \in V(G)$, the set $\{\alpha W; W$ a closed walk emanating from and terminating at $u\}$ constitutes a subgroup of Γ. This subgroup, denoted by Γ_u, is known as the *local group*. It is easy to see that any two local groups Γ_u and Γ_v are conjugate subgroups of Γ.

To explain the way how a new map can be obtained from $M = M(L, T, R)$ using the voltage assignment α let us first assume that the local group Γ_u (and hence each local group Γ_v) is isomorphic to Γ. The new map, called the *derived map*, will be denoted by M^α. Its underlying *derived graph* G^α has vertex set $V(G^\alpha) = V(G) \times \Gamma = \{u_g; u \in V(G), g \in \Gamma\}$ and flag set $F(G^\alpha) = \{x_g; x \in F(G), g \in \Gamma\}$. The incidence in G^α is defined by the following rule: if a flag x in G has initial vertex u and terminal vertex v then, for each $g \in \Gamma$, the flag x_g in G^α emanates from the vertex u_g and terminates at the vertex $v_{g\alpha x}$. The derived map M^α can now be conveniently described by means of three *derived permutations* L^α, T^α and R^α acting on $F(G^\alpha)$: For each flag $x \in F(G^\alpha)$ put

$$L^\alpha x_g = (Lx)_{g\alpha x},$$

$$T^\alpha x_g = (Tx)_g, \text{ and}$$

$$R^\alpha x_g = (Rx)_g.$$

It is a matter of routine to check that L^α, T^α and R^α satisfy the conditions (1)-(4) listed in the preceding section (note that (2) follows from the fact that $\Gamma_u \cong \Gamma$). Thus, the map $M^\alpha = M^\alpha(L^\alpha, T^\alpha, R^\alpha)$ with the underlying graph G^α is well defined.

It is well known (see e.g. [8]) that the projection $p : F(G^\alpha) \to F(G)$ which erases subscripts extends to a regular branched covering $M^\alpha \to M$ which has at most one branch point inside an arbitrary face of M. Conversely, any regular branched covering of two maps can be obtained by means of the just described *lifting technique*, i.e., using a suitable voltage assignment on the underlying graph of the target map.

As already indicated, if Γ_u is a proper subgroup of Γ then the group $< L^\alpha, T^\alpha, R^\alpha >$ does not act transitively on $F(G^\alpha)$. In fact, there are $[\Gamma : \Gamma_u]$ orbits of $< L^\alpha, T^\alpha, R^\alpha >$ on $F(G^\alpha)$ (and hence G^α has $[\Gamma : \Gamma_u]$ connected components). Each of the orbits defines a map; moreover, these maps are mutually isomorphic. In this case M^α will denote any of the $[\Gamma : \Gamma_u]$ maps thus obtained.

Similarly as in Section 2, the last remark will concern orientable maps $M = M(P, Q)$ where P and Q are the rotation and the arc-reversing involution, acting on the set $D(G)$ of arcs of the underlying graph G. For a voltage assignment $\alpha : D(G) \to \Gamma$ we then have $\alpha Qx = (\alpha x)^{-1}, x \in D(G)$. The derived graph G^α is defined analogously as before; in particular, $D(G^\alpha) = D(G) \times \Gamma$. The derived map $M^\alpha = M^\alpha(P^\alpha, Q^\alpha)$ is determined by means of the derived rotation P^α and involution Q^α where, for each arc $x_g \in D(G^\alpha), P^\alpha(x_g) = (Px)_g$ and $Q^\alpha(x_g) = (Qx)_{g\alpha x}$.

4. Lifting map automorphisms

Let $M = M(L, T, R)$ be a map on a surface Σ, with underlying graph G. A *map automorphism* of M is a graph automorphism of G which, in addition, preserves the cell structure of the map M. Map automorphisms can be equally well viewed as self-homeomorphisms of Σ which assign vertices to vertices, edges to edges, faces to faces, and preserve incidence of these elements. Combinatorially, a map automorphism of M can be identified with a bijection $A : F(G) \to F(G)$ which commutes with the three permutations associated with M, i.e., $AL = LA, AT = TA$ and $AR = RA$. Observe that the latter approach enables to reduce questions about automorphisms to algebraic operations. In what follows we shall take advantage of this fact and prefer the combinatorial way of introducing map automorphisms.

Let A be a map automorphism of an orientable map $M = M(L, T, R)$. If A fixes the two orbits of the group $< RL, RT >$ setwise then A is called *orientation-preserving*; otherwise A is a *reflection*. Using the Heffter-Edmonds scheme for M, i.e., representing M as $M(P, Q)$ where P and Q are the rotation and the arc-reversing involution acting on $D(G)$, respectively, then a bijection $B : D(G) \to D(G)$ is an orientation-preserving automorphism of M if $BP = PB$ and $BT = TB$, and a reflection if $BP = P^{-1}B$ and $BT = TB$.

In the previous section we saw how to lift a given map M by means of a voltage assignment to obtain the derived map M^α. Now we are interested in the question of lifting map automorphisms of M to map automorphisms of M^α. To be able to formulate the question precisely, let $M = M(L, T, R)$ be an arbitrary map and G its underlying graph. Consider a voltage assignment $\alpha : F(G) \to \Gamma$ in an arbitrary finite group Γ. Let $p : M^\alpha \to M$ be the natural covering projection of M by the derived map M^α. Further let A and $\tilde{A}$ be map automorphisms of the maps M and M^α, respectively. Then $\tilde{A}$ is said to be a *lift* of A if $Ap = p\tilde{A}$. In other words, $\tilde{A}$ is a lift of A if the following diagram is commutative:

$$\begin{array}{ccc} M^\alpha & \xrightarrow{\tilde{A}} & M^\alpha \\ p \downarrow & & \downarrow p \\ M & \xrightarrow{A} & M \end{array}$$

Our basic question can now be stated in the following way: Under what conditions there exists a lift of an automorphism of M to an automorphism of M^α?

We first present a simple necessary condition in terms of closed walks. Before doing it observe that the action of a map automorphism A of M can be extended to walks $W = x_1 x_2 \ldots x_n$ of the underlying graph by setting $AW = Ax_1 Ax_2 \ldots Ax_n$.

PROPOSITION 1. *Let $p : M^\alpha \to M$ be the natural covering projection of a map M by the derived graph M^α. Let A and $\tilde{A}$ be map automorphisms of M and M^α, respectively such that $\tilde{A}$ is a lift of A. Then for every closed walk S in M it holds that*

(*) $$\alpha S = 1_\Gamma \text{ if and only if } \alpha AS = 1_\Gamma$$

PROOF: Let S be a closed walk in M such that $\alpha S = 1_\Gamma$. This is equivalent to the fact that S lifts to $|\Gamma|$ edge-disjoint closed walks $\tilde{S}_i, 1 \le i \le |\Gamma|$ in M^α (i.e., $p\tilde{S}_i = S$ for each $i \le |\Gamma|$, cf. [8]). Since $\tilde{A}$ is an automorphism of M^α, there are $|\Gamma|$ edge-disjoint images $\tilde{A}\tilde{S}_i, 1 \le i \le |\Gamma|$. Moreover, the relation $Ap = p\tilde{A}$ implies that
$$p\tilde{A}\tilde{S}_i = Ap\tilde{S}_i = AS.$$
Thus, the closed walk AS also lifts to $|\Gamma|$ closed walks $\tilde{A}\tilde{S}_i$. However, the latter is possible if and only if $\alpha AS = 1_\Gamma$. ∎

Surprisingly enough, this obvious necessary condition is also sufficient for the existence of a lift of an arbitrary map automorphism. We state the corresponding result for the case when local groups are isomorphic to the whole voltage group.

THEOREM 2. *Let M be a map with underlying graph G and let A be a map automorphism of M. Let there exist a voltage assignment $\alpha : F(G) \to \Gamma$ in a finite group Γ such that, for every closed walk S containing a fixed vertex u,*

(*) $$\alpha S = 1_\Gamma \text{ if and only if } \alpha AS = 1_\Gamma.$$

Assume further that the local group Γ_u is isomorphic to Γ. Then A lifts to $|\Gamma|$ map automorphisms $A_g(g \in \Gamma)$ of the derived map M^α. Explicitly, these automorphisms can be described by the formula

(**) $$A_g(x_{\alpha W}) = (Ax)_{g\alpha AW}$$

where $x \in F(G)$ and W runs through all walks of G emanating from the vertex u and terminating at the initial vertex of the flag x.

PROOF: Let $M = M(L, T, R)$ and $M^\alpha = M^\alpha(L^\alpha, T^\alpha, R^\alpha)$ be the maps in question. We have to prove that, for each $g \in \Gamma$, A_g is a well-defined bijection on $F(G^\alpha)$ which commutes with L^α, T^α and R^α.

First we show that the mapping A_g is well defined by the formula (**). To see this let W and W' be two walks in G, both emanating from u and terminating at the initial vertex of a flag $x \in F(G)$. Suppose that $\alpha W = \alpha W'$; our aim is to show that $A_g(x_{\alpha W}) = A_g(x_{\alpha W'})$. Now, $S = W'LW$ is a closed walk starting and terminating at u. Moreover,

$$\alpha(S) = \alpha(W'LW) = \alpha(W')\alpha(LW) = \alpha W(\alpha W)^{-1} = 1_\Gamma.$$

By (*), $\alpha AS = 1_\Gamma$. However,

$$\alpha AS = \alpha A(W'LW) = \alpha(AW')\alpha(ALW) = \alpha(AW')(\alpha AW)^{-1},$$

which implies that $\alpha AW = \alpha AW'$. Thus, by (**),

$$A_g(x_{\alpha W}) = (Ax)_{g\alpha AW} = (Ax)_{g\alpha AW'} = A_g(x_{\alpha W'})$$

The fact that A_g is a bijection can be proved similarly. Indeed, suppose that $A_g(x_{\alpha W}) = A_g(y_{\alpha W'})$, i.e., $(Ax)_{g\alpha AW} = (Ay)_{g\alpha AW'}$. Then obviously $Ax = Ay$ and, since A is an automorphism of M, $x = y$. But then both W and W' emanate from u and terminate at the same vertex. From $g\alpha AW = g\alpha AW'$ we have $\alpha AW = \alpha AW'$. Considering again the closed walk $S = W'LW$ at u, the equality $\alpha AW = \alpha AW'$ is equivalent to $\alpha AS = 1_\Gamma$. Applying (*) we have also $\alpha S = 1_\Gamma$, which implies $\alpha W = \alpha W'$. This shows that A_g is a bijection on the (finite) set of flags $F(G^\alpha)$.

It remains to prove that the commutation relations are satisfied. Since A is a map automorphism of M we have $AL = LA, AR = RA$ and $AT = TA$. By virtue of the fact that the flags x, Rx and Tx have the same initial vertex we successively obtain:

$$A_g R^\alpha(x_{\alpha W}) = A_g(Rx)_{\alpha W} = (ARx)_{g\alpha AW} =$$

$$= (RAx)_{g\alpha AW} = R^\alpha(Ax)_{g\alpha AW} = R^\alpha A_g(x_{\alpha W});$$

the relation $A_g T^\alpha = T^\alpha A_g$ can be proved analogously. Finally, employing the definition of L^α we have

$$A_g L^\alpha(x_{\alpha W}) = A_g(Lx)_{\alpha W \alpha x} = A_g(Lx)_{\alpha Wx} = (ALx)_{g\alpha A(Wx)} =$$

$$= (LAx)_{g\alpha(AW)\alpha(Ax)} = L^\alpha(Ax)_{g\alpha AW} = L^\alpha A_g(x_{\alpha W});$$

here we used the fact that the walk Wx terminates at the initial vertex of the flag Lx. We see that A_g commutes with all three derived permutations L^α, T^α and R^α and hence is a map automorphism of M^α. The proof is complete. ∎

It is clear that results analogous to those presented in Proposition 1 and Theorem 2 can be stated and proved for lifts of orientation-preserving automorphisms and reflections of orientable maps. The details are left to the reader.

Note also that the assumption $\Gamma_u \cong \Gamma$ in Theorem 2 was not essential. A careful analysis of the above proof shows that, in the general case, an automorphism of M satisfying (*) lifts to $|\Gamma_u|$ map automorphisms of an arbitrary orbit M^α of the group $<L^\alpha, T^\alpha, R^\alpha>$ acting on $F(G^\alpha)$.

5. The lifted regular maps and their automorphism groups

Intuitively, a map is regular if it admits the largest possible number of symmetries (=map automorphisms). In this section we apply the method of lifting automorphisms to obtain new regular maps from old.

Let M be a map with underlying graph G. The set of all map automorphisms of M constitutes a group called the map automorphism group of M and denoted by $AutM$. It is readily proved (e.g., from Lemma 5.2.5 of [2]) that for each two flags $x, y \in F(G)$ there exists at most one map automorphism $A \in AutM$ such that $Ax = y$. Thus, $|AutM| \leq |F(G)|$. The map M is said to be *regular* if $|AutM| = |F(G)|$. This is equivalent to saying that M is regular if the group $AutM$ acts transitively on the set of flags $F(G)$.

It should be pointed out that the just introduced concept of regularity of a map complies with the one in [20]. For orientable maps M, also a weaker form of regularity is often defined by considering the group Aut^+M of orientation-preserving map automorphisms and regarding its transitive action on the set of arcs of the underlying graph (see e.g. [9]). To avoid confusion we shall call such maps *orientably regular*.

The literature on regular maps is extensive. The results range from classification of regular maps on a given surface ([3,6] and others) through classification of regular maps with a given underlying graph (e.g. [9,13]) to constructions of infinite classes of regular maps ([1,2,16,17] etc.). Our method of lifting map automorphisms, described in Section 4, provides a new tool for constructing new classes of regular maps using voltage assignments in Abelian as well as non-Abelian groups.

THEOREM 3. *Let M be a regular map and let G be its underlying graph. Assume that there is a voltage assignment $\alpha : F(G) \to \Gamma$ such that, for each closed walk S containing a fixed vertex u and each $A \in AutM$,*

$$(***) \qquad \alpha S = 1_\Gamma \text{ if and only if } \alpha AS = 1_\Gamma.$$

Further, let $\Gamma_u \cong \Gamma$. Then the derived map M^α is regular.

PROOF: Let x_g and y_h be flags from $F(G^\alpha)$. We have to show that there exists a map automorphism $\tilde{A} \in AutM^\alpha$ such that $\tilde{A}x_g = y_h$. Since M is regular there is an automorphism $A \in AutM$ such that $Ax = y$. From the fact that $\Gamma_u \cong \Gamma$ it follows that the derived graph G^α is connected. This implies that there exists a walk W in G emanating from u and terminating at the initial vertex of x, such that $\alpha W = g$. Now put $b = h(\alpha AW)^{-1}$ and consider the automorphism A_b of M^α (as defined in Theorem 2). Then,

$$A_b(x_g) = A_b(x_{\alpha W}) = (Ax)_{b\alpha AW} = y_h.$$

Thus we can put $\tilde{A} = A_b$. Theorem 3 follows. ∎

Of course, a similar result holds for the already mentioned orientably regular maps. We leave the formulation and the proof to the reader and concentrate now on the map automorphism group of the derived regular map. First we show that the condition (***) is equivalent to the existence of a natural homomorphism from the group $AutM$ into the Automorphism group $Aut\Gamma$ of the voltage group Γ.

PROPOSITION 4. *Let M be an arbitrary map, with underlying graph G. Let $\alpha : F(G) \to \Gamma$ be a voltage assignment satisfying (***) for each $A \in AutM$. Again, assume that $\Gamma_u \cong \Gamma$. Then, for each $A \in AutM$ the mapping $\phi_A : \Gamma \to \Gamma, \phi_A(\alpha S) = \alpha AS$ (S a closed walk containing u) is an automorphism of the group Γ. Moreover, the assignment $A \mapsto \phi_A$ defines a homomorphism $\phi : AutM \to Aut\Gamma$. Conversely, the existence of ϕ implies (***).*

PROOF: To start with, observe that the condition (***) is equivalent to the following one: For any two closed walks S and S' containing u, $\alpha S = \alpha S'$ if and only if $\alpha AS = \alpha AS'$. This together with $\Gamma_u = \Gamma$ readily implies that ϕ_A is a well-defined bijection $\Gamma \to \Gamma$. The fact that ϕ_A is a group automorphism is clear from the following:

$$\phi_A(\alpha S.\alpha S') = \phi_A(\alpha(SS')) = \alpha A(SS') =$$
$$= \alpha(AS)\alpha(AS') = \phi_A(\alpha S)\phi_A(\alpha S').$$

It remains to show that $\phi : A \mapsto \phi_A$ is a group homomorphism $AutM \to Aut\Gamma$:

$$\phi_{BA}(\alpha S) = \alpha BA(S) = \alpha B(AS) = \phi_B(\alpha AS) = \phi_B(\phi_A(\alpha S)).$$

The fact that the existence of ϕ implies (***) is obvious. Proposition 4 follows. ∎

Now we are able to state and prove how the map automorphism group of a lifted regular map depends on the map automorphism group of the original map and the voltage group.

THEOREM 5. *Let M be a regular map satisfying all assumptions listed in Theorem 3. Then the map automorphism group of the derived regular map M^α is a semidirect product of the map automorphism group of M and the voltage group Γ. Formally,*

$$AutM^\alpha = \Gamma \times_\phi AutM$$

where $\phi : AutM \to Aut\Gamma$ is the homomorphism given by Proposition 4.

PROOF: Let A, B be automorphisms of M, and let g and h be elements of Γ. To be able to determine the composition of the automorphisms B_h and A_g from $AutM^\alpha$ let us represent g in the form $g = \alpha AS$ where S is a suitable closed walk containing the vertex u (this is possible due to the connectedness of the derived graph G^α). Now, by Theorem 2 and Proposition 4,

$$B_h A_g(x_{\alpha W}) = B_h(Ax)_{g\alpha AW} = B_h(Ax)_{\alpha AS.\alpha AW} =$$
$$= B_h(Ax)_{\alpha A(SW)} = (BAx)_{h\alpha BA(SW)} = (BAx)_{h\alpha BA(S).\alpha BA(W)} =$$
$$= (BAx)_{h\phi_B(\alpha AS).\alpha BAW} = (BA)_{h\phi_B(g)}(x_{\alpha W}).$$

Thus, putting $A_g = (g, A)$ and $B_h = (h, B)$ we obtain the following multiplication rule in $AutM^\alpha$:

$$(h, B)(g, A) = (h\phi_B(g), BA).$$

But this is exactly the well known definition of the semidirect product in group theory. Consequently, $AutM^\alpha = \Gamma \times_\phi AutM$, as claimed. ∎

6. Applications

By means of the lifting technique introduced in preceding sections we now construct two infinite classes of regular maps. The first one will be a class of orientable as well as nonorientable maps obtained from voltage assignment in an Abelian group, while the second class will consist of orientable regular maps arising from a non-Abelian voltage group. In both cases the basis of the construction is the group homomorphism ϕ introduced in Proposition 4. Namely, as the reader may already have realized, combining this proposition with Theorem 3 yields the following immediate consequence.

THEOREM 6. *Let α be a voltage assignment on a regular map M in a group Γ. Assume that there exists a group homomorphism $\phi : \text{Aut} M \to \text{Aut}\Gamma$, $A \mapsto \phi_A$ such that*
$$\phi_A(\alpha W) = \alpha AW$$
for every closed walk W in M. Then the derived map M^α is regular. ∎

In our first construction we use a repeated product of cyclic groups as our voltage group Γ. The advantage of this approach is that it applies to an *arbitrary* regular map M and yields both nonorientable as well as orientable derived maps M^α in the case when M is nonorientable.

Let $M = M(L, T, R)$ be an arbitrary regular map with r faces. Let G be the underlying graph of M comprising p vertices and q edges. Consider the group $Z_n \times Z_n \times \cdots \times Z_n$ (q times) where $n \geq 2$. Take now one flag from each orbit of the group $<L, T>$ on $F(G)$, obtaining thereby a sequence of q flags $x_1, x_2, \ldots, x_q$. In order to define a voltage assignment $\alpha : F(G) \to \Gamma$ it is sufficient to define the values $\alpha(x_i), 1 \leq i \leq q$. We set $\alpha(x_i)$ to be the q-tuple of Γ in which all but the i-th entry are zero while the i-th entry is 1. Obviously, $\text{Aut}\Gamma$ contains the symmetric group on q letters as a subgroup and the group $\text{Aut} M$ can also be viewed as a permutation group on q letters ($=$ the q edges of G). Since α is injective, each automorphism $A \in \text{Aut} M$ can be interpreted as a permutation of the Z'_ns in Γ, i.e., as an automorphism from $\text{Aut}\Gamma$. In other words, there exists a canonical homomorphism $\phi : \text{Aut} M \to \text{Aut}\Gamma$ such that $\phi_A(\alpha x) = \alpha Ax$ for every flag $x \in F(G)$. Thus, the assumptions of Theorem 6 are satisfied automatically; it follows that the derived map M^α is regular.

A more detailed analysis shows that the derived graph G^α is disconnected. In fact, this time the local group Γ_u is isomorphic to $Z_n \times Z_n \times \cdots \times Z_n$ ($q-p+1$ times), which implies that G^α consists of n^{p-1} connected components, each containing pn^{q-p+1} vertices. It follows that the derived map M^α has qn^{q-p+1} edges and rn^{q-p} faces. Moreover, observe that every cycle of length l in M lifts to cycles of length nl in M^α. Thus, if M is nonorientable and n is odd, then the derived map M^α is nonorientable, as well. This enables one to construct infinite classes of nonorientable regular maps.

In our second construction we show how regular maps can also be obtained by means of a voltage assignment in a non-Abelian group.

Denote by Γ_n the group with generators a and b and defining relations
$$a^n = b^n = h^n = 1, ha = ah, hb = bh \text{ where } h = aba^{-1}b^{-1}.$$

Thus, Γ_n is a group in which the commutator $h = [a, b]$ belongs to the centre $Z(\Gamma_n)$. One can easily see that Γ_n is not Abelian, $|\Gamma_n| = n^3$, and every element of Γ_n can be expressed in the form $h^p a^q b^r, 0 \leq p, q, r \leq n - 1$. Note also that $a^p b^q = h^{pq} b^q a^p$ for all p and q. As regards group automorphisms of Γ_n, we have the following result.

LEMMA 7. *The mappings* $\phi_i : \Gamma_n \to \Gamma_n (1 \leq i \leq 4)$ *given by*

$$\phi_1 = id$$

$$\phi_2(h^p a^q b^r) = h^p a^{-q} b^{-r}$$

$$\phi_3(h^p a^q b^r) = h^{p+qr} a^{-r} b^q$$

$$\phi_4(h^p a^q b^r) = h^{p+qr} a^r b^{-q}$$

are automorphisms of the group Γ_n.

PROOF: We prove Lemma 7 only for ϕ_3; the remaining cases are similar. Clearly, ϕ_3 is bijection. Moreover,

$$\phi_3[(h^p a^q b^r)(h^k a^l b^m)] = \phi_3(h^{p+q} a^q b^r a^l b^m) =$$
$$= \phi_3(h^{p+k} a^{q+l} b^{r+m} h^{-rl}) = \phi_3(h^{p+k-rl} a^{q+l} b^{r+m}) =$$
$$= h^{p+k-rl+(q+l)(r+m)} a^{-r-m} b^{q+l} = h^{p+k+qr+qm+bm} a^{-r-m} b^{q+l} =$$
$$= h^{p+qr} h^{k+lm} h^{qm} a^{-r} a^{-m} b^q b^l = h^{p+qr} h^{k+lm} a^{-r} b^q a^{-m} b^l =$$
$$= (h^{p+qr} a^{-r} b^q)(h^{k+lm} a^{-m} b^l) = \phi_3(h^p a^q b^r) \phi_3(h^k a^l b^m). \blacksquare$$

Let Σ be an orientable surface; fixing an orientation on Σ makes the surface *oriented*. Let M be an orientably regular map on Σ. Assume that each face of M is bounded by a $4m$-gon and that the underlying graph G has each vertex 4-valent. Consider a voltage assignment α on $D(G)$, the set of arcs of G, in the group Γ_n, defined by the following condition: At each vertex v, the fixed orientation of Σ induces on the arcs emanating from v the cyclic ordering (a, b, a^{-1}, b^{-1}) of elements of Γ_n (see Fig. 3):

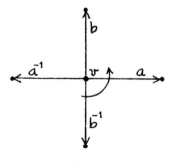

Fig. 3.

It is a matter of routine to check that α is indeed uniquely determined by the above local rule on M. Observe that, for a given vertex u, there is exactly one closed (directed) walk S at u such that all arcs of S carry the voltage a. Let k be the length of S; assume moreover that $g.c.d.(k,n) = g.c.d.(n,m) = 1$. An orientably regular map M satisfying all assumptions listed here will be called a $(k, 4m, \Gamma_n)$-map.

Our next result shows that the just introduced maps are suitable for recursive constructions.

THEOREM 8. *Let M be a $(k, 4m, \Gamma_n)$-map with G as its underlying graph. Then*

(1) *the derived graph G^α is connected,*
(2) *the derived map M^α is orientably regular, and*
(3) *M^α is a $(k', 4m', \Gamma_{n'})$-map for suitable $k' > k, m' > m$ and $n' > n$.*

PROOF: (1) Since M is orientably regular there is an automorphism A of M which rotates the arcs emanating from u by the "angle" $\pi/2$. The way M is defined implies that applying A to the walk S introduced before yields the walk AS at u such that every arc of AS carries voltage b. Thus, $\alpha S = a^k$ and $\alpha AS = b^k$. From the fact that k and n are coprime it follows that a^k and b^k generate the whole group Γ_n. In other words, the local group $(\Gamma_n)_u$ is isomorphic to Γ_n and hence the derived graph G^α is connected.

(2) Let $W = x_1 x_2 \ldots x_t$ be an arbitrary closed walk (x_j arcs of G) and let A be an orientation-preserving map automorphism of M. It is easy to see that there exists an $i, 1 \leq i \leq 4$ such that

$$\alpha AW = \alpha Ax_1 \alpha Ax_2 \ldots \alpha Ax_t = \phi_i(\alpha x_1)\phi_i(\alpha x_2)\ldots \phi_i(\alpha x_t) = \phi_i(\alpha W)$$

where ϕ_i is given by Lemma 7 (it is sufficient to realize how M is constructed). Consequently, $\alpha W = 1$ if and only if $\alpha AW = 1$. By Theorem 6 (and its straightforward modification for orientably regular maps) the derived map M^α is orientably regular.

(3) Clearly, the derived graph G^α is 4-valent. Since m and n are coprime and the net voltage on a boundary of a face of M is $(aba^{-1}b^{-1})^m = h^m$ (or a conjugate of it), each face of M^α has length $4mn$. This enables to define a voltage assignment $\beta : D(G^\alpha) \to \Gamma_{n'} = \Gamma_{n'}(a', b')$ of the same type as described in Fig. 3. (n' will be specified later). Again let k' be the length of a closed walk in G^α whose all arcs carry the voltage a' (note that $k' > k$). Now put $m' = mn$ and take an $n' > n$ which is coprime with both k' and m'. Obviously, this endows M^α with a structure of a $(k', 4m', \Gamma_{n'})$-map, as required. ∎

A routine calculation involving the Euler formula shows that the genus of M^α is larger than the genus of M. Therefore Theorem 8 enables us to construct infinite sequences $M_1, M_2, \ldots, M_t, \ldots$ of orientably regular maps with growing genera, starting from a suitable $(k, 4m, \Gamma_n)$-map M_1. For example, we can take for M_1 a toroidal 4-valent map with underlying graph $G = C_k \Lambda C_k$, the product of two cycles of length k; this is a $(k, 4.1, \Gamma_n)$-map if $g.c.d.(k, n) = 1$.

References

[1] D.Archdeacon, J.Širáň, M.Škoviera, *Regular maps from medial graphs.*. Submitted.
[2] N.L.Biggs, A.T.White, "Permutation groups and combinatorial structures," Cambridge University Press, Cambridge, 1979.
[3] H.S.M.Coxeter, W.O.J.Moser, "Generators and relations for discrete groups (4th ed.)," Springer, Berlin and New York, 1980.
[4] D.Z.Djokovic, *Automorphisms of Graphs and Coverings*, J. Combin. Theory (B) **16** (1974), 243-247.
[5] A.W.M.Dress, *A combinatorial theory of Grunbaum's new regular polyhedra, Part I*, Aequationes Mathematicae **23** (1981), 252-265.
[6] D.Garbe, *Ueber die regulären Zerlegungen geschlossener orientiebarer Flächen*, J. reine angewandte Math **237** (1969), 39-55.
[7] J.L.Gross, *Voltage graphs*, Discrete Math **9** (1974), 239-246.
[8] J.L.Gross, T.W.Tucker, "Topological graph theory," Wiley-Interscience, New York, 1987.
[9] L.D.James, G.A.Jones, *Regular orientable imbeddings of complete graphs*, J. Combin. Theory (B) **39** (1985(3)), 353-367.
[10] G.A.Jones, D.Singerman, *Theory of maps on orientable surfaces*, Proc. London Math. Soc. (3) **37** (1978), 273-307.
[11] G.A.Jones, J.S.Thornton, *Operations on maps, and outer automorphisms*, J. Combin. Theory(B) **35** (1983), 93-103.
[12] W.S.Massey, "Algebraic Topology, An Introduction," Harcourt, Brace & World, Inc., New York, 1967.
[13] R.Nedela, M.Škoviera *Regular maps of the tensor product $G * K_2$*. Submitted.
[14] S.Stahl, *Generalized embedding schemes*, J. Graph Theory **2** (1978), 41-52.
[15] D.B.Surowski, *Lifting Map Automorphisms and MacBeath's Theorem*, J. Combin. Theory (B) **50** (1990), 135-149.
[16] J.Širáň, M.Škoviera, *Regular maps from Cayley graphs 2: Antibalanced Cayley maps*. Submitted.
[17] M.Škoviera, J.Širáň, *Regular maps from Cayley graphs 1: Balanced Cayley maps*. To appear in Discrete Math.
[18] W.T.Tutte, "Graph theory," Addison-Wesley, Reading, Mass., 1984.
[19] S.Wilson, *Riemann surfaces over regular maps*, Can. J. Math. **30** (1978), 763-782.
[20] S.Wilson, *Cantankerous maps and rotary embeddings of K*, J.Combinatorial Theory (B) **47** (1989), 262-273.

Faculty of Mathematics and Physics, Comenius University, 842 15 Bratislava Czechoslovakia

The infinite grid covers the infinite half-grid

BOGDAN OPOROWSKI

1. Introduction

The purpose of this paper is to outline the main ideas that arise in proving the theorem stated in the title. The full details of the proof will appear elsewhere [3]. The remainder of this section introduces the necessary definitions and formally states the main results. Section 2 describes some preliminary lemmas needed for the proofs of these results, while section 3 sketches these proofs.

Graphs in this paper may be finite or infinite and may have loops and multiple edges. We say that a graph H is a *minor* of a graph G, and write $H \leq_m G$, if H can be obtained from a subgraph of G by contracting some of its (possibly infinite) connected subgraphs. If H is a minor of G but G is not a minor of H, then we write $H <_m G$. Two graphs G and H are *minor equivalent*, written $G \cong_m H$, if $G \leq_m H$ and $H \leq_m G$. If H is a subgraph of G, then $G - H$ denotes the subgraph of G induced by the edges of G that are not in H.

It is clear that minor equivalence is indeed an equivalence relation. The equivalence class that contains a graph G will be denoted by $[G]_m$. It is clear that if G is a finite graph, then $[G]_m$ is equal to the isomorphism class containing G.

If G is infinite, then $[G]_m$ may contain graphs from more than one isomorphism class. For example, if G is an infinite clique and H consists of G and a single isolated vertex, then $G \cong_m H$, and thus $[G]_m = [H]_m$, even though G and H are not isomorphic. On the other hand, if G is a two-way-infinite path, then $[G]_m$ is equal to the isomorphism class of G. Another interesting example of a graph whose minor equivalence class coincides with its isomorphism class is a graph that is not isomorphic to any of its proper minors. The existence of such a graph has been shown in [2].

1991 Mathematics Subject Classification: Primary 05C99.

A detailed version [3] of this paper has been submitted elsewhere.

A graph G *covers* another graph H if $H <_m G$ and, for every graph K such that $H \leq_m K \leq_m G$, either $K \cong_m H$ or $K \cong_m G$. It is immediate that

(1.1) *if G covers H, $G \cong_m G'$, and $H \cong_m H'$, then G' covers H'.*

If G is a finite non-null graph, then G covers some graphs, namely those that can be obtained from G by deleting a single edge, contracting a single edge, or deleting a single isolated vertex. Similarly, every finite graph is covered by another graph. However, if G is infinite, then G may not cover, and may not be covered by, another graph. It is easy to verify the following statements:

(1.2) *A graph that has a countably infinite vertex set and no edges covers no graphs.*

(1.3) *A two-way-infinite path covers the disjoint union of two one-way-infinite paths.*

An example of a graph that is covered by no other graph is given in [3].

Let G be a graph and $G \times 2$ be the disjoint union of two copies of G. The graph G is *clonable* if $G \times 2 \leq_m G$. It is also easy to show that

(1.4) *if H and H' are graphs that are both covered by a clonable connected graph G, then $H \cong_m H'$.*

The *countably infinite clique* is a graph whose vertex set is $\mathbf{N}$ and such that every pair of its vertices is joined by an edge. The following two infinite graphs will play a crucial role in this paper. The *full grid*, denoted by $G_{\mathbf{Z} \times \mathbf{Z}}$, has the set $\mathbf{Z} \times \mathbf{Z}$ as its vertex set, and two of its vertices (i,j), (i',j') are joined by an edge if and only if $|i - i'| + |j - j'| = 1$. The *half-grid*, denoted by $G_{\mathbf{Z} \times \mathbf{N}}$, is the graph obtained from $G_{\mathbf{Z} \times \mathbf{Z}}$ by deleting all the vertices whose second coordinate is negative. The $k \times k$-*grid* is the finite subgraph of $G_{\mathbf{Z} \times \mathbf{Z}}$ induced by the subset $\{1, 2, \ldots, k\} \times \{1, 2, \ldots, k\}$ of its vertex set. The main result of this paper is the following.

(1.5) $G_{\mathbf{Z} \times \mathbf{Z}}$ *covers* $G_{\mathbf{Z} \times \mathbf{N}}$.

Two of the many open problems that are related to (1.5) are as follows:

(1.6) Does the half-grid cover any graphs?

(1.7) Does the countably infinite clique cover any graphs?

As both the half-grid and the countably infinite clique are connected and clonable, (1.4) implies that any graph covered by the half-grid and any graph covered by the countably infinite clique would be unique up to minor equivalence. By replacing the minor relation by topological embedding above we obtain several other interesting problems.

In order to outline the proof of (1.5), we need to introduce the concept of an end of an infinite graph. A *ray* is a one-way-infinite path. Two rays

ρ, σ, which are subgraphs of the same graph G, are *equivalent* if, for every finite subgraph H of G, the infinite parts of $\rho - H$ and $\sigma - H$ lie in the same connected component of $G - H$. Halin [1] proved that

(1.8) *two rays are equivalent if and only if there is another ray that meets both of them infinitely often.*

It is easy to verify that the above relation is an equivalence relation on rays which are subgraphs of a fixed infinite graph G. The equivalence classes of this relation are called the *ends* of G. An end is *thick* if it contains infinitely many disjoint rays. A graph is *planar* if it has no minor isomorphic to K_5 or to $K_{3,3}$. We establish the following result which will be used to prove (1.5).

(1.9) *Suppose that G is a planar locally finite graph having exactly one end and that this end is thick. Then $G \cong_m G_{\mathbf{Z} \times \mathbf{Z}}$ or $G \cong_m G_{\mathbf{Z} \times \mathbf{N}}$.*

In the next section we describe the results used to prove (1.5) and (1.9).

2. Auxiliary results

The following assumptions will hold throughout this section:

(2.1) G is a planar locally finite graph that has exactly one end and this end is thick;

(2.2) $\mathcal{F}$ is the set of all subgraphs F of G such that F is the union of a finite number of disjoint rays.

We remark that several results stated in this section can be proved with some of the conditions in (2.1) relaxed. However, we shall not need the more general statements here.

A path P (or a cycle C) of G is *reduced* with respect to a ray ρ if P (or C) either intersects ρ along a path (perhaps consisting of one vertex only) or does not intersect it at all. Path P (or cycle C) is *reduced* with respect to an element of $\mathcal{F}$ if it is reduced with respect to all its components. We show that

(2.3) *for every F in $\mathcal{F}$ and every finite subgraph H of G, if $G - H$ contains a path that joins two of the rays ρ and σ in F, then $G - H$ contains a path that joins ρ and σ and is reduced with respect to F.*

We use (2.1), (2.2), and (2.3) to prove the following result.

(2.4) *G contains a minor isomorphic to the infinite binary tree.*

A path P or a cycle C of G is said to *collate* an element F of $\mathcal{F}$ if it meets all rays in F and is reduced with respect to F. We use (2.1), (2.2), and (2.3) again to show that

(2.5) *for every F in $\mathcal{F}$ and every finite subgraph H of G, there is a path P in $G - H$ that collates F.*

Suppose F is an element of $\mathcal{F}$ and P is an path of G that is reduced with respect to F. Then P induces a linear undirected order on the (possibly empty) subset F_P of F consisting of those rays in F that meet P. The *F-trace* of P is F_P ordered as described. Similarly, if C is a cycle of G that is reduced with respect to F, then C induces a circular undirected order on the subset F_C of F consisting of those rays in F that meet C. The *F-trace* of C is F_C ordered as described.

It follows from an easy application of the infinite version of the pigeon hole principle and from (2.5) that

(2.6) *for every F in $\mathcal{F}$, there is an infinite set R of disjoint paths in G such that each path in R collates F and all the paths in R have the same F-trace.*

We also consider the following property that G may or may not satisfy.

(2.7) For every element F of $\mathcal{F}$ and every finite subgraph H of G, there is a cycle C in $G - H$ that collates F.

It takes another application of the infinite version of the pigeon hole principle to show that

(2.8) *if G satisfies (2.7), then, for every F in $\mathcal{F}$, G contains an infinite set R of disjoint cycles all of which collate F and have the same F-trace.*

We use (2.4), (2.6), and (2.8) to show that

(2.9) $G_{\mathbf{Z} \times \mathbf{N}} \leq_m G$, and

(2.10) *if G satisfies (2.7), then $G_{\mathbf{Z} \times \mathbf{Z}} \leq_m G$.*

Robertson, Seymour, and Thomas [4] proved that

(2.11) *if K is a finite planar graph, then there is an integer k and an isomorphism f from K to a minor of the $k \times k$-grid.*

Two rather technical modifications of this result describe the isomorphism f in (2.11) in more detail. Using these results, it can be shown that

(2.12) $G \leq_m G_{\mathbf{Z} \times \mathbf{Z}}$; and

(2.13) *if G fails to satisfy (2.7), then $G \leq_m G_{\mathbf{Z} \times \mathbf{N}}$.*

3. Proofs of the main results

Observe that (2.9), (2.10), (2.12), and (2.13) imply (1.9). To show (1.5), we must establish that

(3.1) $G_{\mathbf{Z}\times\mathbf{N}} <_m G_{\mathbf{Z}\times\mathbf{Z}}$,

and, for every graph K such that $G_{\mathbf{Z}\times\mathbf{N}} \leq_m K \leq_m G_{\mathbf{Z}\times\mathbf{Z}}$,

(3.2) either $K \cong_m G_{\mathbf{Z}\times\mathbf{N}}$ or $K \cong_m G_{\mathbf{Z}\times\mathbf{Z}}$.

It is clear that $G_{\mathbf{Z}\times\mathbf{N}}$ is a minor of $G_{\mathbf{Z}\times\mathbf{Z}}$. To see that $G_{\mathbf{Z}\times\mathbf{Z}}$ is not a minor of $G_{\mathbf{Z}\times\mathbf{N}}$, observe that if G equals $G_{\mathbf{Z}\times\mathbf{Z}}$, then G satisfies (2.7), whereas if G is a minor of $G_{\mathbf{Z}\times\mathbf{N}}$, it does not. Hence (3.1) follows and we concentrate on proving the other part of the claim.

First we show that

(3.3) if K is a subgraph of $G_{\mathbf{Z}\times\mathbf{Z}}$ such that $G_{\mathbf{Z}\times\mathbf{N}} \leq_m K$, then $K \cong_m G_{\mathbf{Z}\times\mathbf{N}}$ or $K \cong_m G_{\mathbf{Z}\times\mathbf{Z}}$.

We begin the proof of (3.3) by using (1.9) to show that

(3.4) if H is an infinite connected subgraph of $G_{\mathbf{Z}\times\mathbf{Z}}$ and $K = G_{\mathbf{Z}\times\mathbf{Z}} - H$, then $K \leq_m G_{\mathbf{Z}\times\mathbf{N}}$.

Next we consider a subgraph H of $G_{\mathbf{Z}\times\mathbf{Z}}$ all of whose components are finite and such that, for any two distinct components L, M of H, component M is a subgraph of the infinite component of $G_{\mathbf{Z}\times\mathbf{Z}} - L$ and no vertex of L is adjacent in $G_{\mathbf{Z}\times\mathbf{Z}}$ to a vertex of M. Such a subgraph H of $G_{\mathbf{Z}\times\mathbf{Z}}$ will be called *distributed*. We again use (1.9) to show that

(3.5) if H is a distributed subgraph of $G_{\mathbf{Z}\times\mathbf{Z}}$ and $K = G_{\mathbf{Z}\times\mathbf{Z}} - H$, then $K \cong_m G_{\mathbf{Z}\times\mathbf{Z}}$.

Then (3.4) and (3.5) are used to conclude the proof of (3.3).

Result (3.3) states that a deletion minor K of $G_{\mathbf{Z}\times\mathbf{Z}}$ such that $G_{\mathbf{Z}\times\mathbf{N}} \leq_m K$ satisfies (3.2). It remains to show that a contraction minor K of $G_{\mathbf{Z}\times\mathbf{Z}}$ such that $G_{\mathbf{Z}\times\mathbf{N}} \leq_m K$ also satisfies (3.2). To do this, we employ the concept of planar duals. Observe that the full grid is self-dual, and while the half-grid is not, it is minor equivalent to its dual. Using this observation, we prove that

(3.6) if H is an infinite subgraph of $G_{\mathbf{Z}\times\mathbf{Z}}$ and K is obtained from $G_{\mathbf{Z}\times\mathbf{Z}}$ by contracting H, then $K \leq_m G_{\mathbf{Z}\times\mathbf{N}}$; and

(3.7) if H is a distributed subgraph of $G_{\mathbf{Z}\times\mathbf{Z}}$ and K is obtained from $G_{\mathbf{Z}\times\mathbf{Z}}$ by contracting H, then $K \cong_m G_{\mathbf{Z}\times\mathbf{Z}}$.

Finally, since any minor of $G_{\mathbf{Z}\times\mathbf{Z}}$ can be viewed as a contraction minor of a subgraph of $G_{\mathbf{Z}\times\mathbf{Z}}$, we conclude that (1.5) holds.

ACKNOWLEDGEMENTS

This research was partially supported by a grant from the Louisiana Education Quality Support Fund through the Board of Regents. The author is grateful to Robin Thomas for stimulating discussions on the subjects addressed in this paper.

REFERENCES

[1] R. Halin, *Über unendliche Wege in Graphen*, Math. Ann. **157** (1964), 125-137.
[2] B. Oporowski, *A counterexample to Seymour's self-minor conjecture*, J. Graph Theory **14** (1990), 521-524.
[3] B. Oporowski, *Minor equivalence for infinite graphs*, submitted.
[4] N. Robertson, P. Seymour, and R. Thomas, *Quickly excluding a forest*, preprint.

MATHEMATICS DEPARTMENT, LOUISIANA STATE UNIVERSITY, BATON ROUGE, LA 70803, USA
E-mail address: bogdan @ marais.math.lsu.edu

Dominating functions and topological graph minors

REINHARD DIESTEL

An infinite graph G is called dominating if its vertices can be labelled with natural numbers in such a way that for every function $f: \omega \to \omega$ there is a ray in G whose sequence of labels eventually exceeds f. Conversely, G is called bounded if for every labelling of its vertices with natural numbers there exists a function $f: \omega \to \omega$ which eventually exceeds the labelling along any ray in G. This expository paper describes recent classifications of the dominating and the bounded graphs by forbidden topological minors, and indicates some connections of these results to infinite games.

Introduction

If f and g are functions from ω to ω, we say that f *dominates* g if $f(n) \geq g(n)$ for all but finitely many $n \in \omega$. A family $\mathcal{F}$ of $\omega \to \omega$ functions is called a *dominating family* if every function $g: \omega \to \omega$ is dominated by some $f \in \mathcal{F}$. The least cardinality of a dominating family is denoted by $\mathfrak{d}$.

Similarly, a family $\mathcal{F}$ of functions from ω to ω is said to be *bounded* by a function $g: \omega \to \omega$ if g dominates every $f \in \mathcal{F}$; if no such g exists, $\mathcal{F}$ is called *unbounded*. The least cardinality of an unbounded family is denoted by $\mathfrak{b}$.

It is not difficult to see that any unbounded family of functions must be uncountable. Indeed, if $\mathcal{F} = \{ f_n \mid n \in \omega \}$ is a countable family of $\omega \to \omega$ functions, then $g: \omega \to \omega$ defined by $g(n) := \max \{ f_0(n), \ldots, f_n(n) \}$ is a bounding function for $\mathcal{F}$. Thus, $\mathfrak{b} > \omega$. Since any dominating family is clearly unbounded, we further have $\mathfrak{b} \leq \mathfrak{d}$. Finally, the set of *all* $\omega \to \omega$ functions clearly defines a dominating family; since there are exactly continuum many $\omega \to \omega$ functions, we have

$$\omega < \mathfrak{b} \leq \mathfrak{d} \leq 2^\omega.$$

Set theorists have traditionally been interested in the question of when the above inequalities may be strict, and how $\mathfrak{b}$ and $\mathfrak{d}$ compare with other cardinals between ω and 2^ω. In other words, it has been asked just how much the cardinality of a family of functions has to be constrained in order to force it to

---

1991 Mathematics Subject Classification. Primary 5C; Secondary 4.
This paper is in final form and will not be submitted for publication elsewhere.

become bounded or to cease to be dominating. This question depends on the axioms of set theory assumed, and we shall not pursue it further here. Instead, we shall ask how we can force a family of functions to become bounded (or to cease to be dominating) by restricting the independence of its members.

This can be done naturally as follows. We shall label the vertices of an infinite graph with integers, and consider as our family of functions the labellings along one-way infinite paths, or *rays*, in this graph. Depending on how much different rays in the graph intersect, it may turn out that their labellings can never form a dominating or an unbounded family, even when the graph contains continuum many distinct rays.

To be precise, let us say that a graph G is *dominating* if there exists a labelling $\ell: V(G) \to \omega$ of its vertices such that for every $f: \omega \to \omega$ there is a ray $v_0 v_1 \ldots$ in G with $\ell(v_n) \geq f(n)$ for all but finitely many n. Similarly, G is called *bounded* if for every labelling $\ell: V(G) \to \omega$ there exists a function $f: \omega \to \omega$ such that, whenever $v_0 v_1 \ldots$ is a ray in G, we have $\ell(v_n) \leq f(n)$ for all but finitely many n. If G is a graph with a fixed labelling, we shall not always distinguish between a ray in G and its sequence of labels, so that we may speak of functions dominating rays and vice versa.

Note that, by definition, supergraphs of dominating graphs are again dominating, and subgraphs of bounded graphs are again bounded. We may therefore hope to classify the bounded and the dominating graphs by identifying some particular 'minimal' prototypes of unbounded or dominating graphs, and showing that a graph is unbounded or dominating if and only if it contains one of the respective prototypes. In fact, we shall see that a graph is bounded if and only if it contains none of four specified graphs as a topological minor. The dominating graphs will be characterized similarly.

The prototype unbounded or dominating graphs appearing in these characterizations will be discussed in Section 1, together with some other examples. In Section 2 we present the two classification theorems, and take a glance at how they are proved. Section 3 introduces a framework for the definition of infinite games related to the domination of functions arising from rays in labelled graphs; two specific games, the dominating game and the bounding game, are analysed and winning strategies given.

The notation used will be standard; see e.g. [1]. When a graph G contains a subdivision of a graph H as a subgraph, we also say that H is a *topological minor* of G. If P is a path containing vertices u and v, we write uP, Pv and uPv for the obvious subpaths of P starting in u and/or ending in v.

1. Examples

In this section we collect together a few typical examples of graphs that are, or fail to be, bounded or dominating. Note that any dominating graph is automatically unbounded.

We have already seen that any unbounded family of functions must be uncountable. If a graph contains only countably many rays, it is therefore trivially bounded.

For a more interesting example of bounded graphs, consider any connected locally finite graph G, and pick a labelling. For each vertex $v \in G$, we may define a function $f_v: \omega \to \omega$ by setting $f_v(n)$ to be the maximum of all the labels of vertices at distance at most n from v. Clearly, f_v dominates (the labelling along) every ray in G that starts at v. Since G has only countably many vertices, there exists a function f which dominates every f_v with $v \in V(G)$, and hence every ray in G. Therefore G is bounded.

On the other hand, it is not difficult to find graphs that are unbounded or even dominating. For example, the union of $\mathfrak{d}$ disjoint rays is dominating, and the union of $\mathfrak{b}$ disjoint rays is unbounded: just label each ray by a different member of some dominating or unbounded family of functions.

So how about countable unbounded graphs? The complete graph K_ω on a countably infinite set of vertices is clearly dominating. Indeed, consider any labelling ℓ that uses arbitrarily large labels: then, for any $f: \omega \to \omega$, we may find a ray $v_0 v_1 \ldots$ with $\ell(v_n) > f(n)$ for every n. Similarly, the ω-regular tree T_ω is dominating, and hence unbounded: just label its vertices injectively, i.e. so that any two labels are different.

In fact, any subdivision T of T_ω is unbounded. To see this, let again $\ell: V(T) \to \omega$ be any injective labelling, and let $f: \omega \to \omega$ be any given function. We may then choose a ray $v_0 v_1 \ldots$ in T as follows. If $n = 0$ or v_{n-1} is a branch vertex of T, choose v_n so that $\ell(v_n) > f(n)$; this can be done, because ℓ is injective. Otherwise let v_n be any hitherto unused neighbour of v_{n-1}. (This vertex is unique unless $n = 1$.) Since any ray in T contains infinitely many branch vertices, we see that $\ell(v_n) > f(n)$ for infinitely many n, and hence that T is unbounded.

It is an interesting fact that, by contrast, subdivisions of T_ω need not be dominating. Indeed, consider any enumeration $e: \omega \to E(T_\omega)$ of the edges of T_ω, and let T be the tree obtained from T_ω by subdividing $e(n)$ exactly n times for each n.

Proposition 1.1. *T is not dominating.*

Proof. Let $\ell: V(T) \to \omega$ be any labelling of the vertices of T. Let $f: \omega \to \omega$ be an increasing function satisfying $f(n) > \max \{ \ell(v) \mid v \in e(n) \}$ for all $n \in \omega$. We show that, for any ray $R = v_0 v_1 \ldots$ in T and any $i \in \omega$, there exists a $k > i$ with $f(k) > \ell(v_k)$.

Given such R and i, choose $j, k \in \omega$ with $i < j < k$ so that $v_j v_k = e(n)$ for some n. Then R traces out the subdivided edge $e(n)$, so in particular we have $k \geq |v_j R v_k| = n + 2$. Since f is increasing and $f(n) > \ell(v_k)$ by definition of f, this gives $f(k) \geq f(n) > \ell(v_k)$ as desired. □

If, as we have seen, T_ω is dominating but subdivisions of it need not be, can we say exactly which kinds of subdivision of T_ω yield a dominating graph? Indeed we can: as Theorem 2.3 of Section 2 will show, a subdivision of T_ω is dominating if and only if it contains a particularly simple subdivision of T_ω, to be called 'uniform', in which for every branch vertex there exists a bound on how often the edges incident with this vertex in T_ω have been subdivided.

More precisely, let us call a subdivision T of T_ω *uniform* if it has a branch vertex r, called its *root*, such that whenever v is a branch vertex, all the subdivided edges at v that are *not* contained in the unique path from v to r have the same length.

Proposition 1.2. *Uniform subdivisions of T_ω are dominating.*

Proof. Let T be a uniform subdivision of T_ω, with root r say. Let $\ell \colon V(T) \to \omega$ be any injective labelling; we show that for every function $f \colon \omega \to \omega$ there is a ray R in T which dominates f.

Since any ray in T is a concatenation of paths that are subdivided edges of T_ω, we may define $R = v_0 v_1 \ldots$ inductively by choosing these subdivided edges one at a time. Let $v_0 = r$. Suppose that v_n has been defined for every $n \leq m$, and that v_m is a branch vertex of T. Then all the (infinitely many) subdivided edges at v_m that are not contained in the portion of R defined so far have the same length k, and so we can find one of them, say $s_0 \ldots s_k$ where $s_0 = v_m$, such that $\ell(s_i) \geq f(m+i)$ whenever $0 < i \leq k$. Setting $v_{m+i} = s_i$, we obtain $\ell(v_{m+i}) \geq f(m+i)$ for all these i; moreover, v_{m+k} is again a branch vertex of T. This completes the induction step, and hence the construction of R. Since $\ell(v_n) \geq f(n)$ for every $n > 0$, it is clear that R dominates f. □

Surprisingly, there is yet another way to obtain a dominating graph from a subdivision of T_ω: by taking $\mathfrak{b}$ disjoint copies of it. Note that $\mathfrak{d}$ disjoint copies are trivially dominating, since each of them contains a ray; it can be shown, however, that $\mathfrak{b}$ subdivisions of T_ω, which need not be isomorphic, suffice:

Proposition 1.3. [5] *If a graph is the union of $\mathfrak{b}$ disjoint subdivisions of T_ω, then it is dominating.* □

We have now seen all the examples of dominating graphs that will be needed for our classification theorem in Section 2: Theorem 2.3 says that a graph is dominating if and only if it contains a uniform subdivision of T_ω, $\mathfrak{b}$ disjoint subdivisions of T_ω, or $\mathfrak{d}$ disjoint rays.

For the classification of bounded graphs, however, there are two more prototypical unbounded graphs that may occur. One of these is the graph B obtained from a ray $v_0 v_1 \ldots$ by adding, for each $n \in \omega$, a countably infinite set of independent v_{3n+1}–v_{3n+3} paths of length 2 (Fig. 1).

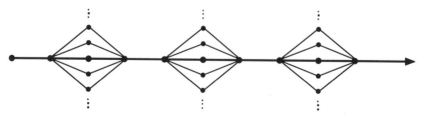

FIGURE 1. The prototype bundle graph B

To show that B is unbounded, let ℓ be any injective labelling of $V(B)$, and assume we are given a function $f\colon \omega \to \omega$. Let us specify a ray $R = u_0 u_1 \ldots$ as follows. Put $u_0 := v_0$ and $u_1 := v_1$. For u_2, we have an infinite choice of neighbours of v_1, all labelled differently. We may thus choose as u_2 a neighbour of v_1 whose label is greater than $f(2)$. We continue with $u_3 := v_3$ and $u_4 := v_4$, where again we have an infinite choice for u_5. Proceeding in this manner, we may choose R in such a way that $\ell(u_n)$ exceeds $f(n)$ for every third value of n, and thus f fails to dominate R.

In a similar way one can show that every subdivision of B is unbounded; any subdivision of B will be called a *bundle graph*.

Our last example is similar to B. Let F be the graph obtained from a ray $V = v_0 v_1 \ldots$ by adding disjoint rays $P_2, P_4, P_6, \ldots$ with $P_k \cap V = \{v_k\}$, and joining v_{2n+1} to all the new vertices of P_{2n+2} for every $n \in \omega$ (Fig. 2).

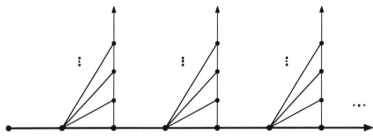

FIGURE 2. The prototype fan graph F

F is again unbounded, the proof being essentially the same as for B. Indeed, given any injective labelling and any function $f\colon \omega \to \omega$, we can easily find a ray R through F whose labels exceed the corresponding values of f again and again. All we have to ensure when defining R is that we start at v_0, and never use an edge of one of the paths P_k in its 'upward' direction. (This would force us to trace out the entire tail of P_k, leaving us unable to return to a vertex of type v_{2n+1} with an infinite choice ahead.)

In a similar way one can show that every subdivision of F is unbounded; any subdivision of F will be called a *fan graph*.

It is perhaps remarkable that F can be made bounded by what would seem to be an inessential change. If the 'fans' in F are flipped horizontally, i.e. if the vertices v_{2n+1} are joined to all the vertices of P_{2n} rather than to those of P_{2n+2} (add a ray P_0), the resulting graph is bounded. This example is due to Halin [6]; the reader may find it amusing to prove its boundedness without using Theorem 2.1 below.

2. The classification theorems

In this section we state the results from [3] and [5] which characterize the bounded and the dominating graphs by forbidden configurations, and give some indications of how these theorems are proved. The full proof of the bounded graph theorem is too complex to be sketched in detail, but we shall give an outline of the main ideas involved. For the dominating graph theorem, we shall be able to give a fairly accurate sketch of what may be the most typical case.

Let us begin with the bounded graph theorem, which was proved in [3]. The result was conjectured by Halin almost 30 years ago, but only few partial results used to be known. Recall that T_ω, B and F denote the ω-regular tree, the prototype bundle graph and the prototype fan graph, respectively, and let $I_\mathfrak{b}$ denote the union of $\mathfrak{b}$ disjoint rays.

Theorem 2.1. (Bounded graph theorem)
A graph is bounded if and only if it contains none of the following graphs as a topological minor: $I_\mathfrak{b}$; T_ω; B; F.

The bounded graph theorem has some fundamental implications for the concept of boundedness for graphs. Let us mention two of these. First, the theorem implies that the boundedness or unboundedness of a graph depends only on its countable subgraphs: unless the graph contains $\mathfrak{b}$ disjoint rays—in which case it is trivially unbounded—it is bounded if and only if all its countable subgraphs are bounded.

Secondly, the translation of the original problem of bounding $\omega \to \omega$ functions to a problem on graphs has been successful, in that the unboundedness of a graph is shown to be a truly structural graph property, not one of the existence of particularly intricate labellings: if a (countable) graph is unbounded then, by the bounded graph theorem, this is witnessed by any injective labelling.

As for the proof of the bounded graph theorem, we have already seen the forward implication: the graphs $I_\mathfrak{b}$, T_ω, B and F and their subdivisions are unbounded, and so they cannot be subgraphs of a bounded graph. For the converse implication, let us first get some intuition for why these four types of subgraph might be natural ones to occur inside an arbitrary unbounded graph G.

If G has $\mathfrak{b}$ or more components each containing a ray, we have $G \supset I_\mathfrak{b}$. If not, it suffices to show that every such component is bounded: by definition of $\mathfrak{b}$, there will then be a bounding function for all of G.

We may thus assume that G is connected. As we saw earlier, locally finite connected graphs are bounded, so G will have vertices of infinite degree. Let v be such a vertex, and assume that G contains infinitely many rays starting at v. (This is not an enormous assumption, given that G contains enough rays to make it unbounded.) Could it be that these rays can be chosen independent, i.e. so that any two of them meet only in v? If so, their union might be viewed as the beginning of a subdivided T_ω: unless v is an atypical vertex of G (in which case the unboundedness of G would not depend on it), there should be another vertex of infinite degree on one of the rays and a similar set of independent rays issuing from it. If this process continues for long enough, it is not unreasonable to assume that the union of all the rays involved contains a subdivision of a graph in which *every* vertex has infinite degree—a graph which is easily shown to contain a copy of T_ω.

Suppose now that there is no infinite set of independent rays starting at v. Then, by Ramsey's theorem, there is an infinite set of rays from v such that every two of them meet also in some other vertex. Now if v is the only vertex that lies on infinitely many of these rays, then it is easy to find a 'fan' in their union. Indeed, let R be one of the rays, and notice that for every finite set of vertices we may find another ray from v avoiding this set. We may thus inductively select an infinite sequence of initial segments of further rays from v, which avoid R except for their two endvertices, and which meet pairwise only in v.

We may therefore assume that there exists a vertex w which lies on infinitely many rays from v. We thus have an infinite set of v–w paths in G whose second vertices are distinct. Using König's infinity lemma, it is not difficult to select an infinite subset of these paths forming either a 'bundle' or a (backward) 'fan' (Fig. 3).

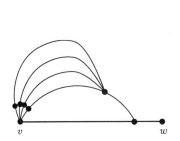 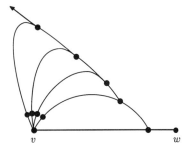

FIGURE 3. Infinitely many v–w paths forming a bundle or a fan

The main purpose of the above considerations was to see that fans and bundles, as well as disjoint or infinitely branching rays, are indeed natural ingredients of unbounded graphs. It must be said, however, that these considerations have not taken us anywhere near a proof of the bounded graph theorem. Indeed, one might well find various fans and bundles in an unbounded graph: the problem is that, in general, they will be far from disjoint. The difficulty in proving the theorem lies in the task of finding some structure in the graph that enables us to construct an infinite sequence of disjoint bundles or fans, as tidily threaded on a ray as in the graphs B and F.

To get to the heart of the proof, let us recall the proofs of the unboundedness of bundle graphs, fan graphs and subdivisions of T_ω, and see what these proofs have in common.

In the case of a subdivision T of T_ω, we just took any injective labelling of T and observed that, for any given function $f: \omega \to \omega$, we could easily find a ray R through T which was not dominated by f. Indeed, no matter how we had chosen an initial segment of R, we would be able to get to another branch vertex of T, where we would have an infinite choice of labels for the next vertex of R. This label could thus be chosen larger than the corresponding term in f.

For a bundle graph or a fan graph, finding such a ray R was hardly more difficult. All we had to make sure of was that the paths P we considered for initial segments of R belonged to a certain family $\mathcal{P}$: a family of paths which could, again and again, be extended to reach unused vertices of infinite degree. In the bundle graph B, $\mathcal{P}$ consisted of the paths from left to right; in the fan graph F, of those towards the right and down the spines of the fans.

The idea of finding such a family of paths in an unbounded graph G is central to the proof of the bounded graph theorem, so let us give it a precise form. Let us call a family $\mathcal{P}$ of finite paths in G a *good family* if

for every path $P \in \mathcal{P}$ there exists an $n \in \omega$ such that P has infinitely many extensions in $\mathcal{P}$ of length n. $\quad(*)$

Note that the existence of a good family of paths in G implies that G is unbounded. Indeed, if in condition $(*)$ we choose n minimal, then there exist infinitely many extensions $P' \in \mathcal{P}$ of P that agree up to their nth ($=$ penultimate) vertex and differ pairwise at their last vertex. Thus if P is viewed as the beginning of a ray R being constructed to elude a given function f (as in our discussion of B and F), then the $(n+1)$th vertex of this ray can be selected from an infinite choice, and therefore in such a way that its label exceeds the corresponding term in f. In this manner we may construct R inductively as the limit of a nested sequence of paths from $\mathcal{P}$.

Moreover, if G contains a bundle graph, a fan graph or a subdivision of T_ω, then clearly G contains a good family of paths. Thus *if* the bounded graph theorem is true, then any graph G not containing $I_\mathfrak{b}$ satisfies the following implications (which in turn imply the bounded graph theorem):

G is unbounded $\quad\Rightarrow\quad$ G contains a good family of paths $\quad\Rightarrow\quad$ G contains T_ω, B or F as a topological minor

It is therefore reasonable to expect that the proof of the bounded graph theorem could be carved up into two chunks, verifying these two implications separately. This is indeed the basic structure of the proof as given in [3]. In practice it turns out to be convenient for the proof of the second implication to strengthen the definition of a good family considerably, but for simplicity we shall here work with the definition given above. The only visible effect of this simplification will be that the first implication will appear to be easier to prove for graphs of cardinality $< \mathfrak{b}$ than for arbitrary graphs. This is not actually the case, i.e. the ideas outlined below for the 'case' of $|G| \geq \mathfrak{b}$ would in fact be essential ingredients even of a proof for countable graphs only.

In order to prove the first implication, let us try to find a good family of paths in G by induction: starting with the family $\mathcal{P}_0$ of *all* finite paths in G, let us recursively discard any paths P from this family that violate the defining condition $(*)$ for a good family, and hope that eventually we will be left with a family that is indeed good. More precisely, let us define subfamilies $\mathcal{P}_\alpha$ of $\mathcal{P}_0$, for all ordinals $\alpha > 0$, as follows.

Suppose first that α is a successor, say $\alpha = \beta + 1$. If $\mathcal{P}_\beta$ contains a path $P =: P_\beta$ which violates $(*)$ (i.e., a path P such that, for each $n \in \omega$, $\mathcal{P}_\beta$ contains only finitely many extensions of P of length n), let $\mathcal{P}_\alpha$ be obtained from $\mathcal{P}_\beta$ by deleting P_β and all its extensions $P' \in \mathcal{P}_\beta$. If $\mathcal{P}_\beta$ contains no such path P, let $\mathcal{P}_\alpha := \mathcal{P}_\beta$. If α is a limit, let $\mathcal{P}_\alpha := \bigcap_{\beta < \alpha} \mathcal{P}_\beta$.

Clearly, there exists an α of cardinality at most $|\mathcal{P}_0|$ such that $\mathcal{P}_{\alpha+1} = \mathcal{P}_\alpha$; let α^* be the least such α, and set $\mathcal{P}_{\alpha^*} =: \mathcal{P}$. Clearly $\mathcal{P}$ satisfies condition $(*)$, and is therefore a good family if and only if it is non-empty. To complete the proof of the first implication, we thus have to show that $\mathcal{P} \neq \emptyset$.

We now prove that $\mathcal{P} \neq \emptyset$, *under the additional assumption that* $\alpha^* < \mathfrak{b}$. The problem of how $\mathcal{P}_0$, and hence α^*, might be reduced to cardinality $< \mathfrak{b}$ without the risk of ending up with $\mathcal{P} = \emptyset$, will be addressed afterwards.

Suppose $\mathcal{P} = \emptyset$; we show that G is bounded, contrary to our assumption. Let $\ell \colon V(G) \to \omega$ be an arbitrary labelling of G. We shall define functions $f_\alpha \colon \omega \to \omega$, one for each $\alpha < \alpha^*$, so that any function f^* that dominates every f_α will bound G; such a function f^* exists by our assumption that $\alpha^* < \mathfrak{b}$.

For each $\alpha < \alpha^*$, let the kth term of f_α be defined as

$$f_\alpha(k) := \begin{cases} \ell(v_k) & \text{if } k \leq |P_\alpha| \\ \max\{\ell(t(P)) \mid P \in \mathcal{P}_\alpha \setminus \mathcal{P}_{\alpha+1} \text{ and } |P| = k\} & \text{if } k > |P_\alpha|, \end{cases}$$

where v_k is the kth vertex of P_α and $t(P)$ is the last vertex of P. Recall that, by definition, P_α has only finitely many extensions $P \in \mathcal{P}_\alpha$ of any given length; since $\mathcal{P}_\alpha \setminus \mathcal{P}_{\alpha+1}$ is precisely the set of P_α and all its extensions in $\mathcal{P}_\alpha$, the maximum used above is therefore just the maximum of a finite set.

To show that f^* bounds G, let $R \subset G$ be any ray. Since every initial segment of R is in $\mathcal{P}_0$, but $\mathcal{P} = \emptyset$ by assumption, each initial segment of R is discarded at some (non-limit) step in the recursive definition of $\mathcal{P}$; let $\alpha < \alpha^*$ be minimal such that $\mathcal{P}_\alpha \setminus \mathcal{P}_{\alpha+1}$ contains an initial segment of R. Then P_α is itself an initial segment of R. Moreover, all the extensions of P_α in R are also in $\mathcal{P}_\alpha$ (by the minimality of α), and were hence discarded together with P_α. Thus, $\mathcal{P}_\alpha \setminus \mathcal{P}_{\alpha+1}$ contains every initial segment of R of length $k \geq |P_\alpha|$. Therefore f_α dominates R, by definition of f_α, and so f^* dominates R as claimed.

To complete the proof of our 'first implication' (i.e. that any unbounded $G \not\supset I_\mathfrak{b}$ contains a good family of paths), let us now see how we can replace $\mathcal{P}_0$ in the above recursion with a smaller starting family. This family should still be large enough to contain a good family (provided that G is unbounded), but small enough to allow the recursion to be completed in fewer than $\mathfrak{b}$ steps.

To achieve this, we make use of a structure theorem for connected graphs not containing a topological K_ω minor; note that we may assume this for G, as otherwise G contains a subdivision of T_ω (and, in particular, a good family of paths). Let $T \subset G$ be a tree, with root r say, and let us call a ray $R \subset T$ *normal* if it starts at r. The tree T will be called a *skeleton* of G if, for each normal ray R, we can assign to every vertex of G a vertex in R, called its R-*height*, so that the following conditions are satisfied:
(i) if t is a vertex on T, then the R-height of t is the vertex of R closest to t in T;
(ii) if v is a vertex of $G \setminus T$ and v is joined to a vertex $t \in T \setminus R$ by a path whose interior avoids T, then the R-height of v equals the R-height of t;
(iii) if a vertex v has R-height x and a vertex w has R-height different from x, then the path rTx separates v from w in G.

Thus, if T is a skeleton of G and $R \subset T$ is a normal ray, and if C is a component of $G \setminus R$ containing the branch B of T, say, then every vertex of C has the same R-height, namely the unique vertex of R to which B is attached in T.

Our structure theorem for graphs without a topological K_ω minor, which builds on work of Jung and Halin and is proved using simplicial decompositions of graphs (see [2]), can now be expressed as follows:

Theorem 2.2. *If G is connected and contains neither $I_\mathfrak{b}$ nor a subdivision of K_ω as a subgraph, then G has a skeleton T of order $< \mathfrak{b}$ such that every ray in G meets some ray in T infinitely often.*

Intuitively, Theorem 2.2 says that G contains a 'small' skeleton, and that the rest of G is 'wrapped around' this skeleton in a bounded sort of way. The reason why this will be useful to us is that, roughly speaking, it allows us to distinguish two cases. Either G is wrapped around T tightly, in which case there is a good family of paths all staying close to the skeleton; since the skeleton itself has order $< \mathfrak{b}$, these paths can be extracted recursively from a family $\mathcal{P}_0$ of order $< \mathfrak{b}$, as desired. Alternatively, the wrapping around T will be bulky in many places of different R-height (for some normal ray R). We may then extract a fan or a bundle at every such place, and combine them into a fan graph or a bundle graph: as the R-heights of different fans or bundles are distinct, they will be pairwise disjoint. (Recall that keeping fans or bundles disjoint is one of the main problems in the entire proof.)

Let us make these ideas more precise. When s and t are comparable vertices of T, i.e. if (say) s lies on the unique r–t path in T, let us say that there is *thick padding* around T at the pair (s,t) if G contains infinitely many (not necessarily independent) s–t paths whose interiors avoid T; otherwise we shall speak of *thin padding* at (s,t). Note that the s–t paths in the padding at (s,t) need not be linked up with each other in $G\backslash T$; in particular, their vertices may have different heights with respect to a normal ray R.

Recall from our introductory discussion that any union of infinitely many s–t paths with distinct second vertices contains a bundle or a fan (Fig. 3). Using similar arguments, it is not difficult to show that if T has thick padding around it at (s,t), then somewhere 'inside this padding' (i.e., using no vertices of T other than s and t), there must be a bundle; note that there cannot be a fan there since, by Theorem 2.2, $G\backslash T$ contains no ray.

We now come to the distinction between the two cases of whether G is 'wrapped around T tightly' or not. Suppose first that, for every normal ray R, the padding around R becomes thin eventually, i.e. there are only finitely many non-overlapping pairs of vertices of R at which there is thick padding. We may then restrict the starting family $\mathcal{P}_0$ for our recursion to paths which leave T only at pairs where the padding is thin; in other words, whenever $P \in \mathcal{P}_0$ is such that sPt is a path of length ≥ 2 and $sPt \cap T = \{s,t\}$, the set of *all* such s–t paths in G is finite. Each path in $\mathcal{P}_0$ is then determined by its sequence of vertices on T and the finite choices of connecting paths between these vertices, so we have $|\mathcal{P}_0| \leq |T| < \mathfrak{b}$ as desired.

We may therefore assume that T contains a normal ray R with thick padding at infinitely many non-overlapping pairs of its vertices. Let us select an infinite sequence of such pairs and extract a bundle from the padding at each of these pairs. Let us call such a bundle *wide* if the set of R-heights of its vertices is finite;

otherwise let it be *tall*. Then there is an infinite subsequence of our pairs such that either each one of the associated bundles is wide, or each of these bundles is tall.

If every bundle is wide, we may select a further subsequence of pairs such that the R-heights of bundles associated with different pairs are disjoint; with a bit of work, these bundles may then be combined into a bundle graph. Suppose finally that every bundle in the sequence is tall. It is not difficult to show that the endvertices of such a bundle (i.e. its two vertices of infinite degree) must be on R. Thus, R has infinitely many vertices—two for each bundle—every one of which has neighbours of arbitrarily large R-height. Choosing appropriate disjoint connecting paths between these neighbours and their respective R-heights on R, we obtain a graph that is easily seen to contain a subdivision of K_ω. This contradicts our assumptions about G, completing the proof of the 'first implication'.

The proof of the second implication, the fact that if G contains a good family of paths then it contains a bundle graph, a fan graph or a subdivision of T_ω, makes up about two thirds of the proof of the bounded graph theorem. The techniques used are largely similar to those outlined above. It is assumed that G is connected and has no topological K_ω minor, and therefore contains a skeleton. The skeleton may then be used to keep different fans and bundles separated by their R-heights, for suitable normal rays R; thus, if sufficiently many fans or bundles can be found, these may be combined to a bundle graph or a fan graph.

Let us then mention briefly how these fans or bundles are generated. Let $\mathcal{P}$ be a good family of paths in G. The essential structure of $\mathcal{P}$ can be represented by a graph $\mathcal{T}$, which looks very similar to a T_ω, and which acts as something like a covering tree for the paths $P \in \mathcal{P}$ in G. If G resembles $\mathcal{T}$ closely, it can be shown to contain a subdivision of T_ω. On the other hand, if the paths of $\mathcal{P}$ intersect a lot more (in G) than do their lifts in $\mathcal{T}$, they generate sufficiently many bundles or fans to make a bundle graph or a fan graph in G. The distinction between the latter two cases depends on the relationship between $\mathcal{T}$ and the skeleton T of G.

To see how $\mathcal{T}$ is obtained, let us first select a simple subfamily of $\mathcal{P}$, which will still be a good family. We start with any path $P_0 \in \mathcal{P}$. By condition (∗), P_0 has infinitely many extensions in $\mathcal{P}$ of some common length n. Among these, select infinitely many that agree up to their penultimate vertices; this can be done if n is first chosen minimal. In the same way, we then find an infinite set of extensions for each of these new paths, again (in each case) of some common length and agreeing up to their penultimate vertices. In ω steps, we have constructed a subfamily $\mathcal{Q}$ of $\mathcal{P}$ which is still good, and which clearly has a structure similar to a subdivision of T_ω: let $\mathcal{Q}'$ be the closure of $\mathcal{Q}$ under taking initial segments, set $V(\mathcal{T}) = \mathcal{Q}'$, and join vertices $P, P' \in \mathcal{Q}'$ by an edge whenever P' is an extension of P by one vertex. The tree $\mathcal{T}$ obtained in this way is a subdivision of T_ω, except for its 'long root' made up of vertices that are proper initial segments of P_0.

Let $\pi\colon V(\mathcal{T}) \to V(G)$ be the map that assigns to each $P \in V(\mathcal{T})$ its last vertex. Then if p is the first vertex of P_0, and the trivial path $\{p\} \in V(\mathcal{T})$ is taken to be the root of $\mathcal{T}$, we see that π maps each path of the form $\{p\}\mathcal{T}P$ to the path $P \subset G$ in a natural way. Conversely, each path $P = v_0 \ldots v_n \in Q'$ lifts uniquely to the path in T with vertices Pv_i, $i = 0, \ldots, n$.

Now let P be a branch vertex of T, with $\pi(P) = v$ say, and consider the subdivided edges at P pointing away from the root $\{p\}$. If these subdivided edges project under π to paths in G that are disjoint except for their common initial vertex v, then these paths might be used in the construction of a subdivided T_ω in G.

On the other hand, it might be the case that G has a vertex w such that every subdivided edge of T at P contains a vertex Q with $\pi(Q) = w$. In this case, the subdivided edges at P have initial segments which project to an infinite set of v–w paths in G. As we saw earlier, the union of such a set of paths makes up a bundle or a fan.

Using an easy Ramsey type argument, we may now assume that all the branch vertices of T behave in the same way; then either we obtain a subdivision of T_ω in G straight away, or else we obtain a bundle or a fan for each branch vertex of T. Using a number of similar Ramsey type arguments, this infinite set of configurations in G (fans or bundles) can then be streamlined in several ways. Eventually we end up with a set of configurations that are not only either all fans or all bundles, but which are compatible (with respect to their position towards the skeleton T) in many other ways too. (In particular, these fans or bundles will be pairwise disjoint.) In this way we eventually obtain a fan graph or a bundle graph in G, completing the proof of the bounded graph theorem. □

We now turn to the dominating graph theorem, proved in [5]. Its proof is much shorter than that of the bounded graph theorem, and we shall be able to give a fairly complete sketch of the most typical case.

Theorem 2.3. (Dominating graph theorem)
A graph G is dominating if and only if it satisfies one of the following three conditions:

(i) *G contains a uniform subdivision of T_ω;*

(ii) *G contains $\mathfrak{b}$ disjoint subdivisions of T_ω;*

(iii) *G contains $\mathfrak{d}$ disjoint rays.*

We prove the theorem under the set theoretic assumption that $\mathfrak{b} = \mathfrak{d}$. This is a comparatively weak assumption, much weaker, say, than Martin's axiom or even the continuum hypothesis. Note that case (ii) of the theorem is now redundant, since $\mathfrak{d}$ subdivisions of T_ω contain $\mathfrak{d}$ disjoint rays.

We have already seen that uniform subdivisions of T_ω and unions of $\mathfrak{d}$ disjoint rays are dominating; it remains to prove that if G is dominating then G contains one of these two types of subgraph.

The basic idea of the proof is similar to the way in which we obtained a good family of paths for the bounded graph theorem. We recursively define a rank function ρ on some or all of the vertices of G, with the following property. If any vertex remains unranked, i.e. if the recursion ends before ρ is defined on all of $V(G)$, then G contains a uniform subdivision of T_ω; if ρ gets defined for every vertex, then either G contains $\mathfrak{d}$ disjoint rays or G is not dominating.

Set $\rho(v) = 0$ for all vertices v that have finite degree in G. Now let $\alpha > 0$ be given, and suppose that for every $\beta < \alpha$ we have assigned to some vertices the ρ-value β. Set $\rho(v) = \alpha$ for all those vertices v for which
- $\rho(v)$ is still undefined, and
- whenever $\mathcal{P}$ is a set of finite paths from v, pairwise disjoint except for v, all of the same length, and each ending in a vertex which has not been given any ρ-value $\beta < \alpha$, then $\mathcal{P}$ is finite.

If there is no such vertex v, we terminate the recursive definition of ρ, and leave ρ undefined for any remaining vertices of G.

It is not difficult to see that if ρ remains undefined for some vertices $v \in G$, then G contains a uniform subdivision of T_ω. Indeed, the definition of ρ implies that for every such v there exists an infinite set of paths from v, pairwise disjoint except for v, all of the same length, and each ending in a vertex for which ρ is also undefined. It is now easy to use all these paths as subdivided edges to build a uniform subdivision of T_ω, choosing them inductively in ω steps and so that the portion of the tree constructed remains connected at all times: then at each point of the construction only finitely many vertices have been used, but there is always an infinite set of disjoint paths from which the next subdivided edge can be chosen.

Let us therefore assume that $\rho(v)$ is defined for all $v \in V(G)$, and that G contains no union of $\mathfrak{d}$ disjoint rays; we shall show that G is not dominating. Let $\ell: V(G) \to \omega$ be any labelling; we now have to find a function $f: \omega \to \omega$ which is not dominated by any ray in G.

Let a path P from u to w in G be called *upward* if

$$\rho(w) = \max\{\rho(v) : v \in P\}.$$

We first show the following.

(2.3.1) *For each $u \in V(G)$ and each integer m, there are only finitely many vertices w such that G contains an upward path of length m from u to w.*

Suppose the contrary, and consider a vertex u, an integer m, and an infinite set $\{w_n : n \in \omega\}$ such that for each n there is an upward path P_n of length m from u to w_n. Choose $k \leq m$ maximal so that there exist a vertex v and an infinite set $\mathcal{P} \subset \{P_n : n \in \omega\}$ such that v is the kth vertex in every $P \in \mathcal{P}$. (Note that k exists, because every P_n starts at u.) We shall now select an infinite sequence $\{P_{n_i} : i \in \omega\}$ of paths from $\mathcal{P}$ so that any two of these paths are disjoint after v; since each P_n is an upward path, and hence $\rho(v) \leq \rho(w_n)$ for every n, this will contradict the definition of $\rho(v)$.

Let P_{n_0} be any path from $\mathcal{P}$. Now suppose $P_{n_0}, \ldots, P_{n_i}$ have been chosen, and let U be the union of their vertex sets. By the maximality of k, there are at most finitely many paths in $\mathcal{P}$ that contain a vertex from U after v; let $P_{n_{i+1}}$ be any other path from $\mathcal{P}$. It is then clear that the full sequence $\{P_{n_i} : i \in \omega\}$ has the required disjointness property. This completes the proof of (2.3.1).

We now define the function f which will show that G is not dominating. Let U be the vertex set of the union of some maximal set of disjoint rays in G. By assumption there are fewer than $\mathfrak{d}$ rays in this set, so $|U| < \mathfrak{d}$. Note also that every ray in G meets U in infinitely many vertices. Using (2.3.1) we may define, for each $u \in U$, a function $f_u : \omega \to \omega$ such that $f_u(m) > \ell(w)$ for every $m \in \omega$ and every vertex w to which u can be linked by an upward path of length m. By our hypothesis that $\mathfrak{b} = \mathfrak{d} > |U|$, there exists an $\omega \to \omega$ function which dominates each of the functions f_u; let f be such a function. Redefining $f(n)$ as $\max\{f(k) : k \leq n\}$ if necessary, we may assume that f is increasing.

Now let $R = v_0 v_1 \ldots$ be any ray in G; we have to show that $f(k) > \ell(v_k)$ for infinitely many $k \in \omega$. Using the fact that the ordinal sequence $\rho(v_0), \rho(v_1), \ldots$ cannot contain an infinite decreasing sequence, it is not difficult to see that we can find an infinite increasing sequence $\{k_i : i \in \omega\}$ such that $\rho(v_{k_i}) \leq \rho(v_{k_{i+1}})$ for each i, and $\rho(v_j) < \rho(v_{k_i})$ whenever $k_i < j < k_{i+1}$. Now pick $k^* \geq k_0$ so that $u := v_{k^*} \in U$. Note that, for each $k_i > k^*$, the path uRv_{k_i} is an upward path of length $k_i - k^*$.

Since f dominates f_u, there is some $K \in \omega$ such that $f_u(k) \leq f(k)$ for all $k \geq K$. But then

$$\ell(v_{k_i}) < f_u(k_i - k^*) \leq f(k_i - k^*) \leq f(k_i)$$

for all i with $k_i - k^* \geq K$, by definition of $f_u = f_{v_{k^*}}$. Thus R fails to dominate f, as required. □

3. Domination games

As is well known, Adam and Eve used to play the following game. For a given graph G, Adam first chooses a labelling $\ell : V(G) \to \omega$. Then the two players move alternately: Eve, who moves first, plays a natural number, Adam a vertex of G. In this way, Eve creates a function $e : \omega \to \omega$, while Adam creates an ω-sequence of vertices.

In the *bounding game*, Adam tries to escape domination by Eve: he wins if and only if he succeeds in constructing a ray $A \subset G$ that is not dominated by Eve's function e. Thus, playing Adam's part in the bounding game is like an interactive version of trying to prove that G is unbounded. (The difference is that now Adam does not know what Eve will play in future moves; for a proof that G is unbounded, it would be sufficient to be able to construct an undominated ray with respect to any $\omega \to \omega$ function given complete at the start.)

It is clear that if G is bounded, then Eve has a winning strategy: once Adam has chosen his labelling of G, she just plays a bounding function with respect to this labelling, without ever paying attention to what Adam is doing. On the other hand, Eve may still have a winning strategy when G is unbounded. For example, Eve has a winning strategy for the graph $I_\mathfrak{b}$: she plays any number in her first move, waits to see from which component of $I_\mathfrak{b}$ Adam picks his first vertex, and then plays a winning strategy for that component. (Recall that each component of $I_\mathfrak{b}$ is a ray, and is therefore bounded.) Since Adam can only construct a ray if he stays in that component, Eve will win the game.

The following result from [4] shows that the example of I_b does indeed mark the difference between the graphs that are unbounded and those for which Adam has a winning strategy in the bounding game. Its proof in [4] gives an explicit winning strategy for either Adam or Eve, as appropriate.

Theorem 3.1. *Adam has a winning strategy in the bounding game on G if and only if G contains one of the graphs T_ω, B and F as a topological minor.*

The *dominating game* is defined like the bounding game, except that now Adam tries to construct his ray A in such a way that it actually dominates the function e created by Eve (and is not just not dominated by it). Again, it is clear that if Adam has a winning strategy in the dominating game on G then G must be a dominating graph. However, unlike the similarity between the bounding game and the bounded graph theorem, it turns out to be much harder for Adam to win the dominating game on G than it is to prove that G is a dominating graph: the following result from [4] implies that the dominating game can be won by Eve on 'most' subdivisions of T_ω, including uniform ones, as well as on disjoint unions of these (cf. Theorem 2.3).

Theorem 3.2. *Adam has a winning strategy on a graph G if and only if $G \supset T_\omega$. Otherwise Eve has a winning strategy.*

Proof. It is clear that Adam has a winning strategy if $G \supset T_\omega$: he chooses a labelling that is injective on this T_ω, and is then able to beat Eve in every move. We shall assume that $G \not\supset T_\omega$, and show that Eve has a winning strategy.

As in the proof of the dominating graph theorem, we start by recursively defining a rank function ρ on some or all of the vertices of G. For each ordinal α, give rank $\alpha =: \rho(v)$ to all vertices v such that all but finitely many neighbours of v have rank $< \alpha$. If any vertex remains unranked, then each of these vertices has infinitely many unranked neighbours, and we may construct a $T_\omega \subset G$ from these vertices by induction (in ω steps, as in the proof of Theorem 2.3).

Thus, since $G \not\supset T_\omega$ by assumption, ρ gets defined for every vertex of G. We may now choose a winning strategy for Eve as follows. Let ℓ be the labelling of G chosen by Adam at the start of the game, and let Eve's first move be arbitrary. Later, if Adam's last chosen vertex is v, let Eve play the number

$$1 + \max \{ \ell(w) \mid w \text{ is a neighbour of } v \text{ and } \rho(w) \geq \rho(v) \};$$

by definition of $\rho(v)$, the maximum above is just one over a finite set.

Now consider a run of the game in which Eve plays the above strategy. If Adam fails to construct a ray, then Eve wins by definition. So assume that Adam does indeed construct a ray $A \subset G$. Since there is no infinite descending sequence of ordinals, A has infinitely many vertices whose rank is at most that of its successor on A. But Eve beats Adam on all these successors, so A fails to dominate her sequence e. Thus, Eve's strategy is indeed a winning strategy. □

References

[1] B. Bollobás, *Extremal Graph Theory*, Academic Press, London 1978.

[2] R. Diestel, *Graph Decompositions—a study in infinite graph theory*, Oxford University Press, Oxford 1990.

[3] R. Diestel and I. Leader, A proof of the bounded graph conjecture, *Inventiones Math.* (to appear).

[4] R. Diestel and I. Leader, Domination games on infinite graphs, in preparation.

[5] R. Diestel, J. Steprans and S. Shelah, Characterizing dominating graphs, submitted.

[6] R. Halin, Bounded graphs, in: (R. Diestel, Ed.) *Directions in infinite graph theory and combinatorics* (Annals of Discrete Mathematics), to appear.

FACULTY OF MATHEMATICS (SFB 343), BIELEFELD UNIVERSITY,
P.O. BOX 8640, D-4800 BIELEFELD, GERMANY

E-mail: diestel@ibm9370.hrz.uni-bielefeld.de

Notes on rays and automorphisms of locally finite graphs

H.A. Jung

ABSTRACT. A well-known equivalence relation on the set of rays in a graph X leads to the notion of an end of X. Several alternate equivalence relations on the set of rays of X are discussed and related to some basic fixed-point results for infinite graphs.

In this note we discuss several equivalence relations on the set $\mathcal{R}(X)$ of rays (= 1-way infinite paths) in a given connected graph X.

Two rays R,R' in X are called *end-equivalent* if there is an infinity of pairwise disjoint paths each connecting a vertex on R to a vertex of R'. It is easy to see that end-equivalence is an equivalence relation on $\mathcal{R}(X)$. A corresponding equivalence class is called an *end* of X. The concept of end is a natural and useful concept. Via Cayley graphs it is linked to the notion of an end in a finitely generated (infinite) group.

By now, there are quite a few "fixed-point" results in the literature involving ends (see [1] - [8]). Most of them degenerate to trivialities in the case of graphs with only one end. By studying appropriate finer equivalence relations on $\mathcal{R}(X)$ one obtains meaningful results for arbitrary locally finite graphs X.

1991 Mathematics Subject Classification. Primary 05C25, 20F32
This paper is a final version and will not be published elsewhere.

An element σ of the automorphism group AutX of the connected graph X is called a *translation* of X if $\sigma(F) \neq F$ for every finite non-empty $F \subseteq V(X)$. The following result was proved by R. Halin in [1] (Theorem 7).

Proposition 1. *Let τ be a translation of the connected graph X. There is a unique end $\mathfrak{D}_\tau$ of X such that $\tau^m(R) \subset R$ for some $m > 0$ and some $R \in \mathfrak{D}_\tau$.*

For a ray $R = v_1 v_2 ...$ we denote the ray $v_n v_{n+1} ...$ by $R[v_n, \infty)$ and $R(v_{n-1}, \infty)$.

We call a sequence $v_1, v_2, ...$ of vertices on a ray R *bounded on* R if there exists a real M such that v_{n+1} is on $R(v_n, \infty)$ and $d_R(v_n, v_{n+1}) \leq M$, for all positive integers n. Two rays R, R' in X are called b-*equivalent* in X if there exist a real M, a bounded sequence $v_1, v_2, ...$ on R and a bounded sequence $w_1, w_2 ...$ on R' such that $d(v_n, w_n) \leq M$ for $n = 1, 2, ...$; here $d(v,w)$ denotes the minimum length of paths from v to w in X. We will show that b-equivalence defines an equivalence relation on $\mathfrak{R}(X)$ and call the corresponding equivalence classes b-*fibers*.

Lemma 2. *Let $x_1, x_2, ...$ and $y_1, y_2, ...$ be bounded sequences on a ray R. There exist a real M, a sequence $m(1) < m(2) < ...$ and a sequence $n(1) < n(2) < ...$ of natural numbers such that $d_R(x_{m(i)}, y_{n(i)}) \leq M$, $m(i+1) - m(i) \leq M$ and $n(i+1) - n(i) \leq M$, $(i = 1, 2...)$.*

Proof. By definition, there exists a real M such that $d_R(x_n, x_{n+1}) \leq M$ and $d_R(y_n, y_{n+1}) \leq M$ for $n = 1, 2, ...$. Set $m(0) = n(0) = 1$ and assume that $m(i)$ and $n(i)$ are constructed for some non-negative integer i. Let $m(i+1)$ be the minimum integer m such that x_m is on $R(y_{n(i)}, \infty)$, and let $n(i+1)$ be the minimum integer n such that y_n is on $R[x_{m(i+1)}, \infty)$.

By construction, $d_R(x_{m(i)}, x_{m(i+1)}) \leq 2M$ and $d_R(y_{n(i)}, y_{n(i+1)}) \leq 2M$ for $i = 1, 2,...$. □

To show that b-equivalence is an equivalence relation on $\mathfrak{R}(X)$ consider two pairs R,.R' and R', R" of b-equivalent rays in X. There exist a real M, sequences $v_1, v_2, ...; x_1, x_2, ... ; y_1, y_2, ...$ and $w_1, w_2, ...$, bounded on respectively R, R', R' and R" such that $d(v_n, x_n) \leq M$ and $d(y_n, w_n) \leq M$ for $n = 1, 2, ...$. We may assume that M is also a common upper bound for all $d_R(v_n, v_{n+1})$, $d_{R'}(x_n, x_{n+1})$, $d_{R'}(y_n, y_{n+1})$ and $d_{R''}(w_n, w_{n+1})$. By Lemma 2 there exist integer sequences $0 < m(1) < m(2) < ...$ and $0 < n(1) < n(2) < ...$ such that $m(i+1) - m(i)$, $n(i+1) - n(i)$ and $d_{R'}(x_{m(i)}, y_{n(i)})$ are bounded by some real M'. It follows that $d(v_{m(i)}, w_{n(i)}) \leq 2M + M'$, $d_R(v_{m(i+1)}, v_{m(i)}) \leq MM'$ and $d_{R''}(w_{n(i+1)}, w_{n(i)}) \leq MM'$. Hence indeed R and R" are b-equivalent. □

Clearly each end is the union of disjoint b-fibers.

For convenience and in view of Theorem 3 below we include a proof of Proposition 1.

Proof of Proposition 1. Pick a vertex x and a path P_0 in G from x to $\tau(x)$. Let P be a minimal subpath of P_0 (with respect to the subpath relation) such that $\tau^m(P) \cap P \neq \emptyset$ for some $m > 0$, say $\tau^m(x) \in V(P)$ for the vertex x on P. Then P has the form $P = P[x, \tau^m(x)]$. If $\tau^i(P) \cap \tau^j(P) \neq \emptyset$ and $i < j$, then $\tau^{j-i}(P) \cap P \neq \emptyset$ and $x \neq \tau^{j-i}(x)$ yield $P = P[x, \tau^{j-i}(x)]$ and consequently $j = m + i$. Therefore for each integer j the graph $R_j = \bigcup (\tau^{im+j}(P): i \geq 0)$ is a ray in X with $\sigma^m(R_j) \subseteq R_j$. Note that moreover $D_j = \bigcup (\tau^{im+j}(P): i \in \mathbb{Z})$ is a double-ray (2-way infinite path), D_0, $D_1, ..., D_{m-1}$ are pairwise disjoint, $\sigma(D_j) = D_{j+1}$ ($j = 0,...$) and $D_m = D_0$.

For the uniqueness part assume that R, R' are rays in X and m, n are natural numbers such that $\tau^m(R) \subset R$ and $\tau^n(R') \subset R'$. Pick $x_0 \in V(R)$, $y_0 \in V(R')$ and abbreviate $x_i = \tau^{mni}(x_0)$, $y_i = \tau^{mni}(y_0)$. Clearly $x_0, x_1, ...$ and $y_0, y_1, ...$ are bounded on R and R', respectively. Since $d(x_i, y_i) = d(x_0, y_0)$ it follows that the rays R, R' are b-equivalent. □

While $\mathcal{B}_\tau$ and $\mathcal{B}_{\tau^{-1}}$ may coincide the following holds.

Theorem 3. *Let τ be a translation of the locally finite connected graph X. There is a unique b-fiber $\mathcal{B}_\tau$ in X such that $\tau^m(R) \subset R$ for some $R \in \mathcal{B}_\tau$ and some $m > 0$. Moreover $\mathcal{B}_\tau \neq \mathcal{B}_{\tau^{-1}}$.*

Proof. By the preceding proof it remains to show $\mathcal{B}_\tau \neq \mathcal{B}_{\tau^{-1}}$. There exist rays $R \in \mathcal{B}_\tau$, $R' \in \mathcal{B}_{\tau^{-1}}$ and positive integers p,q such that $\tau^p(R) \subset R$ and $\tau^{-q}(R') \subset R'$. Assuming $\mathcal{B}_\tau = \mathcal{B}_{\tau^{-1}}$ we can find a real M, a bounded sequence $x_0, x_1, ...$ on R and a bounded sequence $y_0, y_1, ...$ on R' such that $d(x_n, y_n) \leq M$ ($n = 1, 2, ...$). Abbreviate $\tau^{pq} = \sigma$, $u_n = \sigma^n(x_0)$ and $w_n = \sigma^{-n}(y_0)$. Employing Lemma 2 one can find a real M', a bounded sequence $u_{m(1)}, u_{m(2)}, \ldots$ on R and a bounded sequence $w_{n(1)}, w_{n(2)}, \ldots$ on R' such that $d(u_{m(i)}, w_{n(i)}) \leq M'$ for $i = 1, 2, \ldots$. Now $d(w_{n(i)}, u_{m(i)}) = d(y_0, \sigma^{n(i)+m(i)}(x_0))$. Since $\sigma^{n(i)+m(i)}(x_0)$ ($i = 1, 2, \ldots$) are distinct, one contradicts the fact that X is locally finite. □

We say that the rays $R = v_0, v_1, \ldots$ and $R' = w_0, w_1, \ldots$ in X are 1-*equivalent* in X if there exists an upper bound for $d(v_n, w_n)$ ($n = 0, 1, \ldots$). Obviously 1-equivalence is another equivalence relation on $\mathcal{R}(X)$. The corresponding equivalence classes are called 1-*fibers* of X. A 1-fiber $\mathcal{F}$ is *essential* for the automorphism σ of X if $\sigma^m(R) \subset R$ for some $R \in \mathcal{F}$ and some $m > 0$.

Clearly each b-fiber of X decomposes into disjoint 1-fibers and any $\sigma \in \text{Aut} X$ permutes the set of 1-fibers of X. If the 1-fiber $\mathcal{F}$ is essential for $\sigma \in \text{Aut} X$ and X is locally finite and connected, then σ is a translation since some $\langle\sigma\rangle$-orbit and hence all $\langle\sigma\rangle$-orbits on $V(X)$ are infinite; furthermore $\mathcal{F} \subseteq \mathcal{B}_\sigma$.

Theorem 4. *Let τ be a translation of a locally finite connected graph X. Then each 1-fiber $\mathcal{F} \subseteq \mathcal{B}_\tau$ is fixed by τ, and the set of essential 1-fibers for τ is countable.*

Proof. Pick $R \in \mathcal{B}_\tau$ and $p > 0$ such that $\tau^p(R) \subset R$. Let x_0 be the initial vertex of R.

Consider an element R' of $\mathcal{B}_\tau$. By Lemma 2 there exist a bounded sequence $\tau^{n(1)p}(x_0), \tau^{n(2)p}(x_0), \ldots$ on R and a bounded sequence $y_1, y_2, \ldots$ on R' such that $d(\tau^{n(i)p}(x_0), y_i)$ is bounded. Abbreviating $x_i = \tau^{n(i)p}(x_0)$ one obtains $d(y_i, \tau(y_i)) \leq d(y_i, x_i) + d(x_i, \tau(x_i)) + d(\tau(x_i), \tau(y_i)) \leq 2d(y_i, x_i) + d(x_0, \tau(x_0))$ for $i = 1, 2, \ldots$.

Each y on $R'[y_1, \infty)$ is on some $R'[y_i, y_{i+1}]$, and thus $d(y, \tau(y)) \leq d(y, y_i) + d(y_i, \tau(y_i)) + d(\tau(y_i), \tau(y)) \leq 2d(y_i, y_{i+1}) + d(y_i, \tau(y_i))$. Therefore $d(y, \tau(y))$ is bounded for $y \in V(R')$. Hence R' and $\tau(R')$ are in the same 1-fiber $\mathcal{F}$ of X, that is $\tau(\mathcal{F}) = \mathcal{F}$.

Now assume in addition that $\tau^q(R') \subset R'$ and let y_0 be the initial vertex of R'. As $\mathcal{B}_\tau \neq \mathcal{B}_{\tau^{-1}}$ by Theorem 3, necessarily $q > 0$. Abbreviate $\sigma = \tau^{pq}$, $a = |R[x_0, \sigma(x_0)]| - 1$ and $b = |R'[y_0, \sigma(y_0)]| - 1$. We have $d_R(x_0, \sigma^{bi}(x_0)) = abi = d_{R'}(y_0, \sigma^{ai}(y_0))$ and $d(\sigma^{bi}(x_0), \sigma^{ai}(y_0)) = d(y_0, \sigma^{(b-a)i}(x_0))$. If $a = b$, then clearly R and R' are in the same 1-fiber. If $a \neq b$, then, as X is locally finite, the set $\{\sigma^{(b-a)i}(x_0) : i = 0, 1, ...\}$ is infinite and consequently $d(\sigma^{bi}(x_0), \sigma^{ai}(y_0))$ is not bounded. This shows that R, R' are in the same 1-fiber if and only if for $x \in V(R)$ and $y \in V(R')$ arbitrary, $(|R[x, \tau^p(x)]| - 1)/p = (|R'[y, \tau^q]| - 1)/q$.

The rationals $(|R[x, \tau^p(x)]| - 1)/p$ form a "system of invariants" for the essential 1-fibers for τ. □

By a straightforward argument it can be shown that the set of essential 1-fibers for τ as in Theorem 4 is infinite unless X is a double ray.

We call a connected subgraph Y of a graph X a *metric subgraph* of X if there is a real M such that $d_Y(y, y') \leq M d_X(y, y')$ for all $y, y' \in V(Y)$ (that is if $d_Y(.,.)$ and $d_X(.,.)$ are equivalent metrics on $V(Y)$).

Lemma 5. *Let R, R' be b-equivalent rays in X. If R is a metric subgraph of X, then so is R'.*

Proof. There are bounded sequences $x_1, x_2, ...$ on R and $y_1, y_2, ...$ on R' such that $d(x_n, y_n)$ is bounded. Pick a real M such that $d_R(x_n, x_{n+1})$, $d_{R'}(y_n, y_{n+1})$ and $d(x_n, y_n)$ are bounded by M from above, and moreover $d_R(x, x') \leq Md(x, x')$ for all $x, x' \in V(R)$. Without loss of generality assume $R = R[x_1, \infty)$ and $R' = R'[y_1, \infty)$.

Consider distinct elements y, y' of $V(R')$. There exist y_i and y_j such that $d_{R'}(y, y_i) \leq M$ and $d_{R'}(y', y_j) \leq M$. Then $d(x_i, x_j) \leq 4M + d(y, y')$ and

$$d_{R'}(y, y') \leq 2M + d_{R'}(y_i, y_j) \leq 2M + |i - j| M$$
$$\leq 2M + Md_R(x_i, x_j) \leq 2M + M^2 d(x_i, x_j)$$
$$\leq 2M + 4M^3 + M^2 d(y, y')$$
$$\leq (2M + 4M^3 + M^2) d(y, y').$$
□

Theorem 6. *Let τ be a translation of the connected graph X, and suppose τ has more than one fixed end. Then $\mathcal{D}_\tau \neq \mathcal{D}_{\tau^{-1}}$ and each $R \in \mathcal{B}_\tau$ is a metric subgraph of X.*

The claim $\mathcal{D}_\tau \neq \mathcal{D}_{\tau^{-1}}$ was proved by R. Halin [1]. In fact the first part of the following proof is essentially Halin's proof for the fact that τ has at most two fixed ends.

Proof. Let l, l' be different fixed ends of τ. There exist a finite set $F \subseteq V(X)$ and distinct components C and C' of $X - F$ such that C and C' contain respectively rays $R_0 \in l$ and $R'_0 \in l'$. W. l. o. g. assume that F induces a connected subgraph of X. Since τ is a translation there exists some $n > 0$ such that $\tau^i(F) \cap F = \emptyset$ for all $i \geq n$. Therefore $\tau^n(F) \subseteq C_1$ and $\tau^{-n}(F) \subseteq C_2$ for certain components C_1 and C_2 of $X - F$.

If $C_1 \neq C$, then $C \cup F$ induces a connected subgraph of $X - \tau^n(F)$. From $F \subseteq \tau^n(C_2)$ one deduces $C \cup F \subseteq \tau^n(C_2)$, and equivalently $\tau^{-n}(C \cup F) \subseteq C_2$. But $\tau^{-n}(l) = l$ and so $\tau^{-n}(C \cup F)$ contains all but finitely many vertices of R_0. We infer $\tau^{-n}(C \cup F) \subseteq C$. In this case $C_2 = C$ holds, and a symmetric argument yields $\tau^n(C' \cup F) \subseteq C'$. If $C_1 = C$, then $C_1 \neq C'$ and we similarly obtain $\tau^{-n}(C' \cup F) \subseteq C'$ and $\tau^n(C \cup F) \subseteq C$. W. l. o. g. assume the latter case.

Since there is a path $P_0 = P_0[x, \tau^n(x)]$ in C there exist also a path $P \subseteq P_0$ and $m > 0$ such that $R = \bigcup (\tau^{mni}(P) : i \geq 0)$ is a ray (see proof of Proposition 1).

Notice that $R \subseteq C$ and $R \in \mathcal{B}_\tau \subseteq \mathcal{D}_\tau$. Also C' contains a ray R^- such that $\tau^{-kn}(R^-) \subseteq R^-$ for some $k > 0$. In particular $\mathcal{D}_\tau \neq \mathcal{D}_{\tau^{-1}}$, and no finite set can separate l from both $\mathcal{D}_\tau$ and $\mathcal{D}_{\tau^{-1}}$. This shows that τ fixes no end other than $\mathcal{D}_\tau$ and $\mathcal{D}_{\tau^{-1}}$.

Let $R = R[y, \infty)$ and $P_1 = R[y, \tau^{mn}(y)]$. Since $\cap (\tau^{mni}(C) : i \geq 0) = \emptyset$, one can find some $p > 0$ such that $y \notin \tau^{mnp}(C)$. Abbreviate $\tau^{nmp} = \sigma$, $P_2 = R[y, \sigma(y)]$ and $M' = |P_2| - 1$. Any path in X which joins y to $\sigma^i(y)$ $(i > 0)$ intersects F, $\sigma(F)$, ..., $\sigma^i(F)$ and therefore $d(y, \sigma^i(y)) \geq i$.

Now consider distinct vertices v and w on R, with say w on $R(v, \infty)$. As $R = \bigcup (\sigma^i(P_2) : i \geq 0)$ there exist integers j, i such that $j \geq i \geq 0$, $d_R(v, \sigma^i(y)) \leq M$ and $d_R(w, \sigma^j(y)) \leq M$. The estimates

$d_R(v, w) \leq 2M + d_R(\sigma^i(y), \sigma^j(y)) = 2M + M'(j - i)$

$\qquad \leq 2M + M'd(\sigma^j(y), \sigma^i(y))$

$\qquad \leq 2M + M'(d(v, w) + 2M)$

$\qquad \leq (2M + M' + 2MM')d(v, w)$

yield that R is a metric subgraph of X. Hence the claim by Lemma 5. □

Corollary 7. *Let σ, τ be translations of a connected graph X such that $\mathcal{D}_\sigma \neq \mathcal{D}_{\sigma^{-1}}$ and $\mathcal{D}_\tau = \mathcal{D}_\sigma$. Then $\mathcal{D}_\tau \neq \mathcal{D}_{\tau^{-1}}$ and $\mathcal{B}_\sigma = \mathcal{B}_\tau$.*

Proof. As shown in the preceding proof there exist a finite set F, a ray R^+ and distinct components C, C' of X − F such that $\sigma^m(C \cup F) \subseteq C$, $\sigma^{-m}(C' \cup F) \subseteq C'$ and $\sigma^m(R^+) \subset R^+ \subseteq C$ for some $m > 0$.

Pick $R \in \mathcal{B}_\tau$ and $n > 0$ such that $\tau^n(R) \subset R$. By Lemma 5 one can determine $M > 0$ such that $d_{R^+}(x, x') \leq M d(x, x')$ and $d_R(y, y') \leq M d(y, y')$ for all $x, x' \in V(R^+)$ and y, y' on $V(R)$. Let $R = R[x, \infty)$, $P = R[x, \tau^n(x)]$ and $R^- = \bigcup (\tau^{ni}(P): i < 0)$.

First assume that for each $j > 0$ the ray R^- intersects $\sigma^{jm}(C \cup F)$. Then R^- contains infinitely many elements of $\sigma^{jm}(C)$. Pick $j_0 > 0$ such that the double ray $D = R^- \cup R = \bigcup (\tau^{mi}(P): i \in \mathbb{Z})$ is not contained in $\sigma^{j_0 m}(C)$. For each $j \geq j_0$ determine the first and last vertex x_j and y_j respectively of D in $\sigma^{jm}(F)$. Note that $D[x_j, y_j] \subseteq D(x_{j+1}, y_{j+1})$ and therefore $d_D(x_j, y_j) \geq 2(j - j_0)$ On the other hand $d(x_j, y_j) \leq \text{dia } \sigma^{jm}(F) = \text{dia}(F)$. For some $k > 0$ the vertices $v_j = \tau^{kn}(x_j)$ and $w_j = \tau^{kn}(y_j)$ are on R. As $d_D(x_j, y_j) = d_D(v_j, w_j) \geq 2(j - j_0)$ and $d(x_j, y_j) = d(v_j, w_j)$ it follows that R is not metric in X. But $R \in \mathcal{D}_\tau = \mathcal{D}_\sigma$, contrary to Lemma 5.

This shows that for some $j > 0$ the ray R^- does not intersect $\sigma^{jm}(C \cup F)$. Now R, R^+ and $\sigma^{jm}(R^+)$ are in $\mathcal{D}_\sigma$ and $\sigma^{jm}(F)$ separates R^- from $\sigma^{jm}(R^+)$. From $R^- \in \mathcal{B}_{\tau^{-1}}$ we infer $\mathcal{D}_\tau \neq \mathcal{D}_{\tau^{-1}}$.

Pick $j_0 > 0$ such that neither R nor R^+ are contained in $\sigma^{j_0 m}(C)$. For each $j \geq j_0$ determine the last vertex x_j on R^+ and the last vertex y_j on R in $X - \sigma^{jm}(C)$. Then $x_j, y_j \in \sigma^{jm}(F)$ and therefore $d(x_j, y_j) \leq \text{dia } \sigma^{jm}(F) = \text{dia}(F)$. Abbreviating $F_j = \sigma^{jm}(F)$ we obtain

$$d(x_j, x_{j+1}) \leq \text{dia}(F_j) + d(F_j, F_{j+1}) + \text{dia}(F_{j+1}) = 2\text{dia}F + d(F, \sigma^m(F)) =: M'.$$

Similarly $d(y_j, y_{j+1}) \leq M'$ for $j \geq j_0$. Hence indeed R, R^+ are b-equivalent, that is $\mathcal{B}_\tau = \mathcal{B}_\sigma$. □

The part $\mathcal{D}_\tau \neq \mathcal{D}_{\tau^{-1}}$ in the Corollary also follows from results in [1].

REFERENCES

[1] R. HALIN, *Automorphisms and endomorphisms of infinite locally finite graphs*, Abh. Math. Sem. Univ. Hamburg **39** (1973), 251-283

[2] H.A. JUNG, *On finite fixed sets in infinite graphs*, submitted

[3] C. NEBBIA, *Groups of isometries of a tree and the Kunze-Stein phenomenon*, Pac. J. Math. **133** (1988), 141-149

[4] C. NEBBIA, *Amenability and Kunze-Stein property for groups acting on a tree*, Pac. J. Math. **135** (1988), 371-380

[5] N. POLAT AND G. SABIDUSSI, *Fixed elements of infinite trees*, Discrete Applied Math., to appear

[6] J. TITS, *Sur le groupe des automorphismes d'un arbre*, Essays on Topology and Related Topics (Mémoires dédiés à G. de Rham), Springer (1970), 188-211

[7] J. TITS, *A "theorem of Lie-Kolchin" for trees*, Contributions to Algebra (a collection of papers dedicated to Ellis Kolchin), Academic Press, New York-London (1977), 377-388

[8] W. WOESS, *Amenable group actions on infinite graphs*, Math. Ann. **284** (1989), 251-256

Author's address: Technische Universität Berlin
                      Fachbereich 3 - Mathematik
                      Straße des 17. Juni 136
                      1000 Berlin 12

Quasi–Ordinals and Proof Theory

L. GORDEEV

1. Introduction

One of the crucial goals in traditional proof theory is to characterize provability in formal systems which extend **PA** (Peano arithmetic). Since Gentzen proved the consistency of **PA** within **PA** extended by tranfinite induction along (the canonical well-order of the type) ϵ_0, results along these lines are usually presented in the analogous form:

An arithmetical sentence is provable in a given formal recursively axiomatizable theory S iff it is provable in PA extended by a suitable transfinite induction axiom.

In particular, within **PA**, this reduces the question of the consistency of **S** to the one about the well ordering property, $\text{WO}[P]$, for a suitable linear order P. In fact, a primitive recursive P can always be chosen, and the consistency of **S** reduces to the corresponding primitive recursive assertion, $\text{PRWO}[P]$, expressing in $\mathscr{L}_{\mathbf{PA}}$ (the language of **PA**) that P has no infinite descending primitive recursive branch.

So consider a recursively axiomatizable formal theory $\mathbf{S} \supseteq \mathbf{PA}$. It is also assumed that arithmetical theorems of **S** are true in the standard model of **PA**. We are looking for a suitable primitive recursive well-order P with the property required above

(1.1) $\forall \varphi \in \mathscr{L}_{\mathbf{PA}}\colon (\mathbf{S}\ proves\ \varphi) \Leftrightarrow (\mathbf{PA}+\text{TI}[<P]\ proves\ \varphi).$

where $\text{TI}[<P]$ is the arithmetical schema of transfinite induction for all initial segments of P. Since, by Gödel incompleteness, it is hopeless to try "better" proofs of $\text{PRWO}[<P]$, it remains to look for "better" structures P.

1991 Mathematics Subject Classification. Primary 06A10, 03F15.
This paper is in preliminary form and the detailed version of it will be published elsewhere.

At least one might expect P to be easily expressible in familiar mathematical terms.

It turns out that for any **S** in question, (1.1) always has a solution in form of a primitive recursive well-order (w.o.) P_S whose order-type is ω^ω, i.e. $P_S \in \omega^\omega$. This easily follows from an observation of G.Kreisel's that an arithmetical sentence φ is true iff its canonical infinitary cut free proof-search tree is well founded while having order type $< \omega^n$ for some $n < \omega$ (n corresponds to the logical complexity of φ). [Hence the required P_S results in the effective linear supremum of a primitive recursive enumeration, for all $\varphi \in \mathscr{L}_{PA}$ provable in **S**, of these trees – which are well founded since the φ are true.]

Does this mean that $\omega^\omega < \epsilon_0$ is the desired "universal" proof-theoretical ordinal?

The answer is NO, of course, since the structure of P_S is more complicated than provability in **S** itself. In fact, arguing in ordinal-theoretical terms, it is often assumed that "truly proof-theoretical" ordinal of **S** should majorize all ordinals whose primitive recursive representatives are well- ordered, provably in **S**. This leads to the following familiar definition.

(1.2)   *The proof-theoretical ordinal of* **S**, $\varrho(\mathbf{S})$, *is the least recursive ordinal α such that for any primitive recursive well-order (p.r.w.o.) $P \in \alpha$, $\mathrm{PRWO}[<P]$ is not provable in* **S**.

[If **S** admits $2nd$-order formulas, one can just as well replace $\mathrm{PRWO}[<P]$ by $\mathrm{WO}[<P]$. In fact, by H.Friedman's observation, this also makes sense for $1st$-order intuitionistic(!) **S**.]

Now clearly $\omega^\omega \neq \varrho(\mathbf{S})$, since canonical Cantor Normal Form representatives of ordinals below ϵ_0 are known to be well-ordered provably in **PA**, and hence also in **S**.

Having defined the proof theoretical ordinal, it seems plausible to expect primitive recursive representatives of $\varrho(\mathbf{S})$ to fulfill the basic assertion (1.1).

(1.3)   Is it true that (1.1) holds for all p.r.w.o. $P \in \varrho(\mathbf{S})$, i.e.
$\forall \varphi \in \mathscr{L}_{PA}$: (**S** *proves* φ) $\Leftrightarrow$ (**PA**+TI$[<P]$ *proves* φ) ?

In fact, (1.3) fails, as easily follows from the previously mentioned observation of G.Kreisel. [Take $\varphi := \mathrm{Con}(\mathbf{S})$, and let P_s be the canonical proof-tree w.o. of φ. Hence φ is true and **PA**+TI$[P_s] \vdash \varphi$, although $\neg(\mathbf{S} \vdash \varphi)$. But P_s is the initial segment of some $P \in \varrho(\mathbf{S})$, since the order type of P_s is smaller than $\epsilon_0 \leq \varrho(\mathbf{S})$.] Note that this pathology is no exception; actually, (1.3) has both infinitely many solutions and counterexamples P.

Hence "real" proof theory (theory of proofs) is not about proof-theoretical ordinals. Instead, it deals with certain solutions of (1.3), which are called systems of ordinal notations (s.o.n.).

(1.4)   *Proof-theoretical s.o.n. of* **S** *is any p.r.w.o.* $P \in \varrho(\mathbf{S})$ *such that*
$\forall \varphi \in \mathscr{L}_{PA}$: (**S** *proves* φ) $\Leftrightarrow$ (**PA**+TI$[<P]$ *proves* φ).

Of course, this is merely a compromise between the set-theoretical concept $\varrho(S)$, and the pragmatic assertion (1.1); this definition also admits various "pathological" s.o.n. of the above sort. In practice, proof theory deals with some special s.o.n. which are often called canonical.

However, for some S, there are known different "canonical" s.o.n.. Is it possible to pick "truly canonical" ones?

In some cases, the answer is undoubtedly YES – for example, the Cantor Normal Form presentation of ordinals below ϵ_0 seems "truly canonical" indeed. Yet more advanced proof theoretical ordinals are rather involved. When it comes to exhaustive explanations, one argues that the correlated s.o.n. arise as recursively defined term models of certain set, or category theoretical, structures whose real content can only be seen within the framework of the corresponding strong theory (which, anyway, is stronger than the subsystem of Analysis $\Pi_2^1 CA$, whose proof theory, in turn, has not yet been developed). In other words, the ordinal-theoretical background seems more involved than the theory being investigated – clearly not a comfortable position. Actually, this applies to the very idea to first describe $\varrho(S)$ in set, or category theoretical, terms, and afterwards to derive the correlated s.o.n. of S. The "reverse" approach is to look for suitable familiar structures (in simplest cases: finite graphs, trees, etc.) which in certain ways characterize a desired s.o.n. whose order type, in turn, will legitimately present the proof theoretical ordinal. Below, I illustrate this approach in connection with well-quasi-orders (w.q.o.) of finite labeled trees.

2. Basic Notions

Denote by **O** the countable ordinals, i.e. the factor space of countable w.o. $P = \langle M, \preceq \rangle$ under the order of the homomorphic embeddability "$\leq ^o$"

(2.1) $\quad \langle M_1, \preceq_1 \rangle \leq ^o \langle M_2, \preceq_2 \rangle :\Leftrightarrow (\exists f: M_1 \to M_2)(\forall x, y \in M_1)(x \preceq_1 y \to f(x) \preceq_2 f(y))$.

Recall that a w.q.o. is any structure $Q = \langle M, \leq \rangle$, "$\leq$" being countable reflexive, transitive relation having neither an infinite descending branch, nor an infinite antibranch (i.e. an infinite sequence of pairwise incomparable elements). Clearly any w.o. is w.q.o..

Let **Q** denote the factor space of countable w.q.o. under the order of the inverse homomorphic embeddability "$\leq ^q$"

(2.2) $\quad \langle M_1, \leq_1 \rangle \leq ^q \langle M_2, \leq_2 \rangle :\Leftrightarrow (\exists f: M_1 \to M_2)(\forall x, y \in M_1)(f(x) \leq_2 f(y) \to x \leq_1 y)$.

Let me define the appropriate generalized structure **QO** of quasi-ordinals. The underlying idea is as follows. Let $Q = \langle M, \leq \rangle$ be any w.q.o.. Note that "$\leq$" can be extended in an order preserving way to a well-ordered relation "$\preceq$". Usually there are different w.o. extensions, except that Q itself is w.o.. In fact, there is a maximal w.o. that uniquely determines the ordinal $\varrho(Q)$ of Q, due to D.DeJongh and P.Parikh (see [1]), but its definition is not constructive. So a given w.o.-extension $P = \langle M, \preceq \rangle$ can be viewed as an approximation of that maximal w.o.. This leads to the following definition.

Let $Q = \langle M, \leq \rangle$ and $P = \langle M, \preceq \rangle$ be respectively a w.q.o. and a w.o. such that "$\preceq$" extends "$\leq$", i.e. $(\forall x, y \in M)(x \leq y \to x \preceq y)$. The correlated structure $D = \langle M, \leq, \preceq \rangle$ is called *double well-quasi-order* (d.w.q.o.). The space of *quasi-ordinals*, **QO**, is the factor space of all countable d.w.q.o. under the following "double" order "$\leq^d$".

(2.3) $\quad \langle M_1, \leq_1, \preceq_1 \rangle \leq^d \langle M_2, \leq_2, \preceq_2 \rangle :\Leftrightarrow (\exists f: M_1 \to M_2)(\forall x, y \in M_1)$

$((f(x) \leq_2 f(y) \to x \leq_1 y) \land (x \preceq_1 y \to f(x) \preceq_2 f(y)))$.

3. Basic Operations

Let $\langle M, \preceq \rangle$ and $\langle M, \leq, \preceq \rangle$ be any w.o., and any d.w.q.o., respectively. Let

(3.1) $\quad \mathbf{i}(\langle M, \preceq \rangle) := \langle M, \preceq \rangle$, $\mathbf{D}(\langle M, \preceq \rangle) := \langle M, \preceq, \preceq \rangle$, $\mathbf{P_o}(\langle M, \leq, \preceq \rangle) := \langle M, \preceq \rangle$,

$\mathbf{P_q}(\langle M, \leq, \preceq \rangle) := \langle M, \leq \rangle$.

Then let $\mathbf{i}: \mathbf{O} \to \mathbf{Q}$, $\mathbf{D}: \mathbf{O} \to \mathbf{QO}$, $\mathbf{P_o}: \mathbf{QO} \to \mathbf{O}$ and $\mathbf{P_q}: \mathbf{QO} \to \mathbf{Q}$ be the correlated factorizations. Clearly $\mathbf{i}$, $\mathbf{D}$, $\mathbf{P_o}$ and $\mathbf{P_q}$ are homomorphisms; $\mathbf{P_o}$ and $\mathbf{P_q}$ are epimorphisms, while $\mathbf{i}$ and $\mathbf{D}$ are monomorphisms. Since $\mathbf{O}$ is embeddable in $\mathbf{QO}$, ordinals are quasi-ordinals. In particular, this yields finite quasi-ordinals $\mathbf{0}, \mathbf{1}, \ldots, \mathbf{n}, \ldots$ and $\boldsymbol{\omega}$, whose canonical representatives are familiar to everyone. Let me define on (or specify to) $\mathbf{QO}$ the standard operations "sum" and "product" denoted by $\oplus$ and $\otimes$, respectively, which are induced in the factor space of d.w.q.o. by the corresponding homomorphic operations $\oplus$ and $\otimes$ on d.w.q.o.. Another simple homomorphism $\mathbf{L}: \mathbf{QO} \to \mathbf{O}$ will imitate the Kleene-Brouwer linearization.

(3.2) $\quad$ Set $\langle M_1, \leq_1, \preceq_1 \rangle \oplus \langle M_2, \leq_2, \preceq_2 \rangle = \langle M, \leq, \preceq \rangle$, where

$M := (M_1 \times \{1\}) \cup (M_2 \times \{2\})$,

$\langle x, i \rangle \leq \langle y, j \rangle :\Leftrightarrow i = j \land x \leq_i y$,

$\langle x, i \rangle \preceq \langle y, j \rangle :\Leftrightarrow i < j \lor (i = j \land x \preceq_i y)$.

Let $\oplus: \mathbf{QO} \times \mathbf{QO} \to \mathbf{QO}$ be the corresponding factorized operation.

(3.3) $\quad$ Set $\langle M_1, \leq_1, \preceq_1 \rangle \otimes \langle M_2, \leq_2, \preceq_2 \rangle = \langle M, \leq, \preceq \rangle$, where

$M := M_1 \times M_2$,

$\langle x_1, x_2 \rangle \leq \langle y_1, y_2 \rangle :\Leftrightarrow x_1 \leq_1 y_1 \land x_2 \leq_2 y_2$,

$\langle x_1, x_2 \rangle \preceq \langle y_1, y_2 \rangle :\Leftrightarrow x_2 \prec_2 y_2 \lor (x_2 = y_2 \land x_1 \preceq_1 y_1)$.

Let $\otimes: \mathbf{QO} \times \mathbf{QO} \to \mathbf{QO}$ be the corresponding factorized operation.

[Note that the corresponding linear components $\langle M, \preceq \rangle$ provide the standard representatives of the ordinal-sum $\langle M_1, \preceq_1 \rangle + \langle M_2, \preceq_2 \rangle$ and ordinal-product $\langle M_1, \preceq_1 \rangle \cdot \langle M_2, \preceq_2 \rangle$, respectively.]

(3.4)   Let $D = \langle M, \leq, \preceq \rangle$ be any d.w.q.o. Set $L(D) = \langle M', \preceq' \rangle$, where

$$M' := \{\langle x_1,...,x_n \rangle : (\forall i < j \leq n)(x_i \not\preceq x_j)\},$$

$\langle x_1,...,x_n \rangle \preceq' \langle y_1,...,y_m \rangle :\Leftrightarrow$ either $(m \leq n \wedge (\forall i \leq m)(x_i = y_i))$

or $(\exists i \leq m,n)(x_i \prec y_i \wedge (\forall j < i)(x_j = y_j))$.

Let $\mathbf{L}: \mathbf{QO} \to \mathbf{O}$ be the corresponding factorized operation.

As $\oplus$, $\otimes$, $\mathbf{L}$ are homomorphisms, they preserve the underlying orders on d.w.q.o and w.o.. Hence $\oplus$, $\otimes$ and $\mathbf{L}$ are weakly monotone on each argument.

Let me introduce an auxiliary nonconstructive epimorphism $\mathrm{o}: \mathbf{Q} \to \mathbf{O}$ such that, for any $\alpha \in \mathbf{Q}$, $\mathrm{o}(\alpha)$ is the (uniquely determined, see above) ordinal $\mathrm{o}(Q)$ for $Q = \langle M, \leq \rangle \in \alpha$. ["$\mathrm{o}$" is unimportant for proof theoretical applications.] The following diagram clarifies the situation.

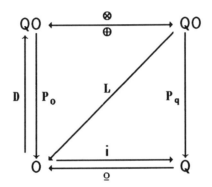

(3.4)   THEOREM. *The following hold for $\alpha, \beta, \tau \in \mathbf{QO}$ where $\mathbf{P_0}(\tau)$ is limit.*

(a) $\mathbf{P_0}(\alpha \oplus \beta) = \mathbf{P_0}(\alpha) + \mathbf{P_0}(\beta)$ *and* $\mathrm{o} \circ \mathbf{P_q}(\alpha \oplus \beta) = \mathrm{o} \circ \mathbf{P_q}(\alpha) \# \mathrm{o} \circ \mathbf{P_q}(\beta)$.

(b) $\mathbf{P_0}(\alpha \otimes \beta) = \mathbf{P_0}(\alpha) \cdot \mathbf{P_0}(\beta)$ *and* $\mathrm{o} \circ \mathbf{P_q}(\alpha \oplus \beta) = \mathrm{o} \circ \mathbf{P_q}(\alpha) \divideontimes \mathrm{o} \circ \mathbf{P_q}(\beta)$.

(c) $\mathbf{P_0}(\alpha) \leq \mathrm{o} \circ \mathbf{P_q}(\alpha) \leq \mathbf{L}(\alpha)$ *and* $\mathbf{L}(\tau) \leq \exp(\mathbf{P_0}(\tau), \mathrm{o} \circ \mathbf{P_q}(\tau) + 1)$.

["+", "·", "#", "$\divideontimes$" and $\exp(\alpha,\beta)$ are the familiar ordinal theoretic operations "sum", "product", "symmetric sum", "symmetric product" and "exponentiation" α^β, respectively.]

PROOF: Straightforward by also using [1]. □

The theorem shows that the operations just defined describe a very elementary ordinal fragment. In order to reach large proof theoretical ordinals, stronger operations are needed.

4. Stronger Operations and Results

I employ labeled trees and/or intervals quasi ordered by the appropriate strong relation of homeomorphic embeddability. Actually, it suffices to deal with binary-branching trees, while finite intervals can be viewed as trees without branchings. Recall that a (finite) tree is a finite nonempty partial order $T = \langle S, < \rangle$ on $\mathbb{N}$, say, (the vertex set S is also denoted by $|T|$) which has a minimal element $r(T) \in S$ (called the root), such that for any $x \in S$, the set of all predecessors $\{y \in S : y < x\}$ is linearly ordered. A given $z \in S$ lies between $x \in S$ and $y \in S$ iff either $x < z < y$ or $y < z < x$; $z \in S$ lies under $x \in S$ iff $z < x$. Note that any $x, y \in S$ uniquely determine their infimum $\inf(x,y) \in S$. A tree T has no branching, i.e. is an interval, iff it is linearly ordered. A tree T is called binary-branching iff every $x \in S$ having a successor has exactly two different immediate successors $y, z \in \text{Suc}(x)$. Now for any two trees T_1 and T_2, let $f \in \text{HEM}: T_1 \to T_2$ express that f is the homeomorphic embedding of T_1 into T_2, i.e. a monomorphism preserving $\inf(-,-,)$ as well as the order of each branching (with respect to the natural order on $\mathbb{N}$ restricted to $\text{Suc}(-)$). In the case of intervals this simply means f is the order preserving embedding.

Let me define homomorphisms $\mathbf{B}: \mathbf{O} \to \mathbf{QO}$ and $\mathbf{J}: \mathbf{O} \to \mathbf{QO}$. $\mathbf{B}(\alpha)$ and $\mathbf{J}(\alpha)$ are quasiordinals of binary-branching trees and intervals, respectively, labeled below α. The underlying operations B and J generalize the analogous Friedman-type definitions applicable for $\alpha < \omega$ (see [2],[3];[5]).

Let $P = \langle M, \preceq \rangle$ be any countable w.o.. Let $B[P]$ be a q.o. of all pairs $\langle T, \ell \rangle$ under the following homeomorphic embeddability "$\leq$", T being a binary-branching finite tree and $\ell : |T| \to M$ a labeling function on T.

(4.1) $\langle T_1, \ell_1 \rangle \leq \langle T_2, \ell_2 \rangle :\Leftrightarrow (\exists f \in \text{HEM}: T_1 \to T_2)$

$(\forall x \in |T_1|)(\forall y \in \text{Suc}(x) \text{ in } T_1)(\forall z \in |T_2|)$

(a) $\ell_1(x) \preceq \ell_2(f(x))$, and

(b) if z lies in T_2 between $f(x)$ and $f(y)$ then $\min\{\ell_1(x), \ell_1(y)\} \preceq \ell_2(z)$, and

(c) if z lies in T_2 under $f(r(T_1))$ then $\ell_1(r(T_1)) \preceq \ell_2(z)$.

That $B[P]$ is w.q.o., as well as the analogous quasi-order of arbitrarily-branching labeled trees, is proved in [3]. In particular, the analogous sub-quasi-order of labeled intervals, $J[P]$, is w.q.o. (cf. [2]). The correlated d.w.q.o.-extensions $B(P)$ and $J(P)$ of $B[P]$ and $J[P]$ arise by adding the lexicographical linear order "$\underline{\preceq}$" on trees and/or intervals, respectively (see below). [Note that in the case of intervals, "$\leq$" is a sort of Higman's w.q.o., and "$\underline{\preceq}$" is the ordinary lexicographical w.o..]

So let $P = \langle M, \preceq \rangle$ and $B[P]$ be as above. Let $B(P)$ be a d.w.q.o. that extends $B[P]$ by the following lexicographical well order "$\underline{\preceq}$". [Below $[\![R, S]\!]$ denotes a tree whose immediate subtrees are R and S, whereas $\#(T_i)$ denotes the cardinality of $|T_i|$.]

(4.2) $\langle T_1, \ell_1 \rangle \preceq \langle T_2, \ell_2 \rangle :\Leftrightarrow$ either $\#(T_1) < \#(T_2)$

or $\#(T_1) = \#(T_2)$ and

either $\ell_1(\mathrm{r}(T_1)) \prec \ell_2(\mathrm{r}(T_2))$

or $\ell_1(\mathrm{r}(T_1)) = \ell_2(\mathrm{r}(T_2))$ and one of the following holds.

(a) $\#(T_1) = \#(T_2) = 1$.

(b) $T_1 = [\![R_1, S_1]\!]$, $T_2 = [\![R_2, S_2]\!]$ and either $\langle R_1, \ell_1 \restriction R_1 \rangle \prec \langle R_2, \ell_2 \restriction R_2 \rangle$

or $\langle R_1, \ell_1 \restriction R_1 \rangle = \langle R_2, \ell_2 \restriction R_2 \rangle$ and $\langle S_1, \ell_1 \restriction S_1 \rangle \prec \langle S_2, \ell_2 \restriction S_2 \rangle$.

Now for any w.o. $P = \langle M, \preceq \rangle$, let $B(P)$ and $J(P)$ be the above d.w.q.o.-extension of $B[P]$ and its sub-d.w.q.o. of labeled intervals, respectively. Let $\mathbf{B}: \mathbf{O} \to \mathbf{QO}$ and $\mathbf{J}: \mathbf{O} \to \mathbf{QO}$ be the corresponding factorizations. Note that $\mathbf{B}$ and $\mathbf{J}$ are homomorphisms. It turns out that by combining $\oplus$, $\otimes$ and $\mathbf{L}$ with $\mathbf{J}$ and $\mathbf{B}$, respectively, one gets crucial proof theoretical ordinals of the predicative and impredicative hierarchy of subsystems of Analysis. In particular, consider the following operations.

(4.3)   For any $n > 0$, define $\mathbf{J_n}: \mathbf{O} \to \mathbf{QO}$ and $\mathbf{B_n}: \mathbf{O} \to \mathbf{QO}$ recursively by setting $\mathbf{J_1} := \mathbf{J}$ and $\mathbf{B_1} := \mathbf{B}$ and $\mathbf{J_{n+1}} := \mathbf{J} \circ \mathbf{L} \circ \mathbf{J_n}$ and $\mathbf{B_{n+1}} := \mathbf{B} \circ \mathbf{L} \circ \mathbf{B_n}$.

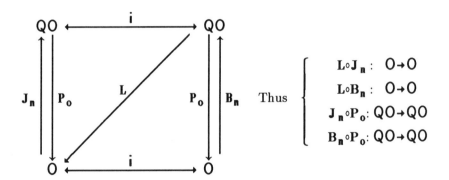

The homomorphisms $\mathbf{L} \circ \mathbf{J_n}: \mathbf{O} \to \mathbf{O}$, $\mathbf{L} \circ \mathbf{B_n}: \mathbf{O} \to \mathbf{O}$ and $\mathbf{J_n} \circ \mathbf{P_0}: \mathbf{QO} \to \mathbf{QO}$, $\mathbf{B_n} \circ \mathbf{P_0}: \mathbf{QO} \to \mathbf{QO}$ are weakly monotone operations on ordinals and quasi-ordinals, respectively. In fact, these operations are the desired fast growing functions on ordinals and quasi-ordinals, because they majorize the proof-theoretical ordinals of the predicative ("jump") and impredicative ("hyperjump") hierarchy of subsystems of Analysis, respectively. Particularly, this yields the following result. [**ITR$_0$** below denotes **ACA$_0$** extended by the axiom of Π_1^1-transfinite recursion.]

(4.4)   THEOREM. *Let $\Sigma_n\{\alpha_n\}$ denote the sum of a sequence $\{\alpha_n\}$, $n<\omega$, i.e. the order-type of the correlated ω-iteration of the operation "sum" on w.o. and/or d.w.q.o.. Then $\Sigma_n\{L\circ J(n)\}$, $\Sigma_n\{L\circ J_2(n)\}$, $\Sigma_n\{L\circ J_n(1)\}$, $\Sigma_n\{L\circ B(n)\}$, $\Sigma_n\{L\circ B\circ L\circ J(n)\}$, $\Sigma_n\{L\circ B_n(1)\}$ are proof-theoretical ordinals of ACA_0, $\Delta_1^1 CA$, ATR_0, $\Pi_1^1 CA_0$, $\Delta_2^1 CA$, ITR_0, respectively, and their canonical representatives are the corresponding proof theoretical s.o.n..*

PROOF SKETCH: That ACA_0, $\Delta_1^1 CA$, ATR_0, $\Pi_1^1 CA_0$, $\Delta_2^1 CA$ and ITR_0 prove the well-foundness of $L\circ J(n)$, $L\circ J_2(n)$, $L\circ J_n(1)$, $L\circ B(n)$, $L\circ B\circ L\circ J(n)$ and $L\circ B_n(1)$ follows from [2], [3] and (3.4) above. As for the reverse, the sufficiency of binary-branching follows from [4]. Since $L\circ J(n)$ converges to $\epsilon_0 = \varrho(PA) = \varrho(ACA_0)$, both $L\circ J_2(n) = L\circ J(\alpha_n)$ and $L\circ B\circ L\circ J(n) = L\circ B(\alpha_n)$ hold for α_n converging to ϵ_0, which corresponds to the familiar equations

$$\varrho(\Delta_1^1 CA) = \varrho(\Pi_0^1 CA_{<\epsilon_0}), \quad \varrho(\Delta_2^1 CA) = \varrho(\Pi_1^1 CA_{<\epsilon_0}) = \varrho(ID_{<\epsilon_0}).$$

On the other hand,

$$\varrho(\Pi_1^1 CA_0) = \varrho(ID_{<\omega}),$$

while $\varrho(ATR_0)$ and $\varrho(ITR_0)$ are the least (positive) fixed points of ordinal-operations $\alpha \mapsto \varrho(\Pi_0^1 CA_{<\alpha})$ and $\alpha \mapsto \varrho(\Pi_1^1 CA_{<\alpha})$, respectively. □

Observe that all s.o.n. involved are w.o. $\Sigma_n\{P_n\}$ for $P_n = \langle M_n, \preceq_n \rangle \in \alpha_n$, where M_n contains objects of types arising recursively as follows: (1) finite strings of binary-branching trees, or intervals, whose nodes are natural numbers; (2) finite strings of ... (as above) whose nodes are finite strings of binary-branching trees, or intervals, whose nodes are natural numbers; (3) etc.. The correlated order "$\preceq_n$" is defined by combinatorial recursion on the type complexity. For any typed object $b \in M$, let $wht(b)$ be the total weight of the expression b, where $wht(n) := n$ for natural numbers n. With respect to this measure, the notions of primitive recursive and linearly growing infinite sequences in P are specified accordingly.

Let S be a theory as in (4.4). If $\Sigma_n\{P_n\}$, $P_n = \langle M_n, \preceq_n \rangle \in \alpha_n$, is s.o.n. of S then $\forall n PRWO[P_n]$ is not provable in S. By H.Friedman's approach, even the analogous weaker assertion $\forall n LWO[P_n]$ about linearly growing sequences, is not provable in S. This yields the following result.

(4.5)   COROLLARY. *The following is true but not provable in S. For every $n>0$ and $k>0$, there is an m so large that for any $b_0, b_1, ..., b_m \in M_n$ with $(\forall i \leq m)(wht(b_i) \leq k+i)$ there are $i<j\leq m$ such that $b_i \preceq_n b_j$.*

By standard proof theoretical arguments, (4.5) extends to the following proposition.

(4.6)   *For sufficiently large n and k, the correlated mapping $F(n,k) := m$ majorizes every recursive function which is everywhere defined provably in S.*

5. Remarks

(5.1) A more exhaustive fragment of quasi-ordinals arises by extending the domain of the (unary) operations **J** and **B**. Namely, let $\mathbf{J}^*: O \times \mathbf{QO} \to \mathbf{QO}$ and $\mathbf{B}^*: O \times \mathbf{QO} \to \mathbf{QO}$ be the analogous binary homomorphisms defined as follows. As before, let $\mathbf{J}^*$ be the restriction of $\mathbf{B}^*$ to non-branching trees. Now for any $P = \langle M, \preceq \rangle \in \alpha \in \mathbf{Q}$ and $D = \langle M', \preceq', \leq' \rangle \in \beta \in \mathbf{QO}$, let $B^*[P, D]$ be a q.o. of all triples $\langle T, \ell, \varrho \rangle$ for $\langle T, \ell \rangle$ being as above and $\varrho: |T| \to M'$ being another ("quasi"-)labeling function on T. Now $\langle T_1, \ell_1, \varrho_1 \rangle \leq \langle T_2, \ell_2, \varrho_1 \rangle$ is defined as in (4.1) above, except extending (a) by $\varrho_1(x) \leq' \varrho_2(f(x))$. The correlated lexicographical linear extension "$\preceq$" of "$\leq$" generalizes (4.2) in an obvious way by using "$\preceq'$". According to [2,3], the resulting operations are well-defined. By combining $\oplus$, $\otimes$, **L**, $\mathbf{J}^*$ and $\mathbf{B}^*$ in the same manner as above, one obtains more quasi-ordinals, whose ordinal-projections range over virtually all known proof-theoretical ordinals before $\varrho(\mathbf{ITR_0})$.

(5.2) W.q.o. statements are of interest from a foundational, or constructive, viewpoint as they enable us to replace $2nd$-order existential axioms of the sort "*For every set X there exists a suitable set Y*" by intuitionistically neutral statements. Namely, $\mathbf{ATR_0}$ and $\mathbf{ITR_0}$ have the same arithmetical theorems as $\mathbf{ACA_0}$ extended for example by the "reverse" universal axioms **UI**: "*For any w.o. P, finite intervals labeled in P are w.q.o. under the embeddability of (4.1)*" and **UT**: "*For every w.o. P, finite trees labeled in P are w.q.o. under the embeddability of (4.1)*", respectively. This follows from [2],[3] and (4.4) above.

(5.3) It is an open problem whether **QO** is w.q.o..

6. Appendix

As mentioned above, the crucial statement "$B[P]$ is w.q.o." generalizes H.Friedman's variant of the Kruskal theorem in which labels are bounded natural numbers. Its proof-theoretical strength is the one of $\Pi_1^1\mathbf{CA_0}$ (cf. [5]). In (4.1) above I replace H.Friedman's asymmetric gap-condition (see [5]) by the modified symmetric gap-condition (b) in order to get the adequate proof theoretical characterizations of theories up to $\mathbf{ITR_0}$. On the other hand, it is possible to get another generalization by replacing (b) by a stronger asymmetric gap-condition due to M.Okada

(b^a)   *if z lies in T_2 between $f(x)$ and $f(y)$ then $\ell_1(y) \preceq \ell_2(z)$*.

Call "asymmetric embeddability" the resulting modification of the embeddability (4.1). Let **UTA** denote the correlated "asymmetric" universal axiom: "*For every w.o. P, finite trees labeled in P are w.q.o. under the asymmetric embeddability*". Obviously **UTA** implies **UT**. Moreover, **UTA** is provable in $\Pi_2^1\mathbf{CA}$ according to the proof of Theorem 1.7 from [6], since the above conditions (a), (b^a) and (c) together are equivalent to the asymmetric gap-condition (1.2.5) from [6].

I claim that in the proof-theoretic sense **UTA** is not stronger than **UT**, i.e. **UTA** is in fact provable in **ITR$_0$**. This result is proved by the appropriate modification of [3] and will be published elsewhere.

On the other hand, according to [6], there is a clear correspondence between vertex-labeled trees under the asymmetric embeddability and edge-labeled trees under the homeomorphic embeddability with the symmetric gap condition (e) of H.Friedman (see below). That is, consider finite trees with edges labeled by ordinals (these trees are referred to as edge-labeled trees). Thus an edge-labeled tree is a pair $\langle T, \ell \rangle$ with $\ell : |T|_E \to O$, $|T|_E$ being the set of all edges of T. If $\langle T_1, \ell_1 \rangle$ and $\langle T_2, \ell_2 \rangle$ are any two edge-labeled trees then set

(6.1) $\langle T_1, \ell_1 \rangle \leq \langle T_2, \ell_2 \rangle :\Leftrightarrow (\exists f \in \text{HEM}: T_1 \to T_2)(\forall x \in |T_1|_E)(\forall y \in |T_2|_E)$

(e) *if y lies in T_2 in the path $f(x)$ then $\ell_1(x) \leq \ell_2(y)$.*

In words, (6.1) expresses that $\langle T_1, \ell_1 \rangle$ is homeomorphically embeddable into $\langle T_2, \ell_2 \rangle$ such that every edge is mapped onto a path with greater-or-equal labels. Let **UTE** be the axiom expressing that all edge-labeled trees are well-quasi-ordered by the latter embeddability of (6.1). Since **UTE** easily follows from **UTA** (cf. [6]), the above claim implies that **UTE** is provable in **ITR$_0$** as well.

REFERENCES

1. D.DeJongh and P.Parikh: *Well-partial-orderings and hierarchies*, Indagationes Mathematicae 39:195–207, 1977
2. L.Gordeev: *Generalizations of the one-dimensional version of the Kruskal-Friedman theorems*, Journal of Symbolic Logic 54(1):100–121, 1989
3. L.Gordeev: *Generalizations of the Kruskal-Friedman theorems*, Journal of Symbolic Logic 55(1):157–181, 1990
4. L.Gordeev: *Systems of iterated projective ordinal notations, etc.*, Archive for Mathematical Logic 29:29–46, 1989
5. S.G.Simpson: *Nonprovability of certain combinatorial properties of finite trees*, in Harvey Friedman's Research on the Foundations of Mathematics, Elsevier, North-Holland, 1985
6. I.Kriz: *Well-quasiordering finite trees with gap-conditions. Proof of Harvey Friedman's conjecture*, Annals of Mathematics 130:215–226, 1989

Minor Classes: Extended Abstract

DIRK VERTIGAN

ABSTRACT. This extended abstract is an informal and partial review of the author's M.Sc. thesis "Minor classes" (1988, University of Tasmania). The thesis introduces the concept of a minor class, develops some theory and applies it to examples from the literature. A paper is in preparation, and is to appear.

1. Introduction

Statements such as "the minor class of directed graphs (with deletion and contraction of edges) has six natural excluded minors of size two and 4890 of size three (and no others)" are informally presented and explained here. Proofs and formal definitions are to be found in the full paper. Terms appearing in this introduction are explained later.

The main aim of this extended abstract is to convey minor class concepts and the nature of the results obtained. It is unfortunately not possible to present all the results, together with the necessary definitions, both briefly and formally. Instead, the formal definition of minor classes in Section 2 is followed by an intuitive discussion of results using some convenient visualisations.

There are several examples of minor classes in the literature, for example, any minor closed class of graphs, digraphs, matroids or chain groups, with the usual definitions of deletion, contraction and isomorphism. By taking a few common properties of these, one obtains the axioms for a minor class. Loosely speaking, a *minor class* consists of a class of *structias* (pronounced, structures) each having a *ground set*, together with a certain axiomatised idea of *minors* and *isomorphism*. A minor of a structia is obtained by removing elements of its ground set in certain *manners* (for example, deletion and contraction). An *isomorphic* structia is obtained simply by renaming the ground set elements. The

---

1991 *Mathematics Subject Classification*. Primary 05, 08.

Key words and phrases. minor classes, minors, graphs, matroids, chain-groups, many-sorted unary algebras.

The detailed version of this paper is to be published elsewhere.

axioms essentially say that element removals can be done in any order giving the same result, and that isomorphism behaves as expected. A minor class contains the information of not only *when* one structia is isomorphic to a minor of another structia, but also *how* it is so obtained. Thus a minor class is richer than just the underlying quasi-order relation 'is isomorphic to a minor of'. Definitions are in Section 2.

When the structias in a minor class are, say, graphs or matroids, there is a concrete way to visualise them. However, in the process of abstraction, we 'forget' what the original structias actually are. A structia is merely an element of a minor class, and it attains its 'structure' by its place in the minor class—by how it is related to other structias by minors or isomorphism.

It is natural to seek results showing how any 'abstract' minor class can be given a concrete description in terms of concrete structias with a natural definition of minors. Such results show how structias in a minor class can be visualised. It also gives a unified way of describing the various minor classes in the literature.

Section 3 defines certain minor classes where the structias on n elements are n-dimensional coloured grids, and minors are certain subgrids (inheriting the colouring). Section 4 presents a result stating that any minor class is contained in some 'coloured grid' minor class. Although this gives a universal way of viewing structias, it does not give a convenient or unique way of describing a minor class.

Section 5 does. It discusses how any minor class (in which ground sets are finite) can be uniquely described by its so-called ψ-*description* given by its ψ-*structias* and its *natural excluded minors*. Section 6 summarises the ψ-description of several minor classes from the literature. (A partial example is the strange, but true, statement at the beginning of this introduction.)

The ψ-description of a minor class gives rise to the following visualisation of its structias, here described in the case that there are two manners of element removal, say deletion and contraction. In this case, structias can be depicted as hypercubes with patterns on some sub-hypercubes. A minor is any sub-hypercube, inheriting the patterns in the natural way. The structias can be depicted by drawing appropriate patterns on the sub-hypercubes corresponding to ψ-structias, subject to the exclusion of the patterned hypercubes corresponding to the natural excluded minors.

For example directed graphs, with deletion and contraction of edges, can be depicted as hypercubes (where the dimension is the number of edges) with patterns on certain 1- and 2-dimensional hypercubes, subject to the exclusion of six patterned hypercubes of dimension 2 and 4890 of dimension 3.

Some notation is as follows. Let A^B denote the set of all functions of the form $f : B \to A$. Let $A - B$ denote set difference ($A - B = \{x | x \in A, x \notin B\}$). For a function $g : A \to B$ and $C \subseteq B$, denote $g|_{g^{-1}(C)}$ by $g|^C$. A one element set $\{q\}$ is often abbreviated simply to q.

2. Definitions

We develop the definition of minor class in stages, accompanied by the familiar example of finite directed graphs (with no isolated vertices) with deletion and contraction of edges. (In this discussion, digraphs can just as well be replaced by graphs, matroids or chain groups.) The defining steps are *structia class*, *isomorphism-structia class* and *minor class*. The first two are of little interest in their own right, but are useful for developing the definition of minor class and illustrating certain points.

Throughout this paper, $\mathcal{Q}$ is a set of sets of the form $\{Q | Q \subseteq \mathcal{Q}_\cup, |Q| < c\}$ for some set $\mathcal{Q}_\cup$ and some cardinal $c \leq |\mathcal{Q}_\cup|$. Also K is a set such that $K \cap \mathcal{Q}_\cup = \emptyset$. The elements of $\mathcal{Q}$ are the allowed ground sets for structias, while K is the set of *manners* of element removal.

For the digraph example let $\mathcal{Q}$ be the set of all finite subsets of some infinite set and let $K = \{delete, contract\}$.

A *$\mathcal{Q}$-structia class* is a family of disjoint sets $\mathcal{S} = (\mathcal{S}_Q | Q \in \mathcal{Q})$. (It is permitted that some $\mathcal{S}_Q = \emptyset$.) The elements of each $\mathcal{S}_Q$ are *structias*. The *ground set* of a structia $S \in \mathcal{S}$, denoted $gs(S)$, is the unique $Q \in \mathcal{Q}$ such that $S \in \mathcal{S}_Q$. Sometimes a structia S with $gs(S) = Q$ is denoted by the pair (S, Q). Note that a structia $S \in \mathcal{S}$ has no properties apart from $gs(S)$.

For example we can define a $\mathcal{Q}$-structia class $\mathcal{G}$ where $\mathcal{G}_Q$ is the set of all digraphs with edge set Q. (As a technicality, vertices are unlabelled, and two digraphs are *equal* if they are identical with respect to the edge labelling.)

Now consider isomorphism. (Isomorphism is a straightforward but irremovable aspect of minor classes.) For a digraph G with edge set Q (or any 'concrete' structia on ground set Q) and a bijection $\omega : Q \to Q'$ one can naturally define $\widehat{\omega} : \mathcal{G}_Q \to \mathcal{G}_{Q'}$ where $\widehat{\omega}(G)$ is obtained from G simply by renaming each edge $q \in Q$ as $\omega(q) \in Q'$. (It is allowed that $q = \omega(q)$ and an edge is 'renamed' to itself.)

In a structia class, structias are devoid of any 'structure' with which to define isomorphism. Instead isomorphism must be axiomatised. The definition of *$\mathcal{Q}$-isomorphism-structia class* below shows how this can be done.

For sets $Q, Q' \in \mathcal{Q}$, let $I_{Q,Q'}$ be the set of all bijections from Q to Q'.

A *$\mathcal{Q}$-isomorphism-structia class* is a pair $(\mathcal{S}, \mathcal{I})$ where $\mathcal{S}$ is a $\mathcal{Q}$-structia class and $\mathcal{I}$ is a family of functions $(\widehat{\omega} : \mathcal{S}_Q \to \mathcal{S}_{Q'} \mid \omega \in I_{Q,Q'}\ Q, Q' \in \mathcal{Q})$ satisfying:
(IC1) If $1_Q \in I_{Q,Q}$ is the identity then $\widehat{1}_Q \in \mathcal{I}_{Q,Q}$ is the identity.
(IC2) If $\omega \in I_{Q,Q'}$ and $\omega' \in I_{Q',Q''}$ then

$$\widehat{\omega'} \circ \widehat{\omega} = \widehat{\omega' \circ \omega}.$$

These are immediately satisfied when isomorphism is defined naturally for 'concrete' structures. On the other hand (IC1) and (IC2) are irredundant; they would not generally hold if they were not imposed. An isomorphism-structia class tells not only when two structias are isomorphic, but also via which bijections they are isomorphic; there is more than just the relation $\cong$. Also a structure $S \in \mathcal{S}_Q$ has an *automorphism* group $Aut(S) = \{\omega | \omega \in I_{Q,Q}, \widehat{\omega}(S) = S\}$. (Conditions (IC1) and (IC2) guarantee that $Aut(S)$ forms a group.) Thus structias in an isomorphism-structia class gain some 'structure' from the functions in $\mathcal{I}$ that they didn't have in the structia class $\mathcal{S}$ alone.

Let us return to K, the set of *manners* of element removal. For any structure (S, Q) in a $(\mathcal{Q}, K)$-minor class (to be defined below) each element $q \in Q$ can be removed in some manner $l \in K$ or else q can be renamed (possibly to q itself). For many 'concrete' structias (digraphs, etc.) there are various explicit and natural definitions of taking minors. But since structias in a structia class have no 'structure' of their own, we must axiomatise minors—impose rules that element removals should satisfy. (These rules are incorporated with the rules for element renamings in an isomorphism-structia class.)

To find the appropriate rules, we consider the example of digraphs. Any edge can be deleted or contracted from a digraph. These edge removals can be performed in any order without affecting the result. It suffices to specify which edges are deleted and which are contracted; they can be considered to be removed simultaneously. Also deletion and contraction respect isomorphism— any statement that one digraph is a minor of another via specified deletions and contractions, remains true if all edges are consistently renamed in the statement. (Of course, distinct edges are not allowed to be given the same name.) After a sequence of removals and renamings, all that matters is how each edge was finally removed or renamed. These general properties also hold for graphs, matroids and chain groups and are to be formalised as axioms for minor classes.

First some notation. For $Q, Q' \in \mathcal{Q}$ let $F_{Q,Q'}$ be the set of all functions of the form $f : Q \cup K \to Q' \cup K$ where $f|_K$ is the identity and $f|^{Q'}$ (see note on notation at end of §1) is a bijection. These f's play the same role for minor classes as did bijections ω for isomorphism-structia classes. The interpretation of the associated $\widehat{f}$'s is given after the definition.

DEFINITION 2.1. A $(\mathcal{Q}, K)$-*minor class* is a pair $(\mathcal{S}, \mathcal{I})$ where $\mathcal{S}$ is a $\mathcal{Q}$-structia class and $\mathcal{I}$ is a family of functions $(\widehat{f} : \mathcal{S}_Q \to \mathcal{S}_{Q'} | f \in F_{Q,Q'} \; Q, Q' \in \mathcal{Q})$ satisfying:
(MC1) If $1_Q \in F_{Q,Q}$ is the identity then $\widehat{1}_Q \in \mathcal{I}_{Q,Q}$ is the identity.
(MC2) If $f \in F_{Q,Q'}$ and $f' \in F_{Q',Q''}$ then

$$\widehat{f'} \circ \widehat{f} = \widehat{f' \circ f}.$$

The succinctness of the conditions suggests that this is a natural definition. Readers familiar with (many-sorted) universal algebra will see that minor classes are algebras and that the definitions and theorems of [1] automatically apply; others can ignore this sentence. Note that a $(\mathcal{Q}, \emptyset)$-minor class is precisely a $\mathcal{Q}$-isomorphism-structia class.

For each $Q, Q' \in \mathcal{Q}$ and $f \in F_{Q,Q'}$ the *isominor operation* $\widehat{f} : \mathcal{S}_Q \to \mathcal{S}_{Q'}$ is interpreted as follows: the structia $\widehat{f}(S) \in \mathcal{S}_{Q'}$ is obtained from $S \in \mathcal{S}_Q$ by
(i) when $f(q) \in K$, *remove* element $q \in Q$ in *manner* $f(q)$,
(ii) when $f(q) \in Q'$, *rename* element $q \in Q$ to $f(q) \in Q'$.

In this interpretation, condition (MC2) says that if you do a bunch of element removals and renamings and then do another bunch, then it is the same as doing them in a single step. Let $q \in Q$, $q' \in Q'$, $q'' \in Q''$, $l \in K$, and consider the following three cases of $q \mapsto f(q) \mapsto f'(f(q))$:
(i) for $q \mapsto q' \mapsto q''$ (MC2) says that renaming q to q' then q' to q'' is the same as renaming q to q''
(ii) for $q \mapsto q' \mapsto l$ (MC2) says that renaming q to q' then removing q' in manner l is the same as removing q in manner l
(iii) for $q \mapsto l \mapsto l$ (MC2) says that if q is removed in manner l then it stays removed in manner l.
Condition (MC1) says that doing nothing to a structia leaves it unchanged. Thus the natural conditions (MC1) and (MC2) capture the desired properties (and ensure that the interpretation makes sense).

A 'concrete' minor class, such as digraphs with deletion and contraction, can be abstracted to a minor class $(\mathcal{S}, \mathcal{I})$ by defining the isominor operations according to the above interpretation and then 'forgetting' what the structias are. (Of course you don't literally forget anything.) As shown in the following sections, these structias attain 'structure' from their place in the minor class, and can be given concrete representations. Typically, the original structias can also be recovered.

To check that a concrete candidate for a minor class really is a minor class, one must first check that it fits into the minor class framework; for example, structias must have ground sets and any element must be removable in any of the given manners. Secondly one must check that (MC1) and (MC2) hold; the main property being that a minor does not depend on the order of element removals. Of course, minor classes do not capture all definitions of 'minor' in the literature but they do capture many examples of interest.

Here are some definitions with examples.

The structia $\widehat{f}(S)$ is an *isominor* of S. If $f|^{Q'}$ is the identity (so that the element renaming part of the operation is trivial) then $\widehat{f}(S)$ is a *minor* of S. If $f^{-1}(K) = K$ (so that no elements are removed) then $\widehat{f}(S)$ is *isomorphic to* S, denoted $\widehat{f}(S) \cong S$. (Note that S' is an isominor of S if and only if S' is isomorphic to a minor of S.)

A *sub minor class* of $(\mathcal{S}, \mathcal{I})$ is a minor class of the form $(\mathcal{S}', \mathcal{I}')$ where $\mathcal{S}'_Q \subseteq \mathcal{S}_Q$ for all $Q \in \mathcal{Q}$, where $\mathcal{S}'$ is closed under isominors and the isominor operations in $\mathcal{I}'$ are those in $\mathcal{I}$ restricted to the $\mathcal{S}_Q$'s. For example the minor class of planar graphs, with deletion and contraction, is a sub minor class of the minor class of graphs with deletion and contraction.

A *homomorphism* from a $(\mathcal{Q}, K)$-minor class $(\mathcal{S}, \mathcal{I})$ to a $(\mathcal{Q}, K)$-minor class

$(\mathcal{S}', \mathcal{I}')$ is a function $\alpha : \mathcal{S} \to \mathcal{S}'$—partitioned into $\alpha_Q : \mathcal{S}_Q \to \mathcal{S}'_Q$ for $Q \in \mathcal{Q}$—which *respects* the isominor operations. If $K = \{delete, contract\}$ then (in conventional notation) this means $\alpha(S\backslash A/B) = (\alpha(S))\backslash A/B$ (and α respects structia isomorphism). More formally, for all $Q, Q' \in \mathcal{Q}$, $f \in F_{Q,Q'}$, $S \in \mathcal{S}_Q$,

$$\alpha_{Q'}(\widehat{f}_S(S)) = \widehat{f}_{S'}(\alpha_Q(S)),$$

where the subscript on the $\widehat{f}$'s indicates which minor class the isominor operation is in. If α is surjective then the minor class $\mathcal{S}'$ is a *homomorphic image* of $\mathcal{S}$. If α is bijective then α is an *isomorphism* and the minor class $\mathcal{S}$ is *isomorphic* to $\mathcal{S}'$. If α is injective then α is an *embedding* and the minor class $\mathcal{S}$ is *embeddable* in $\mathcal{S}'$ (or equivalently $\mathcal{S}$ is isomorphic to a sub minor class of $\mathcal{S}'$). Examples of minor class homomorphisms are the maps sending a digraph to its underlying graph, or the map sending a graph to its graphic matroid, or the map sending a chain group to its coordinatisable matroid.

In the following sections, most formal definitions are omitted and replaced by more intuitive visual descriptions. Also some statements and arguments are specialised to simplify discussions.

3. The 'coloured grid' minor classes

Let K, C, B be sets with $K \cap C = \emptyset$, and let $A = K \cup C$, for the duration of this section. Let $\mathcal{Q}$ be as above. In this section, a certain $(\mathcal{Q}, K)$-minor class, denoted $\mathcal{F}^{\mathcal{Q}}(K, C, B)$, is defined. This is quite a general example because any $(\mathcal{Q}, K)$-minor class can be embedded in a minor class of the form $\mathcal{F}^{\mathcal{Q}}(K, C, B)$ for some sets C and B.

The definition is developed in stages, starting with $\mathcal{F}^{\mathcal{Q}}(K, C, B)$ as a structia class. The structias in $\mathcal{F}^{\mathcal{Q}}(K, C, B)$ with ground set $Q \in \mathcal{Q}$ are all the pairs (g, Q) where $g \in B^{(A^Q)}$, that is, g is a function from A^Q to B. Note that the elements of A^Q are themselves functions from Q to A, and can be thought of as vectors $x = (x_q | q \in Q)$ or $(x(q) | q \in Q)$ where $x_q = x(q) \in A$ for all $q \in Q$.

These structias, that is, functions of the form $g : A^Q \to B$, can be conveniently visualised in the case that Q is finite and $A \subseteq \mathbf{R}$ and B is a set of colours (although the visualisation can be extended to general Q, $A = K \cup C$ and B). Consider the $|Q|$-dimensional Euclidean space $\mathbf{R}^Q$, with coordinate axes labelled by the elements of Q. Then A^Q is the subset of $\mathbf{R}^Q$ consisting of those points whose coordinates are all in A, and these points form an $|A| \times |A| \times \cdots \times |A|$ ($|Q|$-dimensional) *grid* in this space. To each point $x \in A^Q$, in this grid, the colour $g(x) \in B$ is assigned, creating a *coloured grid* in $\mathbf{R}^Q$.

Isomorphism for these structias is defined naturally. For a structia $g \in B^{(A^Q)}$ and a bijection $\omega : Q \to Q'$, the isomorphic structia $\widehat{\omega}(g) \in B^{(A^{Q'})}$ is described, in terms of the coloured grid visualisation, simply by renaming the q-axis to be the $\omega(q)$-axis for each $q \in Q$.

To make $\mathcal{F}^{\mathcal{Q}}(K, C, B)$ a minor class, it is necessary to define element removal. The visualisation of element removal is more illuminating than the formal definition which is omitted. Recall the coloured grid in $\mathbf{R}^Q$, associated with $g : A^Q \to B$. For any element $q \in Q$ and any manner $l \in K$, consider the $|A| \times |A| \times \cdots \times |A|$ (($|Q|-1$)-dimensional) coloured subgrid consisting of those gridpoints with q^{th} coordinate l. (This is the 'cross section' of the coloured grid taken by the hyperplane of $\mathbf{R}^Q$, perpendicular to the q-axis and intersecting this axis at coordinate l). This coloured subgrid can be projected into $\mathbf{R}^{Q-q}$ in the natural way, by discarding the qth coordinate. With each such gridpoint retaining its colour, this yields a coloured grid sitting in $R^{Q-\{q\}}$, which depicts the structia on ground set $Q - q$ (a function from A^{Q-q} to B) obtained from g by removing element q in manner l.

For minors in general, consider $f \in F_{Q,Q'}$ with $f|^{Q'}$ being the identity. Then $\widehat{f}(g)$ on ground set Q' is obtained from g by removing each element $q \in Q - Q'$ in manner $f(q)$. Consider the $|A| \times |A| \times \cdots \times |A|$ ($|Q'|$-dimensional) coloured subgrid consisting of those gridpoints with q^{th} coordinate $f(q)$ for all elements $q \in Q - Q'$. (This is the intersection of all the subgrids associated with removing a single element $q \in Q - Q'$ in manner $f(q)$. Since intersection is independent of order, so is element removal.) This coloured subgrid can be projected into $\mathbf{R}^{Q'}$ in the natural way, by discarding the qth coordinate for all $q \in Q - Q'$. With each such gridpoint retaining its colour, this yields a coloured grid sitting in $\mathbf{R}^{Q'}$, which depicts the structia $\widehat{f}(g)$.

It is routine to check that $\mathcal{F}^Q(K,C,B)$ is a (Q,K)-minor class. The 'coloured grid' visualisation is useful for providing insight.

Consider the case when $K = \{delete, contract\}$ and identify delete and contract with, respectively, 0 and 1, so that $K = \{0,1\}$. Suppose C is the open interval $(0,1)$ so that $A = K \cup C$ is the closed interval $[0,1]$. In this case, the structias in $\mathcal{F}^Q(K,C,B)$ can be visualised as hypercubes $[0,1]^Q$ whose points are coloured by the elements of B. In this visualisation, minors correspond to subhypercubes (obtained by fixing some coordinates to 0 or 1) inheriting the colour pattern.

4. Two embedding theorems

This section presents two embedding theorems for minor classes. The first of these shows that structias in minor classes can always be visualised as coloured grids (see §3).

EMBEDDING THEOREM 1. *For any (Q,K)-minor class $(S,\mathcal{I})$ there exist sets C and B such that $(S,\mathcal{I})$ is embeddable in $\mathcal{F}^Q(K,C,B)$.*

EMBEDDING THEOREM 2. *For any (Q,K)-minor class $(S,\mathcal{I})$ with $|K| \geq 2$ there exists a set B such that $(S,\mathcal{I})$ is embeddable in a homomorphic image of $\mathcal{F}^Q(K,\emptyset,B)$.*

For the purpose of illustration we discuss Embedding Theorem 1 in the case where ground sets are finite and $K = \{delete, contract\}$, visualising in terms of colour-patterned hypercubes (see §3) rather than coloured grids. (This discussion also leads into the next section.) The general proof describes an embedding explicitly (with $|C| = \min\{c|c \geq |Q|, \forall Q \in \mathcal{Q}\}$ and $|B| = |\mathcal{S}/\cong|$, by no means minimising these usually infinite cardinalities). In the case where ground sets are finite so that the isominor ordering on $\mathcal{S}$ has no infinite descending chains, the embedding $\alpha : \mathcal{S} \to \mathcal{F}^{\mathcal{Q}}(K, C, B)$ can be built up inductively. Suppose that α has been defined on all proper isominors of some structia $S \in \mathcal{S}_Q$, say. Thus in the patterned hypercube $[0,1]^Q$ to represent $\alpha(S)$, the colouring of the points on its boundary is already determined leaving its interior $(0,1)^Q$ to be coloured. (Consider the interior of a 0-dimensional hypercube (a point) to be itself.) Defining $\alpha(S)$ immediately determines $\alpha(S')$ for any $S' \cong S$ so there are subtleties of isomorphism and automorphism to consider. One could associate a new colour with S (in fact with $(S \cong) = \{S'|S' \cong S\}$) and colour all of $(0,1)^Q$ with this one colour. However, doing this, it could occur that $Aut(\alpha(S)) \neq Aut(S)$ which contradicts α being an embedding. Instead, a genuine embedding α is defined if one colours $(0,1)^Q$ with two new colours such that the automorphisms (permutations of the coordinate axes) of the colouring of $(0,1)^Q$ are the same as the automorphisms of S. This idea can be extended to give a general construction for Embedding Theorem 1, for arbitrary $(\mathcal{Q}, K)$.

When all ground sets are finite, a minor class $\mathcal{S}$ can be described by specifying sets C and B and the excluded minors of $\mathcal{S}$ in $\mathcal{F}^{\mathcal{Q}}(K, C, B)$. However this is not really satisfactory, as there is much redundancy in $\mathcal{F}^{\mathcal{Q}}(K, C, B)$ and for known minor classes from the literature the list of excluded minors is infinite (though describable). The next section develops a canonical way of describing a minor class, which for many known minor classes is a finite description in this canonical form.

5. ψ-descriptions; ψ-structias and natural excluded minors.

In this section all ground sets are finite and $\mathcal{Q} = \{Q|Q \subseteq \mathcal{Q}_\cup, Q < \aleph_0\}$ for some infinite set $\mathcal{Q}_\cup$. Assume $K = \{delete, contract\}$, although all of this section, rephrased appropriately, applies to arbitary K. (Some parts apply more generally than minor classes.) Conventional deletion and contraction notation is used where appropriate. Recall the patterned hypercube visualisation which applies to the $|K| = 2$ case.

Let $(\mathcal{S}, \mathcal{I})$ be a $(\mathcal{Q}, K)$-minor class. Structias $S, S' \in \mathcal{S}$ are ψ-equivalent, denoted $S\psi S'$, if $gs(S) = gs(S') = Q$ (say) and for all $q \in Q$, $S\backslash q = S'\backslash q$ and $S/q = S'/q$. If $S\psi S'$ and $S \neq S'$ then S and S' are ψ-structias. Thus S is a ψ-structia if and only if it is not uniquely determined by specifying all its proper minors—specifying not only what the proper minors are, but also how each is obtained. Note that $S\psi S'$ and $S \neq S'$ neither implies $S \cong S'$ nor implies $S \not\cong S'$.

The *core* of S, denoted S^ψ, is the sub minor class of S consisting of all isominors of ψ-structias. In many well known minor classes there are just a few ψ-structias and they are easy to find.

For a patterned hypercube on $[0,1]^Q$ depicting a structure (S,Q), the pattern on the interior $(0,1)^Q$ is the *inpattern* for S and the pattern on the exterior $[0,1]^Q - (0,1)^Q$ is the *outpattern* for S.

A motivation for these definitions comes from the previous section. Consider building up the patterning of hypercubes inductively. Suppose structia (S,Q) has not yet been depicted as a patterned hypercube, but that all its proper minors have—so its outpattern is determined, but its inpattern is not yet determined. If S is not a ψ-structia then the outpattern already determines S and the inpattern can simply be chosen to be blank, or given some default colour, say white. However, if S is a ψ-structia, then it is necessary to choose some inpattern for S, so as to distinguish S from any other S' with $S'\psi S$. Also this must be done so that the patterned hypercube has the same automorphisms as S, and the pattern for any structia isomorphic to S is consistent, as discussed in §4.

The minor class S can be extended in a canonical way as now defined. Consider patterned hypercubes where the only (non-blank) patterns on interiors of subhypercubes are inpatterns for ψ-structias, subject to the following compatibility condition:

—For any subhypercube, its exterior has the outpattern for a ψ-structia S if and only if its interior has the inpattern for some S' such that $S'\psi S$. (Note that if $S\psi S'$ then S' and S have the same outpattern.)

Let $\overline{S}$ be the minor class whose structias are exactly the patterned hypercubes of this form. Now S is embeddable in $\overline{S}$ and $\overline{S}$ is called the *completion* of S. In fact $\overline{S}$ is the unique (up to isomorphism of minor classes) maximal minor class with the same core, S^ψ, as S (and S^ψ is of course the unique minimal one).

The *natural excluded minors* of S are its excluded minors in $\overline{S}$. These are the minor minimal patterned hypercubes in $\overline{S}$ which do not depict structias in S. The *ψ-description* of S is a list of its ψ-stuctures and its natural excluded minors. This uniquely determines (and is uniquely determined by) S since the ψ-structias determine S^ψ and hence $\overline{S}$ and the natural excluded minors then determine S. The minor class S has a *finite ψ-description* if there are finitely many ψ-stuctures and natural excluded minors (up to isomorphism of structias).

Consider the patterned hypercube for a natural excluded minor (S,Q) where $|Q| = n$. While $S \notin S$, its immediate minors $S \backslash q$ and S/q are in S for all $q \in Q$. These immediate minors are the $n-1$ dimensional subhypercubes. They intersect (if at all) in $n-2$ dimensional subhypercubes, and they must agree on these intersections. A characterisation, internal to S, of natural excluded minors in terms of their immediate minors is as follows. Let $Q \in \mathcal{Q}$ and for each $q \in Q$

let $S^q, S_q \in S_{Q-q}$. Assume that

$$(S^q)\backslash r = (S^r)\backslash q$$
$$(S^q)/r = (S_r)\backslash q$$
$$(S_q)/r = (S_r)/q.$$

If there is no $S \in S_Q$ such that $S^q = S\backslash q$ and $S_q = S/q$ for all $q \in Q$, then the S^q's and S_q's are the immediate minors of a natural excluded minor. (On the other hand if there is more than one structure satisfying $S^q = S\backslash q$ and $S_q = S/q$, then such structures are ψ-equivalent.) For any minor class S' with sub minor class S, each excluded minor of S in S' is either ψ-equivalent to a structia in S or else it has the same immediate minors as some natural excluded minor of S.

A 'concrete' minor class can be described in various ways so there is no general method for finding its ψ-description, and a variety of techniques are useful. The next section gives ψ-descriptions for some well known minor classes. If S is a subminor class of S' and the ψ-description of S' is known, then finding the ψ-description of S is much the same problem as finding the excluded minors of S in S'.

6. ψ-descriptions of some well known minor classes.

In this section, all ground sets are finite. The *size* of a structia is the cardinality of its ground set. This section summarises the ψ-descriptions of several well known minor classes. Some are given explicitly, while for others, just the numbers (up to isomorphism) of ψ-structures and natural excluded minors of given size are stated.

For $|K| = 2$ we have the patterned hypercube visualisation as described earlier. For arbitrary K, structias can be depicted such that the $|K|$ ways of removing an element q correspond to $|K|$ 'cross sections' perpendicular to the q-axis. For $|K| = 1$ (say $K = \{delete\}$) one can say minors are only those sub-hypercubes containing the origin. Alternatively, for $|K| = 1$, structias of size n can be depicted as patterned n-vertex simplices, with minors being subsimplices (disregarding size 0 strictias).

It is worth comparing (and contrasting!) the ψ-description of a minor class with its image under some homomorphism. Some homomorphisms are mentioned after the ψ-descriptions. For example, the map which sends a digraph to its underlying graph yields a homomorphism from each minor class of digraphs (three examples) to the corresponding minor class of graphs.

6.1. Minor classes of graphs.
After some explanation, Table 1 summarises the ψ-description of six minor classes of graphs.

Edge digraphs are digraphs where the ground set is the edge set, and vertices are unlabelled. Two edge digraphs are *equal* (not merely isomorphic) if they are identical, with respect to the edge labelling. Define *edge graphs, vertex digraphs, vertex graphs* analogously. The six minor classes in the table are respectively, simple vertex digraphs and graphs with vertex deletion, edge digraphs and graphs with edge deletion only, edge digraphs and graphs with both edge deletion and contraction. (Edge digraphs and graphs with edge contraction only, also form minor classes but their ψ-descriptions are infinite.)

The second row refers to structia size. The natural excluded minors are counted up to isomorphism. The numbers in the ψ-structias column are the cardinalities of the ψ-equivalence classes, mentioning each ψ-equivalence class once up to isomorphism.

	$\psi-$	structias			nat.	exc.	min's		
size	0	1	2	3	> 3	1	2	3	4	> 4
$\mathcal{G}(SVD)$			3			0	0	0	0	0
$\mathcal{G}(SVG)$			2			0	0	0	0	0
$\mathcal{G}(ED,1)$		2	2,3,7			0	0	78	0	0
$\mathcal{G}(EG,1)$		2	2,3,3	2		0	0	7	8	0
$\mathcal{G}(ED,2)$		2	2,2,3,5			0	6	4890	0	0
$\mathcal{G}(EG,2)$		2	2,2,2			0	6	125	6	0

Table 1

6.2. Matroids and closure operators with deletion and contraction.
Let $\mathcal{M}$ be the minor class of matroids with deletion and contraction. (See [5] for definitions. For those unfamiliar with matroids, a new visualisation and axiomatisation of matroids arises here.) On a ground set $\{q\}$, say, there are two matroids, called *loop* and *coloop*, which are ψ-equivalent. These are the only ψ-structias. Thus the structias in $\overline{\mathcal{M}}$, the completion of $\mathcal{M}$, can be depicted by hypercubes with just the 1-dimensional faces labelled by L for loop or C for coloop. (The rest of the hypercube remains blank.) It is instructive to look at the 2-point structias in $\overline{\mathcal{M}}$, say those on the 2-element ground set $\{q,r\}$. These are depicted by squares in $\mathbf{R}^{\{q,r\}}$ with their four 1-dimensional edges (corresponding to the 1-point minors obtainable by deleting or contracting q or r) labelled by either L or C. The labelling scheme is given for a structia σ on ground set $\{q,r\}$ in Figure 1. Note that the only possible automorphisms of σ, and the corresponding depiction, are that which fixes q and r (and fixes the corresponding square) and that which swaps q and r (and reflects the square along the dotted line, as in Figure 1). No other symmetries of the square correspond to structia automorphisms.

There are, of course, $2^4 = 16$ structias on ground set $\{q,r\}$ in $\overline{\mathcal{M}}$, but only 10 up to isomorphism. Ten representatives have been named $2a, 2b, \cdots, 2j$ as in Figure 1. (Observe that 2a,2e,2h and 2j have the bijection which swaps q and r, as an automorphism.) By observation, only 2a,2e,2f and 2j are matroids, so that 2b,2c,2d,2g,2h and 2i are natural excluded minors of $\overline{\mathcal{M}}$. These six, called *the matroid six* are the only natural excluded minors of $\mathcal{M}$. Thus, a matroid can be

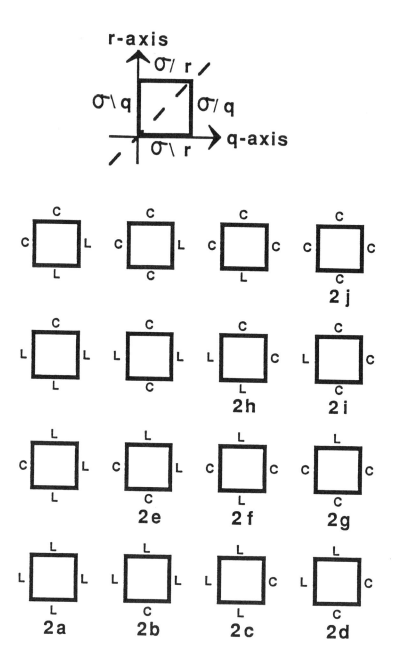

Figure 1: The 2-point structures of $\mathcal{M}$.

depicted as a hypercube with 1-dimensional faces labelled by L or C, such that no 2-dimensional face is one of the matroid six.

Let $\mathcal{M}^C$ be the minor class of closure operators with deletion and contraction. This has the same ψ-structias as $\mathcal{M}$ and only five natural excluded minors, namely 2b, 2c, 2d, 2h and 2i.

6.3. Chain groups with deletion and contraction.

Let R be a field. A *chain group*, S, (see [3]) on ground set Q is a subspace of the vector space R^Q. Loosely speaking, for $q \in Q$, $S \backslash q$ is obtained by intersecting S with R^{Q-q} and S/q is obtained by projecting S onto R^{Q-q}. Let $\mathcal{D}(R)$ be the minor class of chain groups with deletion and contraction.

Let $\mathcal{M}_R$ be the minor class of matroids coordinatisable over R. The motivation for studying $\mathcal{D}(R)$ is that it has $\mathcal{M}_R$ as a homomorphic image. While the ψ-description of $\mathcal{D}(R)$ given below seems natural and aesthetic, the excluded minors of $\mathcal{M}_R$ in $\mathcal{M}$, and hence the ψ-description of $\mathcal{M}_R$, remain unknown for $|R| \geq 4$.

Let chain groups loop(q), coloop(q) be respectively the 1-dimensional, 0-dimensional subspaces of $R^{\{q\}}$. The homomorphism $\alpha : \mathcal{D}(R) \to \mathcal{M}$ is unquely determined by specifying that it sends loop to loop and coloop to coloop (see §6.4). It also extends uniquely to a homomorphism from $\overline{\mathcal{D}(R)}$ to $\overline{\mathcal{M}_R}$.

For $|Q| = n$ let $U_m^n(Q)$ be the uniform matroid of rank m on ground set Q. Let + denote direct sum of matroids.

For $a \in R - \{0\}$ define the chain group $a(q,r)$, called a *slope*, on ground set $\{q,r\}$ where $x \in a(q,r)$ if and only if $x(r) = a \times x(q)$. Note that $\alpha(a(q,r)) = U_1^2(q,r)$.

The only non-trivial ψ-equivalences are:
(i) loop(q)ψcoloop(q),
(ii) $a(q,r)\psi a'(q,r)$, for distinct $a, a' \in R - \{0\}$.

(For $|R| = 2$ case (ii) vanishes.) Thus any structia in $\mathcal{D}(R)$ can be depicted as a hypercube with all 1-faces labelled L or C, as for matroids, and all 2-faces (squares) which look like 2e in Figure 1 (that is U_1^2), are labelled to indicate which slope the corresponding two element minor is. The patterned hypercube depicting $\alpha(S)$ is obtained from that depicting S simply by blanking out the 2-faces.

The natural excluded minors are:
(i) the matroid six,

(For any other natural excluded minor, S, $\alpha(S)$ is a matroid and this can be used to describe S.)

(ii) S where $\alpha(S) = U_1^1(q) + U_1^2(p,r)$ and $S\backslash q = a(p,r)$ and $S/q = b(q,r)$ for some distinct non-zero $a, b \in R$,

(iii) S where $\alpha(S) = U_0^1(q) + U_1^2(p,r)$ and $S\backslash q = a(p,r)$ and $S/q = b(q,r)$ for some distinct non-zero $a, b \in R$,

(iv) S where $\alpha(S) = U_1^3(p,q,r)$ and the slopes $S\backslash p$, $S\backslash q$ and $S\backslash r$ are respectively

(a, q, r), (b, r, p) and (c, p, q) where non-zero $a, b, c \in R$ are such that $abc \neq -1$,
(v) S where $\alpha(S) = U_2^3(p, q, r)$ and the slopes $S/p, S/q, S/r$ are respectively $(a, q, r), (b, r, p), (c, p, q)$ where non-zero $a, b, c \in R$ are such that $abc \neq 1$.
(vi) S where $\alpha(S) = U_4^2(1, 2, 3, 4)$ (for some 4 element ground set $\{1,2,3,4\}$, say) and S does not have a minor as in case (iv) or (v) but S is not in $\mathcal{D}(R)$. More explicitly: for distinct $k, l, m, n \in \{1, 2, 3, 4\}$ let non-zero $a_{mn}^{kl} \in R$ be such that $S \backslash k/l = (a_{mn}^{kl}, m, n)$. (Note that $a_{mn}^{kl} = 1/a_{nm}^{kl}$.) Let $a = a_{23}^{14}$, $b = a_{31}^{24}$, $c = a_{12}^{34}$, $d = a_{43}^{21}$, $e = a_{24}^{31}$, $f = a_{32}^{41}$, $g = a_{41}^{32}$, $h = a_{13}^{42}$, $i = a_{34}^{12}$, $j = a_{21}^{43}$, $k = a_{42}^{13}$, $\ell = a_{14}^{23}$. Then S is a natural excluded minor provided that $1 = abc = def = ghi = jkl$ and $-1 = aik = bdl = ceg = fhj$, but $afi \neq i - ef$.

For any field R (not necessarily finite) $\mathcal{D}(R)$ has ψ-structias of size 1 and 2 and has natural excluded minors of size 2, 3 and 4 (except for $|R| = 2$ where some of the cases vanish). The ψ-description is finite for finite fields. However the ψ-description of $\mathcal{M}_R$ (which is the ψ-description of $\mathcal{M}$ together with the excluded minors of $\mathcal{M}_R$ in $\mathcal{M}$) are known only for $|R| = 2$, [3] and $|R| = 3$, [2]. The question remains open whether $\mathcal{M}_R$ has finitely many excluded minors for each finite R with $|R| \geq 4$. This is a motivation for the study of minor class homomorphisms.

6.4. Some examples of homomorphisms.

The homomorphism in §6.3 is an example of the following situation. There is a surjective homomorphism $\alpha : S' \to S$, where S' has a finite ψ-description and S has a finite core. It is a theorem that α is uniquely determined by $\alpha|S^\psi$ which can be described finitely (listing the finitely many structias in S' sent to the ψ-structias of S). Some homomorphisms, such as the α in §6.3, correspond to simply blanking out some patterns in the patterned hypercubes. This is not true in general, but for reasons omitted here, one can easily reduce to such homomorphisms. However, there are no results for finding the natural excluded minors of S or even characterising when there are finitely many. This is an important problem in minor classes.

Here are some examples of such a homomorphism where a finite list of natural excluded minors of S is known (but seem to bear no obvious relation to those of S'). The only obvious observation from these few examples is that the homomorphic image seems to have fewer but larger natural excluded minors.

—The map which sends a digraph to its underlying graph is a surjective homomorphism from each minor class of digraphs in Table 1 (three examples) to the corresponding minor class of graphs.

—The map sending a digraph or graph to its graphic matroid is a surjective homomorphism from $\mathcal{G}(ED, 2)$ or $\mathcal{G}(EG, 2)$ to $\mathcal{M}_G$, the minor class of graphic matroids. The ψ-description of $\mathcal{M}_G$ is the ψ-description of $\mathcal{M}$ together with the excluded minors of $\mathcal{M}_G$ in $\mathcal{M}$ which (see [4]) have sizes 4, 7, 7, 9, 10.

—The map sending a digraph or graph to its line graph is a surjective homomorphism from $\mathcal{G}(ED, 1)$ or $\mathcal{G}(EG, 1)$ to the minor class of line graphs with vertex deletion. Its excluded minors in $\mathcal{G}(SVG)$ have sizes 4, 6, 6, 6, 6, 7, 7.

—For the field R with $|R| = 3$, the map sending a chain group over R to its ternary matroid is a surjective homomorphism from $\mathcal{D}(R)$ to $\mathcal{M}_3$, the minor class of ternary matroids. The ψ-description of $\mathcal{M}_3$ is the ψ-description of $\mathcal{M}$ together with the excluded minors of $\mathcal{M}_3$ in $\mathcal{M}$ which (see [2]) have sizes 5, 5, 7, 7.

REFERENCES

1. P.J.Higgins, Algebras with a scheme of operators, *Math. Nachr.* 27 (1963), 115-132.
2. P.D.Seymour, Matroid representation over GF(3), *J. Combinatorial Theory* (B) 26 (1979), 159-173.
3. W.T.Tutte, A homotopy theorem for matroids II, *Trans. Amer. Math. Soc.* 88 (1958), 161-174.
4. W.T.Tutte, Lectures on matroids, *J. Res. Nat. Bur. Stand.* 69B (1965), 1-48.
5. D.J.A.Welsh, "Matroid Theory", Academic Press, London, 1976.

DIMACS CENTER, RUTGERS UNIVERSITY, P.O.BOX 1179, PISCATAWAY, NJ 08855-1179, USA
E-mail address: vertigan@dimacs.rutgers.edu

WELL-QUASI-ORDERING FINITE POSETS
(EXTENDED ABSTRACT[1])

JENS GUSTEDT

ABSTRACT. We show that the set of finite posets is a well-quasi-order with respect to a certain relation $\preceq$, called chain minor relation. As a consequence we get that every property which is hereditary with respect to $\preceq$ has a test in $O\left(|P|^c\right)$ where c depends on the property. This test has an easy parallelization with the same costs. On a parallel machine (CRCW PRAM) it may be implemented in such a way that it runs in constant time and needs $O\left(|P|^c\right)$ processors.

1. INTRODUCTION

In the last years algorithmic aspects of well-quasi-orders (wqo's) brought great progress in algorithmic graph theory. In a series of papers Robertson and Seymour (see [RS83], [RS86] ...) showed that a set of graphs together with the graph minor relation forms a wqo. This can be used to show the existence of polynomial time algorithms for a wide class of problems. These problems are those which are hereditary with respect to the graph minor relation.

A similar theory for finite posets was not known until now. In this paper we investigate the chain minor relation between finite posets. This relation was introduced recently by Möhring and Müller in [MM91] to generalize certain approaches in the theory of scheduling stochastic project networks.

We show that this relation indeed leads to a wqo, and that a test for hereditary properties can be done in polynomial sequential time (or in constant time with the same cost on a parallel machine model).

To show polynomiality we give a brute force algorithm to test whether or not two posets are related in the chain minor relation.

2. BASIC NOTATION AND FACTS

$P = (V, <)$ is called a partially ordered set, **poset** for short, or **order** if "<" is a transitive irreflexive relation on the set V. It is **finite** if V is finite, and to denote this we write $|P| = |V|$.

$Q = (V, \leq)$ is called a quasi-order or **qo** if "$\leq$" is transitive and reflexive. A sequence of elements (v_i) in Q is called **good** if there are $i < j$ such that $v_i \leq v_j$. It is **bad** if it is not good and it is **perfect** if $v_i \leq v_j$ for all $i \leq j$.

1991 *Mathematics Subject Classification*. 06A07, 68R99, 68Q22.
Supported by the Deutsche Forschungsgemeinschaft
[1]This is an extended abstract of [Gus91]

Q is a well-quasi-order or **wqo** if it is a qo and if every sequence is good. Observe that in a wqo every sequence has a perfect subsequence and that every antichain is finite. For an overview and bibliography on wqo's refer to the articles of Milner [Mil85] and Pouzet [Pou85] in [GO85] and for a historical overview see for example [Kru72].

For our purposes the difference between the definitions of posets and qo's is not very important. If we define an equivalence relation $\cong$ in a qo $Q = (V, \leq)$ by

$$(1) \qquad (v \cong w) \iff \Big((v \leq w) \wedge (w \leq v) \Big)$$

then $Q/\!\cong\, = (V/\!\cong, <)$ is a poset.

We use the distinction between posets and qo's in the following way: posets will be finite and qo's will represent infinite sets of finite posets. The relations in these qo's will be given by the existence of certain morphisms between the posets.

Let $P = (V, <)$ and $P' = (V', <')$ be posets. We say P is a **chain minor** of P', $P \preceq P'$, if there is a partial mapping $\rho\colon V' \to V$ which has the following property:

> For every chain C in P there is a chain C' in P' such that $\rho\big|_{C'}$ is an isomorphism of chains.

ρ is then called a **chain morphism**. Here $\rho\big|_{C'}$ denotes the partial mapping induced on $P'\big|_{C'}$, the order restricted to the groundset of C'. Observe that every chain morphism is onto and that $\preceq$ defines a qo on any set of posets.

Our first aim is the following theorem.

Theorem 2.1. *Any set of finite posets is a wqo with respect to $\preceq$.*

A property E of the elements of a wqo Q is **hereditary** if the subset of elements with that property forms a lower ideal, that is if

$$(2) \qquad E(w) \wedge (v \leq w) \implies E(v).$$

There is a one-to-one correspondence between hereditary properties in $Q/\!\cong$ and the antichains of $Q/\!\cong$: Every such property defines an antichain, its set of **minimal obstructions**, which are just the minimal elements of the corresponding upper ideal.

Vice versa every antichain $\{[v_1], \ldots, [v_l]\}$ in $Q/\!\cong$ defines a hereditary property in $Q/\!\cong$ (and thus in Q) by

$$(3) \qquad \neg\Big(\big([v_1] \leq [v]\big) \vee \cdots \vee \big([v_l] \leq [v]\big) \Big)$$

For our special relation $\preceq$ we will get

Theorem 2.2. *Every property of finite posets which is hereditary with respect to $\preceq$ has a decision algorithm which runs*
- *in sequential polynomial time*
- *in constant time on a CRCW PRAM and uses polynomially many processors.*

3. Showing the well-quasi-ordering property

We give a sketch of a proof for theorem 2.1. We have to show that any infinite sequence (P_i) of finite posets is good.

First we can easily exclude two cases. The first is that the given sequence of posets (P_i) has members with arbitrary large height. Then we find a large chain $C = \{c_1, \ldots, c_h\}$ in P_l, say, such that $h = |C| = |P_1|$. If $\{v_1, \ldots, v_h\}$ is a linear extension of P_1 the partial mapping ρ given by $\rho(c_i) = v_i$ is a chain morphism. So (P_i) is good in that case.

For the second we assume that height (P_i) is globally bounded. We may therefore assume that height (P_i) is fixed independent of i, by choosing an infinite subsequence. In addition we assume that the number of disjoint chains of maximum length is not globally bounded. Then we find P_l, say, which has more disjoint maximum chains than P_1 has maximal chains. But then clearly $P_1 \preceq P_l$ and (P_i) is good in that case, too. See figure 1.

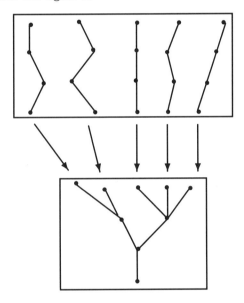

FIGURE 1. Many disjoint chains

So from now on we may assume that height (P_i) is equal to h, say, and that in each of the P_i no $c+1$ maximum chains are mutually disjoint. We choose a set $K_i = \{C_i^1, \ldots, C_i^c\}$ of maximum chains in P_i such that each maximum chain in P_i intersects one of the chains in K_i. We have for all i that

$$(4) \qquad \left| \bigcup_{j=1}^{c} C_i^j \right| \leq c \cdot h \,.$$

So there are only finitely many isomorphism types if we restrict our posets to these sets. Choose a subsequence with isomorphic restrictions, see figure 2. We may now identify all these isomorphic restrictions with one single poset. Denote it with R. All maximal chains in all posets are now classified by their intersection with R. For example, one special class is formed by the set of maximal chains which do not have maximum length and do not intersect R.

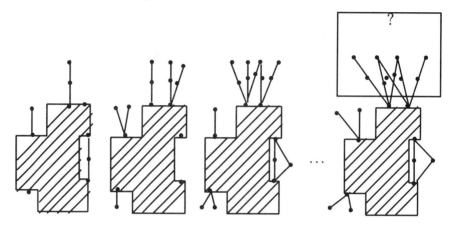

FIGURE 2. Subsequence with isomorphic restrictions

There are only finitely many classes and the chains in each class are necessarily shorter than h if we omit R. For such a class K and $C \in K$ let h_K be $h - |C \cap K|$. We define

$$(5) \qquad s_K(i) = |\{C \mid C \subseteq P_i,\, C \text{ is a chain in } K \text{ with length } h_K\}|.$$

If we now carefully apply an analogous recursive procedure to all the suborders induced by all these classes we end up with a subsequence and restriction R as above with two additional properties. The first is that for each class $s_K(.)$ is either increasing and unbounded or it is 1 for all i. (If it is 1 the corresponding chain belongs to R.) The second is that every two chains in the same class intersect exactly in R.

Now there exists P_l, say, such that each of its unbounded classes has more maximal chains than P_1 has maximal chains, say. But then $P_1 \preceq P_l$ and the sequence is good, too.

4. Algorithmic Aspects

We now prove theorem 2.2, that is we want to show that there are polynomial algorithms for hereditary properties on posets. For that let

$$(6) \qquad c(P) = \sum_{\substack{C \text{ maximal} \\ \text{chain in } P}} |C|$$

Lemma 4.1. *Let $P_1 \preceq P_2$ be posets and $c = c(P_1) \geq 3$. Then there is a poset P_0 which fulfills:*

(1) $P_1 \preceq P_0$
(2) P_0 *is an induced suborder of* P_2
(3) $|P_0| \leq c$.

Proof. Let ρ be a chain morphism which gives $P_1 \preceq P_2$ and let $C_1^1, \ldots, C_1^k$ be the maximal chains of P_1. There are chains $C_2^1, \ldots, C_2^k$ in P_2 such that $\rho\big|_{C_2^i}$ is an isomorphism of chains for all i. Set $V_0 = \bigcup C_2^i$, $P_0 = P_2\big|_{V_0}$ and $\rho_0 = \rho\big|_{V_0}$. Then P_0 and ρ_0 obviously have all the desired properties. $\square$

Lemma 4.2. *Let P_1 and P_2 be finite posets and $c = c(P_1) \geq 3$. Then there is a constant l depending only on P_1 and an algorithm to decide whether or not $P_1 \preceq P_2$ holds that runs*

— *in $O\left(c^2 \cdot |P_2|^c + l\right)$ sequential time*
— *in $O\left(c^2 + l\right)$ time on a CRCW PRAM with $O\left(|P_2|^c\right)$ processors.*

We omit the proof of this lemma. It is based on the observation that with lemma 4.1 we only have to test small subsets of P_2.

Proof (of Theorem 2.2). Let E be a property of finite posets which is hereditary with respect to $\preceq$. By theorem 2.1 we know that the set of minimal obstructions for E is finite, $\{P_1, \ldots, P_l\}$ say. Set c_{max} to the maximum of the constants $c(P_i)$. With lemma 4.2 we know that the test for

(7) $$(P_1 \preceq P) \vee \cdots \vee (P_l \preceq P)$$

can be done in $O\left(|P|^{c_{max}}\right)$ time sequentially and in constant time with $O\left(|P|^{c_{max}}\right)$ processors. $\square$

References

[GO85] Graphs and Orders (Ivan Rival, ed.), D. Reidel Publishing Company, Dordrecht, 1985.
[Gus91] Jens Gustedt, *Well-quasi-ordering finite posets and formal languages*, Tech. Report 290, Technische Universität Berlin, 1991, submitted.
[Kru72] J. B. Kruskal, *The theory of well-quasi-ordering: a frequently discovered concept*, J. Combin. Theory Ser. A **13** (1972), 297–305.
[Mil85] E. C. Milner, *Basic wqo- and bqo-theory*, [GO85], 1985, pp. 487–502.
[MM91] Rolf H. Möhring and Rudolf Müller, *A combinatorial approach to obtain bounds for stochastic project neworks*, Tech. report, Technische Universität Berlin, 1991.
[Pou85] M. Pouzet, *Applications of well quasi-ordering and better quasi-ordering*, [GO85], 1985, pp. 503–519.
[RS83] Neil Robertson and Paul Seymour, *Graph minors I, excluding a forest*, J. Combin. Theory Ser. B **35** (1983), 39–61.
[RS86] Neil Robertson and Paul Seymour, *Graph minors II, algorithmic aspects of tree-width*, J. Algorithms **7** (1986), 309–322.

Technische Universität Berlin, Fachbereich Mathematik, Sekr. MA 6-1, Strasse des 17. Juni 135, W-1000 Berlin 12, Germany
E-mail address: jens@combi.math.tu-berlin.de

The Immersion Relation on Webs

Guoli Ding

Abstract: A web is a cyclicly ordered finite set N together with a partition of N into two-element subsets. We introduce a containment relation on webs called immersion and we prove that webs are well-quasi-ordered under the immersion.

1. Introduction

Let N be a finite set and let Ω be a cyclic permutation of N. Let N be partitioned into two-element subsets and let the set of these subsets be E. Then we call the triple $W = (N, \Omega, E)$ a *web*. The members of N and E are called *nodes* and *edges* of W respectively. Figure 1 gives an example of a web.

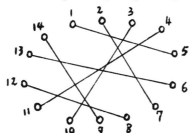

Figure 1. A web W

Webs are combinatorial objects which are very similar to graphs. We shall investigate in this paper a containment relation called "immersion" on webs.

Let $W = (N, \Omega, E)$ and $W' = (N', \Omega', E')$ be two webs. We say W' is a *subweb* of W if $N' \subseteq N$, $E' \subseteq E$ and $\Omega' = \Omega|_{N'}$, where Ω' is the natural restriction of Ω to N', that is, for every $x \in N'$, $\Omega'(x)$ is the first term of the sequence $\Omega(x), \Omega(\Omega(x)), ..., x$ which is in N'. If there are two distinct edges $e_1 = \{x_1, y_1\}$ and $e_2 = \{x_2, y_2\}$ of W such that $\Omega(y_1) = y_2$, $N' = N \setminus \{y_1, y_2\}$, $\Omega' = \Omega|_{N'}$ and $E' = (E \setminus \{e_1, e_2\}) \cup \{e\}$, where $e = \{x_1, x_2\}$, then we say that W' is obtained from W by an *elementary contraction* operation. We call W' a *contraction* of W if W' can be obtained from W by a sequence of elementary

---

AMS Subject classification: 05C55, 05C75 and 05C99.

This is an extended abstract of my thesis [2] which was finished in May 1991 under the supervision of Dr. P. Seymour. It is in final form and no version of it will be submitted for publication elsewhere.

contraction operations. An *immersion* in W is a contraction of a subweb of W. We denote the binary relations "is isomorphic to a subweb of, a contraction of and an immersion in" by $\preceq_s$, $\preceq_c$ and $\preceq_i$ respectively.

Our main question is the following. Given a web W, let $\mathcal{F}(W)$ be the class of all webs U such that none of the immersion in U is isomorphic to W. Then what can we say about the structure of the webs in $\mathcal{F}(W)$ and how can we recognize the members of $\mathcal{F}(W)$? We start with the following definition. A set $\mathcal{G}$ of webs is called a *generator* if for every web W, there exists $G \in \mathcal{G}$ such that $W \preceq_c G$. Let $\mathcal{G}^i = \{G_n^i : n = 1, 2, ...\}$ for $i = 1, ..., 12$, where $G_n^1, ..., G_n^{12}$ are the webs shown in Figure 2. Then our first result is

Theorem 1.1: $\mathcal{G}^1, ..., \mathcal{G}^{12}$ are a generators.

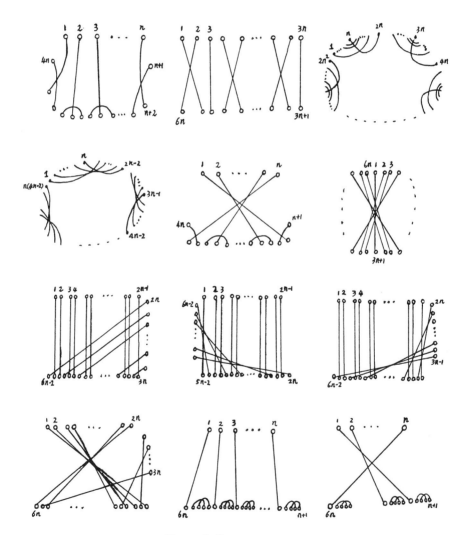

Figure 2. Some generators

Because of this theorem, in order to study the structure of the members of $\mathcal{F}(W)$ for a general web W, we only need to study, for every positive integer n, the structure of the webs without $G_n^1, ..., G_n^{12}$ as subwebs. We shall present the structure of these webs in the next section. In section 3, we present some results concerning well-quasi-orders, and in section 4, we discuss some web algorithms. Finally, in section 5, we point out the connection between the immersion relation on webs and the minor relation on graphs.

2. Excluding the generators

We begin this section by introducing two classes of webs, planar webs and webs of bounded width. It turns out that all members of $\mathcal{F}(W)$ can be built up from these webs by certain elementary operations.

Let $W = (N, \Omega, E)$ be a web and let X, Y be two disjoint subsets of N. We shall say that X *crosses* Y if there are two distinct nodes $x_1, x_2 \in X$ such that $\Omega'(x_1), \Omega'(x_2) \in Y$, where $\Omega' = \Omega|_{Y \cup \{x_1, x_2\}}$. It is not difficult to see that X crosses Y if and only if Y crosses X. An edge of W is called *planar* if it does not cross any other edge. If all the edges of a web are planar, then we call the web a *planar web*. The class of all planar webs will be denoted by $\mathcal{P}$. It is clear that no member of $\mathcal{P}$ has a subweb isomorphic to G_2^i for any i.

Let $W = (N, \Omega, E)$ be a web and let $x, y \in N$. We define $\Omega(x, y)$ to be the closed interval from x to y, that is, $\Omega(x, y)$ is the set of nodes $z \in N$ with $\Omega'(y) = x$, where $\Omega' = \Omega|_{\{x,y,z\}}$. If $x \neq y$, we define $Cr(x, y)$ to be the set of edges $e \in E$ such that $e \cap \Omega(\Omega(x), y) \neq \emptyset \neq e \cap \Omega(\Omega(y), x)$. For convenience, we also define $Cr(x, x) = \emptyset$ for all $x \in N$. To visualize $Cr(x, y)$, we look at two ordered pairs of consecutive nodes $(x, \Omega(x))$ and $(y, \Omega(y))$ of W. Think of these as two pointers. Then $Cr(x, y)$ is the set of edges crossing this pair of pointers. Now we define the *width* $w(W)$ of W as follows. If $N(W) = \emptyset$, then $w(W) = 0$. If $|N(W)| > 0$, then $w(W) = max\{|Cr(x, y)| : x, y \in N\}$. For every positive integer k, the class of all webs of width at most k is denoted by Δ_k. Clearly no member of Δ_k has a subweb isomorphic to G_{k+1}^i for any i.

To generate more webs without any G_n^i subwebs, we define the following composition operation on webs. Let $W' = (N', \Omega', E')$, $W'' = (N'', \Omega'', E'')$ be webs with $N' \cap N'' = \emptyset$ and let $x' \in N'$, $x'' \in N''$. By *sticking* W'' on W' at (x', x'') we get a new web $W = (N, \Omega, E)$ such that $N = N' \cup N''$, $E = E' \cup E''$ and Ω is given by

$$x', \Omega''(x''), \Omega''(\Omega''(x'')), ..., x'', \Omega'(x'), \Omega'(\Omega'(x')), ..., x'.$$

Let W, U be webs and let $\mathcal{W}$ be a class of webs. If there exist a subset X of $N(W)$ and webs $W_x \in \mathcal{W}$ with $y_x \in N(W_x)$ for all $x \in X$, such that U is obtained by sticking W_x on W at (x, y_x) for all $x \in X$, then we say that U is obtained from W by $\mathcal{W}$-*sticking*. For positive integers s and t, we define $\mathcal{S}(s, t)$ to be the class of all webs W such that W can be obtained from a web in $\mathcal{S}(s, t-1)$ by $\mathcal{P} \cup \Delta_s$-sticking, where $\mathcal{S}(s, 0)$ is $\mathcal{P} \cup \Delta_s$. It is not difficult to see that no member of $\mathcal{S}(s, t)$ has a subweb isomorphic to G_{2st}^i for any $i \leq 11$.

To understand more about the class $\mathcal{S}(s, t)$ of webs, we introduce some more notation. Let $W = (N, \Omega, E)$ be a web and let $d(x, y)$ be the number of nonplanar edges in $Cr(x, y)$, where $x, y \in N$. We define the *depth* $d(W)$ of W to be $max\{d(x, y) : x, y \in N\}$. If $N = \emptyset$, we define $d(W) = 0$. We also define webs H_n^0 and H_n^1, as illustrated in Figure 3, for all positive integers n. Then we have

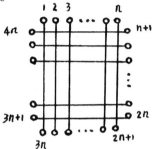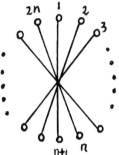

Figure 3. Webs H_n^0 and H_n^1

Theorem 2.1: *For every positive integer n, there exist two numbers δ and γ such that if W is a web with no subweb isomorphic to G_n^1, G_n^2, G_n^3, H_n^0 or H_n^1, then $d(W') \leq \gamma$ for some subweb W' of W with $|E(W) - E(W')| < \delta$.*

Now we are ready to describe the structure of the members of $\mathcal{F}(W)$. Let $W = (N, \Omega, E)$ be a web and let $e = \{x, y\} \in E$. Suppose that N is disjoint from $V(H_n^1) = \{1, 2, ..., 2n\}$. Then we define the *replacement* of e by H_n^1 to be a new web $W' = (N', \Omega', E')$ such that $N' = (N \backslash e) \cup \{1, ..., 2n\}$, $E' = (E \backslash \{e\}) \cup E(H_n^1)$, and Ω' is given by

$$\Omega^{-1}(x), 1, ..., n, \Omega(x), \Omega(\Omega(x)), ..., \Omega^{-1}(y), n+1, ..., 2n, \Omega(y), ..., \Omega^{-1}(x),$$

where $\Omega^{-1}(x)$ is the inverse image of x under Ω. We also define the *replacement* of e by H_n^2 as above except that

$$E' = (E \backslash \{e\}) \cup \{\{1, 2n\}, \{2, 2n-1\}, ..., \{n, n+1\}\}.$$

Let θ be a positive integer. We shall call a web W a θ-web if W can be obtained from a web $U = (N, \Omega, E)$ by replacing every edge $e \in E$ by either $H_{n(e)}^1$ or $H_{n(e)}^2$ such that
(i) $|E| \leq \theta$;
(ii) let E_1 be the set of edges $e \in E$ such that e is replaced by $H_{n(e)}^1$. If $e_1 = \{x_1, y_1\}$, $e_1' = \{x_1', y_1'\} \in E_1$ are crossing edges with $\Omega'(x_1) = x_1'$, where $\Omega' = \Omega|_{\{x_1, y_1, x_1', y_1'\}}$, then both $|\Omega(x_1, x_1')| + |\Omega(y_1, y_1')| > 4$ and $|\Omega(x_1', y_1)| + |\Omega(y_1', x_1)| > 4$;
(iii) none of the edges in $E_2 = E \backslash E_1$ is of the form $\{x, \Omega(x)\}$; and for any two non-crossing edges $e_2 = \{x_2, y_2\}$, $e_2' = \{x_2', y_2'\} \in E_2$, if $\Omega'(x_2) = x_2'$ where $\Omega' = \Omega|_{\{x_2, y_2, x_2', y_2'\}}$, then $|\Omega(x_2, x_2')| + |\Omega(y_2', y_2)| > 4$.

For any two positive integers θ and γ, we denote by $\Sigma(\theta, \gamma)$ the class of all webs obtained from a θ-web by $\mathcal{S}(\gamma, \gamma)$-sticking.

Theorem 2.2 *For every positive integer n, there exist three numbers θ, γ and q such that if W is a web with no subweb isomorphic to G_n^i for any $i = 1, ..., 10$, then $W' \in \Sigma(\theta, \gamma)$ for some subweb W' of W with $|E(W) - E(W')| < q$.*

3. Well-quasi-orientation

A *quasi-orientation* $\mathcal{Q} = (Q, \preceq)$ consists of a set Q and a reflexive binary relation $\preceq$ on Q. We call it a *well-quasi-orientation* if for every infinite sequence $q_1, q_2, q_3, ...$ of elements of Q, there exist $i < j$ such that $q_i \preceq q_j$. Clearly, every well-quasi-order is a well-quasi-orientation and conversely, every transitive well-quasi-orientation is a well-quasi-order. Our first result in this section is about planar webs.

Theorem 3.1: *Both $(\mathcal{P}, \preceq_s)$ and $(\mathcal{P}, \preceq_c)$ are well-quasi-orders.*

The following sequence of webs shows that $(\Delta_4, \preceq_s)$ is not a well-quasi-order.

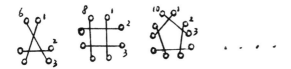

Nevertheless, we are able to show that

Theorem 3.2: (i) $(\Delta_3, \preceq_s)$ *is a well-quasi-order; and*

(ii) $(\Delta_k, \preceq_c)$ *is a well-quasi-order for all $k \geq 0$.*

As we have seen that $(\Delta_k, \preceq_s)$ is not a well-quasi-order in general. However, by weakening the binary relation $\preceq_s$ a little, we are able to prove a result very similar to 3.1. We define for every positive integer k a new binary relation $\propto_k$ on webs as follows. Let W and W' be webs. We say $W \propto_k W'$ if $W \preceq_c W''$ for some subweb W'' of W' with $|N(W'')| < 11(k+1)!|N(W)|$. Then we have

Theorem 3.3: $(\Delta_k, \propto_k)$ *is a well-quasi-orientation for all $k \geq 1$.*

In general, we have

Theorem 3.4: *For all positive integers θ and γ, $(\Sigma(\theta, \gamma), \propto_\gamma)$ is a well-quasi-orientation and $(\Sigma(\theta, \gamma), \preceq_c)$ is a well-quasi-order.*

As consequences, we have the main results of this paper.

Theorem 3.5: *For every web W, there exist finitely many webs $W_1, ..., W_k$ such that for every web U, $W \preceq_m U$ if and only if $W_i \preceq_s U$ for at least one W_i.*

Theorem 3.6: *Webs are well-quasi-ordered by $\preceq_i$.*

4. Algorithms

We shall summarize in this section our algorithms concerning webs. As before, we shall not go into any detail. If a web $W = (N, \Omega, E)$ is the input of an algorithm, we always assume that it takes unit time to compute $\Omega(x)$, $\Omega^{-1}(x)$ and $e(x)$ (the node of W such that $\{e(x), x\} \in E$) for every $x \in N$, and it takes unit time to compute x and y with $\{x, y\} = e$ for every $e \in E$. Moreover, we also assume that it takes unit time to decide if $x \in \Omega(y, z)$ for every choice of $x, y, z \in N$.

With these assumptions, it is clear that it takes constant time to decide if two edges are crossing. It follows that it takes linear time to compute $Cr(x, y)$ for every choice of nodes x, y. Therefore, it takes time $O(|E|)$ to compute the set of edges which cross a given edge (and hence to decide if a given edge is planar), and it takes time $O(|E|^2)$ to compute the width of a web.

It follows from 3.5 that to test if a web U is a member of $\mathcal{F}(W)$ we only need to test if U has a subweb isomorphic to one of $W_1, ..., W_k$. Thus we conclude that there exists an algorithm with running time $O(|E|^n)$, where n is $max\{|E(W_1)|, ..., |E(W_k)|\}$, which decides if U is a member of $\mathcal{F}(W)$. However, by using the technique introduced in [1], we can prove that for every fixed web W, there exists an algorithm with running time $O(|E|^3)$ which decides if U is a member of $\mathcal{F}(W)$.

We first observe that there is a linear time algorithm for every fixed web W_0 in $\mathcal{P}$ (or in Δ_k) which decides if a web in $\mathcal{P}$ (or in Δ_k) has a subweb isomorphic to W_0. Then based on the proof of 2.2 one can show that there is an algorithm with running time $O(|E|^3)$ which, for every fixed web W_0 and fixed positive integer n, decides if a web has a subweb isomorphic to one of W_0 and G_n^i for $i = 1, ..., 12$. Therefore, combined with 3.5, we have proved that there exists an algorithm with running time $O(|E|^3)$ which determines the membership of $\mathcal{F}(W_0)$ for every fixed web W_0. Unfortunately, we do not know how to construct this algorithm because we do not know how to derive the list of webs $W_1, ..., W_k$ described in 3.5.

5. Webs and graphs

In this section, we point out a few connections between webs and graphs. We start with graph ideals. A class $\mathcal{G}$ of graphs is called an *ideal* if for every graph $G \in \mathcal{G}$, all the minors of G are also in $\mathcal{G}$. It has been conjectured that ideals are well-quasi-ordered by the inclusion relation. This conjecture is still open except for the following case. We say an ideal $\mathcal{G}$ has tree width at most w if every graph in $\mathcal{G}$ has tree width at most w. It has been shown [4] that for every fixed positive integer w, graphs of tree width at most w are better-quasi-ordered by the minor relation. This result implies that the conjecture holds for the class of ideals of tree width at most w. In the following, we present a class of ideals with unbounded tree width which is well-quasi-ordered by the inclusion relation.

Let G be a planar graph which is embedded in the plane and let $N \subseteq V(G)$ have even cardinality such that each $x \in N$ is on the boundary of the infinite face F of G. Suppose that the boundary of F is a simple closed curve. Then this curve cyclicly permutes N in the natural way. Let Ω be this cyclic permutation of N (clockwise, say) and let $U = (N, \Omega, E)$ be a web. Let $H(U, G)$ be the graph obtained from G by adding the edges xy for all $\{x, y\} \in E$. Finally we define for every web W an ideal $\mathcal{G}(W)$ of graphs such that every graph in $\mathcal{G}(W)$ is a isomorphic to a minor of some $H(U, G)$, where U is isomorphic to W. Obviously, every $\mathcal{G}(W)$ has an unbounded tree width. Moreover, it is not difficult to see that for every $\gamma \geq 0$, there exists a web W such that $\mathcal{G}(W)$ has genus at least γ.

We observe that for any two webs W and W', $W \preceq_i W'$ implies that $\mathcal{G}(W) \subseteq \mathcal{G}(W')$. Thus from 3.6 we deduce that the class of ideals $\mathcal{G}(W)$, for all webs W, is well-quasi-ordered by the inclusion relation.

Next, we consider the disjoint paths problem, that is, given a graph G and k pairs of vertices of G, do there exist k mutually vertex-disjoint paths of G joining the pairs? It is well known that this problem is NP-complete if k is part of the input. For fixed k, however, this problem is solvable in polynomial time [3]. We shall present in the following an algorithm to solve the disjoint paths problem (with fixed k) for a special class of graphs.

Let us assume that every input graph has the form $G = H(W, G')$ for some graph G' and web $W = (N, \Omega, E)$. Moreover we assume that the k pairs of vertices $s_1, r_1, ..., s_k, r_k$ are all distinct and they are members of N. Let $N_0 = \{s_1, r_1, ..., s_k, r_k\}$, $\Omega_0 = \Omega|_N$, $E_0 = \{\{s_i, r_i\} : i = 1, ..., k\}$ and $W_0 = (N_0, \Omega_0, E_0)$. It is easy to see that if G has the required paths, then W_0 is an immersion in W. It follows from 3.5 that there exists an integer f depending only on W_0 such that $W_0 \preceq_c W'$ for some subweb W' of W with $|N(W')| \leq f$. Suppose that x is a vertex of G' such that the (vertex-region) distance, in G', from x to the infinite face of G' is longer than f. It is not difficult to see that if the required paths exist in G, then the required paths exist in $G \backslash x$. This observation enable us to reduce G to a graph of bounded tree width without changing the problem. Then we may use the technique suggested in [1] to find the required paths.

As a matter of fact, this algorithm is exactly the one presented in [3]. The only reason we present it here is that we use a different argument to show the algorithm is correct.

Finally, we point out a common generalization of webs and graphs. Let $G = (V, E)$ be a graph (G may have loops or multiple edges). For any edge $e = xy$ of E, let us view e as two half-edges which are associated with (e, x) and (e, y). It is clear that there are $2|E|$ half-edges in total. We denote by $N_G(x)$ (or sometimes simply $N(x)$ if there is no confusion), for every $x \in V$, the set of half-edges incident with x. Let Ω_x be a cyclic permutation of $N(x)$ for every $x \in V$, and let $\Omega = \{\Omega_x : x \in V\}$. We shall call the triple $R = (V, E, \Omega)$ a *rotated graph*. To visualize a rotated graph, we may think of G as a graph drawn on an orientable surface and each Ω_x is the natural cyclic permutation of $N(x)$ determined by the drawing (say, clockwise). As a matter of fact, it is well known that for every rotated graph, there exists an orientable surface and a drawing of G on this surface such that each Ω_x is precisely the natural cyclic (clockwise) permutation of $N(x)$ determined by this drawing.

Let $R' = (V', E', \Omega')$ be another rotated graph. We call R' a *sub-rotated graph* of R if $V' \subseteq V$, $E' \subseteq E$ and $\Omega' = \{\Omega_x|_{N_{G'}(x)} : x \in V'\}$. Let $e_1 = xy_1$, $e_2 = xy_2$ be two distinct edges of R and let $h_1 = (e_1, x)$, $h_2 = (e_2, x)$ be half-edges with $\Omega_x(h_1) = h_2$. Then we say R' is obtained from R by *splitting off* e_1, e_2 from x if $V' = V$, $E' = (E \setminus \{e_1, e_2\}) \cup \{e\}$, where $e = y_1 y_2$ is a new edge, and $\Omega' = \{\Omega_x|_{N_{G'}(x)} : x \in V'\}$. If R' is isomorphic to a rotated graph obtained from R by a sequence of splitting off operations, then we write $R' \leq R$. We say R' is an *immersion* in R, and denote it by $R' \leq^* R$, if $R' \leq R''$ for a sub-rotated graph R'' of R. Finally, we call $R = (V, E, \Omega)$ *Eulerian* if each connected component of $G = (V, E)$ is an Eulerian graph.

With the terminology above, we have the following two conjectures:

(1) Rotated graphs are well-quasi-ordered by $\leq^*$.
(2) Eulerian rotated graphs are well-quasi-ordered by $\leq$.

It is obvious that Nash-Williams' graph immersion conjecture is a special case of (1) and 3.6 says that (1) is true for rotated graphs with one vertex. We do not know much about (2) even for webs (that is, Eulerian rotated graphs with one vertex). The only thing we know is that it implies Wagner's graph minor conjecture.

REFERENCES

[1] S. Arnborg and A. Proskurowski, Linear time algorithms for NP-hard problems on graphs embedded in k-trees, *Discrete Applied Math.*, 23(1989) 11-24.

[2] G. Ding, *The immersion relation on webs*, Ph.D thesis, Rutgers University, 1991.

[3] N. Robertson and P. Seymour, Graph minors XIII. The disjoint paths problem, *Manuscript* 1986.

[4] R. Thomas, Well-quasi-ordering infinite graphs with forbidden finite planar minor, *Trans. American Math. Society*, 312 (1989) 279-313.

MATHEMATICS DEPARTMENT, LOUISIANA STATE UNIVERSITY, BATON ROUGE, LA 70803, USA
E-mail address: ding @ marais.math.lsu.edu

STRUCTURAL DESCRIPTIONS OF LOWER IDEALS OF TREES

NEIL ROBERTSON, P. D. SEYMOUR and ROBIN THOMAS

ABSTRACT. A *lower ideal* of trees is a set $\mathcal{I}$ of finite trees such that if $T \in \mathcal{I}$ and T topologically contains S then $S \in \mathcal{I}$. We prove that every lower ideal of trees $\mathcal{I}$ has a structural description, a finite set of rules which describes how to construct an arbitrary element of $\mathcal{I}$.

1. INTRODUCTION

Trees in this paper are non-null, finite, rooted, directed away from the root, and for technical reasons the vertices are assumed to be integers. More precisely, a *tree* is a triple $T = (V, E, r)$, where V, the set of *vertices*, is a non-empty finite set of integers, E, the set of *edges*, is a subset of $V \times V$, and r, the *root* of T, is a member of V, such that for every $t \in V$ there is a unique directed walk from r to t. (A sequence $t_0, t_1, \ldots, t_n$ is a *directed walk from t_0 to t_n* if $(t_{i-1}, t_i) \in E$ for all $i = 1, 2, \ldots, n$.) We write $V(T) = V$, $E(T) = E$, and $\text{root}(T) = r$. The *height* of T is the maximum number of edges in a directed walk in T. For $s, t \in V(T)$, let $s \wedge t$ denote the last vertex of the directed walk from $\text{root}(T)$ to s which belongs to the directed walk from $\text{root}(T)$ to t. We say

---

1991 Mathematics Subject Classification. Primary 05C05, 05C75.

Research of the first and third author was performed under a consulting agreement with Bellcore. The first author was supported by NSF under Grant No. DMS-8903132. The third author was was supported by NSF under Grant No. DMS-8903132.

This paper is a final version, and will not be submitted elsewhere.

that T_2 *topologically contains* T_1, or that T_1 is *topologically contained* in T_2 if there exists a 1-1 mapping $f : V(T_1) \to V(T_2)$, called a *tree-embedding*, with the property that $f(s \wedge t) = f(s) \wedge f(t)$ for every two elements $s, t \in V(T_1)$. Let $\mathcal{I}$ be a set of trees such that if $T \in \mathcal{I}$ and T topologically contains S then $S \in \mathcal{I}$. We say that $\mathcal{I}$ is a *lower ideal of trees*, or a *tree ideal* for short. A tree ideal $\mathcal{I}$ is *proper* if some tree does not belong to $\mathcal{I}$. A tree ideal $\mathcal{I}$ is *coherent* if $\mathcal{I} \neq \emptyset$ and for every $T_1, T_2 \in \mathcal{I}$ there exists $T \in \mathcal{I}$ such that T topologically contains both T_1 and T_2. For notational convenience we introduce a formal symbol Γ, called the *null tree*. We define $V(\Gamma) = E(\Gamma) = \emptyset$, and if T is a tree or Γ we say that Γ is topologically contained in T.

If $T_1, T_2, \ldots, T_n$ ($n \geq 0$) are vertex-disjoint trees or Γ we define a new tree $\text{Tree}(T_1, T_2, \ldots, T_n)$. Its vertices are $V(T_1) \cup V(T_2) \cup \cdots \cup V(T_n) \cup \{t_0\}$, where $t_0 \notin V(T_1) \cup V(T_2) \cup \cdots \cup V(T_n)$ is a new vertex. $\text{Tree}(T_1, T_2, \ldots, T_n)$ has root t_0 and edges $E(T_1) \cup E(T_2) \cup \cdots \cup E(T_n) \cup \{(t_0, \text{root}(T_i)) : 1 \leq i \leq n, T_i \neq \Gamma\}$.

We say that $B = (\mathcal{I}_1, \ldots, \mathcal{I}_n; k; \mathcal{I}_0)$ is a *bit* if $n, k \geq 0$ are integers, $\mathcal{I}_1, \ldots, \mathcal{I}_n$ are coherent tree ideals and $\mathcal{I}_0$ is a tree ideal. The tree ideals $\mathcal{I}_0, \mathcal{I}_1, \ldots, \mathcal{I}_n$ will be called the *lower ideals* of B, and k will be called the *width* of B. Two bits are considered identical if they differ only by a permutation of $\mathcal{I}_1, \ldots, \mathcal{I}_n$. Let $\mathcal{B}$ be a set of bits. We define $I(\mathcal{B})$ to be the intersection of all tree ideals $\mathcal{I}$ that satisfy the following condition:

(*) If $(\mathcal{I}_1, \ldots, \mathcal{I}_n; k; \mathcal{I}_0) \in \mathcal{B}$, $T_i \in \mathcal{I}_i$ for $i = 1, 2, \ldots, n$, $T_{n+i} \in \mathcal{I} \cup \{\Gamma\}$ for $i = 1, 2, \ldots, k$, $j \geq 0$ and $T_{k+n+1}, \ldots, T_{k+n+j} \in \mathcal{I}_0$, then $\text{Tree}(T_1, \ldots, T_{k+n+j})$ belongs to $\mathcal{I}$.

We remark that $I(\mathcal{B})$ is a tree ideal satisfying (*), and that if $\mathcal{B} \neq \emptyset$ then every one-vertex tree belongs to $I(\mathcal{B})$. We offer the following examples. Let $\mathcal{J}$ be the tree ideal consisting of all one-vertex trees. Then $\mathcal{J} = I(\{(; 0; \emptyset)\})$. Further, $I(\{(; 2; \emptyset)\})$ is the tree ideal of all trees with out-degree at most two, and $I(\{(\mathcal{J}; 1; \emptyset)\})$ is the tree ideal of all trees obtained from a directed walk by gluing a directed edge to some of its vertices. As a last example we consider $I(\{(; 2; \emptyset), (; 0; \mathcal{J})\})$. Every element of this tree ideal is obtained from an arbitrary tree with out-degree at most two by gluing stars onto a set of vertices of out-degree zero.

Let $B = (\mathcal{I}_1, \mathcal{I}_2, \ldots, \mathcal{I}_n; k; \mathcal{I}_0)$, $B' = (\mathcal{I}'_1, \mathcal{I}'_2, \ldots, \mathcal{I}'_{n'}; k'; \mathcal{I}'_0)$ be bits and let $\mathcal{I}$ be a tree ideal. We say that B is $\mathcal{I}$-*dominated* by B' if there exist a set $S \subseteq \{1, 2, \ldots, n\}$ and a mapping $f : \{1, 2, \ldots, n\} - S \to \{0, 1, \ldots, n'\}$ such that

(D1) $|S| \leq k' - k$ (and hence $k \leq k'$),

(D2) $\mathcal{I}_i \subseteq \mathcal{I}$ for every $i \in S$,

(D3) for $i, j \in \{1, 2, \ldots, n\} - S$, if $f(i) = f(j) > 0$ then $i = j$, and

(D4) $\mathcal{I}_0 \subseteq \mathcal{I}'_0$, and $\mathcal{I}_i \subseteq \mathcal{I}'_{f(i)}$ for every $i \in \{1, 2, \ldots, n\} - S$.

Let $\mathcal{I}$ be a tree ideal. A *name* of $\mathcal{I}$ is a finite set $\mathcal{B}$ of bits such that

(N1) if $(\mathcal{I}_1, \ldots, \mathcal{I}_n; k; \mathcal{I}_0) \in \mathcal{B}$ then $\mathcal{I}_i \not\subseteq \mathcal{I}_0$ for all $i = 1, 2, \ldots, n$ and $\mathcal{I}_i$ is a proper subset of $\mathcal{I}$ for all $i = 0, 1, \ldots, n$,

(N2) $I(\mathcal{B}) = \mathcal{I}$,

(N3) no bit of $\mathcal{B}$ is $\mathcal{I}$-dominated by a bit of $\mathcal{B}$ other than itself, and

(N4) if $B \in \mathcal{B}$ has width zero and $I(\{B\}) \subseteq \mathcal{I}'$ for a lower ideal $\mathcal{I}'$ of a bit $B' \in \mathcal{B}$, then $B = B' = (\mathcal{I}'; 0; \emptyset)$.

The following is the main result of this paper.

(1.1) *Every proper lower ideal of trees has a unique name.*

There is a trivial difficulty that the lower ideal of all trees does not have a name. This could be overcome for instance by allowing k to be ∞. However, we chose not to do so. A related result, in terms of finite automata, for lower ideals in the minor containment relation was obtained by Gupta [1].

We now put this result into the context of well-quasi-ordering. A *quasi-ordering* is a reflexive and transitive relation. Let Q be a set and let $\leq$ be a quasi-ordering on Q. We say that Q is *quasi-ordered*. We say that Q is *well-quasi-ordered (wqo)* if for every infinite sequence $q_1, q_2, \ldots$ of elements of Q there are indices i, j such that $i < j$ and $q_i \leq q_j$. The following is a theorem of Kruskal [3]; for a simple proof see [4].

(1.2) *The set of all trees quasi-ordered by topological containment is well-quasi-ordered.*

A *lower ideal* in Q is a set $I \subseteq Q$ such that if $q \in I$ and $q' \leq q$ then $q' \in I$. We say that a sequence $q_1, q_2, \ldots$ of elements of Q is *nondecreasing* if $q_i \leq q_j$ for all $i < j$. The following is easy to see.

(1.3) *Let Q be wqo. Then*

(i) every infinite sequence of elements of Q contains an infinite nondecreasing subsequence, and

(ii) there is no infinite strictly decreasing sequence $I_1 \supset I_2 \supset \cdots$ of lower ideals in Q.

It should be noted that if $B = (\mathcal{I}_1, \mathcal{I}_2, \ldots, \mathcal{I}_n; k; \mathcal{I}_0)$ belongs to the name of a lower ideal of trees, then each $\mathcal{I}_i$ also has a name, expressed in terms of finitely many smaller lower ideals. Thus the name of a lower ideal of trees can be regarded as a finitely branching structured tree. It follows from (1.2), (1.3ii) and König's lemma that this structured tree is finite. This is the *structural description* we were referring to in the title and in the abstract. Given the structural description of two lower ideals of trees $\mathcal{I}_1, \mathcal{I}_2$ one would like to be able to test if $\mathcal{I}_1 \subseteq \mathcal{I}_2$. It follows from (2.3) and (2.6) that this can be done. We remark that conditions (N3) and (N4) are only needed to guarantee uniqueness.

Another remark that needs to be made is that it follows from (1.2) that for every lower ideal of trees $\mathcal{I}$ there exists a finite set of trees Ω such that $T \notin \mathcal{I}$ if and only if T topologically contains some member of Ω. Thus both Ω and the name are "finite descriptions" of $\mathcal{I}$. However, there is a difference in how Ω and the name describe $\mathcal{I}$; namely, Ω describes the structure of non-members of $\mathcal{I}$, whereas the name describes the structure of members of $\mathcal{I}$.

In the rest of this section we state several lemmas. Let Q be quasi-ordered. A lower ideal I in Q is said to be *coherent* if $I \neq \emptyset$ and for every $q, q' \in I$ there exists $q'' \in I$ such that $q \leq q''$ and $q' \leq q''$. Thus $\emptyset$ is a lower ideal, but not a coherent lower ideal. We need the following lemma.

(1.4) *Let Q be wqo and let I be a lower ideal in Q. Then there exists a unique finite set $\{I_1, I_2, \ldots, I_n\}$ of coherent lower ideals such that $I_1 \cup I_2 \cup \cdots \cup I_n = I$ and $I_i \not\subseteq I_j$ for all $i, j = 1, 2, \ldots, n$ with $i \neq j$.*

Proof. We first prove that every lower ideal in Q can be represented as a finite union of coherent lower ideals. Suppose to the contrary that there is a lower ideal I in Q which is not expressible as a finite union of coherent lower ideals. By (1.3ii) we may choose I in such a way that every proper lower subideal of I is expressible as a finite union of coherent lower ideals. Since, in particular, I is not coherent, there are $q_1, q_2 \in I$ such that there is no $q \in I$ with $q_1 \leq q$ and $q_2 \leq q$. For $i = 1, 2$ let I_i be the set of all $q \in I$ such that $q_i \not\leq q$. Then $I_1 \cup I_2 = I$, and both I_1 and I_2 are proper lower subideals of I. Since both I_1, I_2 can be expressed as a finite union of coherent lower ideals, so can I, a contradiction.

To prove uniqueness we assume that there are coherent lower ideals $I_1, I_2, \ldots, I_n, I'_1, I'_2, \ldots, I'_{n'}$ such that $I_1 \cup I_2 \cup \ldots \cup I_n = I'_1 \cup I'_2 \cup \ldots \cup I'_{n'} = I$ and $I_i \not\subseteq I_j$ for all $i, j = 1, 2, \ldots, n$ with $i \neq j$ and $I'_i \not\subseteq I'_j$ for all $i, j = 1, 2, \ldots, n'$ with $i \neq j$. Let $i \in \{1, 2, \ldots, n\}$. Since I_i is coherent there exists an integer $\pi(i) \in \{1, 2, \ldots, n'\}$ such that $I_i \subseteq I'_{\pi(i)}$. Similarly there exists an integer $j \in \{1, 2, \ldots, n\}$ such that $I'_{\pi(i)} \subseteq I_j$, and so $i = j$ and thus $I_i = I'_{\pi(i)}$. We deduce that $n = n'$, and that π is a permutation of $\{1, 2, \ldots, n\}$, as desired. □

If $I_1, I_2, \ldots, I_n$ are as in the above lemma we say that $\{I_1, I_2, \ldots, I_n\}$ is the *coherent lower ideal decomposition* of I.

Let $Q_1, Q_2, \ldots, Q_n$ be quasi-ordered. The Cartesian product $Q_1 \times Q_2 \times \ldots \times Q_n$ is quasi-ordered by the coordinatewise quasi-ordering, that is, $(q_1, q_2, \ldots, q_n) \leq (q'_1, q'_2, \ldots, q'_n)$ if $q_i \leq q'_i$ in Q_i for every $i = 1, 2, \ldots, n$. The following lemma follows easily from (1.3i).

(1.5) *If $Q_1, Q_2, \ldots, Q_n$ are wqo, then $Q_1 \times Q_2 \times \ldots \times Q_n$ is wqo.*

Let Q be quasi-ordered. We denote by $Q^{<\omega}$ the set of all finite sequences of elements of Q quasi-ordered by the rule that $(q_1, q_2, \ldots, q_n) \leq (q_1', q_2', \ldots, q_{n'}')$ if there exists a strictly increasing function $f : \{1, 2, \ldots, n\} \to \{1, 2, \ldots, n'\}$ such that $q_i \leq q_{f(i)}'$. The following lemma is due to Higman [2].

(1.6) *If Q is wqo then $Q^{<\omega}$ is wqo.*

We say that a set $I \subseteq Q$ is *generated* by a sequence $q_1, q_2, \ldots$, or that the sequence $q_1, q_2, \ldots$ *generates* I if I is the set of all $q \in Q$ for which there exists an integer $i \geq 1$ with $q \leq q_i$. The following is easy to see.

(1.7) *Let Q be a countable quasi-ordered set and let $I \subseteq Q$. Then I is a coherent lower ideal if and only if I is generated by a nondecreasing sequence.*

2. UNIQUENESS

In this section we prove that every proper lower ideal of trees has at most one name. Let T be a tree, and let $\mathcal{B}$ be a set of bits. We say that T *conforms to* $\mathcal{B}$ if there exists a bit $B = (\mathcal{I}_1, \mathcal{I}_2, \ldots, \mathcal{I}_n; k; \mathcal{I}_0) \in \mathcal{B}$ such that either $T \in \mathcal{I}_0 \cup \mathcal{I}_1 \cup \ldots \cup \mathcal{I}_n$, or there exist an integer $m \geq 0$ and trees $T_1, \ldots, T_{n+k+m}$ such that $T_i \in \mathcal{I}_i \cup \{\Gamma\}$ for $i = 1, 2, \ldots, n$, $T_{n+i} \in I(\mathcal{B}) \cup \{\Gamma\}$ for $i = 1, 2, \ldots, k$, $T_{n+k+i} \in \mathcal{I}_0$ for $i = 1, 2, \ldots, m$ and such that T is isomorphic to $\text{Tree}(T_1, T_2, \ldots, T_{k+n+m})$. We also say that T *conforms to B in $\mathcal{B}$*. We omit the (easy) proof of the following lemma.

(2.1) *Let $\mathcal{B}$ be a set of bits, and let T be a tree. Then T belongs to $I(\mathcal{B})$ if and only if it conforms to some B in $\mathcal{B}$.*

We say that a set $\mathcal{B}$ of bits is *coherent* if $\mathcal{B}$ is nonempty, finite, and either $\mathcal{B}$ contains at most one bit of width zero or $\mathcal{B}$ contains a bit of width at least two. We need the following lemma.

(2.2) *Let $\mathcal{B}$ be a set of bits. If $\mathcal{B}$ is coherent, then $I(\mathcal{B})$ is a coherent lower ideal of trees.*

Proof. Let $\mathcal{B}$ be coherent. Since $\mathcal{B}$ is non-empty we see that $I(\mathcal{B})$ is non-empty. Suppose for a contradiction that $T, T' \in I(\mathcal{B})$ are such that there is no tree in $I(\mathcal{B})$ that topologically contains both T, T', and, subject to that, the sum of the heights of T and T' is minimum. Since $\text{Tree}(T, T')$ topologically contains both T and T', we deduce that

(1) $\quad \text{Tree}(T, T') \notin I(\mathcal{B})$.

From (1) we deduce that $\mathcal{B}$ contains no bit of width two or more, and therefore contains at most one bit of width zero.

(2) If $B = (\mathcal{I}_1, \mathcal{I}_2, \ldots, \mathcal{I}_k; k; \mathcal{I}_0) \in \mathcal{B}$ and $T \in \mathcal{I}_0 \cup \mathcal{I}_1 \cup \cdots \cup \mathcal{I}_n$, then $T' \notin \mathcal{I}_0 \cup \mathcal{I}_1 \cup \cdots \cup \mathcal{I}_n$ and $k = 0$.

This follows from (1) and the fact that $\mathcal{I}_i$ is a coherent tree ideal for all $i = 1, 2, \ldots, n$.

Let $B = (\mathcal{I}_1, \mathcal{I}_2, \ldots, \mathcal{I}_n; k; \mathcal{I}_0), B' = (\mathcal{I}'_1, \mathcal{I}'_2, \ldots, \mathcal{I}'_n; k'; \mathcal{I}'_0) \in \mathcal{B}$ be such that T conforms to B in $\mathcal{B}$ and T' conforms to B' in $\mathcal{B}$. Then $k, k' \leq 1$. Let $T = \text{Tree}(T_1, T_2, \ldots, T_{n+k+m})$, where $m \geq 0$, and for $i = 1, 2, \ldots, n+k+m$, T_i is a tree or Γ in such a way that either $T \in \mathcal{I}_0 \cup \mathcal{I}_1 \cup \cdots \cup \mathcal{I}_n$, or $T_i \in \mathcal{I}_i \cup \{\Gamma\}$ for $i = 1, 2, \ldots, n$, $T_{n+i} \in I(\mathcal{B}) \cup \{\Gamma\}$ for $i = 1, 2, \ldots, k$, and $T_{n+k+i} \in \mathcal{I}_0$ for $i = 1, 2, \ldots, m$. Let $T' = \text{Tree}(T'_1, T'_2, \ldots, T'_{n'+k'+m'})$ similarly.

If $k = 1$, then $T \notin \mathcal{I}_0 \cup \mathcal{I}_1 \cup \cdots \cup \mathcal{I}_n$ by (2). By the choice of T and T' there exists a tree $T'' \in I(\mathcal{B})$ which topologically contains both T_{n+1} and T'. Then

$$\text{Tree}(T_1, T_2, \ldots, T_n, T'', T_{n+2}, T_{n+3}, \ldots, T_{n+k+m})$$

belongs to $I(\mathcal{B})$ and topologically contains both T and T', a contradiction. Because of the symmetry between k and k' we may therefore assume that $k = k' = 0$. Then $B = B'$, and we deduce from (2) that $\{T, T'\} \not\subseteq \mathcal{I}_0 \cup \mathcal{I}_1 \cup \cdots \cup \mathcal{I}_n$. From the symmetry we may assume that $T \notin \mathcal{I}_0 \cup \mathcal{I}_1 \cup \cdots \cup \mathcal{I}_n$. We wish to define T''_i for every $i = 1, 2, \ldots, n$. We choose $T''_i \in \mathcal{I}_i$ in such a way that if $R \in \{T_i, T'_i, T'\} \cap \mathcal{I}_i$, then T''_i topologically contains R. Such a choice is clearly possible, because $\mathcal{I}_i$ is a coherent tree ideal for every $i = 1, 2, \ldots, n$. Let $T_0 = T'$ if $T' \in \mathcal{I}_0$, and let $T_0 = \Gamma$ otherwise. If $T' \in \mathcal{I}_0 \cup \mathcal{I}_1 \cup \cdots \cup \mathcal{I}_n$ then

$$\text{Tree}(T''_1, T''_2, \ldots, T''_n, T_{n+1}, T_{n+2}, \ldots, T_{n+k+m}, T_0)$$

belongs to $I(\mathcal{B})$ and topologically contains both T and T', and if $T' \notin \mathcal{I}_0 \cup \mathcal{I}_1 \cup \cdots \cup \mathcal{I}_n$ then

$$\text{Tree}(T''_1, T''_2, \ldots, T''_n, T_{n+1}, T_{n+2}, \ldots, T_{n+k+m}, T'_{n'+1}, T'_{n'+2}, \ldots, T'_{n'+k'+m'})$$

belongs to $I(\mathcal{B})$ and topologically contains both T and T', a contradiction in both cases. □

(2.3) *Let $\mathcal{B}, \mathcal{B}'$ be finite sets of bits such that $\mathcal{B}$ is coherent. Then $I(\mathcal{B}) \subseteq I(\mathcal{B}')$ if and only if for every bit $B \in \mathcal{B}$ there exists a bit $B' \in \mathcal{B}'$ such that either*
(i) *B is $I(\mathcal{B}')$-dominated by B', or*
(ii) *$I(\{B\}) \subseteq \mathcal{I}'$ for some lower ideal $\mathcal{I}'$ of B', and if B has positive width then $I(\mathcal{B}) \subseteq \mathcal{I}'$.*

Proof. We first prove the "if" part. Let $T \in I(\mathcal{B})$, and assume that no tree of smaller height belongs to $I(\mathcal{B}) - I(\mathcal{B}')$. By (2.1) T conforms to some B in

$\mathcal{B}$; let $B = (\mathcal{I}_1, \mathcal{I}_2, \ldots, \mathcal{I}_n; k; \mathcal{I}_0)$. Then either $T \in \mathcal{I}_0 \cup \mathcal{I}_1 \cup \ldots \cup \mathcal{I}_n$, or T has the form $\text{Tree}(T_1, T_2, \ldots, T_{n+k+m})$, where $T_i \in \mathcal{I}_i \cup \{\Gamma\}$ for $i = 1, 2, \ldots, n$, $T_{n+i} \in I(\mathcal{B}) \cup \{\Gamma\}$ for $i = 1, 2, \ldots, k$, and $T_{n+k+i} \in \mathcal{I}_0$ for $i = 1, 2, \ldots, m$. By the minimality of the height of T it follows that $T_{n+i} \in I(\mathcal{B}') \cup \{\Gamma\}$ for $i = 1, 2, \ldots, k$. By the hypothesis there exists a bit $B' \in \mathcal{B}'$ such that (i) or (ii) holds.

We assume first that (i) holds. Then there exist a set S and mapping f satisfying (D1)-(D4). It is now routine to verify that $T \in I(\mathcal{B}')$. So we may assume that (ii) holds. Let $\mathcal{I}'$ be as in (ii). If B has positive width, then $T \in I(\mathcal{B}) \subseteq \mathcal{I}' \subseteq I(\mathcal{B}')$, and if B has width zero then $T \in I(\{B\}) \subseteq \mathcal{I}' \subseteq I(\mathcal{B}')$. This completes the proof of the "if" part.

To prove "only if" let $B = (\mathcal{I}_1, \mathcal{I}_2, \ldots, \mathcal{I}_n; k; \mathcal{I}_0) \in \mathcal{B}$. For $j = 1, 2, \ldots, n$ let, by (1.7), $\{T_j^i\}_{i \geq 1}$ be a nondecreasing sequence that generates $\mathcal{I}_j$, for $j = n+1, n+2, \ldots, n+k$ let, by (1.7) and (2.2), $\{T_j^i\}_{i \geq 1}$ be a nondecreasing sequence that generates $I(\mathcal{B})$, let $T_{n+k+1}^1, T_{n+k+2}^1, \ldots$ be a sequence that contains infinitely many isomorphic copies of every element of $\mathcal{I}_0$, for $i \geq 2$ and $j \geq n+k+1$ let $T_j^i = T_j^1$, and for $i \geq 1$ let $T^i = \text{Tree}(T_1^i, T_2^i, \ldots, T_{n+k+i}^i)$. Then $T^i \in I(\mathcal{B}) \subseteq I(\mathcal{B}')$, and hence T^i conforms to $\mathcal{B}'$. Since $\mathcal{B}'$ is finite, by (2.1) we may assume (by taking a subsequence) that there exists a bit $B' \in \mathcal{B}'$ such that T^i conforms to B' in $\mathcal{B}'$ for all $i = 1, 2, \ldots$. Let $B' = (\mathcal{I}'_1, \mathcal{I}'_2, \ldots \mathcal{I}'_{n'}; k'; \mathcal{I}'_0)$. There are two cases.

We assume first that $T^i \in \mathcal{I}'_0 \cup \mathcal{I}'_1 \cup \cdots \cup \mathcal{I}'_{n'}$ for all $i \geq 1$. Since the sequence $T^1, T^2, \ldots$ is nondecreasing, we deduce that there exists $j \in \{0, 1, \ldots, n'\}$ such that $T^i \in \mathcal{I}'_j$ for all $i \geq 1$. Since the sequence $\{T^i\}_{i \geq 1}$ generates an ideal that contains $I(\{B\})$ it follows that $I(\{B\}) \subseteq \mathcal{I}'_j$. Moreover, if $k > 0$ then $\{T^i\}_{i \geq 1}$ generates $I(\mathcal{B})$, and hence $I(\mathcal{B}) \subseteq \mathcal{I}'_j$. Thus (ii) holds.

We now assume that (ii) does not hold. In particular, $T^i \notin \mathcal{I}'_0 \cup \mathcal{I}'_1 \cup \cdots \cup \mathcal{I}'_{n'}$ for some $i \geq 1$. By taking a subsequence we may assume that $T^i \notin \mathcal{I}'_0 \cup \mathcal{I}'_1 \cup \cdots \cup \mathcal{I}'_{n'}$ for all $i \geq 1$. We claim that

(1) $\quad \mathcal{I}_0 \subseteq \mathcal{I}'_0$.

Indeed, let $T \in \mathcal{I}_0$ and let i be so big that at least $n+k+1$ of the trees $T_{n+k+1}^i, T_{n+k+2}^i, \ldots, T_{n+k+i}^i$ are isomorphic to T. Since T^i conforms to B' in $\mathcal{B}'$ and $T^i \notin \mathcal{I}'_0 \cup \mathcal{I}'_1 \cup \ldots \cup \mathcal{I}'_{n'}$, we deduce that $T \in \mathcal{I}'_0$. This proves (1).

Since T^i conforms to B' in $\mathcal{B}'$ and $T^i \notin \mathcal{I}'_0 \cup \mathcal{I}'_1 \cup \cdots \cup \mathcal{I}'_{n'}$ there exists a function $f_i : \{1, 2, \ldots, n+k\} \to \{-1, 0, 1, \ldots, n'\}$ such that

(a) $|f_i^{-1}(-1)| \leq k'$,
(b) $T_j^i \in I(\mathcal{B}')$ for all $j = 1, 2, \ldots, k+n$ with $f_i(j) = -1$,
(c) for all $j, j' = 1, 2, \ldots, k+n$, if $f_i(j) = f_i(j') > 0$ then $j = j'$, and
(d) $T_j^i \in \mathcal{I}'_{f_i(j)}$ for all $j = 1, 2, \ldots, k+n$ with $f_i(j) \geq 0$.

By taking a subsequence we may assume that $f_i(j)$ is constant for every fixed $j = 1, 2, \ldots, n + k$. Let $f(j)$ denote this common value.

(2) $f(j) = -1$ for $j = n + 1, n + 2, \ldots, n + k$.

Indeed, let $f(j) \geq 0$ for some $j \in \{n+1, n+2, \ldots, n+k\}$. Since $\{T_j^i\}_{i \geq 1}$ generates $I(\mathcal{B})$, it follows from (d) that $I(\mathcal{B}) \subseteq \mathcal{I}'_{f(j)}$, contrary to our assumption that (ii) does not hold. This proves (2).

Let S be the set of all integers j such that $1 \leq j \leq n$ and $f(j) = -1$. From (1), (2) and (a)-(d) it follows that S and the restriction of f to $\{1, 2, \ldots, n\} - S$ satisfy (D1)-(D4), because $\{T_j^i\}_{i \geq 1}$ generates $\mathcal{I}_j$ for $j = 1, 2, \ldots, n$. Thus (i) holds. □

(2.4) *Let $\mathcal{B}, \mathcal{B}'$ be names of some tree ideals, let $B \in \mathcal{B}$, $B' \in \mathcal{B}'$ and let $\mathcal{I}$ be a tree ideal. If B is $\mathcal{I}$-dominated by B' and B' is $\mathcal{I}$-dominated by B then $B = B'$.*

Proof. Let $B = (\mathcal{I}_1, \mathcal{I}_2, \ldots, \mathcal{I}_n; k; \mathcal{I}_0)$ and $B = (\mathcal{I}'_1, \mathcal{I}'_2, \ldots, \mathcal{I}'_{n'}; k; \mathcal{I}'_0)$. From the fact that B is $\mathcal{I}$-dominated by B' there exist a set $S \subseteq \{1, 2, \ldots, n\}$ and a mapping $f : \{1, 2, \ldots, n\} \to \{0, 1, \ldots, n'\}$ such that (D1)-(D4) hold. Similarly, from the fact that B' is $\mathcal{I}$-dominated by B there exist a set $S' \subseteq \{1, 2, \ldots, n'\}$ and a mapping $f' : \{1, 2, \ldots, n'\} \to \{0, 1, \ldots, n\}$ such that (D1)-(D4) hold with B, S, f replaced by B', S', f'. Thus $k = k'$, $\mathcal{I}_0 = \mathcal{I}'_0$, $S = S' = \emptyset$, $f(i) > 0$ for every $i = 1, 2, \ldots, n$ (because otherwise $\mathcal{I}_i \subseteq \mathcal{I}'_0 \subseteq \mathcal{I}_0$, contrary to (N1)), $f'(i) > 0$ for every $i = 1, 2, \ldots, n'$ and $n = n'$. Since bits that differ by a permutation of $\mathcal{I}_1, \mathcal{I}_2, \ldots, \mathcal{I}_n$ are considered equal, it follows that $B = B'$, as desired. □

(2.5) *Let $\mathcal{B}, \mathcal{B}'$ be names of a tree ideal $\mathcal{I}$, and assume that both $\mathcal{B}, \mathcal{B}'$ are coherent. Then $\mathcal{B} = \mathcal{B}'$.*

Proof. Let $\mathcal{B}_1$ denote the subset of $\mathcal{B}$ consisting of all bits of $\mathcal{B}$ of positive width, and let $\mathcal{B}'_1$ be defined similarly. We first prove the following.

(1) $\mathcal{B}_1 = \mathcal{B}'_1$.

Indeed, let $B \in \mathcal{B}$ have positive width. By (2.3) there exists $B' \in \mathcal{B}'$ such that either B is $\mathcal{I}$-dominated by B', or $I(B)$ is a subset of one of the lower ideals of B', say $\mathcal{I}'$. The latter case is impossible, because then $\mathcal{I} = I(\mathcal{B}) \subseteq \mathcal{I}' \subseteq I(\mathcal{B}') = \mathcal{I}$, and hence $\mathcal{I}' = \mathcal{I}$, contrary to (N1). Thus B is $\mathcal{I}$-dominated by B', and, in particular, B' has positive width. Similarly there exists $B'' \in \mathcal{B}$ such that B' is $\mathcal{I}$-dominated by B''. Since $\mathcal{I}$-domination is transitive we see that B is $\mathcal{I}$-dominated by B'', and hence $B = B''$ by (N3). Thus B is $\mathcal{I}$-dominated by B' and B' is $\mathcal{I}$-dominated by B, and so $B = B'$ by (2.4). This proves (1).

(2) *For every bit $B \in \mathcal{B}$ of width zero there exists a bit $B' \in \mathcal{B}'$ as in (2.3) of width zero.*

Indeed, let $B \in \mathcal{B}$ have width zero, let $B' \in \mathcal{B}'$ and suppose that B' has positive width. Then $B' \in \mathcal{B}$ by (1). From (N3) we deduce that B is not $\mathcal{I}$-dominated by B', and from (N4) it follows that $I(\{B\})$ is a subset of no lower ideal of B'. Hence B' satisfies neither (i) nor (ii) of (2.3). Thus if $B' \in \mathcal{B}'$ is as in (2.3) we see that B' has width zero. By (2.3) at least one such bit exists, and (2) follows.

From the symmetry we deduce that

(3) *For every bit $B' \in \mathcal{B}'$ of width zero there exists a bit $B'' \in \mathcal{B}$ as in (2.3) of width zero.*

(4) *For every bit $B \in \mathcal{B}$ there exists a bit $B' \in \mathcal{B}'$ such that B is $\mathcal{I}$-dominated by B'.*

Indeed, let $B \in \mathcal{B}$. We may assume that B has width zero, because otherwise the result follows from (1). Let $B' \in \mathcal{B}'$ be as in (2); we may assume that $I(\{B\})$ is a subset of one of the lower ideals of B', because otherwise we are done. Let B'' be as in (3). Since B'' has width zero, it follows that $I(\{B\})$ is a subset of one of the lower ideals of B'', say $\mathcal{I}_1$, and so $I(\{B\}) = \mathcal{I}_1$, and $B'' = B = (\mathcal{I}_1; 0; \emptyset)$ by (N4). Similarly $B' = (\mathcal{I}_1; 0; \emptyset)$, and so B is $\mathcal{I}$-dominated by B', as desired. This proves (4).

It follows similarly that

(5) *for every bit $B' \in \mathcal{B}'$ there exists a bit $B'' \in \mathcal{B}$ such that B' is $\mathcal{I}$-dominated by B''.*

Now we are finally ready to finish the proof of (2.5). Let $B \in \mathcal{B}$, let $B' \in \mathcal{B}'$ be as in (4), and let $B'' \in \mathcal{B}$ be as in (5). Then B is $\mathcal{I}$-dominated by B'', and so $B = B''$ by (N3). Consequently, B is $\mathcal{I}$-dominated by B' and B' is $\mathcal{I}$-dominated by B, and hence $B = B'$ by (2.4). Thus $\mathcal{B} = \mathcal{B}'$, as desired. $\square$

Let $\mathcal{B}$ be a finite set of bits. We define a set $\{\mathcal{B}_1, \mathcal{B}_2, \ldots, \mathcal{B}_m\}$ of finite sets of bits, called the *coherent decomposition* of $\mathcal{B}$, as follows. If $\mathcal{B}$ is coherent then we define the coherent decomposition to be $\{\mathcal{B}\}$. Otherwise we let $B_1, B_2, \ldots, B_m$ be all the elements of $\mathcal{B}$ of width zero, and for $i = 1, 2, \ldots, m$ we put $\mathcal{B}_i = (\mathcal{B} - \{B_1, B_2, \ldots, B_m\}) \cup \{B_i\}$ and define the coherent decomposition of $\mathcal{B}$ to be $\{\mathcal{B}_1, \mathcal{B}_2, \ldots, \mathcal{B}_m\}$.

(2.6) *Let $\mathcal{B}$ be a name of a tree ideal $\mathcal{I}$, and let $\{\mathcal{B}_1, \mathcal{B}_2, \ldots, \mathcal{B}_m\}$ be the coherent decomposition of $\mathcal{B}$. Then $I(\mathcal{B}_1) \cup I(\mathcal{B}_2) \cup \cdots \cup I(\mathcal{B}_m)$ is a coherent lower ideal decomposition of $\mathcal{I}$.*

Proof. It follows from the definition that each $\mathcal{B}_i$ is coherent, and hence each $I(\mathcal{B}_i)$ is a coherent lower ideal by (2.2). To prove that $I(\mathcal{B}_i) \not\subseteq I(\mathcal{B}_j)$ for i, j with $1 \leq i, j \leq m$ and $i \neq j$ we suppose the contrary and note that (since then $m > 1$) $\mathcal{B}_i$ contains exactly one bit of width zero, say B and that $B \notin \mathcal{B}_j$. Then

$I(\{B\}) \subseteq I(\mathcal{B}_i) \subseteq I(\mathcal{B}_j) \subseteq \mathcal{I}$. By (2.3) there exists a bit $B' \in \mathcal{B}_j \subseteq \mathcal{B} - \{B\}$ such that either B is $I(\mathcal{B}_j)$-dominated by B', or $I(\{B\})$ is contained in some lower ideal of B'. The latter is impossible by (N4), and so B is $\mathcal{I}$-dominated by B', contrary to (N3). This proves that $I(\mathcal{B}_i) \not\subseteq I(\mathcal{B}_j)$ for $i \neq j$.

Finally, we must show that $\mathcal{I} = \mathcal{I}'$, where $\mathcal{I}' = I(\mathcal{B}_1) \cup I(\mathcal{B}_2) \cup \cdots \cup I(\mathcal{B}_m)$. Since obviously $\mathcal{I}' \subseteq \mathcal{I}$, it remains to prove that $\mathcal{I} \subseteq \mathcal{I}'$. We may assume that $m > 1$, for otherwise the equality $\mathcal{I} = \mathcal{I}'$ is trivial. Then $\mathcal{B}$ contains only bits of width zero and one. Let $T \in \mathcal{I}$ and assume that no tree of smaller height belongs to $\mathcal{I} - \mathcal{I}'$. Since $T \in \mathcal{I} = I(\mathcal{B})$ it conforms to some B in $\mathcal{B}$; let $B = (\mathcal{I}_1, \mathcal{I}_2, \ldots, \mathcal{I}_n; k; \mathcal{I}_0)$, where $k = 0$ or 1. Then either $T \in \mathcal{I}_0 \cup \mathcal{I}_1 \cup \cdots \cup \mathcal{I}_n$, or T has the form $\text{Tree}(T_1, T_2, \ldots, T_{n+k+p})$, where $p \geq 0$, $T_i \in \mathcal{I}_i \cup \{\Gamma\}$ for $i = 1, 2, \ldots, n$, $T_{n+k} \in \mathcal{I} \cup \{\Gamma\}$ and $T_{n+k+j} \in \mathcal{I}_0$ for $j = 1, 2, \ldots, p$. We define $l \in \{1, 2, \ldots, m\}$ as follows. If either $k = 0$ or $T \in \mathcal{I}_0 \cup \mathcal{I}_1 \cup \cdots \cup \mathcal{I}_n$ we let l be such that $B \in \mathcal{B}_l$. Otherwise $T_{n+1} \in \mathcal{I}' \cup \{\Gamma\}$ by the induction hypothesis, and we let l be such that $T_{n+1} \in I(\mathcal{B}_l) \cup \{\Gamma\}$. We remark that $B \in \mathcal{B}_l$, because if $k = 1$ then B belongs to every $\mathcal{B}_i$ for $i = 1, 2, \ldots, m$. In either case we see that T conforms to B in $\mathcal{B}_l$, and hence $T \in \mathcal{I}'$, as desired. □

(2.7) *Every lower ideal of trees $\mathcal{I}$ has at most one name.*

Proof. Let $\mathcal{B}, \mathcal{B}'$ be two names of a lower ideal of trees $\mathcal{I}$, and let $\{\mathcal{B}_1, \mathcal{B}_2, \ldots, \mathcal{B}_k\}$, $\{\mathcal{B}'_1, \mathcal{B}'_2, \ldots, \mathcal{B}'_{k'}\}$ be their respective coherent decompositions. From (1.4) and (2.6) we deduce that $k = k'$ and that we may choose our notation so that $I(\mathcal{B}_i) = I(\mathcal{B}'_i)$ for all $i = 1, 2, \ldots, k$. By (2.5) $\mathcal{B}_i = \mathcal{B}'_i$ for all $i = 1, 2, \ldots, k$, and hence $\mathcal{B} = \mathcal{B}'$, as desired. □

3. EXISTENCE

In this section we complete the proof of (1.1) by showing that every proper lower ideal of trees has at least one name. We need the following lemma.

(3.1) *Let $\mathcal{B}$ be a set of bits, let $B \in \mathcal{B}$ have width zero and assume that $I(\{B\}) \subseteq \mathcal{I}'$ for a lower ideal $\mathcal{I}'$ of a bit $B' \in \mathcal{B}$. If $B = B'$ let $\mathcal{B}' = (\mathcal{B} - \{B\}) \cup \{(\mathcal{I}'; 0; \emptyset)\}$, and otherwise let $\mathcal{B}' = \mathcal{B} - \{B\}$. Then $I(\mathcal{B}) = I(\mathcal{B}')$.*

Proof. Assume first that $B \neq B'$. Then clearly $I(\mathcal{B}') \subseteq I(\mathcal{B})$. To prove the converse let $T \in I(\mathcal{B})$ and assume that every tree in $I(\mathcal{B})$ of smaller height belongs to $I(\mathcal{B}')$. By (2.1) T conforms to some B_1 in $\mathcal{B}$. If $B_1 \neq B$ it follows that $T \in I(\mathcal{B}')$; if $B_1 = B$ then (since B has width zero) $T \in I(\{B\}) \subseteq \mathcal{I}' \subseteq I(\{B'\}) \subseteq I(\mathcal{B}')$, because $B \neq B'$. This completes the proof in the case when $B \neq B'$.

We may therefore assume that $B = B'$. We first show that $I(\mathcal{B}') \subseteq I(\mathcal{B})$. Let $T \in I(\mathcal{B}')$ and assume that every tree in $I(\mathcal{B}')$ of smaller height belongs to

$I(\mathcal{B})$. By (2.1) T conforms to some B_2 in $\mathcal{B}'$. If $B_2 \neq (\mathcal{I}'; 0; \emptyset)$ then it follows that $T \in I(\mathcal{B})$; otherwise $T \in I(\{(\mathcal{I}'; 0; \emptyset)\}) \subseteq I(\{B'\}) = I(\{B\}) \subseteq I(\mathcal{B})$. This proves that $I(\mathcal{B}') \subseteq I(\mathcal{B})$. To prove the converse inequality let $T \in I(\mathcal{B})$, and assume that every tree in $I(\mathcal{B})$ of smaller height belongs to $I(\mathcal{B}')$. By (2.1) T conforms to some B_3 in $\mathcal{B}'$. If $B_3 \neq B$ then it follows that $T \in I(\mathcal{B}')$; otherwise (since B has width zero) $T \in I(\{B\}) = I(\{B'\}) \subseteq I(\{(\mathcal{I}'; 0; \emptyset)\}) \subseteq I(\mathcal{B}')$, as desired. □

We say that $(T_1, T_2, \ldots, T_n; k; M)$ is a *germ* if $n, k \geq 0$ are integers, $T_1, T_2, \ldots, T_n$ are trees, and M is a finite set of trees. If $g = (T_1, T_2, \ldots, T_n; k; M)$ is a germ and $\mathcal{I}$ is a lower ideal of trees, we denote by $H(g, \mathcal{I})$ the set of all trees isomorphic to $\mathrm{Tree}(T'_1, T'_2, \ldots, T'_{n+k+m})$, where $m \geq 0$, T'_i is Γ or a tree topologically contained in T_i for $i = 1, 2, \ldots, n$, $T'_{n+i} \in \mathcal{I} \cup \{\Gamma\}$ for $i = 1, 2, \ldots, k$, and T'_{n+k+i} is a tree topologically contained in some member of M for $i = 1, 2, \ldots, m$. A germ g is *a germ of a lower ideal* of trees $\mathcal{I}$ if $H(g, \mathcal{I}) \subseteq \mathcal{I}$.

We order germs as follows: $(T_1, T_2, \ldots, T_n; k; M) \leq (T'_1, T'_2, \ldots, T'_{n'}; k'; M')$ if there exists a set $S \subseteq \{1, 2, \ldots, n\}$ and a mapping $f : \{1, 2, \ldots, n\} - S \to \{0, 1, \ldots, n'\}$ such that

(i) $|S| \leq k' - k$,

(ii) every element of M is topologically contained in some element of M',

(iii) for $i, j \in \{1, 2, \ldots, n\} - S$, if $f(i) = f(j) > 0$ then $i = j$,

(iv) for every $i \in \{1, 2, \ldots, n\} - S$, if $f(i) > 0$ then T_i is topologically contained in $T'_{f(i)}$, and if $f(i) = 0$ then T_i is topologically contained in some member of M'.

It follows from (1.5) and (1.6) that germs are well-quasi-ordered. We say that a germ $g = (T_1, T_2, \ldots, T_n; k; M)$ of a lower ideal of trees $\mathcal{I}$ is *reduced* if there is no germ $g' = (T'_1, T'_2, \ldots, T'_{n'}; k'; M')$ of $\mathcal{I}$ with $g \leq g'$ and $n > n'$. Clearly for every germ g of $\mathcal{I}$ there exists a reduced germ g' of $\mathcal{I}$ with $g \leq g'$.

(3.2) *Every proper lower ideal of trees has at least one name.*

Proof. Let $\mathcal{I}$ be a proper lower ideal of trees and let $\mathcal{G}$ be the set of all germs of $\mathcal{I}$. Let $\mathcal{G} = \mathcal{G}_1 \cup \mathcal{G}_2 \cup \cdots \cup \mathcal{G}_p$ be the coherent lower ideal decomposition of $\mathcal{G}$. Let us fix $l \in \{1, 2, \ldots, p\}$. By (1.7) $\mathcal{G}_l$ is generated by a nondecreasing sequence, say $g_1, g_2, \ldots$, where $g_i = (T^i_1, T^i_2, \ldots, T^i_{n_i}; k_i; M_i)$. We may assume without loss of generality that each g_i is reduced. (To see this, we may assume by taking a subsequence that $g_1 \notin \bigcup_{i \neq l} \mathcal{G}_i$ and then consider $\tilde{g}_i = (\tilde{T}^i_1, \tilde{T}^i_2, \ldots, \tilde{T}^i_{\tilde{n}_i}; \tilde{k}_i; \tilde{M}_i) \in \mathcal{G}$ with $g_i \leq \tilde{g}_i$ and subject to that with $\tilde{n}_i$ minimum.) Let S_i, f_i be a set and a mapping witnessing that $g_i \leq g_{i+1}$. The sequence $\{k_i\}$ is bounded (because $\mathcal{I}$ is proper), and so we may assume (by taking a subsequence) that $\{k_i\}$ is constant; let k denote the common value. It follows that $S_i = \emptyset$ for every $i \geq 1$, and since each g_i is reduced we deduce that $f_i(n) > 0$ for all $i = 1, 2, \ldots$ and all $n = 1, 2, \ldots, n_i$. We may therefore assume that each f_i is the identity. We now claim that

(1) *the sequence $\{n_i\}_{i \geq 1}$ is bounded.*

To prove (1) suppose that the sequence $\{n_i\}_{i \geq 1}$ is unbounded. By taking a subsequence we may assume that $n_1 \geq 1$, that $\{n_i\}_{i \geq 1}$ is strictly increasing and that, by (1.2) and (1.3i), $\{T_{n_i}^i\}_{i \geq 1}$ is a nondecreasing sequence of trees. We claim that $g = (T_1^1, T_2^1, \ldots, T_{n_1-1}^1; k; M_1 \cup \{T_{n_1}^1\})$ is a germ of $\mathcal{I}$. Indeed, let $T \in H(g, \mathcal{I})$. Then $T = \text{Tree}(T_1, T_2, \ldots, T_{n_1+k+m+t})$, where $m, t \geq 0$, for $j = 1, 2, \ldots, n_1 - 1$, T_j is Γ or a tree topologically contained in T_j^1, $T_{n_1} = \Gamma$, $T_{n_1+j} \in \mathcal{I} \cup \{\Gamma\}$ for $j = 1, 2, \ldots, k$, T_{n_1+k+j} is topologically contained in $T_{n_1}^1$ for $j = 1, 2, \ldots, m$ and $T_{n_1+k+m+j}$ is topologically contained in some member of M_1 for $j = 1, 2, \ldots, t$. Let $j \in \{1, 2, \ldots, m\}$ and put $n(j) = n_1 + k + j$. In the sequence $T_{n_1+k+j}, T_{n_1}^1, T_{n(j)}^{n(j)}, T_{n(j)}^{n(m)}$ each term is topologically contained in the next (the last containment holds because $g_{n(j)} \leq g_{n(m)}$ and all the f_i's are assumed to be the identity). It follows that $T \in H(g_{n(m)}, \mathcal{I})$, and hence $T \in \mathcal{I}$. Thus g is a germ of $\mathcal{I}$, contrary to the fact that g_1 is reduced, since obviously $g_1 \leq g$. This proves (1).

By (1) we may assume by choosing a subsequence that $\{n_i\}_{i \geq 1}$ is constant; let n denote the common value. We have thus arrived at a sequence $g_1, g_2 \ldots$ of elements of $\mathcal{G}_l$. This sequence will be later referred to as a *fundamental sequence* of $\mathcal{G}_l$, and the notation $g_i = (T_1^i, T_2^i, \ldots, T_n^i; k; M_i)$ will be assumed. Let $\mathcal{I}_0$ be the lower ideal generated by $\bigcup_{i \geq 1} M_i$, and for $j = 1, 2, \ldots, n$ let $\mathcal{I}_j$ be the coherent lower ideal generated by $T_j^1, T_j^2, \ldots$.

(2) $\mathcal{I}_0, \mathcal{I}_1, \ldots, \mathcal{I}_n$ *are proper lower subideals of $\mathcal{I}$, and $\mathcal{I}_i \not\subseteq \mathcal{I}_0$ for all $i = 1, 2, \ldots, n$.*

The proof of (2) is very similar to the proof of (1), and so we just sketch it. If $\mathcal{I}_i = \mathcal{I}$ for some $i = 1, 2, \ldots, n$, say for $i = 1$, then it can be shown that $(T_2^1, T_3^1, \ldots, T_{n_1}^1; k+1; M_1)$ is a germ of $\mathcal{I}$, contrary to the fact that g_1 is reduced. Similarly, if $\mathcal{I}_i \subseteq \mathcal{I}_0$ for some $i = 1, 2, \ldots, n$, say for $i = 1$, then it can be shown that $(T_2^1, T_3^1, \ldots, T_{n_1}^1; k; M_1 \cup \{T_1^1\})$ is a germ of $\mathcal{I}$. Finally, if $\mathcal{I}_0 = \mathcal{I}$, then it can be shown that every tree belongs to $\mathcal{I}$, contrary to our assumption that $\mathcal{I}$ is a proper lower ideal of trees. This completes the sketch of the proof of (2).

We now define $B_l = (\mathcal{I}_1, \mathcal{I}_2, \ldots, \mathcal{I}_n; k; \mathcal{I}_0)$ and put $\mathcal{B}_0 = \{B_1, B_2, \ldots, B_p\}$. We will modify $\mathcal{B}_0$ to obtain a name of $\mathcal{I}$. We need the following three claims.

(3) *For $l, l' \in \{1, 2, \ldots, p\}$, if B_l is $\mathcal{I}$-dominated by $B_{l'}$ then $\mathcal{G}_l \subseteq \mathcal{G}_{l'}$.*

To prove (3) let $B_l = (\mathcal{I}_1, \mathcal{I}_2, \ldots, \mathcal{I}_n; k; \mathcal{I}_0)$ and let $B_{l'} = (\mathcal{I}_1', \mathcal{I}_2', \ldots, \mathcal{I}_{n'}'; k'; \mathcal{I}_0')$. By the assumption there exist a set $S \subseteq \{1, 2, \ldots n\}$ and a mapping

$f : \{1, 2, \ldots, n\} \to \{0, 1, \ldots, n'\}$ such that conditions (D1)-(D4) hold. We may assume that there are integers m, t such that $f(i) = i$ for $i = 1, 2, \ldots, m$, $f(i) = 0$ for $i = m+1, m+2, \ldots, t$, and $\{t+1, t+2, \ldots, n\} = S$. Let $g_1, g_2, \ldots$ be a fundamental sequence of $\mathcal{G}_l$; we must show that $g_i \in \mathcal{G}_{l'}$ for every integer $i \geq 1$. To this end let $i \geq 1$ be an integer. Then $g_i = (T_1^i, T_2^i, \ldots, T_n^i; k; M_i)$, where $T_j^i \in \mathcal{I}_j$ for $j = 1, 2, \ldots, n$ and $M_i \subseteq \mathcal{I}_0$. It follows that $T_j^i \in \mathcal{I}_j'$ for $j = 1, 2, \ldots, m$, $\{T_{m+1}^i, T_{m+2}^i, \ldots, T_t^i\} \cup M_i \subseteq \mathcal{I}_0'$ and $k + n - t \leq k'$. We deduce that $g_i \in \mathcal{G}_{l'}$, as desired. This proves (3).

(4) $\quad \mathcal{I} \subseteq I(\mathcal{B}_0)$.

We prove (4) by induction on height. Let $T \in \mathcal{I}$, and assume that every tree in $\mathcal{I}$ of strictly smaller height belongs to $I(\mathcal{B}_0)$. Let $T = \text{Tree}(T_1, T_2, \ldots, T_t)$. Then $g = (T_1, T_2, \ldots, T_t; 0; \emptyset)$ is a germ of $\mathcal{I}$, and so $g \in \mathcal{G}_l$ for some $l = 1, 2, \ldots, p$. Let $g_1, g_2, \ldots$ be a fundamental sequence of $\mathcal{G}_l$. Then $g \leq g_i$ for some integer $i \geq 1$. Thus there exist a set $S \subseteq \{1, 2, \ldots, t\}$ and a mapping $f : \{1, 2, \ldots, t\} - S \to \{0, 1, \ldots, n\}$ satisfying (i)-(iv). Since $T_j^i \in \mathcal{I}_j$ for $j = 1, 2, \ldots, n$, $M_i \subseteq \mathcal{I}_0$ and $T_j \in I(\mathcal{B}_0)$ for $j \in S$ by the choice of T, we deduce that $T \in I(\mathcal{B}_0)$, as desired. This proves (4).

(5) $\quad I(\mathcal{B}_0) \subseteq \mathcal{I}$.

Again, we prove (5) by induction on height. Let $T \in I(\mathcal{B}_0)$, and assume that every tree in $I(\mathcal{B}_0)$ of strictly smaller height belongs to $\mathcal{I}$. Let $l \in \{1, 2, \ldots, p\}$ be such that T conforms to B_l in $\mathcal{B}_0$, and let $B_l = (\mathcal{I}_1, \mathcal{I}_2, \ldots, \mathcal{I}_n; k; \mathcal{I}_0)$. Then either $T \in \mathcal{I}_0 \cup \mathcal{I}_1 \cup \ldots \cup \mathcal{I}_n$, or T has the form $\text{Tree}(T_1, T_2, \ldots, T_{n+k+m})$, where $m \geq 0$, $T_i \in \mathcal{I}_i \cup \{\Gamma\}$ for $i = 1, 2, \ldots, n$, $T_{n+i} \in I(\mathcal{B}_0) \cup \{\Gamma\}$ for $i = 1, 2, \ldots, k$ and $T_{n+k+i} \in \mathcal{I}_0$ for $i = 1, 2, \ldots, m$. In the former case $T \in \mathcal{I}$, and so we assume the latter. Let $g_1, g_2, \ldots$ be a fundamental sequence of $\mathcal{G}_l$. Then there exists an integer $i \geq 1$ such that T_j is topologically contained in T_j^i for $j = 1, 2, \ldots, n$ and T_{n+k+j} is topologically contained in some member of M_i for $j = 1, 2, \ldots, m$. By the induction hypothesis $T_{n+j} \in \mathcal{I} \cup \{\Gamma\}$ for $j = 1, 2, \ldots, k$. Thus $T \in H(g_i, \mathcal{I}) \subseteq \mathcal{I}$, as desired. This proves (5).

It follows from (2), (3), (4), (5) that $\mathcal{B}_0$ satisfies (N1), (N2) and (N3). Let us call a bit B *proper* if B has width zero and is not of the form $(\mathcal{I}_1; 0; \emptyset)$, where $\mathcal{I}_1 = I(\{B\})$. We choose a finite set of bits $\mathcal{B}$ such that

(i) $\mathcal{B}$ satisfies (N1), (N2) and (N3),

(ii) subject to (i), $|\mathcal{B}|$ is minimum, and

(iii) subject to (i) and (ii), the number of proper bits in $\mathcal{B}$ is minimum.

Such a choice is possible, because $\mathcal{B}_0$ satisfies (i). We claim that $\mathcal{B}$ satisfies (N4). Indeed, let $B \in \mathcal{B}$ have width zero, and let $I(\{B\}) \subseteq \mathcal{I}'$ for some ideal $\mathcal{I}'$ of a bit $B' \in \mathcal{B}$. We must show that $B = B'$ and that B is not proper. If $B \neq B'$ we put $\mathcal{B}' = \mathcal{B} - \{B\}$; then $I(\mathcal{B}') = \mathcal{I}$ by (3.1), and hence $\mathcal{B}'$ satisfies (N1), (N2) and (N3), contrary to (ii). Thus $B = B'$. Now $\mathcal{I}' \subseteq I(\{(\mathcal{I}';0;\emptyset)\}) \subseteq I(\{B'\}) = I(\{B\}) \subseteq \mathcal{I}'$; hence equality holds throughout and thus $(\mathcal{I}';0;\emptyset)$ is not proper. If B is is proper let $\mathcal{B}' = (\mathcal{B}-\{B\})\cup\{(\mathcal{I}';0;\emptyset)\}$, and again, using (3.1) we see that $\mathcal{B}'$ satisfies (i), contrary to (iii). This proves that $B = B'$ and that B is not proper, and hence $\mathcal{B}$ satisfies (N4), as desired. □

References

1. A. Gupta, *Constructivity Issues in Graph Minors*, PhD thesis, University of Toronto, 1990.

2. G. Higman, Ordering by divisibility in abstract algebras, *Proc. London Math. Soc.* (3)2 (1952), 326-336.

3. J. Kruskal, Well-quasi-ordering, the tree theorem, and Vázsonyi's conjecture, *Trans. Amer. Math. Soc.* 95 (1960), 210-225.

4. C. St. J. A. Nash-Williams, On well-quasi-ordering finite trees, *Math. Proc. Cambridge Phil. Soc.* 59 (1963), 833-835.

DEPARTMENT OF MATHEMATICS, OHIO STATE UNIVERSITY, COLUMBUS, OH 43210, USA

E-mail: robertso@function.mps.ohio-state.edu

BELLCORE, 445 SOUTH STREET, MORRISTOWN, NJ 07962, USA

E-mail: pds@bellcore.com

SCHOOL OF MATHEMATICS, GEORGIA INSTITUTE OF TECHNOLOGY, ATLANTA, GA 30332, USA

E-mail: thomas@math.gatech.edu

Finite Automata, Bounded Treewidth and Well-Quasiordering

KARL ABRAHAMSON AND MICHAEL FELLOWS

ABSTRACT. Some aspects of the finite-state point of view on bounded treewidth graph properties are exposited, and several new results relating finite-state recognizability to well-quasiordering are presented. The point of view presented emphasizes (1) a division of labor between structural parsing of graphs on the one hand, and standard tools of finite automata theory on the other, and (2) a Myhill-Nerode perspective on the description and handling of finite automata. The main new result is a straightforward necessary and sufficient condition (*cutset regularity*) for a family of graphs to be recognizable from structural parse trees by finite-state tree automata. Based on this main result, a second necessary and sufficient condition for finite-state recognizability is developed, which states that a family of graphs F is finite-state recognizable for bounded treewidth if and only if a certain quasiorder induced by the family is a well-quasiorder. We obtain a similar well-quasiordering characterization of regular formal languages. The criteria are applied to a number of natural graph families. For example, it is shown by a direct Myhill-Nerode argument that for $k \geq 2$ and $w \geq 1$ the family of graphs of bandwidth at most k is not finite-state recognizable from parse trees for graphs of treewidth at most w. It is also shown that for every fixed k, the family of graphs of cutwidth at most k is finite-state recognizable for graphs of bounded treewidth. The main theorem also allows us to give a straightforward Myhill-Nerode style proof of the important theorem of Courcelle relating finite-state recognition to second-order monadic description of graph properties. Finally, it is shown that a quasiorder on graphs of bounded treewidth satisfying a weak technical condition is a well-quasiorder if and only if every lower ideal is finite-state.

1991 *Mathematics Subject Classification*. Primary 05C55, 68Q25; Secondary 05C15, 68R10;.

The first author is supported in part by the U.S. Office of Naval Research under contract N00014-88-K-0456, by the U.S. National Science Foundation under grant MIP-8919312, and by the National Science and Engineering Research Council of Canada.

This is the final version of the results announced in the unpublished, but widely circulated manuscript [FA]. The new results announced in that manuscript, as well as additional material surveying algorithmic aspects of bounded treewidth, were presented in a survey talk at the Graph Minors Workshop, as well as a similar survey talk at the WOBCATS meeting in Banff, July, 1990.

Thanks to Bruno Courcelle for helpful correspondence, and to Hans Bodlaender for valuable discussions. The results of Section 3 benefited from the hospitality of Italo Dejter and the Mathematics Department of the University of Puerto Rico.

1. Introduction

In the setting of bounded treewidth, there are strong and beautiful connections between well-quasiordering and finite-state recognition. These connections have emerged in a body of recent work by several authors and "schools" [ALS] [AP] [Bo1] [BLW] [BPT] [CK] [Co2] [EC] [He] [MP] [Se] [Wi] [WHL] [WPLHH]. It has gradually become clear that almost all of the work that has been done on algorithms for bounded treewidth can be systematically viewed in terms of the standard mathematical objects: finite-state tree automata. The classical theorem of Myhill and Nerode characterizing regular languages can be seen to generalize elegantly to the graph-theoretic setting and to provide a useful characterization of graph families that are easy to recognize, for input restricted to graphs of a given treewidth bound.

The main purpose of this paper is to provide an exposition of this point of view, to present this graph-theoretic generalization of the Myhill-Nerode theorem and to demonstrate its use on concrete problems. We also prove a new characterization of regular formal languages in terms of well-quasiordering, and an analogous well-quasiordering characterization of finite-state graph properties.

The ideal reader is thoroughly grounded in classical finite automata theory including the Myhill-Nerode theorem (as may be found in a textbook such as [DW] or [HU]), is familiar with the basics of graph minor theory, bounded treewidth and well-quasiordering (as may be found in the papers of Robertson and Seymour), and has at least some superficial familiarity with the extensive literature on graph algorithms for graphs of bounded treewidth (a comprehensive bibliography on this subject is maintained by Steve Hedetniemi of Clemson University; see also [Bo3]). The new results presented here are in many respects a sequel to those of [FL], which describe a general method for computing the obstruction sets for lower ideals in the minor order. Tree automata are surveyed usefully and concretely in [Th], and [Mi] can be consulted for background in well-quasiordering theory. Despite these prerequisites to a full appreciation of the paper, we have included examples and illustrations to try to make the main results, and the finite-state point of view on this subject, accessible to readers without them.

A *quasiorder* is a reflexive and transitive relation. A quasiordered set $(S, \leq)$ is *well-quasiordered* (or a wqo) if, for every infinite sequence $(x_1, \ldots)$ of elements of S, there are indices $i < j$ such that $x_i \leq x_j$. (Such a sequence is termed *good*, and a sequence without such a pair of indices is termed *bad*.) To the reader interested in graph minor theory for its applications in computational complexity, wqo's may at first seem unfamiliar and exotic. At least in the setting of bounded treewidth, however, we will show that well-quasiordering is intimately connected to familiar ideas of finite-state automata.

Let $L \subseteq \Sigma^*$ be any formal language. Define the canonical quasiorder induced by L: $x \leq_L y$ if and only if $\forall u, v \in \Sigma^*$, $uyv \in L$ implies $uxv \in L$.

(1.1). *L is finite-state if and only if $\leq_L$ is a wqo.*

(1.1) should be thought of as representing another step of abstraction beyond the Myhill-Nerode Theorem [**HU**] that it half-resembles. Although we do not pursue this further here, the well-quasiordering characterization (1.1) can be used to provide nonconstructive proofs that some formal languages are regular, providing an analogue in the familiar setting of formal languages to some of the nonconstructive consequences of the Graph Minor Theorem. The proof of (1.1) can be found in section 3.

DEFINITION 1.1. A *lower ideal* in a quasiorder $(S, \leq)$ is a subset $A \subseteq S$ such that $x \leq y$ and $y \in A$ imply $x \in A$.

DEFINITION 1.2. A quasiordered set $(Q, \leq)$ is *Noetherian* if there are no properly infinitely descending chains $(x_1 > x_2, > \ldots)$ in Q, where $x > y$ denotes that $x \geq y$ and not $y \geq x$.

DEFINITION 1.3. A quasiorder $\leq$ on Σ^* is *well-behaved* if it is (1) Noetherian, and (2) a congruence with respect to concatenation, that is, $x \leq x'$ and $y \leq y'$ imply $xy \leq x'y'$.

(1.2). *A well-behaved quasiorder $\leq$ on Σ^* is a wqo if and only if $\leq_L$ is a wqo for every lower ideal L of $(\Sigma^*, \leq)$.*

What does it mean for a graph family to be finite-state recognizable? For bounded tree-width we can answer this in an elegant and powerful way by means of (1) structural parse trees to represent the graphs, and (2) finite-state tree automata.

The structure of a graph of bounded treewidth can be represented by a rooted labeled tree. The labels represent a finite set of structural primitives that describe how the graph is built up. (A concrete example of such a parsing formalism can be found in the next section.) Such a *parse tree* of a graph is, in general, not unique. If T is a parse tree, we let $G(T)$ denote the graph that it represents. There are a variety of possible ways to settle the details of the parsing formalism. Any reasonable choice will do, and makes no difference to the theory we present. In the next section we give as an illustrative example a parsing formalism for graphs of treewidth 2.

DEFINITION 1.4. Let t be a positive integer. A family F of graphs is *t-finite state* if there is a leaf-to-root finite state tree automaton [**Th**] that accepts a parse tree T (of a graph of treewidth at most t) if and only if $G(T) \in F$. We say that a family of graphs F is *finite state* if F is t-finite state for every t.

The main results that we describe establish necessary and sufficient conditions for a family of graphs to be finite-state. We apply these characterizations to some natural graph families to obtain both "positive" and "negative" results concerning finite-state recognizability for bounded treewidth.

In this "graph-theoretic" exposition we employ the notion of a t-boundaried graph, as developed in [**FL**] (see also [**Wi**]). A t-boundaried graph is just an ordinary graph equipped with t distinguished vertices labeled 1 through t. Formally, we have:

DEFINITION 1.5. A *t-boundaried graph* $G = (V, E, B, f)$ is an ordinary graph $G = (V, E)$ together with (1) a distinguished subset B of the vertex set V, $|B| = t$, and (2) a bijection $f : B \to \{1, \ldots, t\}$.

A fundamental operation (denoted $\oplus$) on t-boundaried graphs that we consider is that of gluing them together along their boundaries by identifying like-labeled vertices.

DEFINITION 1.6. If $G = (V, E, B, f)$ and $G' = (V', E', B', f')$ are t-boundaried graphs, then $G \oplus G'$ denotes the t-boundaried graph obtained from the disjoint union of the graphs $G = (V, E)$ and $G' = (V', E')$ by identifying each vertex $u \in B$ with the vertex $v \in B'$ for which $f(u) = f'(v)$. (In some situations which will be clear from the context, such as the next definition, we consider $G \oplus G'$ to be an ordinary graph, by "forgetting" the boundary.)

We are concerned with at least two *universes* of t-boundaried graphs: (1) the *large universe* U^t_{large} of all t-boundaried graphs, and (2) the *small universe* U^t_{small} of t-boundaried graphs that arise in the parsing of graphs of treewidth at most t. The universe U^2_{small} is concretely described in the next section.

DEFINITION 1.7. Let F be any graph family, and let U^t be either U^t_{small} or U^t_{large}. Define the canonical quasiorder induced by F: $X \leq_F Y$ if and only if $(\forall Z \in U^t)(Y \oplus Z \in F \to X \oplus Z \in F)$.

We have the following analogues of (1.1) and (1.2).

(1.3). *A graph family F is t-finite-state if and only if $\leq_F$ is a well-quasiorder on U^t_{small}.*

DEFINITION 1.8. A quasiorder $\leq$ on U^t_{small} is *well-behaved* if
(1) $\leq$ is Noetherian, and
(2) $\leq$ is a congruence with respect to $\oplus$, that is, $X \leq Y$ and $X' \leq Y'$ imply $X \oplus X' \leq Y \oplus Y'$.

(1.4). *A well-behaved quasiorder $\leq$ on U^t_{small} is a well-quasiorder if and only if $\leq_F$ is a well-quasiorder on U^t_{small} for every lower ideal F of $\leq$.*

(1.3) and (1.4) follow from the basic result (1.5) stated below. The proofs of (1.1)-(1.5) are presented in the next section. A family of graphs F induces an equivalence relation on a universe U^t (large or small) of t-boundaried graphs as follows.

DEFINITION 1.9. If F is a family of graphs then the *F-canonical congruence* is defined for t-boundaried graphs X, Y by $X \sim_F Y$ if and only if for every t-boundaried graph $Z \in U^t$, $X \oplus Z \in F \leftrightarrow Y \oplus Z \in F$.

The following definition captures an essential feature of the complexity of the "information flow" across a bounded-size cutset necessary to determining membership in a graph family.

DEFINITION 1.10. A graph family F is *t-fully cutset regular* if the F-canonical congruence on t-boundaried graphs has a finite number of equivalence classes (*finite index*) with respect to the large universe U^t_{large} of all t-boundaried graphs. We say that F is *fully cutset regular* if it is t-fully cutset regular for every t.

DEFINITION 1.11. A graph family F is *t-cutset regular* if $\sim_F$ has finite index on U^t_{small}. We say that F is *cutset regular* if t-cutset regular for every t.

Note that if a graph family is t-fully cutset regular then it is t-cutset regular. It is not presently known if these notions are distinct for families of graphs of treewidth bounded by t. (For interesting recent progress on this question see [CL].)

(1.5). *A family of graphs F is t-finite-state if and only if F is t-cutset regular.*

The main interest for algorithms research in RS posets (for bounded treewidth) is the result, based on the finite obstruction sets, that all lower ideals are finite-state, and therefore recognizable in time $O(n \log n)$, (and NC-recognizable). But the collection of t-finite-state graph families is far more extensive than the collection of such lower ideals. For example, it is easy to show directly that for the family F of Hamiltonian graphs, $\sim_F$ has finite index on U^t_{large} (and therefore also U^t_{small}).

RS-posets are defined by two distinct properties: well-quasiordering and feasible order tests. Robertson and Seymour develop more-or-less separate arguments for these two properties for the minor order on general graphs [RS1] [RS2] [RS3]. (1.4) shows that, for bounded treewidth, (well-behaved) well-quasiordering *implies* fast order tests. (Interestingly, this implication is nonconstructive.) (1.2) and (1.4) may open the way to automata-theoretic proofs of well-quasiordering results (such as the Graph Minor Theorem) for graphs of bounded treewidth.

In [Co1] Courcelle establishes a sufficient condition, termed *recognizability* for a graph family to be finite-state recognizable for bounded treewidth. In [Co3] it is shown that *recognizability* coincides with the property we have termed here *full cutset regularity*.

The next section presents the basic machinery of the finite-state point of view on bounded treewidth graph problems, and proves our central result (1.5). Section 3 presents the well-quasiordering characterizations (1.1)–(1.4) of finite-state recognizability. Section 4 describes various applications. Section 5 gives a proof of Courcelle's important theorem on second-order monadic logic in this framework. Section 6 concludes with a discussion of open problems.

2. Bounded Treewidth and Finite-State Automata

We must first describe how a graph of bounded treewidth can be represented as a labeled binary *parse tree*.

DEFINITION 2.1. A *tree-decomposition* of a graph $G = (V, E)$ is a tree T together with a collection of subsets T_x of V indexed by the vertices x of T that satisfies:
  1. For every edge uv of G there is some x such that $\{u, v\} \subseteq T_x$.
  2. If y is a vertex on the unique path in T from x to z then $T_x \cap T_z \subseteq T_y$.

The *width* of a tree decomposition is the maximum over the vertices x of the tree T of the decomposition of $|T_x| - 1$. A graph G has *treewidth* at most k if there is a tree decomposition of G of width at most k. Alternatively, a graph has treewidth at most k if and only if it is a subgraph of a k-tree (see [Ar,AP]), and this provides perhaps the most accessible concrete picture of what bounded treewidth is all about.

EXAMPLE. 2-TREES.
  (1) K_3 is a 2-tree.
  (2) If G is a 2-tree and uv is an edge of G (uv is a subgraph isomorphic to K_2) then the graph G' consisting of G together with a new vertex x and new edges xu and xv is a 2-tree. (Thus the K_2 subgraph is augmented to K_3).
  (3) *Nothing else.*

A recipe for generating (all) k-trees, for k arbitrary, differs from the above in "starting" (1) with the complete graph on $k+1$ vertices, and in (2) by augmenting a complete K_k subgraph of G to K_{k+1}. Figure 1 shows some examples of 2-trees.

At the heart of the automata-theoretic approach to bounded treewidth is the representation of graphs of bounded treewidth as labelled trees, where the labels are taken from a finite alphabet and represent structural primitives and operations for building the graph. Such a labelled tree we refer to as a *parse tree* for the graph. A general notion useful for such representation schemes is that of a *composition operator* for boundaried graphs.

The idea of a boundaried graph composition operator is quite simple: the operator is described by a fixed *operator graph* that has several boundaries (not necessarily disjoint). E.g., for a binary operator, the operator graph is equipped with 3 boundaries, two for the arguments to the operator, and one to be the boundary of the resulting graph. Arguments to the operator (which are boundaried graphs), are attached to the operator graph in the same way as with $\oplus$, by vertex identification along the appropriate argument boundary.

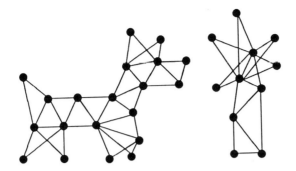

FIGURE 1. SOME 2-TREES.

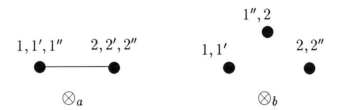

FIGURE 2. OPERATORS FOR PARSING TREEWIDTH 2.

DEFINITION 2.2. An *n-ary t-terminal composition operator* $\otimes$ is defined by the data:

1. A t-boundaried graph $T_\otimes = (V_\otimes, E_\otimes, B_\otimes, f_\otimes)$.
2. Injective maps $f_i : \{1, \ldots, t\} \to V_\otimes$ for $i = 1, \ldots, n$.

For the binary case, if G_i for $i = 1, 2$ is a pair of t-boundaried graphs $G_i = (V_i, E_i, B_i, f_i)$ then $G_1 \otimes G_2$ is defined to be the t-boundaried graph for which the ordinary underlying graph is formed from the disjoint union of G_1, G_2 and $T_\otimes$ by identifying each vertex u of B_i (for $i = 1, 2$) with its image $f_i(u)$ in $V_\otimes$. The boundary set and the labeling for $G_1 \otimes G_2$ is given by $B_\otimes$ and $f_\otimes$.

In this exposition we focus for notational convenience on binary composition operators. All of the theory goes through if one considers composition operators of any fixed arity. Graphs of treewidth at most t can be parsed using a small number of t-ary operators of boundary size t; they can also be parsed with binary operators of size $t + 1$.

Illustration. Parsing graphs of treewidth 2, and the universe U^2_{small}.

The following operators are sufficient to parse graphs of treewidth 2: $\oplus$ and the operators $\otimes_a$ and $\otimes_b$ illustrated in Figure 2. Our notational convention is that the first argument boundary of an operator graph is labelled $\{1', 2'\}$, the second argument boundary is labelled $\{1'', 2''\}$, and the boundary of the resulting graph is labelled $\{1, 2\}$. (Note that these boundaries are not disjoint; a vertex may play several roles.)

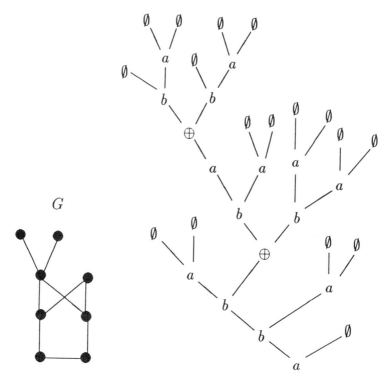

FIGURE 3. AN EXAMPLE OF PARSING.

DEFINITION 2.3. The small universe U^2_{small} is described:
(1) The "empty" graph $\emptyset$ consisting of two isolated vertices and the only possible boundary is in U^2_{small}.
(2) If G_1 and G_2 are boundaried graphs (necessarily having boundary size 2) in U^2_{small} then $G_1 \oplus G_2$, $G_1 \otimes_a G_2$ and $G_1 \otimes_b G_2$ are in U^2_{small}.
(3) Nothing else.

It is straightforward to verify that the underlying (ordinary) graph of any boundaried graph in U^t_{small} has treewidth 2.

Conversely, one can show that if G is any graph of treewidth at most 2 (and having at least 2 vertices) then G can be parsed using the set of labels $\Omega = \{\emptyset, \oplus, a, b\}$ in such a way that each leaf is labeled $\emptyset$. We illustrate with a parse for the graph G shown in Figure 3. The label a denotes the operator $\otimes_a$ and the label b denotes the operator $\otimes_b$ shown in Figure 2.

How fast can a parse tree be found, starting from scratch? One of the more accessible fundamental results of Robertson and Seymour showed that in time $O(n^2)$ one can find a tree decomposition (and therefore a parse tree, by a linear-time translation) of width $t = 5w$ for any graph of treewidth at most w [**RS2**]. This has been improved, parallelized, and randomized in a number of ways (see

[**ACPS**], [**La**] and [**Re**]), culminating in the $t = w$, linear-time algorithm due to Bodlaender [**Bo2**].

The following results are useful.

(2.1). *For an arbitrary graph family F, the canonical equivalence relation $\sim_F$ is a congruence with respect to all binary composition operators.*

PROOF. Let $\otimes$ denote an arbitrary binary composition operator and suppose $X_1 \sim_F X_2$ and $Y_1 \sim_F Y_2$. Suppose there is a boundaried graph Z such that $(X_1 \otimes Y_1) \oplus Z \in F$ and $(X_2 \otimes Y_2) \oplus Z \notin F$. But $X_1 \sim_F X_2$ implies $(X_2 \otimes Y_1) \oplus Z \in F$, and $Y_1 \sim_F Y_2$ further implies $(X_2 \otimes Y_2) \oplus Z \in F$, a contradiction.

(2.2). *An equivalence relation $\sim$ on t-boundaried graphs is a congruence with respect to all composition operators if and only if it is a congruence with respect to all unary composition operators.*

PROOF. The proof is similar to that for (2.1).

We next formally define the finite-state recognition mechanism for sets of binary trees that are vertex-labeled from a finite alphabet. Such a set we will term a *language* of trees.

DEFINITION 2.4. A *finite-state tree automata* is a 5-tuple $M = (\Sigma, Q, q_0, A, \delta)$ where Σ is a finite alphabet, Q is a finite set of *states*, $q_0 \in Q$ is a distinguished *start state*, $A \subseteq Q$ is a set of *accept states* and $\delta : Q \times Q \times \Sigma \to Q$ is a *transition function*.

In order to describe the set of Σ-labeled binary trees *recognized* by a finite-state tree automata M we must first recursively define a function *eval* on these objects. In the natural way, we refer to the two labeled binary trees T_1 and T_2 of a labeled binary tree T as the *children* of T, where if T has root r then T_1 and T_2 are rooted at the children of r in T. To denote this relationship between T_1, T_2 and T in the case where the root r of T is labeled by $a \in \Sigma$ we write $T = T_1 \otimes_a T_2$, and we will refer to $\otimes_a$ as a *tree composition operator*.

Since we are concerned with the situation where the letters in the finite alphabet Σ denote the operators that describe the parsing of a graph of bounded treewidth, it will be convenient to use the notation $\oplus$ to denote the tree composition operator that corresponds to the operator $\oplus$ defined on boundaried graphs.

In the case that T consists of just the single vertex r, we regard the children of T to be $T_1 = T_2 = \emptyset$, the tree with no vertices.

Where Σ is a finite alphabet, we write Σ^{**} to denote the set of all Σ-labeled binary trees.

DEFINITION 2.5. If $M = (\Sigma, Q, q_0, A, \delta)$ is a finite-state tree automata then the function $eval_M : \Sigma^{**} \to Q$ is defined by

(1) $eval_M(\emptyset) = q_0$
(2) $eval_M(T_1 \otimes_a T_2) = \delta(eval(T_1), eval(T_2), a)$

DEFINITION 2.6. The language of trees $L(M)$ *recognized* by a finite-state tree automata M is the set of all labeled trees $T \in \Sigma^{**}$ for which $eval_M(T)$ is in the set of accept states A of M.

DEFINITION 2.7. A language of binary trees $L \subseteq \Sigma^{**}$ is *finite-state* if and only if there is a finite-state tree automata M such that $L(M) = L$.

DEFINITION 2.8. An equivalence relation $\sim$ on Σ^{**} is a *congruence* if and only if $\forall a \in \Sigma$, $T_0 \sim T_0'$ and $T_1 \sim T_1'$ imply $T_0 \otimes_a T_1 \sim T_0' \otimes_a T_1'$.

DEFINITION 2.9. The canonical congruence on Σ^{**} induced by $L \subseteq \Sigma^{**}$ is defined $T \sim_L T'$ if and only if $(\forall a \in \Sigma)(\forall T_0 \in \Sigma^{**})(T \otimes_a T_0 \in L \leftrightarrow T' \otimes_a T_0 \in L)$.

The following standard theorem is the tree automata analogue of the familiar Myhill-Nerode theorem concerning regular formal languages.

(2.3). *A language $L \subseteq \Sigma^{**}$ is finite-state if and only if $\sim_L$ has finite index, and this is true if and only if there is an equivalence relation $\sim$ on Σ^{**} that satisfies:*

(1) $\sim$ *is a congruence*
(2) $\sim$ *has finite index*
(3) $T \sim T'$ *and* $T \in L$ *imply* $T' \in L$

We next prove our central theorem relating finite-state tree automata to bounded treewidth graph families.

(1.5). *A graph family F is t-finite state if and only if F is t-cutset regular.*

PROOF. We argue the case for "only if" by contraposition. Let $X_1, X_2, \ldots$ be an infinite set of representatives of the equivalence classes of $\sim_F$. Let $L = T(F)$ denote the set of all labeled binary trees that are parse trees of graphs in F. We claim that the canonical equivalence relation $\sim_L$ must also have infinite index. For each X_i choose a parse tree T_i. For each pair of indices i, j with $i \neq j$ choose a graph Z_{ij} such that (without loss of generality) $X_i \oplus Z_{ij}$ is in F and $X_j \oplus Z_{ij}$ is not in F, and choose a parse tree T_{ij} of Z_{ij}. Then $T_i \oplus T_{ij}$ is a parse tree for $X_i \oplus Z_{ij} \in F$ and so $T_i \oplus T_{ij} \in T(F)$. By a similar chain of reasoning $T_j \oplus T_{ij} \notin T(F)$. Since the graphs X_i are all distinct, the trees T_i are also distinct, so that $\sim_L$ has infinite index and $T(F)$ is not finite-state.

The converse is argued as follows. Define an equivalence relation $\sim$ on Σ^{**} by $T \sim T'$ if and only if $G(T) \sim_F G(T')$. It suffices to argue that $\sim$ satisfies the three conditions of (2.3) with respect to the language $L = T(F)$ of parse trees of graphs in F. That L is a union of equivalence classes of $\sim$ is immediate. That $\sim$ has finite index follows from the hypothesis that $\sim_F$ has finite index.

It remains only to argue that $\sim$ is a congruence with respect to binary tree composition operators. Suppose $T_1 \sim T_1'$ and $T_2 \sim T_2'$. Let $a \in \Sigma$. Then by the definition of $\sim$ and the fact (2.1) that $\sim_F$ is a congruence with respect to all binary composition operators, we have $G(T_1) \otimes_a G(T_2) \sim_F G(T_1') \otimes_a G(T_2')$ where

$\otimes_a$ denotes in this expression the graph composition operator corresponding to $a \in \Sigma$. By the definition of our parse tree representations, we have $G(T_1 \otimes_a T_2) = G(T_1) \otimes_a G(T_2)$ and $G(T_1' \otimes_a T_2') = G(T_1') \otimes_a G(T_2')$ so that $G(T_1 \otimes_a T_2) \sim_F G(T_1' \otimes_a T_2')$ and therefore $T_1 \otimes_a T_2 \sim T_1' \otimes_a T_2'$.

A useful variant of the above theorem goes as follows. (The proof is straightforward.)

(2.4). *Let F be a family of graphs for which by some means we can decide membership. If for every t a decision algorithm is known for an equivalence relation $\sim$ on t-boundaried graphs that satisfies*

(1) *$\sim$ has finite index,*
(2) *$\sim$ is a congruence with respect to unary composition operators, and*
(3) *$X \sim Y$ and $X \in F$ implies $Y \in F$,*

then F is cutset regular and for every positive integer bound w a finite-state tree automaton can be computed that recognizes the parse trees of the graphs in F of treewidth at most w.

Where F in (2.4) is a lower ideal in a well-quasiorder, and a bound on the obstruction treewidth is known, the automata that can be computed provides a means for computing the obstruction set for F (see [**FL**] and [**APS**] for details).

To illustrate the main idea in (2.4), consider the simpler case of a formal language $L \subseteq \Sigma^*$ for which we know (1) a decision algorithm for L, and (2) a decision algorithm for the Myhill-Nerode congruence $\sim_L$ on Σ^*, and assume that $\sim_L$ has finite index. We wish to compute a finite-state automata that recognizes L. The states of this automata will correspond to the equivalence classes of $\sim_L$, the *start* state will be the equivalence class of the empty word, and the *accept* states will be those equivalence classes that are subsets of L. On a letter $a \in \Sigma$ the transition function takes us from a state $[x]$ to the state $[xa]$. We compute the automata by computing representative words for the states, by a greedy procedure that begins with the empty word representing the *start* state. For a representative x, we use algorithm (2) to determine whether xa represents a *new* state, or one for which we already know a representative. In this way, beginning with the *start* state, we gradually elucidate the states and transitions of the automata. We use algorithm (1) to identify the *accept* states.

(1.5) and (2.4) provide easy to use graph-theoretic tools for showing that graph families are (or are not) finite-state for bounded treewidth. They are easy to use because they focus on a simple and elegant property of the graph family: the amount of information flow required across a bounded sized cutset in order to to decide membership in the family. Some concrete applications are explored in Section 4.

3. Finite Automata and Well-Quasiordering.

In this section we prove our main new results, (1.1) – (1.4). From the applications point of view, one of the main consequences of the Graph Minor Theorem is that every minor order lower ideal is finite-state for bounded treewidth. By a cardinality argument, we have the following converse. (Note that the particulars of the quasiorder play essentially no role in the argument.)

(3.1). *Let $\leq$ be a Noetherian quasiorder on U^t_{small} (on Σ^*). If every lower ideal of $\leq$ is finite-state then $\leq$ is a well-quasiorder.*

PROOF. If $\geq$ is not a well-quasiorder then there is an infinite antichain A. Each (possibly infinite) subset S of A determines a distinct lower ideal $I_S = \{x : x \leq a \in S\}$. If S, T are subsets of A with $S \neq T$ then $I_S \neq I_T$. Thus A yields a collection of uncountably many distinct lower ideals. Since there are only countably many finite-state tree automata (linear automata), there must be a lower ideal F_S that is not finite-state.

We will use the Myhill-Nerode Theorem and (1.5) to prove (1.1) and (1.3), after the following preliminaries.

(3.2). *If F is a family of graphs for which $\leq_F$ is a well-quasiorder on U^t_{large} (U^t_{small}) then $\leq_{F^c}$ is a well-quasiorder on U^t_{large} (U^t_{small}), where F^c denotes the complement of F.*

PROOF. If the statement is false, then we may take $(X_1, X_2, \ldots)$ to be a bad $\leq_{F^c}$ sequence which is $\leq_F$ ascending. For $i < j$ let $Z_{i,j}$ be a choice of evidence for the badness of the sequence: $X_i \oplus Z_{i,j} \in F$ and $X_j \oplus Z_{i,j} \in F^c$. Consider the sequence of t-boundaried graphs $(Z_{1,2}, Z_{3,4}, Z_{5,6}, \ldots)$. Since $\leq_F$ is a well-quasiorder, we must obtain the situation (reindexing for convenience): $X_1 \leq_F X_2 \leq_F X_3 \leq_F X_4$, $X_1 \oplus Z_{1,2} \in F$, $X_2 \oplus Z_{1,2} \notin F$, $X_3 \oplus Z_{3,4} \in F$, $X_4 \oplus Z_{3,4} \notin F$, and $Z_{1,2} \leq_F Z_{3,4}$. But this is a contradiction, since $X_3 \oplus Z_{3,4} \in F$ implies $X_3 \oplus Z_{1,2} \in F$, and this implies $X_2 \oplus Z_{1,2} \in F$.

We have the following similar result for formal languages.

(3.3). *If $L \subseteq \Sigma^*$ is a formal language for which $\leq_L$ is a well-quasiorder on Σ^*, then $\leq_{L^c}$ is also a well-quasiorder on Σ^*.*

PROOF. If the statement is false, then we can find a bad $\leq_{L^c}$ sequence $(x_1, x_2, \ldots)$ that is $\leq_L$ ascending. By considering the associated sequence of evidence as in the proof of (3.2), we obtain the situation (renaming for convenience): $x_1 \leq_L x_2 \leq_L x_3 \leq_L x_4$, $ux_1 v \in L$, $ux_2 v \notin L$, $u'x_3 v' \in L$, $u'x_4 v' \notin L$, and (using that $\leq_L \times \leq_L$ is a wqo) $u \leq_L u'$ and $v \leq_L v'$. This is a contradiction, because $u'x_3 v' \in L$ implies $ux_3 v' \in L$, which implies $ux_3 v \in L$, and in turn $ux_2 v \in L$.

PROOF OF (1.1) AND (1.3). We present the argument for (1.3), the proof of (1.1) is similar. First note that for t-boundaried graphs X and Y, $X \sim_F Y$ if and only if $X \leq_F Y$ and $X \leq_{F^c} Y$, by definition. If $\leq_F$ is a well-quasiorder then by (3.2) $\leq_{F^c}$ is a well-quasiorder. The Cartesian product of $\leq_F$ and $\leq_{F^c}$ is then a well-quasiorder, and therefore $\sim_F$ has finite index. By (1.5) we are done.

PROOF OF (1.2). By (1.1) and (3.1) it is enough to argue that if L is a lower ideal of a well-behaved quasiorder $\leq$ on Σ^*, then $\leq_L$ is a well-quasiorder. If not,

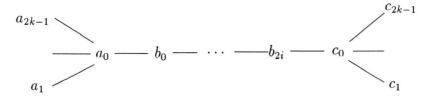

FIGURE 4. REPRESENTATIVES OF $\sim_{F_k}$.

then there is a bad $\leq_L$ sequence $(x_1, x_2, \dots)$. But then, since $\leq$ is wqo, there are indices $i < j$ such that $x_i \leq x_j$. Since it is not the case that $x_i \leq_L x_j$ there are words $u, v \in \Sigma^*$ such that $ux_i v \in L$ and $ux_j v \notin L$. This contradicts that L is a lower ideal, since $x_i \leq x_j$ implies $x_i v \leq x_j v$ which implies $ux_i v \leq ux_j v$.

PROOF OF (1.4). Similarly to (1.2), by (1.3) and (3.1) it remains only to argue that if $\leq$ is a well-behaved quasiorder on U^t_{small} and F is a lower ideal, then $\leq_F$ is wqo. From the definitions, if $X \leq Y$ then $X \leq_F Y$.

4. Some Concrete Applications of Cutset Regularity

The results of Section 2 (especially (2.4)) can be used as the basis for arguments that certain graph families are *not* finite-state. The method of argument is essentially as in textbook applications of the Myhill-Nerode theorem in proving that formal languages are not finite-state. We demonstrate for the k-Bandwidth problem [GJ].

DEFINITION 4.1. A *layout* of a graph $G = (V, E)$ is a one-to-one function $l : V \to \{1, \dots, |V|\}$. The *bandwidth* of layout l of G is the $\max_{uv \in E} |l(u) - l(v)|$. The *bandwidth* of G is the minimum bandwidth of a layout of G.

(4.1). *The family F_k of graphs having bandwidth at most k is not t-cutset regular if $t \geq 1$ and $k \geq 2$.*

PROOF. It suffices to prove the theorem for $t = 1$, since additional boundary vertices can easily be added. Our proof consists simply of exhibiting an infinite set of representatives of distinct equivalence classes of the equivalence relation $\sim_{F_k}$. The representatives are $(X_i, i \geq 1)$, where X_i is the graph illustrated in Figure 4.

The single boundary vertex of X_i is b_i. The stars at either end of X_i strongly constrain the layouts of bandwidth at most k. They must look like Figure 5, up to permutation of the layouts of the stars.

Let Y_i be a path of $2i+1$ vertices, whose single boundary vertex is the centroid of the path. Suppose $n \geq m$. The reader can easily check that $X_m \not\sim_{F_k} X_n$ by observing the following.

(1) $X_n \oplus Y_{kn-n} \in F_k$
(2) $X_m \oplus Y_{kn-n} \notin F_k$

The layout needed to verify assertion (1) is constructed by placing $k - 1$ vertices of Y_{kn-n} between each pair (b_i, b_{i+1}), for $i = 0, \dots, 2n - 1$. There is not

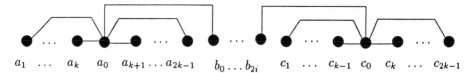

FIGURE 5. THE CONSTRAINED LAYOUT.

enough room in X_m to do the same thing, so some edge will have to be stretched to length more than k to accommodate Y_{kn-n} within the layout for X_m.

So far, the argument has established that k-Bandwidth is not fully cutset regular. Noticing that the representatives X_i are in fact trees, we complete the proof of theorem.

Theorem (4.1) leaves only the case $k = 1$ undetermined. A graph G has a layout of bandwidth 1 if and only if G is a disjoint union of paths. So each equivalence class is determined by where the boundary vertices lie on paths: whether they are isolated, at the end of a path or somewhere in the interior of a path. Hence, for t-boundaried graphs, there are 3^t equivalence classes.

Bruno Courcelle has pointed out that this outcome for the Bandwidth problem can also be obtained indirectly from an undecidability result of Wanke and Weigers concerning context-free graph grammars [**WW**].

Another interesting example of a problem which can be shown to be non-finite-state for bounded treewidth is the Perfect Phylogeny problem of computational biology. Background and details can be found in [**BFW**] where it is shown that for a fixed number k of characters, with $k \geq 4$, the k-Perfect Phylogeny problem is not finite-state for treewidth ≥ 4.

The finite-state characterization theorems of Sections 2 and 3 are also useful for showing that graph families F are finite-state. We wish to point out a useful general method of argument that these characterizations provide. The *method of test sets* proceeds by the following steps.

Step 1: Identify a finite set S of *tests* T to be performed on a t-boundaried graph X. (For example, a test might consist of a t-boundaried graph Y for which we determine whether $X \oplus Y$ is in F.)

Step 2: Argue that the equivalence relation $X \sim Y$ defined: $X \sim Y$ if and only if $S_X = S_Y$ where $S_X = \{T \in S : X \text{ passes test } T\}$ has the properties (1) $X \sim Y$ and $X \in F$ implies $Y \in F$, and (2) $\sim$ is a congruence with respect to unary composition operators.

We demonstrate the method of test sets on the problem of determining whether a graph has a layout of cutwidth at most k.

DEFINITION 4.2. A *layout* of a graph $G = (V, E)$ is a one-to-one function $l : V \to R$, where R is the real line.

DEFINITION 4.3. The *value of a cut at* $\alpha \in R$ is the number of edges uv of G with $l(u) < \alpha$ and $l(v) > \alpha$. The *cutwidth* of a layout l of a graph G is the maximum value of a cut at any point $\alpha \in R$. The *cutwidth* of a graph G is the minimum cutwidth of a layout of G.

(4.2). *For every k and for every t, the problem of deciding whether a graph has a layout of cutwidth at most k is t-finite state.*

PROOF. Fix k and t, and let F denote the family of graphs of cutwidth at most k. Let $w : R \to N$ be a non-negative integer valued function which we will refer to as a *weighting* of the real line R. We will consider the cutwidth of a graph relative to a weighting of R in the following way.

If l is a layout of a graph $G = (V, E)$, the *value of a cut at $\alpha \in R$ with respect to w* is the sum of $w(\alpha)$ and the number of edges uv of G with $l(u) < \alpha$ and $l(v) > \alpha$. The *cutwidth of a layout l of G with respect to w* is the maximum value of a cut at any point $\alpha \in R$. The *cutwidth of G with respect to w* is the minimum cutwidth of a layout of G with respect to w.

A *test* $T = (\pi, S)$ of size n consists of the following: (1) a map $\pi : \{1, \ldots, t\} \to \{1, \ldots, n\}$, and (2) a sequence S of non-negative integers $S = (S(0), \ldots, S(n))$.

To a test T of size n we associate a weighting w_T which consecutively assigns the intervals of R: $(0, 1), (1, 2), \ldots, (n, n+1)$ the values $S(0), \ldots, S(n)$. Precisely, $w_T(\alpha) = S(i)$ if $i < \alpha < i+1$ and $w_T(\alpha) = 0$ otherwise.

We say that a t-boundaried graph $G = (V, E, B, f)$ *passes* a test T of size n if there is a layout l of G of cutwidth at most k relative to w_T that further satisfies the conditions:

(1) For all $v \in V$, $0 < l(v) < n+1$.
(2) If u is a boundary vertex of G with label j, $f(u) = j$, then $l(u) = \pi(j)$. Thus the boundary vertices are laid out in the order and position described by π.
(3) If u is not a boundary vertex, $u \in V - B$, then $l(u) \notin N$. Thus non-boundary vertices of G are assigned layout positions in the interiors of the weighted intervals of R.

Let G and H be t-boundaried graphs and define $G \sim H$ if and only if G and H pass exactly the same set of tests. We will establish the following.

Claim 1. There is a finite set of *reduced tests* such that G and H agree on all tests if and only if they agree on the set of reduced tests.

Claim 2. The relation $\sim$ is a congruence with respect to all unary t-boundary operators, and thus by (2.2) it is a congruence with respect to all t-boundary operators.

Claim 3. If $G \in F$ and $H \sim G$ then $H \in F$. From these three claims it follows by (2.4) that F is t-finite-state.

Note that Claim 3 is trivial, since given a layout l for G, we can describe a test T_l where the weightings of the intervals of R are zero, which G passes. Since H also passes T_l, there is a cutwidth k layout of H, which furthermore places the boundary vertices in the same order.

Claim 2 is also easily seen. Let $\otimes$ denote a unary operator. We must argue that if $H \sim G$ and T is test passed by $\otimes(H)$ then T is also passed by $\otimes(G)$. Let

l be a layout of $\otimes(H)$ that witnesses the fact that H passes $T = (\pi, S)$. Let S' be the modification of S obtained by considering S together with the restriction of l to the graph X of $\otimes$.

The graph X of $\otimes$ thus has two boundaries, one of which is attached to H, and the other is "attached" by l to the weighted line in the manner prescribed by π. We modify S by considering the weighting of R described by S to be subdivided by the images of the vertices of X, and with the weights of the resulting intervals adjusted to reflect the presence of any edges between vertices of X.

Let π' denote the map obtained by (1) normalizing the intervals of the modified weighting of R described above to unit length (by topological deformation, which does not affect the cutwidth combinatorics), and (2) recording the image of the boundary between X and H that is identified in forming $\otimes(H)$.

This gives a test $T' = (\pi', S')$ that is passed by H, and therefore also by G. Let l' be a layout of G passing T'. By extending l' to a layout of $\otimes(G)$ in the same way that the restriction of l extends to l for H, we see that $\otimes(G)$ passes T.

Lastly, we turn to Claim 1. We define an equivalence relation on tests, with $T \sim T'$ if and only if every t-boundaried graph that passes T passes T' and *vice versa*. We will argue that there are a finite number of equivalence classes of tests.

Call the sequence of integer weights *between two consecutive images* of $\{1, \ldots, t\}$ under the map π of a test T a *load pattern* of T. Each test $T = (\pi, S)$ thus has $t + 1$ load patterns between the images of the boundary vertices prescribed by π.

We will use the following notation. The symbols s, s', s_i denote sequences of integers taken from the set $J = \{0, \ldots, k\}$. Letters such as a, b, c will be used to denote single particular values in J. Two load patterns s and s' are termed *equivalent* if $T \sim T'$ for any test T for which s is a load pattern, where T' is the test obtained by replacing s with s'. Write $s \geq s'$ if every t-boundaried graph which passes a test T for which s is a load pattern, also passes the test T' where the load pattern s has been replaced by s'. Thus for example we have $s = (5, 4, 1, 3, 2, 7) \geq (5, 1, 1, 1, 1, 7) \geq (5, 1, 7) = s'$ and $(5, 1, 7) \geq (5, 5, 1, 7, 7, 7) \geq (5, 4, 1, 3, 2, 7)$, and therefore $s \sim s'$, since decreasing the weight of an interval only makes finding a cutwidth k layout easier, and consecutively repeating weights (or deleting such repetitions) makes no difference.

We have the following reduction rules for load patterns.

(R1) If $s = s_1 abc s_2$ with $a \leq b \leq c$ (or $a \geq b \geq c$) then $s \sim s_1 ac s_2$.

(R2) If $s = s_1 a s_2 b s_3$ where each element of s_2 is greater than or equal to the maximum of a and b, and $c = \max(s_2)$, then $s \sim s_1 acb s_3$.

(R3) If $s = s_1 a s_2 b s_3$ where each element of s_2 is at most the minimum of a and b, and $c = \min(s_2)$, then $s \sim s_1 acb s_3$.

(R1) is shown by the sequence of implications: $s_1 abc s_2 \geq s_1 aac s_2 \geq s_1 ac s_2 \geq s_1 acc s_2 \geq s_1 abc s_2$. To see (R2) let $m = \max(a, b)$, and suppose c is in the $i+1$ position of s_2, and that $|s_2| = i + j + 1 = q$. We have the sequence of implications: $s_1 a s_2 b s_3 \geq s_1 a m^i c m^j b s_3 \geq s_1 a^{i+1} c b^{j+1} s_3 \geq s_1 acb s_3 \geq s_1 ac^q b s_3 \geq s_1 a s_2 b s_3$.

The argument for (R3) is similar.

Note that since we are concerned with cutwidth k where k is fixed, every load pattern is trivially equivalent to one where the largest integer weight occurring is at most k. Define a load pattern s to be *reduced* if it contains only integer values in the range $\{0,\ldots,k\}$ and none of the reduction rules (R1)-(R3) can be applied to s. Define a test T to be *reduced* if each of the $t+1$ load patterns of T is reduced.

We next argue that a reduced load pattern s over $\{0,\ldots,k\}$ has length at most $2k+2$. Otherwise, s must contain some value a at least 3 times. Consider the factorization: $s = s_1 a s_2 a s_3 a s_4$ where a does not occur in s_2 or s_3. Neither of the subsequences s_2 or s_3 can be empty, or s is not reduced. Let b be the rightmost integer in s_2 and let c be the leftmost integer in s_3. If $b < a < c$ or $b > a > c$ then (R1) can be applied. So both of b, c must be greater than a, or both must be less than a. Suppose the latter. (The other case is handled similarly.)

Let d denote the rightmost occurrence in the subsequence as_2 of a value greater than or equal to a. Let e denote the leftmost occurrence in the subsequence s_3a of a value greater than or equal to a. There are at least 3 integers in s properly between d and e, and each intervening value is at most a. By (R3) s is reducible.

It follows that there are finitely many reduced tests, completing the proof.

Note that our argument shows that for every k, the family of graphs of cutwidth at most k is fully cutset regular.

There is a difference in the style of the above proof from previous work on algorithms for bounded treewidth, such as can be found for example in [**WHL**] and [**Wi**]. The difference is that in those approaches one essentially describes the finite-state automaton explicitly at the level of states and state-transitions. The test set method, in the "abstract" spirit of the Myhill-Nerode theorem, gives only an *implicit* description of the automaton. In the above proof, a finite test set is constructively described, and from this we can construct the relevant finite automaton by the standard greedy procedure discussed in Section 2. In the next section we further illustrate method of test sets with a proof of Courcelle's theorem concerning second-order monadic logic and bounded treewidth.

5. Courcelle's Theorem

The important theorem discussed in this section was first proved by Courcelle in a universal algebra framework. Similar results were obtained independently by Boric, Parker and Tovey [**BPT**]. (An extension of the theorem to handle some kinds of search problems can be found in [**ALS**].) The "graph-theoretic" proof given below is based on (1.5) and the method of test sets. The theorem is widely useful for establishing that graph decision problems are finite-state for bounded treewidth, since with a little practice it is often quite straightforward to write down a second-order monadic expression that describes the family of graphs of interest. An example of such a description for Hamiltonian graphs is given below. It is interesting to note, however, that some kinds of graph properties have proven to be difficult to express directly in second-order monadic logic, even though it is known by other means (for example, the Graph Minor Theorem) that second-order monadic descriptions of the property exist. Notable in this regard have been the "graph layout" problems, such as the Graph Cutwidth problem addressed the last section.

The syntax of the second-order monadic logic of graphs includes the logical connectives $\wedge, \vee, \neg, \leftarrow, \rightarrow$, variables for vertices, edges, sets of vertices and sets of edges, the quantifiers $\forall, \exists$ that can be applied to these variables, and the five binary relations:

(1) $u \in U$ where u is a vertex variable and U is a vertex set variable

(2) $d \in D$ where d is an edge variable and D is an edge set variable

(3) $inc(d, u)$ where d is an edge variable, u is a vertex variable, and the interpretation is that the edge d is incident on the vertex u

(4) $adj(u, v)$ where u and v are vertex variables and the interpretation is that u and v are adjacent vertices

(5) Equality for vertices, edges, sets of vertices and sets of edges.

Example. Hamiltonian graphs. It is obvious that a graph G is Hamiltonian if and only if the edges of G can be partitioned into two sets, say *red* and *blue*, such that (1) each vertex of G has exactly two incident red edges, and (2) the subgraph induced by the red edges spans G. By progressive refinement of our task, we show to express this in second-order monadic logic. (We use lower case letters for vertex or edge variables, upper case letters for variables representing sets of vertices or edges; early letters of the alphabet for edge variables, late letters of the alphabet for vertex variables.) We first have, by the above observation, and in notation corresponding to the above discussion, with E_1 representing the *red* edges, and E_2 the *blue* edges :

$$\text{Hamiltonian} \leftrightarrow \exists E_1 \exists E_2 \forall u \forall v \, (part(E_1, E_2) \wedge (deg(u, E_1) = 2) \wedge span(u, v, E_1))$$

We can write $part(E_1, E_2)$ as:

$$\forall e \, ((e \in E_1) \vee (e \in E_2)) \wedge \neg((e \in E_1) \wedge (e \in E_2))$$

The expression for $deg(u, E_1) = 2$ can be expanded as:

$$(\exists e_1 \exists e_2 \, (e_1 \neq e_2) \wedge inc(e_1, u) \wedge inc(e_2, u) \wedge (e_1 \in E_1) \wedge (e_2 \in E_1)) \wedge$$

$$\neg(\exists e_1 \exists e_2 \exists e_3 \, (e_1 \neq e_2) \wedge (e_1 \neq e_3) \wedge (e_2 \neq e_3) \wedge inc(e_1, u) \wedge inc(e_2, u) \wedge inc(e_3, u) \wedge$$
$$(e_1 \in E_1) \wedge (e_2 \in E_1) \wedge (e_3 \in E_1))$$

The connectivity requirement $span(u, v, E_1)$ can be expressed (note the neat trick):

$$\forall V_1 \, \forall V_2 \, \forall u \, \forall v \, ((u \in V_1) \wedge (v \in V_2)) \rightarrow$$
$$(\exists e \exists x \exists y \, inc(e, x) \wedge inc(e, y) \wedge (x \in V_1) \wedge (y \in V_2) \wedge (e \in E_1))$$

(5.1) (COURCELLE [Co1] + [Co3]). *If F is a family of graphs described by a sentence in second-order monadic logic, then F is fully cutset regular.*

PROOF. Let ϕ be the sentence that describes the graph family F, and write $\sim_\phi$ for $\sim_F$. We prove the theorem by induction on formulas, for t an arbitrary positive integer describing the boundary size. For convenience, we may write $\partial(X)$ to denote the boundary of a t-boundaried graph X. By the *interior* of X we refer to the vertices in $V(X) - \partial(X)$.

To each formula ϕ (possibly with a nonempty set $Free(\phi)$ of free variables) we associate an equivalence relation $\sim_\phi$ on the set of all t-boundaried graphs that are partially equipped with distinguished vertices, edges, sets of vertices and sets of edges corresponding to the free variables in the set $Free(\phi)$. (The details follow shortly.)

Two partially equipped t-boundaried graphs X and Y are defined to be $\sim_\phi$ related if and only if they have the "same" partial equipment and for every compatibly equipped t-boundaried graph Z, the formula ϕ is true for $X \oplus Z$ if and only if it is true for $Y \oplus Z$.

The compatibility condition on Z insures that the partial equipment of Z agrees with that of X and Y on the boundary vertices and edges, and further insures that all of the free variables have interpretations as the distinguished elements and sets in the product graphs $X \oplus Z$ and $Y \oplus Z$. The formalization of what we mean by "partial equipment" is tedious but unproblematic; it can be accomplished as follows.

Let $fr(0), fr(1), Fr(0)$ and $Fr(1)$ denote, respectively, sets of variables for vertices, edges, sets of vertices and sets of edges. Let $Free$ denote the disjoint union of these four sets. Allow S^2 to denote the set of all 2-element subsets of a set S. Define a *partial equipment signature* σ for $Free$ to be given by the following data.
(1) Disjoint subsets $int_0(\sigma)$ and $\partial_0(\sigma)$ of $fr(0)$, and a map $f_\sigma^0 : \partial_0(\sigma) \to \{1,\ldots,t\}$.
(2) Disjoint subsets $int_1(\sigma)$ and $\partial_1(\sigma)$ of $fr(1)$, and a map $f_\sigma^1 : \partial_1(\sigma) \to \{1,\ldots,t\}^2$.
(3) For each vertex set variable U in $Fr(0)$ a subset $U_\sigma \subseteq \{1,\ldots,t\}$.
(4) For each edge set variable D in $Fr(1)$ a subset $D_\sigma \subseteq \{1,\ldots,t\}^2$.

Let $Free(\sigma)$ denote the union of the sets $int_0(\sigma), \partial_0(\sigma), int_1(\sigma), \partial_1(\sigma), Fr(0)$ and $Fr(1)$. We will say that a t-boundaried graph $X = (V, E, B, f)$ is *σ-partially equipped* if it has distinguished vertices, edges, sets of vertices and sets of edges corresponding to the variables in $Free(\sigma)$, where the correspondence is compatible with the data that describes ϕ. More precisely, the following conditions must be satisfied by these distinguished elements and sets.
(1) If u is a vertex variable, $u \in int_0(\sigma)$, then the distinguished vertex in X corresponding to u must be in the interior of X.
(2) If u is a vertex variable, $u \in \partial_0(\sigma)$, then the distinguished vertex in X corresponding to u must be the unique vertex $x \in V$ for which $f(x) = f_\sigma^0(u)$.

(3) If d is an edge variable, $d \in int_1(\sigma)$, then the distinguished edge in X corresponding to d must have at least one endpoint in the interior of X.

(4) If d is an edge variable, $d \in \partial_1(\sigma)$, then the distinguished edge in X corresponding to d must have as endpoints the pair of vertices x, y of $\partial(X)$ for which $f_\sigma^1(d) = \{f(x), f(y)\}$.

(5) If U is a vertex set variable, $U \in Fr(0)$, then the distinguished set of vertices V_U in X corresponding to U must satisfy $f(V_U \cap \partial(X)) = U_\sigma$.

(6) If D is an edge set variable, $D \in Fr(1)$ then the distinguished set of edges E_D in X corresponding to D must satisfy $\{\{f(x), f(y)\} : \{x, y\} \in E_D \cap \partial(X)^2 = D_\sigma$.

All of this is just as expected. We say that partial equipment signatures σ and $\bar{\sigma}$ for $Free$ are *complements* with respect to $Free$ if the data describing them is in agreement with respect to the boundary and if together they provide for a complete interpretation of the variables of $Free$. More precisely, $\bar{\sigma}$ must satisfy the next listed (symmetric) conditions with respect to σ. (Note that $\bar{\sigma}$ is completely determined by σ.)

(1) $\partial_0(\bar{\sigma}) = \partial_0(\sigma)$ and $f_{\bar{\sigma}}^0 = f_\sigma^0$.
(2) $int_0(\bar{\sigma}) = fr(0) - \partial_0(\sigma) - int_0(\sigma)$.
(3) $\partial_1(\bar{\sigma}) = \partial_1(\sigma)$ and $f_{\bar{\sigma}}^1 = f_\sigma^1$.
(4) $int_1(\bar{\sigma}) = fr(1) - \partial_1(\sigma) - int_1(\sigma)$.
(5) For each vertex set variable U in $Fr(0)$, $U_{\bar{\sigma}} = U_\sigma$.
(6) For each edge set variable D in $Fr(1)$, $D_{\bar{\sigma}} = D_\sigma$.

The important thing is that if partial equipment signatures σ and $\bar{\sigma}$ are complements with respect to $Free$ and if X and Z are t-boundaried graphs with X σ-partially equipped and Z $\bar{\sigma}$-partially equipped then every variable in $Free$ has a consistent interpretation in $X \oplus Z$.

If σ is a partial equipment signature for $Free(\phi)$ and X, Y are σ-partially equipped t-boundaried graphs then we define $X \sim_\phi Y$ if and only if for every $\bar{\sigma}$-partially equipped t-boundaried graph Z: $X \oplus Z \models \phi$ if and only if $Y \oplus Z \models \phi$.

The induction claim may now be stated precisely. In what follows we will avoid mentioning signatures when they can be deduced from the context.

Claim. For every second-order monadic formula ϕ, $\sim_\phi$ has finite index.

We must first show that the claim holds for atomic formulas. Since vertex adjacency is easy to express as the existence of an edge incident on both vertices, it suffices to assume that the only atomic formulas are $d \in D$, $u \in U$ and $inc(e, u)$. Our consideration breaks up into cases according to the possible partial equipment signatures. All of these are easy, and are left entirely to the reader. For example, let $\phi = inc(e, u)$ and suppose σ is described by $e \in int_1(\sigma)$ and $u \in \partial_0(\sigma)$. The relation $\sim_\phi$ has index 2. For another example, let ϕ be the formula $d \in D$ where d is an edge variable and D is an edge set variable, and suppose $d \in \partial_1(\sigma)$ and $D_\sigma = \emptyset$. The equivalence relation $\sim_\phi$ then has index 1.

For the induction step we may suppose (without loss of generality) that ϕ is a formula obtained from simpler formulas in one of the following ways.

(1) $\phi = \neg \phi'$

(2) $\phi = \phi_1 \wedge \phi_2$
(3) $\phi = \exists u \phi'$ where u is a vertex variable free in ϕ'
(4) $\phi = \exists d \phi'$ where d is an edge variable free in ϕ'
(5) $\phi = \exists U \phi'$ where U is a vertex set variable free in ϕ'
(6) $\phi = \exists D \phi'$ where D is an edge set variable free in ϕ'

We treat each of these six cases separately, giving a complete argument for (1),(2),(3) and (6). An appropriate equivalence relation is defined for cases (4) and (5) and the reader will have no difficulty adapting the arguments given for the others to these two cases.

Case 1. $\phi = \neg \phi'$

It is enough to argue that $X \sim_{\phi'} Y$ implies $X \sim_\phi Y$. By contraposition, if $\exists Z, X \oplus Z \models \phi$ and $Y \oplus Z \models \neg \phi$ then immediately $X \oplus Z \models \neg \phi'$ and $Y \oplus Z \models \neg \phi \equiv \phi'$. The converse is just as easy, and in fact $\sim_\phi = \sim_{\phi'}$.

Case 2. $\phi = \phi_1 \wedge \phi_2$

Here we have $Free(\phi) = Free(\phi_1) \cup Free(\phi_2)$. Let σ be a partial equipment signature for $Free(\phi)$. Our strategy is to define a convenient equivalence relation $\sim$ on the σ-equipped t-boundaried graphs, and then show two things: that $\sim$ has finite index, and that it refines $\sim_\phi$.

The sets of variables in $Free(\phi_i), i = 1, 2$, are subsets of the sets of variables in $Free(\phi)$, so the partial equipment signature σ for $Free(\phi)$ induces partial equipment signatures σ_1 for $Free(\phi_1)$ and σ_2 for $Free(\phi_2)$ just by forgetting the unneeded equipment. It is with respect to these induced signatures that we define $\sim$ to be $\sim_{\phi_1} \cap \sim_{\phi_2}$. Since it is, by the induction hypothesis, the intersection of equivalence relations of finite index, $\sim$ has finite index.

It remains to argue that $X \sim Y$ implies $X \sim_\phi Y$. By contraposition, if $\exists Z, X \oplus Z \models \phi$ and $Y \oplus Z \models \neg \phi$ then either (i) $X \oplus Z \models \phi_1$ and $Y \oplus Z \models \neg \phi_1$, or (ii) $X \oplus Z \models \phi_2$ and $Y \oplus Z \models \neg \phi_2$. So either (i) X and Y are not $\sim_{\phi_1}$ equivalent, or (ii) X and Y are not $\sim_{\phi_2}$ equivalent. In either case it follows that X and Y are not $\sim$ equivalent.

Case 3. $\phi = \exists u \phi'$ where u is a vertex variable free in ϕ'

For $i \in \{1, \ldots, t\}$ let X_i denote the partially equipped t-boundaried graph X further equipped with the boundary vertex with label i in correspondence with the variable u.

Let $\alpha(X, Z)$ denote the statement: There is a vertex x in the interior of X such that $X_u \oplus Z \models \phi'$, where X_u is X additionally equipped with the vertex x in correspondence with the variable u.

Define $X \sim_{\exists u} Y$ if and only if for all Z, $\alpha(X, Z) \leftrightarrow \alpha(Y, Z)$.

This is an equivalence relation of finite index, since, by the commutativity of $\oplus$ and the definition of α, the conditions $\alpha(X, Z)$ and $Z \sim_{\phi'} Z'$ together imply $\alpha(X, Z')$, so that $\alpha(X, Z)$ depends in the second argument only on the equivalence class of Z.

Define $X \sim Y$ if and only if
(a) $X \sim_{\phi'} Y$

(b) $X_i \sim_{\phi'} Y_i$ for all $i \in \{1,\ldots,t\}$
(c) $X \sim_{\exists u} Y$

The relation $\sim$ is an equivalence relation of finite index since it is the intersection of finitely many such relations.

It remains only to argue that $X \sim Y$ implies $X \sim_\phi Y$. If not, then $\exists Z$ with $X \oplus Z \models \phi$ and $Y \oplus Z \models \neg\phi$. Choose an instantiation of u to a vertex v in $X \oplus Z$ making ϕ true.

Case (a). The vertex v is in the interior of Z. Let Z_u denote Z further equipped with the vertex v in correspondence with the variable u. Then $X \oplus Z_u \models \phi'$ and by (a), $Y \oplus Z_u \models \phi'$ which implies $Y \oplus Z \models \phi$, a contradiction.

Case (b). The vertex v is a boundary vertex of X. Then for some $i \in \{1,\ldots,t\}$ we have $X_i \oplus Z_i \models \phi'$ which by (b) implies $Y_i \oplus Z_i \models \phi'$ and therefore $Y \oplus Z \models \phi$, a contradiction.

Case (c). The vertex v belongs to the interior of X. Then (c) implies that there is a vertex v' in the interior of Y such that $Y_u \oplus Z \models \phi'$ where Y_u is Y further equipped with the vertex v' in correspondence with the variable u. This implies $Y \oplus Z \models \phi$, a contradiction.

Case 4. $\phi = \exists d \phi'$, where d is an edge variable free in ϕ'

For each $\{i,j\} \in \{1,\ldots,t\}^2$ let X_{ij} denote X further equipped with the edge ij in correspondence with the variable d.

Let $\beta(X,Z)$ be the statement: There is an edge d' with at least one interior endpoint in X such that $X_d \oplus Z \models \phi'$ where X_d denotes X equipped with d' in correspondence with d.

Define $X \sim_{\exists d} Y$ if and only if for all Z, $\beta(X,Z) \leftrightarrow \beta(Y,Z)$. This is an equivalence relation of finite index.

Define $X \sim Y$ if and only if:
(a) $X_{ij} \sim_{\phi'} Y_{ij}$ for all $\{i,j\} \in \{1,\ldots,t\}^2$
(b) $X \sim_{\phi'} Y$
(c) $X \sim_{\exists d} Y$

The relation $\sim$ is an equivalence relation of finite index since it is the intersection of finitely many such relations. The verification that $X \sim Y$ implies $X \sim_\phi Y$ is straightforward.

Case 5. $\phi = \exists U \phi'$ where U is a vertex set variable free in ϕ'

For each subset $S \subseteq \{1,\ldots,t\}$ let $\gamma_S(X,Z)$, defined for Z equipped with $V_U = S$, be the statement: There is a subset U_0 of the interior of X such that $X_U \oplus Z \models \phi'$ where X_U is X equipped with the vertex set $U_0 \cup S$ in correspondence with the variable U.

Define $X \sim_S Y$ if and only if for all Z, $\gamma_S(X,Z) \leftrightarrow \gamma_S(Y,Z)$.

It follows easily from the commutativity of $\oplus$ and the definition of γ_S that $\gamma_S(X,Z)$ and $Z \sim_{\phi'} Z'$ imply $\gamma_S(Y,Z)$, and from this and the induction hypothesis that $\sim_S$ is an equivalence relation of finite index.

Define $X \sim Y$ if and only if for all $S \subseteq \{1,\ldots,t\}$, $X \sim_S Y$.

The relation $\sim$ is an equivalence relation of finite index, since it is the intersection of finitely many such relations. The verification that $X \sim Y$ implies $X \sim_\phi Y$ is straightforward.

Case 6. $\phi = \exists D \phi'$ where D is an edge set variable

For each subset $S \subseteq \{1,\ldots,t\}^2$ let $\delta_S(X,Z)$ be the statement: There is a set of edges D_0, each having at least one interior endpoint in X, such that $X_D \oplus Z \models \phi'$ where X_D is X equipped with the edge set $D_0 \cup S$ in correspondence with the variable D.

Define $X \sim_S Y$ if and only if $\forall Z \delta_S(X,Z) \leftrightarrow \delta_S(Y,Z)$. As in the other cases, this is an equivalence relation of finite index, by the induction hypothesis and the fact that $\delta_S(X,Z)$ depends in the second argument only on the equivalence class of Z with respect to $\sim_{\phi'}$.

Define $X \sim Y$ if and only if $\forall S \subseteq \{1,\ldots,t\}^2 X \sim_S Y$.

The relation $\sim$ is an equivalence relation of finite index since it is the intersection of finitely many such relations. It remains only to show that $X \sim Y$ implies $X \sim_\phi Y$. If not, then $\exists Z$ such that (without loss of generality) $X \oplus Z \models \phi$ and $Y \oplus Z \models \neg\phi$. Fix an instantiation of the variable D to a set of edges E_D in $X \oplus Z$ making ϕ true. Let $S = E_D \cap \partial(X)$. Let Z_D denote Z additionally equipped with the set of edges of E_D in Z in correspondence with the variable D. Thus we have $\delta_S(X, Z_D)$. Since $X \sim Y$, we have also $\delta_S(Y, Z_D)$, which implies $Y \oplus Z \models \phi$, a contradiction.

6. Conclusions and Open Problems

The most intriguing open problems concerning the automata-theoretic point of view on bounded treewidth concern the possibilities for extending this "program." For example, (1.4) raises the possibility of general automata-theoretic methods for proving well-quasiordering results (such as the graph minor theorem for bounded treewidth) by arguments that counterpose the parsing operations, and the operations defining the order.

It would also be interesting to know if some analogue of (1.4) holds in the large universe. It is conceivable that study of the information flow across bounded size cutsets for lower ideals F may provide an avenue for well-quasiordering results in the more general setting.

References

[ACPS] S. Arnborg, B. Courcelle, A. Proskurowski and D. Seese, *An algebraic theory of graph reduction*, Technical Report 90-02, Laboratoire Bordelais de Recherche en Informatique, Universite de Bordeaux I, January 1990.

[ALS] S. Arnborg, J. Lagergren and D. Seese, *Problems easy for tree-decomposable graphs (extended abstract)*, Proc. 15th Int. Coll. Automata, Languages and Programming, Lecture Notes in Computer Science, Vol. 317 (T. Lepisto and A. Salomaa,, eds.), Springer, Berlin, 1988, pp. 38–51.

[Ar] S. Arbborg, *Efficient algorithms for combinatorial problems on graphs with bounded decomposability — a survey*, BIT **25** (1985), 2-23.

[AP] S. Arnborg and A. Proskurowski, *Linear time algorithms for NP-hard problems restricted to partial k-trees*, Disc. Appl. Math. **23** (1989), 11–24.

[BFW] H. Bodlaender, M. Fellows and T. Warnow, *Two strikes against perfect phylogeny*, Proceedings of the 19th International Colloquium on Automata, Languages and Programming, Lecture Notes in Computer Science, Vol. 623 (W. Kuich, eds.), Springer-Verlag, 1992, pp. 273–283.

[Bo1] H. Bodlaender, *Dynamic programming on graphs with bounded treewidth*, Proceedings of the 15th International Colloquium on Automata, Languages and Programming, Lecture Notes in Computer Science, Vol.317, Springer-Verlag, 1988, pp. 105–119.

[Bo2] H. Bodlaender, *A linear time algorithm for finding tree-decompositions of small treewidth*, Manuscript, August 1992.

[Bo3] H. Bodlaender, *A tourist guide through treewidth*, Technical report RUU-CS-92-12, March 1992.

[BLW] M. Bern, E. Lawler and A. Wong, *Linear time computation of optimal subgraphs of decomposable graphs*, J. of Algorithms **8** (1987), 216-235.

[BPT] R. Borie, R. Parker and C. Tovey, *Automatic generation of linear algorithms from predicate calculus descriptions of problems on recursively constructed graph families*, Manuscript, Georgia Institute of Technology, 1988.

[CK] D. G. Corneil and J. M. Keil, *A dynamic programming approach to the dominating set problem on k-trees*, SIAM J. Alg. Disc. Meth. **8** (1987), 535–543.

[CL] B. Courcelle and J. Lagergren, *Recognizable sets of graphs of bounded tree-width*, Manuscript, April, 1992.

[Co1] B. Courcelle, *Recognizability and second-order definability for sets of finite graphs*, Technical Report I-8634, Universite de Bordeaux, 1987.

[Co2] B. Courcelle, *Graph rewriting: an algebraic and logical approach*, In J. van Leeuwen, ed., *Handbook of Theoretical Computer Science, Vol. B*. North-Holland, Amsterdam, 1990, Chapter 5.

[Co3] B. Courcelle, *Recognizable sets of graphs: equivalent definitions and closure properties*, Technical Report 92-06, Bordeaux 1 University, 1992.

[DW] M. Davis and E. Weyuker, *Computability, complexity, and languages*, Academic Press, New York, 1983.

[EC] E. S. El-Mallah and C. J. Colbourn, *Partial k-tree algorithms*, Congressus Numerantium **64** (1988), 105–119.

[FA] M. Fellows and K. Abrahamsom, *Cutset regularity beats well-quasi-ordering for bounded treewidth*, Manuscript, 1989.

[FL] M. Fellows and M. Langston, *An analogue of the Myhill-Nerode theorem and its use in computing finite basis characterizations*, Proc. Symp. Foundations of Comp. Sci. (FOCS) (1989), pp. 520-525.

[GJ] M. Garey and D. Johnson, *Computers and intractability: a guide to the theory of NP-completeness*, W.H. Freeman and Co., 1979.

[He] S. T. Hedetniemi, *Bibliography of algorithms on partial k-trees*, Manuscript, Clemson University, 1989.

[HU] J. E. Hopcroft and J. D. Ullman, *Introduction to automata theory, languages and computation*, Addison-Wesley, Reading, Mass., 1979.

[La] J. Lagergren, *Algorithms and minimal forbidden minors for tree-decomposable graphs*, Dissertation, Department of Numerical Analysis and Computing Science, Royal Institute of Technology, Stockholm, Sweden, March 1991.

[Mi] E. Milner, *Basic WQO and BQO theory*, in Graphs and Orders (I. Rival, eds.), Reidel, Amsterdam, 1985.

[MP] S. Mahajan and J. Peters, *Algorithms for regular properties in recursive graphs*, Technical Report 87-7, School of Computing Science, Simon Fraser University, 1987.

[Re] B. Reed, *Finding approximate separators and computing treewidth quickly*, Proceedings of the ACM Symposium on Theory of Computing, 1992.

[RS1] N. Robertson and P. D. Seymour, *Graph minors IV. Treewidth and well quasi-ordering*, J. Comb. Theory Ser. B (to appear).

[RS2] N. Robertson and P. D. Seymour, *Graph minors XIII. The disjoint paths problem* (to appear).

[RS3] N. Robertson and P. D. Seymour, *Graph minors XV. Wagner's conjecture* (to appear).

[Se] D. Seese, *Tree-partite graphs and the complexity of algorithms*, Technical Report P-Math-08/86, Karl-Weierstrass-Institut fur Mathematik, Akademie der Wissenschaften der DDR, 1986.

[Th] J. W. Thatcher, *Tree automata: an informal survey*, Currents in the Theory of Computing (A. V. Aho, eds.), Prentice-Hall, 1973., p. 143–172.

[WW] Wanke and Wiegers, Information Processing Letters (1989).

[Wi] T. V. Wimer, *Linear algorithms on k-terminal graphs*, Ph.D. dissertation, Clemson University, 1987.

[WHL] T. V. Wimer, S. T. Hedetniemi and R. Laskar, *A methodology for constructing linear graph algorithms*, Congressus Numerantium 50 (1985), 43–60.

[WPLHH] T. V. Wimer, K. Peters, R. Laskar, S. T. Hedetniemi, and E. O. Hare,, *Linear time computability of combinatorial problems on generalized series-parallel graphs*, Perspectives in Computing, Volume 15, Academic Press, 1987, pp. 437–457..

DEPARTMENT OF ELECTRICAL ENGINEERING AND COMPUTER SCIENCE, WASHINGTON STATE UNIVERSITY, PULLMAN, WASHINGTON 99164-1210, U.S.A.

DEPARTMENT OF COMPUTER SCIENCE, UNIVERSITY OF VICTORIA, VICTORIA, BRITISH COLUMBIA V8W 3P6, CANADA

Graph Grammars, Monadic Second-Order Logic And The Theory Of Graph Minors

Bruno COURCELLE

ABSTRACT. We survey the relationships between the descriptions of sets of graphs by formulas of *monadic second-order logic*, by context-free *hyperedge* and *vertex replacement* graph grammars and by *forbidden minors*.

Introduction

Sets of graphs can be specified in different ways: by characteristic graph properties (in particular by forbidden minors or forbidden subgraphs), by recursive formation rules called *graph grammars*, and by *reduction*, i.e., roughly speaking, by formation rules used in the reverse direction.

A vast project consists in comparing these various types of specifications at a general level, and of course in the framework of precise definitions. One may expect to obtain results of the following form: for every graph property expressible in a certain logical language, there exists a grammar of a certain type that generates the (finite) graphs satisfying this property and only them, or vice versa. It is of course desirable to have effective constructions, i.e., to have algorithms that build grammars from logical formulas or vice versa. In this paper, we survey the main results in this direction that concern the relationships between several descriptions of sets of graphs: (i) by characteristic properties expressed in *monadic second-order* (MS) *logic*, (ii) by *context-free graph grammars*, and (iii) by *forbidden minors*. Some of these results also provide effective constructions of efficient graph algorithms.

We now review the main notions and results. In Section 1, we introduce a set $\mathbb{F}$ of *graph operations*, making it possible to define graphs from smaller graphs, typically by gluing two graphs at specified vertices. A *graph expression* is a well-

---

1991 Mathematics Subject Classification. Primary 05C02, 05C75, 68R10, 68R50.
Supported by ESPRIT Basic Research Working Group 3299.
This paper is in final form and no version of it will be submitted for publication elsewhere.

formed algebraic expression built with the operations of $\mathbb{F}$ and nullary symbols denoting basic graphs. It evaluates to a unique graph: we say that the expression *defines* or *denotes* this graph. Graph expressions offer linear notation for the graphs they define, but also tree-structurings of these graphs since every well-formed expression is in some sense a tree. These tree-structurings correspond closely to *tree-decompositions;* see (1.1).

In Section 2, we define *graph grammars* as systems of mutually recursive equations defining sets of graphs, of which one takes least solutions. These equations are written with set union, the operations of $\mathbb{F}$ and nullary symbols denoting fixed graphs. For example, the set of series-parallel graphs, characterized as the least set of graphs containing the one-edge graph e and closed under two operations called *series-* and *parallel-composition*, respectively denoted by • and //, is thus the least solution of the equation S = {e} ∪ S•S ∪ S//S, that we consider as a grammar with three formation rules.

These grammars will be called *HR grammars*; they define the *HR sets* of graphs (HR stands for "Hyperedge Replacement" and refers to an equivalent definition formulated in terms of rewritings of hypergraphs). They have a limited generative power: the HR sets of graphs have bounded tree-width (this follows from (1.1) and the fact that a grammar must be finitely written). A positive counterpart is that they generate graphs together with graph expressions denoting them, and hence with tree-decompositions of these graphs. This fact is interesting because many graph properties (in particular those expressible in monadic second-order logic, see (3.2)) and many functions on graphs (see Section 7) can be tested or computed efficiently on graphs given with tree-decompositions of bounded width.

In Section 3, we explain how a graph can be represented by a logical structure. It follows that every logical language, the formulas of which are meaningful in the structure representing a graph, can be used to write formally properties of this graph. *Monadic second-order logic* (MS logic), namely the extension of first-order logic with quantified variables denoting sets of elements of the domain, is both useful (because many basic graph properties can be expressed in this language) and manageable in the sense that several general complexity and decidability results hold: see (3.2), (3.4), (6.2) and (6.4). Theorem (3.4), stating that the intersection of an HR set of graphs and the set of graphs satisfying an MS property is HR is a powerful tool for constructing new HR grammars from the ones generating the sets listed in (2.3).

In Section 4, we examine what MS logic and HR grammars can bring to the theory of excluded minors. Theorem (4.1) states that the minor-closed sets of graphs of bounded tree-width can be described by HR grammars. We also explain how the minimal forbidden minors can be effectively constructed in certain cases by an algorithm that uses HR grammars and MS logic.

In Section 5 we introduce new graph operations, that are in some sense more powerful than those of Section 1. In particular, we introduce an operation that adds to a graph, "in a single stroke", all edges linking a vertex with label *a* and a vertex with label *b*. Hence, this new operation (together with a few auxiliary ones) constructs a graph as a gluing of cliques and complete bipartite graphs as opposed to a gluing of single edges as do the operations of $\mathbb{F}$. Systems of equations written with these operations give rise to the class of *VR grammars* (where VR stands for "Vertex Replacement" and refers to an alternative definition in terms of graph rewriting). These grammars are strictly more powerful than the HR ones. In particular, they can generate the sets of all cliques and of all complete bipartite graphs, whereas the HR grammars cannot.

The minor-closed sets of graphs of bounded tree-width can be described by HR grammars by (4.1). It would be nice to have a grammatical description of those of *unbounded tree-width,* and VR grammars might appear as good candidates. Unfortunately, they are no good at all for this purpose: if a set of graphs generated by a VR grammar is minor-closed, then it is HR, and hence has bounded tree-width: see (5.3).

In Section 6 we define *MS_1 logic*, a restriction of MS logic, where set variables cannot denote sets of edges, but only sets of vertices. This logic fits with VR grammars as well as MS logic does with HR grammars: Theorems (6.2) and (6.4) are exact counterparts of Theorems (3.2) and (3.4). The MS_1 logic is *strictly less* powerful than the MS logic, whereas VR grammars are *strictly more* powerful than HR ones. However, for graphs of degree at most some fixed integer k, and for graphs that do not contain a fixed graph as a minor, MS-logic is *no more* powerful than MS_1-logic and HR grammars are *no less* powerful than VR-ones.

Section 7 contains historical remarks, comments on references and some pointers to further developments.

1. Graph operations and graph expressions.

Graphs are finite and undirected; they may have loops and multiple edges. We denote by V_G the set of vertices of a graph G and by E_G its set of edges.

Let $\mathbb{C}$ be a fixed countable set of *labels*. An *s-graph* (or a *graph with a possibly empty set of $\mathbb{C}$-labeled distinguished vertices called sources*) is a pair H = <G,f> consisting of a graph G (called the *underlying graph of* H) and a one-to-one mapping f : C ⟶ V_G, where C is a finite subset of $\mathbb{C}$. We say that f(C) is the *set of sources* of H, that f(c) is the *c-source* of H (for c in C), and that c is the *label* of f(c). A vertex that is not a source is called an *internal vertex*. We call C the *type* of H and denote it by τ(H). We denote by $\mathbb{G}$ the class of all s-graphs and by $\mathbb{G}(C)$ the class of s-graphs of type C. (Hence, $\mathbb{G}(\emptyset)$ is the class of all graphs.) Any two isomorphic graphs or s-graphs (for the obvious notion of isomorphism preserving source labels) are considered as equal. In other words, we shall deal with *abstract* graphs and s-graphs, i.e., with isomorphism classes of *concrete* graphs and s-graphs, without being formal about this. The sources are just distinguished vertices; they are frequently called "terminals" but we call them "sources" because the word "terminal" has another well-established meaning in the theory of grammars.

We now define a few operations on s-graphs.

Parallel composition: Let G ∈ $\mathbb{G}$(C) and G' ∈ $\mathbb{G}$(C'). Their *parallel composition* is the abstract s-graph G//G' in $\mathbb{G}$(C∪C') defined as the isomorphism class of a concrete s-graph K defined as follows. We let H and H' be concrete s-graphs isomorphic to G and G' respectively such that $V_H \cap V_{H'} = \emptyset$ and $E_H \cap E_{H'} = \emptyset$. We let K be the s-graph obtained from the union of H and H' by fusing the a-source of H and the a-source of H' for each a in $C \cap C'$. Note that G//G is not equal to G (except in the degenerate case where G has no edge and no internal vertex). Parallel composition is associative and commutative.

Source restriction: Let C be a finite subset of $\mathbb{C}$. We let $\mathbf{rest}_C$ be the mapping on s-graphs such that $\mathbf{rest}_C(<G,f>) = <G,g>$ where g is the restriction of f to C. Hence the sources of H having a label not in C exist as vertices of $\mathbf{rest}_C(H)$ but are no longer sources. They are made internal. The type of $\mathbf{rest}_C(H)$ is thus $\tau(H) \cap C$. We shall use the notations $\mathbf{rest}_a$ and $\mathbf{rest}_{a,b}$ for $\mathbf{rest}_{\{a\}}$ and $\mathbf{rest}_{\{a,b\}}$ respectively.

Source renaming: Let C be a finite subset of $\mathbb{C}$ and h be a bijection: $C \longrightarrow C$. We let $\mathbf{ren}_h$ be the mapping : $\mathbb{G} \longrightarrow \mathbb{G}$ such that $\mathbf{ren}_h(<G,f>) = <G,f \circ h'>$ where h' is the bijection: $\mathbb{C} \longrightarrow \mathbb{C}$ that extends h, being the identity outside of C. It follows that the type of $\mathbf{ren}_h(<G,f>)$ is $h^{-1}(\tau(<G,f>)) \cup (\tau(<G,f>)-C)$.

We let $\mathbb{F}$ denote the set of operations //, $\mathbf{rest}_C$, $\mathbf{ren}_h$.

Remark : We have no operation making an internal vertex into a source. Any two sources may be distinguished (and designated) by their labels, whereas we have no way to distinguish (and designate) internal vertices. We do not wish to deal with nondeterministic operations like : "make any internal vertex of this graph into the a-source" where a is some label. □

Basic s-graphs: We let $\mathbb{B}$ be the set of *basic s-graphs*. A basic s-graph can be of three types: (1) the graph a with no edge consisting of a unique vertex that is the a-source, (2) the graph ab consisting of an edge linking the a-source and the b-source (where $a \neq b$), (3) the graph a^ℓ consisting of a loop incident with a unique vertex that is the a-source, where in the three cases, a and b are any members of $\mathbb{C}$ (and $a \neq b$).

We let $\mathbb{E}$ be the set of finite, algebraic, well-formed expressions built with $\mathbb{F}$ and $\mathbb{B}$. We call them *graph expressions*. The set of those written only with parallel composition, source restrictions and basic graphs with labels in a finite subset C of $\mathbb{C}$ is denoted by $\mathbb{E}_0(C)$. The set of those written only with parallel composition, source restrictions, source renamings $\mathbf{ren}_h$ with h: $C \longrightarrow C$ and basic graphs with labels in C (where C is a finite subset of $\mathbb{C}$) is denoted by $\mathbb{E}(C)$.

Every expression t in $\mathbb{E}$ evaluates to an s-graph denoted by $\mathbf{val}(t)$ and called its *value*. We shall say that t *defines* or *denotes* $\mathbf{val}(t)$. It is clear that every graph with n vertices is $\mathbf{val}(t)$ for some $t \in \mathbb{E}_0(C)$ where $\mathbf{Card}(C) \leq n$. For example, the graph

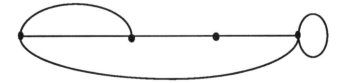

is the value of the expression **rest**$_\emptyset$(ab // ab // bc // cd // d$^\ell$ // ad) and also of the following expression that is written with only three labels, namely a, b, c:

$$\text{rest}_\emptyset(\text{rest}_{a,c}(\text{ab// ab// bc}) \text{ // } \text{rest}_{a,c}(\text{cb // } b^\ell \text{// ab })).$$

It is easy to see that n labels are necessary to define a clique with n vertices. The following result relates graph expressions and tree-decompositions. Let us recall that a *tree-decomposition* of a graph G is a pair (T,f) where T is a tree and f is a mapping that associates with every node u of T a set of vertices of G such that, (i) every vertex of G belongs to some f(u), (ii) every nonloop edge of G has its two ends in some f(u) and (iii) for every vertex x of G, the set of nodes u of T such that x belongs to f(u) induces a connected subgraph of T. The *width* of (T,f) is the maximum cardinality of a set f(u) minus 1, and the *tree-width* of G is the minimum width of a tree-decomposition of G. We denote it by **twd**(G).

(1.1) *For every graph G* :
  (i)   **twd**(G) + 1 = **Min**{**Card**(C) / t $\in$ $\mathbb{E}_0$(C), **val**(t) = G}.
  (ii) **twd**(G) + 1 = **Min**{**Card**(C) / t $\in$ $\mathbb{E}$(C), **val**(t) = G}.

Proof sketch: From a graph expression t in $\mathbb{E}$(C) denoting a graph G, one can construct a tree-decomposition of G as follows. We make a few preliminary observations. The syntactic tree of the expression t is a rooted and directed tree, say T, the nodes of which are labeled by graph operations and basic graphs. Here we let G be a concrete graph. For every node u of T, we let T/u denote the subtree of T rooted at u as well as the corresponding graph expression. There corresponds to each such u a subgraph G_u of G that is isomorphic to **val**(T/u). If u and v are distinct nodes, then G_u and G_v are distinct, even if T/u and T/v are isomorphic subtrees. We take T as the tree of the desired tree-decomposition. The set of vertices f(u) is the set of sources of the concrete s-graph G_u, and (T,f) is the desired tree-decomposition.

Conversely, one can convert as follows a tree-decomposition (T,f) of a concrete graph G of width k into an expression in $\mathbb{E}_0$(C) that denotes G, where C is chosen of cardinality k+1. Without loss of generality, we can assume that each node of T is of degree either 1 or 3. We choose a root r in the tree T and we direct its edges so that every node is reachable from r by a directed path. Hence each node of T either has two successors or no successor (and is a leaf). We color the vertices of G with the elements of C considered as a set of colors in such a way that no two vertices in a same set f(u) have the same color. We let c be the coloring mapping (from V_G to C). For each node u of T, we choose a set of edges E(u) having their ends in f(u) and such that the sets E(u) form a partition of E_G. We let H(u) be the concrete subgraph of G with set of vertices f(u) and set of edges E(u), and K(u) be the concrete s-graph <H(u), s> where s is the restriction of the mapping c^{-1} to c(f(u)). For each u we construct an expression k(u) in $\mathbb{E}_0$(C) denoting K(u). We also construct an expression g(u) in $\mathbb{E}_0$(C) by the following induction: if u is a leaf of T, then g(u) is defined as k(u); if u has successors v and w, we define g(u) as k(u)//**rest**$_S$(g(v))//**rest**$_{S'}$(g(w)) where S = f(u)$\cap$f(v) and S' = f(u)$\cap$f(w). The expression **rest**$_\emptyset$(g(r)) defines G and belongs to $\mathbb{E}_0$(C). □

Remarks: This proof actually works for s-graphs, if we define a tree-decomposition of an s-graph as a tree-decomposition of the underlying graph such that all sources are in some set f(u). The comparison of assertions (1) and (2) shows that the operations of the form **ren**$_h$ are in some sense superflous. They are actually convenient in the writing of derived graph operations to be defined below. □

Example :   Figure 1 shows the syntactic tree of the expression

$$t = \textbf{rest}_\emptyset(\textbf{rest}_{a,c}(ab \mathbin{/\!/} ab \mathbin{/\!/} bc) \mathbin{/\!/} \textbf{rest}_{a,c}(bc \mathbin{/\!/} b^\ell \mathbin{/\!/} ab)$$

in $\mathbb{E}_0(\{a,b,c\})$ already considered above, the corresponding tree-decomposition of the graph **val**(t), and the graph **val**(t) itself. The vertices 1,2,3,4 of the graph appear as sources with respective labels a, b, c, b in the s-graphs defined by the subexpressions of t. □

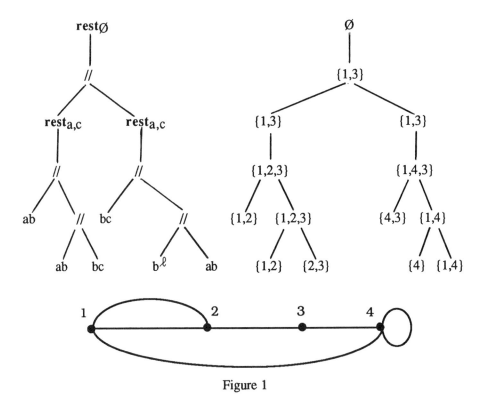

Figure 1

Graph expressions can be used as linear notations for graphs. Unfortunately, they tend to be long and unreadable, even for relatively small graphs. A partial remedy to that consists of defining new graph operations in terms of those of $\mathbb{F}$. Hence, graph expressions written with these new operations will tend to be more concise. We take as an example the *series-composition* of graphs with two sources, respectively labeled by a and b. It can be defined as follows for G and G' in $\mathbb{G}(\{a,b\})$:

$$G \bullet G' = \text{rest}_{a,b}(\text{ren}_h(G) // \text{ren}_{h'}(G'))$$

where h maps c to b, and b to c and h' maps c to a and a to c. Let us note that G•G' is well-defined for any two s-graphs G and G', but we are only interested here in using the operation • for s-graphs in $\mathbb{G}(\{a,b\})$. It follows that the series-parallel graph

can be denoted by the expression :

$$(((ab // ab) \bullet ab) // ab) \bullet (ab \bullet (ab // ab // ab))$$

which is more readable than the one we would obtain by replacing • by its definition.

Here is a formal definition. A *derived graph operation* is a mapping f: $\mathbb{G}^n \to \mathbb{G}$ defined by a well-formed expression t written with the operations of $\mathbb{F}$, the nullary symbols of $\mathbb{B}$ (that denote basic graphs), and variables $x_1,...,x_n$ such that each of them occurs once and only once in t. An algebraic expression t written with the symbols of $\mathbb{F}$ and $\mathbb{B}$, and symbols denoting derived graph operations, will be called an *extended graph expression*. Its value is an s-graph denoted by val(t), as when t is in $\mathbb{E}$.

In many cases, we are only interested in the restriction of a derived operation f to $\mathbb{G}(C_1) \times ... \times \mathbb{G}(C_n)$ where $C_1,...,C_n$ are finite subsets of $\mathbb{C}$ of interest for the intended use of f. Note that, for a given operation f and sets $C_1,...,C_n$, there is a unique set C such that f maps $\mathbb{G}(C_1) \times ... \times \mathbb{G}(C_n)$ into $\mathbb{G}(C)$. We shall say that f *has the profile* $C_1 \times ... \times C_n \to C$. A derived operation has infinitely many profiles. The series-composition operation • has in particular the profile: $\{a,b\} \times \{a,b\} \to \{a,b\}$.

2. Hyperedge replacement graph grammars

Before introducing *Hyperedge Replacement graph grammars* (*HR grammars* for short), we consider as an example the set S of series-parallel graphs with two "ends", formally defined as sources labeled by a and b. Any graph G in S with m edges can be constructed from m copies of the basic graph ab by using m-1 times the series and parallel composition operations defined in Section 1. It follows that S is the least subset of $\mathbb{G}(\{a,b\})$ satisfying the equation:

$$S = \{ab\} \cup S//S \cup S \bullet S,$$

where the binary operations // and • on $\mathbb{G}(\{a,b\})$ are extended to sets in a standard way, that is, $S//S' := \{G//G' / G \in S, G' \in S'\}$, $S \bullet S' := \{G \bullet G' / G \in S, G' \in S'\}$.

More generally, we shall consider systems of equations. An *HR system of equations* is a system Γ of the form :

$$\Gamma = < S_1 = p_1(S_1,...,S_n), ... , S_n = p_n(S_1,...,S_n) >$$

where each S_i is an *unknown*, intended to denote a subset of $\mathbb{G}$, and each p_i is a *polynomial*, i.e., a finite expression of the form $m_1 \cup m_2 \cup ... \cup m_k$, where $m_1,...,m_k$ are monomials. A *monomial* is either an expression denoting an s-graph or an expression of the form $f(S_{i_1},...,S_{i_m})$ where f is a derived graph operation. An *HR grammar* is a pair (Γ,S) consisting of an HR system and one of its unknowns.

Every HR system Γ with unknowns $S_1,...,S_n$ has a least solution in $(\mathcal{P}(\mathbb{G}))^n$ that we shall denote by $(L(\Gamma,S_1),...,L(\Gamma,S_n))$. (We denote by $\mathcal{P}(\mathbb{G})$ the power set of $\mathbb{G}$.) We say that $L(\Gamma,S)$ is the set of s-graphs *generated by the grammar* (Γ,S) where S is some S_i. A set of graphs is *HR* iff it is generated by some HR grammar.

A system Γ as above also has a least solution in the set of extended graph expressions built with the basic graphs and the derived graph operations used to form it. We shall denote by $(T(\Gamma,S_1),...,T(\Gamma,S_n))$ the n-tuple of sets of extended expressions forming this least solution. For example, the least set of extended expressions satisfying the equation defining series-parallel graphs contains in particular the expressions ab, ab//ab, ab•ab, ab//(ab//ab), ab•(ab//ab).

(2.1) *For every* HR *grammar* (Γ,S), *we have* : $L(\Gamma,S) = \{val(t) / t \in T(\Gamma,S)\}$.

This is actually a result of Universal Algebra. It does not depend on the specific domain of s-graphs nor on the specific graph operations introduced in Section 1. An analogous statement holds for other notions of graphs and other graph operations, and the associated systems of equations and graph grammars, and in particular for those we shall consider in Section 5. See also [12] for a treatment of grammars in a Universal Algebra setting.

(2.2) *Every HR set of s-graphs has bounded tree-width, and its graphs have all their source labels in a finite subset of* $\mathbb{C}$.

Proof : The expressions forming $T(\Gamma,S)$ are written with finitely many different derived graph operations. Each of these operations is itself formed with finitely many basic graphs and finitely many operations of $\mathbb{F}$. Hence, by (2.1), every graph G in $L(\Gamma,S)$ is denoted by an expression in $\mathbb{E}(C)$ for some finite set C. Hence, **twd**(G) $\leq$ **Card**(C)-1 by (1.1), and τ(G) is a subset of C. □

Another consequence of (2.1) is that every graph generated by a HR grammar is defined by an extended graph expression from which, by the construction of (1.1), a tree-decomposition of (uniformly) bounded width can be constructed. This fact is important for algorithmic purposes: see (3.2) and Section 7. It follows also from (2.2) that the set of all graphs, the set of all planar graphs, the set of all square grids, and the set of all cliques, just to take a few examples, are not HR. We now list a number of useful HR sets.

(2.3) *For each integer k, the following sets of graphs are HR:*

(i)   *the set of graphs of tree-width at most k,*
(ii) *the set of graphs of branch-width at most k,*
(iii) *the set of graphs of path-width at most k,*
(iv) *the set of graphs of bandwidth at most k,*
(v)   *the set of k-trees.*

Proof : (i) follows from the proof of (1.1). The constructions for the other cases are not difficult. (*Branch-width* is a variant of tree-width introduced in [58]; *path-width* is introduced in [56]; a graph G has *bandwidth at most k* if there exists a bijection h of V_G onto an interval of the set of integers such that $|h(x) - h(y)| \leq k$ for any two adjacent vertices x and y; see [40] or [63].) See [54] for k-trees. □

As an example we show the grammar generating the set of graphs of tree-width at most k. For fixed k, we let $C = \{c_0, c_1,..., c_k\}$, and $C' = \{c_1,..., c_k\}$. The set of graphs of tree-width at most k is $L(\Gamma,U)$, where Γ is the system consisting of the following two equations:

$$U = \mathbf{rest}_\emptyset(T), \qquad T = T//T \cup S(T,T,...,T) \cup B.$$

This grammar is adapted from [1]: we denote by B the set of s-graphs c, $c//c_i^\ell$, and $c//c_i c_j$ for all $1 \leq i < j \leq k$, where c is the s-graph $c_1//c_2//c_3...//c_k$, and by S, the *generalized k-ary series-composition*, a derived operation defined as follows:

$$S(x_1,x_2,...,x_k) = \mathbf{rest}_{C'}(\mathbf{ren}_{h_1}(x_1)//\mathbf{ren}_{h_2}(x_2)//...//\mathbf{ren}_{h_k}(x_k))$$

where h_i exchanges c_0 and c_i. We illustrate as follows the operation S for $k = 3$. The graph $G = S(G_1,G_2,G_3)$ is shown in Figure 2. □

We now list the main properties of HR sets of graphs.

(2.4) *(i) The union of two HR sets is HR.*
(ii) The intersection and the difference of two HR sets is not HR in general.
(iii) The set of all subgraphs and the set of all minors of all graphs of an HR set is HR.
(iv) One can decide in polynomial time whether the set $L(\Gamma,S)$ generated by a given HR grammar (Γ,S) is empty.
(v) For every HR grammar (Γ,S), one can decide whether a given graph belongs to $L(\Gamma,S)$. There exist grammars (Γ,S) for which this problem is NP-complete.
(vi) Given (Γ,S), one can decide whether the set $L(\Gamma,S)$ is finite. If it is, one can enumerate it explicitly.
(vii) Every HR set of simple graphs is sparse.

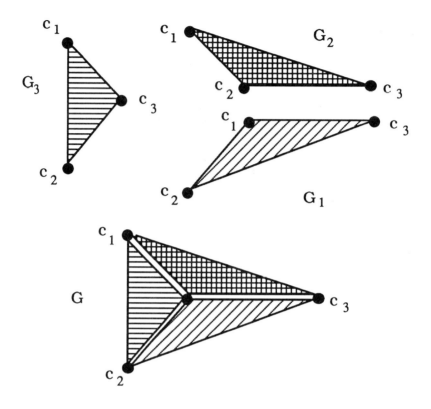

Figure 2

Proof: (i) and (iv) follow easily from (2.1); (ii) is proved in [16, Prop. (3.6.3)]; (iii) follows from the main result of [27] but can also be established directly, without using this difficult theorem; (v) follows from results of [61] and [36]: it is proved in [61] that the membership problem in sets of graphs generated by a so-called BNLC grammar is NP, and in [36] that every HR grammar can be translated (via a linear-time coding of graphs) into a BNLC one; hence, it follows that the membership problem for HR sets is NP; the set of graphs of cycle bandwidth at most 2 is both HR (an HR grammar is constructed in [42, Chap. IV]) and is NP-complete [51]; (vi) is proved in [42, Chap. IV]. We finally consider (vii). A set of graphs is *sparse* if the number of edges is linearly bounded by the number of vertices; since every simple graph generated by a HR grammar is a partial k-tree, hence is obtained from a k-tree by the deletion of edges, the result follows from the observation that $\mathbf{Card}(\mathbf{E}_G) \leq k\mathbf{Card}(\mathbf{V}_G)$ for every k-tree G. □

3. Monadic second-order logic

In this section we explain how graph properties can be expressed in logical languages, and why *monadic second-order logic* is of special interest in this respect.

With every finite set $C \subseteq \mathbb{C}$, we associate a set of relational symbols $\mathbb{R}(C) :=$ {**inc**, $s_c / c \in C$} where **inc** is binary and s_c is unary. With every graph G in $\mathbb{G}(C)$, we associate the $\mathbb{R}(C)$-structure $\|G\| := \langle D_G, \text{inc}_G, (s_{c_G})_{c \in C}\rangle$ such that :

$D_G = V_G \cup E_G$ (we assume that $V_G \cap E_G = \emptyset$),
$\text{inc}_G(x,y) :\Leftrightarrow x [E_G , y [V_G$ and y is incident with x,
$s_{c_G}(x) : \Leftrightarrow$ x is the c-source of G.

It is clear that $\|G\|$ is isomorphic to $\|G'\|$ iff $G = G'$.

Every logical formula φ of a logical language expressing properties of $\mathbb{R}(C)$-structures *expresses* the property P_φ of graphs in $\mathbb{G}(C)$ such that $P_\varphi(G) :\Leftrightarrow \|G\| \vDash \varphi$, and *defines* the subset L_φ of $\mathbb{G}(C)$ such that $L_\varphi := \{G \in \mathbb{G}(C) / \|G\| \vDash \varphi\}$. (The notation $\|G\| \vDash \varphi$ means that the formula φ is true in the structure $\|G\|$.)

For expressing graph properties, first-order logic is rather weak: it expresses only "local conditions" like conditions on degrees of vertices. Second-order logic where quantifications on binary relations are possible is too powerful: nothing can be said about the set of all graphs satisfying a property expressible in second-order logic. *Monadic second-order logic*, namely the extension of first-order logic with quantifications on monadic (unary) relations, i.e., on sets, appears to be both powerful and manageable. We now introduce this language.

We shall use lower case variables x,y,z,... to denote vertices or edges, and upper case variables X,Y,Z,... to denote sets of vertices and edges. The atomic formulas are $x = y$, $x \in X$, **inc**(x,y) and $s_c(x)$ for c in $\mathbb{C}$. The formulas are formed from atomic formulas with Boolean connectives $\wedge, \vee, \neg, \Rightarrow$, *object quantifications* $\exists x, \forall y$, and *set quantifications* $\exists X, \forall Y$. If W is a set of lower and uppercase variables, and if C is a finite subset of $\mathbb{C}$, we denote by $\mathfrak{F}(C,W)$ the set of monadic second-order formulas the free variables of which are in W, and that do not contain occurrences of the relation symbols of the form s_c if c is not in C.

A property P of graphs in $\mathbb{G}(C)$ is *monadic second-order (MS)* if it is of the form P_φ for some formula φ in $\mathfrak{F}(C,\emptyset)$. A set of graphs $L \subseteq \mathbb{G}(C)$ is *monadic second-order (MS) definable* if membership in L is a MS property. One says that the *monadic theory* of a set of graphs L is decidable if there exists an algorithm deciding for every MS formula φ whether $L \subseteq L_\varphi$, i.e., whether all graphs in L satisfy P_φ.

(3.1) (i) *The following properties of a graph G are MS:*

G *is k-connected, (for any fixed k),*
G *is a tree,*

G contains H as a minor (for any fixed H),
G is planar,
G has tree-width at most k (for any fixed k),
G is Hamiltonian,
G is k-colorable (for any fixed k),
G is a square grid.

(ii) *The following properties of a graph G are not MS :*

G has bandwidth at most 3,
G has a nontrivial automorphism.

Proof : (i) See [17-19]. Most of these proofs are based on the existence of a formula $\varphi \in \mathcal{B}(\emptyset, \{x, y, X\})$ such that :

$(\|G\|, \overline{x}, \overline{y}, \overline{X}) \models \varphi$ iff $\overline{x}$ and $\overline{y}$ are two distinct vertices
and $\overline{X}$ is the set of edges of a path in G linking x to y.

(We write $(\|G\|, \overline{x}, \overline{y}, \overline{X}) \models \varphi$ to mean that φ holds true in the structure $\|G\|$ when its free variables x, y, X take the respective values $\overline{x}$, $\overline{y}$, and $\overline{X}$ with $\overline{x}, \overline{y} \in V_G \cup E_G$ and $\overline{X} \subseteq V_G \cup E_G$.)

(ii) The case of bandwidth at most 3 follows from the main result of [66]. The case of graphs with nontrivial automorphisms can be obtained from a similar result for labeled graphs proved in [17]. □

The families **HR** of HR sets of graphs and **MS** of MS definable sets are incomparable: the set of planar graphs is in **MS** but not in **HR**, whereas the set of graphs of bandwidth at most 3 is in **HR** but not in **MS**. See the end of Section 4 for a comparison diagram.

At this point, the following problems are quite natural :

Problem 1 : *How can one decide whether a given graph G satisfies a given MS formula?*

Problem 2 : *Can one decide whether all graphs of a (possibly infinite) set L of (finite) graphs satisfy a given MS formula? Equivalently, is there an algorithm deciding the monadic theory of L?*

For the first problem one can verify in a straightforward (brute force) way whether a formula holds in the structure $\|G\|$, because this structure is finite. The only interesting question is the complexity of this verification. Quite a lot of NP-complete graph problems can be expressed by MS formulas: 3-vertex colorability and the existence of a Hamiltonian circuit are two examples of such. A detailed list can be found in [2]. There exist efficient algorithms for graphs of bounded tree-width as stated in the next proposition.

(3.2) *Let φ be an MS formula, let K be a finite set of graph operations, derived graph operations and basic graphs. For every extended graph expression t written with K, one can decide in time $O(|t|)$ whether the graph* **val**(t) *satisfies property* P_φ.

We refer the reader to [17] for the proof; we only present the main tool. Let f be a derived graph operation with profile $C_1 \times ... \times C_m \to C$ of particular interest. We say that a finite set of graph properties $\mathcal{P}$ is *inductive with respect to* the operation f if for every property P in $\mathcal{P}$, there exist properties $P_{j,i}$ in $\mathcal{P}$ such that for all graphs $G_1,...,G_m$ respectively in $\mathbb{G}(C_1),...,\mathbb{G}(C_m)$:

(3.2.1) $\qquad P(f(G_1,...,G_m)) = \bigvee_{1 \le j \le k} P_{j,1}(G_1) \wedge ... \wedge P_{j,m}(G_m).$

(We denote by P(G) the Boolean value **true** if P(G) holds and **false** otherwise.) In words this means that the validity of a property P for a graph defined as $f(G_1,...,G_m)$ where $G_1,...,G_m$ are of respective types $C_1,...,C_m$, can be determined from the knowledge of the validity of some properties of $\mathcal{P}$ for the graphs $G_1,...,G_m$. It is proved in [17] that for every MS formula φ, one can find finitely many auxiliary formulas $\varphi_1,...,\varphi_m$ such that $\{P_\varphi, P_{\varphi_1},...,P_{\varphi_m}\}$ is inductive with respect to the given set K. (It should be noted that the construction of $\varphi_1,...,\varphi_m$ depends on K and that it is essential that K be finite.) Then, in order to decide whether **val**(t) satisfies φ, one computes the tuple of Boolean values $<P_\varphi(\mathbf{val}(t'))$, $P_{\varphi_1}(\mathbf{val}(t')),...,P_{\varphi_m}(\mathbf{val}(t'))>$ for all subexpressions t' of t. This can be done in time $O(|t|)$ by means of the equalities (3.2.1). (The integer m and the constant of the time bound depend on the size of the formula and on the width of the tree-composition associated with the graph expression via a tower of exponentials of height equal to the number of nested quantifications.)

We now consider the second problem. Seese has proved in [62] that the monadic theory of the set of all square grids, and of any set of graphs containing infinitely many square grids as minors, is undecidable. It follows then from [57] and (2.2):

(3.3) *If a set of graphs has a decidable monadic theory, then its tree-width is bounded. Hence, it is a subset of some HR set of graphs.*

Theorem (3.4) will state that every HR set of graphs has a decidable monadic theory. Note that interesting graph theoretic conjectures or theorems can be stated in logical terms. For instance, if we knew an algorithm deciding the monadic theory of the set of planar graphs, then this algorithm would prove the 4-color theorem: it would suffice to run it for the MS formula expressing that a graph is 4-colorable and wait, perhaps very long, for the answer. Unfortunately (?) no such algorithm exists by (3.3).

Equalities (3.2.1) also form the basic tool for proving the following theorem:

(3.4) *Let C be a finite subset of $\mathbb{C}$, let Γ be an HR grammar such that $L(\Gamma) \subseteq \mathbb{G}(C)$ and let φ be a formula in $\mathcal{B}(C,\emptyset)$. One can construct an HR grammar Γ_φ*

generating $L(\Gamma) \cap L_\varphi$. *One can decide whether* $\|G\| \vDash \varphi$ *for some graph* G *in* $L(\Gamma)$. *The monadic theory of* $L(\Gamma)$ *is decidable.*

Proof sketch: See [17] for the construction of Γ_φ. For every HR grammar Γ', one can decide whether $L(\Gamma') \neq \emptyset$. This test can be applied to Γ_φ and says whether there exists a graph in $L(\Gamma)$ such that $\|G\| \vDash \varphi$. For deciding the monadic theory of $L(\Gamma)$, note that $\|G\| \vDash \varphi$ for every G in $L(\Gamma)$ iff $L(\Gamma_{\neg \varphi}) = \emptyset$, which is decidable. □

This theorem provides us with systematic methods for constructing HR grammars from others. In particular, the sets of graphs defined as the restrictions of those listed in (2.3) to graphs satisfying MS properties (typically planarity, Hamiltonicity or connectedness to take a few examples) are HR, and HR grammars can be constructed to generate them.

4. Minor-closed sets of graphs

By using the results of Robertson and Seymour [57,59], we show how MS logic and HR grammars can help to describe minor-closed sets of graphs. We then discuss the problem of the effective construction of their sets of obstructions. We shall denote by **OBST**(L) the set of obstructions of a minor-closed set L, that is, the set of minimal graphs not in L, where minimality is understood w.r.t. minor inclusion.

The Graph Minor Theorem establishes that every minor-closed set of graphs has a finite set of obstructions (see [59] and the related papers). This is the case in particular, for every k, of the set of graphs of tree-width at most k. The corresponding sets of obstructions are known for k = 1,2,3 (see [3]). In principle, these sets can be computed by the algorithm of [49]. However, this algorithm is intractable, and the sets of obstructions are still unknown for k ≥ 4.

(4.1) *(i) Every minor-closed set of graphs is MS-definable.*
(ii) Every minor-closed set of graphs of bounded tree-width is also HR.

Proof ([19]) : (i) It is an easy exercise to construct, for every fixed graph H, an MS formula that defines the set of graphs that do not contain H as a minor. The result then follows from the Graph Minor Theorem.
(ii) If L is minor-closed and has tree-width ≤ k, then $L = L \cap L_k$, where L_k is the set of graphs of tree-width at most k. Since L_k is HR and L is MS definable by the first part, it follows that L is also HR by (3.4). □

The second assertion of (4.1) yields *internal descriptions* of minor-closed sets of graphs of bounded tree-width. We say that they are "internal" because grammars describe the graphs of some set in terms of smaller graphs belonging to the same set or to finitely many auxiliary sets, described in the same way. By contrast, forbidden configurations are *external* to the sets they characterize.

The proof of (4.1) is effective in the following way: from **OBST**(L) one can construct an MS formula defining L and also a grammar for L in the case where **OBST**(L) contains a planar graph. However, this construction may involve huge constants. The bound on the tree-width of the set of graphs excluding a planar graph as a minor given in [57] has several levels of exponentiation; a smaller one is given in [60].

No algorithm is known by which one could construct **OBST**(L) from a formalized proof that L is minor-closed. In many cases, obstruction sets are unknown. The following result from [19] gives conditions making it possible to construct **OBST**(L) from an MS formula defining L.

(4.2) *Let L be a minor-closed set of graphs. From an MS formula φ defining L, one can construct an MS formula ψ defining* **OBST**(L). *If in addition, either we know that L has bounded tree-width, or we know an upper bound on* **twd**(**OBST**(L)), *then we can construct* **OBST**(L).

Let us recall that the knowledge of an MS formula ψ together with the fact that L_ψ is finite, is not sufficient to construct L_ψ effectively. This can be seen as follows. From a deterministic Turing machine M given with its initial configuration, one can construct an MS formula ψ such that L_ψ is the set of square grids such that a terminated computation of M can be written on this grid, with one symbol per vertex, in such a way that the successive lines of the grid encode the successive configurations of the unique computation, and this cannot be done on any strictly smaller grid. Hence, the set L_ψ is empty or consists of a single grid. If one could compute it from ψ, one could solve the halting problem for Turing machines.

We sketch the proof. From an upper bound on **twd**(**OBST**(L)), one can construct by (3.4) an HR grammar generating the finite set **OBST**(L); from assertion (vi) of (2.4), one can construct **OBST**(L). If we know that the tree-width of L is finite (we need not know any upper bound), then, by enumerating all square grids and testing whether they satisfy φ, one can find the smallest one not in L, whence by [57] a bound, say k, on **twd**(L) so that **twd**(**OBST**(L)) $\leq$ k+1, and one can obtain **OBST**(L) as before. This result can also be obtained by the method of Fellows and Langston [39]. They use *effectively given* congruences with finitely many classes on an appropriate algebra of graphs. Every MS definable set of graphs saturates such a congruence, as proved in [17]. This congruence relation can be effectively determined from a given logical formula; hence, the main theorem of [39] also yields (4.2), but the proof we have sketched is simpler.

Theorems (4.1) and (4.2) show that, in order to define a minor-closed set of graphs of bounded tree-width, it is equivalent to have the obstruction set or an MS formula, because either of them can be constructed from the other. Both of them can be used to construct an HR grammar. However, it is not presently known how to construct an MS formula or the obstruction set *from a given grammar*, though we have no theoretical result saying that this is impossible. The result of Fellows and Langston [38] (see also Van Leeuwen [63]) showing that one cannot construct **OBST**(L) from a membership algorithm for L does not apply if L is given by a grammar. There may exist an algorithm that takes an HR grammar Γ and produces a finite set of graphs A(Γ) such that, whenever **L**(Γ) is minor-closed the set A(Γ) is precisely **OBST**(**L**(Γ)). The existence of such an algorithm, and *a fortiori* an explicit construction of it, is an open problem. There exists such an algorithm in the case of context-free grammars (generating words, i.e., linear, directed, labeled graphs) where subword inclusion is the relevant restriction of minor inclusion; see [26].

Here are two examples of minor-closed sets, the obstructions of which are unknown. Let k-**PI** (where k is some fixed integer) denote the set of graphs that are subgraphs of some planar graph with diameter at most k. It is not very hard to prove that this set is minor-closed and that it does not contain large square grids. Its definition is not MS, so that we cannot use (4.2) to construct its obstruction set. *There is* some equivalent MS definition by (4.1.i), i.e., basically, by the Graph Minor Theorem, but to write it, one needs to know the obstruction set. This alternative definition is not known yet and neither is the set **OBST**(k-**PI**). (This example is due to M. Fellows.)

Here is the second example (by R. Thomas): a graph is *apex* if one can make it planar by deleting at most one vertex (and the incident edges). It is not hard to prove that every minor of an apex graph is apex, and to construct an MS formula characterizing apex graphs. No bound on the tree-width of the graphs of the obstruction set is known, and all grids are apex. Here again, one cannot apply (4.2), and the obstruction set of the set of apex graphs is unknown.

An important case is that of graphs of tree-width at most k. There exists an algorithm for computing the corresponding obstruction sets [49] (this algorithm uses [39]); however it is intractable and the sets are not known for $k \geq 4$.

We conclude this section by presenting a diagram comparing the classes of sets of graphs **HR**, **MS** (see Section 3), the class **B** of sets of graphs of bounded tree-width, and the class **MC** of minor-closed sets.

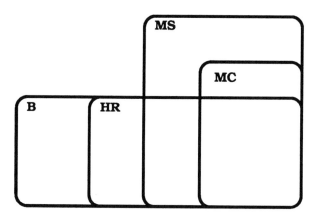

This diagram shows in particular that $B \cap MS = HR \cap MS$ and that $B \cap MC = HR \cap MC$.

5. Vertex replacement graph grammars

We introduce new graph operations. Systems of recursive equations written with these operations yield the *VR grammars* which are strictly more powerful than the HR ones for generating simple graphs. These new operations will concern *simple graphs*, that is, graphs without loops and multiple edges; these graphs will still be finite and undirected.

We let $\mathbb{P}$ be a fixed, countable set of *labels*. A *p-graph (or a simple graph with a possibly empty set of $\mathbb{P}$-labeled vertices called ports)* is a pair H = <G,w> consisting of a simple graph G and a partial mapping w: $V_G \to \mathbb{P}$. We say that {v $\in V_G$ / w(v) is defined} is the set of *ports* of H, and that v is a *p-port* if w(v) = p. Note the difference with the notion of source: a graph with sources has at most one c-source for each c $\in \mathbb{C}$ but a graph with ports may have several p-ports. As for s-graphs, any two isomorphic p-graphs are considered as equal.

We denote by $\mathbb{H}(P)$ the set of p-graphs with ports labeled in P where P is a finite subset of $\mathbb{P}$. (If G $\in \mathbb{H}(P)$, then every port of G is a p-port for some p $\in$ P, but G may have no q-port for some q in P.) We denote by $\mathbb{H}$ the set of all p-graphs. We now define a few operations combining p-graphs.

Disjoint union : If G $\in \mathbb{H}(P)$ and G' $\in \mathbb{H}(P')$, we let G$\oplus$G' be the disjoint union of G and G' (i.e., the union of disjoint copies of G and G'). Clearly, G$\oplus$G' $\in \mathbb{H}(P \cup P')$, because a p-port in G (resp. in G') remains a p-port in G$\oplus$G'. This operation is associative and commutative.

Edge creation : For any two labels a and b in $\mathbb{P}$, and any graph G in $\mathbb{H}$, we let $\eta_{a,b}(G)$ be the graph in $\mathbb{H}$ consisting of G augmented with all edges linking an a-port to a b-port. This definition applies to the case where a = b. We do not create multiple edges and loops. Hence, an edge linking v to w is actually created only if w ≠ v and no edge already links v and w. If G has no a-port or no b-port, then no edge is created.

Port redefinition: Let z be a partial mapping: $\mathbb{P} \to \mathbb{P}$ which is the identity outside of a finite subset of $\mathbb{P}$. We let **redef**$_z$ be the mapping: $\mathbb{H} \to \mathbb{H}$ such that **redef**$_z$(<G,w>) = <G,z∘w>. Hence, **redef**$_z$ maps $\mathbb{H}(P)$ into $\mathbb{H}(z(P))$ for any finite subset P of $\mathbb{P}$.

Port restriction: For every finite subset P of $\mathbb{P}$, we let **rest**$_P$ be the mapping on p-graphs such that **rest**$_P$(<G,w>) = <G,w'> where w' is the restriction of w to $w^{-1}(P)$. Hence, the ports of <G,w> having a label not in P exist as vertices of **rest**$_P$(<G,w>) but are no longer ports. We shall use the notations **rest**$_a$ and **rest**$_{a,b}$ for **rest**$_{\{a\}}$ and **rest**$_{\{a,b\}}$ respectively, as for the similar operations of $\mathbb{F}$.

We denote by $\mathbb{F}'$ the set of these operations.

Basic p-graphs: For each a in $\mathbb{P}$, the graph consisting of a single vertex that is the a-port is a basic graph denoted by a.

These operations and basic p-graphs can be used as those of Section 1 to form algebraic expressions denoting p-graphs, and a complexity measure of simple graphs can be defined in a natural way: the complexity of G is the least number of port labels necessary to form an expression denoting this graph. All cliques are of complexity 1: the n-clique K_n is the value of the expression **rest**$_\emptyset(\eta_{a,a}(a \oplus a \oplus ... \oplus a))$ with n-occurrences of a. Infinitely many source labels are necessary to define all cliques with the operations of $\mathbb{F}$ by (1.1). Hence, these new operations are more powerful than those of $\mathbb{F}$.

Little is known about this new complexity measure (an essentially equivalent notion is introduced in [65]). In particular, it is not known whether the set of graphs of complexity at most k is MS definable, whereas the set graphs of tree-width at most k is by (4.1). (By (1.1), tree-width is the complexity measure on graphs naturally associated with the operations of $\mathbb{F}$.) Let us recall that (4.1) rests upon the Graph Minor Theorem, and that we have no similar tool in the present case.

Systems of equations written with the operations of $\mathbb{F}'$ can be built as those of Section 2 are built with the operations of $\mathbb{F}$. We obtain the notions of a *VR system*, of a *VR grammar* and of a *VR set of p-graphs*. (Let us recall that all graphs in a VR set are simple.)

Example: A *diameter-critical* graph is a connected simple graph such that the addition of any edge (that is not parallel to any existing edge) decreases the diameter. Ore has proved that the diameter-critical graphs with at least three vertices can be described in terms of chains of cliques (see [47]). A *chain of cliques* is a graph of the following form: it consists of n cliques $A_1,...,A_n$, with edges between every vertex of A_i and every vertex of A_{i+1} for i=1,...,n-1, and no edge between A_i and A_j if |i-j| ≥ 2. Such a graph is denoted by $[A_1,...,A_n]$. The result of [47] states that the diameter-critical graphs with at least three vertices are the chains of cliques $[A_1,...,A_n]$ with n ≥ 3 where A_1 and A_n are reduced to single vertices. It follows that the VR grammar (Γ,U) generates the set of diameter-critical graphs with at least three vertices, where Γ is the following system of equations :

$$U = \text{rest}_\emptyset(\eta_{b,d}(V \oplus d)) \cup \text{rest}_\emptyset(\eta_{c,d}(W \oplus d))$$
$$V = \text{rest}_b(\eta_{a,b}(a \oplus K_b)) \cup \text{rest}_b(\eta_{c,b}(W \oplus K_b))$$
$$W = \text{rest}_c(\eta_{b,c}(V \oplus K_c))$$
$$K_b = b \cup \eta_{b,b}(b \oplus K_b)$$
$$K_c = c \cup \eta_{c,c}(c \oplus K_c).$$

Here are typical examples of VR and nonVR sets of graphs.

(5.1) *The following sets of graphs are VR :*

(i)   every HR set of simple graphs,
(ii) the set of all cliques,
(iii) the set of all complete bipartite graphs.

The following sets of graphs are not VR :

(iv) the set of all planar graphs,
(v)   the set of all square grids,
(vi) the set of all chordal graphs.

Proof : (i) is proved in [36, 28]; (iv) and (v) follow from (6.4) below and the fact that the MS_1 theory of these sets is undecidable; (vi) is similar by the results of [24]. □

The following proposition lists the main properties of VR sets of graphs. It should be compared with (2.4) that concerns HR sets similarly.

(5.2) *(i)   The union of two VR sets is VR.*
(ii)   The intersection and the difference of two VR sets is not VR in general.
(iii)   The set of all induced subgraphs and the set of edge complements of all graphs of a VR set is VR.
(iv)   One can decide in polynomial time whether $L(\Gamma,S)$ is empty for any given VR grammar (Γ,S).
(v)   For every VR grammar (Γ,S) the membership problem in $L(\Gamma,S)$ is decidable.

Proof : For (i), (ii), (iv), the proofs are as for the corresponding assertions in (2.4); (iii) is a consequence of [34] and [28], and see also [27]; and for (v), see [8] on the complexity of the membership problem. □

The borderline between the class of HR sets and that of VR sets that are not HR is quite well-known :

(5.3) *Let L be a VR set of graphs. The following conditions are equivalent :*
(i)   L is HR,
(ii)   L has bounded tree-width,
(iii)   L excludes some graph as a minor,
(iv)   L excludes some complete bipartite graph $K_{n,n}$ as a subgraph,
(v)   L is sparse.
These conditions are decidable from any VR grammar generating L.

Proof : The implications (i) ⇒ (ii) ∧ (iii) ∧ (iv) ∧ (v), (ii) ⇒ (iii) ∧ (iv) ∧ (v) are evident from previous results. The implication (iii) ⇒ (v) follows from a result by Mader saying that if G is simple and $\mathbf{Card}(E_G) \geq 2^k \cdot \mathbf{Card}(V_G)$, then it contains the k-clique K_k as a minor (see [52]). Implication (iii)∨(ii) ⇒ (i) is a difficult theorem proved in [27]. Implication (iv)∨(v) ⇒ (i) and the decidability results are proved in [25]. □

The equivalence of (i), (ii) and (iii) shows that a minor-closed set of graphs is VR iff it has bounded tree-width iff it is HR. Hence, the only minor-closed sets of graphs that we know presently how to describe by graph grammars (either VR or HR) are those of bounded tree-width.

The following diagram collects some these results. It shows in particular that **VR**∩**B** = **HR** and **VR**∩**MC** = **HR**∩**MC** = **B**∩**MC**. We denote by **VR** the class of VR sets of graphs. See the diagram of Section 4 for the other notation.

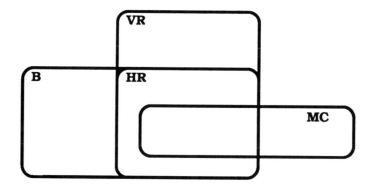

6. Monadic second-order logic without edge quantifications

Results similar to (3.2) and (3.4) hold for VR grammars but only for a restricted form of monadic second-order logic where the variables cannot denote sets of edges. Rather than to a modification of the logical language, this restriction corresponds to a modification *of the representation of a graph by a logical structure*, that we now describe.

For every finite subset P of $\mathbb{P}$, we let $\mathbb{R}'(P)$ be the finite set of relational symbols $\{\mathbf{adj}, \mathbf{p_a} / a \in P\}$ where **adj** is binary and $\mathbf{p_a}$ is unary.

For every graph G in $\mathbb{H}(P)$ we let $|G|$ be the $\mathbb{R}'(P)$-structure $<V_G, \mathbf{adj}_G, (\mathbf{p_a}_G)_{a \in P}>$ where $\mathbf{adj}_G(x,y)$ holds iff there is an edge in G linking x and y, and $\mathbf{p_a}_G(x)$ holds iff x is an a-port of G. Since all graphs in $\mathbb{H}(P)$ are simple, it is clear that for any two of them G and G', $|G| = |G'|$ iff G = G'.

The edges are no longer present as elements of the domain of the structure. They are represented by a basic binary relation. Since in MS logic one cannot quantify over binary relations, one cannot quantify over sets of edges. (One can actually quantify over individual edges because "there exists an edge e....." can be written "there exist vertices x and y.....".)

We refer by MS_1 to the monadic second-order logic as a language for expressing graph properties via the representation of a graph G by the structure $|G|$. Hamiltonicity is an example of a property that is expressible in MS but not in MS_1. However, MS and MS_1 have the same expressive power in certain cases, as proved in [22]:

(6.1) *The same properties of simple graphs*

   (i)   *either that have degree at most some fixed* **k**,
   (ii) *or that do not contain a fixed graph* H *as a minor,*
are expressible in MS and MS_1.

The proof consists in constructing for each k (respectively H) a translation **tr**: $\mathfrak{I}(\mathbb{R}(\emptyset),\emptyset) \to \mathfrak{I}(\mathbb{R}'(\emptyset),\emptyset)$ such that, for every simple graph G of degree at most k (respectively that does not contain H as a minor):

$$|G| \models \mathbf{tr}(\varphi) \quad \text{iff} \quad \|G\| \models \varphi \, .$$

The following theorem is fully analogous to (3.2) (and these two results are actually two instances of a single one; see [23]).

(6.2) *Let φ be a MS_1 formula; let K be a finite set of graph operations from $\mathbb{R}'$, of derived graph operations constructed with $\mathbb{R}'$ and of basic graphs. For every graph expression t over K, one can decide in time $O(|t|)$ whether the graph* **val**(t) *satisfies property* P_φ.

We have no theorem comparable to (3.3), but only a conjecture.

(6.3) Conjecture: *If a set of simple graphs has a decidable MS_1 theory, then it is a subset of some VR set.*

For working with VR grammars, we have the following result, fully analogous to (3.4) (and they have a common proof [23]).

(6.4) *Let P be a finite subset of $\mathbb{P}$; let Γ be a VR grammar such that $L(\Gamma) \subseteq \mathbb{H}(P)$, and let φ be a formula in $\mathfrak{I}(\mathbb{R}'(P),\emptyset)$. One can construct a VR grammar generating $L(\Gamma) \cap L_\varphi$. One can decide whether $|G| \models \varphi$ for some graph G in $L(\Gamma)$. The MS_1 theory of $L(\Gamma)$ is decidable.*

7. Guide to the literature

Introduction

We have called a *graph grammar* what should be called more precisely a *context-free graph grammar*. Many notions of (general) graph grammars have been defined, with different motivations ranging from biology to parallel computing, which explains the large variety of definitions. The reader will find these definitions in the proceedings of the four international workshops on graph grammars ([11, 31, 32, 30]). Most of the definitions are based on graph-rewriting rules by which, according to the case, one can replace a vertex, an edge, a hyperedge or a subgraph by a graph. Some of these grammars can be called *context-free* by reference to the notion of a context-free grammar generating a context-free language. This is the case if rewriting sequences can be represented by *derivation trees* in such a way that any two sequences having the same tree produce the same graph. (This means that rewriting steps on different branches of the tree can be permuted.) It should be noted here that such a tree defines a tree-structuring of the produced graph, that is not always a tree-decomposition. Another characteristic property of context-free graph grammars is the possibility of representing them by systems of mutually recursive definitions. See Courcelle [15] on the context-freeness of graph grammars. In this paper, we define graph grammars directly as systems of equations, so that they are necessarily context-free.

Graph reduction as a tool for recognizing graphs is presented in [1]. A graph is recognized as a member of the specified set if it can be reduced to one of finitely many "accepting graphs" by means of fixed graph-rewriting rules that reduce the sizes of the graphs to which they apply. This paper establishes that every MS definable set of graphs of bounded tree-width has a linear membership algorithm.

Section 1

In order to concentrate on the main aspects relevant to the Theory of Graph Minors, we have restricted our exposition to undirected, unlabeled graphs. The adaptation to directed and/or labeled graphs is straightforward. Operations on directed, labeled hypergraphs have been introduced by Bauderon and Courcelle [4, 13, 16]. Other graph operations are used in [1, 23, 25].

The operations of $\mathbb{F}$ are different from those of [4, 13]; they are closer to those of [1] or [67]. The main difference is that these operations are defined on graphs of all types. This makes it possible to avoid equipping operations with *sorts* and simplifies considerably the presentation. (This complicates proofs only slightly.) The equivalence is not hard to establish (the basic ideas for this equivalence are in [27]) and proves that the HR sets are indeed the sets generated by the context-free graph grammars of [4] or the hyperedge replacement grammars of [42, 43]. That we deal here with undirected graphs, whereas [4, 42, 43] deal with directed hypergraphs, is a minor technical detail.

Section 2

Complexity properties of sets of graphs generated by grammars of various types are discussed in [8, 9]. Whereas the set of graphs of cycle bandwidth at most 2 is NP-complete [51], it can be decided in time $O(n^k)$ whether a graph has bandwidth at most k. See [40]. Conditions ensuring that a graph grammar has a polynomial parsing algorithm are considered in [9, 50, 64] just to cite a few papers. Efficient algorithms have been given in [6, 48] to construct tree-decompositions. These algorithms can be considered as parsing algorithms relative to HR grammars generating the sets of graphs of tree-width at most k.

Section 3

Monadic second-order logic is popular among logicians because it is a relatively powerful extension of first-order logic enjoying many decidability results. (See the survey by Gurevich [41]; the central result is the very powerful and difficult result by Rabin [55].) Its use for characterizing sets of finite and infinite graphs (in the context of the theory of graph grammars) is the subject of the series of papers [17-23] to which [15] and the survey [16] can be added. The definitions of these papers which deal with directed hypergraphs have been simplified and adapted to undirected graphs in the present paper. Using logical formulas for defining sets of graphs is also fruitful in the theory of complexity. See the surveys [45] and [47].

Monadic second-order logic can also be used to define transformations from graphs to graphs (see [21-24]) and functions from graphs to integers, reals, sets of integers (see [29]).

Theorem (3.2) (established in 1986, see [14]) has been proved in a different way and extended to the so-called Extended Monadic Second-Order logic, making possible some arithmetical computations, by Arnborg et al. [2]. Other extensions, which answer questions raised in [5], can be found in [7, 29, 44]. The algorithms constructed from these papers take as inputs tree-decompositions or graph expressions of the graphs they compute on. The result of [1] improves (3.2) in that the construction of the tree-decomposition is not necessary, so that the resulting algorithm is linear. It is not clear yet whether the technique of [1] can be adapted to the situations of [2, 7, 29, 44].

Section 5

Although VR grammars are more difficult to study than HR ones, they were introduced earlier. They stem from the NLC grammars ([46, 61, 37]) defined in terms of graph rewritings by which a vertex is replaced by a graph. They are not always context-free (see [15]); the ed-NCE grammars are of the same style but more powerful. The C-edNCE ones (see [33]) are context-free: systems quite similar to VR systems are introduced in [28] and it is proved that the least solutions of these systems and the sets generated by C-edNCE grammars coincide. (Let us make a technical remark: in [28], a vertex may be simultaneously a p-port and q-port where p and q are different; the equivalence of the VR systems and those of [28] follows from Lemma 1 of [35].) The definition we gave in terms of VR systems is the simplest one. However, it comes as the result of quite hard work.

Section 6

It is proved in [10, 35] that if a VR set of graphs has bounded degree, then it is HR. It follows from (5.3.iii) and (6.1) that the same conditions on a class of simple graphs, namely, that it has bounded degree or excludes some graph as a minor, have two different effects: they make MS and MS_1 equally expressive (for the graphs of this class) and they make HR and VR grammars equally powerful (for generating graphs satisfying these conditions). The second of these effects is actually a consequence of the first and the logical characterizations of the classes **HR** and **VR** given respectively in [27] and [34], which state that a set of graphs is HR (resp. VR) iff it is the image of a recognizable set of finite trees under a function from trees to graphs specified by MS formulas (resp. by MS_1 formulas). These results indicate how close are the links between monadic second-order logic and context-free graph grammars.

Acknowledgements: I thank N. Robertson and P. Seymour who invited me to write this survey. I thank J. Engelfriet and the referee for helpful comments, and K. Callaway who helped me once again with English style.

REFERENCES

L.N.C.S. means: "Lecture Notes in Computer Science, Springer Verlag, Heidelberg, Berlin, New-York".

1. ARNBORG S., COURCELLE B., PROSKUROWSKI A., SEESE D., An algebraic theory of graph reduction, Research Report 90-02 (revision 91-36), Bordeaux-1 University, see also short version in [30] pp. 70-83.
2. ARNBORG S., LAGERGREN J., SEESE D., Easy problems for tree-decomposable graphs, J. of Algorithms **12** (1991) 308-340.
3. ARNBORG S., PROSKUROWSKI A., CORNEIL D., Forbidden minors characterization of partial 3-trees, Discrete Mathematics **80** (1990) 1-19.
4. BAUDERON M., COURCELLE B., Graph expressions and graph rewritings, Mathematical Systems Theory **20** (1987) 83-127.
5. BERN M., LAWLER E., WONG A., Linear time computation of optimal subgraphs of decomposable graphs, J. Algorithms **8** (1987) 216-235.
6. BODLAENDER H., KLOKS T., Better algorithms for path-width and tree-width of graphs, L.N.C.S. **510** (1991) 544-555.
7. BORIE R., PARKER R., TOVEY C., Automatic generation of linear algorithms from predicate calculus descriptions of problems on recursively constructed graph families, Algorithmica, to appear.
8. BRANDENBURG F.-J., On partially ordered graph grammars, in [32] pp.99-111.
9. BRANDENBURG F.-J., On polynomial time graph grammars, L.N.C.S. **294** (1988) 227-236.
10. BRANDENBURG F.-J., The equivalence of boundary and confluent graph grammars on graph languages with bounded degree, L.N.C.S. **488** (1991) 312-322.
11. CLAUS V., EHRIG H., ROZENBERG G., *First International Workshop on Graph Grammars and their Applications to Computer Science and Biology*, L.N.C.S. **73** (1979).
12. COURCELLE B., Equivalences and transformations of regular systems; applications to recursive program schemes and grammars, Theoret. Comput. Sci. **42** (1986) 1-122.
13. COURCELLE B., A representation of graphs by algebraic expressions and its use for graph rewriting systems, in [32], pp. 112-132.
14. COURCELLE B., On context-free sets of graphs and their monadic second-order theory, in [32], pp. 133-146.
15. COURCELLE B., An axiomatic definition of context-free rewriting and its application to NLC graph grammars, Theoret. Comput. Sci. **55** (1987) 141-181.
16. COURCELLE B., Graph rewriting: An algebraic and logic approach, in *Handbook of Theoretical Computer Science, Volume B*, J. Van Leeuwen ed., Elsevier,1990, pp.193-242.
17. COURCELLE B., The monadic second-order logic of graphs I, Recognizable sets of finite graphs, Information and Computation **85** (1990) 12-75.
18. COURCELLE B., The monadic second-order logic of graphs II, Infinite graphs of bounded width, Mathematical Systems Theory **21** (1989) 187-221.
19. COURCELLE B., The monadic second-order logic of graphs III, Tree-decompositions, minors and complexity issues, Informatique Théorique et Applications, to appear.
20. COURCELLE B., The monadic second-order logic of graphs IV, Definability properties of equational graphs, Annals Pure Applied Logic **49** (1990) 193-255.
21. COURCELLE B., The monadic second-order logic of graphs V: On closing the gap between definability and recognizability, Theoret. Comput. Sci. **80** (1991) 153-202.
22. COURCELLE B., The monadic second order logic of graphs VI: On several representations of graphs by relational structures, Report 89-99, Bordeaux-I University, Discrete Applied Mathematics, to appear (see also "Logic in Computer Science", 1990, Philadelphia).
23. COURCELLE B., The monadic second order logic of graphs VII: Graphs as relational structures, Theoret. Comput. Sci., in press, Research Report 91-40, see also short version in [30] pp. 238-252.

24. COURCELLE B., Monadic second-order definable graph transductions, Proceedings of CAAP'92, Rennes, February 1992, L.N.C.S., to appear.
25 COURCELLE B., On the structure of vertex replacement sets of graphs, Research Report 91-44, Bordeaux-1 University, December 1991.
26. COURCELLE B., On constructing obstruction sets of words, Bulletin of EATCS **44**, June 1991, pp. 178-185.
27. COURCELLE B., ENGELFRIET J., A logical characterization of the sets of hypergraphs generated by hyperedge replacement grammars, Research Report 91-41, Bordeaux-1 University, 1991.
28. COURCELLE B., ENGELFRIET J., ROZENBERG G., Handle-rewriting hypergraph grammars, Report 90-84, Bordeaux-I University, J. Comput. Syst. Sci., to appear, see also short version in [30] pp. 253-268.
29. COURCELLE B., MOSBAH M., Monadic second-order evaluations on tree-decomposable graphs, Theoret. Comput. Sci., to appear, Research Report 90-110, Bordeaux-1 University, short version in the Proceedings of WG'91, L.N.C.S. **570**, (1992).
30. EHRIG H., KREOWSKI H.-J., ROZENBERG G., *Fourth International Workshop on Graph Grammars and their Applications to Computer Science*, L.N.C.S. **532**, 1991.
31. EHRIG H., NAGL M., ROZENBERG G., *Second International Workshop on Graph Grammars and their Applications to Computer Science*, L.N.C.S. **153**, 1983.
32. EHRIG H., NAGL M., ROZENBERG G., ROSENFELD A., *Third International Workshop on Graph Grammars and their Applications to Computer Science*, L.N.C.S. **291**, 1987.
33. ENGELFRIET J., Context-free NCE graph grammars, L.N.C.S. **380** (1989) 148-161.
34. ENGELFRIET J., A characterization of context-free NCE graph languages by monadic second-order logic on trees, in [30] pp. 311-327.
35. ENGELFRIET J., HEYKER L., Hypergraph languages of bounded degree, Research Report 91-01, University of Leiden, 1991.
36. ENGELFRIET J., ROZENBERG G., A comparison of boundary graph grammars and contextfree hypergraph grammars, Information and Computation **84** (1990) 163-206.
37. ENGELFRIET J., ROZENBERG G., Graph grammars based on node rewriting: an introduction to NLC graph grammars, in [30] pp. 12-23.
38. FELLOWS M., LANGSTON M., On search decision and the efficiency of polynomial time algorithms, Proceedings ACM-STOC 1989, pp. 501-512.
39. FELLOWS M., LANGSTON M., An analogue of the Myhill-Nerode theorem and its use in computing finite basis characterizations, 30th Annual IEEE Symposium on Foundations of Computer Science, 1989, pp. 520-525.
40. GUARI E., SUDBOROUGH I., Improved dynamic programming algorithms for bandwidth minimization and the mincut linear arrangement problem, J. Algorithms **5** (1984) 531- 546.
41. GUREVICH Y., Monadic second-order theories, in *Model theoretic logic*, J. Barwise and S. Feferman eds., Springer, Berlin, 1985, pp. 479-506.
42. HABEL A., Hyperedge replacement: grammars and languages, Doctoral dissertation, University of Bremen, 1989.
43. HABEL A., KREOWSKI H.-J., May we introduce to you: Hyperedge replacement?, in [32], pp.15-26.
44. HOHBERG W., REISCHUK R., A framework to design algorithms for optimization problems on graphs, Research Report, Technische Hochschule Darmstadt, Germany, October 1990.
45. IMMERMAN N., Languages that capture complexity classes, SIAM J. Comput.**16** (1987) 760-777.
46. JANSSENS D., ROZENBERG G., A survey of NLC grammars, L.N.C.S. **159** (1983) 131-151.
47. KANNELLAKIS P., Elements of relational data base theory, in *Handbook of Theoretical Computer Science, Volume B*, J. Van Leeuwen ed., Elsevier, 1990, pp. 1073-1156.

48. LAGERGREN J., Efficient parallel algorithms for tree-decomposition and related problems, Proceedings of IEEE Symp. on F.O.C.S., 1990, pp. 173-182.
49. LAGERGREN J., ARNBORG S., Finding minimal forbidden minors using a finite congruence, L.N.C.S. **510** (1991) 532-543.
50. LAUTEMANN C., The complexity of graph languages generated by hyperedge replacement, Acta Informatica **27** (1990) 399-421.
51. LEUNG J., WITTHOF J., VORNBERGER O., On some variants on the bandwidth minimization problem, SIAM J. on Comput.**13** (1984) 650-667.
52. MADER W., Homomorphieeigenschaften und mittlere Kanten dichte von Graphen, Math. Ann. **174** (1967) 265-268.
53. ORE O., Diameters in graphs, J. of Combinatorial Theory **5** (1968) 75-81.
54. PROSKUROWSKI A., Recursive graphs, recursive labellings and shortest paths, SIAM J. Comput. **10** (1981) 391-397.
55. RABIN M., Decidability of second-order theories and automata on infinite trees, Trans. Am. Math. Soc. **141** (1969) 1-35.
56. ROBERTSON N., SEYMOUR P., Graph minors I, Excluding a forest, J. Comb. Theory Ser. B, **35** (1983) 39-61.
57. ROBERTSON N., SEYMOUR P., Graph minors V, Excluding a planar graph, J. Comb. Theory Ser. B, **41** (1986) 92-114.
58. ROBERTSON N., SEYMOUR P., Graph minors X, Obstructions to tree-decomposition, J. Comb. Theory Ser. B, **52** (1991) 153-190.
59. ROBERTSON N., SEYMOUR P., Graph minors XX, Wagner's conjecture, Preprint, September 1988.
60. ROBERTSON N., SEYMOUR P.,THOMAS R., Quickly excluding a planar graph , Submitted
61. ROZENBERG G., WELZL E., Boundary NLC grammars, Basic definitions, normal forms and complexity, Information and Control **69** (1986) 136-167.
62. SEESE D., The structure of the models of decidable monadic theories of graphs, Annals of Pure Applied Logic **53** (1991) 169-195.
63. VAN LEEUWEN J., Graph algorithms, in *Handbook of Theoretical Computer Science, Vol. A*, J. Van Leeuwen ed., 1990, Elsevier, pp. 523-631.
64. VOGLER W., Recognizing edge replacement graph languages in cubic time, in [30] pp. 676-687.
65. WANKE E., Algorithms for graph problems on k-NLC trees, Discrete Applied Mathematics, to appear.
66. WANKE E., WIEGERS M., Undecidability of the bandwidth problem on linear graph languages, Inf. Proc. Letters **33** (1989) 193-197.
67. WIMER T., HEDETNIEMI S., LASKAR R., A methodology for constructing linear graph algorithms, Congressus Numerantium **50** (1985) 43-60.

LaBRI (CNRS Laboratory n° 1304),BORDEAUX-I University, TALENCE, France
Current address: Université BORDEAUX-I, LaBRI, 351, Cours de la Libération, 33405 TALENCE, France
E-mail address: courcell@geocub.greco-prog.fr

Graph reductions, and techniques for finding minimal forbidden minors

Andrzej Proskurowski

Abstract

Knowing that a class of graphs has a finite set of minimal forbidden minors is one thing, knowing what they are is another. We present an account of some techniques used to find small sets of minimal forbidden minors for a few classes of graphs with treewidth at most 3.

1 Introduction

A finite representation of an infinite class of objects constitutes a very attractive tool and an elegant result. For graphs, there have been a number of forbidden substructure characterizations, the most famous being the Planar Graphs Theorem of Kuratowski:

> A graph is planar if and only if it does not contain a subgraph homeomorphic to either K_5 or $K_{3,3}$.

1991 Mathematics Subject Classification. Primary 05C10, 05C75, 68C05.

This is a preliminary presentation. The expanded version of this paper will be submitted for publication elsewhere.

Wilf [21] gives an introduction to the notion of obstructions in his column that introduces work of Robertson and Seymour. There, one can find Kuratowski's theorem stated in terms of minors (Thomassen [19] attributes it to Harary and Tutte):

> A graph is planar if and only if it does not contain a minor isomorphic to either K_5 or $K_{3,3}$.

As a reminder, we state few basic definitions. A graph H is a *minor* of a graph G if contracting some edges of a subgraph of G ('minor-taking') would give a graph isomorphic to H (we will consider only *simple* graphs, without multiple edges). For a class $\mathcal{C}$ of graphs closed under minor-taking, F is a *minimal forbidden minor* if it is not in $\mathcal{C}$, but every minor of F is in $\mathcal{C}$. Henceforth, we will call the minimal forbidden minors a little loosely *obstructions*.

For a fixed positive integer k, the complete graph on k vertices, K_k is a k-tree and every k-tree with $n > k$ vertices can be constructed from a k-tree with $n-1$ vertices by adding to it a vertex adjacent to all vertices of a subgraph isomorphic to K_k. A graph that can be embedded in a k-tree is called *partial k-tree*, or alternatively, it is said to have *treewidth* at most k.

In the study of graphs with bounded treewidth (partial k-trees), there are obvious characterizations for $k = 1, 2$ by forbidden subgraphs homeomorphic to K_3 and K_4, respectively. Although K_5 is likewise forbidden for $k = 3$, the set of obstructions for partial 3-trees is obviously larger. Before discussing the tools used in the discoveries of that set, we briefly present approaches to determining obstruction sets for some smaller graph classes.

For *partial 1-trees* (forests), the completeness of $\{K_3\}$ as the set of obstructions follows directly from the definition (the acyclicity of the graphs).

Partial 2-trees can be recognized by iterating degree 0, 1 and 2 vertex reduction (where a degree 2 vertex is replaced by an edge incident with its neighbors). A proof of $\{K_4\}$ as their obstruction set follows from considering the reduction rule: An obstruction must be cubic and biconnected, and end vertices of any edge must have a common neighbor. K_4 is the only graph that fits this description.

K_4 is also an obstruction for the *outerplanar graphs*. By the definition, so is the graph $K_{2,3}$. To see that these two graphs constitute the set of obstructions for outerplanar graphs, consider any plane embedding of a series-parallel non-outerplanar graph. It must

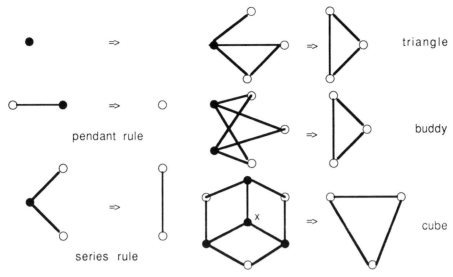

Figure 1: Reduction rules for recognition of partial 3-trees

have an interior vertex and at least one vertex between its attachments to the cycle constituting the boundary of the exterior mesh in that embedding, in each direction around that cycle. Such a graph has a subgraph homeomorphic to $K_{2,3}$ and thus has $K_{2,3}$ as a minor. (Thomassen [19] traces this characterization to Chartrand and Harary.)

In this note, we intend to illustrate the concept of graph reduction in the search for a complete set of obstructions. For this purpose, we give a short survey of the different approaches that lead to the discovery of the set of obstructions for partial 3-trees. We will start with a presentation of one of the tools (vertex reduction system), then describe the search for obstructions, and conclude with a brief description of continued efforts to find a general (constructible) algorithm paradigm for obstruction sets.

2 Vertex reductions for partial k-trees

Recognition of forests by checking the irreducible result of repeated 'pruning of leaves' (removal of degree 1 vertices) and discarding isolated vertices has been taken for granted for a long time. Recognition of partial 2-trees by, in addition, contracting an edge incident with a degree 2 vertex has been proposed by Wald and Colbourn [20]. Arnborg and Proskurowski [3] give a complete set of confluent vertex reduction rules for partial 3-trees. (This means that a graph is a partial 3-tree if and only if any sequence of applications of these rules reduces it to the empty graph; if one does, so does any other.) There, the three reduction rules mentioned above are augmented by three more reductions of degree 3 vertices (see Figure 1). These mimic pruning 3-leaves (degree 3 vertices) of an embedding 3-tree, but also indicate that not all degree 3 vertices in a partial 3-tree are such 3-leaves.

Independently, the same set of reduction rules was derived by Kajitani *et al.* [11] who discovered the necessity of certain configurations in 3-connected partial 3-trees following a very similar line of reasoning.

A fairly natural implementation of these rules leads to an $\mathcal{O}(n \log n)$ algorithm; Matoušek and Thomas [15] noticed that the rules can be modified to yield a linear algorithm.

As these reduction rules constitute an important tool in the investigations of 'small' properties of partial k-trees, the following result of Lagergren [13] is quite discouraging.

> There is no complete set of confluent vertex reduction rules that reduce a graph to the empty graph if and only if the graph is a partial k-tree, for $k > 3$.

Yet, it turns out that there exist more general graph reduction systems that decide membership in classes of partial k-trees. Namely, Arnborg *et al.* show in [1] that this is the case for any subclass of partial k-trees (fixed k) definable by the *Monadic Second Order Logic (MSOL)* (*cf.*, for instance, Courcelle [7]).

> For any class of graphs of bounded treewidth that can be described by an expression in the *MSOL* formalism there is a finite terminating graph rewriting system with the following property: Repeated applications of the rewrite rules lead to an irreducible graph that is a member of a finite accepting set of graphs if and only if the original graph is a member of the graph class in question.

Such a graph rewriting system can be implemented as a linear time (although space intensive) algorithm. As usual, however, constructing such a system might be a difficult task, even though the existence proof (of the above result) is constructive. More importantly for the subject of this note, such a system gives little insight

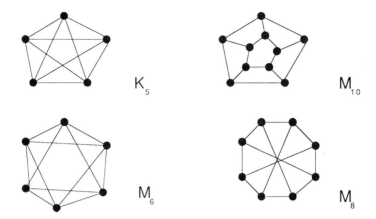

Figure 2: The set of obstructions for partial 3-trees

into construction of the set of obstructions. Yet, it might provide some computational help, see section 4.

3 Obstructions for partial 3-trees

The sets of obstructions for *partial 3-trees* and for *planar partial 3-trees* have been discussed independently, even though the former – in conjunction with the minors form of the Planar Graph Theorem – implies the latter.

El Mallah and Colbourn [10] state the following characterization of planar partial 3-trees.

> A planar graph is a partial 3-tree if and only if it does not have a minor isomorphic to either M_6 or M_{10} in Figure 2.

In their proofs, they exploit the duality between Δ-Y and Y-Δ reductions (replacing triangle K_3 by star $K_{1,3}$ and *vice-versa*, respectively) and properties of geometric duals of the graphs defined with help of these reductions. The proof of a similar result presented by Dai and Sato [9] is based on Tutte's characterization of planar 3-connected graphs. Both papers rely on the presence of the lefthand-sides of vertex reduction rules used in recognition of partial 3-trees in edge-contracted planar graphs.

Arnborg, Proskurowski and Corneil characterize the class of partial 3-trees in [4].

A graph is a partial 3-tree if and only if it does not have a minor isomorphic to any of the graphs in Figure 2.

Their proof depends very heavily on the small complete set of confluent reduction rules for this class of graphs (see above). The minimality of the investigated minors implies 3-connectivity, and the reduction rules (and the fact that they are vertex-reducing) imply that any vertex degree can be only 3 or 4. Investigation of cases of possible neighborhood configurations of contracted or extracted edge in any obstruction (those 'configurations' must admit vertex reductions) completes their proof.

Theirs was just one of several independent investigations that ended with similar results. The approach based on the same set of reduction rules for partial 3-trees was used by Borie, Parker and Tovey [6].

Satyanarayana and Tung [17] do not use the reduction rules in their proofs, but they rediscover (in fact) the properties of 3-connected components of obstructions implied by those reductions. The flow of their proofs follows a similar path of discovering cubic obstructions, then 4-regular such graphs, and then showing that minimum vertex degree 3 implies 3-regularity of an obstruction.

A recent paper by Satyanarayana and Politoff [16] gives an alternative proof of the obstruction set for partial 3-trees. In their discussion of *quasi 4-connected graphs* (that have no 3-vertex separators except for those that separate several degree 3 vertices) they find that only few graphs are 'responsible' for this property. Namely, a non-planar quasi 4-connected graph has a K_5 minor or is a 'small graph'. A planar quasi 4-connected graph has M_6 as a minor or is some other 'small graph'. These 'small graphs' are M_8 and M_{10} from Figure 2, and some partial 3-trees. Since no partial 3-tree, except for some small ones with only trivial 3-separators, is quasi 4-connected, every large enough quasi 4-connected graph has a minor from the set of obstructions for partial 3-trees. Analysis of those small partial 3-trees and the 'small graphs' in their lemmata implies the desired characterization of partial 3-trees. Although tediously relying on case analysis, the proofs are somewhat shorter than case analyses in the previously published proofs.

4 Other tools for finding obstructions

Graph reduction rules are of some help in constructing the obstruction set, but they are by no means the only tool available. The computational power of modern computers and the skill of their programmers can go a long way in searching for obstructions, especially among subclasses of partial k-trees, for small values of k. An example of such a result is the set of 110 obstructions for graphs with pathwidth 2 constructed by Kinnersley [12]. Similar result concerning acyclic such minors for $k = 2, 3$ and 4 is presented by Takahashi *et al.* [18], who actually provide the (rather long) lists of trees.

Another approach, yet to be implemented, is to construct the obstruction set using raw computational power for searching a finite list of graphs among which all such graphs are guaranteed to be found. Arnborg *et al.* [5] describe the translation process of a *MSOL* formula defining a subclass of bounded treewidth graphs into a *tree automaton*. The number of states in the resulting automaton can be used to determine a bound on the number of vertices in an obstruction for that class.

Using an encoding of tree decompositions of width k, Lagergren and Arnborg [14] find a finite congruence in a *graph algebra* that defines the class of partial k-trees. Subsequently, they describe how to obtain the set of irreducibles that contains the obstruction set by a procedure similar to the construction algorithm for the corresponding graph reduction system of [1] (*cf.* Section 2).

Given a *graph grammar* defining a class of graphs, Courcelle and Proskurowski [8] use formal linquistic tools to derive another graph grammar that generates a finite superset of obstructions for the original class. Graph grammars, upon which we will not elaborate here, provide another way of defining classes of graphs with interesting algorithmic properties.

Using a finite characterization, a class of graphs with bounded treewidth can be defined by:

(i) a set of obstructions,
(ii) an *MSOL* description,
(iii) a graph reduction system.

These have been shown equivalent by Lagergren and Arnborg [14]. While we have some developed understanding of the conceptual relationships between these description methods, constructive proofs of their equivalence remain still an important research topic. This is due in part to the potentially gigantic size of any such finite characterization.

We have chosen a small example of the class of partial 3-trees to illustrate some of the notions used in discovering the corresponding obstruction set. Vertex reduction rules used in this process can not be directly generalized for graphs with larger treewidth, but more general graph reduction systems might bring some assistance in the search for minimal forbidden minors by bounding the set of candidate graphs.

Disclaimer: It is not within the scope of this note to present a complete historical and methodological survey of this extensive and exciting area. The author readily accepts the blame for omissions of references to and timing of any independent work.

References

[1] S. Arnborg, B. Courcelle, A. Proskurowski, D. Seese, An algebraic theory of graph reduction, (to appaer in *JACM*), Report No. 90-02, LaBRI, Université de Bordeaux I (1990).

[2] S. Arnborg, J. Lagergren and D. Seese, Problems easy for tree-decomposable graphs, *J. of Algorithms* 12 (1991), 308-340.

[3] S. Arnborg and A. Proskurowski, Characterization and recognition of partial 3-trees, *SIAM J. Alg. and Discr. Methods* 7, 305-314 (1986).

[4] S. Arnborg, A. Proskurowski and D.G. Corneil, Minimal forbidden minor characterization of a class of graphs, *Colloquia Mathematica Societatis János Bolyai* 52 (1987), 49-62.

[5] S. Arnborg, A. Proskurowski and D. Seese, Monadic second order logic: tree automata and forbidden minors, *UO-CIS-TR-90/23* (1990).

[6] R. Borie, R.G. Parker and C.A. Tovey, The regular forbidden minors of partial 3-trees (manuscript) (1988).

[7] B. Courcelle, Some applications of logic of universal algebra, and of category theory to the theory of graph transformations, *Bulletin of EATCS 36* (1988), 161-218.

[8] B. Courcelle and A. Proskurowski, in preparation (1991).

[9] W.W-M. Dai and M. Sato, Minimal forbidden minor characterization of planar partial 3-trees and application to circuit layout, *IEEE Int'l Symp. on Circuits and Systems* (1990), 2677-2681.

[10] E.S. El Mallah and C.J. Colbourn, On two dual classes of planar graphs, *Discrete Mathematics 80* (1990), 21-40.

[11] Y. Kajitani, A. Ishizuka, and S. Ueno, Characterization of partial 3-trees in terms of three structures, *Graphs and Combinatorics 2* (1986), 233-246.

[12] N. Kinnersley, Obstruction set isolation for layout permutation problems, PhD. Thesis, Washington State University (1989).

[13] J. Lagergren, Nonexistence of reduction rules giving an embedding into a k-tree, (manuscript) (1990)

[14] J. Lagergren and S. Arnborg, Finding minimal forbidden minors using a finite congruence, *LNCS: Proceedings of ICALP'91*, Springer-Verlag (1991).

[15] J. Matoušek and R. Thomas, Algorithms finding tree-decompositions of graphs, *J. of Algorithms 12* (1991), 1-22.

[16] A. Satyanarayana and T. Politoff, A characterization of quasi 4-connected graphs (manuscript) (1991).

[17] A. Satyanarayana and L. Tung, A characterization of partial 3-trees, *Networks 20* (1990), 299-322.

[18] A. Takahashi, S. Ueno, and Y. Kajitani, Minimal acyclic forbidden minors for the family of graphs with bounded path-width, *SIGAL 91-19-3* (1991).

[19] C. Thomassen, Embeddings and minors, in *Handbook of Combinatorics*, R.L Graham, M. Grötschel and L. Lovász, eds., North Holland (to appear).

[20] J.A. Wald and C.J. Colbourn, Steiner trees, partial 2-trees, and minimum IFI networks, *Networks 13* (1983), 159-167.

[21] H.S. Wilf, Finite list of obstructions (The Editor's Corner), *Mathematics Monthly* (March 1987), 267-271.

Author's address:

Department of Computer and Information Science
University of Oregon, Eugene, Oregon 97403, USA

An Upper Bound on the Size of an Obstruction

JENS LAGERGREN

ABSTRACT. We prove constructively that every minor closed family which is recognized by a finite congruence and has an obstruction set that is of bounded tree-width has a finite number of obstructions. Our proof gives a general bound on the size of these obstructions. They cannot have more than $2^{O(c^{w+1}w!)}$ edges where w is the tree-width and c is the index of the congruence. This general bound is our first main result. It applies in particular to minor closed families of graphs which are of bounded tree-width and recognized by a finite congruence.

We define explicitly a finite congruence for graphs of tree-width at most w, and bound its index. So for graphs of tree-width at most w, the general bound implies an upper bound on the number of edges in an obstruction. This bound, which is triply exponential in w^4, is our second main result.

1. Introduction

In this note, we investigate the relationships between two ways of characterizing a family of graphs. The first is by excluding substructures, and more precisely *minors*. The second is by a *finite congruence*. We introduce a number of operations that construct new graphs from given graphs. A congruence is defined as an equivalence relation between graphs such that the operations induced by the graph operations on the equivalence classes are well defined. A congruence *recognizes*, or is a congruence *for*, a family of graphs F if each of its classes either contains no graph from F or only graphs from F.

Each minor-closed family of graphs has by the graph minor theorem (GMT), [7], a finite set of obstructions. It follows from [4] that given the obstructions for a family F it is possible to construct a congruence for F. Moreover, given the obstructions an algorithm that *decides* the congruence (that is, decides whether

---

1991 *Mathematics Subject Classification.* 05C75.

Key words and phrases. Universal algebra, graphs, minors, congruence, obstruction, bounded tree-width.

The final version of this paper will be submitted for publication elsewhere

or not two given graphs are congruent) can be constructed. But, the proof of GMT is non-constructive. From a formalized proof that a family F is minor-closed one does not know how to construct the obstructions or bound the size of them. Of course, there are families F with arbitrarily large obstructions. So, a bound on the size of an obstruction must in some way depend on the family F, or some characterization of F.

In [5], it was shown, using the non-constructively proved GMT, that given a bound on the tree-width for a minor-closed family of graphs F and an algorithm that decides a congruence for F it is possible to find the obstructions for F. So, for families F of bounded tree-width there is an algorithm that produces the obstructions from a congruence. In [6], the same result was proved constructively without using the Graph Minor Theorem. In this note, we take advantage of the constructive techniques used there to obtain a general upper bound on the number of edges in an obstruction. This bound is $2^{O(c^{w+1}w!)}$ where w is the tree-width and c is the index of the congruence. Actually, the condition that F is of bounded tree-width can be relaxed. It is enough that there is a bound w on the tree-width of the *obstructions* of F. Also, in this case we obtain a bound $2^{O(c^{w+1}w!)}$ on the *size* of an obstruction, but here w is the bound on the tree-width of the obstructions.

The most interesting family of graphs to apply the above general method to is that of graphs of tree-width at most w. For $w = 1, 2, 3$, the obstructions have been known for some time, although found using other techniques, see [2]. In[6], the first construction of a congruence for graphs of tree-width at most w and the first algorithm to decide such a congruence appeared. Hence, the first algorithm to find the obstructions for graphs of tree-width at most w was obtained. Also, the results there together with results from [1] showed how to construct a linear time recognition algorithm for graphs of tree-width at most w.

In this note, we use almost exactly the same construction of a congruence for graphs of tree-width at most w as in [6]. But, we also obtain an upper bound on the size of the congruence. This and the general bound, mentioned earlier, implies an upper bound triply exponential in w^4 on the number of edges of an obstruction for graphs of tree-width at most w.

2. Preliminaries

We consider undirected simple graphs. We denote the vertices of a graph G by $V(G)$ and the edges by $E(G)$. A graph H is a *minor* of another graph G if H can be obtained from G by a sequence of deletions, contractions of edges, and deletion of isolated vertices. The graph H is a *proper* minor of G if H is a minor of G but not equal to G. A *minor minimal* graph in a set S of graphs is a graph G such that G is in S but no proper minor of G is. A family of graphs F is *minor closed* if $G \in F$ and H is isomorphic to a minor of G implies that $H \in F$. An *obstruction* for a minor closed family F of graphs is a graph G

such that G is not in F but every proper minor of G is (and hence G is a minor minimal graph in the complement of F). A *separation* of a graph G is a pair of graphs (G_1, G_2) such that $V(G_1) \cup V(G_2) = V(G)$, $E(G_1) \cup E(G_2) = E(G)$, and $E(G_1) \cap E(G_2) = \emptyset$. If G is a graph, $A \subseteq V(G)$, $B \subseteq V(G)$, $Z \subseteq V(G)$, and each path in G from a vertex in A to a vertex in B contains a vertex in Z then Z is said to be a A, B-*separator*.

A graph G has *tree-width at most w* if there is a pair (X, T) such that T is a tree and $X = \{X_t\}_{t \in V(T)}$ a family of subsets of $V(G)$, called *bags*, such that: those nodes t in T whose bags X_t contain a given vertex in G induce a subtree of T, every pair of adjacent vertices in G share membership of some bag X_t, every vertex in G is in some bag X_t, and $|X_t| \leq w + 1$ for all $t \in V(T)$. Such a pair (X, T) is called a *tree-decomposition of G of width at most w*. The family of partial k-trees is exactly the family of graphs of tree-width at most k (since we deal with simple graphs). We consider a path to be a connected graph where exactly two vertices, called endvertices, have degree 1 and the rest degree 2. We say that a number of paths $P_1, \ldots, P_k$ are totally vertex disjoint if P_i and P_j have no vertex in common unless $i = j$. By ε, we denote the empty tree, that is, the tree without vertices and edges. We consider ε to be a subtree of every tree. For every $n \in \mathbb{N}$, we denote by $[n]$ the set $\{1, \ldots, n\}$ (with $[0] = \emptyset$).

A concrete i-sourced graph is a graph G given together with an injection $S_G : [i] \to V(G)$. The image of S_G is called the set of *sources* of G. An i-sourced graph is an isomorphism class of a concrete i-sourced graph (an isomorphism of two concrete i-sourced graphs must take source j of one graph into source j of the other, for $1 \leq j \leq i$). The *underlying graph* of an i-sourced graph is the corresponding graph without sources. We identify the family of graphs with the family of 0-sourced graphs. By a *minor of an i-sourced graph* we will mean the obvious generalization of a minor of an ordinary graph with the restriction that contracting an edge between two sources is not allowed. Hence, a minor of an i-sourced graph is i-sourced. A tree-decomposition of an i-sourced graph is a tree-decomposition of its underlying graph as defined above.

Algebras of i-sourced graphs have been defined in [1, 3]. We will consider the following variant. The algebra M has sorts $\{g_0, g_1, \ldots\}$. The domain corresponding to sort g_i is the family $\mathbf{G}_i$ of i-sourced graphs. The signature contains two constant symbols 0 of sort g_0 and $e^{(2)}$ of sort g_2. Moreover, for each $i \in \mathbb{N}$, there are unary operation symbols $l^{(i,j)}$, for each $j \in [i]$, of profile $g_{i-1} \to g_i$ and $r^{(i)}$ of profile $g_i \to g_{i-1}$, and one binary operation symbol $/\!/^{(i)}$ of profile $g_i \times g_i \to g_i$. For each $i \in \mathbb{N}$, we denote this set of operation symbols D_i. We denote the set of all these operation symbols by L. Corresponding to the constant symbols there are two constants: 0_M the empty graph and $e_M^{(2)}$ the edge with two sources as end-vertices. Corresponding to the operation symbols $l^{(i,j)}$, $r^{(i)}$, and $/\!/^{(i)}$ there are operations on i-sourced graphs $l_M^{(i,j)}$, $r_M^{(i)}$, and $/\!/_M^{(i)}$, respectively. These operations are defined as follows:

: $/\!/_M^{(i)} : \mathbf{G}_i \times \mathbf{G}_i \to \mathbf{G}_i$; the parallel composition of two i-sourced graphs.

It is obtained by fusing corresponding sources of the two i-sourced graphs and deleting one edge in any pair of multiple edges that may appear (to make sure the resulting graph is simple). $/\!/_M^{(0)}$ is the special case of disjoint union of two graphs. $/\!/_M^{(i)}$ is associative and commutative.

: $l_M^{(i,j)}$: $\mathbf{G}_{i-1} \to \mathbf{G}_i$, for $1 \leq j \leq i$; the lifting of an $(i-1)$-sourced graph to an i-sourced graph by insertion of a new isolated vertex. For $l = j, \ldots, i-1$ the l:th source becomes the $(l+1)$:th source and the new isolated vertex becomes the j:th source.

: $r_M^{(i)}$: $\mathbf{G}_i \to \mathbf{G}_{i-1}$, $i \geq 1$; removes the i:th source from the source set of an i-sourced graph (but keeps the vertex as a vertex in the graph).

We denote by ν the unique homomorphism between the term algebra and M. That is, if f is an L expression, $\nu(f)$ is the graph denoted by f, or $\nu(f)$ is the *value* of f.

We denote by M_w the subalgebra of M consisting of the sorts $g_0, \ldots, g_{w+1}$, corresponding domains $\mathbf{G}_0, \ldots, \mathbf{G}_{w+1}$, operation symbols

$$L_w = \{0, e_2\} \cup \bigcup_0^{w+1} D_i,$$

and the correspondence between operation symbols and operations as above. For M_w, we denote the operation and the unique homomorphism between the term algebra and M as above. This is justified by the fact that M_w is a subalgebra of M.

By a L_w (L) *context of argument sort* g_i, we mean a well-sorted expression over the operation symbols in L_w (L) with at most one variable x of sort g_i occurring at most once. We write a context $f[x]$. We call the operation associated in the classical way with a context $f[x]$ its *derived operation* and denote it by $f_M(x)$. A *congruence* on M_w (M) is an equivalence relation $\approx$ over objects in M_w (M) such that

: (i) If $m \approx m'$, then for some i both m and m' belong to $\mathbf{G}_i$.
: (ii) If $m \approx m'$, where m and m' belongs to G_i then for every L_w (L) context $f[x]$ of argument sort g_i, $f_M(m) \approx f_M(m')$

The equivalence classes of this relation are called *congruence classes*. The *index* of the congruence is its number of congruence classes. A congruence is *finite* if for each i it has finitely many congruence classes containing elements from the domain $\mathbf{G}_i$. A family F of graphs is *recognizable* over M_w if it is the union of a number of congruence classes of a finite congruence on M_w.

3. The general bound

The theorem below follows from GMT.

THEOREM 3.1 (ROBERTSON, SEYMOUR [7]). Every minor-closed family of graphs has a finite number of obstructions.

The proof of GMT is non-constructive. Until now, no general methods to bound the number of obstructions or the size of them has been suggested. It is a result of this paper that the size of an obstruction can be bounded for certain restricted minor-closed families; that is, for minor-closed families of graphs which are recognizable and whose obstructions have bounded tree-width, and, in particular, for minor-closed families of graphs which are recognizable and of bounded tree-width. The most interesting and important family that our methods apply to is that of graphs of tree-width at most w. The proof of the general bound is for the case of recognizable minor-closed graph families whose obstructions have bounded tree-width an elementary and constructive proof of (3.1). For this reason we will not use (3.1) here.

We have omitted the proofs of the following two theorems. The first is easily proved, and the second is technical but seems to be well known.

THEOREM 3.2. *Let G be a graph, $A, B \subseteq V(G)$, (X, T) a tree-decomposition of width at most w of G with A a subset of one bag and B a subset of one bag. Let Z be a minimum A, B-separator and (G_1, G_2) a separation of G such that $V(G_1) \cap V(G_2) = Z$, $A \subseteq V(G_1)$, and $B \subseteq V(G_2)$. Then there is a tree-decomposition of width at most w of G_1 with A as one bag and Z as one bag.*

THEOREM 3.3. *Let G be an i-sourced graph with at least one vertex and $r \geq 1$. There is a tree-decomposition (X, T) of G of width at most w where $\{t_1, \ldots, t_r\} \subseteq V(T)$ and X_{t_1} is the set of sources of G iff there is an L_w expression f_1 such that $\nu(f_1) = G$, and f_1 has subexpressions $f_2, \ldots, f_r$ where $\nu(f_i)$ has source set X_{t_i}, moreover, f_i is a subexpression of f_j if and only if t_j lies on the unique t_1, t_i-path in T.*

THEOREM 3.4. *Let $f[f']$ be an L_w expression, A the source set of $\nu(f[f'])$, A' the source set of $\nu(f')$, and Z a minimum A, A'-separator in $\nu(f[f'])$. Then there are two L_w contexts $f_1[x]$ and $f_2[x]$ such that $\nu(f_1[f_2[f']]) = \nu(f[f'])$ and $\nu(f_2[f'])$ has source set Z.*

Proof. By (3.3), there is a tree-decomposition (X, T) of $\nu(f[f'])$ with A and A' as two bags. Since Z is an A, A'-separator in $\nu(f[f'])$, there is a separation (G_1, G_2) of $\nu(f[f'])$ such that $V(G_1) \cap V(G_2) = Z$, $A \subseteq V(G_1)$, and $A' \subseteq V(G_2)$. By (3.2), there is a tree-decomposition (X^1, T^1) of G_1 with A and Z as two bags. Also by (3.2), there is a tree-decomposition (X^2, T^2) of G_2 with A' and Z as two bags. Choose a vertex t from T^1 such that $X_t = Z$ and a vertex t' from T^2 such that $X_{t'} = Z$. Let T be the tree obtained taking the disjoint union of T^1 and T^2 and then making t adjacent to t'. Let (X, T) be the tree-decomposition defined by $X_t = X_t^1$ for all $t \in V(T^1)$, and $X_t = X_t^2$ for all $t \in V(T^2)$. Obviously, (X, T) is a tree-decomposition of $\nu(f'[f])$ where T has three vertices s, s', s'' such that $X_s = A$, $X_{s'} = Z$, and $X_{s''} = A'$ in that order on a path. Hence, by (3.3), there exist three L_w contexts $h_1[x]$, $h_2[x]$, and h_3 such that: $\nu(h_2[h_3])$ has source set $X_{s'} = Z$, $\nu(h_3)$ has source set $X_{s''} = A'$, and $\nu(h_1[h_2[h_3]]) = \nu(f[f'])$. Hence the theorem is satisfied if we let $f_1[x]$ be $h_1[x]$ and $f_2[x]$ be $h_2[x/\!/^{(i)}h_3]$ where $i = |A'|$. □

LEMMA 3.5. Let $a_1, \ldots, a_w$ and a be positive numbers. Let L be a sequence of length $\sum_{i=1}^{w} a^i \prod_{j=1}^{i} a_j$ where each element is a set of cardinality at least 1 and at most w. Then for some k, $1 \leq k \leq w$, there is a subsequence L' of consecutive elements of L that contains at least aa_k sets of cardinality k and no set of cardinality less than k.

Proof. Assume the opposite. Consider the following property:

: (P) L' is a subsequence of consecutive elements of L of length

$$\sum_{i=k}^{w} a^{i+1-k} \prod_{j=k}^{i} a_j$$

and all its elements are sets of cardinality at least k.

Let L' be a sequence that satisfies (P) with $k = l$ where l is the greatest number such that for $k = l$ there is a sequence with property (P). For $k = 1$, L is a sequence with property (P), and hence $l \geq 1$. Assume $l = w$. Since $\sum_{i=w}^{w} a^{i+1-w} \prod_{j=w}^{i} a_j = aa_w$, L' satisfies the theorem for $k = w$. Hence, we have reached a contradiction

Assume $l < w$. By the original assumption that the theorem is false, there are fewer than aa_l sets of cardinality l in L'. Hence, there are at most aa_l different maximal subsequences of consecutive elements of L' without a set of cardinality l. By a standard average argument, at least one such subsequence contains at least

$$\frac{\sum_{i=l}^{w} a^{i+1-l} \prod_{j=l}^{i} a_j - aa_l}{aa_l} = \frac{\sum_{i=l+1}^{w} a^{i+1-l} \prod_{j=l}^{i} a_j}{aa_l} = \sum_{i=l+1}^{w} a^{i+1-(l+1)} \prod_{j=l+1}^{i} a_j$$

sets, all of cardinality at least $l + 1$. But this contradicts the choice of l. That is, we know that $1 \leq l \leq w$, and both the assumptions $l = w$ and $l < w$ leads to contradictions. Hence, we have a contradiction to our original assumption that the theorem is false. □

We say that a nesting of contexts $f_1[f_2[\ldots[f_r]\ldots]]$ is *without repetition* if

$$\nu(f_i[f_{i+1}[\ldots[f_r]\ldots]]) = \nu(f_j[f_{j+1}[\ldots[f_r]\ldots]])$$

implies $i = j$. Aided by the above lemmas, we can state and prove the following technical result from which the general bound follows.

THEOREM 3.6. Let $\sim$ be a congruence on M_w with index c. If G is a minor minimal graph in a congruence class and generated by L_w then G has fewer than 2^r edges where $r = (c+1)\sum_{i=1}^{w} c^i \prod_{j=1}^{i} (j! + 1)$.

Proof. Assume the opposite, i.e., G has at least 2^r edges. The maximal arity of any operation in M_w is 2. Furthermore, edges can only be introduced by constants and constant symbols are leaves in an L_w expression. If we, moreover, use the facts that G is generated by L_w and that any nesting of contexts can be reduced to one without repetition, we get that there is a context $f[x]$ and a nesting of contexts $f_1[\ldots[f_r]\ldots]$ without repetition such that $f[f_1[\ldots[f_r]\ldots]]$ has G as value. Actually, we can choose this nesting of contexts such that for each $l \in [r]$, $f_l[x]$ is either $l^{(i,j)}(x)$, $r^{(i)}(x)$, or, for some t, $x/\!/^{(i)}t$. We claim that at least a fraction $(c+1)^{-1}$ of these contexts have sort $g_0, \ldots, g_{w-1}$, or g_w. Assume the opposite. Then there is an l such that $f_l[\ldots], \ldots, f_{l+c}[\ldots]$ all have sort g_{w+1}. That implies that for each $m \in \{0, \ldots, c\}$, there is some t such that $f_{l+c}[x]$ is $x/\!/^{(i)}t$. That is, $\nu(f_l[\ldots]), \ldots, \nu(f_{l+c}[\ldots])$ all have the same source set. Since $\sim$ has index c, there are also $m, m' \in \{0, \ldots, c\}$ such that $\nu(f_{l+m}[\ldots]) \sim \nu(f_{l+m'}[\ldots])$ and $m < m'$. We can draw two conclusions. First, $\nu(f[f_1[\ldots f_{l+m-1}[f_{l+m'}[\ldots]]\ldots]])$ is a proper minor of G, since our original nesting was without repetition, and $\nu(f_{l+m}[\ldots])$ and $\nu(f_{l+m'}[\ldots])$ have the same set of sources. Second, $\nu(f[f_1[\ldots f_{l+m-1}[f_{l+m'}[\ldots]]\ldots]])$ and G belong to the same congruence class, since $\nu(f_{l+m}[\ldots])$ and $\nu(f_{l+m'}[\ldots])$ do. But this contradicts the given fact that G is minor minimal in its congruence class. We conclude that at least a fraction $(c+1)^{-1}$ of the contexts $f_1[x], \ldots, f_r[x]$ have sort $g_0, \ldots, g_{w-1}$, or g_w.

These $\sum_{i=1}^{w} c^i \prod_{j=1}^{i}(j!+1)$ context of sort $g_0, \ldots, g_{w-1}$, and g_w among

$$f_1[x], \ldots, f_r[x]$$

are important to us for the following reason. They, clearly, imply the existence of a context h and a nesting of contexts $h_1[\ldots[h_s]\ldots]$ such that: the value of $h[h_1[\ldots]]$ is G, $s \geq \sum_{i=1}^{w} c^i \prod_{j=1}^{i}(j!+1)$, and if A_i is the source set of $\nu(h_i[\ldots])$ then $1 \leq |A_i| \leq w$. We can without loss of generality assume that $h_1[x], \ldots, h_s[x]$ minimizes $\sum_{i=1}^{s} |A_i|$ w.r.t. the above conditions.

According to (3.5) (with $a = c$ and $a_k = (k!+1)$), there is a subsequence $A_i, \ldots, A_m$ of $A_1, \ldots, A_s$ that contains at least $c(k!+1)$ sets of cardinality k and no set of cardinality less than k. We can without loss of generality assume that $|A_i| = |A_m| = k$. We want to prove that there are at least k totally vertex disjoint A_i, A_m-paths in G. Assume the opposite. Let j be the least number such that there are k totally vertex disjoint A_j, A_m-paths and let $P_1, \ldots, P_k$ be such paths satisfying $|V(P_l) \cap A_j| = 1$. Let $x_l \in V(P_l) \cap A_j$ and $X = \{x_1, \ldots, x_k\}$. Let Z_1 be a minimum A_{j-1}, X-separator in $H = \nu(h_{j-1}[\ldots]) - (\nu(h_j[\ldots]) - A_j)$. We see that $|Z_1| < k = |X|$. Because by Menger's Theorem, if $|Z_1| \geq k$ then there would be k totally vertex disjoint A_{j-1}, X-paths in H, and hence k totally vertex disjoint A_{j-1}, A_m-paths in G which would contradict the choice of j.

Let $Z = Z_1 \cup (A_j - X)$. That $|Z_1| < |X|$ and $X \subseteq A_j$ imply $|Z| < |A_j|$. Furthermore, Z is an A_{j-1}, A_{j+1}-separator in G. Because, otherwise there is an A_{j-1}, A_{j+1}-path P in $G - Z$. The path P must contain a vertex from A_j, since

A_j is an A_{j-1}, A_{j+1}-separator in G. Let v be a vertex on P that belongs to A_j, as well. Since P is a path in $G - Z$ we have $v \notin Z$, that is, $v \notin A_j - X$. We conclude that $v \in X$. But then there is, obviously, a subpath of P which is an A_{j-1}, X-path in $H - Z_1$, since $V(P) \cap Z = \emptyset$ and $Z_1 \subseteq Z$. This contradict the choice of Z_1 as a A_{j-1}, X-separator in H. We conclude that Z is an A_{j-1}, A_{j+1}-separator in G.

Let Z' be a minimum A_{j-1}, A_{j+1}-separator in G. Then by (3.4), we can find two L_w contexts $g_1[x]$ and $g_2[x]$ such that

$$\nu(g_1[g_2[h_{j+1}[\ldots[h_s]\ldots]]]) = \nu(h_{j-1}[\ldots[h_s]\ldots])$$

and $\nu(g_2[h_{j+1}[\ldots[h_s]\ldots]])$ has source set Z'. Hence,

$$\nu(h[h_1[\ldots h_{j-2}[g_1[g_2[h_{j+1}[\ldots[h_s]\ldots]]]]]\ldots]]) = G.$$

This contradict the choice of $h_1[x], \ldots, h_s[x]$, since we get

$$|Z'| + \sum_{\substack{1 \leq i \leq s \\ i \neq j}} |A_i| < \sum_{1 \leq i \leq s} |A_i|$$

from $|Z'| \leq |Z| < |A_j|$ and $h_1[x], \ldots, h_s[x]$ were chosen to minimizes this sum. We conclude that there are k totally vertex disjoint A_i, A_m-paths $Q_1, \ldots, Q_k$ in G. By a standard average argument there is a congruence class C such that there are $(k! + 1)$ numbers j with the following properties: $i \leq j \leq m$, A_j has cardinality k, and $\nu(h_j[\ldots])$ belongs to C. Since the order of the permutation group on k elements is $k!$, there are two such integers j and j' such that: $i \leq j < j' \leq m$, and Q_l contains the l:th source of $\nu(h_j[\ldots])$ and the l:th source of $\nu(h_{j'}[\ldots])$. Let $H' = \nu(h_j[\ldots]) - (\nu(h_{j'}[\ldots]) - A_{j'})$. If we delete all edges in $E(H) - \cup_{i=1}^{k} E(Q_i)$ from G and contract all the edges in $\cup_{i=1}^{k} E(Q_i)$, then we obtain $\nu(h[h_1[\ldots h_{j-1}[h_{j'}[\ldots]]]\ldots]])$. Since $h[h_1[\ldots]]$ is without repetition this is a proper minor of G. Also, $\nu(h_j[\ldots])$ and $\nu(h'_j[\ldots])$ are congruent, so $\nu(h[h_1[\ldots h_{j-1}[h_{j'}[\ldots]]]\ldots]])$ is congruent to G. That is, $\nu(h[h_1[\ldots h_{j-1}[h_{j'}[\ldots]]]\ldots]])$ is both congruent to G and a minor of G. We have once again reached a contradiction, but this time to our original assumption that the theorem is false. This finishes the proof. □

The above theorem implies the general bound, our first main result, below.

THEOREM 3.7. *Let F be a minor-closed family of graphs recognized by a congruence $\sim$ on M_w with index c whose obstructions are of tree-width at most w. If G is an obstruction for F then G has fewer than 2^r edges where $r = (c+1) \sum_{i=1}^{w} c^i \prod_{j=1}^{i} (j! + 1)$.*

Proof. The obstructions of F have tree-width at most w. Hence, they are generated by L_w. Also, each obstruction for F must, naturally, be a minor minimal graph in the congruence class that it belongs to. Hence, the general bound above follows from (3.6). □

In particular, if the family F has tree-width at most w then the obstructions have tree-width at most $w+1$. Hence we can apply the above theorem.

4. Graphs of tree-width at most w

In this section, we will construct a finite congruence on M_k for graphs of tree-width at most w. Also, a bound doubly exponential in $\mathcal{O}(k^4)$ on its index will be obtained. The bound on the index of the congruence together with the general bound, from the previous section, are the most important ingredients in the proof of our second main result below.

THEOREM 4.8. *If G is an obstruction for graphs of tree-width at most w then $|E(G)|$ is at most triply exponential in $\mathcal{O}(w^4)$*

Proof. According to (4.14), yet to be stated and proved, there is a congruence on M_{w+1} of size doubly exponential in $\mathcal{O}(w^4)$ for graphs of tree-width at most w. Also, as easily proved, the obstructions for tree-width at most w have tree-width at most $w+1$. Thus the general bound, (3.7), implies that each obstruction has at most a number of edges that is triply exponential in $\mathcal{O}(w^4)$. □

4.1. Encodings of tree-decompositions of i-sourced graphs.

In this subsection, we introduce encodings of tree-decompositions of i-sourced graphs. These encodings will be used to define the congruence for graphs of tree-width at most w. The most important property of this congruence is that given two i-sourced graphs G_1 and G_2 it is possible to determine from the congruence classes of G_1 and G_2 whether or not $G_1 /\!/_M^{(i)} G_2$ has tree-width at most w. Of course, similar statements hold for $r_M^{(i)}$ and $l_M^{(i+1,j)}$, as well, but $/\!/_M^{(i)}$ is more interesting to us. The reason for this is that $r_M^{(i)}(G)$ and $l_M^{(i+1,j)}(G)$ always has the same tree-width as G, while the tree-width of $G_1 /\!/_M^{(i)} G_2$ is not determined by the tree-width of G_1 and the tree-width of G_2.

By definition, if the i-sourced graph $G_1 /\!/_M^{(i)} G_2$ has tree-width at most w, there is a tree-decomposition (X, T) of $G_1 /\!/_M^{(i)} G_2$. The tree-decomposition (X, T) can be made into a tree-decomposition (X^1, T) of G_1 by letting X^1 be the family

$$\{X_t^1 = X_t \cap V(G_1)|\ X_t \text{ belongs to } X\}.$$

In the same way we can find a tree-decomposition (X^2, T) of G_2 by letting X^2 be the family

$$\{X_t^2 = X_t \cap V(G_2)|X_t \text{ belongs to } X\}.$$

These two tree-decompositions (X^1, T) and (X^2, T) can obviously be fitted together to form a tree-decomposition of $G_1 /\!/_M^{(i)} G_2$ by taking

$$\{X_t = X_t^1 \cup X_t^2 \mid X_t^1 \text{ belongs to } X^1 \text{ and } X_t^2 \text{ to } X^2\}$$

as the family of bags and T as the tree. Of course, the tree-decomposition we get is just (X, T). The important observation here is, however, that if $G_1 /\!/_M^{(i)} G_2$ has tree-width at most w, then we can always find a tree-decomposition of G_1 and one of G_2 that can be fitted together to form a tree-decomposition of $G_1 /\!/_M^{(i)} G_2$. That is, the set of all tree-decompositions of width at most w of an i-sourced graph G_1 and the set of all tree-decompositions of width at most w of an i-sourced graph G_2 completely determine whether $G_1 /\!/_M^{(i)} G_2$ has tree-width at most w. This makes the set of all tree-decompositions of width at most w of an i-sourced graph G look like a good candidate to determine the congruence class of G. However, if we do not refine this approach, it gives us a congruence with infinite index. If there existed only finitely many tree-decompositions of i-sourced graphs this definition would suffice, but since there actually are infinitely many it does not.

Our refinement can intuitively be thought of as using finitely many encodings of tree-decompositions instead of tree-decompositions. Three factors cause the existence of infinitely many tree-decompositions. First, there are infinitely many possible bags. We will avoid this by replacing each bag by a number and a set; the *number of non-sources* in the bag and the set of all *source numbers* of the sources in the bag. Since the sources are the only vertices that can appear in the two graphs that we are taking the parallel composition of, this is enough information to determine whether or not two tree-decompositions can be fitted together. We call the tree-decomposition-like structure obtained by replacing bags in this way an *i-profile*.

For any integer i we define an *i-profile* to be a triple (Z, Y, T) where T is a tree, Z is a family of numbers, and Y is a family of subsets of $[i]$ where both Z and Y are indexed by the vertices of T. If T' is a subtree of T then the profile

$$(\{z_t \mid t \in V(T')\}, \{Y_t \mid t \in V(T')\}, T')$$

is called the profile *induced* by T' and denoted $(Z, Y, T)[T']$. Of course, for any i-profile, the i-profile induced by the ε is the i-profile $(\emptyset, \emptyset, \varepsilon)$. We will frequently refer to vertices of T as vertices of (Z, Y, T). The *width* of a profile (Z, Y, T) is the maximum of $z_t + |Y_t| - 1$ where t is any vertex in T. If (X, T) is a tree-decomposition of an i-sourced graph G then the G profile of (X, T) is the i-profile (Z, Y, T) where the members of Z and Y are defined by

$$z_t = |X_t - S_G([i])| \text{ and } Y_t = \{j \mid S_G(j) \in X_t\}$$

respectively. That is, z_t is the number of non-sources in X_t, and Y_t is the set of source numbers of sources of G belonging to X_t. We call Y_t the *source bag* for t. The *set of sources* of an i-profile (Z, Y, T) is defined to be $\cup_{t \in V(T)} Y_t$. This is a slight abuse of terminology, since the set of sources of an i-sourced graph G is $S_G([i])$, not $[i]$. Clearly, by using i-profiles instead of tree-decompositions we have achieved what we wanted; there is a bounded number of possible z_t and Y_t when t varies over all vertices in all i-profiles.

Second, a tree in a tree-decomposition can have an unbounded number of leaves. But to obtain a tree-decomposition of $G_1 /\!\!/_M^{(i)} G_2$, it is enough to fit together a subtree of the tree T^1 of a tree-decomposition of G_1 with a subtree of the tree T^2 of a tree-decomposition of G_2. As long as for each source number j, there is some node t_1 in T^1 such that X_{t_1} contains the source with number j of G_1 that is merged with some node t_2 in T^2 such that X_{t_2} contains the source with number j in G_2, we will obtain a tree-decomposition of $G_1 /\!\!/_M^{(i)} G_2$. In particular, if X_{t_1} contains all sources of G_1 and X_{t_2} all sources of G_2, it is enough to fit together t_1 and t_2. That is, to identify t_1 and t_2, call the vertex obtained t, and set $z_t = z_{t_1} + z_{t_2}$ and $Y_t = Y_{t_1}$ (hence $Y_t = Y_{t_1} \cup Y_{t_2}$).

This leads us to the following definitions. A *kernel* of an i-profile (Z, Y, T) is a minimal subtree K of T such that: for each j in $[i]$ there is a vertex t in K such that $j \in Y_t$. If (X, T) is a tree-decomposition of an i-sourced graph G, (Z, Y, T) the G profile of (X, T), and K a kernel of (Z, Y, T), then we also say that K is a G kernel of (X, T).

We will call an i-profile (Z, Y, T) *singular* if there is a vertex t such that $[i] \subseteq Y_t$. Notice that unless (Z, Y, T) is singular, there is a unique kernel K of (Z, Y, T). Let us prove this. Observe, since (Z, Y, T) is the i-profile of a tree-decomposition, the set of vertices in T whose source bags contain a given source $j \in [i]$ induce a subtree of T. Assume that there are two kernels K and K' of (Z, Y, T). If K and K' have a non-empty common subtree K'' then clearly K'' is a kernel, as well. Since K and K' are minimal, we get $K = K'' = K'$. If K and K' do not intersect, let P be a path with one endvertex in K and the other in K' and no internal vertex in K or in K'. By the above observation, for each vertex t on P, $[i] \subseteq Y_t$, a contradiction. This concludes the proof.

In a similar way, one can prove that if there is a kernel which contains exactly one vertex then all kernels contains exactly one vertex. We conclude that an i-profile either has a unique kernel or each of its kernels contains exactly one vertex.

DEFINITION 4.9. Let (X, T) be a tree-decomposition of an i-sourced graph G, (Z, Y, T) the G profile of (X, T), and K a kernel of (Z, Y, T). Then the i-profile $(Z, Y, T)[K]$ is an *encoding* of (X, T) with respect to G.

We also say that $(Z, Y, T)[K]$ is the encoding of (X, T) with respect to G and K. Notice, if (X, T) is a tree-decomposition of width at most w and E is an encoding of (X, T) then the width of E is at most w. Whenever we say that E is an encoding of (X, T) with respect to G, we actually mean also that (X, T) is a tree-decomposition of G of width at most w. Also, when we say that E is an encoding of (X, T) with respect to a kernel K and G, we actually mean also that: (X,T) is a tree-decomposition of G of width at most w and K is a G kernel of (X,T). We say that (Z,Y,K) is an i-sourced encoding if (Z,Y,K) is an encoding of a tree-decomposition (X,T), of an i-sourced graph G, with respect to G. As easily seen, this is equivalent to (Z,Y,K) has source set $[i]$ and width at most w, and the set of vertices in K whose source bags contain a given source $j \in [i]$ induce a subtree of K. Clearly, if l is a leaf in an i-sourced encoding E then there is some $j \in [i]$ in the source bag of l that is not in any other source bag in E. Hence, there are at most i leaves in an i-sourced encoding.

We call the i-sourced encoding (Z,Y,T) where T has one single vertex, say, t, $z_t = 0$, and $Y_t = [i]$ *the minimum singular i-sourced encoding*. We call $(\emptyset, \emptyset, \varepsilon)$ the *empty encoding*; it is the only 0-sourced encoding.

Third, a path in the tree of a tree-decomposition can have unbounded length. To solve this problem, we introduce an equivalence relation, $\sim_D$, *between encodings* (this is not our congruence). That two encodings are equivalent under $\sim_D$ should intuitively be interpreted as that they are encodings of two tree-decompositions that are equally hard to fit together with any third tree-decomposition. To be able to define $\sim_D$ we first introduce subdivisions of encodings and a quasi-order $\leq_D$. That $E \leq_D F$ should intuitively be interpreted as that E is an encoding of a tree-decomposition that can be fitted together with any tree-decomposition that can be fitted together with the tree-decomposition that F is an encoding of. The quasi-order $\leq_D$ and the equivalence relation $\sim_D$ are defined below. In (4.3), we actually solve this third problem by showing that: In each equivalence class of $\sim_D$ there is at least one "non-redundant encoding"; and that a path in a tree of a non-redundant encoding has bounded length. This gives us a bound on the number of i-sourced non-redundant encodings and the number of equivalence classes of $\sim_D$.

DEFINITION 4.10. Let (Z,Y,K) be an encoding, e an edge in K, and t a non-leaf vertex incident to e. Then (Z',Y',K') is said to be a *subdivision* of (Z,Y,K), with respect to e and t if the following conditions are satisfied:
- the tree K' is obtained by inserting a new vertex t' on the edge e
- $z'_s = z_s$ and $Y'_s = Y_s$ for all vertices $s \in V(K)$
- $z'_{t'} = z_t$ and $Y'_{t'} = Y_t$

The reason for not allowing t to be a leaf is that we want a subdivision of an encoding to be an encoding. If there is a sequence of encodings $E_1, \ldots, E_{r+1}$ such that E_{i+1} is a subdivision of E_i then we also say that E_{r+1} is a subdivision of E_1.

An i-sourced encoding (Z,Y,K) is said to be *directly dominated* ($\leq_{DD}$) by another i-sourced encoding (Z',Y',K') if there is an isomorphism f from K to K' such that for each vertex t in K, $Y_t \subseteq Y'_{f(t)}$ and $z_t \leq z'_{f(t)}$. The isomorphism f is said to be an *isomorphism of the direct dominance* $(Z,Y,K) \leq_{DD} (Z',Y',K')$. We say that f *respects sources* if $Y_t = Y'_{f(t)}$ for all vertices t in K. An i-sourced encoding E is said to be *dominated* ($\leq_D$) by another i-sourced encoding

F if there is a subdivision E' of E and a subdivision F' of F such that E' is directly dominated by F'. Two encodings that directly dominate each other are considered to be equal. We consider the empty encoding to directly dominate the empty encoding.

Let $E_1, \ldots, E_r$ be a sequence of encodings where E_{i+1} is a subdivision of E_i with respect to (t_i, t'_i) and t_i. Let F_1 be an encoding such that f_1 is an isomorphism from the tree of E_1 to the tree of F_1. Then, the subdivision of F_1 with respect to $(f_1(t_1), f_1(t'_1))$ and $f_1(t_1)$, call it F_2, has a tree isomorphic to the tree of E_2. Moreover, if $E_1 \geq_{DD} F_1$ ($E_1 \leq_{DD} F_1$) and f_1 is an isomorphism of this direct dominance then $E_2 \geq_{DD} F_2$ ($E_2 \leq_{DD} F_2$).

We can, naturally, continue this process recursively and, for each $1 \leq i \leq r-1$, let F_{i+1} be the subdivision of F_i with respect to $(f_i(t_i), f_i(t'_i))$ and $f_i(t_i)$ where f_i is an isomorphism from the tree of E_i to the tree of F_i. The encoding F_r is said to be a subdivision of F_1 according to the E_1 to E_r scheme (the isomorphism f_1 is assumed to be clear from the context). Notice, the encoding E_r has, obviously, a tree isomorphic to the tree of F_r. Moreover, if $E_1 \geq_{DD} F_1$ ($E_1 \leq_{DD} F_1$) and we in each step choose f_i to be an isomorphism of the direct dominance $E_i \geq_{DD} F_i$ ($E_i \leq_{DD} F_i$), then $E_r \geq_{DD} F_r$ ($E_r \leq_{DD} F_r$). Similar statements hold if the tree of F_1 is isomorphic to a subtree of the tree of E_1, or if the tree of E_1 is isomorphic to a subtree of the tree of F_1.

As easily seen, $\leq_D$ is reflexive. Actually, $\leq_D$ is transitive as well and, hence, a quasi-order. We will not prove this. But it follows, basically, from the transitivity of $\leq_{DD}$ and the fact that two different subdivisions of an encoding always have a common subdivision.

THEOREM 4.11. $\leq_D$ is a quasi-order.

Given $\leq_D$ we can define, as always when given a quasi-order, an equivalence relation $\sim_D$; $E \sim_D F$ if and only if $E \leq_D F$ and $F \leq_D E$. (Remember, this is not the congruence relation that we are aiming to define.) Define $C(G)$, the *set of encodings*, of an i-sourced graph G to be the set of all encodings of all tree-decompositions of width at most w of G, i.e.

$$C(G) = \{E | E \text{ is an encoding of } (X, T) \text{ w.r.t. } G\}.$$

The following lemma is easily proved. We have omitted the proof.

LEMMA 4.12. $C(G) = \{E' | E' \text{ is a subdivision of some } E \in C(G)\}$

For any set C of encodings we define $\overline{C}$ to be the closure of C upwards w.r.t. $\leq_D$, that is,

$$\overline{C} = \{E' | E' \text{ is an encoding of width at most } w \text{ and } E' \geq_D E \text{ for some } E \in C\}.$$

4.2. The congruence relation $\sim$.
We are ready to define our congruence relation, which we denote by $\sim$.

DEFINITION 4.13. $G_1 \sim G_2$ iff $\overline{C(G_1)} = \overline{C(G_2)}$.

We will spend this and the next subsection proving the following theorem.

THEOREM 4.14. For every $r \in \mathbb{N}$ the relation $\sim$ is a finite congruence on M_r for graphs of tree-width at most w. Moreover if $r \geq w$, the index of $\sim$ is at most doubly exponential in $\mathcal{O}(r^4)$.

In the next subsection, (4.3), the index of $\sim$ will be dealt with, see (4.23). The rest of this subsection is used to prove that $\sim$ actually is a congruence, that is, (4.15) below. The theorem above, (4.14), follows immediately from (4.15) and (4.23).

LEMMA 4.15. For every $r \in \mathbb{N}$ the relation $\sim$ is a congruence relation on M_r for graphs of tree-width at most w.

It is trivial to see that a graph has tree-width at most w iff $\overline{C(G)}$ is non-empty. Hence, if $\sim$ is a congruence then it is a congruence on M_r for graphs of tree-width at most w. To prove that $\sim$, actually, is a congruence we need to show that it is stable with respect to the operations $/\!/_M^{(i)}$, $l_M^{(i,j)}$, and $r_M^{(i)}$. That is, that the congruence class of the result of an operation is uniquely determined by the congruence class of the argument (or the arguments) to the operation. To show that this is true we shall define three operations on sets of encodings: $\oplus^{(i)}$, $\lambda^{(i,j)}$, and $\rho^{(i)}$. The operations are defined so that each of them will correspond to exactly one of the graph operations: $\oplus^{(i)}$ corresponds to $/\!/_M^{(i)}$, $\lambda^{(i,j)}$ to $l_M^{(i,j)}$, and $\rho^{(i)}$ to $r_M^{(i)}$. Of course, $\oplus^{(i)}$ is a precise formulation of the vaguely defined operation previously referred to as fitting together.

Then a number of lemmas (4.16)-(4.22) are stated, which together imply that $\sim$ is a congruence on M_r, that is, (4.15). Let us, for example, prove using these lemmas that the congruence class of $G_1 /\!/_M^{(i)} G_1'$ is uniquely determined by the congruence classes of G_1 and G_1'. The important lemmas about $/\!/_M^{(i)}$ and $\oplus^{(i)}$ are (4.16) and (4.20). They say that $\overline{C(G/\!/_M^{(i)}G')} = \overline{C(G)\oplus^{(i)}C(G')}$ and $\overline{C(G)\oplus^{(i)}C(G')} = \overline{\overline{C(G)}\oplus^{(i)}\overline{C(G')}}$ for any two i-sourced graphs G and G'. Now, assume that $G_1 \sim G_1'$ and $G_2 \sim G_2'$, that is, $\overline{C(G_1)} = \overline{C(G_1')}$ and $\overline{C(G_2)} = \overline{C(G_2')}$. Then

$$\overline{C(G_1/\!/_M^{(i)}G_2)} = \overline{C(G_1)\oplus^{(i)}C(G_2)} = \overline{\overline{C(G_1)}\oplus^{(i)}\overline{C(G_2)}} =$$
$$= \overline{\overline{C(G_1')}\oplus^{(i)}\overline{C(G_2')}} = \overline{C(G_1')\oplus^{(i)}C(G_2')} = \overline{C(G_1'/\!/_M^{(i)}G_2')}.$$

Hence $C(G_1/\!/_M^{(i)}G_2) \sim C(G_1'/\!/_M^{(i)}G_2')$, that is, $\sim$ is stable with respect to $/\!/_M^{(i)}$. The proofs that $\sim$ is stable with respect to $l_M^{(i,j)}$ and $r_M^{(i)}$ are analogous.

We start by defining the operations $\oplus^{(i)}$, $\lambda^{(i,j)}$, and $\rho^{(i)}$ as operations on encodings, but we will use them as operations on sets of encodings, as well; the value of the operation applied to a set C is defined to be the union of the values of the operation applied to the members of C.

Let (Z, Y, K) and (Z', Y', K') be two i-sourced encodings such that there is an isomorphism f from K to K'. Then (Z, Y, K) and (Z', Y', K') are said to be *compatible because of f* if for each $j \in [i]$ there is a vertex t in K such that $j \in Y_t \cap Y'_{f(t)}$. If (Z, Y, K) and (Z', Y', K') are compatible because of f then $+(f, (Z, Y, K), (Z', Y', K'))$ is defined to be the i-profile (Z'', Y'', K) where $z''_t = z_t + z'_{f(t)}$ and $Y''_t = Y_t \cup Y'_{f(t)}$. For any pair of i-sourced encodings (Z, Y, K) and (Z', Y', K'), we define $(Z, Y, K) \oplus^{(i)} (Z', Y', K')$ as follows: $E \in (Z, Y, K) \oplus^{(i)} (Z', Y', K')$ if and only if either (i) E, (Z, Y, K), and (Z', Y', K') all are the empty encoding or (ii) (Z, Y, K) and (Z', Y', K') are compatible because of an isomorphism f, $+(f, (Z, Y, K), (Z', Y', K'))$ has width $\leq w$ and $E = +(f, (Z, Y, K), (Z', Y', K'))[K'']$ where K'' is a kernel of

$$+(f, (Z, Y, K), (Z', Y', K')).$$

Let (Z, Y, T) be an i-profile. Then (Z', Y', T) is said to be (Z, Y, T) with i *forgotten* if for all vertices t in T: $z'_t = z_t + |Y_t \cap \{i\}|$ and $Y'_t = Y_t - \{i\}$. Let F be an i-sourced encoding; then $E \in \rho^{(i)}(F)$ if and only if F' is F with i forgotten and either F' is singular and E is the minimum singular $(i-1)$-sourced encoding or $E = F'[K]$ where K is a kernel of F'.

Let F be an $(i-1)$-sourced encoding. Then (Z, Y, T) is said to be F *with sources renamed from j* if (Z, Y, T) can be obtained from F by increasing the source number by one for all the sources $j, \ldots, i-1$, that is, source r becomes source $r+1$. Let F be an $(i-1)$-sourced encoding. We define $\lambda^{(i,j)}$ by considering two cases. If $i = 1$, and hence also $j = 1$, then $E \in \lambda^{(i,j)}((Z, Y, K))$ if and only if (Z, Y, K) is the empty encoding and E is the minimum singular 1-sourced encoding. If $i > 1$, then $E \in \lambda^{(i,j)}((Z, Y, K))$ iff E can be obtained from (Z, Y, K) by renaming the sources from j and then either

- setting $Y_t = Y_t \cup \{j\}$ for some vertex t in T; or
- adding two new vertices t' and t'' to T; making t' adjacent to some vertex t in T; making t'' adjacent to t'; and setting $z_{t'} = z_{t''} = 0$, $Y_{t'} = \emptyset$, and $Y_{t''} = \{j\}$

LEMMA 4.16. *If G_1 and G_2 are two i-sourced graphs then*

$$\overline{C(G_1 /\!/_M^{(i)} G_2)} = \overline{C(G_1) \oplus^{(i)} C(G_2)}.$$

Proof. Assume that $(Z, Y, K) \in C(G_1 /\!/_M^{(i)} G_2)$. Let (Z, Y, K) be an encoding of (X, T) with respect to $G_1 /\!/_M^{(i)} G_2$. Then (X^i, T) where $X_t^i = X_t \cap V(G_i)$ is a tree-decomposition of G_i with K as a G_i kernel. Let (Z^i, Y^i, K) be the encoding of (X^i, T) with respect to K and G_i, and f be the identity mapping on $V(K)$. Then K is the kernel of $+(f, (Z^1, Y^1, K), (Z^2, Y^2, K))$ and

$$(Z, Y, K) = +(f, (Z^1, Y^1, K), (Z^2, Y^2, K))[K] \in (Z^1, Y^1, K) \oplus^{(i)} (Z^2, Y^2, K),$$

i.e.

$$(Z, Y, K) \in C(G_1) \oplus^{(i)} C(G_2).$$

Assume that $(Z, Y, K) \in C(G_1) \oplus^{(i)} C(G_2)$. Then there is a tree-decomposition (X^i, T^i) of G_i with (Z^i, Y^i, K^i) as an encoding with respect to G such that

$$(Z, Y, K) = +(f, (Z^1, Y^1, K^1), (Z^2, Y^2, K^2))[K].$$

We can without loss of generality assume that sources with corresponding source numbers in G_1 and G_2 are identical, that $V(T^1) \cap V(T^2) = V(K)$, and that f is the identity mapping on $V(K)$. Let $T = T^1 \cup T^2$ and let X be defined by

$$X_t = \begin{cases} X_t^1 \cup X_t^2 & \text{if } t \in V(K) \\ X_t^1 & \text{if } t \in V(T^1) - V(K) \\ X_t^2 & \text{if } t \in V(T^2) - V(K) \end{cases}$$

then (X, T) is a tree-decomposition of $G_1 /\!\!/_M^{(i)} G_2$. Since K is a

$$+(f, (Z^1, Y^1, K^1), (Z^2, Y^2, K^2))$$

kernel, K is a $G_1 /\!\!/_M^{(i)} G_2$ kernel of (X, T), as well. The encoding of (X, T) with respect to K and $G_1 /\!\!/_M^{(i)} G_2$ is (Z, Y, K), i.e. $(Z, Y, K) \in C(G_1 /\!\!/_M^{(i)} G_2)$. □

LEMMA 4.17. *If G is an i-sourced graph then $\overline{C(r_M^{(i)}(G))} = \overline{\rho^{(i)}(C(G))}$.*

Proof. Assume $E \in C(r_M^{(i)}(G))$. Let E be an encoding of a tree-decomposition (X, T) with respect to a kernel K and $r_M^{(i)}(G)$. Let (Z, Y, T) be the $r_M^{(i)}(G)$ profile of (X, T) and (Z', Y', T) the G profile of (X, T). Clearly, (Z, Y, T) is (Z', Y', T) with i forgotten. We consider the two cases: (Z, Y, T) is not singular and (Z, Y, T) is singular. First, if (Z, Y, T) is not singular then neither is (Z', Y', T) and hence both have unique kernels and, as easily seen, the G kernel K' of (X, T) satisfies $K \subseteq K'$. By definition $(Z', Y', T)[K']$ belongs to $C(G)$. Let F be $(Z', Y', T)[K']$ with i forgotten. Since the following hold K is a (Z, Y, T) kernel, (Z, Y, T) is (Z', Y', T) with i forgotten, and $K \subseteq K'$ we get that K is an F kernel, as well. Hence $E = F[K] \in \overline{\rho^{(i)}(C(G))}$ as was to be proved. Second, if (Z, Y, T) is singular then E is a singular $(i-1)$-sourced encoding. Now choose a minimal subtree K' of T' such that some vertex s in K' satisfies $[i-1] \subseteq Y_s'$ and some vertex s' in K' satisfies $i \in Y_{s'}'$. It follows from the properties of encodings that K' is a G kernel of (Z', Y', T). Thus $(Z', Y', T)[K']$ belongs to $C(G)$. Let F be $(Z', Y', T)[K']$ with i forgotten. Since F is singular, the minimum singular $[i-1]$-sourced encoding belongs to $\rho^{(i)}(C(G))$ and hence we get that $E \in \overline{\rho^{(i)}(C(G))}$ as was to be proved.

Assume that $E \in \overline{\rho^{(i)}(C(G))}$. Let $E \in \rho^{(i)}(F)$ where F is an encoding of a tree-decomposition (X, T) with respect to a kernel K and G. Let (Z, Y, T) be the G profile of (X, T) and (Z', Y', T) be (Z, Y, T) with i forgotten. Clearly, (Z', Y', T) is the $r_M^{(i)}(C(G))$ profile of (X, T). If $E = F'[K']$ where F' is F

with i forgotten and K' is a kernel of F', then K' is an $r_M^{(i)}(C(G))$ kernel of (Z', Y', T). Hence E is the encoding of (X, T) with respect to K' and $r_M^{(i)}(G)$, that is, $E \in C(r_M^{(i)}(G))$. If E is a singular $(i-1)$-sourced encoding then (Z', Y', T) is singular. Hence, there is some other tree-decomposition (X', T') of G where some bag X'_t is *exactly* the source set of $r_M^{(i)}(G)$. Thus, t is a $r_M^{(i)}(G)$ kernel of (X', T'). The encoding of (X', T') with respect to t and $\overline{r_M^{(i)}(G)}$ is the singular i-sourced encoding which is dominated by E. Hence $E \in \overline{C(r_M^{(i)}(G))}$. □

We have omitted the proof of the next lemma, since it, like the last lemma, consist of rather technical verification.

LEMMA 4.18. *If G is an i-sourced graph then $\overline{C(l_M^{(i,j)}(G))} = \overline{\lambda^{(i,j)}(C(G))}$*

That an encoding E is directly dominated by another encoding F and an isomorphism of this direct dominance respects sources is denoted $E \leq_{DD} F$. Let G be an i-sourced graph. Assume that $(Z, Y, K) \in C(G)$ and (Z', Y', K') are i-sourced encodings such that $(Z, Y, K) \leq_{DD} (Z', Y', K')$. Then there is an i-sourced encoding $F \in C(G)$ such that $F \leq_{DD} (Z', Y', K')$. Let us prove this. Let f be an isomorphism of the direct dominance $(Z, Y, K) \leq_{DD} (Z', Y', K')$. Let (X, T) be a tree-decomposition that has (Z, Y, K) as an encoding with respect to G. Define another tree-decomposition (X', T) of G by: $X'_t = X_t \cup S_G(Y'_{f(t)})$ for all $t \in V(K)$ and $X'_t = X_t$ for all $t \in V(T) - V(K)$. It is easy to see that K is a G kernel of (X', T). Moreover, if F is the encoding of (X', T) with respect to G and K then $F \in C(G)$ and $F \leq_{DD} (Z', Y', K')$. The encoding F is said to be obtained by *augmenting* (Z, Y, K) with the sources of (Z', Y', K'). Notice, if $F \leq_{DD} E$, then any encoding compatible with E is also compatible with F.

LEMMA 4.19. *If $E \leq_{DD} E'$ then $\overline{\lambda^{(i,j)}(E')} \subseteq \overline{\lambda^{(i,j)}(E)}$ and $\overline{\rho^{(i)}(E')} \subseteq \overline{\rho^{(i)}(E)}$. Furthermore, every encoding F compatible with E' is compatible with E, and $\overline{E' \oplus^{(i)} F} \subseteq \overline{E \oplus^{(i)} F}$.*

Proof. We only give the proof for $\rho^{(i)}$. The proofs of the statements regarding $\oplus^{(i)}$ and $\lambda^{(i,j)}$ are analogous.

Let B be E with i forgotten and B' be E' with i forgotten. It is easy to see that any B kernel K is an B' kernel, as well, and that $B[K] \leq_{DD} B'[K]$. Also, B is a singular i-profile iff B' is. Hence, $\overline{\rho^{(i)}(E')} \subseteq \overline{\rho^{(i)}(E)}$. □

LEMMA 4.20. *Let G_1 and G_2 be two i-sourced graphs then*

$$\overline{C(G_1) \oplus^{(i)} C(G_2)} = \overline{\overline{C(G_1)} \oplus^{(i)} \overline{C(G_2)}}.$$

Proof. It is easy to see that $\overline{C(G_1) \oplus^{(i)} C(G_2)} \subseteq \overline{\overline{C(G_1)} \oplus^{(i)} \overline{C(G_2)}}$. Also, since $\oplus^{(i)}$ is commutative, we only have to show that for any set of encodings C that is closed under subdivision $\overline{C(G_1) \oplus^{(i)} C} \subseteq \overline{C(G_1) \oplus^{(i)} C}$ to be able to conclude that the lemma holds.

Assume that $F \in \overline{\overline{C(G_1)} \oplus^{(i)} C}$, and let $F \in E_1 \oplus^{(i)} E_2$ where $E_1 \in \overline{C(G_1)}$ and $E_2 \in C$. Then there is an encoding $D_1 \in C(G_1)$ such that $D_1 \leq_D E_1$. Let D_1' and E_1' be subdivisions of D_1 and E_1, respectively, such that $D_1' \leq_{DD} E_1'$. Let E_2' be the subdivision of E_2 according to the E_1 to E_1' scheme. Let D_1'' be D_1' augmented with the sources of E_1'. By (4.12) and the properties of augmented encodings, $D_1'' \in C(G_1)$ and $D_1'' \leq_{DD} E_1'$. By (4.19), $\overline{E_1' \oplus^{(i)} E_2'} \subseteq \overline{D_1'' \oplus^{(i)} E_2'}$. Moreover, since F subdivided according to the E_1 to E_1' scheme belongs to $E_1' \oplus^{(i)} E_2'$, we have $F \in \overline{E_1' \oplus^{(i)} E_2'}$. Hence, $F \in \overline{D_1'' \oplus^{(i)} E_2'} \subseteq \overline{C(G_1) \oplus^{(i)} C}$. □

LEMMA 4.21. *If G is an i-sourced graph then $\overline{\rho^{(i)}(\overline{C(G)})} = \overline{\rho^{(i)}(C(G))}$.*

Proof. Clearly, $\overline{\rho^{(i)}(\overline{C(G)})} \supseteq \overline{\rho^{(i)}(C(G))}$ holds, since $\rho^{(i)}$ is monotone. Assume that $F \in \overline{\rho^{(i)}(\overline{C(G)})}$, and let $F \in \rho^{(i)}(E)$ where $D \in C(G)$ and $E \geq_D D$. Let E' and D' be subdivisions of E and D, respectively, such that $D' \leq_{DD} E'$. Let D'' be D' augmented with the sources of E'. By (4.12) and the properties of augmented encodings, $D'' \in C(G_1)$ and $D'' \leq_{DD} E'$. By (4.19), $\overline{\rho^{(i)}(E')} \subseteq \overline{\rho^{(i)}(D'')}$. Moreover, since F subdivided according to the E to E' scheme belongs to $\rho^{(i)}(E')$, $F \in \overline{\rho^{(i)}(D'')}$. Hence, $F \in \overline{\rho^{(i)}(C(G))}$. □

We have omitted the proof of the next lemma, since it is analogous to the proof of the last lemma.

LEMMA 4.22. *If G is an i-sourced graph then $\overline{\lambda^{(i,j)}(\overline{C(G)})} = \overline{\lambda^{(i,j)}(C(G))}$.*

4.3. The index of $\sim$. In this subsection, we will give an upper bound on the index of our congruence relation $\sim$. We start by defining non-redundant profiles. Directly after the definition, their importance is motivated. A profile (Z, Y, T) is said to be *redundant* if there are vertices t and t' in T such that the unique path P from t to t' in T satisfies:
- $z_t \leq z_s \leq z_{t'}$ and $Y_t = Y_s = Y_{t'}$ for all vertices s in P,
- all internal vertices in P have degree 2 in T
- P has at least one internal vertex.

The internal vertices in a path P satisfying these three conditions are called *redundant*. By shortcutting redundant vertices we mean: deleting the internal vertices in a path $t_0, \ldots, t_n$ where $t_1, \ldots, t_{n-1}$ are redundant, and then making t_0 and t_n adjacent.

The interesting thing about non-redundancy is that each equivalence class of $\sim_D$ that contains i-sourced encodings contains at least one non-redundant such. This is because, if we shortcut some redundant vertices in an encoding E then the resulting encoding is, clearly, equivalent to E under $\sim_D$. So, we can always obtain a non-redundant encoding from the equivalence class that E belongs to by recursively shortcutting redundant vertices until no more remains. Hence, if b is a bound on the number of non-redundant i-sourced encodings then it is a bound on the number of equivalence classes under $\sim_D$ that contains i-sourced encodings, as well. So, 2^b is a bound on the number of congruence classes with elements from the domain $\mathbf{G}_i$, of our congruence $\sim$ for graphs of tree-width at most w. The rest of this section will be used to show, the way indicated above, that the number of equivalence classes of $\sim_D$ containing i-sourced encodings is at most $2^{O(i^3 w + i^2 w \log w)}$. This is stated in the last lemma of this subsection, (4.28). But, we already conclude the following from it.

LEMMA 4.23. *If $r \geq w$ then $\sim$ has at most $2^{2^{O(r^4)}}$ congruence classes containing graphs from the domains $G_0, \ldots, G_r$.*

If (Z, Y, T) is a profile such that T is a path, and $Y_t = Y_{t'}$ holds for all vertices t and t' in T then (Z, Y, T) is called a *source homogeneous path profile*. If (Z, Y, P) is a source homogeneous path profile and t a vertex in P such that $z_t > z_{t'}$ for all other vertices t' in P or $z_t < z_{t'}$ for all other vertices t' in P then t is said to be an extreme vertex for (Z, Y, P).

LEMMA 4.24. *Every non-redundant source homogeneous path profile (Z, Y, P) with more than two vertices has an extreme vertex.*

Proof. Assume that $P = v_1, \ldots, v_m$. Let $A = \{u \in V(P) | z_u \leq z_{u'}$ for all $u' \in V(P)\}$ and $B = \{u \in V(P) | z_u \geq z_{u'}$ for all $u' \in V(P)\}$. Assume that the lemma is false. That is $|A|, |B| \geq 2$. Notice, both A and B are non-empty. Let $u_1, \ldots, u_l$ be the vertices of $A \cup B$ in the order that they appear in P. Assume that there is an $i \in [l-1]$ such that u_i and u_{i+1} both belong to the same set, say, A. Then the following holds.

If there is a vertex $u_k \in B$ after u_{i+1} in P then the internal vertices of the subpath of P between u_i and u_k are redundant and we have at least one such vertex—u_{i+1}. If there is a vertex $u_k \in B$ before u_i in P then the internal vertices of the subpath of P between u_k and u_{i+1} are redundant and we have at least one such vertex—u_i. The case when u_i and u_{i+1} both belong to B is completely analogous.

Hence, if u_1 belongs to A then u_4 belongs to B, and if u_1 belongs to B then u_4 belongs to A. In both cases, the internal vertices of the path between u_1 and u_4 are redundant and both u_2 and u_3 are such vertices. □

LEMMA 4.25. *Let (Z, Y, P) be a non-redundant source homogeneous path profile with more than two vertices and an extreme endvertex v. Then also the single neighbor of v is an extreme vertex.*

Proof. Assume that $P = v_1, \ldots, v_m$, where $v = v_1$. If $z_{v_1} > z_{v_i}$ for all $i \in [m]$ then $z_{v_2} < z_{v_i}$ must hold for all $i \in [m]$. Let $A = \{v_i | z_{v_i} \leq z_{v_j}$ for all $j \in [m]\}$;

since v_k is the last vertex in P that belongs to A, and $k \neq 2$ it follows that $v_2, \ldots, v_{k-1}$ are redundant. The case when $z_{v_1} < z_{v_i}$ for all $i \in [m]$ is analogous. □

Equipped with this lemma it is easy to prove, by induction, that if (Z, Y, P) is a non-redundant source homogeneous path profile of width at most w with an extreme endvertex then P has at most $w + 2$ vertices. By (4.24), we can break any non-redundant source homogeneous path profile into two such, both with an extreme endvertex. This gives the following theorem.

THEOREM 4.26. *If (Z, Y, P) is a non-redundant source homogeneous path profile of width at most w then P has at most $2w + 3$ vertices.*

We shall now bound the size of the tree of a non-redundant i-sourced encoding (Z, Y, K) by a constant, depending on w and i. First a bound on the *diameter* of K, that is, the length of a maximum length subpath P of T is obtained. Clearly, the maximum length subpath P is a path between two leaves. Let us assume that (Z, Y, K) has k leaves. The number of *boundary edges* (that is, edges (s, t) such that $Y_s \neq Y_t$) in P can not exceed $2(i - k) + 2$. Let us prove this. Direct each edge (s, t) of P so that P becomes a directed path. Mark a directed edge (s, t) with a $-$ if there is some source j that appears in the source set of its tail but not in the source set of its head, that is, $j \in Y_s - Y_t$. Mark a directed edge (s, t) with a $+$ if it is not marked $-$ and there is some source j that appears in the source set of its head but not in the source set of its tail, that is, $j \in Y_t - Y_s$. The fact that (Z, Y, K) is an encoding implies the following two observations. First, each source can contribute to at most two marks, since a source belongs to the source bags for a set of vertices that induce a subtree of K. Second, with each leaf in K except the two endvertices of P, we can associate one source that will not contribute to any mark. Different sources for different leaves. This is because, for each leaf l in K, there is at least one source that belong to the source bag of l, but not to any other source bag in (Z, Y, K). Hence, except for the first and last edge we have at most $2(i - k)$ marks to use. Since each boundary edge must have a mark, there are at most $2(i - k) + 2$ boundary edges.

We can also see that except for the two endvertices of P there are at most $2(i - k) + 1$ maximal source homogeneous subpaths of P. Since the encoding moreover is non-redundant, (4.26) tells us that each such path has at most $2w + 3$ vertices. So, P has at most $(2(i-k)+1)(2w+3)+2$ vertices. Hence, we can pick a vertex v such that no path from v to a leaf has more than $(i-k)(2w+3)+w+2$ vertices except from v. So, K has at most $k((i-k)(2w+3)+w+2)+1$ vertices. That is, we have at most $\mathcal{O}(i^2 w)$ vertices.

Thus we have proved the following theorem.

THEOREM 4.27. *If (Z, Y, K) is a non-redundant i-sourced encoding then K has at most $\mathcal{O}(i^2 w)$ vertices.*

The number of trees with at most n vertices is $\mathcal{O}(4^n)$. There are clearly at most 2^i different source sets and at most $w+1$ possible z-values to associated with vertices in an encoding. Hence, for some c the number of non-redundant i-sourced encodings is at most $(2^i(w+1))^{ci^2w} 4^{ci^2w}$, which is $2^{O(i^3 w + i^2 w \log w)}$. As noted above, this implies the following theorem.

LEMMA 4.28. $\sim_D$ has at most $2^{O(i^3 w + i^2 w \log w)}$ equivalence classes containing i-sourced encodings.

5. Conclusions

It is interesting to note that we have only used what is equivalent to a bounded number of inductive definitions to prove that there is a finite number of obstructions for the family of graphs of tree-width at most w. It is known to be necessary and sufficient to use an arbitrarily large finite number of iterations of inductive definitions to prove bounded GMT (stating that the family of graphs of bounded tree-width is well-quasi-ordered under minor taking). Bounded GMT is easily seen to imply that there is a finite number of obstructions for the family of graphs of tree-width at most w. We conclude that bounded GMT, in a metamathematical sense, is the stronger of these two theorems.

For the case of graphs of tree-width at most 1, at most 2, and at most 3 the obstruction sets are known. They have size 1, 1 and 4, respectively, see [2]. We conclude that the bounds we obtain here are not sharp.

6. Acknowledgment

I am grateful Stefan Arnborg for his support, the ideas he has shared with me, and the many discussions. I would also like to thank the anonymous referee for his creative criticism and good suggestions.

REFERENCES

1. S. Arnborg, B. Courcelle, A. Proskurowski, and D. Seese. An algebraic theory of graph reduction. Technical Report LaBRI-90-02, Universite de Bordeaux, Jan. 1990.
2. S. Arnborg and A. Proskurowski. Forbidden minors characterization of partial 3-trees. *Discrete Math.*, 80:1–19, 1990.
3. M. Bauderon and B. Courcelle. Graph expressions and graph rewritings. *Mathematical Systems Theory*, 20:83–127, 1987.
4. B. Courcelle. The monadic second order logic of graphs I: Recognizable sets of finite graphs. *Information and Computation*, 85:12–75, March 1990.
5. M. Fellows and M. Langston. An analogue of the Myhill-Nerode theorem and its use in computing finite basis characterizations. In *IEEE FoCS*, pages 520–525, 1989.
6. J. Lagergren and S. Arnborg. Finding minimal forbidden minors using a finite congruence. In *18 th ICALP*, pages 532–543, 1991.
7. N. Robertson and P.D. Seymour. Graph minors. XX. Wagner's conjecture. manuscript in preparation, 1988.

ROYAL INSTITUTE OF TECHNOLOGY, NADA/KTH, S-100 44 STOCKHOLM, SWEDEN
E-mail address: jensl@nada.kth.se

An obstruction-based approach to layout optimization

MICHAEL A. LANGSTON

ABSTRACT. Fast obstruction tests have potential as practical VLSI design tools. In this brief review, ongoing efforts to develop such tests are discussed. The emphasis is on providing compact layouts for circuits under representative metrics such as pathwidth and cutwidth.

1. Background

$\mathcal{N}P$-complete graph width problems arise at the heart of a number of VLSI layout styles. Well-known examples include pathwidth [12] (aka gate matrix layout) and cutwidth [7] (aka min cut linear arrangement). Accordingly, vast assortments of heuristic algorithms have been proposed to deliver approximate but not-necessarily-optimal solutions.

It is not at all clear, however, that optimality must be sacrificed for the sake of $\mathcal{N}P$-completeness. Many of these problems can be solved in low-order polynomial time whenever the width is fixed [5].

Pathwidth, for example, can in principle be decided in $O(n^2)$ time for any fixed width using a finite set of minor tests. Cutwidth, also amenable to this approach, can be decided analogously for any fixed width with a finite number of immersion tests.

Remarkably, these results rely on nonconstructive techniques [13, 14]. That is, the promised sets are not provided in the proof of polynomial-time complexity. All that is guaranteed is that each set contains a finite number of elements, henceforth termed *obstructions*, and that testing for each obstruction can be

---

1991 *Mathematics Subject Classification*. Primary 68C25, 68C40, 68E10.

This research has been supported in part by the National Science Foundation under grant MIP-8919312 and by the Office of Naval Research under contract N00014-90-J-1855

This paper summarizes material presented at the Workshop of Graph Minors held in Seattle, Washington, in June, 1991. It is intended solely as a survey on ongoing work, not as a preliminary version of a paper to be submitted elsewhere for final publication

layout instance	number of nets	number of gates	K_4 present	pathwidth two
1	5	7	yes	no
2	7	9	yes	no
3	8	8	yes	no
4	8	8	yes	no
5	8	9	yes	no
6	11	10	yes	no
7	17	16	yes	no
8	5	5	no	yes
9	5	8	no	yes
10	6	6	no	yes
11	6	6	no	yes
12	8	8	no	yes

TABLE 1. Sample circuits from the literature

accomplished in polynomial time. Constructivization strategies have recently been proposed [6], but even these cannot provide complete obstruction sets.

How then does one translate the knowledge of such "technically efficient" methods into anything resembling a practical algorithm? In this brief survey, we discuss ongoing efforts to resolve this foundational question, exploring the notion that not all obstructions are equally important.

2. Pathwidth

For every fixed k, the family of graphs with pathwidth at most k is minor-closed. But only sketchy information is known about pathwidth obstructions and their underlying structure [8]. One obstruction exists for pathwidth zero. Two obstructions suffice for pathwidth one. For pathwidth two, 110 obstructions make up the set. For pathwidth three, at least 122 million obstructions exist.

Recognizing such a rapid growth rate, we began this effort by studying layouts for real circuits from the literature. (Circuits are modeled as graphs, with vertices corresponding to nets and edges representing net incompatibilities.) Interestingly, every instance we could find had the property that, if the pathwidth exceeded k, then the instance contained as a minor K_{k+2}, an obstruction to pathwidth k. A dozen such instances are listed in Table 1, where we show the situation for $k = 2$.

Therefore, despite the immense number of possible obstructions for even modest values of k, it seems plausible that K_{k+2} alone is an excellent discriminator for deciding whether a circuit has pathwidth k. To explore this possibility beginning with $k = 2$, we have generated great numbers of instances at random.

number of vertices in graph	number of graphs generated	graphs with obstruction	graphs without obstruction	
			pathwidth two or less	pathwidth three or more
6	1000	40	960	0
7	1000	141	857	2
8	1000	271	711	18
9	1000	532	454	14
10	1000	757	221	22
11	1000	880	95	25
12	1000	975	17	8
13	1000	992	6	2
14	1000	1000	0	0

TABLE 2. Typical effectiveness of K_4 test

(Of course we make no pretense that a randomly generated graph is particularly likely to represent a useful circuit. Rather, we only wish to gather as much data as possible.) For each instance, we performed the easy K_4 topological test of [9] first, thereby eliminating non-series-parallel graphs (and, presumably, almost all graphs with pathwidth exceeding two). For those instances without K_4, we used the dynamic programming formulation of [4] to find an optimal path decomposition.

Sample results are displayed in Table 2. In this particular set of experiments, each edge was chosen with probability 1/3. We observe that the single obstruction K_4 was able to screen out instances with pathwidth three or more roughly 98% of the time.

Although these empirical results are encouraging, one must be cautious not to attach too much significance to them. They rely in part on the fact that real circuits are not arbitrarily sparse. In particular, the bigger, more tree-like obstructions tend to occur only in the presence of a K_4 minor unless the edge probability is exceedingly small. (It has been observed that, if one can make the edge probability arbitrarily small and the number of vertices arbitrarily large, then any tree can be made more likely than any clique of size three or more in a random graph.)

In order to account for more relatively dense obstructions, we have developed a practical, linear-time test for K_4 and the five other obstructions depicted below.

The improvement obtained with this method is illustrated in Table 3.

number of vertices in graph	number of graphs generated	graphs with obstruction	graphs without obstruction	
			pathwidth two or less	pathwidth three or more
6	1000	40	960	0
7	1000	143	857	0
8	1000	289	711	0
9	1000	545	454	1
10	1000	778	221	1
11	1000	904	95	1
12	1000	982	17	1
13	1000	994	6	0
14	1000	1000	0	0

TABLE 3. Effectiveness of six-obstruction test

These tests may be attractive from a practical standpoint because series-parallel circuits occur frequently (a restriction to series-parallel connections is common in the design of CMOS cells [15]). Even circuits that are not series-parallel are often decomposable into series-parallel subcircuits, for which pathwidth two layouts can be useful.

Testing for other complete graphs is a next logical step. An $O(n^2)$-time K_5 minor test has recently been reported in [10]. We are hopeful that this algorithm or methods based on it may prove practical to implement and run.

3. Cutwidth

For every fixed k, the family of graphs with cutwidth at most k is immersion-closed. Even less is known about cutwidth obstructions. Two obstructions exist for cutwidth zero. Two suffice for cutwidth one as well. For cutwidth two, sixteen obstructions are needed. And for cutwidth three, eighty-five obstructions are known, although many more are thought likely to exist.

Again, complete graphs are obstructions. Based on this and our experience with pathwidth, we have considered fast tests to determine whether a complete graph is immersed in an arbitrary graph. Testing for K_1 and K_2 are of course trivial. Testing for K_3 is easy: K_3 is immersed in any graph of order three or more unless the graph is a tree with no pair of multiple edges incident on a common vertex.

The first really difficult test is that for K_4. Unlike the aforementioned simple K_4 topological test, no genuinely practical test was previously known for deciding whether K_4 is immersed in a graph. (For graphs of maximum degree three, topological containment is equivalent to minor containment, but only sufficient and not necessary for immersion containment.) It is possible in principle to use the method sketched in [16] to obtain a tree decomposition of width two, and then to use the dynamic programming formulation of [13] on the decomposition. Although this two step procedure runs in $O(n)$ time, the resultant constant of proportionality is prohibitively high.

One problem is that multiple edges cannot be ignored, making immersion tests much more "slippery" than minor tests. Thus we have developed a linear-time K_4 test from scratch [2], which we now sketch. Our method requires the following three technical lemmas.

(3.1) *K_4 is immersed in G if and only if it is immersed in some three-edge-connected component of G.*

(3.2) *Each three-edge-connected component of a series-parallel graph is series-parallel.*

(3.3) *Any series-parallel graph contains at least two vertices with at most two neighbors.*

Our K_4 test, algorithm **immerse**, proceeds in three steps. We first invoke algorithm **decompose**, which finds a series-parallel decomposition of G if any exist. If **decompose** is successful, we next invoke algorithm **components**, which breaks G into three-edge-connected components. Finally, we invoke algorithm **test** on each component until either a K_4 is encountered or all components have been eliminated (3.1).

Algorithm **test** is the heart of our method, and is described in the pidgin Algol that follows. Its input is series-parallel (3.2). We proceed by examining vertices with at most two neighbors (3.3). At each iteration, some such vertex is selected. The vertex is deleted (after deleting or lifting its incident edges) if we can determine that it is not contained in every copy of K_4 should K_4 be present. Otherwise, the vertex is marked. We assume all vertices are initially unmarked. As the algorithm progresses, a vertex may be marked then later unmarked again as its neighborhood changes.

<u>algorithm</u> **test**
delete all but three copies of edges incident on vertices with one neighbor
<u>if</u> any articulation point has degree seven or more
    <u>then</u> report "yes" and halt
<u>while</u> there are unmarked vertices with fewer than three neighbors <u>do</u>
    <u>if</u> there are fewer than four vertices
        <u>then</u> report "no" and halt
    <u>while</u> there is a vertex v with exactly one neighbor w <u>do</u>
        delete v and every copy of the edge vw
        unmark w if it is marked
        <u>end while</u>

```
if there is an unmarked vertex v with exactly two neighbors u and w
    then
        assume the multiplicity of uv is no less than that of vw
        if u or v is an articulation point
            then
                lift all possible pairs of edges uv and vw,
                delete any remaining copies of uv, and delete v
            else
                if there is only one copy of edge vw
                    then
                        lift uv and vw
                        delete any remaining copies of uv
                        delete v
                        if the degree of u exceeds three
                            then report "yes" and halt
                            else unmark u if it is marked
                        end if
                    else mark v
                end if
        end if
    end if
end while
if there is a vertex with degree five or more
    then report "yes" and halt
end test
```

(3.4) *Algorithm* **immerse** *runs in $O(n)$ time and correctly decides whether K_4 is immersed in an input graph.*

This algorithm is fast, reasonably simple to code, and easy to modify if one wants to locate a model of an immersed K_4 when any exist. It may also serve to simplify other obstruction tests, because all other obstructions (and any input graph that fails the test) must be series-parallel. In addition to cutwidth, a variety of other load factor problems can be decided by finite sets of immersion tests that include K_4.

4. Related Efforts

On a more general front, new approaches of undetermined practical potential offer asymptotic improvements over the $O(n^2)$ method of [13]. These include an $O(n \log^2 n)$ pathwidth scheme [3], an $O(n \log n)$ tree decomposition algorithm [11], and an $O(n)$ graph rewriting technique [1]. It will be interesting to see how well these and other new ideas can make the transition from theory to practice.

References

1. S. Arnborg, B. Courcelle, A. Proskurowski and D. Seese, *An algebraic theory of graph reduction*, Technical Report, Laboratoire de Recherche en Informatique Bordeaux, 1990.
2. H. D. Booth, R. Govindan, M. A. Langston and S. Ramachandramurthi, *Fast sequential and parallel algorithms for detecting an immersed K_4*, to appear.
3. H. L. Bodlaender and T. Kloks, *Better algorithms for the pathwidth and treewidth of graphs*, Technical Report, University of Utrecht, 1990.
4. N. Deo, M. S. Krishnamoorthy and M. A. Langston, *Exact and approximate solutions for the gate matrix layout problem*, IEEE Transactions on Computer-Aided Design **6**, (1987), 79–84.
5. M. R. Fellows and M. A. Langston, *On well-partial-order theory and its application to combinatorial problems of VLSI design*, SIAM J. Discrete Math. **5**, (1992), 117–126.
6. _____, *On aearch, decision and the efficiency of polynomial-time algorithms*, J. Comput. Systems Sci., to appear.
7. F. Gavril, *Some $\mathcal{N}P$-complete problems on graphs*, Proceedings, 11th Annual Conference on Information Sciences and Systems, Baltimore, MD, 1977, 91–95.
8. N. G. Kinnersley and M. A. Langston, *Obstruction set esolation for the gate matrix layout problem*, Technical Report, University of Tennessee, 1991.
9. P. C. Liu and R. C. Geldmacher, *An $O(max(m,n))$ algorithm for finding a subgraph homeomorphic to K_4*, Congr. Numer. **29**, (1980), 597–609.
10. P. J. McGuinness and A. E. Kezdy, *An algorithm to find a K_5 Minor*, Technical Report, University of Illinois, 1991.
11. B. Reed, private communication.
12. N. Robertson and P. D. Seymour, *Graph minors I. Excluding a forest*, J. Combin. Theory Ser. B **35**, (1983), no. 1, 39–61.
13. _____, *Graph Minors XIII. The disjoint paths problem*, to appear.
14. _____, *Graph Minors XVI. Wagner's conjecture*, to appear.
15. C. C. Su and M. Sarrafzadeh, *Optimal gate matrix layout of CMOS functional cells*, Integration, the VLSI Journal **9**, (1990), 3–23.
16. J. A. Wald and C. J. Colbourn, *Steiner trees, partial 2-trees and minimum IFI networks*, Networks **13**, (1983), 159–167.

Department of Computer Science, University of Tennessee, Knoxville, TN 37996
E-mail address: langstoncs.utk.edu

Decomposing 3-Connected Graphs

COLLETTE R. COULLARD   DONALD K. WAGNER

ABSTRACT. This note describes ongoing work by the authors on the decomposition of 3-connected graphs. Section 1 describes the history of the problem and our progress to date. Section 2 outlines our current approach. Section 3 discusses applications of this work.

1. History and Progress

Cunningham and Edmonds [5] proved a generalization of Tutte's theorem (Chapter 11 of [10]) that a 2-connected graph G has a unique minimal decomposition into graphs, each of which is either 3-connected, a bond, or a polygon. They define the notion of a good split, and first prove that G has a unique minimal decomposition into graphs, none of which has a good split, and second prove that the graphs that do not have a good split are precisely 3-connected graphs, bonds and polygons.

Coullard, Gardner, and Wagner [3] proved an analogue of the first result above for 3-connected graphs, and an analogue of the second for minimally 3-connected graphs. Following the basic strategy of Cunningham and Edmonds, they defined an appropriate notion of good split. They proved first that if G is a 3-connected graph, then G has a unique minimal decomposition into graphs none of which has a good split. Then they proved that the minimally 3-connected graphs that do not have a good split are precisely cyclically 4-connected graphs, twirls, and wheels. Thus, it follows that if G is a minimally 3-connected graph,

---

[0]1991 Mathematics Subject Classification. Primary 05C40.
[0]This paper is a preliminary version and the detailed version will be published elsewhere.

then G has a unique minimal decomposition into graphs, each of which is either cyclically 4-connected, a twirl or a wheel.

Robertson and Shih also have a unique decomposition for 3-connected graphs; their work is unpublished. Their decomposition is different from ours in that the objects obtained in the decomposition are allowed to be hypergraphs, some edges of which are incident to three vertices.

The result of Coullard et al falls short of a complete analog of the Tutte uniqueness result for general 3-connected graphs. We still hope to find that complete analog, as described in the following section.

2. Proposed Approach

As pointed out in the previous section, Coullard et al have a unique decomposition for 3-connected graphs. The reason this result falls short is that we have not been able to completely describe the general graphs with no good split, according to the definition of good split that yields the unique decomposition. That seemed to be the next step. Now we are taking a different approach.

In order to describe the new approach, we must give a little background. A split of a graph can be defined many ways; indeed, one of the main contributions of [3] is a definition that does yield a unique decomposition. In [3], a split is defined as a special tripartition of the edge set of the graph. (For brevity, the precise definition is omitted here.) These are called *edge splits*. An alternate approach is to define a split as a special pair of subsets of the vertices of the graph, the union of which is the entire vertex set. (Again, we omit the detailed definition.) These are called *vertex splits*. Vertex splits are introduced in the latter part of [3], where it is shown that vertex splits and edge splits are equivalent for the class of minimally 3-connected graphs. Then we proceed to characterize the minimally 3-connected graphs with no good vertex splits.

Recently we have been able to characterize the 3-connected graphs with no good vertex splits, extending the above result to general graphs. Thus, if vertex splits and edge splits were equivalent for general graphs, we would have the desired analog of the Tutte uniqueness result.

Unfortunately, vertex splits and edge splits are not equivalent for general graphs. Thus, a unique decomposition that uses good vertex splits, instead of good edge splits must be defined. There are several possibilities to be investigated.

3. Applications

While direct applications of the *unique* decomposition of 3-connected graphs seem unlikely, the uniqueness issue has enjoyed enough interest in the past (see for example [1], [5], [6], [7], [10]) that researchers in the area consider it important in its own right. On the other hand, decomposing graphs (not necessarily uniquely) has played an important role in optimization for quite some time. (see [2], [4], [8], [9]) We intend to continue our thrust in the direction of finding decomposition-based solution techniques for optimization problems on graphs.

REFERENCES

1. R.E. Bixby, *Composition and decomposition of matroids and related topics.* Ph.D. thesis, Cornell University, 1972.

2. G. Cornuejols, D. Naddef and W.R. Pulleyblank, "Halin graphs and the traveling salesman problem." *Mathematical Programming* **26** (1983), 287-294.

3. C.R. Coullard, L.L. Gardner, and D.K. Wagner, "Decomposition of 3-connected graphs," 1990, to appear in *Combinatorica*.

4. C.R. Coullard and W.R. Pulleyblank, "On Cycle Cones and Polyhedra," *Linear Algebra and Its Applications* **114/115** (1989), 613-640.

5. W.H. Cunningham and J. Edmonds, "A combinatorial decomposition theory," *Canadian Journal of Mathematics* **32** (1980), 734-765.

6. J.E. Hopcroft and R.E. Tarjan, "Dividing a graph into triconnected components," *SIAM Journal on Computing* **2** (1973), 135-158.

7. S. MacLane, "A structural characterization of planar combinatorial graphs," *Duke Mathematics Journal* **3** (1937), 460-472.

8. A. Rajan, *Algorithmic implications of connectivity and related topics in matroid theory*, Ph.D. thesis, Northwestern University, 1986.

9. K. Truemper, "A decomposition theory for matroids, I. General results," *Journal of Combinatorial Theory (B)* **39** (1985), 43-76.

10. W.T. Tutte, *Connectivity in Graphs*, University of Toronto Press, Toronto, 1966.

Northwestern University
e-mail address: coullard@iems.nwu.edu

Office of Naval Research
e-mail address: dwagner@ocnr-hq.navy.mil

Graph Planarity and Related Topics

A.K. KELMANS

ABSTRACT. We describe different results on graphs containing or avoiding subdivisions of some special graphs, and in particular, different refinements of Kuratowski's planarity criterion for 3-connected and quasi 4-connected graphs. Some results on non-separating circuits in a graph are presented. In particular some more refinements of Kuratowski's theorem and graph planarity criteria in terms of non-separating circuits are given for 3-connected and quasi 4-connected graphs. An ear-like decomposition for quasi 4-connected graphs is described similar to that of for 3-connected graphs, and is shown to be a very useful tool for investigating graph planarity and some other problems for quasi 4-connected graphs. Refinements of different kinds are given for Whitney's graph planarity criterion. Some results on Dirac's conjecture and Barnette's conjecture are also presented.

CONTENTS

1. Introduction

2. The main concepts and notation

3. Some classical results

4. Simple reductions of the graph planarity problem

5. Subdivisions of K_5, $K_{3,3}$, and L in a graph

6. Subdivisions of $K_{3,3}$ in a 3-connected graph with some edges not subdivided

7. A vertex in a matroid and the corresponding notion and dual notion for graphs

---

1991 *Mathematics Subject Classification.* Primary 05C10, 05C70.

Key words and phrases. graph, matroid, planarity, connectivity, circuit, cocircuit, non-separating circuit, Hamiltonian circuit, graph minors.

Support by the National Science Foundation under Grant NSF–STC88–09648 and the Air Force Office of Scientific Research under Grants AFOSR–89–0512 and AFOSR–90–0008 to Rutgers University

This paper is in final form and no version of it will be submitted for publication elsewhere

8. More about non-separating circuits in a graph

9. Triangle and 3-cut reductions of the graph planarity problem

10. Subdivisions of K, M, and N in quasi 4-connected graphs

11. An ear-like decomposition for quasi 4-connected graphs

12. Non–separating circuits in quasi 4-connected graphs

13. Some refinements of Whitney's planarity criterion

14. On Dirac's conjecture

15. On Barnette's conjecture

REFERENCES

1. Introduction

Graph planarity theory is one of the classical fields of graph theory [4, 6, 13]. It is a part of theory of graph embeddings into surfaces of different types. Whitney's planarity criterion illustrates the interconnection between graph planarity theory and matroid theory [55]. MacLane's criterion gives an interpretation of graph planarity in terms of linear algebra. Kuratowski's planarity criterion shows that this theory is also a natural part of graph minors theory [38]. Steinitz's theorem shows the interconnection between planar graphs and convex polytopes in 3-dimensional space [11]. In the 60's W.T. Tutte found very interesting and important results concerning graph planarity and related topics [49, 51, 52, 50]. Many interesting results in this direction were found since then by D.S. Archdeacon, D.W. Barnette, R.E. Bixby, W.H. Cunningham, G.A. Dirac, J. Edmonds, J.-C. Fournier, B. Grünbaum, L. Lovász, F. Jaeger, H. Jung, N. Robertson, P. Seymour, C. Thomassen, W.T. Tutte, (e.g. [1, 3, 5, 8, 9, 10, 14, 15, 35, 37, 38, 39, 40, 41, 43, 44, 45, 46, 47, 53]) and many others. In the 70's by considering the above mentioned problems from the point of view of matroid theory, we rediscovered some of Tutte's results (with different proofs) and found some more results in this direction. We also found simple proofs, natural generalizations and strengthenings of the classical results of Kuratowski and Whitney on planarity and circuit isomorphism of graphs, some results on Dirac's and Barnette's conjectures, etc.

The main part of this paper is an outline of some results on graph planarity and related topics which have been obtained by the author. Some of these results have never been published and some have only been published in Russian. The results described in the paper show different directions in which this classical theory can be naturally developed.

2. The main concepts and notation

All the concepts about graphs and matroids used but not defined here can be found in [4, 6, 13, 55].

We consider *undirected graphs* which may have loops and parallel edges. A graph without loops or parallel edges is called a *simple graph*. The sets of *vertices* and *edges* of a graph G are denoted by $V(G)$ and $E(G)$, respectively.

Given a set X of edges of G let $G \setminus X$ and G/X denote the graph obtained from G by deleting and contracting the edges in X, respectively. For a subgraph F of G we write G/F instead of $G/E(F)$.

A graph G is called *2-connected* if G has no loops and at least two vertices, for $|V(G)| = 2$ G has at least two parallel edges, and for $|V(G)| \geq 3$ the graph obtained from G by deleting any vertex is connected. A graph G is called *k-connected*, $k \geq 3$, if G is simple, $|V(G)| \geq k+1$, and the graph obtained from G by deleting any $k-1$ vertices is connected. Sometimes it is convenient to treat the loop as a 2-connected graph and the graph with two vertices and three parallel edges as a 3-connected graph.

Let K_n denote the complete graph with n vertices and $K_{n,m}$ denote the complete bipartite graph with two parts of n and m vertices. Let L denote the graph consisting of two disjoint triangles connected by three disjoint edges so that L is the cubic (3-connected) simple graph on 6 vertices distinct from $K_{3,3}$ (see Fig. 1).

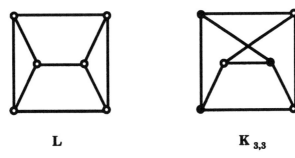

L   **K**$_{3,3}$

FIGURE 1. The cubic simple graphs on 6 vertices.

For $i \in \{0, 1, 2, 3\}$, and $n \geq 1$, we denote by $K_{3,n}^i$ the graph obtained from $K_{3,n}$ by adding i edges between the vertices of a 3-vertex part of $K_{3,n}$ (and so $K_{3,n}^0 = K_{3,n}$).

A *circuit-graph* is a connected graph with each vertex of degree 2 – it is sometimes called a *polygon*. A *cocircuit-graph* is a connected graph with two vertices and some (parallel) edges – it is sometimes called a *bond*.

A *wheel* is a graph obtained from a circuit-graph by adding a new vertex and all edges between the new vertex and the vertices of the circuit.

Paths and *circuits* in a graph will be assumed to be *non-self-intersecting*. A path P will be assumed to have at least two vertices and will sometimes be denoted by xPy to identify x and y as the *end-vertices* of P. If P has exactly one edge then x and y are the *end-vertices* of the edge. A vertex of xPy distinct from x and y is an *inner vertex* of P. A path P of G is called a *thread* of G if all inner vertices of P are of degree 2 and the end-vertices are of degree $\neq 2$ in G. A thread with no inner vertices (i.e. with one edge) is *an edge-thread*. Given a path xPy and a circuit C in G, we call xPy a *path-chord of* C if $P \cap C = \{x, y\}$. A path-chord of C with one edge is an *edge-chord* of C.

A set $\mathcal{C}$ of circuits of G is called *independent* if for every nonempty subset $\mathcal{A}$ of $\mathcal{C}$ the symmetric difference of the circuits in $\mathcal{A}$ is not empty. A maximal independent set of circuits of G is called *a cycle basis of G*. It is easy to see that every cycle basis of G has $|E(G)| - |V(G)| + c(G)$ circuits where $c(G)$ is the number of components of G.

A set X of edges of a graph G is called an *edge cut* of G if $G \setminus X$ has more components than G. An edge cut of G minimal under inclusion is called a *cocircuit* of G. Associated with each vertex in G is a *vertex star* which is the set of edges in G incident to that vertex. A set S of vertices of a graph G is called a *vertex cut* of G if $G \setminus S$ has more components than G.

A graph G is called *planar* if there exists an embedding of G into the sphere (or into the plane) such that the vertices are points and the edges are segments of Jordan curves, a vertex is an end-vertex of an edge if and only if the corresponding point is an end-point of the corresponding segment, and any common point of two segments corresponds to a common end-vertex of the corresponding two edges of G.

A *face-circuit* of a planar graph embedded into the sphere is a circuit which bounds a face of the embedded graph.

A graph F is called a *subdivision* of G, written $F = TopG$, if F can be obtained from G by *subdividing* some edges of G, i.e. by a sequence of operations of replacing an edge $e = (x, y)$ of G by a path xPy having exactly two vertices $\{x, y\}$ in common with the current graph.

The *topological length* or *top length* $l_G(P)$ of a path (a circuit) P in G is the number of inner vertices (respectively, vertices) of P of degree at least 3 in G.

Given two disjoint graphs G_1 and G_2, a graph G is called a *2-sum of G_1 and G_2* if G is obtained from G_1 and G_2 by identifying an edge e_1 from G_1 and an edge e_2 from G_2 with a new edge e and by deleting the new edge e from the resulting graph.

A graph G is *combinatorial dual* to G^* if there exists a one-to-one mapping $e : E(G) \to E(G^*)$ of the edge set of G onto the edge set of G^* such that C is a circuit of G if and only if $e(C)$ is a cocircuit of G^*.

Given a matroid M let $E(M) = E$, $\mathcal{C}(M)$, and $\mathcal{C}^*(M)$ denote the ground set, the set of circuits, and the set of cocircuits of M, respectively. Given $X \subseteq E(M)$ let $M \setminus X$ and M/X denote the matroids obtained from M by *deleting* and *contracting* X, respectively, which means that they are defined on the ground set $E \setminus X$ and $\mathcal{C}(M \setminus X) = \{C : C \subseteq E \setminus X, C \in \mathcal{C}(M)\}$, and $\mathcal{C}^*(M/X) = \{C^* : C^* \subseteq E \setminus X, C^* \in \mathcal{C}^*(M)\}$.

Given $x, y \in E(M)$ we write xCy (xC^*y) if $x = y$ or there is a circuit (respectively, a cocircuit) of M which contains both x and y. It is easy to see that $xCy \Leftrightarrow xC^*y$ and that C is an equivalence relation. The equivalence classes under C are called the *components* of M. A matroid is *connected* if it has one component. A partition (X_1, X_2) of $E(M)$ is called a *k-separation of M* if $|X_1|, |X_2| \geq k$ and $\rho(X_1) + \rho(X_1) \leq \rho(M) + k - 1$ where ρ is the rank function of M. A matroid is *k-connected* if it has no l-separation for $l < k$. It is easy to see that M is k-connected if and only if M^* is k-connected. We are only concerned with the cases $k \leq 3$. It is easy to show that (1) every matroid is 1-connected, (2) a matroid is 2-connected if and only if it is connected, (3) a graph with no isolated vertices and with at least two edges is 2-connected if and only if its circuit matroid is connected, and (4) a simple graph with at least 4 edges is 3-connected if and only if its circuit matroid is 3-connected.

A matroid is *binary* if $|C \cap C^*|$ is even for all $C \in \mathcal{C}(M)$ and $C^* \in \mathcal{C}^*(M)$.

3. Some classical results

The following three planarity criteria are classical.

3.1. Kuratowski's planarity criterion [33]

A graph G is planar if and only if it does not contain a subdivision of K_5 or $K_{3,3}$ (see Fig. 2).

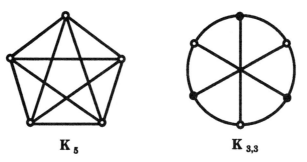

FIGURE 2. The Kuratowski graphs.

3.2. Whitney's planarity criterion [56]

A graph G is planar if and only if it has a combinatorial dual graph G^*.

3.3. MacLane's planarity criterion [36]

A graph G is planar if and only if it has a cycle basis such that each edge of G belongs to at most two circuits of the basis.

4. Simple reductions of the graph planarity problem

It is well-known and is easy to show that the graph planarity problem can be refined by simple reductions described in **4.1**, **4.2**, and **4.3**.

4.1 *A graph is planar if and only if each of its components is planar.*

Therefore we may consider connected graphs.

4.2. *A connected graph is planar if and only if each of its blocks is planar.*

Therefore we may consider 2-connected graphs.

Let a 2-connected graph G have a vertex 2-cut $X = \{x_1, x_2\}$ so that $G = F_1 \cup F_2$ and $F_1 \cap F_2 = X$. Let G_i, $i = 1, 2$, be obtained from F_i by adding a new edge $e_i = (x_1, x_2)$ so that G is the 2-sum of G_1 with the specified edge e_1 and G_2 with the specified edge e_2 (see Fig. 3).

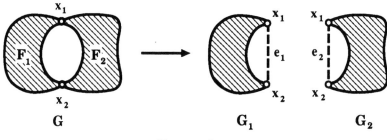

FIGURE 3.

4.3. *A 2-connected graph G is planar if and only if G_1 and G_2 are planar.*

Therefore 3-connected graphs can be considered as the *main bricks in the graph planarity problem*. Also 3-connected planar graphs are of special interest because of the following reason. It is very easy to see that

4.4. *The skeleton-graph of a convex polytope in 3-dimensional space is a planar graph.*

One can easily prove (by using linear programming arguments) that

4.5. *For all $k \geq 1$ the skeleton-graph of a convex polytope in k-dimensional space is a k-connected graph.*

Therefore

4.6. *The skeleton-graph of a convex polytope in 3-dimensional space is a planar 3-connected graph.*

It turns out that the reverse is also true. Steinitz's theorem says [3] that

4.7. *A graph is the skeleton-graph of a convex polytope in 3-dimensional space if and only if it is planar and 3-connected.*

5. Subdivisions of K_5, $K_{3,3}$, and L in a graph

It is well-known that K_5 and $K_{3,3}$ play different roles in Kuratowski's planarity criterion; namely, it is easy to prove that [12, 54]:

5.1. *A 3-connected graph G distinct from K_5 is planar if and only if it does not contain a subdivision of $K_{3,3}$.*

Given a graph F and an edge $e = (x,y)$ of F, let $\mathcal{T}(F,e)$ denote the set of graphs obtained from F by subdividing e (once) and adding a new edge $p = (m,s)$ connecting the "middle point" m of e with the "middle point" s of some other edge of F or with a vertex s of F distinct from x and y. The proof of **5.1** follows immediately from Kuratowski's planarity criterion and the fact that if F is K_5 then any graph in $\mathcal{T}(F,e)$ contains a subdivision of $K_{3,3}$.

Moreover, a recursive description can be given of 2-connected non-planar graphs which contain no $TopK_{3,3}$ (and consequently contain $TopK_5$). A *two-pole* is a graph with two distinguished *pole-vertices* as follows. A two-pole is *planar* if the graph obtained from it by adding a new edge between the poles is planar. A K_n-*two-pole* is obtained from K_n by distinguishing any two pole-vertices and by deleting the edge connecting the poles. It is easy to prove that [16, 54]:

5.2. *A 2-connected non-planar graph contains no subdivision of $K_{3,3}$ if and only if it can be obtained recursively from K_5 by replacing an edge by either a planar two-pole or a K_5-two-pole.*

In other words

5.2'. *A 2-connected graph contains no subdivision of $K_{3,3}$ if and only if it can be obtained by 2-sums from planar graphs and K_5.*

The last result can be generalized. Let us consider K_n, $n \geq 5$, instead of K_5, and $H(K_{n-1})$ instead of $K_{3,3}$ where $H(K_{n-1})$ is obtained from K_{n-1} by adding a new edge between the "middles" of two non-incident edges so that $H(K_4)$ is $K_{3,3}$. We say that G is *n-planar* if it contains no subdivision of K_n or $H(K_{n-1})$. Thus *5-planar* means *planar*. Then it is easy to give a recursive description of non-n-planar graphs containing no subdivision of $H(K_{n-1})$ similar to that of in **5.2**. [**16**]:

5.3. Let $n \geq 5$. A 2-connected non-n-planar graph contains no subdivision of $H(K_{n-1})$ (and so it contains $TopK_n$) if and only if it can be obtained recursively from K_n by replacing an edge by either an n-planar two-pole or the K_n-two-pole.

In other words

5.3'. Let $n \geq 5$. A 2-connected graph contains no subdivision of $H(K_{n-1})$ if and only if it can be obtained by 2-sums from n-planar graphs and K_n.

This splitting effect has been described in a much more general case (not only for graphs but also for matroids) by P. Seymour in his paper on decomposition of regular matroids [**40**].

It is easy to prove that "almost" all 3-connected graphs contain two disjoint circuits. More precisely [**7, 34**] (see also [**35**], page 377–379)

5.4. A 3-connected graph G with at least 6 vertices has no two disjoint circuits if and only if G is either a wheel or $K_{3,n}^i$ for $i \in \{0,1,2,3\}$ and $n \geq 3$.

By Menger's theorem a 3-connected graph contains $TopL$ if and only if it contains two disjoint circuits. Therefore we have from **5.4**:

5.5. A 3-connected graph G with at least 6 vertices does not contain $TopL$ if and only if G is a wheel or $K_{3,n}^i$ for $i \in \{0,1,2,3\}$ and $n \geq 3$.

From **5.5** we have:

5.6. A 3-connected graph G with at least 6 vertices contains neither $TopL$ nor $TopK_{3,3}$ if and only if G is a wheel.

From **5.5** it is easy to obtain a description of 2-connected graphs which contain no $TopL$. This description is similar to that of **5.2'** and **5.3'**.

5.7. A 2-connected graph contains no $TopL$ if and only if it can be obtained by 2-sums from circuit-graphs, cocircuit-graphs, wheels, K_5 and $K_{3,n}^i$, $i \in \{0,1,2,3\}$ and $n \geq 3$.

6. Subdivisions of $K_{3,3}$ in a 3-connected graph with some edges not subdivided

It turns out that some deeper strengthenings of Kuratowski's planarity criterion can be obtained for 3-connected graphs.

Let $F = K_{3,3}$ and let $e \in E(F)$. Then it is easy to see that any graph in $\mathcal{T}(F, e)$ has a new subdivision F' of $K_{3,3}$ such that it contains the new edge $p = (m, s)$, and one of the two edges (x, m) and (y, m) (obtained from $e = (x, y)$ by dividing it by the middle point m in two parts) as a thread of F'. From this fact it follows immediately that

6.1. *A non-planar 3-connected graph G distinct from K_5 contains a $TopK_{3,3}$ at least one edge-thread.*

In 1975 we proved that a stronger result is true:

6.2. *A non-planar 3-connected graph G distinct from K_5 contains a $TopK_{3,3}$ with at least two edge-thread.*

The following construction shows that the above statement is not true if we replace "two edges not subdivided " by "four edges not subdivided"; it gives infinitely many non-planar 3-connected graphs which have no $TopK_{3,3}$ having four edge-threads. *To replace a cubic vertex x in G by a triangle* means to construct a new graph G' from G by adding a new edge connecting the "middles" of some two edges with a common end-vertex x. Let G be a cubic non-planar 3-connected graph and let $G^{\triangle}$ be the graph obtained from G by replacing every vertex of G by a triangle. Obviously $G^{\triangle}$ is also a non-planar 3-connected graph G distinct from K_5. By **5.1**, $G^{\triangle}$ contains $TopK_{3,3}$. Since $TopK_{3,3}$ does not have a triangle, in every subgraph $H = TopK_{3,3}$ of G at least one of any two incident edges is subdivided. Therefore any subset of edge-threads of H induces a matching and so at most three threads of H are edges.

A natural conjecture arises about the existence of $TopK_{3,3}$ with three edge-threads in any non-planar 3-connected graph G distinct from K_5. This conjecture was published in 1981 in [**21**]. Later we proved this conjecture by using ear–decompositions of 3-connected graphs (presented at the Moscow Seminar on Discrete Mathematics in December, 1981, published in [**22**], see also [**23**]). Thus we have the following strengthening of Kuratowski's planarity criterion:

6.3. *A 3-connected graph G distinct from K_5 is non-planar if and only if it has a special subdivision of $K_{3,3}$, namely a $TopK_{3,3}$ with three three disjoint edge-threads (or, in other words it has a circuit with three crossing edge-chords).*

In [**46**] C. Thomassen used a similar approach to prove the above conjecture.

7. A vertex in a matroid and the corresponding notion and dual notion for graphs

It turns out that it is natural to consider the planarity problem for 3-connected graphs (and also some other graph-theoretical problems and results) from the point of view of matroid theory [17, 18, 19]. We introduce the concept of a *vertex* in a matroid. This concept is helpful not only in that it facilitates understanding and obtaining simple proofs of certain classical results but also in that it reveals new knowledge about 3-connected graphs. Thus, consideration of a vertex in the matroid which is dual to the cycle matroid of a graph leads to the very natural and very useful concept of a non-separating circuit of a graph.

Given a connected matroid M a cocircuit C^* of M is called *a vertex* or *a non-separating cocircuit of* M if $M \setminus C^*$ is also a connected matroid. Let $\mathcal{N}^*(M)$ denote the list of non-separating cocircuits of M.

By matroid duality, we have a notion of a covertex. Given a connected matroid M a circuit C of M is called *a covertex* or *a non-separating circuit of* M if M/C is also a connected matroid. Let $\mathcal{N}(M)$ denote the list of non-separating circuits of M.

(Compare: *A point x of a convex set S is an extreme point of S if $S \setminus x$ is also a convex set*).

A graph is *cyclically connected* if the circuit matroid of the graph is connected, i.e. if any two edges belong to a common circuit. It is easy to see that a graph G with at least two edges is cyclically connected if and only if the graph obtained from G by deleting all isolated vertices is 2-connected.

Now we have the notions of a non-separating cocircuit and a non-separating circuit of a the cycle matroid of a graph. Given a cyclically connected graph G a cocircuit (i.e. a minimal edge cut) C^* of G is called *a matroid vertex* or *a non-separating cocircuit* of G if $G \setminus C^*$ is also a cyclically connected graph. Let $\mathcal{N}^*(G)$ denote the list of non-separating cocircuits of G.

Given a cyclically connected graph G a circuit (more exactly the edge set of a circuit) C of G is called *a covertex* or *a non-separating circuit* of G if G/C is also a cyclically connected graph. A *separating circuit* of G is a circuit of G which is not non-separating. Let $\mathcal{N}(G)$ denote the list of non-separating circuits of G.

It is easy to see that

7.1. *A non-separating cocircuit of G is a vertex star of G.*

For 3-connected graphs the reverse is obviously true.

7.2. *If G is 3-connected then the list $\mathcal{N}^*(G)$ of non-separating cocircuits of*

G is exactly the list $\mathcal{S}(G)$ of vertex stars of G.

A graph G is uniquely defined by its set $\mathcal{S}(G)$ of vertex stars, and the set $\mathcal{N}^*(G)$ of non-separating cocircuits of G is uniquely defined by the list $\mathcal{C}(G)$ of the edge sets of circuits of G. Therefore we proved [17, 18] the following well-known circuit isomorphism theorem for 3-connected graphs due to Whitney [57]:

7.3. *Let G be a 3-connected graph, G' be a graph without isolated vertices, and $e : E(G) \to E(G)'$ be a one-to-one mapping of the edge set of G onto the edge set of G' such that C is a circuit of G if and only if $e(C)$ is a circuit of G'. Then G and G' are isomorphic and there exists an isomorphism of G onto G' which induces e.*

P. Seymour informed me recently that J. Edmonds told him the same simple proof of the above theorem in 1976. Some simple proofs of Whitney's circuit isomorphism theorem for 2-connected graphs can be found in [29, 30, 48].

By using the notions of a non-separating circuit and a non-separating cocircuit of a graph, it is also very easy to prove [17, 18] the following theorem of Whitney on unique embedding of 3-connected planar graphs into the sphere.

Indeed let G_e be any embedding of a 3-connected planar graph G into the sphere. Let G_e^* be the embedded graph geometrically dual to G_e. Then G_e and G_e^* have dual matroids and so the list $\mathcal{N}(G)$ of non-separating circuits of G_e corresponds to the list $\mathcal{N}^*(G_e^*)$ of non-separating cocircuits of G_e^*. It is easy to see that if G_e is 3-connected then G_e^* is also 3-connected, and that any face-circuit of G_e corresponds to a vertex star of G_e^* and vice versa. By **7.2**, the list $\mathcal{N}^*(G_e^*)$ of non-separating cocircuits of G_e^* is exactly the list $\mathcal{S}(G_e^*)$ of vertex stars of G_e^*. Therefore we have :

7.4. *Let G be 3-connected and let G_e be any embedding of a 3-connected planar graph G into the sphere. Then the list $\mathcal{F}(G_e)$ of circuits-faces of G_e is exactly the list $\mathcal{N}(G)$ of non-separating circuits of G_e.*

We say that a *planar graph is uniquely embedded in the sphere* if every embedding of the graph in the sphere has the same set of circuits-faces. From **7.4** we have the following theorem of Whitney [58]:

7.5. *A 3-connected planar graph is uniquely embedded in the sphere.*

From **7.4** we also have the following theorem of W.T. Tutte [50]:

7.6. *Every edge of a 3-connected planar graph belongs to exactly two non-separating circuits.*

By Menger's theorem, for every edge $e = (x,y)$ of a 3-connected graph G there exist two paths $xR'y$ and $xS'y$ in $G \setminus e$ having exactly two vertices x and y in common, or, equivalently, there exist two circuits $R = R' \cup e$ and $S = S' \cup e$ of G containing only $\{e, x, y\}$ in common. A circuit C of G is an (R,e)–*circuit* if $R \cap C = \{e, x, y\}$; and so S is an (R,e)-circuit of G. Obviously for any (R,e)-circuit C of G there exists a block $B_R^e(C)$ of G/R containing $(R \cup C)/R$. Let $b_R^e(C)$ denote the number of edges of the block $B_R^e(C)$ of G/R. We say that an (R,e)-circuit Q is *extremal* if $b_R^e(Q) \geq b_R^e(C)$ for any (R,e)-circuit C of G. It is not difficult to prove [**17, 19**] that

7.7. *An extremal (R,e)-circuit of a 3-connected graph G is a non-separating circuit of G (distinct from R).*

Let Q be an extremal (R,e)-circuit of G. Obviously R is a (Q,e)-circuit of G. Therefore by the same reason as above there exists an extremal (Q,e)-circuit, say P, of G. By **7.7**, P is a non-separating circuit of G distinct from Q.

Therefore we have another proof of the following theorem due to Tutte [**50**]:

7.8. *Let G be a 3-connected graph and let $e = (x,y)$ be an edge of G. Then*
(a) *$e = (x,y)$ belongs to at least two non-separating circuits, moreover*
(b) *there exist two non-separating circuits P and Q of G such that $P \cap Q = \{x, e, y\}$.*

Theorem **7.8(a)** is a particular case of **8.2** below.

The above idea to consider a circuit which is extremal in a sense can be used to find some special subgraphs in a graph (see, for example, **8.7**–**8.9** and **12.4** below).

Note that from **7.2** we have the obvious "dual" result:

7.8*. *Every edge of a 3-connected graph belongs to exactly two non-separating cocircuits.*

Now we can formulate a graph planarity criterion in terms of non-separating circuits:

7.9. *Let G be a 3-connected graph. The following two conditions are equivalent:*
(a) *G is a planar graph, and*
(b) *every edge of G belongs to exactly two non-separating circuits.*

In [**17, 19**] a simple proof of this graph planarity criterion is given which does not use any known planarity criteria. As was noticed in [17] the planarity criterion **7.9** can also be obtained easily from **7.6** and **8.6** below due to W.T. Tutte [50] and MacLane's planarity criterion **3.3** [36].

Direct proofs are also given [17] that (b) is equivalent to the conditions (c) and (d) below:

(c) G does not contain a subdivision of K_5 or $K_{3,3}$, (see **3.1**), and
(d) G has a combinatorial dual graph G^* (see **3.2**).

Thus other proofs of Kuratowski's planarity criterion and Whitney's planarity criterion were found which are based on the criterion **7.9**.

7.10. Here is a simple proof [17] of Whitney's planarity criterion for 3-connected graphs [56] (actually a proof of $(b) \Leftrightarrow (d)$) based on **7.9**.

If G is planar then it has a geometrically dual graph G', and obviously G and G' are combinatorial dual graphs. Let us prove that if G is 3-connected, and there exists a graph G^* which is combinatorial dual to G then G is planar. Obviously the list $\mathcal{N}(G)$ of non-separating circuits of G corresponds to the list $\mathcal{N}^*(G^*)$ of non-separating cocircuits of G^*. By **7.1**, a non-separating cocircuit of G^* is a vertex star of G^*. Therefore each edge of G belongs to at most two non-separating circuits. By **7.8**, each edge of G belongs to at least two non-separating circuits. Therefore each edge of G belongs to exactly two non-separating circuits. By **7.9**, G is planar. Thus we proved Whitney's planarity criterion **3.2** for 3-connected graphs.

A direct proof of $(b) \Rightarrow (c)$ is outlined in **8.8**, **8.10**, and **8.11** below. Three non-separating circuits with a common edge in a 3-connected graph G can be used to find a subdivision of K_5 or $K_{3,3}$ in G [17] which gives a direct proof $(c) \Rightarrow (b)$.

The above idea to concider circuits having some extremal properties was also used to find some results for matroids. Let P and Q are distinct circuits of a matroid M. We say that Q *does not separate* P, writing $QnsP$, if there exists a component of M/Q containing $P \setminus Q$. Let us denote this component by $B_P(Q)$. Put $\mathcal{C}_P(M) = \{C \in \mathcal{C}(M) : CnsP, C \neq P\}$. For $Q, R \in \mathcal{C}_P(M)$ we write $Q \succeq^P R$ if $B_P(Q) \cap R = \emptyset$. This relation has the following property [17].

7.11. Let $P \in \mathcal{C}(M)$, and let $Q, R, S \in \mathcal{C}_P(M)$. If $Q \succeq^P R$ and $R \succeq^P S$, then $Q \succeq^P S$ (i.e. the relation $\succeq^P$ is transitive).

Therefore we can introduce the concept of the $\sim^P$-*equivalence relation* and the concept of a $\succeq^P$-*minimal equivalence class* of circuits in $\mathcal{C}_P(M)$. By using these concepts we proved in 1978 the following generalization of theorem **7.8** for matroids.

7.12. *Every element of a 3-connected binary matroid with at least 4 elements belongs to at least two non-separating circuits.*

Uniform matroids show that the above statement is not true for an arbitrary matroid. This theorem is a particular case of **8.2m** below.

Proposition **7.8*** can be reformulated as follows: If a 3-connected matroid M is a circuit matroid of a graph, then every element of M belongs to exactly two non-separating cocircuits of M. We proved that for binary matroids the reverse is also true.

7.13. *A 3-connected binary matroid M is a circuit matroid of a graph if and only if every element of M belongs to exactly two non-separating cocircuits of M.*

The scheme of the proof of **7.13.** is similar to that of for **7.8** [17, 19].

Recently W.H. Cunningham drew my attention on the paper [5] due to R.E. Bixby and W.H. Cunningham where theorems **7.12.** and **7.13.** were proved in another way (by using some concepts and results due to W.T. Tutte [51, 52]). They also investigated for a binary matroid M the properties of so-called avoidance graph of components of $M\setminus C$ where C is a circuit of M. By using these results they gave a recursive procedure for testing a matroid for 3-connectivity.

8. More about non-separating circuits in a graph

Here are some other results on non-separating circuits of a graph found in [17, 18].

For $k \geq 3$, a subdivision of the k-bond (i.e. the graph with two vertices and k parallel edges) is called *a k-necklace*; it has exactly two vertices of degree $k \geq 3$ which are called *essential vertices of the k-necklace*, and k threads connecting the essential vertices. A subgraph N of G is a *proper k-necklace of G* if each thread of N is also a thread of G.

8.1. *Let G be a 2-connected graph distinct from a circuit and a k-necklace, $k \geq 3$. Then*
   (a) *if G has no proper 3-necklaces then G has at least two non-separating circuits whose intersection is either empty or a path,*
   (b) *if G has no non-separating circuits then it has at least two proper 3-necklaces with distinct pairs of essential vertices,*
   and consequently,
   (c) *if G has exactly one proper 3-necklace then it has at least one non-*

separating circuit,

(d) *if G has exactly one non-separating circuit then it has at least one proper 3-necklace.*

It is easy to see that these results are sharp. The proof (**8.7** and **8.8** in [**17**]) of the above theorem provides polynomial algorithms for solving the corresponding problems. The corresponding dual (in the matroid sense) theorem is also given (**8.8*** in [**17**]). The analogous result was also proved for connected binary matroids. Theorem **8.1(b)** was proved in another way in [**45**].

Since any forest with at least one edge has at least two vertices of degree 1, we have from **7.2** the following obvious fact:

8.2*. *Let G be 3-connected. If $A \subseteq E(G)$ includes no circuit, then there exist two non-separating cocircuits of G each of which has exactly one common edge with A. Moreover if $|A| \geq 2$ then there exist two non-separating cocircuits P and Q of G and two edges p and q in A such that $A \cap P = p$ and $A \cap Q = q$.*

It turns out that the "dual result" is also true, namely [**17**]:

8.2. *Let G be 3-connected. If $A \subseteq E(G)$ includes no cocircuit (i.e. $G \backslash A$ is connected), then there exist two non-separating circuits of G each of which has exactly one common edge with A. Moreover if $|A| \geq 2$ then there exist two non-separating circuits P and Q of G and two distinct edges p and q in A such that $A \cap P = p$ and $A \cap Q = q$.*

Note that **7.8(a)** is a particular case of the above theorem.

The result analogous to **8.2*** and **8.2** is not true for an arbitrary matroid but is true for a binary matroid:

8.2m. *Let M be a binary 3-connected matroid with at least 4 elements. If A is an independent set of M, then there exist two non-separating cocircuits of G each of which has exactly one common element with A. Moreover if $|A| \geq 2$ then there exist two non-separating cocircuits P and Q of M and two distinct elements p and q in A such that $A \cap P = p$ and $A \cap Q = q$.*

Note that **7.12** is a particular case of the above theorem.

It is easy to see that in a 2-connected graph whose edges are colored in two colors (so that no color class is empty) there exist at least two two-colored vertex stars. Therefore we have from **7.2** the following simple fact:

8.3*. *In a 3-connected graph with two-colored edges there exist at least two two-colored non-separating cocircuit.*

It turns out that again the "dual result" is also true, namely [**17**]:

8.3. *In a 3-connected graph with two-colored edges there exist at least two two-colored non-separating circuits.*

Again the result analogous to **8.3*** and **8.3** is not true for an arbitrary matroid but is true for a 3-connected binary matroid with at least 4 elements.

P. Seymour noted that **8.3** can be obtained from **8.2** and **8.3*** as follows. Let G be a 3-connected graph with two-colored edges. By **8.3***, there exists a two-colored non-separating cocircuit C^* of G. Let A be the edges of C^* in color one. Let A be the edges of C^* in color one. Since A is a proper subset of the cocircuit C^*, it includes no cocircuit of G. By **8.2**, there are two non-separating circuits, say C_1 and C_2, each with exactly one edge in A. Since $|C_i \cap C^*| \neq 1$, we have $C_i \cap C^* \setminus A \neq \emptyset$, and so the non-separating circuits C_1 and C_2 are two-colored.

Non-separating circuits can be used to characterize subdivisions of 3-connected graphs. Two edges a and b of G are said to be $\mathcal{N}$-*equivalent* if there exists a sequence $a = e_0 C_1 e_1 C_2 ... e_{k-1} C_k e_k = b$ such that $e_0, ..., e_k$ are edges and $C_1, ... C_k$ are non-separating circuits of G and $e_i, e_{i+1} \in C_{i+1}, i = 0, 1, ..., k-1$. One can prove [**17**]:

8.4. *Let G be a graph distinct from a circuit and from a 3-necklace. Then G is a subdivision of a 3-connected graph if and only if every two edges of G are $\mathcal{N}$-equivalent.*

The result analogous to **8.4** also holds for binary matroids.

Whitney's circuit isomorphism theorem **7.3** can be strengthened by replacing in the theorem the set of all circuits by the set of non-separating circuits of a 3-connected graph, namely [**17, 18**]:

8.5. *Let G be a 3-connected graph, G' be a graph without isolated vertices, and $e : E(G) \to E(G')$ be a one-to-one mapping such that C is a non-separating circuit of G if and only if $e(C)$ is a non-separating circuit of G'. Then G and G' are isomorphic and there exists an isomorphism of G onto G' which induces e.*

An outline of some results on refinements, strengthenings, and generalizations of Whitney's circuit isomorphism theorem can be found in [**25, 29**].

Theorem **8.5.** can be generalized to binary matroids:

8.5m. *Let M be a 3-connected binary matroid, M' be a binary matroid, and $e : E(M) \to E(M')$ be a one-to-one mapping such that C is a non-separating circuit of M if and only if $e(C)$ is a non-separating circuit of M'. Then e is an isomorphism of M onto M' (and so C is a circuit of M if and only if $e(C)$ is a* *circuit of M').*

From **8.5** we have the following theorem due to W.T. Tutte [**50**]:

8.6. *A 3-connected graph G has an independent set of $|E(G)| - |V(G)| + 1$ of non-separating circuits (which is a cycle basis of G).*

In 1980 a referee drew my attention to a remarkable paper [**50**] W.T. Tutte published many years before. In this paper a concept of a non-separating circuit of a graph (a so-called *peripheral circuit*) was introduced for another reason and theorems **7.6**, **7.8**, and **8.6** were proved in another way (see **(2.3)**, **(2.5)**, **(3.1)** and **(3.2)** in [**50**]). Different results on non-separating circuits of a graph can also be found, for example, in [**32, 45, 47**]. Recently C. Thomassen obtained interesting results [**47**] showing that non-separating circuits of a graph play an essential role in embeddings of a graph not only into the sphere but also into orientable 2-dimensional surfaces.

Consider a pair (F, D) where F is a graph and D is a circuit of F. We say that (H, C) is a *subdivision* of (F, D), $(H, C) = Top(F, D)$, if H is a subdivision of F and C is a circuit of H which is the corresponding subdivision of D in F. Let $\mathcal{F}(G)$ denote the set of pairs (H, C) such that H is a subgraph of G, and $(H, C) = Top(F, D)$. Suppose that F is a 2-connected graph and D is a non-separating circuit of F. Then there exists a block $B(H, C)$ of G/C containing H/C. Let $b(F, C)$ denote the number of edges of the block $B(F, C)$. A pair (R, Q) from $\mathcal{F}(G)$ is called *extremal* if $b(R, Q) \geq b(H, C)$ for every $(H, C) \in \mathcal{F}(G)$. Given $(H, C) \in \mathcal{F}(G)$, let $A(H, C)$ denote the set of vertices in C which are of degree at least 3 in H. If all vertices of D are of degree at least 3 in F then $|V(D)| = |A(H, C)|$ for every $(H, C) \in \mathcal{F}(G)$. Given a circuit T in G and 4-vertex subset $X = \{x_1, x_2, x_3, x_4\}$ of T (the vertices in X are listed along the circuit), we say that T is an *almost non-separating circuit of G with respect to X* if $G \setminus V(T)$ is connected, T has at most two chord-edges with both end-vertices in X, and any such edge-chord is either (x_1, x_3) or (x_2, x_4). We can prove:

8.7. *Suppose that G is a 3-connected graph, F is a 2-connected graph, G has a subgraph $TopF$, and D is a non-separating circuit of F which has three or four vertices and each of these vertices is of degree 3 in F. Let (R, Q) be an extremal pair from $\mathcal{F}(G)$. Then*
   (a) *if D has three vertices then Q is a non-separating circuit of G,*
   (b) *if D has four vertices then Q is an almost non-separating circuit of G with respect to $A(R, Q)$.*

Now from **5.1** and **8.7** we have the following refinement of Kuratowski's planarity criterion **3.1** [**17**]:

8.8. *A 3-connected graph G distinct from K_5 is non-planar if and only if it contains a subgraph $R = TopK_{3,3}$ such that one of its circuits Q of top length 4 is an almost non-separating circuit of G with respect to $A(R,Q)$.*

From **5.5** and **8.7** we have:

8.9. *Let G be a 3-connected graph with at least 6 vertices distinct from a wheel and from $K_{3,n}^i$, $i \in \{0,1,2,3\}$ and $n \geq 3$. Then G contains a subgraph $R = TopL$ such that one of its circuits with top length three is a non-separating circuit of G.*

We can also prove the following [**17**]:

8.10. *Let G be a 3-connected graph, let $R = TopK_{3,3}$ be a subgraph of G, and let Q be a circuit of R such that it is of top length 4 in R and it is an almost non-separating circuit of G with respect to $A(R,Q)$. Then each edge of C belongs to at least three non-separating circuits of G.*

From **8.8** and **8.10** we have [**17**]:

8.11. *Let G be a 3-connected graph. If G contains a subdivision of K_5 or $K_{3,3}$ then G has at least four edges belonging to at least three non-separating circuits of G.*

9. Triangle and 3-cut reductions of the graph planarity problem

In **4.1**, **4.2**, and **4.3** three simple reductions for the graph planarity problem are described. Reductions **4.2** and **4.3** use vertex 1 and 2-cuts of the graph, respectively. The natural question arises [**24, 26**] whether it is possible to do some more steps in this direction, namely to use vertex 3-cuts, 4-cuts and so on for deeper reductions of the problem.

A vertex 3-cut X of G is called *non-essential* if $G \setminus X$ consists of exactly two components one of which is an isolated vertex, and *essential* otherwise. For example, each vertex 3-cut of $K_{3,3}$ is essential. A 3-connected graph is called *inner 4-connected* if it has no essential 3-cut.

Consider a 3-connected graph G which has a vertex 3-cut $X = \{x_1, x_2, x_3\}$ so that $G = F_1 \cup F_2$ and $F_1 \cap F_2 = X$. Let G_i, $i = 1,2$, be obtained from F_i by adding a new vertex v_i and three new edges (v_i, x_j), $j = 1,2,3$ (see Fig. 4).

Note that this splitting operation can be used to obtain a decomposition of 3-connected graphs similar to that we have for 2-connected graphs.

9.1. *A 3-connected graph G is planar if and only if G_1 and G_2 are planar.*

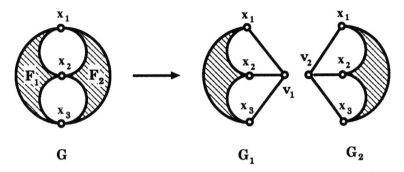

FIGURE 4.

The proof is easy [26] as follows. Since G is 3-connected, for any vertex y of $F_1 \setminus X$ there are three disjoint paths P_1, P_2, and P_3 from y to X in G (actually in F_1). Then the subgraph $F_2 \cup P_1 \cup P_2 \cup P_3$ of G is a subdivision of G_2. By the same reason G has a subgraph which is a subdivision of G_1. Therefore if G is planar then both G_1 and G_2 are planar. Conversely if G_1 and G_2 are planar then G is also planar because two equal triangles T and T' in the plane with the vertex sets $\{x_1, x_2, x_3\}$ and $\{x'_1, x'_2, x'_3\}$, respectively, can be identified by the operations of moving and reflecting the plane such that the vertices x_i and x'_i coincide, $i = 1, 2, 3$.

Note that two equal squares S and S' in the plane with the vertices labeled along the boundaries $\{x_1, x_2, x_3, x_4\}$ and $\{x'_1, x'_3, x'_2, x'_4\}$, respectively, cannot be identified by the operations of moving and reflecting the plane such that the vertices x_i and x'_i coincide, $i = 1, 2, 3, 4$. Therefore if we replace in the above proposition "the vertex 3-cut X" by "a vertex 4-cut" then the planarity of G_1 and G_2 do not imply the planarity of G. Thus the planarity problem for 3-connected graphs can be reduced to the problem for inner 4-connected graphs. But a "4-cut" reduction of the problem (if any) seems to be not as natural as the previous reductions.

Note that some other problems on disjoint paths in a graph can be reduced to the problem for inner 4-connected graphs because at most one path from a set of disjoint paths can go through a vertex 3-cut and come back (see for example **12.4**).

Moreover we can get rid of triangles when considering the planarity problem for inner 4-connected graphs. Suppose that an inner 4-connected graph G contains a triangle T with vertex set $\{t_1, t_2, t_3\}$. If G has a cubic vertex v adjacent to t_1, t_2, t_3 only then put $G' = G \setminus E(T)$. If G has no such cubic vertex then let G' be obtained from G by *replacing the triangle T by a star*, namely by deleting the edges of T and by adding a new vertex v and three new edges (v, t_j), $j = 1, 2, 3$. Clearly the planarity of G' implies the planarity of G. Suppose that G is inner 4-connected and planar. Then if T is a face of G then G' is also planar. Otherwise since G is 3-connected, by **7.8**, T is a separating circuit of G. Since G has no parallel edges, $V(T)$ is a vertex 3-cut of G. Since G is inner 4-connected, it has no essential 3-cut. Therefore $V(T)$ is not an essential 3-cut of G and so there exists a cubic vertex v of G adjacent to t_1, t_2, t_3. Then obviously G' is planar. Thus we proved that [**26**]:

9.2. *An inner 4-connected graph G is planar if and only if G' is planar.*

A graphs G is called *quasi 4-connected* if G is 3-connected and has no essential 3-cut or triangle. Let $\mathcal{Q}$ denote the class of quasi 4-connected graphs. Note that the class of cubic quasi 4-connected graphs is exactly the class of so called *cubic cyclically 4-connected graphs*. The class $\mathcal{Q}$ of quasi 4-connected graphs turns out to be a very natural class of graphs and some results about this class may be interesting in their own right.

By **9.1** and **9.2**, a planarity problem for an arbitrary graph can be easily reduced to the same problem for quasi 4-connected graphs. Thus quasi 4-connected graphs are, even more than 3-connected graphs the *essential bricks for the planarity problem*.

10. Subdivisions of K, M, and N in quasi 4-connected graphs

Let K be the skeleton of the 3-dimensional cube, N be obtained from K by adding a main diagonal of the 3-dimensional cube, and M^k denotes the so called *Möbius k-ladder* or *Wagner's graph*, i.e. M^k is the $2k$-circuit with k main diagonals, $k \geq 4$. Thus $M^4 = M$ is a *a twisted cube* (see Fig. 5). Obviously:

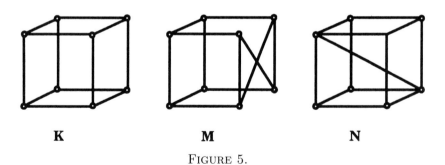

K **M** **N**

FIGURE 5.

10.1. *K and M are quasi 4-connected graphs.*

By **5.4**, any quasi 4-connected graph contains two disjoint circuits of length at least 4. Thus by using Menger's theorem, we have [**24, 26**]:

10.2. *Any quasi 4-connected graph contains a subdivision of K or M (so that K and M are minimal quasi 4-connected graphs).*

"Almost all" quasi 4-connected graphs turn out to contain a subdivision of K. More precisely [24, 26]:

10.3. *A quasi 4-connected graph G does not contain a TopK (and consequently contains a TopM) if and only if G is either a Möbius k–ladder, $k \geq 4$, or the Petersen graph, or is obtained from the Petersen graph by contracting an edge.*

It turns out that a number of results described above can be refined for quasi 4-connected graphs. For instance, the following refinement of Kuratowski's planarity criterion for quasi 4-connected graphs can be proved [24, 26].

10.4. *A quasi 4-connected graph G is planar if and only if it does not contain a subdivision of M or N.*

As we mentioned before K_5 and $K_{3,3}$ play different roles in Kuratowski's theorem (see **5.1** and **5.2**). It turns out that the graphs M and N also play different roles in the above theorem, i.e. "almost all" non-planar quasi 4-connected graphs contain a subdivision of M, namely [24, 26]:

10.5. *A non-planar quasi 4-connected graph G does not contain a TopM (and consequently contains a TopN) if and only if G is a bipartite graph and one of its parts consists of four vertices.*

From **10.3** and **10.5** we have:

10.6. *A quasi 4-connected graph G does not contain a TopM (and so it contains a TopK) if and only if G is either planar or a bipartite graph such that one of its parts consists of four vertices.*

Note that the statements **10.2**, **10.3**, and **10.5** on quasi 4-connected graphs are similar, respectively, to the statements **5.6**, **5.5**, and **5.1** on 3-connected graphs.

11. An ear-like decomposition for quasi 4-connected graphs

The class $\mathcal{Q}$ of quasi 4-connected graphs admits a good constructive description, namely, there exists an *ear-like decomposition* of quasi 4-connected graphs [24, 26]. In order to describe this decomposition we need some more definitions.

Given a path P in G let as before $l_G(P)$ denote the number of inner vertices of P which are of degree at least 3 in G. We define $l_G(x,y) = \min\{l_G(xPy) : xPy$ is a path of G $\}$. Thus $l_G(x,y)$ can be treated as the *topological distance* between the vertices x and y in G. Two distinct vertices are *top adjacent* if they are on topological distance 0.

Let $\mathcal{T}$ denote the set of pairs (G, F) of graphs such that F is a path with the end-vertices x and y, $G \cap F = \{x, y\}$ and $l_G(x, y) \geq 2$; we say that $G' = G \cup F$ is *obtained from G by operation T* (see Fig. 6). If in addition x and y are vertices of degree 2 in G then the set of such pairs (G, F) in $\mathcal{T}$ is denoted by $\mathcal{H}$; we say that $G' = G \cup F$ is obtained from G operation H.

Let $\mathcal{X}$ denote the set of pairs (G, F) of graphs such that F consists of two disjoint paths $x_1 P_1 y_1$ and $x_2 P_2 y_2$, $G \cap F = \{x_1, y_1, x_2, y_2\}$ and G has a 3-necklace with three inner disjoint paths $a X_0^i x_1^i X_1^i x_2^i X_2^i b$ with $x_i^i = x_i$, $i = 1, 2$, and $Y = a Y_2 y_2 Y_1 y_1 Y_0 b$, where Y and X_j^i are threads of G for $i = 1, 2$ and $j = 0, 1, 2$; we say that $G' = G \cup F$ is *obtained from G by operation X* (see Fig. 6).

Let $\mathcal{Y}$ denote the set of pairs (G, F) of graphs such that F consists of three inner disjoint paths xP_1y_1, xP_2y_2 and xP_3y_3 with a single common vertex x (and so $F = TopK_{1,3}$), $G \cap F = \{y_1, y_2, y_3\}$, $l_G(y_i, y_j) = 1$ for any $y_i \neq y_j$, G has no vertex top adjacent to each y_i, and at least one of the vertices y_1, y_2, y_3 is of degree 2 and at least one of them is of degree at least 3 in G (and so there exists a circuit in G of top length 4 or 5 containing all y_i's); we say that $G' = G \cup F$ is *obtained from G by operation Y* (see Fig. 6).

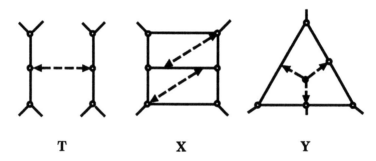

FIGURE 6. Operations T, X, and Y.

In Fig. 6 the degree of every "white" vertex in G is at least the degree in the figure, every "solid" edge in the figure is a thread or a subthread in G, and the "dashed" lines are the threads of F.

Let $\mathcal{E} = \mathcal{T} \cup \mathcal{X} \cup \mathcal{Y}$. A sequence $D = (G_0; T_1, ..., T_k)$ of subgraphs of G is called a $\mathcal{E}$- *sequence of G* if for all $i \in \{0, ..., k-1\}$ we have $(G_i, T_{i+1}) \in \mathcal{E}$ where $G_i = G_{i-1} \cup T_i$. If in addition $G_k = G$ then D is called an $\mathcal{E}$-*decomposition* of the graph G.

We proved the following decomposition theorem for quasi 4-connected graphs (presented at the Tbilisi Workshop on Discrete Mathematics in 1977) [**24, 26**]:

11.1. *Let G be a subdivision of a quasi 4-connected graph, and let F be a graph without parallel edges or triangles and with every vertex of degree at least 3. Suppose that G contains a subgraph $TopF$. Then there exists an $\mathcal{E}$-decomposition $D = (G_0; T_1, ..., T_k)$ of G such that, setting $G_i = G_0 \cup T_1 \cup ... \cup T_i$ for $i = 1, ..., k$:*
  (a) *$G_0/E = TopF$ for some circuit free edge subset E of $E(G_0)$ (in particular if F is cubic then $G_0 = TopF$), and*
  (b) *if G_i is a subdivision of a quasi 4-connected graph, then so are all $G_{i+1}, \ldots G_k$.*

Our proof of this theorem provides a polynomial time algorithm for finding the corresponding $\mathcal{E}$-decomposition of a graph or revealing that the graph has no such decomposition (and therefore is not a subdivision of a quasi 4-connected graph). The algorithm is recursive. The $\mathcal{E}$-decomposition of the current subgraph have to be rebuilded sometimes to make it possible to enlarge the subgraph.

P. Seymour informed me recently that N. Robertson presented a version of the above theorem at the Rome conference in 1972, but did not publish it yet.

The decomposition theorem **11.1** is a very useful tool for investigating the class of quasi 4-connected graphs. We found different strengthenings of this theorem which guarantee the existence of an $\mathcal{E}$-decomposition with some special properties. A strengthened version of theorem **11.1** was used to find some results analogous to the circuit isomorphism theorem due to H. Whitney [26, 29]. One of these strengthenings shows that we can get reed of operation Y, but the algorithm of finding such $\mathcal{E}$-decomposition becomes more complicated (more complicated rebuildings of the current $\mathcal{E}$-sequence are required), and in this case we cannot guarantee some useful properties of an $\mathcal{E}$-decomposition.

Let $\mathcal{E}' = \mathcal{T} \cup \mathcal{X}$. A sequence $D = (G_0; T_1, ..., T_k)$ of subgraphs of G is called a $\mathcal{E}'$-*sequence of* G if for all $i \in \{0, ..., k-1\}$ we have $(G_i, T_{i+1}) \in \mathcal{E}'$ where $G_i = G_{i-1} \cup T_i$. If in addition $G_k = G$ then D is called an $\mathcal{E}'$-*decomposition* of the graph G.

11.1' *Let G be a subdivision of a quasi 4-connected graph, and let F be a graph without parallel edges or triangles and with every vertex of degree at least 3. Suppose that G contains a subgraph $TopF$. Then there exists an $\mathcal{E}'$-decomposition $D = (G_0; T_1, ..., T_k)$ of G such that, setting $G_i = G_0 \cup T_1 \cup ... \cup T_i$ for $i = 1, ..., k$:*
  (a) *$G_0/E = TopF$ for some circuit free edge subset E of $E(G_0)$ (in particular if F is cubic then $G_0 = TopF$), and*
  (b) *if G_i is a subdivision of a quasi 4-connected graph, then so are all $G_{i+1}, \ldots G_k$.*

From **10.1**, **10.2**, and **11.1'** we have [24, 26]:

11.2 *Let G be a subdivision of a quasi 4-connected graph, and let K be the* skeleton-graph of the 3-dimensional cube and M be the Möbius 4-ladder (see Fig 5). Then G can be obtained from K or M by a sequence of operations T or X so that any intermediate graph is also quasi 4-connected. Moreover G can be obtained by these operations from any cubic graph F without parallel edges or triangles if G contains a subdivision of F.

As we have already mentioned the class of cubic quasi 4-connected graphs is exactly the class of cubic cyclically 4-connected graphs. Therefore we have, in particular[24, 26]:

11.3. Let G be a subdivision of a cubic cyclically 4-connected graph, and let F be a simple cubic graph without triangles. Suppose that G contains a subgraph $TopF$. Then there exists an $\mathcal{E}$-decomposition $D = (G_0; T_1, ..., T_k)$ of G such that, setting $G_i = G_0 \cup T_1 \cup ... \cup T_i$ for $i = 1, ..., k$:

(a) $G_0 = TopF$,
(b) $(G_i, T_{i+1}) \in \mathcal{H}$, and
(c) if G_i is a subdivision of a cubic cyclically 4-connected graph, then so are all $G_{i+1}, ..., G_k$.

From **10.1**, **10.2**, and **11.1** (or **11.1'**) we have:

11.4. Let G be a subdivision of a cubic cyclically 4-connected graph. Then G can be obtained from K or M by a sequence of operations H, so that any intermediate graph is also cubic cyclically 4-connected. Moreover G can be obtained by these operations from any simple cubic graph F without triangles if G contains a subdivision of F.

This theorem was proved before in [16].

The decomposition theorem **11.1'** enables us to give the following simple proof of the above refinement **10.4** of Kuratowski's planarity criterion for quasi 4-connected graphs. Let us consider the class $\mathcal{L}$ of non-planar quasi 4-connected graphs, and let $G \in \mathcal{L}$. Then by **5.1**, G contains a subdivision of $K_{3,3}$. By **11.1'**, G can be obtained from $K_{3,3}$ by operations T or X. Let us apply each of these two operations to $K_{3,3}$. The operation T gives the graph M, and the operation X gives the graph N (see Fig. 7). (As to operation Y when applied to $K_{3,3}$ it gives the graph S (see Fig. 7) which can also be obtained from $K_{3,3}$ by two operations T.) Therefore any non-planar quasi 4-connected graph contains a subdivision of M or N. Thus we proved **10.4**. Moreover we proved that

11.5. Any non-planar quasi 4-connected graph can be obtained from M or N by a sequence of operations T or X.

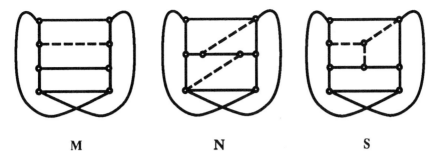

FIGURE 7. Applying operations T, X, and Y to $K_{3,3}$.

12. Non–separating circuits in quasi 4-connected graphs

The planarity criterion **7.13** can also be refined for quasi 4-connected graphs. Let us rewrite **7.13** as follows:

12.0. *Suppose that G is 3-connected. Then*
  (a) *if G is planar then EACH edge of G belongs to exactly two non-separating circuits, and*
  (b) *if G is non-planar then there EXISTS an edge of G belonging to at least three non-separating circuits.*

Now we have for quasi 4-connected graphs [**28**]:

12.1. *Let G be a quasi 4-connected graph or a 4-connected graph. If G is non-planar then EACH edge of G belongs to at least three non-separating circuits.*

To illustrate this result let us consider a 4-connected triangulation T of the plane. Then by **12.0**(a), EACH edge of T belongs to exactly two non-separating circuits which are triangles. Now let us add one more edge between non-adjacent vertices of G. Then the resulting graph G' is non-planar and 4-connected. By **12.1**, EACH edge of G belongs to at least three non-separating circuits. Therefore the local action (adding to G only one edge) results in a global effect.

Theorem **12.1** follows easily from the following stronger result [**28**]. Consider a 3-connected graph G which has a vertex 3-cut $X = \{x_1, x_2, x_3\}$ so that $G = F_1 \cup F_2$ and $F_1 \cap F_2 = X$ (see Fig. 4). Given an edge subset A of F_1, we say that A *is cut off by a vertex 3-cut X in G.*

12.2. *Let G be a 3-connected non-planar graph, and let $e = (x, y)$ be an edge of G. Suppose that G has no essential 3-cut which cut e off. Then e belongs to exactly two non-separating circuits if and only if e belongs to exactly two triangles and one of the end-vertices of e is of degree 3 in G.*

The planarity criterion **8.8** (which is a refinement of Kuratowski's planarity criterion) can also be refined for quasi 4-connected graphs [**26, 28**]:

12.3. *Let G be a non-planar quasi 4-connected graph. Then for any non-separating circuit C of G there exists $H = TopK_{3,3}$ in G such that C is a circuit of H with top length 4 in H.*

Note that **12.1** for quasi 4-connected graph also follows immediately from **7.12**, **8.10**, and **12.3**.

Theorem **8.7** can be used to obtain different results on some special subgraphs with non-separating circuits of a quasi 4-connected graph similar to **8.8** and **8.9** for 3-connected graphs. Here is an example of these results.

12.4. *Any quasi 4-connected graph G contains a subgraph H such that H is either $TopK$ or $TopM$ and some top length 4 circuit of H is a non-separating circuit of G.*

Theorem **12.3** follows easily from the following criterion of existence of two disjoint and crossing path-chords of a circuit in a 3-connected graph (found in 1979 and published in [**20**]). Consider a 3-connected graph G which has a vertex 3-cut $X = \{x_1, x_2, x_3\}$ so that $G = F_1 \cup F_2$ and $F_1 \cap F_2 = X$. Let G_1 be obtained from F_1 by adding a new vertex v_1 and three new edges (v_1, x_j), $j = 1, 2, 3$ (see Fig. 4). Let C be a circuit of length at least 4 in the subgraph F_1 of G, so that C is also a circuit of G_1 and C is cut of by X in G. It is clear that C has two disjoint and crossing path-chords in G if and only if C has two disjoint and crossing path-chords in G_1. Therefore we can consider the corresponding refined problem.

12.5. *Let G be a 3-connected graph and C be a circuit of G of at least four vertices. Suppose that G has no essential vertex 3-cut which cut C off. Then C does not have two disjoint and crossing path-chords in G if and only if G is planar and C is a face-circuit of G.*

In [**39, 41, 42, 44**] a criterion was given for the existence of two disjoint paths in a graph G between two pairs of vertices, say, x_1, y_1 and x_2, y_2. This result follows easily from **12.4** if we add to G four artificial edges (x_1, x_2), (x_2, y_1), (y_1, y_2), (y_2, x_1) and apply criterion **12.4** to the resulting graph G' with the artificial 4-circuit $C = (x_1, x_2, y_1, y_2, x_1)$. By means of **12.5** it is easy to prove the following [**20**]:

12.6. *Let G be an inner 4-connected graph. Then the following conditions are equivalent:*

(a) G is a maximal planar graph (i.e. a triangulation of the sphere) or a non-planar graph,

(b) any circuit C of G with at least four vertices has two crossing chords-paths in G,

(c) any non-separating circuit of G of at least 4 edges is a circuit of top length 4 in a subgraph $H = TopK_{3,3}$ of G,

(d) for any two disjoint edges e_1 and e_2 of G there exists a subgraph $H = TopK_4$ in G and two disjoint threads T_1 and T_2 of H such that $e_1 \in T_1$ and $e_2 \in T_2$,

(e) for any two disjoint edges e_1 and e_2 of G there exists a subgraph $H = TopK_4$ in G and two disjoint threads T_1 and T_2 of H such that $e_1 = T_1$ and $e_2 \in T_2$,

(f) for any two distinct pairs $\{x_1, y_1\}$ and $\{x_2, y_2\}$ of vertices of G there exist two disjoint paths $x_1 P_1 y_1$ and $x_2 P_2 y_2$ in G.

The equivalence $(a) \Leftrightarrow (f)$ above is a strengthening of Jung's theorem on 4-connected graphs [**15**] (a simpler proof of Jung's theorem was given by C. Thomassen [**44**]). I have been recently informed that some of the results on inner 4-connected graphs described here have been found independently by N. Robertson [**37**].

13. Some refinements of Whitney's planarity criterion

Now the question arises whether it is possible to find some natural refinements, strengthenings or generalizations of Whitney's planarity criterion [**56**] (see **3.2**) and related results. Whitney's criterion is based on the concept of a combinatorial or matroid duality of graphs.

We recall that two graphs G and G^* are called *combinatorial dual* or *matroid dual* if there exists a one-to-one mapping $e : E(G) \to E(G^*)$ of the edge set of G onto the edge set of G^* such that C is a circuit of G if and only if $e(C)$ is a cocircuit of G^*. The correspondence $e : E(G) \to E(G^*)$ will be called *a circuit duality of G onto G^**.

The concept of circuit duality can be generalized in different directions [**30, 31**]. Let Λ be an algorithm which for any finite set E of elements and for any set $\mathcal{C} \subseteq 2^E$ of subsets of E gives a set of subsets of E: $\Lambda(E, \mathcal{C}) \subseteq 2^E$. Given a graph G, put $\Lambda(G) = \Lambda(E(G), \mathcal{C}(G))$ and $\Lambda^*(G) = \Lambda(E(G), \mathcal{C}^*(G))$.

We say that G is Λ-*dual* (Λ-*semi-dual*) to G^* if there exists a one-to-one mapping $e : E(G) \to E(G^*)$ of the edge set of G onto the edge set of G^* such that $A \in \Lambda(G) \Leftrightarrow e(A) \in \Lambda^*(G^*)$ (respectively, $A \in \Lambda(G) \Rightarrow e(A) \in \Lambda^*(G^*)$).

A circuit of G without edge-chords is called a *hole* of G. In other words a circuit A of G is a *hole* of G if either A is a loop or G/A has the same set of loops as G. Now the notion of a *cohole* (which is matroid dual to a hole) can be defined. A cocircuit A of G is a *cohole* of G if either A is a coloop (an isthmus) or $G \setminus A$ has the same set of coloops as G. If $\Lambda(G)$ is the set of holes of G, then $\Lambda^*(G)$ is the set of coholes of G. Thus we get the concepts of *hole duality* and *hole semi-duality of graphs*. We can prove [**30**]:

13.1. *Let G be a graph without parallel edges. Then G is hole dual to G^* if and only if G is circuit dual to G^*.*

Therefore we have the following refinement of Whitney's planarity criterion.

13.1'. *A graph G without parallel edges is planar if and only if there exists a graph G^* which is hole dual to G.*

If $\Lambda(G) = \mathcal{N}(G)$ (the list of non-separating circuits of G) then we have the concepts of *$\mathcal{N}$-duality* and *$\mathcal{N}$-semi-duality of graphs*. One can prove [**31**]:

13.2. *Let G be a 3-connected graph. Then G is $\mathcal{N}$-semi-dual to G^* if and only if G is circuit dual to G^*.*

Therefore we have the following refinement of Whitney's planarity criterion for 3-connected graphs.

13.2'. *A 3-connected graph G is planar if and only if there exists a graph G^* which is $\mathcal{N}$-semi-dual to G.*

If $\Lambda(G) = \mathcal{C}(G)$ (the list of the edge sets of circuits of G) then we have the original concept of *circuit duality* and also the concept of *circuit semi-duality* of graphs. The following is true [**31**]:

13.3. *Let G be a 3-connected graph. Then G is circuit semi-dual to G^* if and only if G is circuit dual to G^*.*

It is not difficult to prove the following statement for matroids. We say that M is *semi-isomorphic* to M' if there exists a one-to-one correspondence $e : E(M) \to E(M')$ of the ground set of a matroid M the ground set of a matroid M' such that $A \in \mathcal{C}(M) \Rightarrow e(A) \in \mathcal{C}(M')$ where $\mathcal{C}(M)$ is the set of circuits of M. A circuit of a matroid M is *Hamiltonian* if it contains a base of M.

13.4. *Let M be a matroid. There exists a matroid semi-isomorphic but not isomorphic to M if and only if M has no Hamiltonian circuit.*

From **13.3** and **13.4** we have the following refinement of Whitney's planarity criterion.

13.5. *Let G be a 3-connected or Hamiltonian graph. Then G is planar if and only if there exists a graph G^* which is circuit semi-dual to G.*

If $\Lambda(G) = \mathcal{C}^*(G)$ (the set of the cocircuits of G) then we have the concept of *cocircuit duality* (which is the same as circuit duality), and also the concept of *cocircuit semi-duality* of graphs. The following can be proved [31]:

13.6. *Let G be a 3-connected graph, and let every non-separating circuit of G contain a vertex of degree 3 in G. Then G is cocircuit semi-dual to G^* if and only if G is cocircuit dual (or circuit dual) to G^*.*

From **13.4** and **13.6** we have:

13.7. *Let a graph G have at least one of the following properties: (p1) G is a 3-connected graph G and every non-separating circuit of G contains a vertex of degree 3, and (p2) G has a Hamiltonian cocircuit. Then G is planar if and only if it has a cocircuit semi-dual graph.*

Note that **13.6.** and **13.7.** hold in particular for a cubic graph.

It turns out that in general the theorem analogous to **13.3** is not true for cocircuit semi-duality of graphs. The following theorem can be proved [31].

13.8. *Let G be a triangulation of the projective plane and G^p be the graph geometrically dual to G in the projective plane. Let $e : E(G) \to E(G^p)$ be the corresponding one-to-one edge mapping. Then G is 3-connected, e is a cocircuit semi-duality of G onto G^p, and e is not a circuit duality of G onto G^p.*

From **13.4** and **13.8** we have:

13.9. *A triangulation of the projective plane has no Hamiltonian cocircuit.*

14. On Dirac's conjecture

We know that an n-vertex triangulation T of the plane has exactly $3n - 6$ edges. If we add to T a new edge between non-adjacent vertices, then obviously the resulting graph T' has $3n - 5$ edges and is non-planar. Therefore by Kuratowski's theorem T' contains $TopK_5$ or $TopK_{3,3}$. It is easy to prove that if T' has no parallel edges then it contains $TopK_5$. Dirac conjectured [8] that

14.1. *For $n \geq 3$, any simple undirected graph G with n vertices and at least $3n - 5$ edges contains $TopK_5$.*

C. Thomassen proved [**43**] that

14.2. *For $n \geq 5$, any simple undirected graph G with n vertices and at least $4n - 10$ edges contains $TopK_5$.*

We investigate properties of a minimum counterexample to the Dirac conjecture (if any). Obviously the conjecture is true when $n = 5$. Let D be a simple graph such that

  (a) D has exactly $3|V(D)| - 5$ edges,
  (b) D has no $TopK_5$, and
  (c) if F is a proper subgraph of D with at least $3|V(F)| - 5$ edges, then F has $TopK_5$.

We proved that (presented at the Moscow Seminar on Discrete Mathematics in 1979) [**22**]:

14.3. *Any such graph D has the following properties:*

  (a) *D is 5-connected,*
  (b) *for every vertex x of D the subgraph induced by the set of vertices adjacent to x consists of components which are either isolated vertices or paths,*
  (c) *D contains no subgraph isomorphic to K_4, $K_{3,3}$, L or K_4' where K_4' is obtained from K_4 by replacing an edge by a 2-edge path,*
  (d) *D has at least 13 vertices, and*
  (e) *D has at least 10 vertices of degree 5, and if D has exactly 10 vertices of degree 5 then all the other vertices of D are of degree 6.*

Recently A. Kézdy and P. McGuinness rediscovered **14.3.**(a) and proved the following additional property of the graph D by using 5-connectedness of D (announced at the Seattle Graph Minors conference, July 1991):

14.4. *D does not contain a subgraph obtained from K_4 by deleting an edge. In other words the neighborhood of every vertex of D consists of components of at most 2 vertices.*

15. On Barnette's conjecture

The well-known conjecture of Barnette claims [**2**] that

B_3. *Any planar, cubic, bipartite, and 3-connected graph is Hamiltonian.*

If we replace in Barnette's conjecture *3-connected* by *cyclically 4-connected* then we obtain a weaker conjecture:

B_4. *Any planar, cubic, bipartite, cyclically 4-connected graph is Hamiltonian.*

It turns out that if B_3 or B_4 is true, then a much stronger result would be also true, namely [**27**]:

15.1. *Suppose that B_3 or B_4 is true. Let G be a graph which satisfies the hypotheses of the corresponding true conjecture. Then for any face-circuit C of G and for any two edges x and y of C there exists a Hamiltonian circuit of G which contains x and avoids y, and also there exists a Hamiltonian circuit of G which contains both x and y.*

By using the approach developed in [**27**] we proved that

15.2. *The two conjectures B_3 and B_4 are equivalent.*

Acknowledgment. I am thankful to Paul Seymour for a very careful reading of the paper and very useful remarks.

References

1. D.S. Archdeacon, A Kuratowski theorem for the projective plane, *J. Graph Theory* **5** (1981) 243–246.
2. D.W. Barnette, Conjecture 5, *Recent Progress in Combinatorics*, Ed. by W.T.Tutte, (1969), 343.
3. D.W. Barnette and B. Grünbaum, On Steinitz's theorem concerning convex 3-polytopes and on some properties of planar graphs, In *The Many Facets of Graph Theory. Lecture Notes in Mathematics*, Springer Verlag, New York **110** (1969) 27–40.
4. C. Berge, *Theory of Graphs and its Applications*, Methuen London, 1962.
5. R.E. Bixby and W.H. Cunningham, Matroids, graphs, and 3-connectivity, In *Graph Theory and Related Topics*, Ed. by J.A. Bondy and U.S.R. Murty, Academic Press, NY (1979), 91–103.
6. J.A. Bondy and U.S.R. Murty, *Graph Theory with Applications*, MacMillan Co., New York, 1976.
7. G.A. Dirac, Some results concerning the structure of graphs, *Can. Math. Bull.* **6** (1962) 183–210.
8. G.A. Dirac, Homeomorphism theorem for graphs, *Math. Annalen*, **153** (1964), 69–80.
9. J. Edmonds, On the surface duality of linear graphs, *J. Res. Nat. Bur. Stand.* **69B** (1965) 121–123.
10. J.-C. Fournier, Propriétés combinatoires et algébraic des graphes planaires, *Publ. Journées Comb. et Informatique, Bordeaux* **1** (1975) 153–179.
11. B. Grünbaum, *Convex polytopes*, Wilet-Interscience, New York, 1967.
12. D. W. Hall, A note on primitive skew curves, *Bull. Amer. Math. Soc.* **49** (1943), 935–937.
13. F. Harary, *Graph Theory*, Addison–Wesley, Reading, Mass., 1969.
14. F. Jaeger, Interval matroids and graphs, *Discrete Math.* **27** (1979) 331–336.
15. H.A. Jung, Eine Verallgemeineinerung des n–fachen zusammenhangs für Graphen, *Math. Ann.* **187** (1970), 95–103.
16. A.K. Kelmans, Graph expansion and reduction, In *Algebraic Methods in Graph Theory*, Vol. 1, Colloq. Math. Soc. János Bolyai, (Szeged, Hungary, 1978) North-Holland **25** (1981), 318–343.

17. A.K. Kelmans, The concept of a vertex in a matroid, the non-separating cycles and a new criterion for graph planarity, In *Algebraic Methods in Graph Theory*, Vol. 1, Colloq. Math. Soc. János Bolyai, (Szeged, Hungary, 1978) North–Holland **25** (1981), 345–388.
18. A.K. Kelmans, Concept of a vertex in a matroid and 3-connected graphs. *J. Graph Theory* **4** (1), (1980), 13–19.
19. A.K. Kelmans, A new planarity criterion for 3-connected graphs. *J. Graph Theory* **5** (1981), 259–267.
20. A.K. Kelmans, Finding special subdivisions of K_4 in a graph, In *Finite and Infinite Sets* (Eger, Hungary, 1981), Colloq. Math. Sci. János Bolyai, Vol. 37, North–Holland (1984), 487–508.
21. A.K. Kelmans, A problem on a strengthening of Kuratowski's planarity criterion, In *Finite and Infinite Sets*, Vol. 2, Colloq. Math. Soc. János Bolyai (Eger,Hungary, 1981), North-Holland, **37** (1984), 881–882.
22. A.K. Kelmans, On existence of given subgraphs in a graph, In *Algoritmy Diskretnoy Optimizacii i ih Primeneiya v Vychislitelnih Systemah*, Yarosl. Gos. Universitet, Yaroslavl (1983), 3–20.
23. A.K. Kelmans, A strengthening of the Kuratowski planarity criterion for 3-connected graphs, *Discrete Mathematics* **51** (1984), 215–220.
24. A.K. Kelmans, On planarity of 3-connected graphs without essential 3-cuts or triangles, In *Proceedings of the III All-Union conference methods of solving optimal problems on graphs and networks*, Tashkent 1984, AN SSSR, Sibirskoe otdelenie, Novosibirsk, 1984.
25. A.K. Kelmans, On homeomorphic embedding of graphs with given properties, *Soviet Math. Dokl.* **29** (1984) 130–135.
26. A.K. Kelmans, On 3-connected graphs without essential 3-cuts or triangles, *Soviet Math. Dokl.* **33** (1986), 698–703.
27. A.K. Kelmans, Constructions of cubic, bipartite and 3-connected graphs without Hamiltonian circuits, In *Analiz Zadach Formirovaniya i Vybora Alternativ*, VNIISI, Moscow **10** (1986), 64-72 [in Russian].
28. A.K. Kelmans, Non–separating cycles and planarity of 3-connected graphs without essential 3-cuts or triangles. In *Problemi Diskretnoy Optimizacii i Metodi ih Resheniya*, AN SSSR, CEMI, Moscow (1987), 224–233.
29. A.K. Kelmans, A short proof and strengthenings of the Whitney 2-isomorphism theorem on graphs, *Discrete Mathematics* **64** (1987), 13–25.
30. A.K. Kelmans, Matroids and Whitney theorems on 2-isomorphism and planarity of graphs, *Uspehi Mat. Nauk* **43** (5) (1988), 199–200.
31. A.K. Kelmans, On semi-isomorphisms and semi-dualities of graphs. *RUTCOR Research Report 59-90*, Rutgers University (1990), 1–15.
32. U. Krusenstjerna–Hafstrom and B. Toft, Special subdivisions of K_4 and 4-chromatic graphs, *Monatsh. Math.* **89** (1980), 101–110.
33. K. Kuratowski, Sur le problème des courbes gauches en topologie, *Fund. Math.* 15 (1930), 271–283.
34. L. Lovász, On graphs containing no disjoint circuits, *Mat. Lapok* **16** (1965) 289–299.
35. L. Lovász, *Combinatorial Problems and Exercises*, North-Holand Publishing Company, Amsterdam, New York, Oxford, 1979.
36. S. MacLane, A combinatorial condition for planar graphs, *Fund. Math.* **28**(1937) 22–32.
37. N. Robertson, Private communication, Seattle Graph Minors conference, July 1991.
38. N. Robertson and P.D. Seymour, Graph minors. VIII. A Kuratowski theorem for general surfaces, *J. Combinatorial Theory* **B-48** (1990), 255–288.
39. N. Robertson and P.D. Seymour, Graph minors. IX. Disjoint crossed paths, *J. Combinatorial Theory* **B-49** (1990), 40–77.
40. P.D. Seymour, Decomposition of regular matroids, *J. Combinatorial Theory* **B-28** (1980), 305–359.
41. P.D. Seymour, Disjoint paths in graphs, *Discrete Math.* **29** (1980), 293–309.
42. Y. Shiloach, A polynomial solution to the undirected two paths problem, *J. Assoc. Comput. Mach.* **27**, 445–456.

43. C. Thomassen, Some homeomorphism properties of graphs, *Math. Nachr.* **64** (1974) 119–133.
44. C. Thomassen, 2-linked graphs, *European J. Combinatorics* **1** (1980), 371–378.
45. C. Thomassen and B.Toft, Induced non-separating cycles in graphs, *J. Combinatorial Theory* **B-31** (1981), 199–224.
46. C. Thomassen, A refinement of Kuratowski's theorem, *J. Combinatorial Theory* **B-37** (1984), 245–253.
47. C. Thomassen, Embedding of graphs with no short non-contractible cycles, *J. Combinatorial Theory* **B-48** (1990), 155–177.
48. K. Truemper, On Whitney's 2-isomorphism theorem for graphs, *J. Graph Theory* **4** (1980), 43–49.
49. W.T. Tutte, Convex representations of graphs, *Proc. London Math. Soc.* **10** (1960), 304–320.
50. W.T. Tutte, How to draw a graph, *Proc. London Math. Soc.* **13** (1963), 743–768.
51. W.T. Tutte, Lectures on matroids, *J. Res. Nat. Bur. Stand.* **69B** (1965), 1–47.
52. W.T. Tutte, Connectivity in matroids, *Canad. J. Math.* **18** (1966), 1301–1324.
53. W.T. Tutte, Bridges and Hamiltonian circuits in planar graphs, *Aequatioes Math.* **15** (1977) 1–34.
54. K. Wagner, Über eine Erweiterung eines Satzes von Kuratowski, *Deutsche Math.*, **2** (1937), 280–253.
55. D. Welsh, *Matroid Theory*, Academic Press, London, 1976.
56. H. Whitney, Planar graphs, *Fund. Math.* **21**, (1933), 245–254.
57. H. Whitney, 2-isomorphic graphs, *Amer. J. Math* **55** (1933), 245–254.
58. H. Whitney, A set of topological invariants for graphs, *Fund. Math.* **55** (1933), 231–235.

RUTCOR, RUTGERS UNIVERSITY, NEW BRUNSWICK, NJ 08903, USA
E-mail address: kelmans@rutcor.rutgers.edu

EXCLUDING A GRAPH WITH ONE CROSSING

NEIL ROBERTSON AND PAUL SEYMOUR

ABSTRACT. Let H be a graph with crossing number ≤ 1. We prove that for some integer N, every graph with no minor isomorphic to H may be constructed by 0-, 1-, 2- and 3-sums, starting from planar graphs and graphs of tree-width $\leq N$. We also find the 41 minor-minimal graphs with crossing number ≥ 2.

1. INTRODUCTION

K. Wagner [13] and D. W. Hall [3] proved that a graph has no $K_{3,3}$ minor if and only if it can be constructed by 0-, 1- and 2-sums (defined below), starting from planar graphs and copies of K_5. (All graphs in this paper are finite. H is a *minor* of G if H can be obtained from a subgraph of G by contracting edges.) Wagner [12] also proved that a graph has no K_5 minor if and only if it can be constructed by 0-, 1-, 2- and 3-sums, starting from planar graphs and copies of the four-rung Möbius ladder. It turns out that there is a cause for the similarity of these two theorems, namely, that K_5 and $K_{3,3}$ both can be drawn in the plane with one crossing. We shall see that the same sort of structure is produced by the exclusion of any graph with crossing number one.

In earlier papers we investigated the structure produced by the exclusion of a graph with crossing number zero, that is, a planar graph. Let us say a *tree-decomposition* of a graph G is a pair $(T, (X_t : t \in V(T)))$, where T is a tree and $(X_t : t \in V(T))$ is a family of subsets of $V(T)$, such that

(i) $V(G) = \bigcup(X_t : t \in V(T))$, and for every edge e of G there exists $t \in V(T)$ such that X_t contains both ends of e, and

(ii) if $t, t', t'' \in V(T)$ and t' lies on the path of T between t and t'' then $X_t \cap X_{t''} \subseteq X_{t'}$.

1991 *Mathematics Subject Classification.* Primary 05C10, 57M15.

Research of the first author was performed under a consulting agreement with Bellcore and was supported by NSF under Grant No. DMS-8903132.

This paper is in final form and no version of it will be submitted for publication elsewhere.

We say G has *tree-width* $\leq N$ if a tree-decomposition $(T, (X_t : t \in V(T)))$ of G exists such that $|X_t| \leq N+1$ for all $t \in V(T)$. It was shown in [4] (see also [10] for a better proof) that

(1.1) *For any planar graph H, there is an integer $N \geq 0$ such that every graph with no minor isomorphic to H has tree-width $\leq N$.*

Of course, this theorem is not so exact as the theorems of Hall and Wagner above, because the structure it provides is necessary but not sufficient for the exclusion of H, while Hall's and Wagner's theorem give structure that is necessary and sufficient for the exclusion of the minor. On the other hand, (1.1) is best possible in another sense, because for any non-planar graph H there is no integer N as in (1.1). In that sense, bounded tree-width is the structure resulting from the exclusion of a planar graph as a minor. Our objective in this paper is to describe the structure resulting from the exclusion of a graph with crossing number 1 (or more precisely, of a minor of such a graph).

To describe this structure we need "k-sums". A *separation* of a graph G is a pair (A, B) of subgraphs of G with $A \cup B = G$ and $E(A \cap B) = \emptyset$, and its *order* is $|V(A \cap B)|$. Let (A, B) be a separation of G, of order k. Let A^+ be obtained from A by adding $\binom{k}{2}$ new edges to A joining every pair of vertices in $V(A \cap B)$, and let B^+ be obtained from B similarly. We say that G is the *k-sum* of A^+ and B^+. For example, the graphs which can be obtained from copies of K_1 and K_2 by repeated 1-sums are the trees, and if we permit 0-sums as well we obtain the forests. If G is the k-sum of A^+ and B^+, and A^+ and B^+ are both isomorphic to proper minors of G (every minor of G is *proper* except G itself) we say that G is the *proper k-sum* of A^+ and B^+. We shall only need 0-, 1-, 2- and 3-sums.

Let us say that a graph H is *singly-crossing* if H is isomorphic to a minor of a graph H' which can be drawn in the sphere with ≤ 1 crossing. (Unlike being planar, having crossing number ≤ 1 is a property not always inherited by minors.) Our main result is the following.

(1.2) *For any singly-crossing graph H there is an integer $N \geq 0$ such that every graph with no minor isomorphic to H may be obtained by proper 0-, 1-, 2- and 3-sums, starting from planar graphs and graphs of tree-width $\leq N$.*

(1.2) is a consequence of the following.

(1.3) *For any singly-crossing graph H there is an integer $N \geq 0$ such that every graph with no minor isomorphic to H is either*

(i) *the proper 0-, 1-, 2- or 3-sum of two graphs, or*

(ii) *planar, or*

(iii) *of tree-width $\leq N$.*

Proof of (1.2), assuming (1.3). Let N be as in (1.3); then we claim that (1.2) is satisfied. Let G be a graph with no minor isomorphic to H. We prove, by induction on $|V(G)| + |E(G)|$, that G may be constructed as in (1.2). If G is the proper 0-, 1-, 2- or 3-sum of G_1 and G_2 say, then G_1 and G_2 have no minor isomorphic to H because G_1 and G_2 are themselves isomorphic to minors of G. From the inductive hypothesis, G_1 and G_2, and hence G, can be obtained in the required way. We assume then that G is not the proper 0-, 1-, 2- or 3-sum of two graphs. By (1.3), either G is planar or it has tree-width $\leq N$, and in either case it satisfies the theorem. ∎

It remains then to prove (1.3), and that is the objective of the next section.

Let us observe that (1.2) is indeed best possible in the sense we specified, that for any non-singly-crossing graph H there is no integer N as in (1.2). Let G' be a 4-connected planar triangulation with large tree-width, and let a, b, c and b, c, d be the vertices of two regions with a common edge. Add an edge with ends a, d forming G. Then G is non-planar and has large tree-width, and cannot be expressed as the proper 0-, 1-, 2- or 3-sum of two graphs (since G is 4-connected), and so does not have the structure specified in (1.2). On the other hand, G has no minor isomorphic to H, since G has crossing number 1 and H is non-singly-crossing. Thus, there is no N so that (1.2) holds.

2. The Main Proof

We shall see that (1.3) follows rather easily from results about tangles in the Graph Minors series. A *tangle* in G of *order* θ is a set $\mathcal{T}$ of separations of G, each of order $< \theta$, such that

(i) if (A, B) is a separation of order $< \theta$ then $\mathcal{T}$ contains one of (A, B), (B, A)

(ii) if $(A_1, B_1), (A_2, B_2), (A_3, B_3) \in \mathcal{T}$ then $A_1 \cup A_2 \cup A_3 \neq G$

(iii) if $(A, B) \in \mathcal{T}$ then $V(A) \neq V(G)$.

Tangles were introduced in [6].

Let G be a non-null connected planar graph drawn in a sphere Σ. Let K be another graph drawn in Σ such that

(i) $V(G) \subseteq V(K)$ and every edge of K has one end in $V(G)$ and the other in $V(K) - V(G)$

(ii) every region of G (regions are open discs) contains a unique vertex of K

(iii) no edge of G (edges are open line segments) intersects the drawing of K

(iv) for each $v \in V(G)$, the edges of G and of K incident with v alternate in their cyclic order around v.

Such a graph K is called a *radial graph* of G.

Now let G be a connected planar graph drawn in Σ with $E(G) \neq \emptyset$, and let K be a radial graph of G. Let $\mathcal{T}$ be a tangle in G of order θ. If C is a circuit of K with $|E(C)| < 2\theta$, we define $ins(C)$ to be the unique closed disc Δ bounded by C with

$$(G \cap \Delta, G \cap \overline{\Sigma - \Delta}) \in \mathcal{T},$$

where $G \cap \Delta$ denotes the subgraph of G drawn in Δ. (For $X \subseteq \Sigma$, $\overline{X}$ denotes the closure of X.) If W is a closed walk of K of length $< 2\theta$ we define $ins(W)$ to be the union of $ins(C)$, taken over all circuits C of K', together with all points of the drawing of K', where K' is the subgraph of K formed by the vertices and edges in W. If $u,v \in V(G)$ we define

$$d(u,v) = \begin{cases} 0 & \text{if } u = v \\ k & \text{if } u \neq v \text{ and there is a closed walk } W \text{ of } K \text{ of length } < 2\theta \text{ with} \\ & u,v \in ins(W), \text{ and the shortest such walk has length } 2k \\ \theta & \text{otherwise.} \end{cases}$$

We call d the *metric of* $\mathcal{T}$. (It is indeed a metric, incidentally.) It does not depend on the choice of K, and every connected planar graph G with $E(G) \neq \emptyset$ has a radial graph, and so d is determined by G, as drawn in Σ, and by $\mathcal{T}$.

If G' is a subgraph of G, a G'-*path* in G is a path in G with distinct ends, both in $V(G')$, and with no other vertex or edge in G'. We need the following two lemmas.

(2.1) *For any singly-crossing graph H there is an integer $N_1 \geq 0$ with the following property. Let G' be a subgraph of a graph G, and let there be a G'-path with ends u, v. Let G' be planar and connected, drawn in a sphere, and let $\mathcal{T}$ be a tangle in G' with metric d. If $d(u,v) \geq N_1$ then G has a minor isomorphic to H.*

(2.2) *For any singly-crossing graph H there is an integer $N_2 \geq 0$ with the following property. Let G' be a subgraph of a graph G, and let there be two disjoint G'-paths with ends u_1, v_1 and u_2, v_2 respectively. Let G' be planar and connected, drawn in a sphere, and let $\mathcal{T}$ be a tangle in G' of order $\geq N_2$. Let C be a circuit of G' bounding a region, with $u_1, u_2, v_1, v_2 \in V(C)$ in order. Let there be no $(A,B) \in \mathcal{T}$ of order ≤ 3 with $\{u_1, v_1, u_2, v_2\} \subseteq V(A)$. Then G has a minor isomorphic to H.*

The proofs of (2.1) and (2.2) are similar, and much like the proofs of [5, section 9] or [7, theorems (4.4) and (4.5)], and we omit the details.

A drawing G in a sphere Σ is *rigid* if for every $F \subseteq \Sigma$ homeomorphic to a circle, meeting the drawing only in vertices and with $|F \cap V(G)| \leq 2$, there is a closed disc $\Delta \subseteq \Sigma$ bounded by F such that either

(i) G is drawn entirely in $\overline{\Sigma - \Delta}$, or

(ii) $|F \cap V(G)| = 2$ and $G \cap \Delta$ is a path with ends in F.

Proof of (1.3). Let N_1, N_2 satisfy (2.1), (2.2) respectively. We may assume that $N_1 \geq 12$ and is even. Let $\theta = 12N_1 + 12 + \max(4N_1 + 3, N_2)$. Let N be such that every graph with tree-width $> N$ has a subgraph G' such that

(i) G' is planar, and has a rigid drawing in a sphere, and

(ii) G' has a tangle of order θ.

This is possible by (1.1). Indeed, we may choose G' to be a subdivision of a large piece of the hexagonal lattice, since such graphs have tangles of large order.

We claim that N satisfies the theorem. For let G be a graph with no minor isomorphic to H, and that is not expressible as the proper 0-, 1-, 2- or 3-sum of two graphs. We may assume that G has tree-width $> N$, for otherwise it satisfies the theorem. By the choice of N, G has a subgraph G' with a rigid drawing in a sphere Σ, and G' has a tangle $\mathcal{T}$ of order θ. Since $\theta \geq 2$, it follows that $E(G') \neq \emptyset$ and the metric of $\mathcal{T}$ is defined, d say.

(1) $d(u,v) < N_1$ for all $u,v \in V(G')$ such that some G'-path in G has ends u, v.

For otherwise G has a minor isomorphic to H, by (2.1), a contradiction.

(2) There is no connected subgraph G'' of G', with a tangle T'' of order $\geq \theta - 12N_1 - 12$, such that
  (i) there are two disjoint G''-paths in G with ends u_1, v_1 and u_2, v_2 respectively
  (ii) there is a circuit C of G'' bounding a region of G'', with $u_1, u_2, v_1, v_2 \in V(C)$ in order
  (iii) there is no $(A,B) \in T''$ of order ≤ 3 with $\{u_1, v_1, u_2, v_2\} \subseteq V(A)$.

For otherwise G has a minor isomorphic to H by (2.2), since $\theta - 12N_1 - 12 \geq N_2$, a contradiction.

Since G cannot be expressed as the proper 0-, 1-, 2- or 3-sum of two graphs, it follows easily that

(3) There is no separation (A,B) of G of order ≤ 3 such that A, B both have circuits and $V(A), V(B) \neq V(G)$.

By (1), (2) and [8, theorem (10.1)], it follows that there exists a family $((A_i, B_i) : i \in I)$ of separations of G, a family $(\Delta_i : i \in I)$ of closed discs in Σ, and a function α with domain

$$X = \bigcup (V(A_i \cap B_i) : i \in I)$$

satisfying the following:
  (i) for each $i \in I$, (A_i, B_i) has order ≤ 3, and $(A_i \cap G', B_i \cap G') \in T$
  (ii) $G = \bigcup(A_i : i \in I)$, and $A_i \subseteq B_j$ for all distinct $i, j \in I$
  (iii) for each $i \in I$ and $v \in V(A_i \cap B_i)$, $\alpha(v) \in bd(\Delta_i)$
  (iv) for distinct $i, j \in I$, if $x \in \Delta_i \cap \Delta_j$ then $x = \alpha(v)$ for some $v \in V(A_i \cap A_j)$
  (v) $\alpha(u) \neq \alpha(v)$ for distinct $u, v \in X$.

(4) For each $I \in I$, A_i may be drawn in Δ_i so that each $v \in V(A_i \cap B_i)$ is represented by $\alpha(v)$.

By (i), $(A_i \cap G', B_i \cap G') \in T$, and so $B_i \cap G'$ has a circuit, by [6, theorem (2.10)]; and $V(A_i \cap G') \neq V(G')$ since $(A_i \cap G', B_i \cap G') \in T$. By (3), either $V(B_i) = V(G)$ or A_i has no circuit, and in either case the claim follows.

For each $i \in I$, take a drawing of A_i in Δ_i as in (4); then we obtain a drawing of G in Σ, and so G is planar, as required. ∎

3. Excluded Minors

A second natural question about singly-crossing graphs is, what are the minor-minimal graphs that are not singly-crossing? In a conversation with T. Böhme and R. Thomas, during the conference of this Proceedings, we were surprised to observe that this question can be solved very easily using known results, and it seems worthwhile to record the answer here. We need the following three lemmas.

(3.1) [1] A graph G can be drawn in the projective plane if and only if it has none of 35 specific graphs (say $H_1, \ldots, H_{35}$) as a minor.

(3.2) [2,11] *If G is a graph which can be drawn in the projective plane Σ, then the following are equivalent:*
 (i) *some drawing of G in Σ is not 3-representative*
 (ii) *every drawing of G in Σ is not 3-representative*
 (iii) *G has one of 6 specific graphs (say $H_{36}, \ldots, H_{41}$) as a minor.*

[$H_{36}, \ldots, H_{41}$ are the six graphs different from $K_{4,4}$ minus one edge, that can be obtained by $Y\Delta$ and ΔY exchanges starting from K_6. See [9]. A drawing in Σ is *3-representative* if every non-null-homotopic simple closed curve in Σ which meets the drawing only in vertices, passes through at least 3 vertices.]

(3.3) *A graph G is singly-crossing if and only if it has a drawing in the projective plane which is not 3-representative.*

We omit the proof, which is easy. But now it follows that

(3.4) *A graph is singly-crossing if and only if it has none of $H_1, \ldots, H_{41}$ as a minor.*

Proof. If G is singly-crossing then by (3.3), (3.1) and the equivalence of (i) and (iii) in (3.2), it follows that G has none of $H_1, \ldots, H_{41}$ as a minor. For the converse, suppose G has none of $H_1, \ldots, H_{41}$ as a minor. By (3.1), G has a drawing in the projective plane, and by the equivalence of (ii) and (iii) in (3.2), this drawing is not 3-representative. From (3.3), G is singly-crossing, as required. ∎

One can also check that $H_1, \ldots, H_{41}$ are all distinct, and indeed none of these is a minor of another.

References

1. D. Archdeacon, "A Kuratowski theorem for the projective plane", Thesis, Ohio State University, 1980.
2. D. W. Barnette, "Generating projective plane polyhedral maps", *J. Combinatorial Theory, Ser. B*, 51 (1991), 277-291.
3. D. W. Hall, "A note on primitive skew curves", *Bull. Amer. Math. Soc.* 49 (1943), 935-937.
4. N. Robertson and P. D. Seymour, "Graph minors. V. Excluding a planar graph", *J. Combinatorial Theory, Ser. B*, 41 (1986), 92-114.
5. N. Robertson and P. D. Seymour, "Graph minors. VII. Disjoint paths on a surface", *J. Combinatorial Theory, Ser. B*, 45 (1988), 212-254.
6. N. Robertson and P. D. Seymour, "Graph minors. X. Obstructions to tree-decomposition", *J. Combinatorial Theory, Ser. B*, 52 (1991), 153-190.
7. N. Robertson and P. D. Seymour, "Graph minors. XIV. Distance on a surface", submitted.
8. N. Robertson and P. D. Seymour, "Graph minors. XV. Extending an embedding", submitted.
9. N. Robertson, P. D. Seymour and R. Thomas, "A survey of linkless embeddings", this volume.
10. N. Robertson, P. D. Seymour and R. Thomas, "Quickly excluding a planar graph", *J. Combinatorial Theory, Ser. B*, to appear.

11. R. Vitray, "Representativity and flexibility of drawings of graphs on the projective plane", Thesis, Ohio State University, 1987.
12. K. Wagner, "Über eine Eigenshaft der ebenen Komplexe", *Math. Ann.* 114 (1937), 570-590.
13. K. Wagner, "Über eine Erweiterung eines Satzes von Kuratowski", *Deutsche Math.* 2 (1937), 280-285.

DEPARTMENT OF MATHEMATICS, OHIO STATE UNIVERSITY, 231 W. 18TH AVE., COLUMBUS, OHIO 43210, USA
 E-mail address: robertso@function.mps.ohio-state.edu

BELLCORE, 445 SOUTH ST., MORRISTOWN, NEW JERSEY 07962, USA
 E-mail address: pds@breeze.bellcore.com

Open Problems

NATHANIEL DEAN

ABSTRACT. Graph minors is a field that has motivated numerous investigations in discrete mathematics and computers science, a fact demonstrated by the variety of papers and open problems appearing in this volume. This section summarizes those problems which were submitted by participants of the conference for inclusion in this special section of the proceedings.

1. Introduction

There were many open problems discussed in problem sessions, meals, excursions and other gatherings at the conference. Several of these appear in the other papers of this volume, and a few of the participants submitted problems to be included in this section. This paper is partitioned into several sections addressing a variety of problem areas: path-width and tree-width (Section 2), paths, cycles and independent sets (Section 3), coverings and integer flows (Section 4), well-quasi-ordering (Section 5), geometry and topology (Section 6), logic (Section 7), and disjoint paths (Section 8). Each subsection focuses on presenting a particular problem or group of problems, and the name of the author who contributed that section is given in the title. To enhance the exposition and have a more consistent format, several of the contributions were edited and so I apologize for any errors that may have been introduced.

With only a few exceptions, our notation and terminology is consistent with that of Bondy and Murty [7]. A graph H is said to be a **minor** of a graph G if H can be obtained from a subgraph of G by contracting edges. A graph may have loops and multiple edges unless we explicitly state otherwise.

1991 *Mathematics Subject Classification.* Primary 05C75, 05C10; Secondary 68R10.
Key words and phrases. graph minors, conjectures, problems, algorithms, hamiltonian, cycles, disjoint paths, path-width, tree-width, bandwidth, logic.
This paper is in final form and no version of it will be submitted for publication elsewhere.

2. Path-Width and Tree-Width

Recently, researchers in structural as well as algorithmic graph theory have shown much interest in the notions of path-width and tree-width. A **tree-decomposition** of a graph $G = (V, E)$ is a pair $(\{X_i \mid i \in I\}, T = (I, F))$, with T a tree, and each $X_i \subseteq V$, such that

(i) $\bigcup_{i \in I} X_i = V$,
(ii) for all $vw \in E$, there exists an $i \in I$ with $v, w \in X_i$, and
(iii) for each $v \in V$, $\{i \in I \mid v \in X_i\}$ forms a connected subtree of T.

The **width** of a tree-decomposition is $\max_{i \in I} |X_i| - 1$. The **tree-width** $tw(G)$ of G is the minimum tree-width over all tree-decompositions of G. When $tw(G) \leq k$, G is often called a **partial** k-**tree** or, if G is maximal with respect to this property, a k-**tree**. The terms **path-decomposition** and **path-width** $pw(G)$ are defined similarly, except that T is required to be a path.

2.1. Unavoidable Minors (Paul Seymour).
Let H_1 be a graph with a vertex v such that $H_1 - v$ is a forest. Let H_2 be an outerplanar graph, that is, a graph that can be drawn in the plane so that every vertex is incident with the infinite region.

OPEN PROBLEM 1. Does every 2-connected graph of sufficiently large path-width (or equivalently, containing a sufficiently large uniform depth binary tree as a minor) contain H_1 or H_2 as a minor?

2.2. The Path-Width of k-Trees (Hans L. Bodlaender and Jens Gustedt).
The general problem of determining the path-width of a given graph is known to be **NP**-hard, proven independently by various authors [23]. Many problems become polynomial time solvable, when restricted to graphs whose tree-width is bounded by a constant [3]. Thus, we are interested in the complexity of the problem of determining the path-width of graphs G with tree-width bounded by some constant k. So far, this problem has only been solved in the case $k = 1$ (i.e., G is a forest) by Scheffler [30], among others, who gives a linear time algorithm for the problem. For $k > 1$, the problem is open.

CONJECTURE 2. For any fixed k the problem of calculating $pw(G)$ on the class of graphs G with $tw(G) \leq k$ is solvable in polynomial time.

Here is an indication why Conjecture 2 should hold. Recently Bodlaender and Kloks [6] gave an algorithm to determine whether or not $pw(G) \leq l$ holds in time $O((2l)! \cdot n)$, if a tree-decomposition of G of size $\leq k$ is given, where k is constant. The 'O' hides a term, exponential in k. Since $pw(G) \leq tw(G) \cdot \log n$ (see [5]), this gives a subexponential algorithm which uses time, polynomial in $n^{\log n}$.

Now for every $k > 1$ one of the following three cases must hold

(i) The problem is **NP**-hard for k
(ii) It has a lower bound which is super-polynomial
(iii) It has a polynomial algorithm, that is Conjecture 2 holds.

Of course, (i) would separate **NP** from **EXP** and (ii) would separate **P** from **NP**. So these two cases, if proven, would both imply deep results in complexity theory. This seems very unlikely. On the other hand many people have tried and failed to prove the conjecture in the special case $k = 2$.

CONJECTURE 3. *The problem of calculating $pw(G)$ on the class of partial 2-trees is solvable in polynomial time.*

One reason for this seems to be the large number of cases that result (for example) when several graphs are glued together at a triangle, for full 2-trees, or a cycle, for partial 2-trees. At the moment it seems more likely that Conjecture 2 would be solved in one step, avoiding intensive case analysis.

3. Paths, Cycles and Independent Sets

3.1. Neighbors of an Independent Set (Nathaniel Dean).

CONJECTURE 4. *Let G be a k-connected, nonhamiltonian graph with $k \geq 2$. Then some cycle of G contains k independent vertices and their neighbors.*

CONJECTURE 5. *Let G be a k-connected, nontraceable graph with $k \geq 1$. Then some path of G contains $k + 1$ independent vertices and their neighbors.*

The cases $k = 2, 3$ of Conjecture 4 have already been proved (see [14] and [20]). In fact, the case $k = 2$ of Conjectures 4 and 5 follows easily from a proof given by Dirac [11] (see also Exercise 10.27 in [19]). Further, cases $k = 1, 2$ of Conjecture 5 are easy.

3.2. Dominating an Independent Set (Paul Seymour).

OPEN PROBLEM 6. *Let G be a simple graph with $n + k$ vertices, and let X be an independent set with $|X| = k < V(G)$. Suppose $|E(G)| > kn$ and each vertex v not in X is joined to X by k paths, mutually (vertex) disjoint except at v. Can one always contract edges (not identifying any two vertices in X) to obtain a graph in which there are two vertices both adjacent to every vertex in X?*

Robertson, Seymour and Thomas have proved this for $k \leq 5$. There are examples showing that the condition $|E(G)| > kn$ is best possible. For example, if G is obtained from an $(n + 1)$-vertex path P with one end $x \in X$ by joining each of the remaining $k - 1$ vertices of X to every vertex of $P - x$, then G has kn edges and no such contraction is possible.

3.3. Hamiltonian Cycles (Robin Thomas).

Tutte [41] proved that *every 4-connected planar graph is hamiltonian*, and Wagner [44] characterized the graphs having no K_5-minor. These results together imply that every 4-connected graph with no K_5-minor is hamiltonian. Can this be extended to larger connectivities by excluding larger minors?

CONJECTURE 7. *For every $k > 1$, every k-connected graph with no K_{k+1}-minor is hamiltonian.*

The conjecture is vacuously true for $k < 4$.

4. Coverings and Integer Flows

4.1. Triangles (Andras Sebő).

OPEN PROBLEM 8 (S. POLJAK, A. SEBŐ, P. D. SEYMOUR). Characterize graphs which can be regularly edge-covered by triangles, that is for which there exists a set of triangles covering every edge of the graph the same positive number of times.

Equivalently, when is the all 1's function on the edges of a graph in the cone generated by the edge-characteristic vectors of triangles? More generally, what is the set of linear inequalities describing the cone of triangles (as edge-sets) of a graph? Namely, is the following conjecture true?

CONJECTURE 9 (S. POLJAK, A. SEBŐ, P. D. SEYMOUR). Let $G = (V, E)$ be a graph. Then the cone generated by the edge-characteristic vectors of triangles is $\{x : v_H^T x \geq 0\}$, where $H = (V', E')$ is a triangle free subgraph of G, and

$$v_H(e) = \begin{cases} -1 & \text{if } e \in E(H) \\ 2 & \text{if } e = xy \in E(G) - E(H) \text{ and } x, y \in V' \\ 1/2 & \text{if } e \text{ has exactly one end in } V' \\ 0 & \text{otherwise.} \end{cases}$$

For a reference on cones of circuits see [35], and for more information on triangle covers see [22]. Svatya Poljak notes (personal communication) that for random graphs Problem 8 has already been settled since in this case the all 1's function is in the cone of cuts if and only if there is a regular triangle cover.

OPEN PROBLEM 10 (A. SEBŐ). Let G be a graph. Is it true that for all $R \subseteq E(G)$ satisfying $|E - R| \leq |E \cap R|$ there exist edge-disjoint circuits each containing exactly one edge of R, if and only if there exists no R for which the same inequality holds and in addition the union of cuts for which equality holds contains an odd cut?

This problem comes from the undirected integer multiflow problem where the union of the demand and supply edges is planar [21]. If the cut condition holds a fractional solution always exists (see [37] or for a short proof see [32]). A common generalization of these is proved in [15], and some special cases where the condition stated in the conjecture is also sufficient are stated in [33].

OPEN PROBLEM 11 (A. SEBŐ). Let G be a graph. Is it true, that (i) holds for arbitrary $R \subseteq E(G)$, and adding to G parallel copies of edges in an eulerian way (copies inherit membership in R), if and only if (ii) also holds for all these?

(i) There exist edge-disjoint circuits each containing exactly one edge of R, if and only if there exists a fractional packing of such circuits.

(ii) There exist edge-disjoint circuits each containing exactly one edge of R, if and only if for every $0-1-\infty$ function l on $E(G) - R$ which is even on cycles, $\sum_{r \in R} \mu_l(r) \leq \sum_{e \in E-R} \mu_l(e)$, where $\mu_l(e)$ denotes the shortest path length with edge-weights l, between the ends of e.

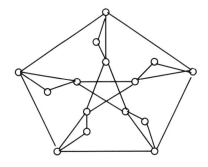 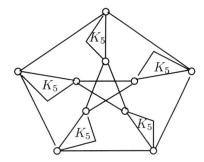

FIGURE 1. 2-connected graphs with no even cycle decomposition

The property that a fractional solution to multiflow problems implies an integer solution provided the data is Eulerian is strictly related to the bipartiteness of the metrics providing a necessary condition for the existence of a fractional flow. Some tools for Problem 11 and a proof for a special case can be found in [**34**] and [**31**].

4.2. Even cycle decomposition (Cun-Quan Zhang). Let G be an eulerian graph. A partition $\mathcal{F}$ of $E(G)$ into cycles is called a **cycle decomposition of** $E(G)$, and the decomposition is **even** if every cycle of $\mathcal{F}$ has even length.

CONJECTURE 12. Every 3-connected eulerian graph $G \neq K_5$ containing an even number of edges has an even cycle decomposition.

Note that K_5 does not have an even cycle decomposition. The conjecture was proved by Seymour [**36**] for 2-connected planar graphs and by Zhang [**47**] for 2-connected graphs with no K_5-minor. Furthermore, the connectivity condition in the conjecture cannot be reduced since there are 2-connected eulerian graphs $\neq K_5$ that contain an even number of edges and have no even cycle decomposition (see Figure 1).

4.3. A Weak 4-Flow Conjecture (Cun-Quan Zhang). The following conjecture of Tutte is a well-known refinement of the Four Color Problem (see [**40**] or [**18**]).

    4-Flow Conjecture: Every 2-connected graph containing no
    Petersen-minor admits a nowhere-zero 4-flow.

The only progress was made by Jaeger [**18**] who proved that *every 4-edge-connected graph admits a nowhere-zero 4-flow*.

A **cycle double cover** of a graph G is a family $\mathcal{F}$ of even subgraphs such that each edge of G is contained in exactly two even subgraphs of F. The relation between the 4-flow problem and the cycle cover problem is evident in the following well-known proposition.

PROPOSITION 4.1. *Let G be a 2-connected graph. The following statements are equivalent:*
  (i) *G has a nowhere-zero 4-flow;*
  (ii) *G has a cycle double cover consisting of at most 4 even subgraphs.*

CONJECTURE 13 (WEAK 4-FLOW CONJECTURE). There is an integer k such that every 2-connected cubic graph containing no subdivision of the Petersen graph has a cycle double cover consisting of at most 4 even subgraphs.

4.4. Disjoint Parity Subgraphs (Cun-Quan Zhang).

Let G be a graph. A subgraph H of G is called a **parity subgraph** of G if $d_H(v) \equiv d_G(v) \bmod 2$ for every vertex v of G. The **odd-connectivity** of G is the smallest number of edges in an edge cut of containing an odd number of edges.

CONJECTURE 14. If the odd-connectivity of a graph G is k_o, then G contains at least $(k_o - 1)/2$ edge-disjoint parity subgraphs.

The following related results are either well-known or obvious.
  (i) Every $2k$-edge-connected graph contains at least k edge-disjoint spanning trees (see [25] or [42]).
  (ii) Every spanning tree of a graph G contains a parity subgraph of G.
  (iii) A graph admits a nowhere-zero 4-flow if and only if it contains at least three edge-disjoint parity subgraphs.
  (iv) In any graph the maximum number of edge-disjoint parity subgraphs is odd.

4.5. Equivalence of SCC and CPP (Cun-Quan Zhang).

The (undirected) Chinese Postman Problem **CPP** is to find a shortest postman tour (i.e. a closed walk using each edge at least once) of a given graph. The Shortest Cycle Cover Problem **SCC** is that of finding a family $\mathcal{F}$ of cycles of a graph G such that each edge of G lies in at least one cycle of $\mathcal{F}$ and the total length of all cycles in $\mathcal{F}$ is as small as possible. It is obvious that an optimum solution of **CPP** cannot be greater than a solution of **SCC**. Further, these two solutions need not be equal; see, for example, the Petersen graph. We say that **CPP** and **SCC** are **equivalent** for G if an optimum solution of **CPP** equals an solution of **SCC**.

A weight function $w : E(G) \to \{1, 2\}$ is called **eulerian** if the total weight of each edge-cut is even. A graph G with an associated weight function w is denoted by (G, w). The set of all $(1, 2)$-eulerian weight functions of G is denoted by $\mathcal{W}_G$. If (G, w) has a family $\mathcal{F}$ of cycles such that each edge e of G is contained in precisely $w(e)$ cycles of $\mathcal{F}$, then $\mathcal{F}$ is called a **cycle w-cover** and G is **cycle w-coverable** (see [2], [18] and [48]). A graph G is said to have the **cycle cover**

property if G is cycle w-coverable for every $w \in \mathcal{W}_G$. It is not hard to see that the **SCC** and the **CPP** are equivalent for G if G has the cycle cover property.

CONJECTURE 15. *Let G be a 3-connected graph. Then **SCC** and the **CPP** are equivalent for G if and only if G has the cycle cover property.*

Note that this conjecture is stronger than Tutte's 4-flow conjecture. Moreover, it is known that if G has no Petersen-minor or admits a nowhere-zero 4-flow, then G has both properties: the cycle cover property and the equivalency of **SCC** and **CPP** (see [1], [2], [18] and [48]). Note that the connectivity condition in the conjecture cannot be reduced since there are 2-connected graphs for which **SCC** and **CPP** are equivalent and which do not have the cycle cover property.

5. Well-Quasi-Ordering

A simple graph H is an **induced minor** of a simple graph G if H can be obtained from an induced subgraph of G by contracting edges (multiple edges that result are deleted). Let us denote by $G \preceq H$ if G is isomorphic to an induced subgraph of H.

5.1. Alternating Double Wheels (Robin Thomas).

An **alternating double wheel** is a graph obtained from an even circuit by adding two non-adjacent vertices and joining one of them to one color class of the circuit and joining the other vertex to the other color class. Let $\mathcal{F}$ be a set of (isomorphism classes) of planar graphs such that if $G \in \mathcal{F}$ and H is an induced minor of G then $H \in \mathcal{F}$.

OPEN PROBLEM 16. *Is it true that $\mathcal{F}$ is well-quasi-ordered in the induced minor ordering if and only if it contains only finitely many alternating double wheels?*

See [39] for a related result.

5.2. Well-Quasi-Orders From Bipartite Graphs (Guoli Ding).

Let P_n be the path on n vertices and let G_n be the class of graphs with no induced P_n and $\overline{P_n}$ (the complement of P_n). It has been shown in [10] that $(G_4, \preceq)$ is a well-quasi-order but $(G_5, \preceq)$ is not. Let Π be the class of permutation graphs [16]. Then we ask:

OPEN PROBLEM 17. *Is $(G_n \cap \Pi, \preceq)$ a well-quasi-order for $n \geq 5$?*

Let $G = (X, Y, E)$ be a connected bipartite graph. We define the **bipartite complement** of G to be the bipartite graph $G^* = (X, Y, X \times Y - E)$. Let B_n be the class of bipartite graphs without induced P_n and P_n^*. It is proved in [11] that $(B_6, \preceq)$ is a well-quasi-order yet $(B_8, \preceq)$ is not.

OPEN PROBLEM 18. *Is $(B_7, \preceq)$ a well-quasi-order?*

6. Geometry and Topology

6.1. Higher Genus Polyhedra (Nathaniel Dean).

A polyhedron P of genus γ is constructed by joining polyhedral faces edge-to-edge so that the re-

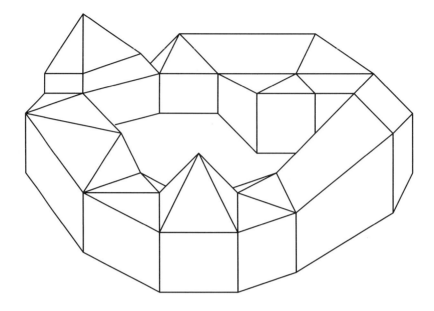

FIGURE 2. A 4-connected toroidal polyhedron

sulting figure is topologically equivalent to a surface of genus γ (see Figure 2). The 1-skeleton of P is a graph of genus γ, but there are toroidal graphs that cannot be realized as a toroidal polyhedra [**46**].

CONJECTURE 19. *Every 4-connected toroidal polyhedron is hamiltonian.*

Grunbaum has made the stronger conjecture that *every 4-connected toroidal graph is hamiltonian.*

6.2. Crossing Number (Detlef Seese). For which minor closed classes $\mathcal{K}$ of graphs does there exist a polynomial time algorithm to determine the crossing number of the elements of $\mathcal{K}$?

OPEN PROBLEM 20. *Is there a polynomial time algorithm to determine the crossing number of partial k-trees, for fixed k?*

6.3. (Robin Thomas). Colin de Verdiere [**8**] introduced an invariant μ defined for every graph G. He showed that $\mu(G) \leq 3$ if and only if G is planar.

OPEN PROBLEM 21. *Is it true that $\mu(G) \leq 4$ if and only if G can be embedded in 3-space in such a way that every circuit of G bounds a disk disjoint from the graph?*

See [**29**] for more information about this kind of spatial embeddings.

6.4. Bandwidth of the Integer Simplex (Douglas West). Consider the problem of labeling the vertices of a graph with integers so as to minimize the maximum of the absolute values of the difference between the labels of adjacent vertices. This minimum over all possible labelings is called the **bandwidth** of the graph. Let $G(k,l)$ be the graph whose vertices are the vectors with $k+1$ non-negative integer coordinates summing to l, putting vectors adjacent when they differ by one in two positions. It is easy to draw $G(2,l)$ in the plane as a triangulation of a large equilateral triangle. Numbering the vertices successively in rows confirms that the bandwidth is at most $l+1$.

CONJECTURE 22. *The bandwidth of $G(2,l)$ equals $l+1$.*

More generally, when the vertices of $G(k,l)$ are assigned distinct $k+1$-dimensional integer vectors, the minimum sum of coordinate differences between adjacent vertices is called the $(k+1)$-**dimensional bandwidth** (see [4] and [45]).

CONJECTURE 23. *The $(k+1)$- dimensional bandwidth of $G(k,l)$ is $l+1$.*

6.5. Pagenumber of the Complete Bipartite Graph (Douglas West).
The **pagenumber** of a graph is the minimum number of pages in a book such that the graph can be embedded without crossings with the vertices in some fixed order on the spine of the book and each edge on a single page. By construction, the pagenumber of $K_{m,n}$ has been shown to be at most $\lceil (2m+n)/4 \rceil$, where $m \geq n$.

CONJECTURE 24. *The pagenumber of $K_{m,n}$ is $\lceil (2m+n)/4 \rceil$, where $m \geq n$.*

Of particular interest is the answer for $K_{n,n}$ which by our conjecture should be $3m/4$ (see [24]).

7. Logic

7.1. Trees and Second Order Theory (Detlef Seese).
CONJECTURE 25. *Let $\mathcal{K}$ be a class of finite graphs with a decidable monadic second order theory. Then there is a class $\mathcal{T}$ of trees such that the monadic second order theory of $\mathcal{K}$ is interpretable into the monadic second order theory of $\mathcal{T}$.*

For the class of planar graphs, graphs of bounded genus and any class of graphs excluding an arbitrary fixed graph as a minor one can show that this conjecture holds (see [38] and [9]).

7.2. Digraphs and Decidability (Dirk Vertigan).
For a directed graph G, let $\overline{G}$ denote the corresponding undirected graph. Let $\mathcal{C}$ be a minor closed class of directed graphs. Let $\overline{\mathcal{C}}$ be the corresponding minor closed class of undirected graphs, that is, $\overline{\mathcal{C}} = \{\overline{G} : G \in \mathcal{C}\}$. Note that $\mathcal{C}$ is not uniquely determined by $\overline{\mathcal{C}}$. By the Robertson-Seymour theorem [27], both $\mathcal{C}$ and $\overline{\mathcal{C}}$ have finitely many excluded minors. How are these excluded minors related? Consider the following problem (see also [43]).

> **Input**: An antichain of directed graphs (being the excluded minors of some $\mathcal{C}$).
> **Output**: The excluded minors of $\overline{\mathcal{C}}$.

Is this problem decidable? (Equivalently, is there a computable bound on the size of excluded minors of $\overline{\mathcal{C}}$, in terms of such a bound for $\mathcal{C}$?) If so, what stronger statements can be made? What about special cases such as when $\mathcal{C}$ has a planar excluded minor, or a single excluded minor? For minor insight, here are some examples:

Input: Acyclically oriented C_3. **Output**: $K_{1,1,2}$.
Input: Cyclically oriented C_3. **Output**: K_4, C_5.
Input: Both oriented C_3's. **Output**: C_3.

8. Disjoint Paths (Nathaniel Dean)

Probably no stucture is more fundamental to graph theory than a path. Results concerning disjoint paths tend to have significant theoretical and practical implications. Consider the following algorithmic problems.

CYCLE THROUGH K VERTICES
INSTANCE: Graph $G = (V, E)$ with integer edge weights and a subset S of V of size k.
QUESTION: Is there a cycle in G that contains S?

SHORTEST CYCLE THROUGH K VERTICES
INSTANCE: Graph $G = (V, E)$ with integer edge weights, an integer w, and a subset S of V of size k.
QUESTION: Is there a cycle in G of weight at most w that contains S?

DISJOINT CONNECTING PATHS
INSTANCE: Graph $G = (V, E)$ and a collection of k disjoint vertex pairs $(s_1, t_1), (s_2, t_2), \ldots, (s_k, t_k)$.
QUESTION: Does G contain k mutually vertex-disjoint paths, one connecting s_i to t_i for each $i = 1, 2, \ldots, k$?

SHORTEST DISJOINT CONNECTING PATHS
INSTANCE: Graph $G = (V, E)$ with integer edge weights, an integer w, and a collection of k disjoint vertex pairs $(s_1, t_1), (s_2, t_2), \ldots, (s_k, t_k)$.
QUESTION: Does G contain k mutually vertex-disjoint paths of total weight at most w, one connecting s_i to t_i for each $i = 1, 2, \ldots, k$?

The DISJOINT CONNECTING PATHS problem is solvable in polynomial time for fixed k (see [**26**] and [**28**])). This implies a polynomial time solution to CYCLE THROUGH K VERTICES for fixed k. The case $k = 3$ can actually be solved in linear time [**13**]. This algorithm is straightforward and appears to be quite fast. For $k > 3$ nothing is known.

OPEN PROBLEM 26. Is SHORTEST CYCLE THROUGH K VERTICES solvable in polynomial time for fixed k?

OPEN PROBLEM 27. Is SHORTEST DISJOINT CONNECTING PATHS solvable in polynomial time for fixed k?

REFERENCES

1. B. Alspach, L. Goddyn and C. Q. Zhang, Graphs with the circuit cover property, *Trans. Amer. Math. Soc.*, submitted.
2. B. Alspach and C. Q. Zhang, Cycle coverings of cubic multigraphs, *Discrete Math.*, to appear.
3. S. Arnborg, Efficient algorithms for combinatorial problems on graphs with bounded decomposability, *BIT* **25** (1985) 2–23.
4. Bandwidth of the integer simplex, *Amer. Math. Monthly* **94** (1987), 997-1000. (Solution to elementary problem.)
5. H. L. Bodlaender, J. R. Gilbert, Hjálmtýr Hafsteinson, and Ton Kloks, *Approximating treewidth, pathwidth and minimum elimination tree height*, in *Graph-Theoretic Concepts in Computer Science* ed. by G. Tinhofer et al, Springer-Verlag (1991) to appear.
6. H. L. Bodlaender and T. Kloks, *Better algorithms for the pathwidth and treewidth of graphs*, in *Proceedings of the 18'th International Colloquium on Automata, Languages and Programming*, Springer-Verlag (1991) 544–555.
7. J. A. Bondy and U. S. R. Murty, *Graph Theory with Applications*, American Elsevier, New York (1976).
8. Y. Colin de Verdiere, Sur un Nouvel Invariant des Graphes et un Critere de Planarite, *J. Combin. Theory Ser. B* **50** (1990) 11–21.
9. B. Courcelle, The monadic second order logic of graphs VI: on several representatiuons of graphs by relational structures, preprint.
10. P. Damaschke, Induced subgraphs and well-quasi-ordering, *J. Graph Theory* **14** (1990) 427–435.
11. G. Ding, Subgraphs and well-quasi-ordering, *Graph Theory*, to appear.
12. G. A. Dirac, Some theorems on abstract graphs, *Proc. London Math. Soc.* **2** (1952) 69–81.
13. H. Fleischner and G. J. Woeginger, Detecting cycles through three fixed vertices in a graph, *Info. Proc. Letters* **41** (1992) 29–33.
14. I. Fournier, Thèse, Université Paris Sud, 91405 Orsay, France (1982).
15. A. Gerards, Fraphs and polyhedra–binary spaces and cutting planes, PhD thesis, Tilburg Univ., Tilburg (1988).
16. M. C. Golumbic, *Algorithmic Graph Theory and Perfect Graphs*, Academic Press, New York, 1980.
17. B. Jackson, Shortest circuit covers and postman tours in graphs with a nowhere zero 4-flow, *SIAM J. Discrete Math.*, to appear.
18. F. Jaeger, Flows and generalized coloring theorems in graphs, *J. Combin. Theory Ser B* **26** (1979) 205–216.
19. L. Lovász, *Combinatorial Problems and Exercises*. North-Holland, New York (1979).
20. Y. Manoussakis, Covering by cycles the vertices of a 3-connected graph, Research Report 219, Labortoire de recherche en Informatique, Université Paris Sud, 91405 Orsay, France (1985).
21. M. Middendorf and F. Pfeiffer, On the complexity of disjoint paths problems, in *Polyhedral Combinatorics* **DIMACS I** AMS, ACM, ed. by W. Cook, P. D. Seymour (1990) 171–178.
22. S. Milici and Z. Tuza, Coverable graphs, technical report CCS-91004, Comp. and Aut. Inst., Hung. Acad. Sci. (1991).
23. Rolf H. Möhring, *Graph problems related to gate matrix layout and PLA folding*, Computational Graph Theory (G. Tinhofer et al., eds.), Springer-Verlag (1990) 17–52.
24. D. J. Muder, M. L. Weaver, and D. B. West, Pagenumber of complete bipartite graphs, *J. Graph Theory* **12** (1988) 469–489.
25. C. St. J. A. Nash-Williams, Edge-disjoint spanning trees of finite graphs, *J. London Math. Soc.* **36** (1961) 445–450.
26. N. Robertson and P. D. Seymour, Graph minors XIII. The disjoint paths problem, manuscript (1986).
27. N. Robertson and P. D. Seymour, Graph minors XX. Wagner's conjecture, manuscript (1988).
28. N. Robertson and P. D. Seymour, Graph minors XXII. Irrelevant vertices in linkage prob-

lems, manuscript (1992).
29. N. Robertson, P. D. Seymour and R. Thomas, A survey of linkless embeddings, this volume.
30. Petra Scheffler, A linear algorithm for the pathwidth of trees, in *Topics in Combinatorics and Graph Theory* ed. by R. Bodendiek and R. Henn, Physica Verlag, Heidelberg (1990) 613–620.
31. W. Schwärzler, A. Sebő, A generalized cut condition for multiflows in matroids, *Discrete Math.*, to appear.
32. A. Sebő, A quick proof of Seymour's theorem on T-joins, *Discrete Math.* **64** (1987) 101–103.
33. A. Sebő, Dual Integrality and multicommodity flows, in *Infinite and Finite Sets* ed. by A. Hajnal, V. T. D-Sós, Eger (Hungary) (1987).
34. A. Sebő, The cographic multiflow problem: an epilogue, in *Polyhedral Combinatorics* ed. by W. Cook, P. D. Seymour **DIMACS I**, AMS, ACM (1990) 203–234.
35. P. D. Seymour, Sums of Circuits, in *Graph Theory and Related Topics* ed. by J. A. Bondy and U. S. R. Murty, Academic Press, New York (1979) 341–355.
36. P. D. Seymour, Even circuits in planar graphs, *J. Combin. Theory Ser B* **31** (1981) 178–192.
37. P. D. Seymour, On odd cuts and planar multicommosdity flows, *Proc. Lond. Math. Soc.* **42** (1981) 178–192
38. D. Seese, The structure of the models of decidable monadic theories of graphs, *Annals of Pure and Applied Logic* **53** (1991) 169-195.
39. R. Thomas, Graphs without K_4 and well-quasi-ordering, *J. Combin. Theory Ser. B* **38** (1985) 240–247.
40. W. T. Tutte, On the embedding of linear graphs in surfaces, *Proc. London Math. Soc., Ser. 2* **51** (1949) 474–489.
41. W. T. Tutte, A theorem on planar graphs, *Trans. Am. Math. Soc.* **82** (1956) 99–116.
42. W. T. Tutte, On the problem of decomposing graphs into n connected factors, *J. London Math. Soc.* **36** (1961) 221–230.
43. D. L. Vertigan, Minor Classes. In preparation (manuscript 1988).
44. K. Wagner, Über ein Eigenschaft der ebenen Komplexe, *Math. Ann.* **114** (1937) 570–590.
45. M. L. Weaver, *Graph Labelings*, Ph.D Thesis, University of Illinois (1987).
46. N. Huy Xuong, *Sur quelques problemes d'immersion d'un graphe dans une surface*, These de Doctorat d'Etat, Grenoble (1977) Chapter III.
47. C-Q. Zhang, On even cycle decomposition of eulerian graphs, (preprint).
48. C-Q. Zhang, Minimum cycle coverings and integer flows, *J. Graph Theory*, **14** (1990) 537–546.

BELLCORE, 445 SOUTH STREET, MORRISTOWN, NJ 07960-1910, USA
E-mail address: nate@bellcore.com

Recent Titles in This Series

(Continued from the front of this publication)

117 **Morton Brown, Editor,** Continuum theory and dynamical systems, 1991
116 **Brian Harbourne and Robert Speiser, Editors,** Algebraic geometry: Sundance 1988, 1991
115 **Nancy Flournoy and Robert K. Tsutakawa, Editors,** Statistical multiple integration, 1991
114 **Jeffrey C. Lagarias and Michael J. Todd, Editors,** Mathematical developments arising from linear programming, 1990
113 **Eric Grinberg and Eric Todd Quinto, Editors,** Integral geometry and tomography, 1990
112 **Philip J. Brown and Wayne A. Fuller, Editors,** Statistical analysis of measurement error models and applications, 1990
111 **Earl S. Kramer and Spyros S. Magliveras, Editors,** Finite geometries and combinatorial designs, 1990
110 **Georgia Benkart and J. Marshall Osborn, Editors,** Lie algebras and related topics, 1990
109 **Benjamin Fine, Anthony Gaglione, and Francis C. Y. Tang, Editors,** Combinatorial group theory, 1990
108 **Melvyn S. Berger, Editor,** Mathematics of nonlinear science, 1990
107 **Mario Milman and Tomas Schonbek, Editors,** Harmonic analysis and partial differential equations, 1990
106 **Wilfried Sieg, Editor,** Logic and computation, 1990
105 **Jerome Kaminker, Editor,** Geometric and topological invariants of elliptic operators, 1990
104 **Michael Makkai and Robert Paré,** Accessible categories: The foundations of categorical model theory, 1989
103 **Steve Fisk,** Coloring theories, 1989
102 **Stephen McAdam,** Primes associated to an ideal, 1989
101 **S.-Y. Cheng, H. Choi, and Robert E. Greene, Editors,** Recent developments in geometry, 1989
100 **W. Brent Lindquist, Editor,** Current progress in hyperbolic systems: Riemann problems and computations, 1989
99 **Basil Nicolaenko, Ciprian Foias, and Roger Temam, Editors,** The connection between infinite dimensional and finite dimensional dynamical systems, 1989
98 **Kenneth Appel and Wolfgang Haken,** Every planar map is four colorable, 1989
97 **J. E. Marsden, P. S. Krishnaprasad, and J. C. Simo, Editors,** Dynamics and control of multibody systems, 1989
96 **Mark Mahowald and Stewart Priddy, Editors,** Algebraic topology, 1989
95 **Joel V. Brawley and George E. Schnibben,** Infinite algebraic extensions of finite fields, 1989
94 **R. Daniel Mauldin, R. M. Shortt, and Cesar E. Silva, Editors,** Measure and measurable dynamics, 1989
93 **M. Isaacs, A. Lichtman, D. Passman, S. Sehgal, N. J. A. Sloane, and H. Zassenhaus, Editors,** Representation theory, group rings, and coding theory, 1989
92 **John W. Gray and Andre Scedrov, Editors,** Categories in computer science and logic, 1989
91 **David Colella, Editor,** Commutative harmonic analysis, 1989
90 **Richard Randell, Editor,** Singularities, 1989
89 **R. Bruce Richter, Editor,** Graphs and algorithms, 1989
88 **R. Fossum, W. Haboush, M. Hochster, and V. Lakshmibai, Editors,** Invariant theory, 1989
87 **Laszlo Fuchs, Rüdiger Göbel, and Phillip Schultz, Editors,** Abelian group theory, 1989

(See the AMS catalog for earlier titles)